ENGINEERING DESIGN GRAPHICS

AutoCAD® Release 12
Eighth Edition

ENGINEERING DESIGN GRAPHICS

AutoCAD® Release 12
Eighth Edition

JAMES H. EARLE

Texas A & M University

Addison-Wesley Publishing Company
Reading, Massachusetts • Menlo Park, California • New York
Don Mills, Ontario • Wokingham, England • Amsterdam • Bonn
Sydney • Singapore • Tokyo • Madrid • San Juan • Milan • Paris

Executive Editor: Michael Payne
Sponsoring Editor: Denise Descoteaux
Senior Production Supervisor: David Dwyer
Senior Production Coordinator: Genevra A. Hanke
Cover Design Supervisor: Peter M. Blaiwas
Senior Manufacturing Manager: Roy Logan
Manufacturing Coordinator: Judy Sullivan
Text Designer: Jean Hammond
Copyeditor: Jerrold C. Moore
Proofreader: Phyllis Coyne
Layout Artist: Julia M. Fair

Cover credits: Courtesy of Trilby Wallace, McDonnell Douglas Space Systems, Kennedy Space Center, Florida, and Intergraph Corporation, Huntsville, Alabama.

Many of the designations used by manufacturers and sellers to distinguish their products are claimed as trademarks. Where those designations appear in this book, and Addison-Wesley was aware of a trademark claim, the designations have been printed in initial caps or all caps.

The programs and applications presented in this book have been included for their instructional value. They have been tested with care, but are not guaranteed for any particular purpose. The publisher does not offer any warranties or representations, nor does it accept any liabilities with respect to the programs or applications.

Library of Congress Cataloging-in-Publication Data
Earle, James H.
    Engineering design graphics : AutoCAD release 12 / James H. Earle.
  -- 8th ed.
        p.   cm.
    Includes index.
    ISBN 0-201-51982-8 (alk. paper)
    1. Engineering design. 2. Engineering graphics. I. Title.
TA174.E23 1994
620'.0042'028566869--dc20   93-15665
                                                                 CIP

5 6 7 8 9 10 – DOC – 979695

Dedicated to my father,
Hubert Lewis Earle,
October 25, 1900–October 22, 1967

Preface

This eighth edition of *Engineering Design Graphics* has been revised to keep pace with the needs of education and industry. It is a significant modification of the seventh edition that retains the classroom-tested sequence of presentation and its tried and true teaching features while adding many new features to help students and teachers. *Engineering Design Graphics* covers the principles of:

- engineering drawing,
- computer graphics,
- descriptive geometry,
- design, and
- problem solving.

Objective

The objective of this book is to support a course in which the student learns:

- ANSI standards and techniques of preparing engineering drawings

- How to improve spatial analysis skills (descriptive geometry)
- How to prepare drawings on the drawing board
- How to prepare drawings by computer
- How to use graphics as a medium of design

Above all, this textbook is designed to help students expand their creative talents and communicate their ideas in an effective manner.

Revision Features

Engineering Design Graphics, Eighth Edition has been made as teachable as possible. All aspects of the book have been designed to help teachers in their presentations and students in learning on their own when necessary.

Some features included to help the student are:

- Several hundred new illustrations
- Almost 1,000 revised or redrawn illustrations

- Use of human figures to show viewpoints

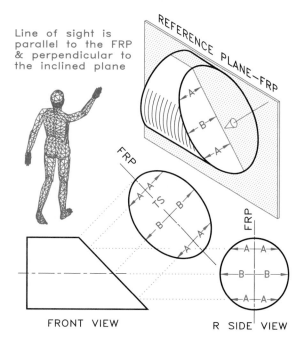

Line of sight is parallel to the FRP & perpendicular to the inclined plane

REFERENCE PLANE-FRP

FRP

FRP

FRONT VIEW

R SIDE VIEW

- Notes within the figures to highlight key points
- Use of rotational art where appropriate to illustrate how revolving a view changes the way objects are seen and drawn
- More accessible page design with clearer separation of art and text; color is used only to enhance the pedagogical effectiveness of figures
- New working drawing problems in Chapter 23
- Updated and expanded AutoCAD® coverage, including Release 12

Format

Engineering Design Graphics, Eighth Edition retains the proven features that have made it a time-tested teaching and learning textbook. Its self-instructional examples enable students to work independently with a minimum of assistance from the

instructor. Clear, teachable examples assist with visualization.

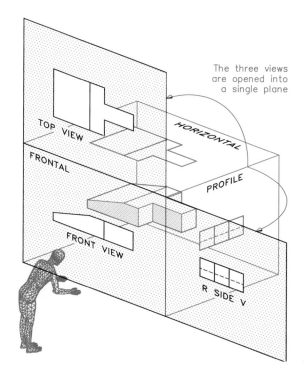

The three views are opened into a single plane

TOP VIEW

HORIZONTAL

FRONTAL

PROFILE

FRONT VIEW

R SIDE V

Most instructional examples are presented in a step-by-step sequence to illustrate how problems are solved.

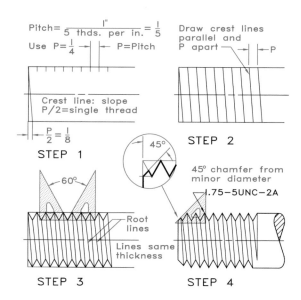

$Pitch = \dfrac{1"}{5\ thds.\ per\ in.} = \dfrac{1}{5}$

Use $P = \dfrac{1}{4}$ ← ← P=Pitch

Draw crest lines parallel and P apart ← ← P

Crest line: slope P/2=single thread

$\dfrac{P}{2} = \dfrac{1}{8}$

STEP 1

STEP 2

60°

45°

Root lines

Lines same thickness

45° chamfer from minor diameter

1.75–5UNC–2A

STEP 3

STEP 4

Industrial applications and illustrations are used to make the examples and problems meaningful to the student.

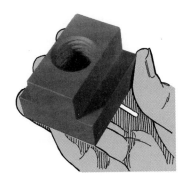

A second color is used to emphasize sequential steps, key points, and explanatory notes in the figures. Computer graphics examples have been boxed in color to identify them on the page; these examples and their supporting text are preceded by a mouse icon.

Key points and terms appear in bold face type. Figure references in the text also are in bold type for easier referencing.

Computer-Aided Design (CAD)

The main purpose of this book is to help the student learn the principles of graphics, whether done on the drawing board or on the computer. *Engineering Design Graphics* has been designed to be as applicable for courses not using CAD as for courses that do. However, students will benefit from reading about computer graphics even if CAD is not used in the course. For courses that do use CAD, this book gives computer-graphics instruction as an integrated part of most chapters.

AutoCAD software is the featured software because it has the largest market share (approximately 75%) and consequently is the software that students are most likely to encounter in industry. Chapter 35 gives a general overview of hardware and software used for computer graphics. Chapter 36 covers AutoCAD Release 12, and Chapter 37

gives an introduction to solid modeling using AutoCAD's AME extension.

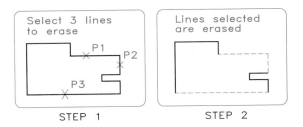

All computer graphics principles have been presented as two-step, three-step, or four-step illustrations. Each step illustrates what the computer user would see on the screen. Commands and prompts for these steps conveniently appear beneath these illustrations.

A Career Reference Book

Some material in this book may not be covered in the course for which it is used due to time limitations or the emphasis of the course by the instructor. Because the course may be the only graphics course that a student will encounter, this book should be retained for reference.

A Teaching System

This book used in combination with the supplements listed below comprises a complete teaching system.

Textbook Problems
Over 500 problems are offered to aid the student in mastering important principles.

Problem Manuals
Nineteen problem books and teachers' guides (with outlines, problem solutions, tests, and test solutions) are available for use with this book, and new problem books will be introduced in the future. Fifteen of the manuals have computer graphics problems on the backs of the problem sheets, which allows the solution of the problems

by both computer and pencil. A listing of these books and their source appears in the endpapers.

Visual Aids

Sixteen modules of *SoftVisuals* are available on disks from which multicolored overhead transparencies can be plotted on transparency film for classroom presentations. Transparency selection can be made from over 500 *SoftVisuals* keyed to this textbook that can be plotted with AutoCAD.

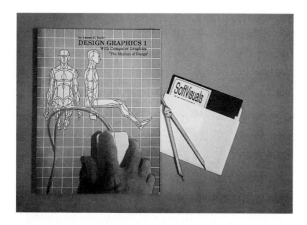

Acknowledgments

We are grateful for the assistance of many who have influenced the development of this volume. Many industries have furnished photographs, drawings, and applications that have been acknowledged in the corresponding legends. The Engineering Design Graphics staff of Texas A & M University have been helpful in making suggestions for the revision of this book.

Professor Tom Pollock provided valuable information on various metals for Chapter 19. Professor Leendert Kersten of the University of Nebraska, Lincoln, kindly provided his descriptive geometry computer programs for inclusion, and his cooperation is appreciated.

We are indebted to Neal Alen, Rodger Payne, and Jimm Meloy of Autodesk, Inc., for their assistance with AutoCAD. We appreciate the assistance and cooperation of Karen Kershaw of MegaCADD, Inc. David Ratner of Biomechanics Corporation was helpful in providing HUMANCAD® software.

Matthew Whiteacre and Larry Tucker were especially helpful in reviewing and critiquing the manuscript.

We are appreciative of the many institutions that have thought enough of our publications to adopt them for classroom use. It is an honor for one's work to be accepted by his colleagues. We are hopeful that this textbook will fill the needs of engineering and technology programs. As always, comments and suggestions for improvement and revision of this book will be appreciated.

College Station, Texas   *Jim Earle*

Brief Contents

Contents

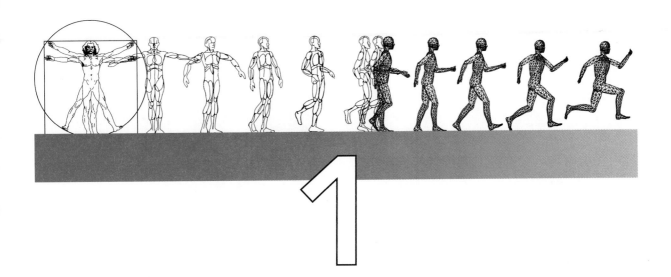

Introduction to Engineering and Technology

1.1 Introduction

This book deals with the field of engineering design graphics and its application to the design process. **Engineering graphics is the primary medium for developing and communicating design concepts.** The solution of most engineering problems requires a combination of organization, analysis, problem-solving principles, graphics, skill, and communication (**Fig. 1.1**).

This book is intended to help you use your creativity and develop your imagination, because innovation is essential to a successful career in engineering and technology. Albert Einstein said, "Imagination is more important than knowledge, for knowledge is limited, whereas imagination embraces the entire world . . . stimulating progress, or, giving birth to evolution" (**Fig. 1.2**).

1.2 Engineering Graphics

Engineering graphics covers the total field of graphical problem solving and has two major areas of specialization: descriptive geometry and

THE ENGINEERING PROBLEM

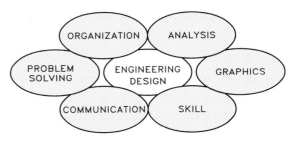

Figure 1.1 This text illustrates the total approach to engineering, with the engineering problem as the focal point.

documentation drawings. Other areas of application are nomography, graphical mathematics, empirical equations, technical illustration, vector analysis, data analysis, and computer graphics.

Graphics is one of the designer's most effective tools for developing design concepts and solving three-dimensional problems. It also is the designer's best means of communicating ideas to others.

1

Figure 1.2 Albert Einstein: "Imagination is more important than knowledge."

Figure 1.3 Gaspard Monge was the "father of descriptive geometry."

Descriptive Geometry

Gaspard Monge (1746–1818), the "father of descriptive geometry" (**Fig. 1.3**), used graphical methods to solve design problems related to fortifications and battlements while a military student in France. His headmaster scolded him for not using the usual long, tedious mathematical process. Only after lengthy explanations and demonstrations of his technique was he able to convince the faculty that graphical methods (now called descriptive geometry) produced solutions in less time.

Descriptive geometry was such an improvement over mathematical methods that it was kept as a military secret for fifteen years before the authorities allowed it to be taught as part of the civilian curriculum. Monge went on to become a scientific and mathematical aide to Napoleon.

Descriptive geometry is the projection of three-dimensional figures on the two-dimensional plane of paper in a manner that allows geometric manipulations to determine lengths, angles, shapes, and other geometric information about the figures.

1.3 Technological Advances

Many of the technological advances of the twentieth century are engineering achievements. Since

TECHNOLOGICAL ADVANCES OF THE TWENTIETH CENTURY

TECHNOLOGICAL ADVANCEMENTS OF THE 20TH CENTURY			
1900	Vacuum cleaner	1950	A−bomb tests
	Airplane		Optical fibers
	Dial telephone		Soviet satelite
	Light bulb		Microchip
	Model T Ford	1960	Commun. satelite
1910	Washing machine		Indus. robot
	Refrigerator		Nuclear reactor
	Wireless phone		Heart transplant
1920	Radio broadcasts		Man on moon
	Telephone service	1970	Silicon chip
	35mm camera		Personal computer
	Cartoons & sound		Videocassette record.
1930	Tape recorder		Supersonic jet
	Atom split		Neutron bomb
	Jet engine	1980	Stealth bomber
	Television		Space shuttle
1940	Elect. computer		Artificial heart
	Missile		Soviet space station
	Transistor	1990	Computer voice recoginition
	Microwave		Artificial intelligence
	Polaroid camera		Space−based assembly plant

Figure 1.4 This chronology lists some of the significant technological advances of the twentieth century.

1900, technology has taken us from the horse-drawn carriage to the moon and back, and more advances are certain in the future.

Figure 1.4 shows a few of the many technological mileposts since 1900. It identifies products and processes that have provided millions of jobs

Figure 1.5 Technological and design team members, with their varying backgrounds and areas of expertise, must communicate and interact with each other. (Courtesy of Honeywell, Inc.)

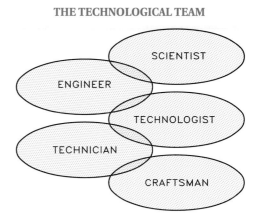

Figure 1.6 This ranking of the typical technological team is from the most theoretical level (scientists) to the least technical level (craftspeople).

Figure 1.7 This geologist is studying seismological charts to determine the likelihood of petroleum deposits. (Courtesy of Texas Eastern; photo by Bob Thigpen.)

and a better way of life for all. Other significant achievements were building a railroad from Nebraska to California that met at Promontory Point, Utah, in 1869 in less than four years; constructing the Empire State Building with 102 floors in a mere thirteen and a half months in 1931; and retooling industry in 1942 for World War II to produce 4.5 naval vessels, 3.7 cargo ships, 203 airplanes, and 6 tanks each day while supporting 15 million Americans in the armed forces.

1.4 The Technological and Design Team

Technology and design have become so broad and complex that teams of specialists rather than individuals undertake most projects (**Fig. 1.5**). Such teams usually consist of one or more scientists, engineers, technologists, technicians, and craftspeople, and may include designers and stylists (**Fig. 1.6**).

Scientists

Scientists are researchers who seek to discover new laws and principles of nature through experimentation and scientific testing (**Fig. 1.7**). They are more concerned with the discovery of scientific principles than with the application of those principles to products and systems. Their discoveries may not find applications until years later.

Figure 1.8 These engineers are discussing geological data and area surveys to arrive at the best placement of exploratory oil wells. (Courtesy of Texas Eastern; photo by Bob Thigpen.)

Figure 1.9 Engineering technologists combine their knowledge of production techniques with the design talents of engineers to produce a product. (Courtesy of Omark Industries, Inc.)

Engineers

Engineers receive training in science, mathematics, and industrial processes to prepare them to apply the findings of the scientists (**Fig. 1.8**). Thus engineers are concerned with converting raw materials and power sources into needed products and services. **Creatively applying scientific principles to develop new products and systems is the design process, the engineer's primary function.** In general, engineers use known principles and available resources to achieve a practical end at a reasonable cost.

Technologists

Technologists obtain backgrounds in science, mathematics, and industrial processes. But, whereas engineers are responsible for analysis, overall design, and research, technologists are concerned with the application of engineering principles to planning, detail design, and production (**Fig. 1.9**). Technologists apply their knowledge of engineering principles, manufacturing, and testing to assist in the implementation of projects and production. They also provide support and act as liaison between engineers and technicians.

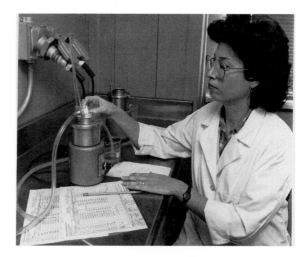

Figure 1.10 This engineering technician performs laboratory tests as a member of the technological and design team. (Courtesy of Texas Eastern; photo by Bob Thigpen.)

Technicians

Technicians assist engineers and technologists at a less theoretical level than technologists and provide

Figure 1.11 This craftsperson, a welder, is skilled in joining metal parts according to prescribed specifications. (Courtesy of Texas Eastern, *TE Today;* photo by Bob Thigpen.)

Figure 1.12 Thomas A. Edison essentially had no formal education, but he gave the world some of its most creative designs.

liaison between technologists and craftspeople (**Fig. 1.10**). They have backgrounds in mathematics, drafting, computer programming, and materials testing. Their work varies from conducting routine laboratory experiments to supervising craftspeople in manufacturing or construction.

Craftspeople

Craftspeople are responsible for implementing designs by fabricating them according to engineers' specifications. They may be machinists who make product parts or electricians who assemble electrical components. Their ability to produce a part according to design specifications is as necessary to the success of a project as engineers' ability to design it. Craftspeople include electricians, welders, machinists, fabricators, drafters, and members of many other occupational groups (**Fig. 1.11**).

Designers

Designers may be engineers, technologists, inventors, or industrial designers who have special talents for devising creative solutions. Designers do not necessarily have engineering backgrounds, especial-

ly in newer technologies where there is little design precedent. Thomas A. Edison (**Fig. 1.12**), for example, had little formal education, but he created some of the world's most significant inventions.

Stylists

Stylists are concerned with the appearance and market appeal of a product rather than its fundamental design (**Fig. 1.13**). They may design an automobile body or the exterior of an electric iron. Automobile stylists, for example, consider the car's appearance, driver's vision, passengers' enclosure, power unit's space requirement, and so on. However, they are not involved with the design of the car's internal mechanical functions, such as the engine, steering linkage, and brakes. Stylists must have a high degree of aesthetic awareness and an instinct for styling that appeals to the consumer.

1.5 Engineering Fields

Recent changes in engineering include the emergence of technologists and technicians and the growing number of women pursuing engineering careers. More than 15% of today's freshman engineering students are women; 15% of master's degrees and 10% of doctor's degrees in engineering are awarded to women.

Figure 1.14 Aerospace engineers design and develop space systems such as this payload assembly. (Courtesy of Trilby Wallace, McDonnell Douglas; Kennedy Space Center, Florida; and Intergraph Corporation, Huntsville, Alabama.)

Figure 1.13 The stylist develops a product's outward appearance to make it as marketable as possible, as illustrated by this body design for GM's Ultralite, a concept car of the future. (Courtesy of General Motors Corporation.)

Aerospace Engineering

Aerospace engineering has progressed from the Wright brothers' first flight at Kittyhawk, North Carolina, in 1903 to the penetration of outer space. It deals with all aspects (speeds and altitudes) of flight. Aerospace engineering assignments range from developing complex vehicles capable of traveling millions of miles into space to hover aircraft that can transport and position large construction components. In the space

exploration branch of this profession, aerospace engineers work on all types of aircraft and space-craft—missiles, rockets, propeller-driven planes, and jet-powered planes (**Fig. 1.14**).

Second only to the auto industry in sales, the aerospace industry contributes immeasurably to national defense and the economy. Specialized areas include aerodynamics, structural design, instrumentation, propulsion systems, materials, reliability testing, and production methods.

Aerospace engineers specialize in one of two major areas: **research engineering** and **design engineering.** Research engineers investigate known principles in search of new ideas and concepts; design engineers translate them into workable applications. The professional society for aerospace engineers is the American Institute of Aeronautics and Astronautics (AIAA).

Agricultural Engineering

Agricultural engineers are trained to serve the world's largest industry—agriculture—in which they deal with the production, processing, and handling of food and fiber.

Figure 1.15 Agricultural engineers of the future will design food production environments and systems for space settlements. (Courtesy of NASA.)

Mechanical Power Agricultural engineers who work with manufacturers of farm equipment are concerned with gasoline and diesel engine equipment, including pumps, irrigation machinery, and tractors. Machinery must be designed for the electrical curing of hay, milk and fruit processing, and heating environments for livestock and poultry. Farm machinery designed by agricultural engineers has been largely responsible for the vastly increased productivity in U.S. agriculture. Today's farmer produces enough food for 120 people, whereas one hundred years ago a farmer was able to feed only 4 people.

Farm Structures The construction of barns, shelters, silos, granaries, processing centers, and other agricultural buildings requires specialists in agricultural engineering. They must understand heating, ventilation, and chemical changes that might affect the storage of crops.

Electrical Power Agricultural engineers design electrical systems and select equipment that will operate efficiently and meet the requirements of many situations. They may serve as consultants or designers for manufacturers or processors of agricultural products.

Soil and Water Control Agricultural engineers are responsible for devising systems to improve drainage and irrigation systems, resurface fields, and construct water reservoirs (**Fig. 1.15**). They may perform activities in conjunction with the U.S. Department of Agriculture, the U.S. Department of the Interior, state agricultural universities, consulting engineering firms, or irrigation companies.

Characteristics Most agricultural engineers are employed in private industry, especially by manufacturers of heavy farm equipment and specialized lines of field, barnyard, and household equipment; by electrical service companies; and by distributors of farm equipment and supplies. Although few agricultural engineers live on farms, they need to understand agricultural problems—farming, crops, animals, and farmers themselves. The professional society for agricultural engineers is the American Society for Agricultural Engineers (ASAE).

Chemical Engineering

Chemical engineering involves the design and selection of equipment used to process and manufacture large quantities of chemicals. Chemical engineers develop and design methods of transporting fluids through ducts and pipelines, transporting solid material through pipes or conveyors, transferring heat from one fluid or substance to another through plate or tube walls, absorbing gases by bubbling them through liquids, evaporating liquids to increase concentration of solutions, distilling mixed liquids to separate them, and handling many other chemical processes. Process control and instrumentation are important specialties in chemical engineering. With the measurement of quality and quantity by instrumentation, process control is fully automatic.

Chemical engineers often utilize chemical reactions of raw products, such as oxidation, hydrogenation, reduction, chlorination, nitration, sulfonation, pyrolysis, and polymerization, in their work (**Fig. 1.16**). They develop and process chemicals such as acids, alkalies, salts, coal-tar products, dyes, synthetic chemicals, plastics, insecticides, and fungicides for industrial and domestic uses. Chemical engineers help develop drugs and medicines, cosmetics, explosives, ceramics, cements, paints, petroleum products, lubricants, synthetic fibers, rubber, and detergents. They also design equipment for food preparation and canning plants.

Approximately 80% of chemical engineers work in manufacturing industries, primarily the chemical industry. The other 20% work for government agencies, independent research institutes, and as independent consultants. New fields requiring chemical engineers are nuclear sciences, rocket fuel development, and environmental pollution control. The professional society for chemical engineers is the American Institute of Chemical Engineers (AIChE).

Civil Engineering

Civil engineering, the oldest branch of engineering, is closely related to virtually all our daily activities. The buildings we live and work in, the transportation we use, the water we drink, and the drainage and sewage systems we rely on are all the results of civil engineering.

Construction Civil engineers manage the workers, finances, and materials used on construction projects. **Structural engineers** design and supervise the construction of buildings, harbors, airfields, tunnels, bridges, stadiums, and other types of structures.

Figure 1.16 The contributions of scientists, chemical engineers, petroleum engineers, structural engineers, and others are required in the construction and operation of refineries that produce the petroleum products and by-products vital to our economy. (Courtesy of Exxon Corporation.)

City Planning Civil engineers working as city planners develop plans for the growth of cities and systems related to their operation. They are involved in street planning, zoning, residential subdivisions, and industrial site development.

Hydraulics Civil engineers work with the behavior of water and other fluids from their conservation to their transportation. Civil engineers design wells, canals, dams, pipelines, flood control and drainage systems, and other methods of controlling and using those resources (**Fig. 1.17**).

Transportation Civil engineers design and supervise construction, modification, and maintenance of railroad and mass transit systems. They design and supervise construction of airport runways, control towers, passenger and freight stations, and aircraft hangars. They are involved in all phases of developing the national systems and local networks of highways and interchanges for moving automobile traffic including the design of tunnels, culverts, and traffic control systems.

Figure 1.17 Civil, structural, hydraulic, mechanical, and electrical engineers work on projects such as the Morrow Point Dam in Colorado, which provides a water reserve and controls downstream flooding. (Courtesy of the U.S. Department of Interior.)

Sanitary Engineering Civil engineers maintain public health by designing pipelines, treatment plants, and other facilities for water purification and water and air pollution control. They also are involved in solid waste disposal activities.

Other Areas of Specialization Civil engineers are involved in **geotechnical engineering,** which deals with the study of soils. They also work in **environmental engineering,** which relates to all aspects of the environment and methods of preserving it.

Characteristics Many civil engineers hold positions in administration and municipal manage-

ment and are associated with federal, state, and local government agencies and the construction industry. Many work as consulting engineers for architectural firms and independent consultants. The remainder work for public utilities, railroads, educational institutions, and various manufacturing industries. The professional society for civil engineers is the American Society of Civil Engineers (ASCE). Founded in 1852, it is the oldest engineering society in the United States.

Electrical Engineering

Electrical engineers are concerned with **power**, which deals with providing the large amounts of energy required by cities and large industries, and **electronics**, which deals with the small amounts of power used for communications and automated operations that have become part of everyday life.

Power Power generation, transmission, and distribution pose many electrical engineering problems from the design of generators for producing electricity to the development of transmission equipment. Power applications are quite numerous in homes for computers, washers, dryers, vacuum cleaners, and other appliances. Only about one-quarter of electric energy is consumed in the home. About half is used by industry for metal refining, heating, motor drives, welding, machinery controls, chemical processes, plating, and electrolysis. **Illumination** (lighting) is required in nearly every phase of modern life. Improving its efficiency is a challenging area for electrical engineers.

Electronics Computers have contributed to the development of a gigantic industry that is the domain of electrical engineers. Used with industrial electronics (such as robotics), computers have changed industry's manufacturing and production processes, resulting in greater precision and less manual labor.

Communications applications are devoted to the improvement of radio, telephone, telegraph, and television systems, the nerve centers of most industrial operations (**Fig. 1.18**).

Figure 1.18 Electrical engineers in cooperation with aerospace engineers develop space outposts, such as this satellite for providing mobile voice and data services to customers beyond the range of standard cellular systems. (Courtesy of BCE, Inc.)

Figure 1.19 Industrial engineers design and lay out automatic conveyor systems such as those used at the Sara Lee plant. (Courtesy of Honeywell, Inc.)

Instrumentation applies systems of electronic instruments to industrial processes. Electrical engineers make extensive use of the cathode-ray tube and the electronic amplifier in industry and nuclear reactors. Increasingly, engineers are using instrumentation for applications in medical diagnosis and therapy.

Military electronics encompasses most weapons and tactical systems ranging from the walkie-talkie to radar networks for detecting enemy aircraft. Remote-controlled electronic systems are used for navigation and interception of guided missiles.

Characteristics There are currently more electrical engineers than any other type of engineer. The increasing need for electrical equipment, automation, and computerized systems is expected to sustain the rapid growth of this field. The professional society for electrical engineers is the Institute of Electrical and Electronic Engineers (IEEE). Founded in 1884, the IEEE is the world's largest professional society.

Industrial Engineering

Industrial engineering, one of the newer engineering fields, differs from other branches of engineering in that it relates primarily to people, their performance, and working conditions. Industrial engineers often manage people, machines, materials, methods, and money in the production and marketing of goods.

Industrial engineers may be responsible for plant layout, development of plant processes, or determination of operating standards that will improve the efficiency of a plant operation. They also design and supervise systems for improved safety of personnel and increased production at lower costs (**Fig. 1.19**).

Areas of industrial engineering include management, plant design and engineering, electronic data processing, systems analysis and design, control of production and quality, performance standards and measurements, and research. Industrial engineers are increasingly involved in implementing automated production systems.

People-oriented areas include the development of wage incentive systems, job evaluation, work measurement, and environmental system design. Industrial engineers are often involved in management–labor agreements that affect operations and production.

More than two-thirds of all industrial engineers are employed in manufacturing industries. Others work for insurance companies, construction and mining firms, public utilities, large businesses, and government agencies. The professional society for industrial engineers is the American Institute of Industrial Engineers (AIIE), which was organized in 1948.

Mechanical Engineering

Mechanical engineering's major areas of specialization are power generation, transportation, manufacturing, power services, and atomic energy.

Power Generation Mechanical engineers develop and design prime movers (machines that convert natural energy into work) to power electric generators that produce electricity. They are involved in the design and operation of steam engines, turbines, internal combustion engines, and other prime movers (**Fig. 1.20**).

Transportation Mechanical engineers participate in the design of trucks, buses, automobiles, locomotives, marine vessels, and aircraft. In aeronautics they develop aircraft engines, controls, and internal environmental systems. Mechanical engineers design marine vessels powered by steam, diesel, or gas-turbine engines, as well as power services throughout the vessels, such as lighting, water, refrigeration, and ventilation.

Manufacturing Mechanical engineers design new products and the equipment and factories needed to manufacture them economically and at a uniformly high level of quality. The professional society for manufacturing engineers is the Society of Manufacturing Engineers (SME).

Power Services Mechanical engineers in this area must have a knowledge of pumps, ventilation

Figure 1.20 Mechanical engineers designed and supervised the building of this 3.0-liter, V-6, two-stroke automobile engine. (Courtesy of General Motors Corporation.)

equipment, fans, and compressors. They apply this knowledge to the development of methods for moving liquids and gases through pipelines and the design and installation of refrigeration systems, elevators, and escalators.

Nuclear Energy Mechanical engineers develop and handle protective equipment and materials and assist in constructing nuclear reactors. The professional society for mechanical engineers is the American Society of Mechanical Engineers (ASME).

Mining and Metallurgical Engineering

Mining engineers are responsible for developing methods of extracting minerals from the earth and preparing them for use by manufacturing industries. Working with geologists to locate ore deposits, which are exploited through the construction of tunnels and underground operations or surface strip mining, mining engineers must understand safety, ventilation, water supply, and communications. Mining engineers who work at mining sites usually are employed near small, out-of-the-way communities, whereas those in research and consulting most often work in large urban areas.

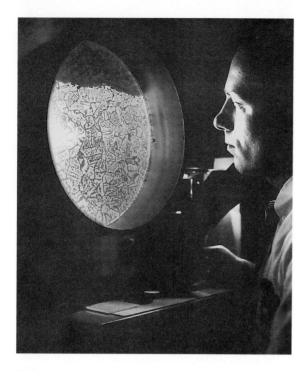

Figure 1.21 A metallograph shows the structure of an alloy that may be used in the construction of a refinery unit. Metallurgical engineers specially develop materials for specific applications for various industrial uses. (Courtesy of Exxon Corporation.)

The two main areas of metallurgical engineering are **extractive metallurgy,** the extraction of metal from raw ores to form pure metals, and **physical metallurgy,** the development of new products and alloys. Many metallurgical engineers work on the development of machinery for electrical equipment and in the aircraft industry. The need to develop new lightweight, high-strength materials for spacecraft, jet aircraft, missiles, and satellites will require more metallurgical engineers in the future (**Fig. 1.21**). The professional society for mining and metallurgical engineers is the American Institute of Mining, Metallurgical, and Petroleum Engineering (AIME).

Nuclear Engineering

The earliest work in nuclear engineering involved military applications. Nuclear power for

Figure 1.22 Nuclear engineers will design the nuclear reactors to economically produce much of the electric power needed in the future.

domestic needs was developed for a time, but the new emphasis in nuclear engineering is for medical applications.

Peaceful applications of nuclear engineering fall into two major areas: radiation and nuclear power reactors. **Radiation** is the propagation of energy through matter or space in the form of waves. In atomic physics, radiation includes fast-moving particles (alpha and beta rays, free neutrons, and so on), gamma rays, and X rays. Nuclear science is closely allied with botany, chemistry, medicine, and biology.

The use of **nuclear energy** to produce mechanical or electric power is a major peaceful application of nuclear engineering (**Fig. 1.22**). In the production of electric power, nuclear energy is the fuel used to produce steam to drive turbine generators.

Most nuclear engineering training focuses on the design, construction, and operation of nuclear reactors. Other areas include the processing of nuclear fuels, thermonuclear engineering, and the use of various nuclear by-products. The professional society for nuclear engineers is the American Nuclear Society.

Petroleum Engineering

The recovery of petroleum and natural gas is the primary concern of petroleum engineers, but they also develop methods for transporting and sepa-

Figure 1.23 Petroleum engineers supervise the operation of offshore platforms used in the exploration for oil. (Courtesy of Texas Eastern; photo by Bob Thigpen.)

Figure 1.24 This drafter is aided by a computer graphics system. (Courtesy of Hewlett-Packard.)

rating various petroleum products. They also are responsible for improving drilling equipment and ensuring its economical operation (**Fig. 1.23**). In exploring for petroleum, petroleum engineers are assisted by geologists and by instruments such as the airborne magnetometer, which indicates uplifts in the earth's subsurfaces that could hold oil or gas.

Petroleum engineers develop equipment to remove oil from the ground most efficiently and supervise oil-well drilling. They also design systems of pipes and pumps to transport oil to shipping points or to refineries. Development, design, and operation of petroleum processing facilities are done jointly with chemical engineers.

The Society of Petroleum Engineers (SPE) is a branch of AIME, which includes mining and metallurgical engineers and geologists.

1.6 Drafting

Drafters help engineers and designers select materials. They also prepare construction documents and specifications and translate designs into working drawings and technical illustrations. **Working drawings** are visual instructions for fabricating products and erecting structures in all fields of engineering. **Technical illustration,** the

most artistic area of engineering design graphics, visually depicts projects and products for operations manuals and presentations.

In addition, drafters prepare **maps, geological sections,** and **plats** from data given them by engineers, geologists, and surveyors. These drawings show the locations of property lines, physical features, strata, rights-of-way, building sites, bridges, dams, mines, and utility lines.

The three levels of certification for drafters are drafters, design drafters, and engineering designers:

- **Drafters** are graduates of a two-year, post–high school curriculum in engineering design graphics.
- **Design drafters** complete two-year programs at an approved junior college or technical institute.
- **Engineering designers** are graduates of a four-year college course in engineering design graphics who can become certified as technologists.

Computer Graphics

Industry, in particular, is embracing computer graphics as a way to improve drafters' productivity. **However, the use of computer graphics systems does not lessen the need for a knowledge of graphics. The principles of graphics remain the same; only the medium—the computer—is different (Fig. 1.24).**

1. Write a report that outlines the specific duties of and relationships among the scientist, engineer, technologist, technician, craftsperson, designer, and stylist in an engineering field of your choice. For example, explain this relationship for an engineering team involved in an aspect of civil engineering. Your report should be supported by factual information obtained from interviews, brochures, or library references.

2. Investigate and write a report on the employment opportunities, job requirements, professional challenges, and activities of your chosen branch of engineering or technology. Illustrate this report with charts and graphs where possible for easy interpretation. Compare your personal abilities and interests with those required by the profession.

3. Arrange a personal interview with a practicing engineer, technologist, or technician in your field of interest. Discuss with that person the general duties and responsibilities of the position to gain a better understanding of this field. Summarize your interview in a written report.

4. Write to the professional society in your field of study for information about it. Prepare a notebook of these materials for easy reference. Include in the notebook a list of books that provide career information for that field.

Addresses of Professional Societies

Publications and information from these societies were used in preparing this chapter.

American Ceramic Society
65 Ceramic Drive, Columbus, OH 43214

The American Institute of Aeronautics and Astronautics
1290 Avenue of the Americas, New York, NY 10019

American Institute of Chemical Engineers
345 East 47th Street, New York, NY 10017

American Institute for Design and Drafting
3119 Price Road, Bartlesville, OK 74003

The American Institute of Industrial Engineers
345 East 47th Street, New York, NY 10017

American Institute of Mining, Metallurgical, and Petroleum Engineering
345 East 47th Street, New York, NY 10017

American Nuclear Society
244A East Ogden Avenue, Hinsdale, IL 60521

American Society of Agricultural Engineers
2950 Niles Road, St. Joseph, MI 49085

American Society of Civil Engineers
345 East 47th Street, New York, NY 10017

American Society for Engineering Education
11 DuPont Circle, Suite 200, Washington, DC 20036

American Society of Mechanical Engineers
345 East 47th Street, New York, NY 10017

The Institute of Electrical and Electronic Engineers
345 East 47th Street, New York, NY 10017

National Society of Professional Engineers
2029 K Street, N.W., Washington, DC 20006

Society of Petroleum Engineers (SPE)
P.O. Box 833836, Richardson, TX 75083

Society of Women Engineers
United Engineering Center, Room 305
345 East 47th Street, New York, NY 10017

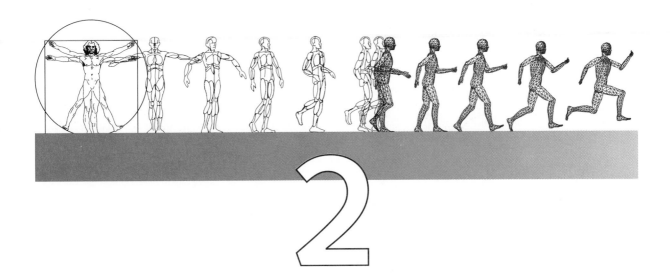

2

The Design Process

2.1 Introduction

The design process is a way of devising innovative solutions to problems that will result in new products or systems. Engineering graphics and descriptive geometry are excellent tools for developing designs from initial concepts to final working drawings. Initially, a design consists of sketches and ultimately becomes precise detail drawings and specifications. They, in turn, become part of the contract documents for the parties involved in funding and constructing a project.

At first glance, the solution of a design problem may appear to involve merely the identification of a need and the application of effort toward its solution, but most engineering designs are more complex than that. The engineering and design efforts may be the easiest parts of a project.

For example, engineers who develop roadway systems must deal with constraints such as ordinances, historical data, human factors, social considerations, scientific principles, budgeting, and politics **(Fig. 2.1)**. Engineers can readily design driving surfaces, drainage systems, overpasses, and

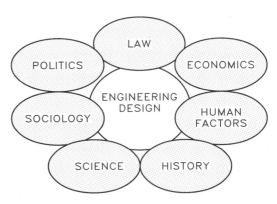

Figure 2.1 An engineering project may involve the interaction of people representing many professions and interests, with engineering design as the central function.

other components of the system. However, adherence to budgetary limitations is essential, and funding is closely related to politics.

Traffic laws, zoning ordinances, environmental impact statements, right-of-way acquisition, and liability clearances are legal aspects of road-

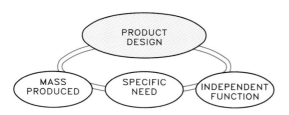

Figure 2.2 Product design seeks to develop a product that meets a specific need, that can function independently, and that can be mass produced.

way design that engineers must deal with. Past trends, historical data, human factors (including driver characteristics), and safety features affecting the function of the traffic system must be analyzed. Social problems may arise if proposed roadways will be heavily traveled and attract commercial development such as shopping centers, fast-food outlets, and service stations. Finally, designers must apply engineering principles developed through research and experience to obtain durable roads, economical bridges, and fully functional systems.

2.2 Types of Design Problems

Most design problems fall into one of two categories: product design and systems design.

Product Design

Product design is the creation, testing, and manufacture of an item that usually will be mass produced, such as an appliance, a tool, or a toy (Fig. 2.2). In general, a product must have sufficiently broad appeal for meeting a specific need and performing an independent function to warrant its production in quantity. Designers of products must respond to current market needs, production costs, function, sales, distribution methods, and profit predictions **(Fig. 2.3)**.

Products can perform one or many functions. For instance, the primary function of an automo-

Figure 2.3 These factors affect product design.

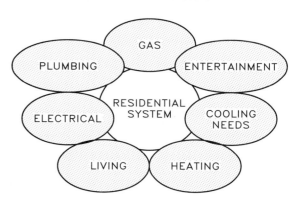

Figure 2.4 The typical residence is a system composed of many components and products.

bile is to provide transportation, but it also contains products that provide communications, illumination, comfort, and safety. Because it is mass produced for a large consumer market and can be purchased as a unit, the automobile is regarded as a product. However, because it consists of many products that perform various functions, the automobile also is a system.

Figure 2.5 The cart for carrying luggage in an airport terminal is a product and part of a system. (Courtesy of Smarte Carte.)

Figure 2.6 Luggage carts are dispensed from coin-operated centers at major entrances of an airport terminal. (Courtesy of Smarte Carte.)

Figure 2.7 The coin-operated gate releases a cart to the customer. (Courtesy of Smarte Carte.)

Systems Design

Systems design combines products and their components into a unique arrangement and provides a method for their operation. A residential building is a system of products consisting of heating and cooling, plumbing, natural gas, electrical power, and others that together form the overall system **(Fig. 2.4)**.

Systems Design Example

Suppose that you were carrying luggage to a faraway gate in an airport terminal. You would soon recognize the need for a luggage cart that you could use and then leave behind for others to use. If you have this need, then others do, too. You could design a cart like the one shown in **Fig. 2.5** to hold luggage—and even a child—and make it available to travelers. The cart is a product.

How could you profit from providing such a cart? First you would need a method of holding the carts and dispensing them to customers, such as

the one shown in **Fig. 2.6**. You would also need a method for users to pay for cart rental, so you could design a coin-operated gate for releasing them **(Fig. 2.7)**.

For customer convenience you could provide a vending machine that will take bills both to make change and to issue cards entitling customers to multiple use of carts at this and other terminals

Figure 2.8 A vending machine makes change for $5, $10, and $20 bills and issues baggage cards that can be used at any other airport terminal having the same cart system. (Courtesy of Smarte Carte.)

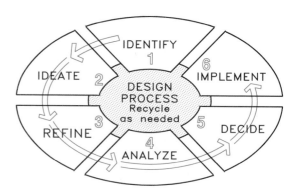

Figure 2.9 The design process consists of six steps. Each step can be repeated as necessary.

(Fig. 2.8). A mechanism to encourage customers to return carts to another, conveniently located dispensing unit would be helpful and efficient. A coin dispenser (see Fig. 2.6) that gives a partial refund on the rental fee to customers returning carts to a dispensing unit at their destination might work.

In combination, these products make up a system. Such a system of products is more valuable than the sum of the products alone.

2.3 The Design Process

Design is the process of creating a product or system to satisfy a set of requirements that has multiple solutions by using any available resources. In essentially all cases, the final design must be completed at a profit or within a budget.

The six steps of the design process **(Fig. 2.9)** are

1. problem identification,
2. preliminary ideas (ideation),
3. refinement,
4. analysis,
5. decision, and
6. implementation.

Designers should work sequentially from step to step but should review previous steps periodically and rework them if a new approach comes to mind during the process.

Problem Identification

Most engineering problems are not clearly defined at the outset and require identification before an attempt is made to solve them **(Fig. 2.10)**. For example, air pollution is a concern, but we must identify its causes before we can solve the problem. Is it caused by automobiles, factories, atmospheric conditions that harbor impurities, or geographic features that trap impure atmospheres?

Another example is traffic congestion. When you enter a street where traffic is unusually congested, can you identify the reasons for the congestion? Are there too many cars? Are the signals poorly synchronized? Are there visual obstructions? Has an accident blocked traffic?

Problem identification involves much more than simply stating, "We need to eliminate air pollution." We need data of several types: opinion surveys, historical records, personal observations,

PROBLEM IDENTIFICATION

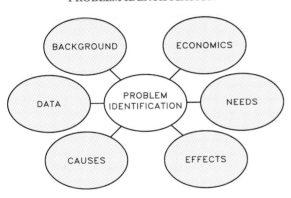

Figure 2.10 Problem identification requires that the designer accumulate as much information about a problem as possible before attempting a solution. The designer also should keep product marketing in mind.

REFINEMENT

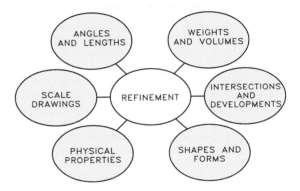

Figure 2.12 Refinement begins with the construction of scale drawings of the best preliminary ideas. Descriptive geometry and graphical methods are used to describe geometric characteristics.

PRELIMINARY IDEAS

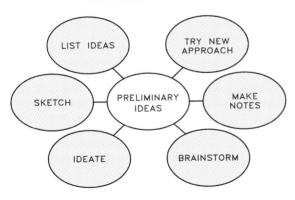

Figure 2.11 The designer develops preliminary ideas after completing the identification step. The designer should list and sketch all these ideas to have a broad selection to work from.

experimental data, and physical measurements from the field.

Preliminary Ideas
The second step of the design process is the development of as many ideas for problem solution as possible (**Fig. 2.11**). A brainstorming session is a good way to collect ideas that are bold, revolutionary, and even wild. Rough sketches, notes, and comments can capture and preserve preliminary ideas for further refinement. The more ideas, the better at this stage.

Refinement
Next, several of the better preliminary ideas can be selected for refinement to determine their merits. Rough sketches should be converted to scale drawings for spatial analysis, determination of critical measurements, and calculation of areas and volumes affecting the design (**Fig. 2.12**). Descriptive geometry aids in determining spatial relationships, angles between planes, lengths of structural members, intersections of surfaces and planes, and other geometric relationships.

Analysis
Analysis is the step at which engineering and scientific principles are used most intensively to evaluate the best designs and compare their merits with respect to cost, strength, function, and market appeal (**Fig. 2.13**). Graphical methods play an important role in analysis. Data can be analyzed graphically, forces analyzed by graphical vectors,

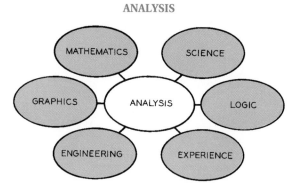

Figure 2.13 The designer should use all available methods, from science to technology to graphics to experience, to analyze designs.

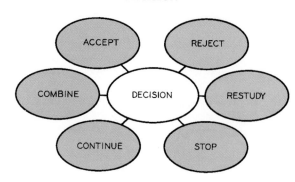

Figure 2.14 Decision involves selecting the best design or design features to be implemented. This step may require a compromise or a rejection of the proposed solution.

and empirical data can be analyzed, integrated, and differentiated by other graphical methods.

Decision

After analysis, a single design, which may be a compromise among several designs, is selected as the solution to the problem **(Fig. 2.14)**. The designer alone, or a team, may make the decision. The outstanding aspects of each design usually lend themselves to graphical comparisons of manufacturing costs, weights, operational characteristics, and other data essential in decision making.

Implementation

The final design must be described in detail in working drawings and specifications from which the project will be built, whether it is a computer chip or a suspension bridge **(Fig. 2.15)**. Workers must have precise instructions for the manufacture of each component, often measured within thousandths of an inch to ensure proper fabrication assembly. Working drawings must be sufficiently explicit to serve as part of the legal contract with the successful bidder on the job.

2.4 Application of the Design Process

The following example illustrates the application of the design process to a simple problem.

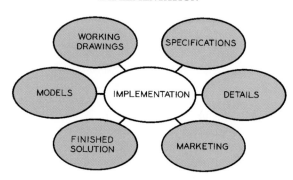

Figure 2.15 Implementation is the preparation of drawings and specifications from which the final product can be made. The product is produced and its marketing is begun.

Swing Set Anchor Problem

A child's swing set can become unstable and overturn from the forces of the swinging motion. This hazardous condition is worsened when two or three swings in the set swing in unison. Design a product that eliminates this hazard and has market appeal to swing set owners.

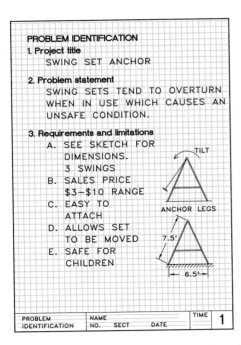

Figure 2.16 This worksheet shows aspects of problem identification for the swing set problem.

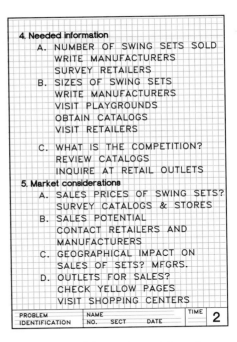

Figure 2.17 This worksheet shows information needed before the designer can proceed, including sources from which it can be obtained.

Problem Identification

First, write a statement of the problem and a statement of need (**Fig. 2.16**). List limitations and desirable features and make descriptive sketches to better identify the requirements (**Fig. 2.17**). Even if much of this information may be obvious, writing statements and making sketches about the problem will help you "warm up" to the problem and begin the creative process. Also, you must begin thinking about sales outlets and marketing methods.

Preliminary Ideas

Brainstorm the problem and possible solutions with others. List the ideas obtained on a worksheet (**Fig. 2.18**), and summarize the best ideas and design features on a separate worksheet (**Fig. 2.19**). Then translate and expand these verbal ideas into rapidly drawn freehand sketches (**Fig.**

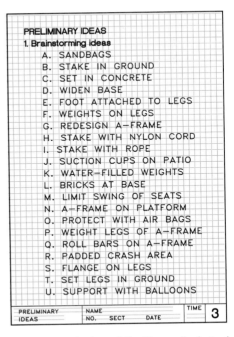

Figure 2.18 A recorder lists all the ideas voiced at a brainstorming session.

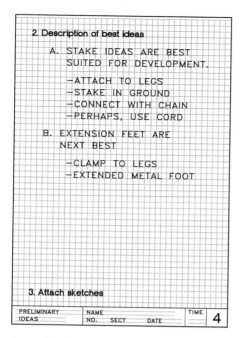

Figure 2.19 The designer selects the best ideas from the brainstorming session to be developed as preliminary ideas.

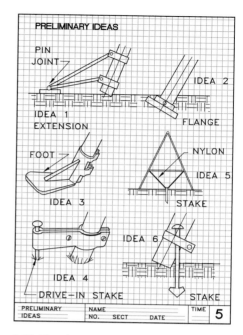

Figure 2.20 The designer sketches preliminary ideas and makes brief notes for problem solution. This step is the most creative part of the design process.

2.20). You should attempt to develop as many ideas as possible because a large number of ideas represents a high level of creativity.

Problem Refinement

Describe the design features of one or more preliminary ideas on a worksheet for comparison (**Fig. 2.21**). Draw the better designs to scale in preparation for analysis; you need to show only a few dimensions at this stage (**Fig. 2.22**).

Use instrument-drawn orthographic projections, computer drawings, and descriptive geometry to refine the designs and ensure precision. Let's say that you select ideas 3 and 5 for analysis. Figure 2.22 depicts orthographic and auxiliary views of the two designs.

Analysis

Use an analysis worksheet to analyze the tubular stake design (**Figs. 2.23–2.26**). If you are considering more than one solution, analyze each design.

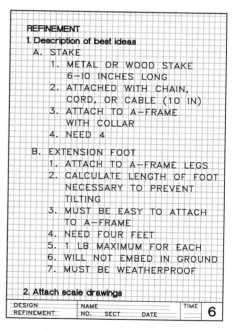

Figure 2.21 Refining preliminary ideas begins with written descriptions of the ideas selected.

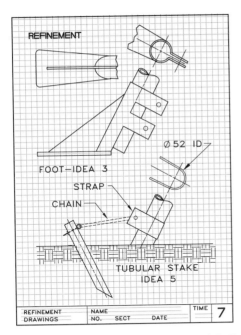

REFINEMENT

Ø 52 ID

FOOT—IDEA 3

STRAP

CHAIN

TUBULAR STAKE
IDEA 5

REFINEMENT DRAWINGS	NAME NO. SECT DATE	TIME	7

Figure 2.22 Scaled drawings refine ideas 3 and 5. Few, if any, dimensions are needed.

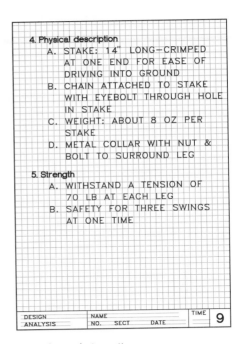

4. Physical description
   A. STAKE: 14" LONG—CRIMPED
      AT ONE END FOR EASE OF
      DRIVING INTO GROUND
   B. CHAIN ATTACHED TO STAKE
      WITH EYEBOLT THROUGH HOLE
      IN STAKE
   C. WEIGHT: ABOUT 8 OZ PER
      STAKE
   D. METAL COLLAR WITH NUT &
      BOLT TO SURROUND LEG

5. Strength
   A. WITHSTAND A TENSION OF
      70 LB AT EACH LEG
   B. SAFETY FOR THREE SWINGS
      AT ONE TIME

DESIGN ANALYSIS	NAME NO. SECT DATE	TIME	9

Figure 2.24 The analysis continues.

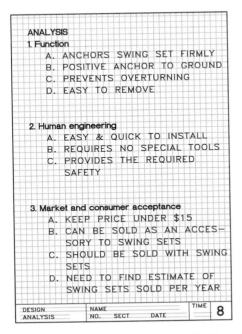

ANALYSIS
1. Function
   A. ANCHORS SWING SET FIRMLY
   B. POSITIVE ANCHOR TO GROUND
   C. PREVENTS OVERTURNING
   D. EASY TO REMOVE

2. Human engineering
   A. EASY & QUICK TO INSTALL
   B. REQUIRES NO SPECIAL TOOLS
   C. PROVIDES THE REQUIRED
      SAFETY

3. Market and consumer acceptance
   A. KEEP PRICE UNDER $15
   B. CAN BE SOLD AS AN ACCES—
      SORY TO SWING SETS
   C. SHOULD BE SOLD WITH SWING
      SETS
   D. NEED TO FIND ESTIMATE OF
      SWING SETS SOLD PER YEAR

DESIGN ANALYSIS	NAME NO. SECT DATE	TIME	8

Figure 2.23 An analysis worksheet shows the important design aspects that must be considered (for idea 5).

6. Production procedures
   A. USE 15 GAUGE GALVANIZED
      IRON PIPE (Ø40 OD) STAKE.
      CUT ON DIAGONAL TO CRIMP
      AND POINT STAKE
   B. 15 GAUGE METAL COLLAR TO
      BE FORMED INTO U—SHAPE &
      DRILLED FOR Ø5 BOLT
   C. STAKE DRILLED Ø8 FOR Ø5
      BOLT FOR ATTACHING CHAIN

7. Economic analysis
   A. COSTS
      1. CHAIN $0.40
      2. STAKE   .20
      3. COLLAR   .20 1.00
      4. EYEBOLT & NUT .10
      5. COLLAR BOLT .10
   B. LABOR   .70
   C. PACKAGING   .30
   D. PROFIT   2.00
   E. WHOLESALE PRICE 6.00
   F. RETAIL PRICE $10.00

DESIGN ANALYSIS	NAME NO. SECT DATE	TIME	10

Figure 2.25 The analysis continues focusing on production and economic considerations.

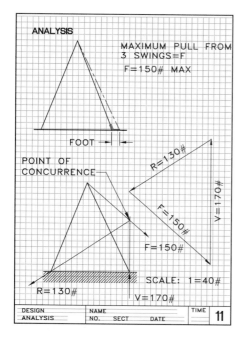

Figure 2.26 The designer uses descriptive geometry and graphics to analyze a solution. Here, reaction *R* is obtained graphically.

DECISION

1. Decision table for evaluation

DESIGN 1: STAKE AND CHAIN
DESIGN 2: EXTENSION FOOT
DEISGN 3:
DESIGN 4:
DESIGN 5:

MAX	FACTORS	1	2	3	4	5
2	FUNCTION	2	1			
2	HUMAN FACT.	1.5	1.5			
1	MARKET ANAL	.5	.2			
1	STRENGTH	1	1			
1	PRODUCTION	.5	.5			
1	COST	.5	.2			
2	PROFITABILITY	1.5	1			
0	APPEARANCE	0	0			
10	TOTALS	8	5.9			

DESIGN DECISION | NAME | NO. SECT DATE | TIME | 12

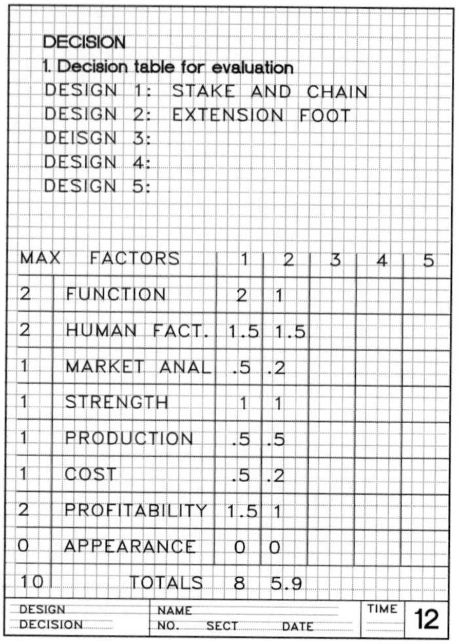

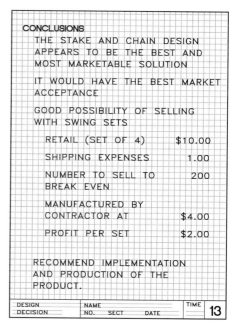

Figure 2.27 A decision table is used to evaluate alternative designs.

Measure or estimate Force *F* at the critical angle as the basis of a vector polygon to determine the magnitude of reaction *R* at the base that must be resisted by the anchor **(Fig. 2.26)**. Again, you are using graphics as a design tool.

Decision

The decision table **(Fig. 2.27)** compares two designs: the stake and the extension foot. Weight the factors to be analyzed by assigning points to them that add up to 10. You can then rank the designs by their overall scores from highest to lowest.

Draw conclusions and summarize the features of the recommended design, along with a projection of its marketability **(Fig. 2.28)**. In this case, let's say that you decide to implement the stake and chain solution (idea 5).

CONCLUSIONS

THE STAKE AND CHAIN DESIGN APPEARS TO BE THE BEST AND MOST MARKETABLE SOLUTION

IT WOULD HAVE THE BEST MARKET ACCEPTANCE

GOOD POSSIBILITY OF SELLING WITH SWING SETS

RETAIL (SET OF 4)	$10.00
SHIPPING EXPENSES	1.00
NUMBER TO SELL TO BREAK EVEN	200
MANUFACTURED BY CONTRACTOR AT	$4.00
PROFIT PER SET	$2.00

RECOMMEND IMPLEMENTATION AND PRODUCTION OF THE PRODUCT.

DESIGN DECISION | NAME | NO. SECT DATE | TIME | 13

Figure 2.28 The designer summarizes the final decision, other conclusions, and recommendations.

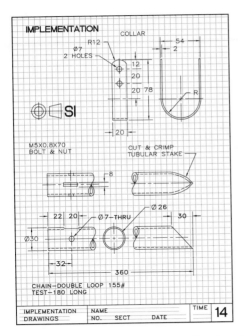

Figure 2.30 The completed swing set anchor is ready to be marketed.

Figure 2.29 The designer makes a working drawing of the design to be implemented.

Implementation

Detail the tubular stake design in a working drawing, graphically describing and dimensioning each individual part (**Fig. 2.29**).

Use notes to specify standard parts—the nuts, bolts, and chain—but do not draw them because they will be bought, not specially fabricated. With this working drawing—and assembly drawing to show how the parts are to be assembled—you now need to build a prototype or model of the product.

Figure 2.30 shows the final product. After determining how to package it for distribution, your next task is to determine how to market the product.

Problems

Most of the following problems are to be solved on 8½-by-11-inch paper, using instruments or drawing freehand as specified. Use paper with a 0.20-inch printed grid or plain paper and an engineer's scale to lay out the problems. The grid provides intervals that can be counted and transferred to other grid paper or scaled onto plain paper.

Endorse each problem sheet with your name and file number, date, and problem number. Letter all points, lines, and planes with ⅛-in. letters, using guidelines.

Letter the answers to essay problems with approved, single-stroke Gothic lettering. (See Chapter 11.)

1. Outline your plan of activities for the weekend. Indicate aspects of your plans that you feel display creativity or imagination. Explain why.

2. Write a short report (not to exceed two pages) on the engineer or engineering achievement that you believe exhibits a high degree of creativity. Justify your selection by outlining the creative aspects of your choice.

3. Test your ability to recognize the need for new designs. List as many improvements as you can think of for the typical automobile. Make suggestions for implementing these improvements. Follow this same procedure for another product of your choice.

4. List as many systems as you can that affect your daily life. Separate several of these systems into their components (subsystems).

5. Subdivide the following items into components: (a) a classroom, (b) a wristwatch, (c) a movie theater, (d) an electric motor, (e) a coffee percolator, (f) a golf course, (g) a service station, and (h) a bridge.

6. Indicate which of the items in Problem 5 are systems and which are products. Explain your answers.

7. Make a list of products and systems that you believe would be necessary for life on the moon.

8. Assume that you have been assigned the responsibility for organizing and designing a skateboard installation on your campus. It must be a self-supporting enterprise. Write a paragraph on each of the six steps of the design process, explaining how you would apply each step to the problem. For example, what action would you take to identify the problem?

9. Suppose that you are responsible for designing a motorized wheelbarrow to be marketed for home use. Write a paragraph on each of the six steps of the design process, explaining how you would apply each step to

the problem. For example, what action would you take to identify the problem?

10. List and explain a sequence of steps that you believe would be adequate for, yet different from, the design process given in this chapter. Your version of the design process may contain any of the steps discussed.

11. Design a simple device for holding a fishing pole in a fishing position while the person fishing rows the boat. Make sketches and notes to describe your design.

12. Design a doorstop that could be used to keep a door from slamming into the wall behind it. Make sketches and notes as necessary to show that you proceeded through the six design steps, labeling each step. Your work should be entirely freehand and rapid. Do not spend more than thirty minutes on this problem. Indicate any information you would need at the decision and implementation steps that you may not have now.

13. List factors that you must consider during the problem identification step of a design project for (a) a skillet, (b) a bicycle lock, (c) a handle for a piece of luggage, (d) an escape from prison, (e) a child's toy, (f) a stadium seat, (g) a desk lamp, (h) an umbrella, and (i) a hot dog stand.

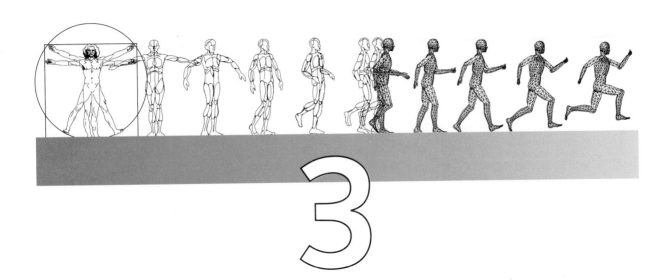

3

Problem Identification

3.1 Introduction

Problem identification is the initial step that a designer takes to solve a problem. It first involves recognizing a need and then proposing design criteria (Fig. 3.1).

Recognizing a need may begin with observation of a problem with or a defect in a product or a system that needs to be corrected. The need may be for an improved automobile safety belt, a system allowing full wheelchair access to public buildings, or a new exercise apparatus. The solution may be a product or a system improvement so that it will be more reliable and perhaps more profitable.

Proposing design criteria follows need recognition. Here the designer proposes the specifications a new product or system must meet.

3.2 Example: Preventing Gutter Damage

When you use a ladder for house repairs, it works well enough when placed against the roof or fascia at the eaves of the house (**Fig. 3.2**). However, if a

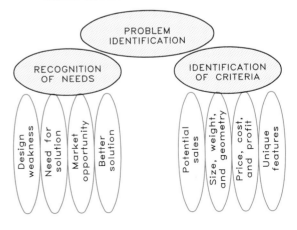

Figure 3.1 Problem identification has two aspects.

gutter is attached to the fascia, the ladder damages the gutter when placed against it. The need is to prevent gutter damage by the ladder.

Now, you must propose design criteria before proceeding with a solution. You need to know the geometry and dimensions of typical gutters, pitch-

27

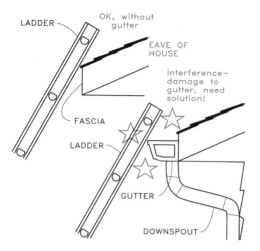

Figure 3.2 A ladder placed against the fascia or roof works fine, but when a gutter is attached to the fascia, the ladder damages the gutter. This problem needs a solution.

es of roofs, angles of ladder placement, and the potential market for the solution.

Later in the design process, you might design a product that attaches to the ladder, allowing it to clear the gutter (**Fig. 3.3**). Other solutions are possible, but they all begin with recognizing the need.

3.3 The Identification Process

Problem identification requires the designer to determine requirements, limitations, and other background information without becoming involved in problem solution. The designer should take the following steps during problem identification (**Fig. 3.4**).

1. **Problem statement.** Describe the problem to begin the thinking process.

2. **Requirements.** List the conditions that the design must meet. Most will be questions to be answered after data are gathered.

3. **Limitations.** List the factors affecting design specifications, such as maximum weight or size.

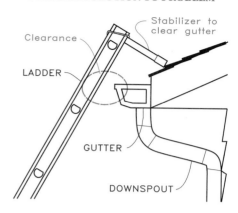

Figure 3.3 The problem recognized in Fig. 3.2 can be solved with a stabilizer attached to the ladder, which helps the ladder clear the gutter.

STEPS OF PROBLEM IDENTIFICATION

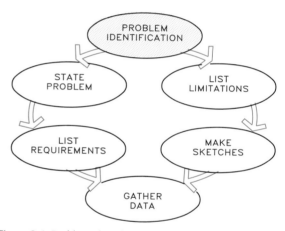

Figure 3.4 Problem identification consists of five steps.

4. **Sketches.** Make sketches and add notes and dimensions to identify geometric and physical characteristics.

5. **Data collection.** Gather data on population trends, related designs, physical characteristics, sales records, and market studies. Graph the data for interpretation, as in **Fig. 3.5**, for easier interpretation than the raw numbers or data tabulation permit.

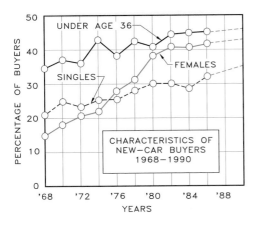

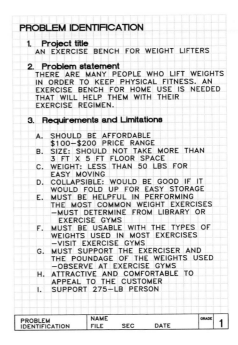

Figure 3.5 Data gathered to identify the changing characteristics of new-car buyers can be interpreted more easily when it has been graphed.

3.4 Design Worksheets

Designers must make numerous notes and sketches on worksheets throughout the design process. They serve to document what has been done and allow periodic review of earlier ideas to avoid overlooking previously identified concepts. Moreover, a written and visual record of design work helps establish ownership of patentable ideas.

The following materials aid in maintaining permanent records of design activities.

1. **Worksheets** (8½ by 11 inches). Sheets can be either grid-lined or plain and should be three-hole punched for a notebook or binder.

2. **Pencils.** A medium-grade pencil (F or HB) is adequate for most purposes.

3. **Binder or envelope.** Keep worksheets in a binder or envelope for reference.

3.5 Example: Exercise Bench

Design an exercise bench that can be used by those who lift weights for body fitness. This apparatus should be as versatile as possible and at the low end of the price range of exercise equipment.

Figure 3.6 Part of the problem identification step is to give the project a title, to describe what the product will be used for, and to list basic requirements and limitations for the design.

The contents of worksheets shown in **Figs. 3.6–3.9** identify the problem in a typical manner.

First, give the title of the project and a brief problem statement to describe the problem better. Then list requirements and limitations and add sketches as necessary. **You will have to list some requirements as questions for the time being, but in all cases make estimates and give sources for the answers (Fig. 3.6).** Make estimates as you go along; for example, it must cost between $100 and $200; it must weigh less than 50 pounds; and it must at least support a person weighing 250 pounds. Give estimates as ranges of prices or weights rather than exact numbers. Use catalogs offering similar products as sources for prices, weights, and sizes.

Next, make a list of the questions that need answers. **Follow each question with a source for its answer and give a preliminary answer.** How many people exercise? What are their ages? Do they buy exercise equipment? What are the most

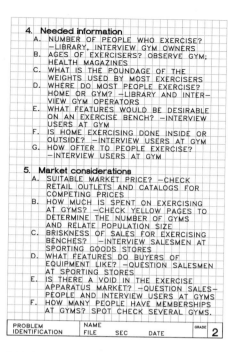

Figure 3.7 Continuation of problem identification for the exercise bench includes gathering information on market potential.

PROBLEM IDENTIFICATION
A. CHECK YELLOW PAGES FOR GYMS
  –8 GYMS FOR POPULATION OF 100,000
  –ONE GYM PER 12,500 PEOPLE
B. GYM MEMBERSHIP FOR STILLMAN'S GYM

YEAR	MEN	WOMEN
1984	75	20
1986	70	36
1988	110	70
1990	175	82
1992	180	120
1994	210	135

C. INTERVIEW RETAILERS OF EQUIPMENT
  –3 DEALERS POSITIVE ABOUT BENCH
  –2 DEALERS NEUTRAL
  –1 DEALER NEGATIVE ABOUT PROSPECTS

D. AGES OF EXERCISERS (BY OBSERVATION)
  –20% UNDER 20
  –40% BETWEEN 20 AND 30
  –30% BETWEEN 30 AND 50
  –10% OVER 50

E. PRICES AT STORES
  1. COMPLEX MULTI–USE $1000
  2. MEDIUM–RANGE EQUIPMENT 700
  3. LIGHTWEIGHT BENCHES   130
  4. WEIGHTS   100

F. TYPICAL BRANDS OF EQUIPMENT ON THE MARKET?
  1. WEIDER
  2. NAUTILUS
  3. BODY WONDERFUL
  4. HEALTH PLUS

Figure 3.8 This worksheet shows the data collected for use in designing the exercise bench.

popular weight exercises? You may obtain this type of information from interviews, product catalogs, the library, and sporting-goods stores. Market considerations include the average income of a typical exerciser. How much does he or she spend on physical fitness per year? The opinions of sporting-goods dealers are helpful, and they should be able to direct you to other sources of information (**Fig. 3.7**).

Record the data that you gather by interviewing gym managers, looking at catalogs, and visiting sporting-goods stores to learn about the people who exercise and the equipment they use (**Fig. 3.8**). Determine the most popular weight exercises of those for whom this bench is to be designed by surveying users, coaches, and sporting-goods outlets. Sketch those exercises on a worksheet (**Fig. 3.9**).

Think about costs and pricing the product even during problem identification. **Figure 3.10** shows how products may be priced from wholesale to

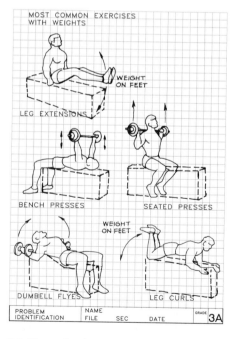

Figure 3.9 These sketches are of the most popular weight exercises.

ECONOMIC ANALYSIS MODEL

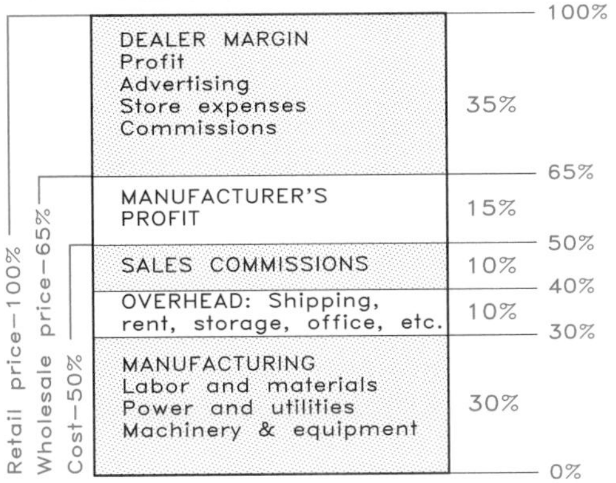

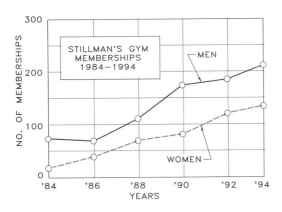

Figure 3.10 This chart shows the breakdown of costs involved in the retail price of a product.

DATA GRAPHED FOR INTERPRETATION

STILLMAN'S GYM
MEMBERSHIPS
1984–1994

Figure 3.11 This graph describes visually the trends in potential customers for the exercise apparatus.

retail, but these percentages vary by product. For instance, profit margins are smaller for food sales than furniture sales. If an item retails for $50, the production and overhead cost cannot be more than about $20 to maintain the necessary margins.

Data are easier to interpret if presented graphically (**Fig. 3.11**). Thorough problem identification includes graphs, sketches, and schematics that improve the communication of your findings.

STEPS OF SCHEDULING

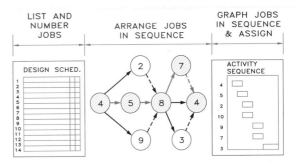

Figure 3.12 Planning and scheduling a project involves three steps.

Problem identification is not complete at this point. However, this example should give you a basic understanding of the process.

3.6 Organization of Effort

The designer should prepare a schedule of required design activities after completing problem identification. **The project evaluation and review technique (PERT), developed by project managers for projects requiring coordination of many activities, aids in scheduling tasks.** PERT is a means of scheduling activities in sequence and reviewing progress made toward their completion.

The critical path method (CPM) of scheduling evolved from PERT and is used with it. The tasks that must be completed before others can begin are critical tasks and should be given priority. The critical path is the sequence of tasks requiring the longest time from start to finish and has the least flexibility. Activities not in the critical path receive less emphasis.

3.7 Planning Design Activities

The chart in **Fig. 3.12** illustrates the steps involved in planning a job (project): (1) list the tasks on a form called a Design Schedule and Progress Record; (2) prepare an Activities Network Diagram,

![Design Schedule & Progress Record](Design Schedule and Progress Record chart)

Design Schedule & Progress Record

TEAM __7__ PROJECT __PRODUCT DESIGN__
WORK PERIODS __11__ MAN-HOURS __70__ FINISHING DATE __4-11__ SECTION __501__

JOB	ASSIGNMENT	ESTIMD HOURS	ACTUAL HOURS	PER CENT COMPLETE 0 50 100
1	WRITE REPORT-ALL	15		
2	BRAINSTORM-ALL	.5	1	
3	WRITE MFGRS-JHE	2		
4	GRAPH DATA-HLE	2		
5	MARKET ANAL-DL	3	4	
6	COLLECT DATA-TT	2		

Figure 3.13 The Design Schedule and Progress Record (DS&PR) shows typical entries for a product design project.

ACTIVITIES NETWORK

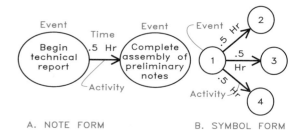

A. NOTE FORM   B. SYMBOL FORM

Figure 3.14 Two ways of preparing an Activities Network are (A) note form and (B) symbol form, both of which graphically arrange activities in sequence.

arranging the tasks in sequence; and (3) prepare an Activity Sequence Chart that graphs the tasks in the sequence shown in the network.

Design Schedule and Progress Record

The designer separates the job into tasks, numbers them, and lists them in the **Design Schedule and Progress Record (DS&PR)** without concern for their sequence (**Fig. 3.13**). Next, the designer estimates the amount of time required for each task and enters it in the third column. The designer adjusts the amount of time for each task so that the total matches the time allotted for the job.

Activities Network

The designer prepares an **Activities Network (AN)** by arranging the tasks from the DS&PR in their proper sequence in either note or symbol form. The note form identifies activities by task name (**Fig. 3.14**); the symbol form identifies activities by task number. Arrows connect activities, showing their sequence. Labels on the arrows indicate the amount of time needed to complete the activities. For example, the first activity in preparing the technical report is to assemble preliminary notes, which requires 0.5 hour (**Fig. 3. 14A**).

Dummy activities simply indicate connections between activities that involve no work and no time, because activities must be connected in the

THE CRITICAL PATH

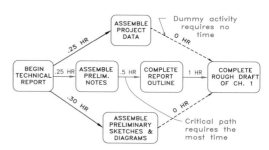

Figure 3.15 The critical path is the path from start to end of the project that requires the most time; shown here for a portion of the product design project.

network. The two dummy activities (dashed lines) in **Fig. 3.15** show sequential connections but no expenditure of time. In other words, the project data and the preliminary sketches and diagrams that were assembled were not needed to complete the draft of Chapter 1 but are available for later use in the project.

The critical path identified in the Activities Network (**Fig. 3.15**) is the longest time path from the project's beginning to its end. For the partial Activities Network shown, 1.75 hours are required to complete the draft of Chapter 1 of the technical report.

Activity Sequence Chart

Next, the designer lists the activities on the **Activity Sequence Chart (ASC)** (**Fig. 3.16**), which shows task numbers and team member assignments from the DS&PR. A bar graph shows the sequence of and the time allocated to each task. For example, 0.5 hour is scheduled for brainstorming, or task 2, from the DS&PR and is to be done first. The ASC also is the overall project schedule.

As work progresses, the designer graphs the status of each activity on the DS&PR (**Fig. 3.13**). The actual hours required to complete tasks appear in column 4, for easy comparison with the original time estimates. When the extra time required

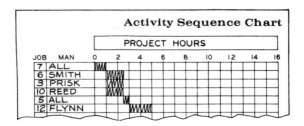

Figure 3.16 The Activity Sequence Chart (ASC) contains tasks, task assignments, and initial time estimates from the DS&PR and activity sequence from the Activity Network.

becomes greater than that scheduled, the designer has to make adjustments in the ASC.

Problems

Problems should be presented on 8½-by-11-inch paper, grid or plain. All notes, sketches, drawings, and graphs should be neat and accurate. Written matter should be typed or lettered on paper with ⅛-in. guidelines.

General

1. Identify a need for a design solution as a short class assignment requiring less than three hours to complete. Submit a proposal outlining this need and your general plan for a design solution. Limit the written proposal to two pages.

2. Assume that you were marooned on an uninhabited island with no tools or supplies. Identify the major problems that you would have to solve. List the factors that you would have to consider before attempting a solution for each problem. For example: There is the need for food. Determine (a) available sources of food on island, (b) methods of gathering/catching, (c) methods of storing a food supply, and (d) method of cooking.

3. While you were walking to class today, what irritants or discomforts did you recognize? Using worksheets, identify the cause of these problems, write a problem statement (including need recognition), and list the requirements for and limitations on solutions.

4. Repeat Problem 3 for your (a) living quarters, (b) classroom, (c) recreation facilities, (d) dining facilities, or (e) another environment with which you are familiar.

Product Design Problems

5. Assume that you are attempting to reconstruct the designer's approach to development of the tab-opening can. Even though the problem has been solved and the solution marketed, identify the need for the product and propose specifications for it. Is the existing solution the most appropriate one, or does your problem identification statement suggest others? Explain your answer.

6. Repeat Problem 5 for a travel iron for pressing clothes.

7. List the problems involved in designing a motorized wheelbarrow.

8. Suppose that you recognize the need for a device that, attached to a bicycle, would allow riding the bicycle over street curbs to sidewalk level. Identify the problems involved in determining the marketability of such a device.

9. Identify the problems you might encounter in developing a portable engineering travel kit to give engineers the capability of making engineering calculations, notes, sketches, and drawings. The kit might include a carrying case, calculator, drawing instruments, paper, reference material, and other products.

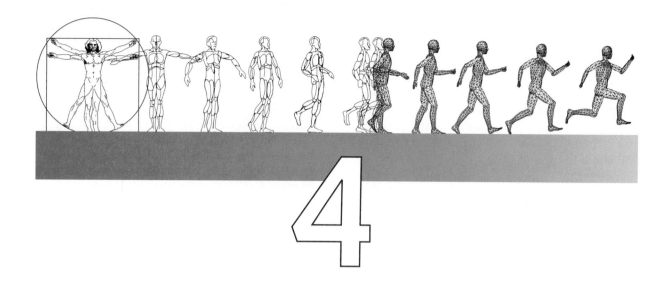

Preliminary Ideas

4.1 Introduction

Creativity is highest during the preliminary idea step of the design process because there are no limitations on being innovative, experimental, and daring. During subsequent steps of the design process, freedom of creativity diminishes and the need for information increases. **Figure 4.1** shows the relationship between creativity and information accumulation during the design process.

An example of a creative solution to a problem is the Wearable Data Terminal (**Fig. 4.2**), which utilizes electronic technology in a unique manner. The device is an optical scanner, worn on the forearm of the user, that reads bar codes. A mirror on the shoulder-mounted model reflects the data from screen to user (**Fig. 4.3**). Another advanced design is the prototype voice/data/fax mobile terminal (**Fig. 4.4**), which is being developed for trucking and transportation companies. A short time ago these ideas were considered futuristic fantasies. Progress in the future will be limited only by our imagination in applying technology.

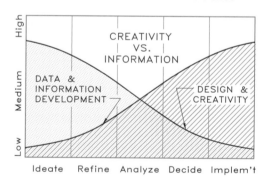

Figure 4.1 Engineering creativity is highest during the early steps of the design process (excluding the problem identification step) and information development increases during the later steps.

4.2 Individual Versus Team

Designers work both as individuals and as members of design teams.

Figure 4.2 This Wearable Data Terminal, an optical scanner that reads bar codes, is worn on the forearm and is an example of what would have been merely a fantasy several years ago. (Courtesy of NEC U.S. Communications Office.)

Figure 4.3 The Wearable Data Terminal transmits data to the screen of the shoulder unit and is reflected onto a mirror, permitting the user to see the reading. (Courtesy of NEC U.S. Communications Office.)

Figure 4.4 This prototype voice/data/fax mobile terminal operates from satellite signals and is being developed for trucking and transportation companies, utilities, and resource companies. (Courtesy of BCE, Inc.)

Individual Approach

Designers working alone must make sketches and notes to communicate first with themselves and then with others. **Their primary goal is to generate as many ideas as possible, because better ideas are more likely to come from long lists than from short lists.** Quick sketches can capture fleeting thoughts that might otherwise be lost during ideation.

Team Approach

The team approach brings diversity and a broad range of ideas to the design process, but along with it comes problems of management and coordination. Groups perform better with a leader they select to guide their activities.

Teams should alternate between individual and group work to take advantage of both approaches. For example, each member could individually develop preliminary ideas, bring them to a team meeting, compare solutions, and return to individual work with a renewed outlook.

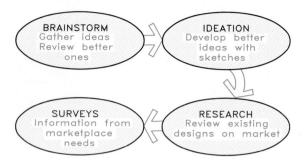

Figure 4.5 A plan of action is needed for the preliminary ideas step of the design process.

4.3 Plan of Action

The following steps are suggested for completing the preliminary ideas step of the design process: (1) hold brainstorming sessions, (2) prepare sketches and notes, (3) gather background data, and (4) conduct surveys (**Fig. 4.5**). Periodically reviewing notes and worksheets made during problem identification ensures that efforts will stay focused on the design objectives.

4.4 Brainstorming

Brainstorming is a problem-solving technique in which members of a group spontaneously contribute ideas.

Rules of Brainstorming
The guidelines for a brainstorming session are:[*]

1. **Criticism is ruled out.** Judgment of ideas must be withheld until later.

2. **"Free-wheeling"** is encouraged. The wilder the idea, the better; it is easier to tame down than to think up.

3. **Quantity is wanted.** The larger the number of ideas, the greater will be the likelihood of useful ideas.

4. **Combination and improvement are sought.** Participants should seek ways of improving the ideas of others.

Organization of a Brainstorming Session
The organization of a brainstorming session involves selecting the panel, becoming familiar with the problem, selecting a moderator and recorder, holding the session, and following up.

Panel Selection The optimum number of participants in a brainstorming session is twelve people. For diversity they should be people both with and without knowledge of the subject. Because supervisors may restrict the flow of ideas, panels should be composed of nonsupervisory professionals of similar status.

Preliminary Work An information sheet about the session should be given to panel members a couple of days before the session to allow ideas to incubate.

The Problem The problem to be brainstormed should be defined concisely. Instead of presenting the problem as "how to improve our campus," it should be presented by category, such as "how to improve student parking," or "how to improve food services."

The Moderator and Recorder The panel selects a moderator to be in charge of the session and a recorder to keep track of the ideas presented.

The Session The moderator identifies the problem and recognizes the first member holding up a hand to respond. The person responding should state an idea as briefly as possible. The moderator then recognizes the next person holding up a hand, and the process continues. A suggestion made by one member often stimulates ideas in

---

[*]From Alex F. Osborn, *Applied Imagination.* New York: Scribner, 1963, p. 156.

Figure 4.6 Designers use sketches and notes to develop preliminary ideas and communicate with themselves and others. (Courtesy of Ford Motor Company.)

Figure 4.7 Designers and stylists used sketches to develop the exterior design of automobiles such as GM's Ultralite. (Courtesy of General Motors Corporation.)

others, who hold up their hands and snap their fingers to signify that they want to "hitchhike" on the previous idea. This interaction is central to a brainstorming session.

Length A session should move at a brisk pace and end when ideas slow to an unproductive rate. An effective session can last from a few minutes to an hour, but twenty minutes is considered about the best length.

Follow-up The recorder should reproduce the list of ideas gathered during the session and distribute them to the participants. As many as a hundred ideas may be gathered during a twenty-minute session. The designer should pare them down to those having the most merit.

4.5 Sketching and Notes

Sketching is the designer's medium (Fig. 4.6). Sketching allows the designer's ideas to take form as three-dimensional pictorials or as two-dimensional views that are visual extensions of the thinking

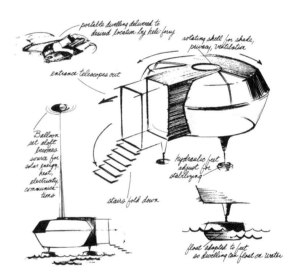

Figure 4.8 These preliminary sketches depict a Transportable Uni-Lodge, a mobile dwelling of the future. (Courtesy of Lippincott and Margulies, Inc., and Charles Bruning Company.)

process. The sketch shown in **Fig. 4.7** was one of many used to develop the styling features of an automobile.

The Transportable Uni-Lodge (**Fig. 4.8**) illus-

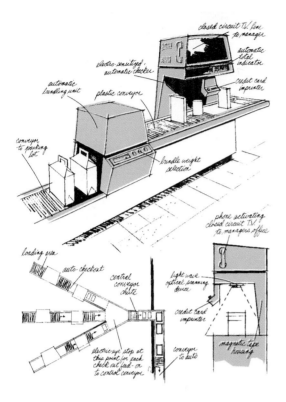

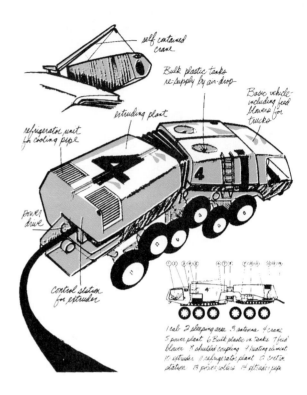

Figure 4.9 This sequence of sketches shows a concept for an automatic checkout/packaging unit for a supermarket. (Courtesy of Lester Beall, Inc., and Charles Bruning Company.)

Figure 4.10 This sketch is a designer's preliminary design of a self-contained pipelayer for laying irrigation pipe in large tracts of desert. (Courtesy of Donald Desky Associates, Inc., and Charles Bruning Company.)

trates the value of sketches in developing and communicating ideas. The Uni-Lodge can be transported by helicopter to previously unreachable areas. Its retractable legs can be equipped with pontoons, allowing it to float on water. Additional ideas are noted on the sketches.

Figure 4.9 shows a sketch of a checkout system for a grocery store. The shopper sets the machine in operation by inserting a credit card in a slot. After the card is scanned, the customer receives an order number tag, and the conveyor moves the items under a scanner and totals the prices. If a question arises, the customer can stop the machine and talk to a store employee by lifting the phone. The conveyor moves the items to a unit where they are packaged in plastic containers and marked with the customer's number. A central conveyor transports large orders to an exterior pickup point near where the customer is parked.

Another concept is a self-contained pipelayer **(Fig. 4.10)** for laying irrigation pipe in desert areas. The tractor has a cab, sleeping accommodations, radio equipment, power plant, and bulk storage tanks for plastic. The van consists of an extrusion machine, a refrigeration unit, and a control station. The pipelayer transports bulk plastic and machinery that can extrude and lay approximately two miles of pipe from each pair of storage tanks. Empty tanks are discarded and replaced by full tanks that are air-dropped to the crew.

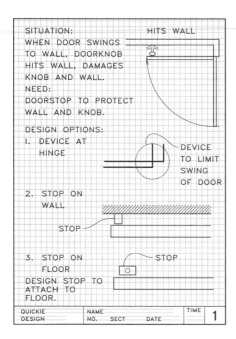

Figure 4.11 We used this worksheet of sketches and notes to identify the need for a device to prevent damage caused by a door swinging into a wall.

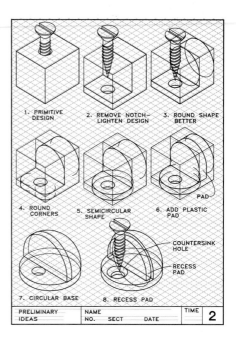

Figure 4.12 This worksheet illustrates the evolution of a doorstop design with a series of sketches.

4.6 Quickie Design

Let's assume that we recognize the need for a product to protect walls and doorknobs where doors swing into walls. We make sketches on a worksheet to identify the problem and its geometry (**Fig. 4.11**). If we decide to develop a floor-mounted doorstop, we might begin with sketches (made by hand or by computer) of a design that begins as basic block (**Fig. 4.12**). We then refine the design with more sketches until we find a suitable solution.

We make more sketches (**Fig. 4.13**) to develop the doorstop's details, which we show to others for discussion and evaluation. **Figure 4.14** shows other solutions to this problem that are products already being marketed. These products were developed in a similar manner, by designers using graphics as the medium of design.

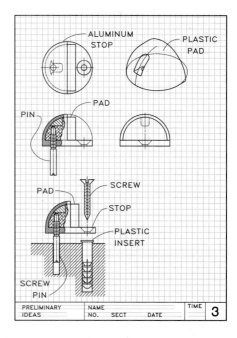

Figure 4.13 We used detailed sketches to refine the design and make it easily understandable for others.

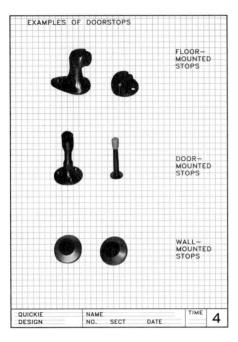

Figure 4.14 These doorstops are examples of those being marketed already.

4.7 Background Information

One way of gathering preliminary ideas is to look for products and designs that are similar to the one being considered. Sources of background information include magazines, patents, and consultants.

Magazines

Articles in both general and technical magazines often explain unique designs, complete with drawings and photographs. Advertisements in these magazines may give helpful information on materials and innovative devices.

Patents

Patents from the U.S. Patent Office (at $1.50 each) illustrate the details of patented designs. The designer can use them to understand competing products better.

Consultants

Complex projects may require specialists in manufacturing, electronics, materials, and instrumentation. Manufacturers' representatives also provide valuable assistance with projects related to their companies' products.

4.8 Opinion Surveys

Designers need to know consumers' attitudes about a new product at the preliminary design stage. Is there a need for the product? Are consumers excited about the design? Would they buy the product? What features do consumers like or dislike? What price would they willingly pay for it? What do retailers think it will sell for? Does size or color matter?

For a survey to be of value, the population for whom the product is being developed should be identified. Are they homeowners, high school males, or single women? A selected sample of the population can be surveyed by a personal interview, telephone survey, or mail questionnaire.

The Personal Interview

A personal interview survey should be organized to obtain and summarize reliable responses quickly and easily. For this reason, questions should be true–false or multiple choice.

Interviewers should introduce themselves, explain the purpose of the interview, and ask for permission to proceed. They should record responses and thank the interviewee for participating.

The Telephone Interview

If the opinions of the general public are desired, interviewers talk to people selected from the telephone book. If opinions from a certain group— say, sporting-goods retailers—are needed, the Yellow Pages provide prospective interviewees.

The Mail Questionnaire

An economical method of contacting a large number of people in many locations is by mail ques-

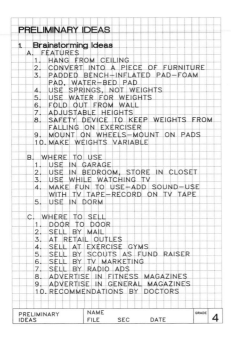

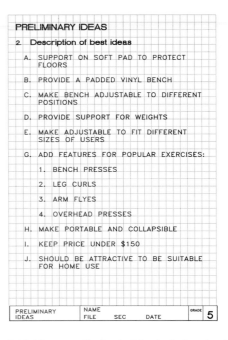

Figure 4.15 This worksheet contains a list of the brainstorming ideas for the exercise bench.

Figure 4.16 These are the best of the brainstorming ideas, selected and summarized from the first list.

tionnaire. To test the suitability of questions to be asked, the questionnaire can be mailed to a small group of people for their response. The final questionnaire should be mailed to at least three times as many people as the number of responses desired. Inserting a self-addressed, stamped envelope will increase responses.

4.9 Preliminary Ideas: Exercise Bench

We introduced the design of an exercise bench in Chapter 3. Recall that it is to be used by those who lift weights to maintain body fitness. This apparatus should be versatile and at the low end of the price range for exercise equipment.

List the brainstorming ideas obtained from a session with team members on a worksheet (**Fig. 4.15**). Then select the better ideas and list their features described on the worksheet (**Fig. 4.16**), even if they have more features than you could possibly use in a single design. The reason is to

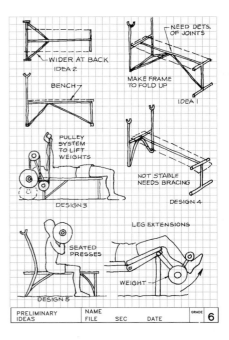

Figure 4.17 These sketches and notes illustrate preliminary ideas for the exercise bench.

make sure that you do not forget or lose any.

Using rapid freehand techniques, such as orthographic views and three-dimensional pictorial sketches, sketch ideas on worksheets. Note on the drawings any ideas or questions that come to mind while you sketch.

In **Fig. 4.17**, note the adaptation of ideas from various types of benches and exercise techniques. A number identifies each idea. Another worksheet (**Fig. 4.18**) shows other ideas and modifications of previous ideas.

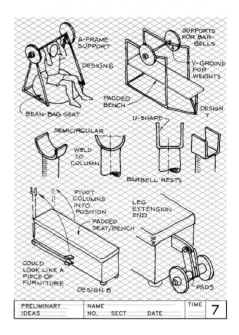

Figure 4.18 These sketches and notes describe additional design concepts for the exercise bench.

Problems

Problems should be presented on 8½-by-11-inch paper, grid or plain. All notes, sketches, drawings, and graphs should be neat and accurate. Written matter should be legibly lettered, using ⅛-in. guidelines.

1. Select one or more of the following items and list as many of their uses as possible: (a) empty vegetable cans (3-in. diameter by 5 in.), (b) two thousand sheets of 8½-by-11-inch bond paper, (c) one cubic yard of dirt, (d) three empty oil drums (24-in. DIA by 36 in.), (e) a load of egg cartons, (f) twenty-five bamboo poles (10 ft. long), (g) ten old tires, or (h) old newspapers.

2. If you were going to select an ideal team to develop an engineering solution to a problem, what characteristics would you look for? Explain.

3. What are the advantages and disadvantages of working independently on a project? Of working as a member of a design team? Explain and give examples of instances in which each approach would have the advantage.

4. Gather background information on one of the following design problems or on one of your own selection: (a) a one-person canoe, (b) a built-in car jack, (c) an automatic blackboard eraser, (d) a built-in coffee maker for an automobile, (e) a self-opening door to permit a pet to leave or enter the house, (f) an emergency fire escape for a two-story building, (g) a rain protector for people attending outdoor spectator activities, (h) a new household appliance, or (i) a home exerciser. You are concerned with costs, methods of construction, dimensions, existing products, estimates of need, and other information that will assist you in understanding the problem and its feasibility as a project. List your references and present your findings.

5. List and describe the type of consulting services required for the following design projects: (a) a zoning system for a city of twenty thousand people, (b) a shopping center, (c) a go-cart operation, (d) a water-purification facility, (e) a hydroelectric system, (f) a nuclear fallout disaster plan, (g) a processing plant for refining petroleum products, and (h) a drainage system for residential and rural areas.

6. Develop a questionnaire to determine the public's attitude toward a particular product of your selection. Explain how you would tabulate the responses to the questionnaire.

7. Organize a group of classmates and hold a brainstorming session to identify problems in need of solutions. Make a list of the ideas.

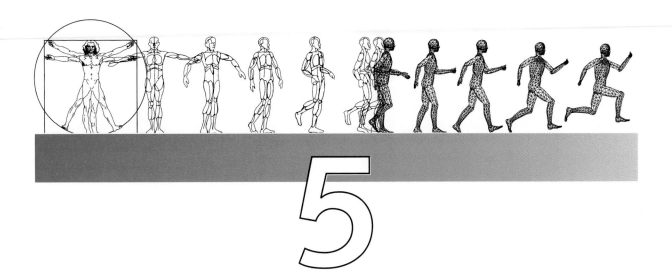

5

Idea Refinement

5.1 Introduction

Refinement of preliminary ideas is the first departure from unrestricted creativity and imagination. The designer must now give primary consideration to function and practicality.

Descriptive geometry has its greatest application in the idea refinement step of the design process. This step calls for the designer to make scale drawings with instruments to check dimensions and geometry that cannot be accurately measured in unscaled sketches.

GEOMETRY OF AUTOMOBILE SEATING

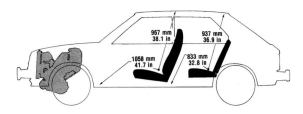

Figure 5.1 This scale drawing gives the dimensions and clearances necessary for a comfortable automobile interior. (Courtesy of Chrysler Corporation.)

5.2 Physical Properties

Important in refining an idea is determining the product's physical properties. An example of a refinement drawing is the profile of an automobile, which shows its basic geometry and seat positions **(Fig. 5.1)**.

HARDWARE AND OPERATING SYSTEM

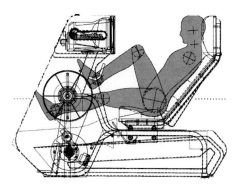

Figure 5.2 This computer drawing was made to refine the geometry of the operation of an exercise apparatus. (Courtesy of CADSHARE Resources.)

43

Figure 5.3 A computer can be used to make three-dimensional refinement drawings that can be inspected on the screen, as for this linkage system. (Courtesy of Intergraph Corporation.)

Figure 5.4 The computer drawing in Fig. 5.3 eventually was translated into a real product. (Courtesy of Intergraph Corporation.)

The computer is a significant aid to the designer. **Figure 5.2** shows a computer-generated scale drawing of an exercise machine that allows its geometry to be studied. A similar refinement drawing shows the linkage design for a front-loading scoop (**Fig. 5.3**). The design is developed and tested on the screen to understand its geometry, and then is translated into the final product (**Fig. 5.4**).

Designers used the computer to develop a three-dimensional image of the body of the HX3 experimental car (**Fig. 5.5**). Once developed, the image can be rotated on the screen to refine the design.

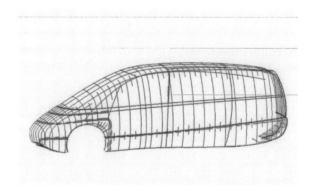

Figure 5.5 The body of this HX3 experimental car has been drawn on the computer screen for three-dimensional refinement. (Courtesy of General Motors Corporation.)

5.3 Application of Descriptive Geometry

Descriptive geometry is the study of points, lines, and surfaces in three-dimensional space, the geometric elements that comprise all forms. **Before descriptive geometry can be applied, the designer must draw orthographic views to scale, from which auxiliary views can be projected.** **Figure 5.6** shows how descriptive geometry is used to determine the clearance between a hydraulic cylinder and the fender of an automobile.

The design of a surgical light requires the application of descriptive geometry. The light fixture is to provide maximum light on the operating area (**Fig. 5.7**). This scale drawing depicts the converging beams of light emitted from the reflectors. The beams are positioned so their narrowest rays are at shoulder level to minimize shadows cast by the surgeon's shoulders, arms, and hands.

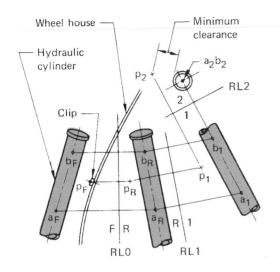

Figure 5.6 Descriptive geometry is an effective way to determine clearances between components, such as the clearance between a hydraulic cylinder and a fender. (Courtesy of General Motors Corporation.)

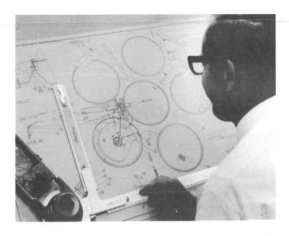

Figure 5.8 By using scale drawings developed in the refinement step of the design process, the designer can study the geometry of a surgical lamp. (Courtesy of Sybron Corporation; photo by Brad Bliss.)

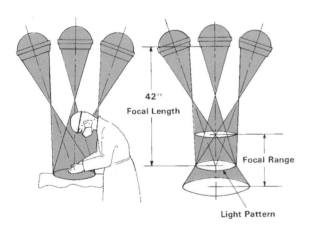

Figure 5.7 A well-designed surgical lamp emits light that passes around the surgeon's shoulders with the minimum of shadow. The focal range of this surgical lamp is between 30 and 60 inches. (Courtesy of Sybron Corporation.)

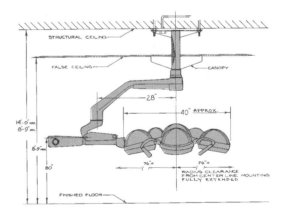

Figure 5.9 Refinement drawing shows the overall dimensions of the final design of the surgical lamp. (Courtesy of Sybron Corporation.)

From these scale drawings, lengths, angles, areas, and other geometric relations can be determined **(Fig. 5.8)**. The critical dimensions of the surgical lamp are shown in the refinement drawing in **Fig. 5.9**.

5.4 Refinement Considerations

In advanced designs, such as that of a new model automobile, numerous features must be refined. The exhaust system shown in **Fig. 5.10** is the result of the many refinement drawings needed to deter-

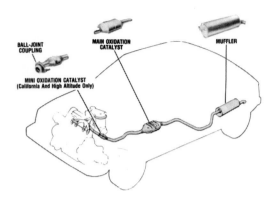

Figure 5.10 Descriptive geometry is useful in determining the lengths and angles of an automobile's exhaust system that are necessary to clear the structural members of the chassis. (Courtesy of Chrysler Corporation.)

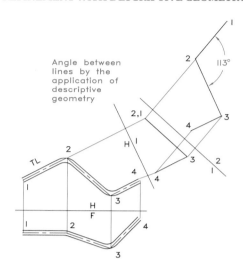

Figure 5.11 Descriptive geometry is applied to find the lengths and angles between exhaust pipe segments.

mine its geometry. Descriptive geometry was used to determine the exhaust pipe's bend angles, its length, and clearances required for fitting it to the chassis (**Fig. 5.11**).

Designers refined a command module by using orthographic drawings and notes. Orthographic views of the module show its overall size, and notes identify other features (**Fig. 5.12**).

5.5 Refinement: Exercise Bench

In Chapter 3 the exercise bench design problem was identified and in Chapter 4 preliminary ideas for it were developed. Now, in this step, we refine the preliminary ideas for the exercise bench with instrument drawings.

First, list the features to be incorporated into the design on a worksheet (**Fig. 5.13**). Then refine a preliminary idea, say, idea 2 from Fig. 4.17 in an orthographic scale drawing of the seat (**Fig. 5.14**). Block in extruded parts, such as the framework members, to expedite the drawing process and omit unneeded hidden lines.

Refinement drawings must be drawn to scale. Use of instruments is important to precisely por-

REFINEMENT DRAWING OF A SPACE CAPSULE

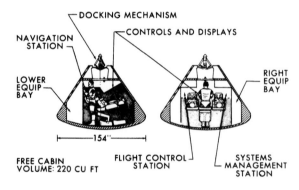

Figure 5.12 This scaled refinement drawing gives information about the orbiter vehicle. (Courtesy of NASA.)

tray the design from which angles, lengths, shapes, and other geometric elements will be obtained. The drawing shows only overall dimensions and several connecting joints are detailed to explain the design. **Figure 5.15** shows additional design

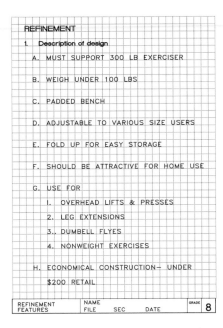

Figure 5.13 This list of desirable features is a refinement of the exercise bench.

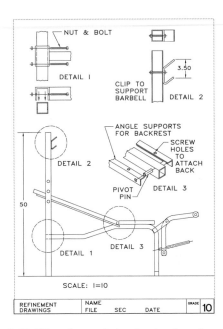

Figure 5.15 This refinement drawing is of another design concept for the exercise bench.

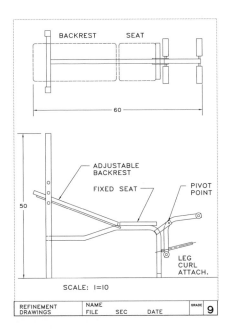

Figure 5.14 This refinement drawing of a preliminary idea for the exercise bench is a scale drawing, with only a few major dimensions shown.

features. These worksheets depict representative types of drawings required to refine a design; additional drawings would be required for a complete refinement of the design.

5.6 Standard Parts

When preparing refinement drawings, specify standard parts that are readily available whenever possible because they are cheaper. Merchandise catalogs, sales brochures, magazine advertisements, newspapers, and similar sources contain specifications for standard parts.

Make a practice of keeping files on stock items—such as wheels **(Fig. 5.16)** and casters **(Fig. 5.17)**—that can be used in developing products and specified in refinement drawings by referring to literature from manufacturers and vendors. You can become a better designer by observing how parts are made and how they function and thinking about how to use them in your designs.

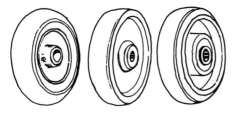

Figure 5.16 These standard types of wheels for light applications are available commercially.

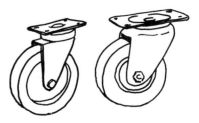

Figure 5.17 These casters are typical of those used for light applications.

Problems

Problems should be presented on 8½-by-11-inch paper, grid or plain. All notes, sketches, drawings, and graphical work should be neatly presented. Written matter should be lettered legibly, using ⅛-inch guidelines.

1. When refining a design for a folding lawn chair, what physical properties would a designer need to determine? What physical properties would be needed for a (a) TV set base, (b) golf cart, (c) child's swing set, (d) portable typewriter, (e) shortwave radio, (f) portable camping tent, and (g) warehouse dolly used for moving heavy boxes?

2. Why should scale drawings rather than freehand sketches be used in the refinement of a design?

3. List five examples of problems involving spatial relationships that could be solved by the application of descriptive geometry. Explain your answers.

4. In the refinement step, how many preliminary designs should be refined? Why?

5. Make a list of refinement drawings that would be needed to develop the installation and design of a 100-foot radio antenna. Make rough sketches of the types of drawings needed, with notes to explain their purposes.

6. If a design is eliminated as a possible solution after refinement drawings have been made, what should be the designer's next step? Explain.

7. For the exercise bench discussed in Section 5.5, what refinement drawings are necessary in addition to those presented? Make freehand sketches of the necessary drawings, with notes to explain what they should show.

Refinement Problems

Figures 5.18–5.28 show products that were developed from preliminary ideas. Use drawing instruments and prepare refinement drawings to scale for each product, from which the product can be made. You may use

a) orthographic views of each individual product part;

b) orthographic views of the product's assemblies and subassemblies; and

c) pictorial representations to explain relationships between parts.

Feel free to improve upon these products by adding your own solutions in making the refinement drawings.

8. Pipe hangers (**Fig. 5.18**). Make scale drawings of one of the pipe hangers, which are used to support overhead plumbing pipes. The base plates are about 6 in. across.

9. Sit-up device (**Fig. 5.19**). This device clamps to an interior door to hold the exerciser's feet while doing sit-ups for fitness.

10. Fireplace caddy (**Fig. 5.20**). This small handcart is to be used for carrying firewood and is to be decorative enough to be kept indoors as a holder for the wood.

11. Cable connector (**Fig. 5.21**). Make orthographic refinement drawings of the parts of the cable connector.

12. Gear puller (**Fig. 5.22**). Make orthographic refinement drawings of the parts of the gear puller.

13. Pointer mount (**Fig. 5.23**). This mount fastens to the top of a drawing table to hold a rotational pencil pointer.

Figure 5.18 Problem 8. Pipe hangers with six-in. plates.

Figure 5.19 Problem 9. Sit-up device.

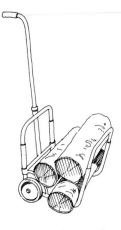

Figure 5.20 Problem 10. Fireplace caddy.

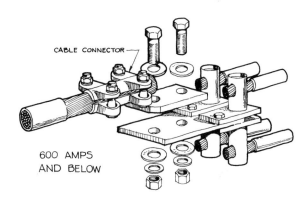

CABLE CONNECTOR

600 AMPS
AND BELOW

Figure 5.21 Problem 11. Cable connector.

14. Sharpener guide (**Fig. 5.24**). This device holds a chisel cutting edge at a constant angle while the user sharpens it on a whetstone.

15. Luggage carrier (**Fig. 5.25**). This portable cart for carrying luggage is to fold up to as small a size as possible.

16. Pipe hanger (**Fig. 5.26**). Make orthographic refinement drawings of the parts of the pipe hanger. It connects to the bottom flange of I-beams that measure from 6 in. to 10 in. across the flange.

17. Plywood table (**Fig. 5.27**). Make orthographic refinement drawings of the parts of the plywood table, which is made of ½-in. plywood.

18. Picture-frame tacker (**Fig. 5.28**). Make orthographic refinement drawings of the parts of the framing aid used to insert nails in a picture frame to hold the picture in place.

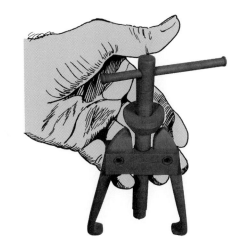

Figure 5.22 Problem 12. Gear puller.

Figure 5.23 Problem 13. Pointer mount.

Figure 5.26 Problem 16. Pipe hanger. (Courtesy of Grinnel Corporation.)

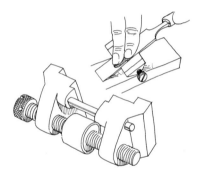

Figure 5.24 Problem 14. Sharpener guide.

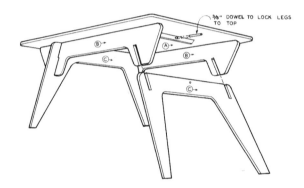

Figure 5.27 Problem 17. Plywood table. (Courtesy of American Plywood Association.)

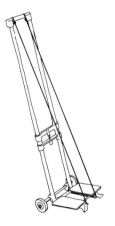

Figure 5.25 Problem 15. Luggage carrier.

Figure 5.28 Problem 18. Picture frame tacker.

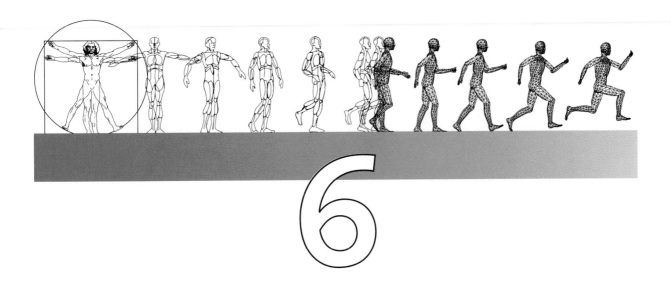

6

Design Analysis

6.1 Introduction

Analysis is the step in the design process most thoroughly covered in engineering courses. Analysis is the evaluation of a proposed design by objective thinking and the application of technology. For example, bridge designs are analyzed for loads, stresses, dimensions, material selection, and member sizes, function, and economy. **Less creativity is needed during analysis than during the previous steps of the design process.**

6.2 Graphics and Analysis

Engineering graphics and descriptive geometry are effective tools for analyzing a design. Empirical data obtained from laboratory experiments and field observation can be transformed into formats suitable for graphical analysis (**Fig. 6.1**).

Graphical calculus is useful in analyzing data that do not fit algebraic equations. **Figure 6.2** shows an example of graphics applied to the analysis of a linkage system to determine clearances and limits. Graphics also is an effective way

to present and analyze technical information, background data, market surveys, population trends, and sales projections (**Fig. 6.3**).

6.3 Types of Analysis

Analysis includes evaluation of

1. function,
2. human factors,
3. product market,
4. physical specifications,
5. strength,
6. economic factors, and
7. models.

Function
Function is the most important characteristic of a design because a product that does not function properly is a failure regardless of its other desirable features (Fig. 6.4).

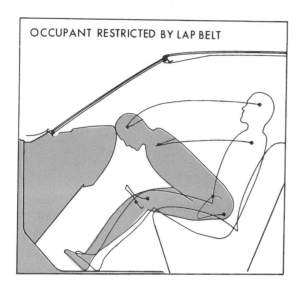

OCCUPANT RESTRICTED BY LAP BELT

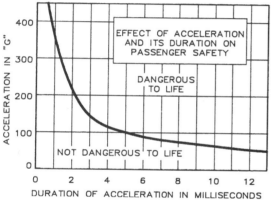

Figure 6.1 Designers analyze experimental data and human factors to determine comfortable and safe automobile designs. Graphics is a helpful tool in this step of the design process. (Courtesy of General Motors Corporation.)

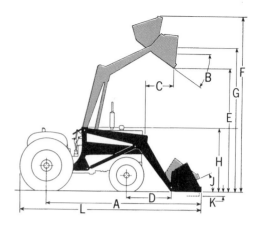

Figure 6.2 Clearance between functional parts and linkage systems can be analyzed efficiently with graphical methods. (Courtesy of Navistar International.)

U.S. CRUDE OIL AND PETROLEUM PRODUCT IMPORTS

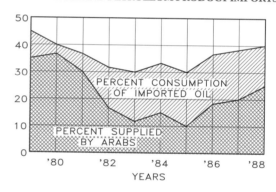

Figure 6.3 Graphs can be used to analyze data obtained from laboratory experiments, marketing information, and many other sources.

Many products serve a narrow, utilitarian purpose, such as the hand-operated clamp used for holding a part during machining or drilling (**Fig. 6.5**). In those cases, function is the designer's main concern. Functional analysis usually involves the optimization of several aspects of the design, including economics, durability, appearance, and marketability. For example, despite high fuel costs, most consumers will not accept cars designed to get good mileage at the expense of comfort, safety, and styling. Buyers may be willing to give up some features, but few would give up air conditioning and safety features for improved gasoline mileage.

Human Factors

Human engineering (ergonomics) is the design of products and workplaces suited to the humans who use and occupy them. Safety and comfort are

Figure 6.4 Designs of products such as the one for this Lincoln Marque-X must be analyzed to determine their physical and operating characteristics, particularly in terms of how well they will perform their intended functions. (Courtesy of Ford Motor Company.)

A.

GRAPHICAL ANALYSIS

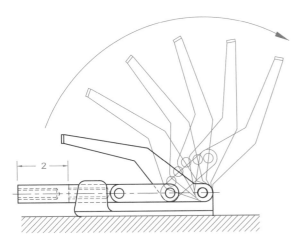

Figure 6.5 Graphical analysis is an efficient way to determine operating limits for a product such as this hand-operated clamping device.

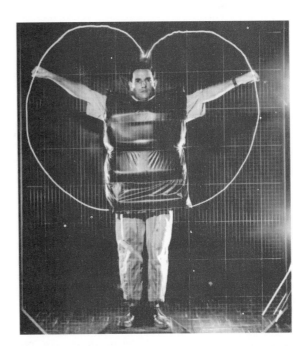

B.

Figure 6.6

A Leonardo da Vinci analyzed body dimensions and proportions in 1473.

B Engineers used his techniques in the twentieth century to analyze motions by astronauts when restricted by radiation protection vests. (Courtesy of General Dynamics Corporation.)

Figure 6.7 Analysis of human factors and living environments was part of designing the interior of this automobile. (Courtesy of General Motors Corporation.)

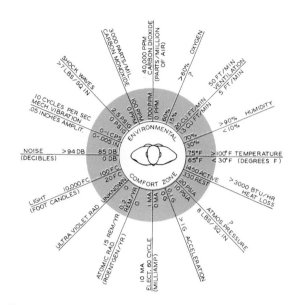

Figure 6.8 The inner circle represents the environmental comfort zone and the outer circle the bearable limit zone of the human environment. (Courtesy of Henry Dreyfuss, *The Measure of Man.*)

essential for efficiency, productivity, and profitability. Therefore the designer must consider the physical, mental, safety, and emotional needs of the user and how to best satisfy them.

Leonardo da Vinci analyzed body dimensions in about 1473 (**Fig. 6.6A**). Nearly five hundred years later NASA performed similar analyses to determine the range of mobility permitted by a radiation protection garment used by astronauts (**Fig. 6.6B**). Analysis of human factors is crucial in the space program because even the simplest, most familiar tasks require training and adaptation when astronauts perform them while in a weightless state. The configuration of spacecraft cabins shares many similarities with an automobile's interior (**Fig. 6.7**), because both must provide comfort and space in which to function.

Dimensions and Ranges A design must take into account the sizes, ranges of movement, senses, and comfort zones of the people using the finished product (**Fig. 6.8**). Variations in people's physical characteristics conform to the normal distribution curve shown in **Fig. 6.9**. Designers must use the body dimensions of the average American man (**Fig. 6.10**) and the average American woman (**Fig. 6.11**) as the bases for industrial designs.

PERCENTILE DISTRIBUTION OF AMERICAN MALES BY HEIGHT

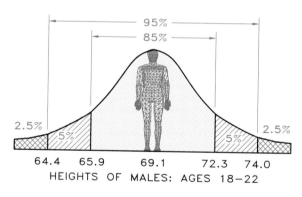

HEIGHTS OF MALES: AGES 18–22

Figure 6.9 This chart shows the distribution of average heights in inches of American men from 18 to 22 years of age. Fifty percent of American men in this age range are taller than 69.1 in., and 50% are shorter. (Courtesy of HumanCAD.)

A child can easily raise the safety bar of the car seat shown in **Fig. 6.12A** and get out of the seat without assistance. The car seat shown in **Fig.**

AVERAGE AMERICAN ADULT MALES
AVERAGE BUILD

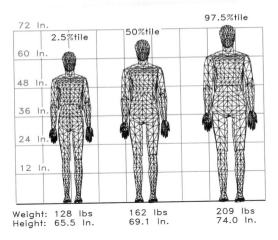

Weight: 128 lbs   162 lbs   209 lbs
Height: 65.5 In.   69.1 In.   74.0 In.

Figure 6.10 Men of average build have the body measurements shown. (Courtesy of HumanCAD.)

AVERAGE AMERICAN ADULT FEMALES
AVERAGE BUILD

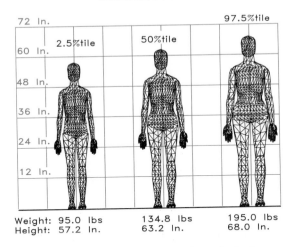

Weight: 95.0 lbs   134.8 lbs   195.0 lbs
Height: 57.2 In.   63.2 In.   68.0 In.

Figure 6.11 Women of average build have the body measurements shown. (Courtesy of HumanCAD.)

6.12B has a safety bar that the child cannot easily lift, making the car seat difficult to get out of without assistance and offering more protection to an unattended child in the back seat.

A.

B.

Figure 6.12
A A child can easily lift the safety bar of this seat, making it better for use in the front seat where the child will be supervised.

B A child has more difficulty raising this seat, making it better for use in the back seat where the child may be unattended.

Motion The study of body motion includes the amount of space required for a person to function comfortably, safely, and efficiently. **Figure 6.13** shows a computer-generated analysis of a grinding

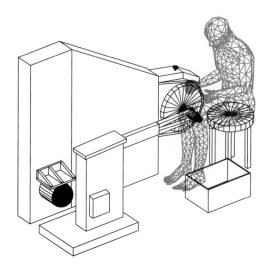

Figure 6.13 Three-dimensional analysis of this grinding wheel workstation helps ensure that it is well adapted to the operator. (Courtesy of HumanCAD.)

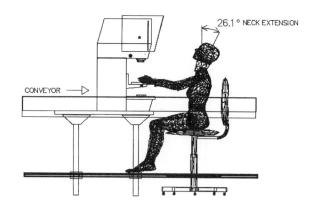

Figure 6.15 Analysis of this workstation revealed that it caused hand, arm, shoulder, neck, and upper back stress, indicating that the station needs to be redesigned. (Courtesy of HumanCAD.)

Figure 6.14 The design of an efficient, safe automobile instrument panel involves analysis of human factors. (Courtesy of Ford Motor Company.)

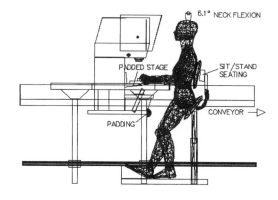

Figure 6.16 The workstation in Fig. 6.15 was redesigned to provide a work surface at stage height, to permit the operator to sit or stand, and to add padding on the edges of the stage. (Courtesy of HumanCAD.)

wheel workstation for the purpose of adapting it to the operator's needs.

Vision Designs that include gauges and controls must provide the most visually effective means of aiding the operator. The automobile dashboard configuration shown in **Fig. 6.14** is a futuristic

design, but it is similar enough to traditional dashboards to make the driver feel at home with it. Lights allow the driver to operate the controls easily and to obtain information from the instruments quickly.

When the computer station shown in **Fig. 6.15** was analyzed to determine why it caused operator fatigue, the designer found that position of the body, line of vision, and amount of clearance caused stress. **Figure 6.16** depicts a station designed to fit the user better.

Sound Sound must be within specified frequencies so as not to adversely affect a person's stress level and productivity.

Environment Working environments may include an entire industrial plant, a particular workstation, or a specialized location, such as the cockpit of an airplane. Important environmental factors are temperature, lighting, color, sound, and comfort.

Product Market

Designers study the market for a product during all stages of product development (Chapters 3–5) and more formally review it during the analysis step. Areas of product analysis are market prospects, retail outlets, sales features, and advertising.

Market Prospects Market information should be collected to learn about the age groups, income brackets, and geographical locations of prospective purchasers of the product. This information is helpful in planning advertising campaigns to reach potential customers.

Retail Outlets The product may be marketed through existing wholesale and retail channels, from newly established dealerships, or by the manufacturer directly. For example, mainframe computers are not suitable for distribution through retail outlets, so manufacturers' technical representatives work with clients individually. However, an exercise machine can be sold effectively through department stores and sporting-goods outlets.

Sales Features The designer should itemize the unique features of a new design that would stimulate interest in the product and attract consumers.

Advertising Manufacturers, wholesalers, and retailers use several media, including personal contact, direct mail, radio, TV, newspapers, and periodicals, to attract potential customers to their products. Advertising costs vary widely and each medium should be analyzed for suitability before one or more is selected.

Physical Specifications

During the refinement step, the designer specified various measurements, such as lengths, areas, shapes, and angles, for the product. During the analysis step, the designer uses the product's geometry and materials to calculate member sizes and dimensions, weights, volumes, capacities, velocities, operating ranges, packaging and shipping requirements, and similar information (**Fig. 6.17**).

Sizes and Dimensions The designer must evaluate product sizes and dimensions to ensure that they meet any standards specified, such as permissible widths, lengths, and weights in automobile design. For products that have moving parts, such as a construction crane, the designer must analyze the size of the product when extended, contracted, or positioned differently, as well as weight and balance requirements.

Ranges Many products have ranges of operation, capacities, and speeds that the designer must analyze before finalizing a design. For example, the designer must determine ranges and maximum limits such as seating capacity, miles per gallon, pounds of laundry per cycle, flows in gallons per minute, or power required.

Packaging and Shipping The designer must also be concerned with product packaging and shipping. Packaging relates both to product protection and consumer appeal: how the product is to be shipped—air, rail, mail, or truck—and whether it is to be shipped one at a time or in quantity are important considerations. Shipping and marketing a product assembled, partially assembled, or dis-

Figure 6.17 Designs must be analyzed to determine their physical properties, including weights, ranges, geometries, and capacities. (Courtesy of Air Technical Industries.)

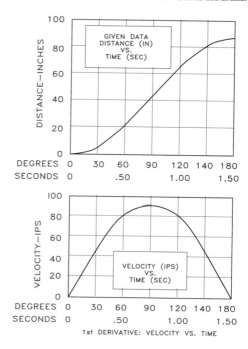

Figure 6.18 This graph shows the distance traveled versus the time required for a part on a conveyor. The designer graphically differentiates the plot of the data given to obtain a graph of velocity versus time as the first step in designing parts of sufficient strength for the conveyor.

assembled requires design attention and analysis, as does the cost of each method.

Strength

Much of engineering is devoted to analyzing a product's strength to support maximum design loads, withstand specified shocks, and endure necessary repetitive motions. **Figure 6.18** illustrates motion analysis for a moving part on a conveyor. The designer plots the data given and then uses graphical calculus to find the conveyor's velocity versus time profile as the first step in determining the strength needed by the part.

Economic Factors

Designs must be economically competitive to have a chance of being successful. Therefore, before releasing a product for production, the designer must analyze its cost and expected profit margin. Two methods of pricing a product are itemizing and comparative pricing.

Itemizing The process of totaling the costs of each part and its related overhead to determine its final cost is the first step in itemizing a product's price. From the working drawings, the designer (or an estimator) can estimate the costs of materials, manufacturing, labor, overhead, and other items to arrive at the total production expense. The wholesale price equals production cost plus profit. Dealer margin plus the wholesale cost equals retail price. An example of an economic model is shown in **Fig. 6.19**; these percentages vary for different areas of manufacturing and retailing.

Comparative Pricing The other method used to estimate the price of a proposed product is to compare it with the prices of similar products. For example, the power tools shown in **Fig. 6.20** are

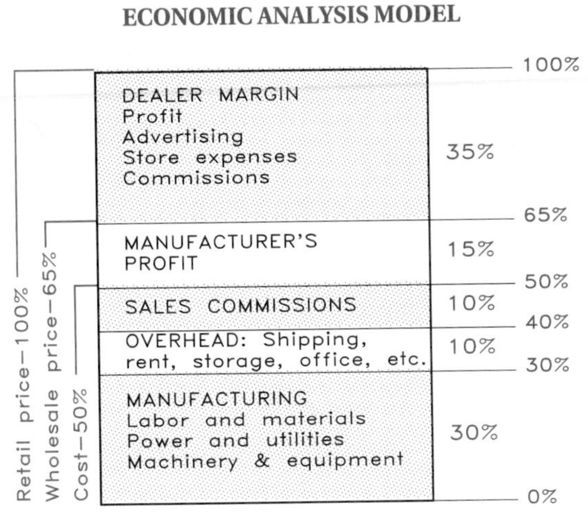

ECONOMIC ANALYSIS MODEL

DEALER MARGIN Profit Advertising Store expenses Commissions	35%	100% 65%
MANUFACTURER'S PROFIT	15%	 50%
SALES COMMISSIONS	10%	40%
OVERHEAD: Shipping, rent, storage, office, etc.	10%	30%
MANUFACTURING Labor and materials Power and utilities Machinery & equipment	30%	 0%

Retail price—100% / Wholesale price—65% / Cost—50%

Figure 6.19 An economic model that can be used as a guide in the economic analysis and pricing of a product.

Figure 6.20 Each of these three products retails for approximately $100. They can be priced comparatively because they are similar in design and have essentially the same market size. (Courtesy of Sears, Roebuck and Company.)

priced at $99 each. These tools are similar: All use the same type of power source, are made from the same materials, have the same styling, and serve an identical market size. Approximately the same number of drills, sanders, and saws are sold; consequently, production costs and retail prices are similar for each.

Another example of comparative pricing are the prices of the hunting seat (**Fig. 6.21**) and the exercise bench (**Fig. 6.22**), both sold by Sears, Roebuck and Company. Both products' manufacturing requirements and market volumes are similar, and both sell for about $90–$100. However, the baby stroller (**Fig. 6.23**) sells for about $50 because of its larger market, even though it is similar in several ways to the hunting seat and the exercise bench.

Manufacturers also use comparative pricing to estimate cost per square foot, per mile, per cubic foot, or per day. These factors yield rough cost estimates as a basis for doing more detailed studies.

Miscellaneous Expenses Various expenses incurred in the development of new products can be easily overlooked and thereby affect a prod-

Figure 6.21 This hunting seat and the exercise bench in Fig. 6.22 can be comparatively priced at between $90 and $100 because both have similar manufacturing requirements and market volume potential. (Courtesy of Baker Manufacturing Company, Valdosta, Georgia.)

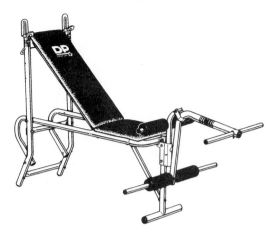

Figure 6.22 Priced at $100, this exercise apparatus is similar in manufacture and market appeal to the hunting seat shown in Fig. 6.21. (Courtesy of Diversified Products Corporation.)

Figure 6.24 A mock-up is a full-size model of a product made to represent its final appearance rather than its operation. (Courtesy of General Motors Corporation.)

Figure 6.23 These baby strollers are priced from $55 to $75, or considerably less than the hunting seat and exercise apparatus, because the stroller market is larger and competition for customers is greater. (Courtesy of Strolee of California.)

Figure 6.25 This system layout model is used to analyze the details of construction of a refinery. (Courtesy of E. I. du Pont de Nemours and Company.)

uct's profitability projections. For example, warehousing and storage costs for finished products must be included in their price. Associated with warehousing are the costs of insurance, temperature control, shelving, forklifts, and employees.

Models

Models are effective aids for analyzing a design in the final stages of its development. Designers use three-dimensional models to study a product's proportion, operation, size, function, and efficien-

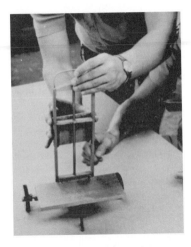

Figure 6.26 A student's model demonstrates how a portable home caddy will fold flat for ease of storage.

cy. Types of models often used are conceptual models, mock-ups, prototypes, and system layout models.

Conceptual Models Designers use rough models to analyze a preliminary design or feature concept.

Mock-ups Designers use full-size dummies of the finished design to demonstrate the product's size, appearance, and component relationships (**Fig. 6.24**). Mock-ups present a visual impression rather than demonstrate its operation.

Prototypes Designers use full-size working models to demonstrate the operation of a final product. Because prototypes are made mostly by hand, materials that are easy to fabricate are used instead of those to be used in final production.

System Layout Models Designers use detailed scale models that show the relationships among components of large manufacturing systems, building complexes, and traffic layouts. Designers usually construct system layout models of refineries to supplement working drawings for contractors and construction supervisors during construction (**Fig. 6.25**).

Model Materials Designers commonly use balsawood in model construction because it is easy to shape and requires few tools. Standard parts such as wheels, tubing, figures, dowels, and other structural shapes can be purchased, rather than made, to save time and effort. Plexiglas can be used to construct models that illustrate both inside and outside design features. Finished models should give a realistic impression of the design, especially when they are used for presentations and displays.

Model Scale A model should be large enough to show the function of the smallest significant moving parts. For example, the student model of a portable home caddy shown in **Fig. 6.26** (made of balsawood) demonstrates a linkage system that permits the wheels to be collapsed for storage. The model's scale is large enough to permit the linkage system to operate as it will in the final product.

Model Testing Using models to test performance is helpful in determining how well a design meets requirements. Aerodynamic characteristics of the rear styling of an automobile can be evaluated by wind tunnel tests (**Fig. 6.27**). Physical relationships and the functional workings of movable components, as for the hatch and storage area of a car, can be tested in a prototype. Designers also use models to test consumer reactions to new products before releasing the design for production (**Fig. 6.28**).

Figure 6.27 A mock-up of a product is useful for testing its resistance to air flow in the laboratory prior to a decision about its final configuration. (Courtesy of General Motors Corporation.)

Figure 6.28 Prototypes are full-size models for demonstrating and testing a design's operation. (Courtesy of General Motors Corporation.)

6.4 Analysis: Exercise Bench

To illustrate a method of analyzing a product design, we return to the exercise bench, which was carried through the first three steps of the design process in Chapters 3–5. The main areas of analysis listed on the worksheets in **Figs. 6.29–6.32** will assist you in analyzing the design. Additional worksheets and large sheet sizes for analysis drawings may be used if more space is needed.

Figure 6.32 shows how graphics is used to determine the range of positions of the backrest. Those positions affect the design of the angle-iron

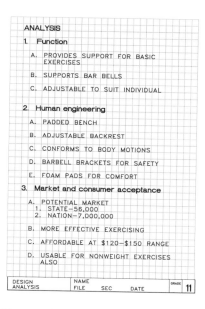

ANALYSIS

1. Function

   A. PROVIDES SUPPORT FOR BASIC EXERCISES

   B. SUPPORTS BAR BELLS

   C. ADJUSTABLE TO SUIT INDIVIDUAL

2. Human engineering

   A. PADDED BENCH

   B. ADJUSTABLE BACKREST

   C. CONFORMS TO BODY MOTIONS

   D. BARBELL BRACKETS FOR SAFETY

   E. FOAM PADS FOR COMFORT

3. Market and consumer acceptance

   A. POTENTIAL MARKET
     1. STATE—56,000
     2. NATION—7,000,000

   B. MORE EFFECTIVE EXERCISING

   C. AFFORDABLE AT $120–$150 RANGE

   D. USABLE FOR NONWEIGHT EXERCISES ALSO

DESIGN ANALYSIS | NAME FILE   SEC   DATE | GRADE 11

Figure 6.29 A worksheet containing an analysis of function, human engineering, and market considerations for the exercise bench.

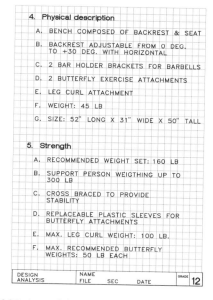

4. Physical description

   A. BENCH COMPOSED OF BACKREST & SEAT

   B. BACKREST ADJUSTABLE FROM 0 DEG. TO +30 DEG. WITH HORIZONTAL

   C. 2 BAR HOLDER BRACKETS FOR BARBELLS

   D. 2 BUTTERFLY EXERCISE ATTACHMENTS

   E. LEG CURL ATTACHMENT

   F. WEIGHT: 45 LB

   G. SIZE: 52" LONG X 31" WIDE X 50" TALL

5. Strength

   A. RECOMMENDED WEIGHT SET: 160 LB

   B. SUPPORT PERSON WEIGTHING UP TO 300 LB

   C. CROSS BRACED TO PROVIDE STABILITY

   D. REPLACEABLE PLASTIC SLEEVES FOR BUTTERFLY ATTACHMENTS

   E. MAX. LEG CURL WEIGHT: 100 LB.

   F. MAX. RECOMMENDED BUTTERFLY WEIGHTS: 50 LB EACH

DESIGN ANALYSIS | NAME FILE   SEC   DATE | GRADE 12

Figure 6.30 A worksheet giving the physical description and strength analysis for the exercise bench.

supports for the backrest and the locations of the semicircular holes in the angle irons for a range of settings of 30°.

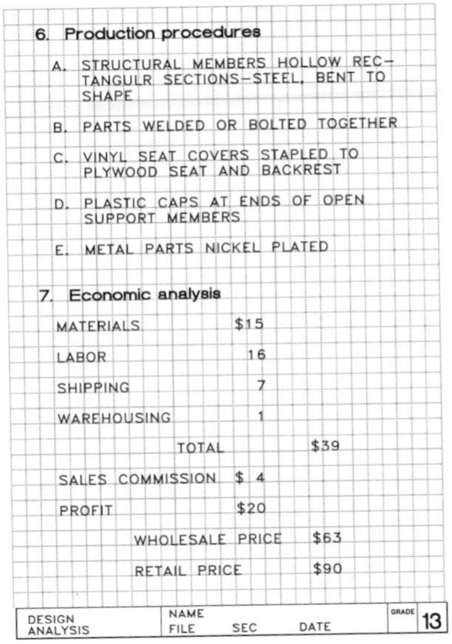

6. Production procedures
   A. STRUCTURAL MEMBERS HOLLOW REC-TANGULR SECTIONS—STEEL, BENT TO SHAPE
   B. PARTS WELDED OR BOLTED TOGETHER
   C. VINYL SEAT COVERS STAPLED TO PLYWOOD SEAT AND BACKREST
   D. PLASTIC CAPS AT ENDS OF OPEN SUPPORT MEMBERS
   E. METAL PARTS NICKEL PLATED

7. Economic analysis
   MATERIALS $15
   LABOR   16
   SHIPPING   7
   WAREHOUSING   1
              TOTAL $39
   SALES COMMISSION $ 4
   PROFIT $20
   WHOLESALE PRICE $63
   RETAIL PRICE $90

   DESIGN ANALYSIS | NAME FILE | SEC | DATE | GRADE 13

Figure 6.31 A worksheet containing an analysis of the production procedures for and economics of the exercise bench.

Figure 6.33 A full-size model of the exercise bench is tested for function and acceptability.

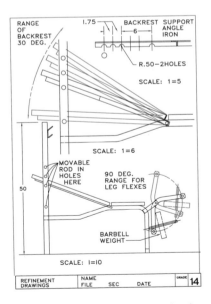

RANGE OF BACKREST 30 DEG.

1.75

BACKREST SUPPORT ANGLE IRON

6

R.50—2HOLES

SCALE: 1=5

SCALE: 1=6

MOVABLE ROD IN HOLES HERE

90 DEG. RANGE FOR LEG FLEXES

50

BARBELL WEIGHT

SCALE: 1=10

REFINEMENT DRAWINGS | NAME FILE | SEC | DATE | GRADE 14

Figure 6.32 A worksheet that shows graphical analysis of the range of movements for the adjustable parts of the exercise bench.

Figure 6.34 This catalog description gives the key features of the exercise bench: Weider® bench with butterfly attachment. It features no-pinch supports, multiposition padded back and leg lift, and tubular steel frame. Total weight capacity is 1000 lb; butterfly capacity, 50 lb; leg lift capacity, 65 lb; overall size, 58" × 45" × 41"; weight, 48 lb; and price, $89.99. (Courtesy of Sears, Roebuck and Company.)

The leg-exercising attachment at the end of the bench is designed to move through a 90° arc, which is sufficient for leg extensions. By determining the maximum loads on the backrest and the leg exerciser, you can select the member sizes and materials that provide the strength required. For further analysis, construct a model and test the design for suitability.

Figure 6.33 illustrates testing of a commercial version of the exercise bench to measure its functional features. The catalog description in Fig. 6.34 lists the physical properties of the Weider® exercise bench to help the consumer understand its features.

Problems

The following problems should be solved on 8½-by-11-inch paper and the solution presented in drawing, note, and text forms. Answers to essay problems can be typed or lettered. All sheets should be placed in a binder or folder.

General

1. Make a list of human factors that must be considered in designing the following items: (a) canoe, (b) hairbrush, (c) water cooler, (d) automobile, (e) wheelbarrow, (f) drawing table, (g) study desk, (h) a pair of binoculars, (i) baby stroller, (j) golf course, and (k) seating in a stadium.

2. What physical quantities have to be determined for the designs in Problem 1?

3. Select one of the items in Problem 1 and outline the steps required to analyze: (a) function, (b) human factors, (c) product market, (d) physical specifications, (e) strength, (f) economic factors, and (g) a prototype model.

Human Engineering

4. Design a drawing table for your body size to meet your own working and comfort needs. Make a drawing indicating the optimum working areas and tilt angle for the board when you sit at the drawing table. The drawing also should show the most efficient positioning of instruments for working.

5. Using the dimensions for the average man and woman given in Figs. 6.10 and 6.11, design stadium benches to meet the optimum needs of spectators. A primary consideration is slope of the stadium seating to allow an adequate view of the playing field. Spectator comfort and provision for traffic along aisles in front of the benches also must be considered.

6. Compare the measurements of the male and female students in your class with the averages given in Figs. 6.10 and 6.11.

7. Design a backpack for use on a week-long camping trip. Determine the minimum number of articles a camper should carry and use their weights and volumes in establishing design criteria. Make sketches of the pack and the method of attaching it to the body to provide mobility, comfort, and capacity.

8. State the dimensions, facilities, and provisions needed for a one-person storm shelter to provide protection for forty-eight hours. Make sketches of the interior in relationship to a person and the supplies.

9. Design a manhole access to an underground facility. Determine the diameter of the manhole required to permit a person to climb a ladder for a distance of ten feet with freedom of movement. Make a sketch of your design and explain your method of solving the problem.

10. Design an observation facility for temporary service in the Arctic. This facility is to be as compact as possible but must provide for the needs of one person during a seventy-two-hour duty watch. Make sketches of your design and explain the items considered essential to human survival in that harsh climate.

11. Design an automobile steering wheel that is different from current designs but that is just as functional. Base your design on human factors such as arm position, grip, and vision. Make sketches of your design and list the factors that you considered.

12. Make sketches to indicate safety features that you would build into your automobile to reduce the severity of personal injury in case of a bad accident. Explain your ideas and the advantages of your designs.

13. Assume that you prefer to alternate between sitting and standing when working at a drafting workstation. Determine the ideal height of the table top for working in each position. Indicate how you would devise the table to permit instant conversion from the height for standing to the height for sitting.

14. Identify some human engineering problems that you believe need to be solved. Present several to your instructor for approval. Solve the approved problems. Make a series of sketches and notes to explain your approach.

Market Analysis

15. Conduct a market analysis for the drill shown in Fig. 6.20, covering the areas mentioned in the text. Assume that this power tool has never been introduced before. Outline the steps you would take in conducting a product market analysis.

16. Make a market analysis of the car seat shown in Fig. 6.12B, following the steps suggested in the text. Determine a reasonable price, potential outlets, and other marketing information for the product.

17. Assume that the costs of producing hunting seats are estimated as: 100 seats, $35 each; 200 seats, $20 each; 400 seats, $10 each; 1000 seats, $8.50 each. Using these figures, determine the price at which you could introduce the seats to consumers on a trial basis and still make some money. Explain your plan.

18. List the unique features of the hunting seat that would be important to a sales campaign, including advertising. Make sketches and notes to explain these features.

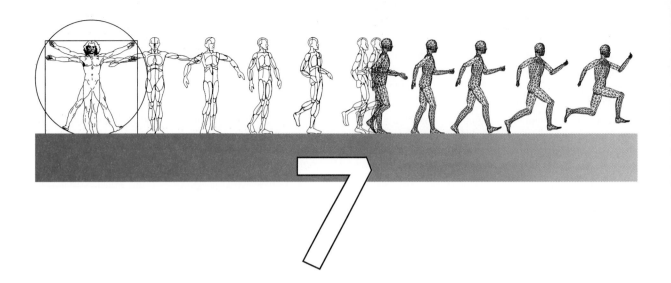

Decision

7.1 Introduction

After the designer has conceived, developed, refined, and analyzed several designs, one must be selected for implementation. The decision process begins with a presentation by the designer (or design team) of all significant findings, features, estimates, and recommendations. The presentation should be organized in an easy-to-follow form and it must communicate the designer's conclusions and recommendations because it is the means of gaining support for the project in order for it to become a reality. A committee usually makes the decision when funding must be obtained. Although decision making is aided by facts, data, and analyses, it is subjective at best.

7.2 Decision

The purpose of oral presentations and written reports is to present the findings of a project so that a decision can be made whether to implement it. One of three types of decisions may be made:

- **Acceptance.** A design may be accepted in its entirety, which indicates success by the designer.

- **Rejection.** A design may be rejected in its entirety, which does not necessarily mean that the designer failed. Changes in the economic climate, moves by competitors, or other factors beyond the designer's control may make the design obsolete, premature, or unprofitable.

- **Compromise.** A design may not be approved in parts, and compromises may be suggested. For example, the initial production run might be increased or decreased, or various features might be eliminated, modified, merged, or added.

7.3 Decision: Exercise Bench

We have used the exercise bench problem introduced in Chapter 3 to illustrate the first four steps of the design process. We continue to use it here to help explain the decision step.

DECISION

1. Decision for evaluation

DESIGN 1   A-FRAME
DESIGN 2   U-FRAME
DESIGN 3   2-COLUMN FRAME
DESIGN 4
DESIGN 5

MAX	FACTORS	1	2	3	4
3.0	FUNCTION	2.0	2.3	2.5	
2.0	HUMAN FACTORS	1.6	1.4	1.7	
0.5	MARKET ANALYSIS	0.4	0.4	0.4	
1.0	STRENGTH	1.0	1.0	1.0	
0.5	PRODUCTION EASE	0.3	0.2	0.4	
1.0	COST	0.7	0.6	0.8	
1.5	PROFITABILITY	1.1	1.0	1.3	
0.5	APPEARANCE	0.3	0.4	0.4	
10	TOTALS	7.4	7.3	8.5	

DESIGN DECISION   NAME   FILE   SEC   DATE   GRADE **16**

Figure 7.1 This worksheet shows the decision table used to evaluate the design alternatives for the exercise bench.

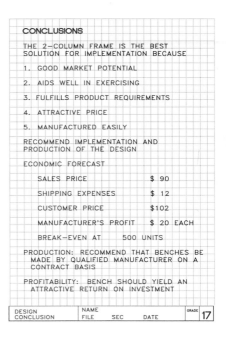

Figure 7.2 This worksheet summarizes the designer's conclusions about the best alternative and recommendations concerning the feasibility of implementing that design.

Decision Table

Use a table like the one shown in **Fig. 7.1** to compare designs, where each idea is listed and given a number for identification. Assign maximum values for each factor of analysis, based on your best judgment, so they total to ten points. Rate each factor for the competing designs by entering points for each.

Sum the columns of numbers to determine the total for each design and compare the scores of each design. Your instincts may disagree with the outcome of this numerical analysis. If so, have enough faith in your judgment to go with your intuition. **The scores from the decision table are meant to be a guide for you and not the absolute final word in your decision.**

Conclusion

After making a decision, state it and the reasons for it clearly (**Fig. 7.2**). Record any additional information, such as number to be produced initially, selling price per unit, profit per unit, estimated sales during the first year, break-even number, and the product's most marketable features, that will help you prepare your presentation.

If you believe that none of your designs are satisfactory, you should recommend that they not be implemented. **A negative recommendation is not a failure of the design process; it means only that your solutions developed so far are not feasible.** Going forward with an inadequate solution could cause both monetary losses and wasted effort.

Presentation

Until now your efforts have been self-directed and mostly free from supervision. The work is your own (or that of your team), you have solved the problem to the best of your ability, and you are ready to make recommendations regarding its implementation.

At this point the project usually involves the input of others besides the designers. These out-

Figure 7.3 A decision may be the outcome of a presentation, where ideas and designs are discussed and sketched informally. (Courtesy of the MITRE Corporation.)

Figure 7.4 Planning cards are useful in preparing the sequence of a presentation by arranging the cards on a planning board (as shown here) or on a table top. (Courtesy of Eastman Kodak.)

siders may be other engineers, managers, administrators, salespeople, company shareholders, investors or bankers who will loan money for the project. You must prepare a presentation suitable for your audience in order to communicate the important features of your design, the data you gathered and analyzed, and the benefits to be gained by implementing your design.

Present your findings, conclusions, and recommendations as objectively as possible so that the group can make a valid decision. At no time should your enthusiasm for the project outweigh an impartial presentation of the facts.

7.4 Types of Presentations

Presentations may be made to groups ranging from a few knowledgeable design associates to a large number of laypeople unfamiliar with the project and its objectives. Presentations of the first type usually are informal; the second type, formal.

Informal Presentations

Informal presentations are made to several associates and perhaps a supervisor. Although formally prepared visual aids are unnecessary for presentations to a small group, the designer nevertheless needs to graph data, draw pictorials, sketch

schematics, and build models to explain design concepts. Ideas and concepts may be sketched on a blackboard and informally discussed (**Fig. 7.3**).

Formal Presentations

Formal presentations usually involve large groups that may include associates, administrators, and/or laypeople, or a combination. They may be clients for whom the project is designed, potential investors, or politicians who will vote to approve or disapprove the design. Function and acceptability of a design are the primary concerns of engineering associates, and profitability is most important to investors.

7.5 Organizing a Presentation

An effective method of planning oral and written reports is to use 3-by-5-inch index cards for the ideas to be presented. Placing the cards on a table or tacking them to a bulletin board (**Fig. 7.4**) allows an easy choice of sequence and rearrangement as needed. Each card (**Fig. 7.5**) should contain the following information:

1. **Number.** The card's position in the presentation sequence.

2. **Illustration.** A sketch of the illustration, if any.

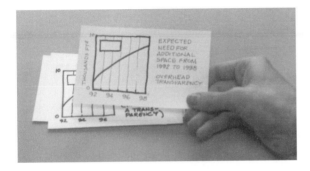

Figure 7.5 This layout on a 3-by-5-inch card, showing a sketch of the visual and its accompanying text, illustrates sound planning.

Figure 7.6 The flip chart is an effective method of presentation to small groups. (Photo by Hazel Hankin/Stock, Boston.)

3. **Text.** A brief outline of the points to be covered for that idea.

7.6 Visual Aids

Visual aids commonly used in presentations are flip charts, photographic slides, overhead visuals, models, and videotapes. The following suggestions apply to the preparation of visual aids:

1. Limit each visual to a single concept or point.

2. Reduce statements to key points to communicate thoughts clearly and concisely.

3. Present tables of data in a simplified graphic form.

4. Make visuals containing text large enough to be readable.

5. Use illustrations, color, and attention-getting devices.

6. Prepare enough visuals so that notes are unnecessary.

Flip Charts

Flip charts consist of bold illustrations drawn on sheets (usually 30 × 36 inches) for presentation to groups no larger than about thirty people (**Fig. 7.6**).

Paper Flip charts may be drawn on brown wrapping paper or white newsprint paper attached to a cardboard backing board. A stand or easel is needed to support the cardboard-backed set of sheets.

Lettering Felt-tipped markers, ink, tempera, or sign paints are fine for lettering. When used correctly, felt-tipped markers can yield bold, visible lines in a variety of colors and with sophisticated effects (**Fig. 7.7**). India ink is an effective medium for lettering and for adding emphasis to a chart (**Fig. 7.8**).

Color Construction paper cutouts mounted with rubber cement are especially effective for adding color to bar graphs. The use of felt-tipped markers and tempera colors also adds color and interest to a chart.

Assembly Flip chart sheets are arranged in order of presentation, with a title page covered by a blank sheet of paper on top to prevent audience anticipation. The sheets are fastened at the top to the backing board.

Presentation Each sheet is flipped in sequence after the presenter has covered the points on it. A pointer should be used to direct the audience's

Figure 7.7 Felt-tipped markers are well-suited to the preparation of flip charts and other types of visual aids.

Figure 7.8 This flip chart sheet is brown wrapping paper on which India ink was applied with a brush. Overlays of colored construction paper or colored felt-tipped markers can be used to highlight topics.

attention to specific items on the sheet. Well-prepared flip charts should require no additional notes.

Photographic Slides

Slides are effective for larger audiences and for showing actual scenes or examples of hardware.

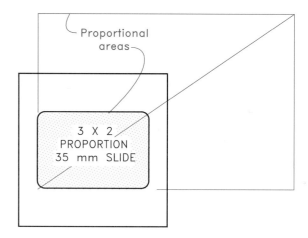

Figure 7.9 Use of this method of proportionately sizing artwork ensures that artwork will properly fill a photographic slide.

Artwork Figure 7.9 shows the method for proportionally sizing artwork for a 35-mm slide. An 8-by-12-inch size is suitable for most slides. The artwork should contain color to make the slides more attractive and effective in maintaining audience interest. Colored construction paper, mat board, and other poster materials should be used in preparing slides.

Allow at least an inch margin on all artwork, so the edges will not show when photographed. Uppercase letters are best for slides, with the space between lines of text equal to the height of the letters. Do not use a white background for slide artwork, because it is tiring to the eyes.

Photographing Layouts To make slides a camera, copy stand, and lights are required. A 35-mm reflex camera with through-the-lens view finder is best because the photographer sees exactly what is being photographed. The copy stand holds the camera steady (**Fig. 7.10**). If all the artwork is uniform in size, the camera can be left in the same

Figure 7.10 Artwork and charts can be reproduced with a 35-mm reflex camera mounted onto a copy stand, which holds the camera in position.

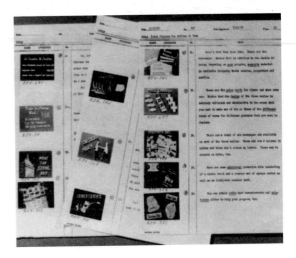

Figure 7.12 A slide script is useful for lengthy slide presentations and those that will be given repetitively. (Courtesy of Eastman Kodak.)

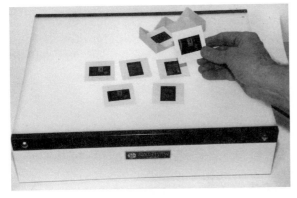

Figure 7.11 Photographic slides should be sorted and arranged in proper sequence and then loaded in a slide tray.

position during photography. Small illustrations can be photographed with a close-up lens. The finished slides are reviewed, arranged in sequence, numbered, and loaded in a tray for showing (**Fig. 7.11**).

Slide Scripts   A slide script is useful when a presentation will be made repeatedly. Photographic copies of the slides attached to the left side of the script serve as prompts for the presenter (**Fig. 7.12**).

Overhead Projector Transparencies

Overhead projector transparencies are reproduced on 8-by-10-inch plastic sheets by the heat-transfer or diazo processes, or plotted by the computer. Tracing paper is the most commonly used drawing surface for preparing art from which transparencies are made. Tracing paper can be used in both the heat-transfer and diazo processes (opaque paper cannot be used in the diazo process). Computer plotting can be done directly onto plastic film with special pens. Diazo transparencies are

OVERHEAD VISUAL

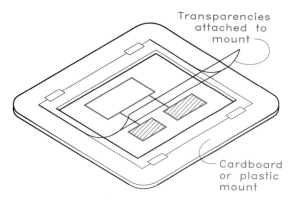

Transparencies
attached to
mount

Cardboard
or plastic
mount

Figure 7.13 A transparency used on an overhead projector consists of an 8½"-×-11" transparency mounted on a 10"-×-12" frame. The projection area within the frame is about 7½"× 9½". To show information sequentially, different-colored overlays can be flipped over, one at a time.

Figure 7.14 Overhead projector transparencies can be generated by computer on plastic film with plotter pens that match the film. (Courtesy of Hewlett Packard Corporation.)

reproduced on plastic film in the same manner in which blue-line prints are made.

Drawings should be made in black India ink. Stick-on shapes and graphing tapes can give the drawing a professional appearance. Lettering should be at least 0.20 in. high.

Color Overlays Several color overlays can be hinged to the transparency mount for presentations with multiple steps (**Fig. 7.13**). Each color overlay requires a separate piece of artwork, which is drawn on tracing paper placed over the basic layout.

Computer-Plotted Transparencies Computer-generated art and text can be plotted directly on plastic film with fiber-tipped pens, which come in many colors and match the film (**Fig. 7.14**). The Romand font of AutoCAD is better suited for large lettering on transparencies than is the Romans font.

Presentation The presenter stands or sits near the projector in a semilighted room and refers to the transparencies while facing the audience. The presenter can emphasize items on the stage of the projector with a pointer, which is projected onto the screen. In the same manner as multiple overlays are hinged to a mount, paper overlays can be attached to mounts in order to block out parts of the transparency to control audience attention.

Models

A model is the most realistic visual aid for showing a final design. Models should be large enough to be seen by all in the audience. A series of photographic close-ups of the model, taken from different angles, can be used to supplement the presentation. Obviously, a full-sized prototype of the completed design (**Fig. 7.15**) provides the most accurate description of the product and demonstrates its operation.

Videotapes

Videotaping presentations or visuals supplemented by a voice-over narration is an effective and sophisticated method. It may include various special effects such as music, sound effects, close-ups, motion, and precise realism. Formal presentations in the future will be multimedia shows using video and the computer in combination.

Figure 7.15 A full-sized prototype model of a design is an effective method of presentation to a small group. (Courtesy of Chrysler Corporation.)

Figure 7.16 A presentation should include graphical aids and models to help the speaker communicate with the audience. (Courtesy of Hewlett Packard Corporation.)

The table below should be completed jointly by the team with only the grade column completed by the instructor who will use the chart on page 4 and the factor "F" that was computed for each member.

Oral Report

Team No. __5__   Project: __TOY MFGR.__

Names No. (N=7)	%Contribution (C)	F=NC	GRADE
1. BROWN, G.	17	119	91
2. PRISK, A.	14.3	100	87
3. SMITH, L.	20	140	95
4. REED, T.	5.7	40	63
5. POTTER, M	14.3	100	87
6. FLYNN, O.	14.3	100	87
7. ROSS, N.	14.4	100	87
8.			

Evaluation by instructor   Max.   Comments:

		Max.	
1.	Introduction of team members	2	2
2.	Proper dress of team members	2	2
3.	Statement of purpose of presentation	5	4
4.	Use of visuals—point to important points, do not block screen, do not fumble, etc.	10	8
5.	Adequate number of visual aids	9	9
6.	Quality of visual aids	15	12
7.	Clear presentation of recommended design	10	8
8.	Presentation of alternate solutions considered	2	2
9.	Consideration of human factors	5	5
10.	Coverage of economics (manufacturing, shipping, packing, overhead, mark-up, etc.)	10	7
11.	Presentation of an effective conclusion	5	4
12.	Continuity of presentation	3	3
13.	Poise and professionalism	2	2
14.	Team participation (perfect score if all participate)	10	10
15.	Use of allotted time	10	9
	TOTAL	100	87

Instructor comments on back of this sheet.

Figure 7.17 This evaluation form was used to grade a team's oral presentation.

7.7 Making and Evaluating the Presentation

You should inspect the room in which the presentation is to be made in advance of the meeting. You also should view projected visuals from various audience locations. Projectors and visual-aid equipment must be positioned and focused before the audience arrives, and remote controls for slide projectors should be readied for use. You should be aware of where to stand so as to not block anyone's view.

Delivery

You should move through the presentation at a moderate pace while emphasizing information on the visual aids with a pointer (**Fig. 7.16**). A positive approach in selling ideas should not yield to deceptive high-pressure salesmanship. The presenter should be frank in pointing out weaknesses in a design and show alternatives that compensate for them.

Conclusions and recommendations should be supported by data and analyses. A recommendation to reject or accept a design should be support-

Project Uniqueness

TEAM NO. __5__ PROJECT __TOY MFGR__

	NAMES	NO. (N = 7)	% CONTRIBU-TION (C)	F = NC	GRADE (G)
1.	BROWN, G.		14.3	100	83
2.	PRISK, A.		14.3	100	83
3.	SMITH, L.		18.0	126	90
4.	REED, T.		14.3	100	83
5.	POTTER, M.		14.3	100	83
6.	FLYNN, O.		14.3	100	83
7.	ROSS, N.		10.5	74	75
8.	_____				

100%

EVALUATION BY INSTRUCTOR: Degree of uniqueness and originality in solution.

		Max. Value	Points Earned
1.	Serves a needed function	10	8
2.	Functions effectively	10	7.5
3.	Needed by consumers	10	9
4.	Simple, uncomplicated solution	10	8.5
5.	Reasonable, attractive price	10	7
6.	An attractive investment for marketing	10	9
7.	Level of imagination and ingenuity	10	8
8.	Aesthetically pleasing and attractive	10	8
9.	Competition of similar products on the market	10	9
10.	Degree of fulfillment of problem statement	10	9
		100	83

Additional comments by the instructor are on the back of this sheet.

Figure 7.18 This evaluation form was used to grade the uniqueness of a team's project.

ed by reasons. A period for questions and answers should follow the presentation for clarification purposes. If available, a technical report (discussed in the next section) should be given to the audience.

Critique

When your team gives a presentation to the class, both your design recommendations and the skill of your presentation will be evaluated and critiqued. The form shown in **Fig. 7.17** is typical of the evaluation form that may be used for your critique. It can also be used as a guide in preparing the presentation. The names of the team members are listed at the top of the sheet. As a group, your team must agree on the percent contribution of each member to the project prior to presentation. The sum of the percent contributions of all team

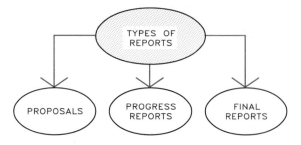

Figure 7.19 The three basic types of written reports.

members must equal 100 percent. The F-factor for each member is the number of members (N) times the contribution of each (C). The chart in Appendix 46 illustrates how an individual's contribution to the project is translated into his or her individual grade by using the F-factor.

The form shown in **Fig. 7.18** is used to evaluate the degree of creativity in the project's solution. The grade given by the instructor for creativity is proportionally divided among the team members using the same technique as for presentations. The chart in Appendix 46 is used to arrive at individual grades by using their F-factors.

7.8 Written Reports

Engineers, technologists, and technicians must know how to prepare a well-written report. The three basic types of written reports are proposals, progress reports, and final reports (**Fig. 7.19**).

Proposals

Proposals are written to establish the need for projects and to obtain authorization of funds and support to pursue them. A proposal outlines data, costs, specifications, time schedules, personnel requirements, completion dates, and other information concerning the project. The purpose of the project is stated, with emphasis on its benefit to the client or organization.

Proposals should reflect the interests and language of the reader. For instance, investors are interested in profits and returns, whereas engineers are more concerned with function and feasibility. Typical elements of a proposal are the following:

- **Statement of the problem.** Identify the problem and its purpose.
- **Method of approach.** Outline procedures for attacking the problem.
- **Personnel needs and facilities.** Itemize requirements for equipment, space, and personnel.
- **Time schedule.** Give estimated completion dates for each phase of the project.
- **Budget.** Itemize the funds required for each phase of the project.
- **Summary.** Review the important points made.

Progress Reports

Progress reports are periodic reports on the status of a project. Usually they are brief and may take the form of a letter or memo. They generally review progress and project the outlook for further progress, including the need for increases or decreases in expenditures or time.

Final Reports

The most comprehensive type of written reports are final reports, which contain five sections:

1. Problem identification.
2. Method.
3. Body.
4. Findings.
5. Conclusions and recommendations.

Some reports present the conclusions at the beginning and others at the end. The order of presentation will vary with the requirements of your instructor or employer.

Report Format Typically, the sequence of a technical report's contents (**Fig. 7.20**) is as follows:

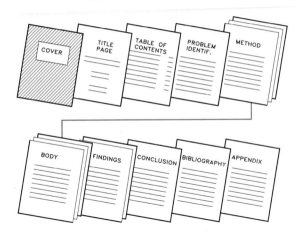

Figure 7.20 The elements of a typical technical report.

1. **Cover.** The report should be inserted in a binder, with its title and author indicated on the cover.
2. **Letter of transmittal (optional).** The second page may be a letter of transmittal, describing briefly the contents of the report and the reasons for the project.
3. **Title page.** The title page contains the title of the report, the name of the person or team that prepared it, and the date.
4. **Table of contents.** The major headings in the report and their page numbers are listed.
5. **Table of illustrations (optional).** A list of the illustrations contained in the report may be included, especially in long, formal reports having many illustrations.
6. **Problem identification.** (Actual heading should be appropriate to the report rather than this general term.) This section explains the importance of and need for a solution to the problem by outlining background information on the problem.
7. **Method.** (Actual heading should be appropriate to the report rather than this general term.) This section should cover the general method used in solving the problem.

8. **Body**. (Actual heading should be appropriate to the report rather than this general term.) This section is the main part of the report. It describes the data collected and analyzed and the steps taken to solve the problem. Subheadings should be used to emphasize the various parts of the report's body.

9. **Findings**. (Actual heading should be appropriate to the report rather than this general term.) Project findings should relate clearly to the data gathered, analyzed, and described in the preceding section.

10. **Conclusions**. Conclusions should be based on the findings; recommendations may be included as necessary.

11. **Bibliography**. The reference materials—books, magazines, brochures, interviews—used in the preparation of the report are listed alphabetically by author.

12. **Appendix**. The appendix includes information (such as drawings, sketches, raw data, brochures, and letters) that supplements (often in more detail) the main sections of the report.

Common Omissions Topics that are often overlooked and omitted from reports by students are sales estimates, advertising methods, shipping costs, product packaging, miscellaneous overhead expenses, and recommendation summaries. Each of these topics must be adequately covered to provide a complete understanding of the total project.

Illustrations Technical reports should be liberally illustrated. Drawings and other illustrations should be numbered and given captions that describe them and relate them to the text. The text should refer to each figure by number and explain the figure's contents. For example, "Figure 6 shows the number of boats sold between 1983 and 1994" identifies the figure being referred to and what it depicts.

Illustrations for the report should be drawn or reproduced on opaque paper rather than tracing

Written Report

TEAM NO. 5   PROJECT TOY MFGR

	NAMES	NO. (N = 7)	% CONTRIBUTION (C)	F = NC	GRADE (G)
1.	BROWN, G.		14.3	100	92
2.	PRISK, A.		14.3	100	92
3.	SMITH, L.		12.0	84	87
4.	REED, T.		16.3	114	93
5.	POTTER, M.		10.0	70	82
6.	FLYNN, O.		8.8	62	79
7.	ROSS, N.		14.3	100	92
8.					

	EVALUATION BY INSTRUCTOR 100%	Max. Value	Points Earned
1.	Use of an appropriate cover	2	2
2.	Inclusion of an evaluation sheet	2	2
3.	Inculsion of a proper letter of transmittal	2	2
4.	Correct title page	2	2
5.	Proper table of contents	2	2
6.	Sufficient introduction to the report	5	4
7.	Thoroughness in identifying the problem	10	8
8.	Continuity and quality of the body the report	10	9
9.	Collection and presentation of background data	5	4.5
10.	Justification of major decisions	5	5
11.	Review of costs, overhead expenses, shipping costs and similar expenses	5	5
12.	Arrival at strong conclusion and recommendation	5	4.5
13.	Sufficient number of graphs and graphics	10	9
14.	Quality of graphics	10	9
15.	Bibliography—form and content	5	5
16.	Use of footnotes	5	5
17.	Appendix—content and form	5	5
18.	Form and appearance of report (spelling, punctuation, margins, typing, neatness)	10	9
		100	92

Figure 7.21 This evaluation form was used to grade a team's written report.

paper; ink illustrations are preferred. Drawings should be positioned on the sheet so that they can be read from the bottom or from the right. Large drawings should be folded to 8½ by 11 inches to fit the binder and they should be easy to unfold for reading.

Evaluating a Written Report A good way to begin learning is to work on a team project, write a technical report covering it, and have your instructor evaluate it. Again, the team decides the contributions of team members, which are needed to determine each person's grade for the project. The evaluation form shown in **Fig. 7.21** will help you and your classmates plan a team project, write the technical report, and know how the report is to be evaluated. The chart in Appendix 46 saves calculating individual grades.

1. Prepare a checklist for evaluating an oral presentation by one of your classmates. List items to consider and develop a point scale for them. Devise a rating system to arrive at an overall evaluation.

2. Use 3-by-5-inch cards to plan a flip-chart presentation that will last no more than five minutes. The subject of your presentation may be of your choosing or one assigned by your instructor. Some examples are (a) your career plans for the first two years after graduation, (b) the role of this course in your overall educational program, (c) the importance of effective communication, (d) the need for a design project that you are proposing, and (e) a comparison of engineering with another profession.

3. Prepare graphical aids for an oral presentation using the methods and materials covered in this chapter.

4. Using the planning cards developed in Problem 2, prepare a five-minute briefing on a technique that you choose or assigned by your instructor. Present this briefing to your class.

5. Assume that you are an engineer responsible for developing a proposal for a project that could result in a sizable contract for your firm. Make a list of instructions to give to your assistants for their help in preparing a presentation for a group of twenty people, ranging in background from bankers to engineers. Your instructions should outline the materials needed, types and number of graphical aids required, method of projection or presentation, assistance needed during the presentation, room seating arrangements, and other factors. Your outline should cover the entire presentation for the length of time you think most desirable. Select a topic or use one assigned by your instructor.

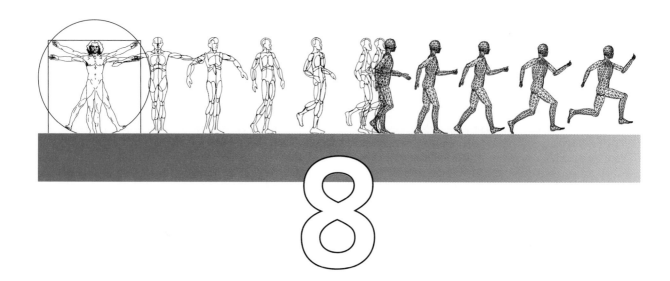

Implementation

8.1 Introduction

Implementation is the final step of the design process, in which the design becomes a reality. The designer details the product in working drawings with specifications and notes for its fabrication. Graphical methods are particularly important during implementation, because all products are manufactured from working drawings and specifications. Implementation also involves the packaging, warehousing, distribution, and sales of the manufactured product.

8.2 Working Drawings

Working drawings, with orthographic views, dimensions, and notes, describe how to make the individual parts of a product. **Figure 8.1** shows a computer-generated working drawing of a single part. Properly executed working drawings ensure that the resulting products will be identical when the instructions on the drawings are followed, regardless of the shop in which they are made.

When making working drawings, designers draw several parts on the same sheet without attempting to arrange them in relationship to each other or in order of their assembly. **The names of the parts, their identifying numbers, the quantity required, and the materials to be used in making them are noted near the views.**

8.3 Specifications

Specifications are written notes and instructions that supplement the information shown in drawings. Specifications may be prepared as separate typed documents that accompany drawings or that stand alone when graphical representation is unnecessary. Instructions such as the following are adequate as written specifications without drawings:

> METALLURGICAL INSPECTION IS
> REQUIRED BEFORE MACHINING.

or

> PAINT WITH TWO COATS OF FLAT BLACK
> PAINT (NO. 780) AFTER FINISHING.

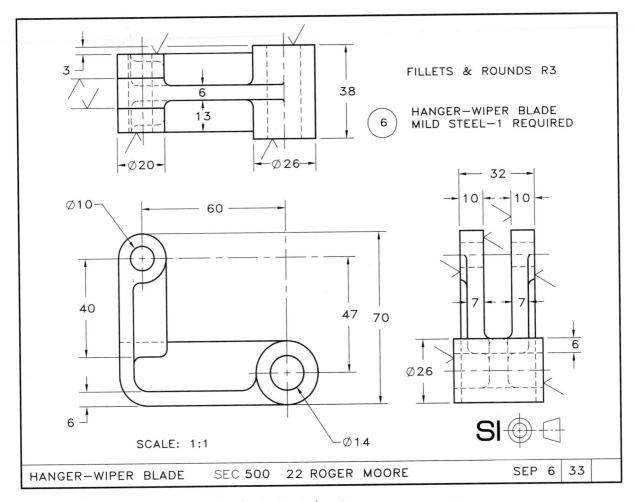

Figure 8.1 This computer-generated working drawing is a single part.

When space permits, specifications should be given on the working drawing rather than in a separate document.

8.4 Assembly Drawings

Assembly drawings illustrate how individual parts are to be put together to become the final product. They can be drawn as three-dimensional pictorials or orthographic views that are fully assembled, fully exploded, or partially exploded. **Figure 8.2** shows a partially exploded orthographic view of an assembly with part numbers in balloons. An assembly drawing usually contains a parts list for easy reference.

8.5 Miscellaneous Considerations

After preparing drawings and specifications, designers must consider other aspects of implementation: product packaging, storage, shipping, and marketing.

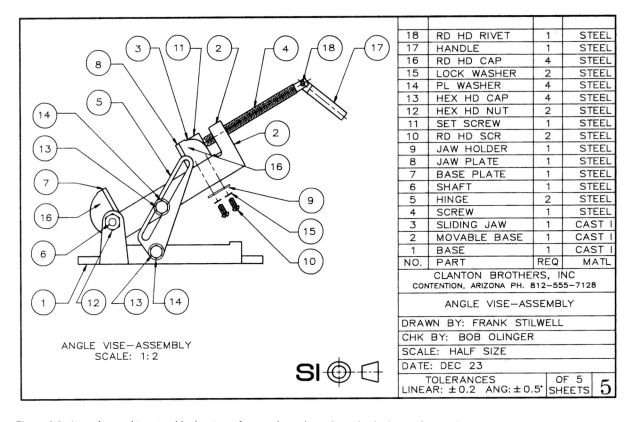

18	RD HD RIVET	1	STEEL
17	HANDLE	1	STEEL
16	RD HD CAP	4	STEEL
15	LOCK WASHER	2	STEEL
14	PL WASHER	4	STEEL
13	HEX HD CAP	4	STEEL
12	HEX HD NUT	2	STEEL
11	SET SCREW	1	STEEL
10	RD HD SCR	2	STEEL
9	JAW HOLDER	1	STEEL
8	JAW PLATE	1	STEEL
7	BASE PLATE	1	STEEL
6	SHAFT	1	STEEL
5	HINGE	2	STEEL
4	SCREW	1	STEEL
3	SLIDING JAW	1	CAST I
2	MOVABLE BASE	1	CAST I
1	BASE	1	CAST I
NO.	PART	REQ	MATL

CLANTON BROTHERS, INC
CONTENTION, ARIZONA PH. 812-555-7128

ANGLE VISE—ASSEMBLY

DRAWN BY: FRANK STILWELL

CHK BY: BOB OLINGER

SCALE: HALF SIZE

DATE: DEC 23

TOLERANCES
LINEAR: ±0.2 ANG:±0.5° OF 5 SHEETS 5

ANGLE VISE—ASSEMBLY
SCALE: 1:2

Figure 8.2 An orthographic assembly drawing of a vise shows how the individual parts fit together.

Packaging

In some industries such as the toy industry, packaging is elaborate and may be as expensive as the product. Designers must be aware of packaging problems as they develop a design because a product that is difficult to package will cost more. Many products are shipped partially disassembled to make packaging easier and cheaper.

Storage

Most manufacturers maintain an inventory of products for shipment. Therefore warehousing costs must be figured into the product's final selling price.

Shipping

Industries that locate warehouse facilities in the middle of their market areas have lower shipping costs than those with warehouses at the edges of their market areas.

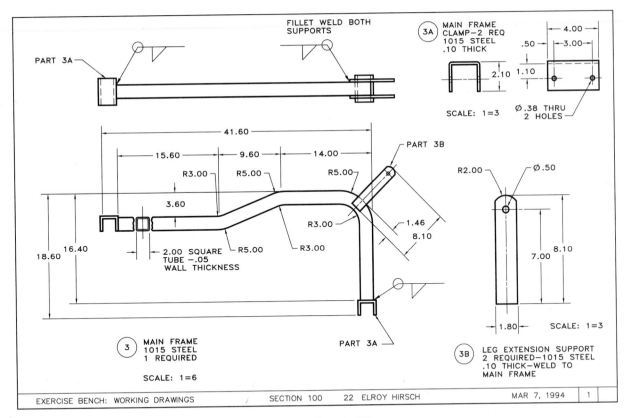

Figure 8.3 This working drawing depicts exercise bench parts (sheet 1 of 5).

Marketing

Designers must be concerned with all aspects of a product after it enters the marketplace, including its marketability and consumer acceptance. Complaints about a product's reliability and function are important to designers, alerting them to design or manufacturing defects that must be overcome in future versions of the product.

8.6 Implementation: Exercise Bench

To illustrate implementation of a product design, we return to the exercise bench, which was introduced in Chapter 3 and has been used to demonstrate the application of each step in the design process.

Working Drawings

The two working drawings shown in **Figs. 8.3** and **8.4** depict some details of the exercise bench design. Additional working drawings are required to show the other parts of the bench, which are dimensioned in decimal inches. Standard parts to be purchased from suppliers are not drawn but are

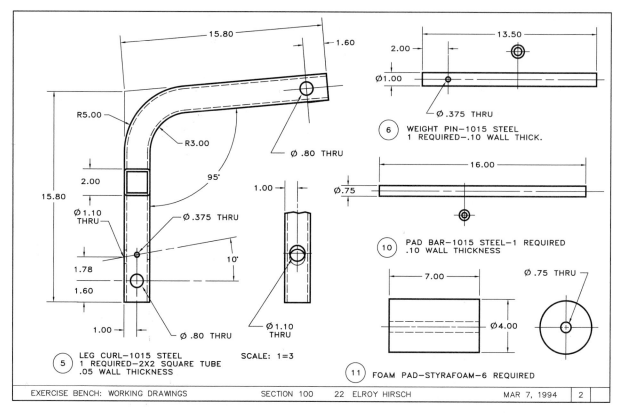

Figure 8.4 This working drawing depicts additional exercise bench parts (sheet 2 of 5).

itemized on the drawing, given part numbers, and listed in the parts list on the assembly drawing.*

Assembly Drawing

Figure 8.5 shows an assembly drawing that illustrates how the parts are to be assembled after they have been made. The assembly is shown pictorially, with the different parts identified by numbered balloons attached to leaders. The parts list identifies each part by number and describes it generally.

*This particular design was developed and patented and is marketed by Weider Health and Fitness, 2100 Erwin Street, Woodland Hills, CA 91367.

Packaging

The Weider exercise bench is packaged in a corrugated cardboard box and weighs approximately 40 pounds (**Fig. 8.6**). It is shipped unassembled so that it will fit into a smaller carton for ease of handling during shipment (**Fig. 8.7**).

Storage

An inventory of benches must be maintained to meet retailer demand. The need to hold inventory increases overhead costs for interest payments, warehouse rent, warehouse personnel, and loading equipment.

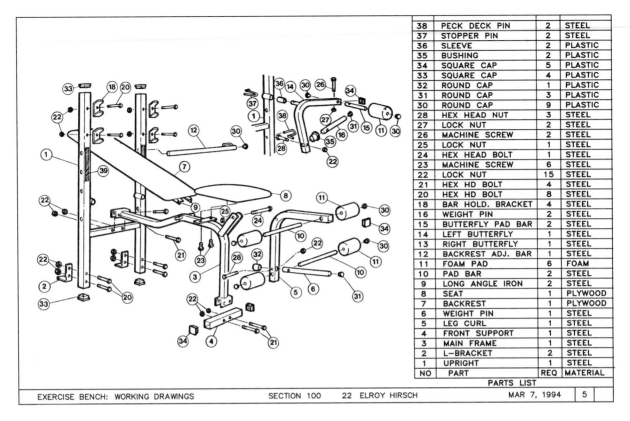

38	PECK DECK PIN	2	STEEL
37	STOPPER PIN	2	STEEL
36	SLEEVE	2	PLASTIC
35	BUSHING	2	PLASTIC
34	SQUARE CAP	5	PLASTIC
33	SQUARE CAP	4	PLASTIC
32	ROUND CAP	1	PLASTIC
31	ROUND CAP	3	PLASTIC
30	ROUND CAP	9	PLASTIC
28	HEX HEAD NUT	3	STEEL
27	LOCK NUT	2	STEEL
26	MACHINE SCREW	2	STEEL
25	LOCK NUT	1	STEEL
24	HEX HEAD BOLT	1	STEEL
23	MACHINE SCREW	6	STEEL
22	LOCK NUT	15	STEEL
21	HEX HD BOLT	4	STEEL
20	HEX HD BOLT	8	STEEL
18	BAR HOLD. BRACKET	4	STEEL
16	WEIGHT PIN	2	STEEL
15	BUTTERFLY PAD BAR	2	STEEL
14	LEFT BUTTERFLY	1	STEEL
13	RIGHT BUTTERFLY	1	STEEL
12	BACKREST ADJ. BAR	1	STEEL
11	FOAM PAD	6	FOAM
10	PAD BAR	2	STEEL
9	LONG ANGLE IRON	2	STEEL
8	SEAT	1	PLYWOOD
7	BACKREST	1	PLYWOOD
6	WEIGHT PIN	1	STEEL
5	LEG CURL	1	STEEL
4	FRONT SUPPORT	1	STEEL
3	MAIN FRAME	1	STEEL
2	L—BRACKET	2	STEEL
1	UPRIGHT	1	STEEL
NO	PART	REQ	MATERIAL
	PARTS LIST		

| EXERCISE BENCH: WORKING DRAWINGS | SECTION 100 | 22 ELROY HIRSCH | MAR 7, 1994 | 5 |

Figure 8.5 This assembly drawing demonstrates how the parts of the Weider exercise bench design are to be assembled (sheet 5 of 5). (Courtesy of Weider Health and Fitness.)

Shipping

Shipping costs for all types of carriers (rail, motor freight, air delivery, and mail services) must be evaluated. The shipping cost for a Weider bench with its accessories is $10–$15, depending on distance, when shipped one at a time by United Parcel Service. The cost per unit is about 50% less when units are shipped in bundles of ten to the same destination.

Accessories

Examples of accessories, or add-ons, are the butterfly attachments for arm exercises (**Fig. 8.8**). Accessories enable buyers to upgrade the basic product in stages, which can increase product marketability and sales.

Prices

The retail price of the Weider bench is about $100. This type of product generally retails for about five or six times the cost of manufacturing them (materials and labor). Retailers receive approximately a 40% margin, distributors earn about 10%, and the remainder of the price represents advertising costs and the other miscellaneous costs mentioned previously. The consumer pays all of these costs (pro-rated to each exercise bench) as part of the purchase price.

Figure 8.6 The Weider exercise bench is shipped in a corrugated cardboard box to the retailer or directly to the consumer.

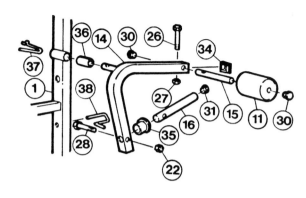

Figure 8.8 The butterfly exercise attachments are examples of accessories designed for use with the Weider exercise bench. (Courtesy of Weider Health and Fitness.)

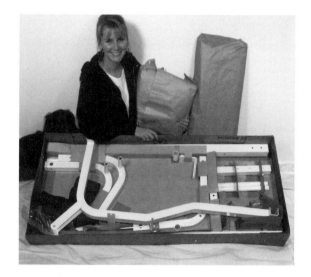

Figure 8.7 The exercise bench is packaged unassembled and flat for ease of packaging and handling during shipment.

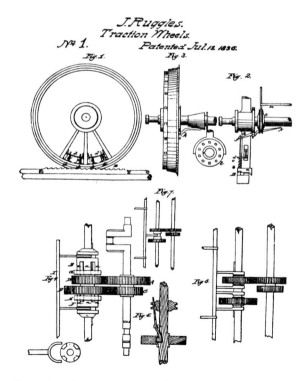

Figure 8.9 The U.S. Patent and Trademark Office issued this first patent in 1836.

Figure 8.10 Thomas Edison received this patent for the electric lamp in 1880.

8.7 Patents

Inventors of original processes or products should investigate the possibility of obtaining patents on them from the U.S. Patent and Trademark Office (PTO) before disclosing the invention. As an item of interest, the first U.S. patent was for traction wheels (**Fig. 8.9**). The patent procedure is outlined in *General Information Concerning Patents,* a publication available from the PTO that was the primary reference for the following material.

What May Be Patented?

Any person who "invents or discovers any new and useful process, machine, manufacture, or composition of matter, may obtain a patent," subject to the conditions and requirements of law. These categories include everything made by humans and the processes for making them (**Fig. 8.10**).

Inventions used for the development of nuclear and atomic weapons for warfare are not patentable because they are not considered "useful." Also, a design for a mechanism that will not operate as described is not patentable. An idea or concept for a new invention is not patentable; it must be designed and described in detail before it can be considered for patent registration.

Who May Apply for a Patent?

Only the inventor may apply for a patent. A patent given to a person who was not the inventor would be void and the recipient subject to prosecution for perjury. However, the executor of a deceased inventor's estate may apply for a patent, and two or more people may apply for a patent as joint inventors.

Patent Rights

An inventor granted a patent has the right to exclude others from making, using, or selling the invention throughout the United States for seventeen years. At the end of that time, anyone may make, use, or sell the invention without authorization from the patent holder.

Application for a Patent

An inventor applying for a patent must provide:

1. a completed form that includes a petition, specification (description and claims), and oath or declaration;

2. a drawing, if a drawing is possible; and

3. the filing fee.

Petition and Oath In the petition and oath (usually on one form) the inventor asks to be given a patent on the invention and declares that he or she is the original inventor of the device described in the application.

Specification The inventor must submit a written specification, describing the invention in detail so

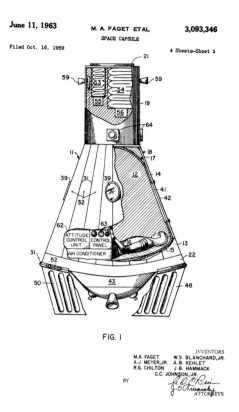

June 11, 1963   M. A. FAGET ETAL   3,093,346
SPACE CAPSULE

Filed Oct. 16, 1959   4 Sheets-Sheet 1

FIG. I

INVENTORS
M.A. FAGET   W.S. BLANCHARD,JR.
A.J. MEYER,JR.   A.B. KEHLET
R.G. CHILTON   J.B. HAMMACK
C.C. JOHNSON, JR.
BY
ATTORNEYS

Figure 8.11 This patent drawing of a space capsule was developed by the National Aeronautics and Space Administration (NASA). (Courtesy of the U.S. Patent and Trademark Office.)

that a person skilled in the field to which the invention pertains can produce the item. Drawings should carry figure numbers and contain part numbers for text references (**Fig. 8.11**).

Claims The inventor's claims are brief descriptions of the invention's features that distinguish it from already patented items. The PTO studies claims to judge the novelty and patentability of an invention.

Fee As part of the application for a patent, the inventor must submit a $170 filing fee. After the PTO has accepted the application, the applicant has three months in which to pay a $280 issue fee. There are other miscellaneous charges for multiple

claims. As a general rule, patents cost the inventor about $2000, excluding attorney's fees.

8.8 Patent Drawings

A booklet, *Guide for Patent Draftsmen* (available from the U.S. Government Printing Office), outlines the required format for patent drawings. If the inventor cannot furnish drawings, the PTO will recommend a drafter who can prepare them at the inventor's expense.

Patent Drawing Standards
Patent drawings must meet the following standards.

Paper and Ink Drawings must be on pure white paper of the thickness of a two- or three-ply Bristol board with a surface that is calendered and smooth to permit erasure and correction. India ink is required for permanence and solid black lines. The use of white pigment to cover errors is not allowed.

Sheet Size and Margins Sheet size must be 8½ by 14 inches (21.6 by 35.6 cm) or 21.0 by 29.7 cm. All sheets in a particular application must be the same size. One of the shorter sides is regarded as the top of the sheet. On 8½-by-14-inch sheets, the top margin is 2 inches and the side and bottom margins are ¼ inch. Margin border lines cannot be drawn on the sheets, but all work must be included within the margins. Sheets may be punched with two ¼-inch holes, with their centerlines ¹¹⁄₁₆ inch below the top edge and 2¾ inches apart and centered from the sides of the sheet. The margins for 21.0-×-29.7-cm sheets are 2.5 cm from the top, 2.5 cm from the left, 1.5 cm from the right, and 1 cm from the bottom.

Character of Lines All lines and lettering must be absolutely black regardless of line thickness. Freehand work is to be avoided.

Hatching and Shading Hatching lines used to shade the surface of an object should be parallel and at least ¹⁄₂₀ inch apart (**Fig. 8.12**). Heavy lines

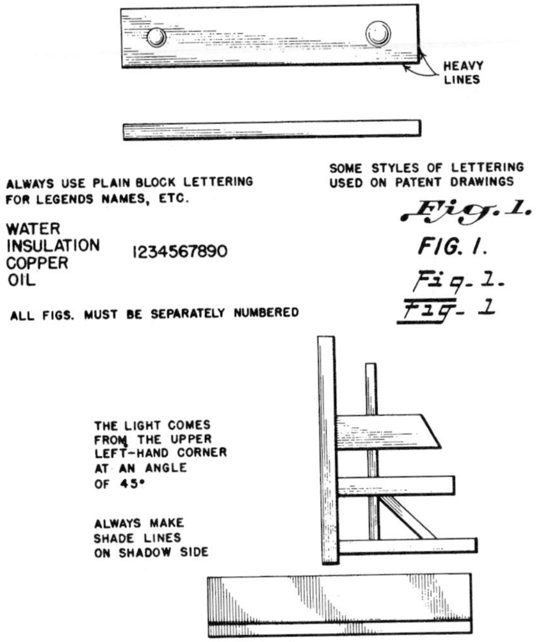

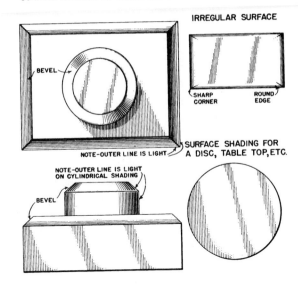

ALWAYS USE PLAIN BLOCK LETTERING
FOR LEGENDS NAMES, ETC.

SOME STYLES OF LETTERING
USED ON PATENT DRAWINGS

WATER
INSULATION
COPPER
OIL

1234567890

Fig. 1.

FIG. 1.

Fig. 1.

Fig. 1

ALL FIGS. MUST BE SEPARATELY NUMBERED

THE LIGHT COMES
FROM THE UPPER
LEFT-HAND CORNER
AT AN ANGLE
OF 45°

ALWAYS MAKE
SHADE LINES
ON SHADOW SIDE

HEAVY
LINES

Figure 8.12 These are typical examples of lines and lettering recommended for patent drawings.

SURFACE SHADING ON BEVEL EDGES

IRREGULAR SURFACE

BEVEL

SHARP
CORNER

ROUND
EDGE

NOTE-OUTER LINE IS LIGHT

SURFACE SHADING FOR
A DISC, TABLE TOP, ETC.

NOTE-OUTER LINE IS LIGHT
ON CYLINDRICAL SHADING

BEVEL

Figure 8.13 Several techniques of representing surfaces and beveled planes may be used on patent drawings.

are used on the shade side of the views if they do not confuse the drawing. The light is assumed to come from the upper left-hand corner at an angle of 45°. **Figure 8.13** depicts several types of surface delineation.

Scale The scale must be large enough to show the mechanism without crowding when the drawing is reduced for reproduction. Portions of the mechanism may be drawn at a larger scale to show details.

Reference Characters The drafter should identify different views of a mechanism by consecutive plain, legible numerals at least ⅛-inch high figure numbers, not encircled, and placed close to their parts (**Fig. 8.14**). A blank space should be provided on hatched surfaces if numbers are to be placed on them. The same part appearing in more than one view on the drawing should be labeled with the same numeral.

Symbols Symbols used to represent various materials in sections, electrical components, and mechanical devices are recommended by the PTO and conform to engineering drawing standards.

Signature and Names The signature or name of the applicant and the signature of the attorney or agent are placed in the lower right-hand corner of each sheet within the marginal lines or below the lower marginal line.

Views Figures should be numbered consecutively in order of their appearance. Figures may be plan, elevation, section, perspective, or detail views. Exploded views may be used to describe an assembly of multiple parts. Large parts may be broken into sections and drawn on several sheets if this approach is not confusing. Removed sections may be used if the cutting plane is labeled to indicate the section by number. All sheet headings and sig-

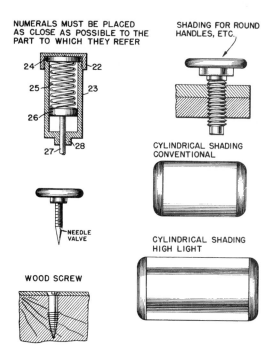

NUMERALS MUST BE PLACED AS CLOSE AS POSSIBLE TO THE PART TO WHICH THEY REFER

SHADING FOR ROUND HANDLES, ETC.

CYLINDRICAL SHADING CONVENTIONAL

NEEDLE VALVE

CYLINDRICAL SHADING HIGH LIGHT

WOOD SCREW

Figure 8.14 Various methods of numbering parts and rendering details may be used for patent drawings.

natures are to be placed in the same position on the sheet whether the drawing is read from the bottom or the right of the sheet. Completed drawings should be sent flat, protected by heavy board, or rolled in a suitable mailing tube.

8.9 Patent Searches

A patent can be granted only after PTO examiners have searched existing patents to verify that the invention has not been patented previously. With more than 4,500,000 patents on record, a search is the most time-consuming part of obtaining a patent. Most inventors employ patent attorneys or agents to do preliminary searches for possible infringement on other patents.

8.10 Questions and Answers About Patents

We used the PTO's pamphlet, *Questions and Answers About Patents,* to answer the following questions.

Nature and Duration of Patents

1. **Q.** *What is a patent?*

   A. A patent is a grant issued by the U.S. Government, giving an inventor the right to exclude all others from making, using, or selling his or her invention within the United States, its territories, and possessions.

2. **Q.** *For how long is a patent granted?*

   A. Seventeen years from the date on which it is issued; except for patents on ornamental designs, which are granted for terms of 3½, 7, or 14 years.

3. **Q.** *May the term of a patent be extended?*

   A. Only by special act of Congress, which occurs rarely and only under exceptional circumstances.

4. **Q.** *Does the person granted the patent have any control over the patent after it expires?*

   A. No. Anyone has the right to use an invention covered in an expired patent so long as they do not use features covered in other unexpired patents.

5. **Q.** *On what subject matter may a patent be granted?*

   A. A patent may be granted to the inventor or discoverer of any new and useful process, machine, manufacture, or composition of matter, or any new and useful improvement thereof, or on any distinct and new variety of plant, or on any new, original, and ornamental design for an article of manufacture.

6. **Q.** *What may not be patented?*

   A. A patent may not be granted on a useless device, on printed matter, on a method of doing business, on an improvement in a device that would be obvious to a person skilled in the art, or on a machine that will not operate, particularly on alleged perpetual motion machines.

Meaning of Words "Patent Pending"

7. **Q.** *What do "patent pending" and "patent applied for" mean?*

   A. They are used by a manufacturer or seller of an article to indicate that a patent application for that article is on file with the U.S. Patent and Trademark Office. Those using these terms falsely to deceive the public can be fined.

Patent Applications

8. **Q.** *I have made some changes and improvements in my invention after my patent application was filed with the PTO. May I amend my patent application by adding a description or illustration of these features?*

   A. No. The law provides that new matter shall not be introduced into a patent application. You should call to the attention of your patent agent any such changes you may make, or plan to make, so steps may be taken for your protection.

9. **Q.** *How does someone apply for a patent?*

   A. By making application to the Commissioner of Patents, Patent and Trademark Office, Washington, DC, 20231.

10. **Q.** *What are the PTO's fees in connection with filing of an application for patent and issuance of the patent?*

    A. A filing fee of $170 plus certain additional charges for claims, depending on their number and the manner of their presentation, are required when the application is filed. An issue fee of $280 plus certain printing charges are required if the patent is to be granted.

11. **Q.** *Are models required as a part of the application?*

    A. Only in the exceptional cases. The PTO has the authority to require that a model be submitted, but rarely exercises it.

12. **Q.** *Is it necessary for me to go to the PTO in Washington to transact business concerning patent matters?*

    A. No. Most business is conducted by correspondence. Interviews regarding pending applications can be arranged with examiners if necessary and often are helpful.

13. **Q.** *Can the PTO give me advice about whether to apply for a patent?*

    A. No. It can only consider the patentability of an invention when an application comes before it.

14. **Q.** *Is there any danger that the PTO will give others information contained in my application while it is pending?*

    A. No. All patent applications are kept secret until the patent is issued. After the patent is issued, the PTO file containing the application and all correspondence leading to its issuance is made available in the Patent Office Search Room to anyone, and copies may be purchased from the PTO.

15. **Q.** *May I write to the PTO about my application after it is filed?*

    A. The PTO will answer your inquiries about the status of the application and indicate whether the application has been rejected, allowed, or is awaiting action. However, you should forward correspondence through your patent attorney or agent.

16. **Q.** *What happens when two inventors apply separately for a patent on the same invention?*

    A. The PTO declares an "interference" and requires that testimony be submitted to determine which inventor is entitled to the patent.

17. Q. *May applications be examined out of their regular order?*

A. No. All applications are examined in the order in which they are filed, except under special conditions.

When to Apply for a Patent

18. Q. *I have been making and selling my invention for the past thirteen months and have not filed any patent application. Is it too late for me to apply?*

A. Yes. A patent may not be obtained if the invention has been in public use or for sale in this country for more than a year prior to application. Your own use and sale of it for more than a year before filing will bar your right to a patent as though someone else had done so.

19. Q. *I published an article describing my invention in a magazine thirteen months ago. Is it too late to apply for a patent?*

A. Yes. The inventor is not entitled to a patent if the invention has been described in a printed publication anywhere in the world more than a year before filing an application.

Who May Obtain a Patent

20. Q. *If two or more people work together on an invention, to whom will the patent be granted?*

A. If each had a share in the ideas forming the invention, they are joint inventors and a patent will be issued to them jointly if an application is filed by them jointly. If one person provided all the ideas and the other has only followed instructions in making the device, the person contributing the ideas is the sole inventor and the patent application and patent should be in his or her name only.

21. Q. *If one person furnishes all the ideas for an invention and someone else employs that person or furnishes the money for building and testing the invention, should the patent application be filed by them jointly?*

A. No. The application must be signed, executed, sworn to, and filed in the name of the inventor, who is the person furnishing the ideas, not the employer or the person furnishing the money.

22. Q. *May a patent be granted if an inventor dies before filing an application?*

A. Yes. The application may be filed by the executor or administrator of the inventor's estate.

23. Q. *While in England this summer, I found an ingenious article that has not been introduced into the United States or patented. May I obtain a U.S. patent on it?*

A. No. A U.S. patent may be obtained only by the inventor, not by someone learning of someone else's invention.

Ownership and Sale of Patent Rights

24. Q. *May the inventor sell or otherwise transfer the right to the patent or patent application to someone else?*

A. Yes. The inventor may sell all or part of the interest in the patent application or patent to anyone by a properly worded legal assignment. However, the application for a patent must be filed in the name of the inventor, not in the name of the purchaser.

25. Q. *Is it advisable to conduct a search of patents and other records before applying for a patent?*

A. Yes. If the device has been patented previously, making application is useless. A patent search avoids the expense of filing a needless application.

Technical Knowledge Available from Patents

26. Q. *May I obtain information through patents of what has been done by others to solve a particular problem?*

A. The patents in the Patent Office Search Room in Washington contain a wealth of technical information. It is organized so that you can easily find and review previous work related to your problem or general field of interest. You may review these patents personally, or hire a patent practitioner to do so and send you copies of patents related to your problem.

27. **Q.** *May I make a search or obtain technical information from patents at locations other than the Patent Office Search Room in Washington?*

A. Yes. Libraries have sets of patent copies arranged numerically in bound volumes that may be used for search or other information purposes, as discussed in the answer to Question 28.

28. **Q.** *How do I find technical information in a library collection of patents arranged in bound volumes in numerical order?*

A. You must first find out from the *Manual of Classification* in the library the PTO classes and subclasses that cover your field of interest. By referring to microfilm reels or volumes of the *Index of Patents* in the library, you can then identify patents in these subclasses and look at them in the bound volumes. Further information on this subject is available in the leaflet *Obtaining Information from Patents,* which is available from the PTO.

Infringement of Others' Patents

29. **Q.** *If I obtain a patent on my invention, will that protect me against the claims of others who assert that I am infringing on their patents when I make, use, or sell my own invention?*

A. No. There may be a patent of a more basic nature on which your invention is an improvement. If your invention is a detailed refinement or feature of a basically protected invention, you may not use it without consent of the patentee, just as no one will have the right to use your patented improvement without your consent.

Enforcement of Patent Rights

30. **Q.** *Will the PTO help me prosecute others if they infringe on the rights granted me by my patent?*

A. No. The PTO has no jurisdiction over questions relating to the infringement of patent rights. If your patent is infringed upon, you may sue the infringer at your own expense.

Patent Protection in Foreign Countries

31. **Q.** *Does a U.S. patent give protection in foreign countries?*

A. No. The U.S. patent protects your invention only in this country. If you want to protect your invention in foreign countries, you must file an application in the patent office of each such country within the time required by law.

Problems

Working Drawings

1. Prepare working and assembly drawings as the implementation step of the design process for one of the refinement problems at the end of Chapter 5. Produce the drawings on 11-by-17-in. sheets of tracing vellum or film.

2. Prepare working and assembly drawings for one of the problems assigned or selected from those at the end of Chapter 23. Produce the drawings on 11-by-17-in. sheets of tracing vellum or film.

Patents

3. Write for a copy of a patent that is of interest to you. List the features used as a basis for obtaining the patent.

4. Suggest modifications to the patent obtained in Problem 3. Sketch innovations that would improve the patented mechanism.

5. Write the U.S. Patent and Trademark Office for patent application forms. Prepare a patent application for a simple invention that has been previously patented, such as a fountain pen, drafting instrument, or similar item. Determine the drawings and materials needed to complete your application.

6. Prepare patent drawings in accordance with the standards presented in Section 8.8 to depict the invention selected in Problem 5.

7. Make a list of ideas for products that you believe to be patentable. These may be ideas that you have developed during work on design problems assigned in class.

8. Write a technical report on the history and significance of the patent system and its role in our industrial society. Consult your library and available government publications on patents.

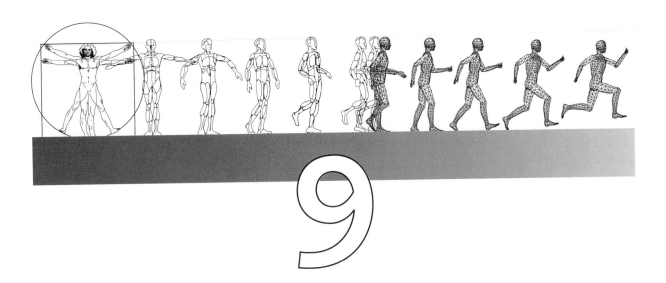

Design Problems

9.1 Introduction

This chapter offers problems that are suitable for both individual assignments and team projects to provide experience in applying the methods of creative problem solving presented in this textbook. Graphics has many applications during the design process: All new products begin with sketches at the preliminary idea step and end with working and assembly drawings at the implementation step.

9.2 The Individual Approach

The solution of short problems (one to two hours) is best suited to students working alone. Although simple design problems may involve fewer details and less depth than comprehensive problems do, the same design steps are involved.

9.3 The Team Approach

An effectively organized team working on a problem represents more talent than the typical individual possesses. However, management of talent becomes as much of the process as solving the problem.

Team Size

Student design teams should have from three to eight members. Three is the minimum number needed for a valid team experience, and four is the number needed to minimize the possibility of domination by one or two members.

Team Composition

In practice, an engineering team often consists of representatives of different departments or even different firms who may be unacquainted. This situation can be advantageous because it reduces the impact of preconceived notions about individuals.

Team Leader

A leader is necessary for teams to function effectively. The leader is responsible for making assignments, ensuring that deadlines are met, and mediating disagreements.

9.4 Selection of a Problem

The best problem for a student design project is one that involves familiar and accessible conditions that can be observed, measured, and inspected. A design for a water-ski rack for an automobile is more feasible than is a design for a support bracket for an airplane. When a student design team selects a design problem, the team should prepare a written proposal identifying the problem and outlining its limits. Assignment of problems by the instructor in the classroom is analogous to assignments by a supervisor in the workplace.

9.5 Problem Specifications

An individual or team may be expected to complete any or all of the following tasks.

Short Problems (One or Two Work Hours)

1. Worksheets that record development of a design procedure (Chapter 2).

2. Freehand sketches of the design for implementation (Chapter 13).

3. Instrument drawings of the solution (Chapters 8 and 23).

4. Pictorial sketches (or drawings made with instruments) illustrating the design (Chapters 3, 13, and 25).

5. Visual aids, flip charts, or other media for presentation to a group (Chapter 7).

Comprehensive Problems (40 to 100 Work Hours)

1. A proposal identifying the problem and outlining an approach for solving it (Chapters 2, 3, and 7).

2. Worksheets documenting the preliminary ideas for a solution (Chapter 4).

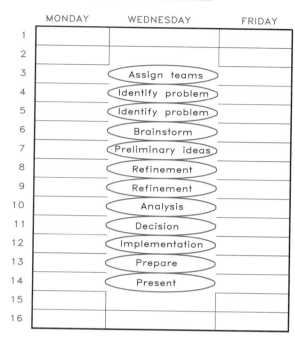

Figure 9.1 A suggested semester schedule shows periods for work on a comprehensive design project.

3. Schematic diagrams, flowcharts, or other graphics to illustrate refinements of the design (Chapter 5).

4. A market survey evaluating the product's possible acceptance and estimated profit (Chapters 3 and 6).

5. A model or prototype for analysis and/or presentation (Chapter 6).

6. Pictorials to illustrate features of the final design solution (Chapters 7 and 25).

7. Dimensioned working drawings and assembly drawings to give details and specifications (Chapters 8 and 23).

8. A written or oral report, illustrated with graphs and diagrams, to explain the method of solution and present conclusions and recommendations (Chapter 7).

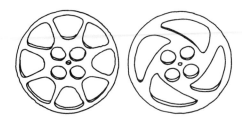

Figure 9.2 Problem 4. A typical movie projector reel and one allowing more finger room.

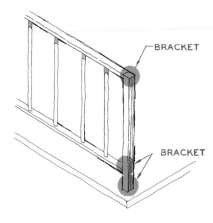

Figure 9.3 Problem 6. A porch railing system.

9.6 Scheduling Team Activities

The semester schedule shown in **Fig. 9.1** is suggested for a comprehensive design project. Spreading design projects over the semester allows time for thinking about the problem, gathering information, and working on the solution. Refer to the exercise bench example in Chapters 3–8 as a guide for carrying out your project.

9.7 Short Design Problems

The following short design problems can be completed in less than two hours.

1. **Lamp bracket.** Design a simple bracket to attach a desk lamp to a vertical wall for reading in bed. It should be removable for use as a conventional desk lamp.

2. **Towel bar.** Design a towel bar for a kitchen or bathroom. Determine optimum size and consider styling, ease of use, and method of attachment.

3. **Pipe aligner for welding.** The initial requirement for joining pipes with a butt weld is to align the pipes. Design a device for aligning 2-to-4-in.-diameter pipes for on-the-job welding.

4. **Film reel (Fig. 9.2).** The film reel used on projectors is difficult to thread because of limited working space. Redesign the typical 12-in.-diameter movie reel shown to allow more space for threading.

5. **Side-mounted mirror.** Design an improved side-mounted rearview mirror for an automobile. Consider aerodynamics, protection from inclement weather, visibility, and other factors.

6. **Railing-post mount.** An ornamental iron railing is to be attached to a wooden porch surface and supported by several 1-in.-square tubular posts, as shown in **Fig. 9.3**. Design one mounting bracket to anchor the posts to the porch surface and a second bracket to connect the railings to the support posts.

7. **Nail feeder.** Design a device that can be attached to a worker's chest for holding roofing nails and that will feed to the worker nails that are lined up and ready for removal and driving.

8. **Paint-can holder.** Paint cans held by their wire bails are difficult to hold and get paint brushes into. Design a holding device that can be attached and removed easily from a gallon-size paint can. Consider weight, grip, balance, and function.

Figure 9.4 Problem 16. A wall-mounted stool.

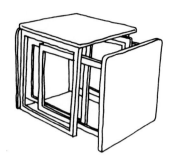

Figure 9.5 Problem 19. Nesting tables.

9. **Self-holding hinge.** Design a hinge that will hold a door in a completely open position to prevent it from remaining slightly ajar.

10. **Audio cassette storage unit.** Design a storage unit for an automobile that will hold several audio cassettes, making them accessible to the driver but not to a thief.

11. **Slide projector elevator.** Design a device for raising a slide projector to the proper angle for projection on a screen. It may be part of the original projector or an accessory to be attached to existing projectors.

12. **Book holder 1.** Design a holder to support a book for reading in bed.

13. **Table leg design.** Do-it-yourselfers build a variety of tables using hollow doors or plywood for the tops and commercially available legs. Determine standard heights for various types of tables and design a family of legs that can be attached to table tops with screws.

14. **Boat rack.** Design an accessory that will enable one person to load a boat (ranging from 14 to 17 ft in length and weighing from 100 to 200 lb) on an automobile, secure it, and later remove it.

15. **Toothbrush holder.** Design a toothbrush holder for a cup and two toothbrushes that can be attached to a bathroom wall.

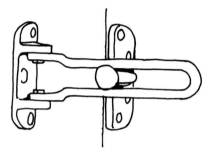

Figure 9.6 Problem 22. A door latch.

16. **Wall-mounted stool (Fig. 9.4).** Design a stool that can be attached to a wall and that will swing out of the way when it is not in use.

17. **Book holder 2.** Design a holder to support a textbook or reference book at a workstation for ease of reading and accessibility.

18. **Clothes hook.** Design a clothes hook that can be attached to a closet door for hanging clothes on.

19. **Nesting tables (Fig. 9.5).** Design three tables that have 18-in.-square table tops and that nest together to form a cube to save space. The tables should be coffee-table height.

20. **Hammock support.** Design a hammock support that will fold up and that will require minimal storage space.

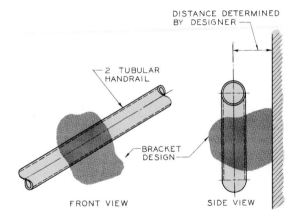

Figure 9.7 Problem 26. A handrail bracket.

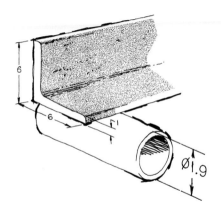

Figure 9.8 Problem 27. A pipe clamp.

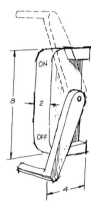

Figure 9.9 Problem 32. A safety lock.

21. **Door stop.** Design a door stop that can be attached to a wall or floor to prevent a door knob from hitting the wall.

22. **Door latch (Fig. 9.6).** Design a door latch that provides security to a home owner or an apartment dweller.

23. **Cup dispenser.** Design a dispenser that can be attached to a vertical wall to hold 2-in.-diameter paper cups that measure 6 in. in height when stacked together.

24. **Drawer handle.** Design a handle for a standard file cabinet drawer.

25. **Paper dispenser.** Design a dispenser that will hold a 6-×-24-in. roll of wrapping paper.

26. **Handrail bracket (Fig. 9.7).** Design a bracket that will support a tubular handrail to be used on a staircase.

27. **Pipe clamp (Fig. 9.8).** A pipe with a 4-in. diameter must be supported by angles that are spaced 8 ft apart. Design a clamp that will support the pipe without drilling holes in the angles.

28. **TV yoke.** Design a yoke to hang from a classroom ceiling that will support a TV set and that will permit it to be adjusted for viewing from various parts of the room.

29. **Flagpole socket.** Design a flagpole socket that is to be attached to a vertical wall.

30. **Cup holder.** Design a holder that will support a soft-drink can or bottle in an automobile.

31. **Gate hinge.** Design a hinge that can be attached to a 3-in.-diameter tubular post to support a 3-ft-wide wooden gate.

32. **Safety lock (Fig. 9.9).** Design a safety lock to hold a high-voltage power switch in the "off" and "on" positions to prevent an accident.

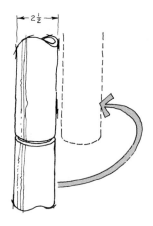

Figure 9.10 Problem 33. A tubular hinge.

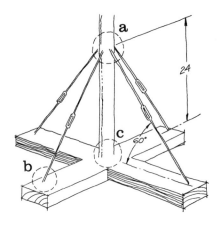

Figure 9.12 Problem 35. Base hardware for a volleyball net.

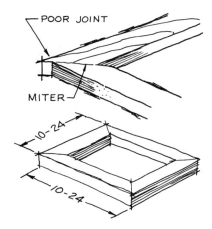

Figure 9.11 Problem 34. A miter fixture.

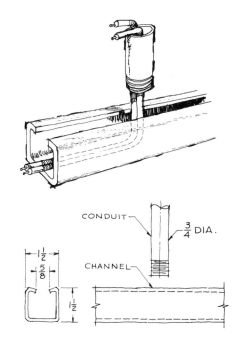

Figure 9.13 Problem 36. A conduit connector hanger.

33. **Tubular hinge (Fig. 9.10).** Design a hinge for portable scaffolding that can be used to hinge 2.5-in. outside diameter, high-strength aluminum pipe in the manner shown.

34. **Miter jig (Fig. 9.11).** Design a jig for assembling mitered wooden frames at 90° angles. The stock used for the frames is rectangular in cross section, varying from 0.75 × 1.5 in. to 1.60 × 3.60 in. and in length from 10 to 24 in.

35. **Base hardware (Fig. 9.12).** Design the hardware needed for points *a, b,* and *c* on a standard volleyball net support. The 7-ft-tall pipes are attached to crossing 2-×-4-in. boards.

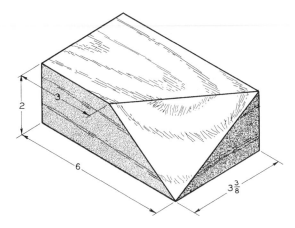

Figure 9.14 Problem 37. A fixture design for sawing a block at an angle.

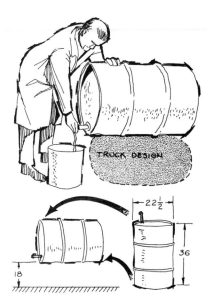

Figure 9.15 Problem 38. A drum truck.

36. Conduit connector hanger (Fig. 9.13). Design an attachment for a 3/4-in. conduit to support a channel used as an adjustable raceway for electrical wiring.

37. Fixture (Fig. 9.14). Design a fixture for a small manufacturer that will permit sawing the corner of blocks as shown.

38. Drum truck (Fig. 9.15). Design a truck that can be used for handling 55-gal drums of turpentine (7.28 lb per gallon) one at a time. Drums are stored in a vertical position and used in a horizontal position. The truck should be useful in tipping a drum into a horizontal position (as shown), as well as for moving the drum.

9.8 Systems Design Problems

Systems problems require analysis of the interrelationship of various components, as well the application of design principles.

39. Multipurpose utility meter. Residences have meters for electricity, water, and gas, which are checked monthly by separate utility companies. Consider the feasibility of combining all the meters into a single unit that one meter-reading service could read. Describe briefly how to organize and implement such a system.

40. Portable bleachers. Design portable bleachers that can be easily assembled, disassembled, and stored. Identify the uses that would justify their production for the general market.

41. Archery range. Determine the feasibility of providing an archery range that can be operated profitably. Investigate the potential market for such a facility and factors such as location, equipment needed, method of operation, utilities, concessions, parking, costs, and fees.

42. Car rental system. Investigate the feasibility of a student-operated car rental system. Determine student interest, cost factors, number of cars needed, prices, personnel needs, storage, maintenance, and so on. Summarize your location, operating cost, and profitability conclusions.

43. Model-airplane field. Investigate the need for a model-airplane field, including space requirements, types of surface needed, sound control, safety factors, and method of operation. Select a site on or near your campus that is adequate for this facility, and evaluate the equipment, utilities, and site preparation required.

44. Overnight campsite. Analyze the feasibility of converting a vacant tract of land near a major highway into sites for overnight campers. Determine the facilities required by the campers and the venture's profitability.

45. Swimming pool study. Determine the cost of building and operating an outdoor swimming pool on your campus. Consider, among other factors, where it could be located, how many students would be likely to use it, and the amounts of equipment and labor necessary to operate it. Would it be financially feasible?

46. Hot-water supply. Your weekend cottage does not have a hot-water supply, but cold water is available from a private well. Design a system that uses the sun's energy in the summer to heat water for bathing and kitchen use. Devise a system for heating the water in the winter by some other source. Determine whether one or both of these systems could be made portable for showers on camping trips and, if so, how.

47. Information center. Design a drive-by information center to help campus visitors find their way around. Determine the best location for it and the informational material needed, such as slides, photographs, maps, sound, and other audiovisual aids.

48. Golf driving range ball-return system. Balls at golf driving ranges are usually retrieved by hand or with a specially designed vehicle. Design a system capable of automatically returning balls to the tee area.

49. Instructional system. Classroom instruction could be improved by providing feedback from the students to the teacher. The teacher could proceed more efficiently by having some idea of whether the class understands the points being made. Devise a system to give students a way to signify understanding, or a lack of understanding, without having to ask questions.

50. Instant motel. Many communities need temporary housing for celebrations and sporting events. Investigate methods of providing an "instant motel" involving the use of tents, vans, trailers, train cars, or other temporary accommodations. Estimate profitability.

51. Mountain lodge. Determine the food supply and other provisions needed for a mountain lodge to accommodate six people who might be snowed-in for two weeks. Identify and explain the features that should be included in the design of the lodge.

52. Campus planning. Assume that your campus is to be used at full capacity, twelve months a year, twenty-four hours a day. Determine the number of students that could be accommodated by the existing classrooms, dormitories, and other facilities. Evaluate teaching schedules, administrative assignments, and service personnel requirements to identify potential problems. Evaluate changes that would occur in parking and pedestrian traffic and the modifications that would be needed in transportation management.

53. Diazo machine layout. A diazo machine will accept drawings along its forty-inch belt at a rate of ten feet per minute. The drawing size used most frequently by your company is 11 by 17 in. The diazo paper must be run through the developing chamber of the machine directly above the intake at a speed of 10 ft per minute. The machine is 60 in.

Figure 9.16 Problem 54. The Texas A&M bonfire.

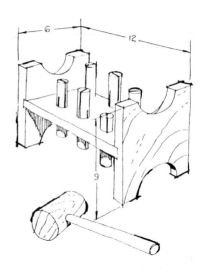

Figure 9.17 Problem 60. An educational toy.

wide, 24 in. deep, and 48 in. high. Determine the equipment and tables required and their positions for the most efficient operation.

What is the cost per drawing of the operator working at peak efficiency? Consider the other duties of the operator, such as stapling sets of drawings together and gathering original drawings to be returned with their prints.

54. **Bonfire (Fig. 9.16).** Develop a plan for building a bonfire to be burned the night before a major football game at your school. A few of the questions that must be answered are: Where will logs be obtained? How many people will be required to build it? Where can it be located? How can it be funded? What hazards must be controlled?

55. **Football stadium expansion.** Study the recent attendance figures for your school's football stadium and predict future attendance. Recommend whether to enlarge the stadium. If you recommend that it be enlarged, determine the quantity of seating needed and prepare a schedule for adding it.

56. **Shopping checkout system.** A problem for food markets and shopping malls is checkout, bagging, payment, and delivery of pur-

chases to customers' automobiles. Develop a system that will improve these conditions.

57. **Injury-proof playground.** Design a playground that permits the greatest degree of participation by children with the least risk of injury.

58. **Modification of an existing facility.** Select a facility on your campus or in your community that is inadequate for the demands made on it, such as a street intersection, parking lot, recreational area, or classroom. Identify its deficiencies and propose improvements to it.

59. **Recreational facility.** Analyze the various recreational activities on your campus that could be improved with the construction of a multipurpose facility to accommodate activities such as movies, plays, sports, meetings, and dances. The facility should be designed as an outdoor installation with the minimum of structures.

60. **Educational toy (Fig. 9.17).** Assume that you are responsible for establishing the pro-

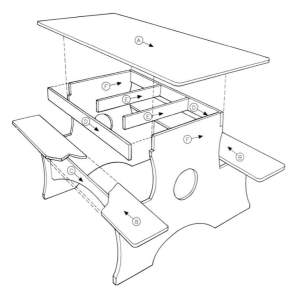

Figure 9.18 Problem 61. A plywood patio table. (Courtesy U.S. Plywood.)

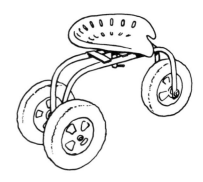

Figure 9.19 Problem 64. A garden seat.

duction system for manufacturing the educational toy shown at a rate of 5000 per month. Determine the space needed for the storage of raw materials, office facilities, manufacturing, and warehousing; the number and types of workstations required; the number and types of employees needed, and their rates of pay; and the quantity of materials needed. Calculate the cost of producing the item and the selling price necessary to provide a profit.

61. Patio table (Fig. 9.18). Proceed as in Problem 60 for production of a plywood patio table. Determine the number to be manufactured per month to break even and the selling price. Establish quantity breaks for selling prices for quantities that exceed the break-even level.

62. Simple product. Select a simple product and determine the production system and

requirements for its manufacture, proceeding as in Problem 60. Your instructor should approve the product before you go ahead.

9.9 Product Design Problems

Product design involves developing a device that will perform a specific function, be mass produced, and be sold to a large number of consumers.

63. Hunting blind. Design a portable hunting blind to house two hunters for hunting geese or ducks. It should be portable so that the two hunters can carry it.

64. Garden seat (Fig. 9.19). Design a mobile seat that can be used for work that requires much bending over, such as work in the garden.

65. Writing table arm. Design a writing table arm that can be attached to a folding chair.

66. Firewood caddy (Fig. 9.20). Design a cart that can be used for carrying firewood inside from the outdoors. It should be easy to handle when the user is climbing steps.

67. Computer mount. Design a device that can be clamped to a desk top for holding a com-

Figure 9.20 Problem 65. A firewood caddy for hauling firewood.

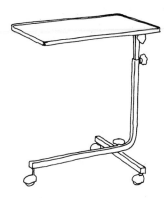

Figure 9.22 Problem 72. A hospital bed table.

Figure 9.21 Problem 68. A portable hauler.

puter, permitting it to be adjusted to various positions while leaving the desk top free to work on.

68. **Portable hauler (Fig. 9.21).** Design a portable, collapsible hauler that can be used for various applications.

69. **Workers' stilts.** Design stilts to give workers access to an 8-ft high ceiling, permitting them to nail 4-×-8-ft ceiling panels into position.

70. **Pole-vault uprights.** Pole-vault uprights must be adjusted for each vaulter by moving them forward or backward 18 in. The cross-

bar must be replaced at heights of over 18 ft by using poles and ladders. Develop a more efficient set of uprights that can be readily adjusted and allow the crossbar to be replaced easily.

71. **Sportsman's chair.** Design a sportsman's chair that can be used for camping, for fishing from a bank or boat, at sporting events, and for other purposes.

72. **Bed table (Fig. 9.22).** Modify the design of a hospital table to permit it to be used for studying, computing, writing, and eating in bed.

73. **Bicycle child carrier.** Design a seat that can be used to carry a small child as a passenger on a bicycle.

74. **Car washer.** Design a garden-hose attachment that can apply water and agitation to wash a car. Suggest other applications for this device.

75. **Power lawn-fertilizer attachment.** The rotary power lawnmower emits a force caused by the rotating blades that might be used to distribute fertilizer during mowing. Design an attachment for a lawnmower that can be used in this manner.

Figure 9.23 Problem 76. A portable sawhorse.

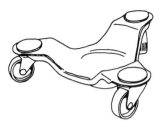

Figure 9.24 Problem 78. A heavy-appliance mover.

76. **Sawhorse (Fig. 9.23).** Design a portable sawhorse for use in carpentry projects that folds up for easy storage. It should be about 36 in. long and 30 in. high.

77. **Projector cabinet.** Design a cabinet to serve as an end table or some other function while housing a slide projector and slide trays ready for use.

78. **Heavy-appliance mover (Fig. 9.24).** Design a device for moving large appliances—stoves, refrigerators, and washers—about the house for the purposes of rearranging, cleaning, and servicing them.

79. **Car jack.** The average car jack does not attach itself adequately to the automobile's frame or bumper, causing a safety problem. Design one that would employ a different method of lifting a car on various types of terrain.

80. **Map holder.** Design a map holder to give the driver a view of the map in a convenient location in the car while driving. Provide a method of lighting the map that will not distract the driver.

81. **Stump remover.** Design an apparatus that can be attached to a car bumper to remove dead stumps by pushing or pulling.

82. **Gate opener.** An annoyance to farmers and ranchers is the necessity of opening and closing gates. Design a manually operated gate that could be opened and closed without the driver having to get out of the vehicle.

83. **Paint mixer.** Design a product for use at paint stores or by paint contractors to mix paint quickly in the store or on the job.

84. **Automobile coffee maker.** Design a device that will provide hot coffee from the dashboard of an automobile. Consider the method of changing and adding water, the spigot system, and similar details.

85. **Baby seat (cantilever).** Design a chair to support a child. It should attach to and be cantilevered from a standard table top. The chair also should be collapsible for ease of storage.

86. **Miniature-TV support.** Design a device that would support TV sets ranging in size from 6×6 in. to 7×7 in. for viewing from a bed. Provide adjustments on the device to allow positioning of the set.

87. **Panel applicator.** A worker applying $4\text{-}\times\text{-}8\text{-}$ft plasterboard to a ceiling needs a helper to hold the panel in place while she nails it. Design a device to hold the panel and eliminate the need for an assistant.

88. **Backpack.** Design a backpack that can be used for carrying camping supplies. Adapt

Figure 9.25 Problem 92. A log splitter.

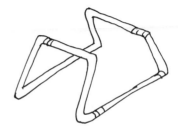

Figure 9.26 Problem 100. A push-up and sit-up exerciser.

the backpack to the human body for maximum comfort for extended periods of time. Suggest other uses for your design.

89. **Automobile controls.** Design driving controls that can be easily attached to the standard automobile to permit a car to be driven without using the legs.

90. **Bathing apparatus.** Design an apparatus that would help a wheelchair-bound person to get in and out of a bathtub without assistance from others.

91. **Adjustable TV base.** Design a base to support full-sized TV sets and allow maximum adjustment up and down and rotation about vertical and horizontal axes.

92. **Log splitter (Fig. 9.25).** Design a device to aid in splitting logs for firewood.

93. **Projector cabinet.** Design a portable cabinet that can remain permanently in a classroom to house a slide projector and/or a movie projector in a ready-to-use position. It should provide both convenience and security from theft.

94. **Projector eraser.** Design a device to erase grease-pencil markings from the acetate roll of a specially equipped overhead projector as the acetate is cranked past the stage of the projector.

95. **Cement mixer.** Design a portable cement mixer that a home owner can operate manually. Such mixers are used only occasionally, so it also should be affordable to make it marketable.

96. **Boat trailer.** Design a trailer from which a boat hangs rather than riding on top of it so that the boat can be launched in very shallow water.

97. **Washing machine.** Design a manually operated washing machine. It may be considered an "undesign" of an electrically powered washing machine.

98. **Pickup truck hoist.** Design a lift that can be attached to the tailgate of a pickup truck for raising and lowering loads from the ground to the bed of the truck without a motor.

99. **Display booth.** Design a portable display booth for use behind or on an 8-ft-long table for displaying your company's name and product information. It must be collapsible so that it can be carried by one person as airplane luggage.

100. **Exerciser (Fig. 9.26).** Design a simple exercise device that can be used for doing push-ups and sit-ups and that would sell for about $20.

Figure 9.27 Problem 101. A patio grill.

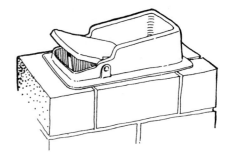

Figure 9.29 Problem 104. A chimney cover.

Figure 9.28 Problem 102. A shop bench.

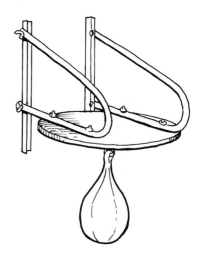

Figure 9.30 Problem 105. A punching bag platform.

101. **Patio grill (Fig. 9.27).** Design a portable charcoal grill for cooking on the patio. Consider how it would be cleaned, stored, and used. Study competing products already on the market.

102. **Shop bench (Fig. 9.28).** Design an adjustable shop bench that is collapsible for easy storage. Design it to accommodate accessories such as vises, anvils, and electrical tools.

103. **Deer-hunting seat.** Design a deer-hunting seat that can be carried to the field and attached to a tree trunk.

104. **Chimney cover (Fig. 9.29).** Design a chimney cover that can be closed from inside the house for repelling rain and reducing temperature loss.

Figure 9.31 Problem 106. A shopping caddy.

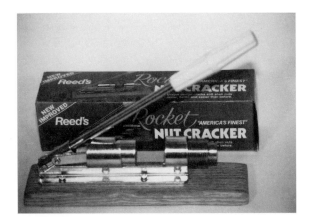

Figure 9.32 Problem 107. A nutcracker.

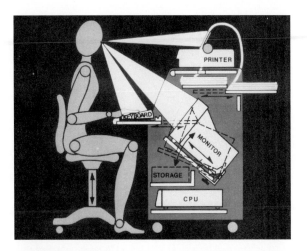

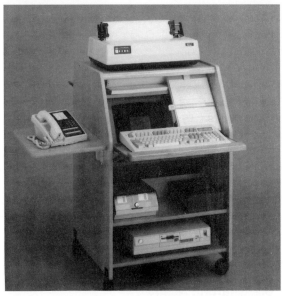

Figure 9.33 Problem 108. The Peanut Ultra View Work-station. (Courtesy of Continental Engineering Group, Inc.)

105. Punching bag platform (Fig. 9.30). Design a platform for a speed punching bag that is adjustable to various heights, portable, and ships in a flat box.

106. Shopping caddy (Fig. 9.31). Design a portable lightweight caddy that a shopper can use for carrying parcels.

107. Nutcracker (Fig. 9.32). Design a nutcracker similar to the one shown.

108. Computer workstation. Figure 9.33 illustrates a unique design for a computer workstation that reduces user discomfort. Components of the computer system are positioned for greatest visibility and comfort. Design a similar workstation, using this example as a guide.

Drawing Instruments

10.1 Introduction

The preparation of technical drawings requires knowledge of and skill in the use of drafting instruments. Skill and productivity increase with practice. Even people with little artistic ability can produce professional technical drawings when they learn to use drawing instruments properly.

10.2 Drafting Media

Pencils

A good drawing begins with the correct pencil grade and its proper use. Pencil grades range from the hardest, 9H, to the softest, 7B (**Fig. 10.1**). The pencils in the medium-grade range, 4H–4B, are used most often for drafting work of the type covered in this textbook.

Figure 10.2A–C shows three standard pencils used for drawing. The leads used in the lead holder shown in Fig. 10.2A (the best all-around pencil of the three) are marked white at their ends. The fine-line leads used in the lead holder in Fig. 10.2B are

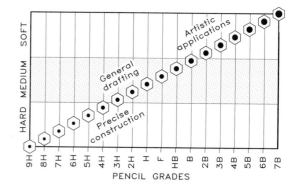

AVAILABLE PENCILS

Figure 10.1 The hardest pencil lead is 9H, and the softest is 7B. The diameters of the hard leads are smaller than those of the soft leads.

more difficult to identify because their sizes are smaller and are not marked. A different fine-line holder must be used for each size of lead. The common sizes are 0.3 mm, 0.5 mm, and .007 mm, which are the diameters of the leads. A disadvan-

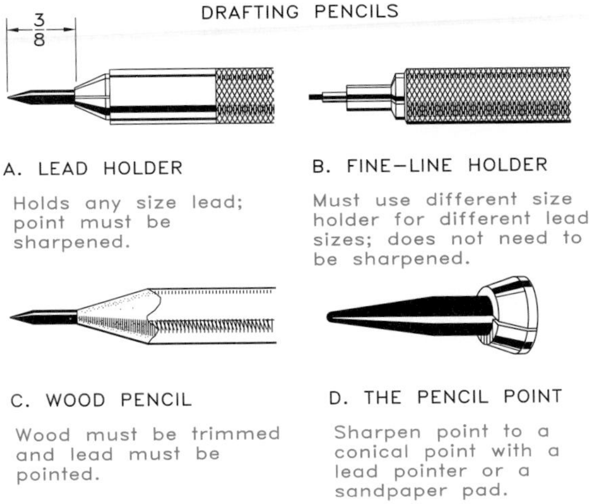

DRAFTING PENCILS

A. LEAD HOLDER

Holds any size lead;
point must be
sharpened.

B. FINE—LINE HOLDER

Must use different size
holder for different lead
sizes; does not need to
be sharpened.

C. WOOD PENCIL

Wood must be trimmed
and lead must be
pointed.

D. THE PENCIL POINT

Sharpen point to a
conical point with a
lead pointer or a
sandpaper pad.

Figure 10.2 Sharpen the drafting pencil to a tapered conical point (not a needle point) with a sandpaper pad or other type of sharpener.

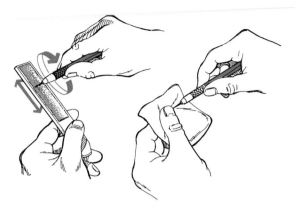

Figure 10.3 Revolve the drafting pencil about its axis as you stroke the sandpaper pad to form a conical point. Wipe away the graphite from the sharpened point with a tissue.

PENCIL POINTER

Figure 10.4 A pencil pointer of this type is useful in sharpening pencils.

tage of the fine-line pencil is the tendency of the lead to snap off when you apply pressure to it.

Although you have to sharpen and point its lead, the wood pencil shown in Fig. 10.2C is very satisfactory for lettering and drawing. The grade of the lead appears clearly on one end of the pencil. Sharpen the opposite end of the pencil so the identity of the grade of lead will be retained.

You must sharpen a pencil point properly, as shown in **Fig. 10.2D**. You may do so with a small knife or a drafter's pencil sharpener, which removes the wood and leaves approximately 3/8 inch of lead exposed. You should then sharpen the point to a conical point with a sandpaper pad by stroking the sandpaper with the pencil point while revolving the pencil about its axis **(Fig. 10.3)**. Wipe excess graphite from the point with a cloth or tissue.

You also may use a pencil pointer **(Fig. 10.4)** to sharpen wood and mechanical pencils. Insert the pencil in the hole and revolve it to sharpen the lead. Other types of small hand-held point sharpeners also are available.

Pens and Ink

Unlike drawings made by pencil, drawings made in ink remain dark and distinct. Ink lines also reproduce better than pencil lines. India ink—a dense, black carbon ink that is much thicker and

Figure 10.5 Ink the pen between the nibs by using the spout on the ink bottle cap.

RULING A LINE

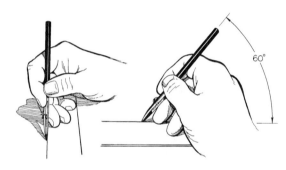

Figure 10.6 Hold the ruling pen in a plane perpendicular to the paper and at 60° to the drawing surface.

faster drying than fountain pen ink—is excellent for engineering drawings.

Ink the ruling pen by using the spout on the cap of the ink bottle **(Fig. 10.5)**. Learning the proper amount of ink to apply to the nibs takes practice. When drawing horizontal lines, hold the ruling pen as you would hold a pencil **(Fig. 10.6)**, maintaining a space between the nibs and straightedge.

An alternative type of pen is the technical ink pen **(Fig. 10.7)**, which comes in sets with pen points of various sizes for drawing lines of differ-

INK PEN

Figure 10.7 The Rapidograph® is one type of technical ink pen.

DRAFTING SHEET SIZES

	ENGINEERS'	ARCHITECTS'		METRIC
A	11" X 8.5"	12" X 9"	A4	297 X 210
B	17" X 11"	18" X 12"	A3	420 X 297
C	22" X 17"	24" X 18"	A2	594 X 420
D	34" X 22"	36" X 24"	A1	841 X 594
E	44" X 34"	48" X 36"	A0	1189 X 841

Figure 10.8 Standard sheet sizes vary by purpose.

ent thicknesses. Lines drawn with this type of pen dry faster than those drawn with ruling pens because the ink is applied in a thinner layer.

Papers and Films

Sizes Sheet sizes are specified by the letters A–E. These sizes **(Fig. 10.8)** are multiples of either the standard 8 1/2-×-11-inch sheet (used by engineers) or the 9-×-12-inch sheet (used by architects). The metric sizes (A4–A0) are equivalent to the 8 1/2-×-11-inch modular sizes.

Detail Paper When drawings are not to be reproduced by the diazo or blue-line process, you may use an opaque paper, called **detail paper**, as the drawing surface. The higher the rag content (cotton additive) of the paper, the better is its quality and durability.

You may draw preliminary layouts on detail paper and then trace them onto the final surface.

Tracing Paper   Tracing paper or tracing vellum is a thin, translucent paper used for making detail drawings. This paper permits light to pass through it, allowing reproduction by the blue-line process. Tracing papers that yield the best reproductions are the most translucent ones. **Vellum** is a chemically treated tracing paper. The treatment improves its translucency, but vellum does not retain its original quality as long as do high-quality, untreated tracing papers.

Tracing Cloth   Tracing cloth is a permanent drafting medium used for both ink and pencil drawings. It is made of cotton fabric that has been covered with a starch compound to provide a tough, erasable drafting surface that yields excellent blue-line reproductions. Tracing cloth does not change shape as much as tracing paper with variations in temperature and humidity. Repeated erasures do not damage the surface of tracing cloth, an especially important feature for ink drawings. An application of a coating of powder or pounce to absorb oily spots that may repel an ink line prepares tracing cloths for inking.

Polyester Film   An excellent drafting surface is polyester film, which is available under several trade names such as **Mylar**. It is more transparent, stable, and tougher than paper or cloth and is waterproof. Mylar film is used for both pencil and ink drawings. A plastic-lead pencil must be used with some films, whereas standard lead pencils may be used with others. Ink will not wash off with water and will not erase with a dry eraser; a dampened hand-held eraser is required.

When you use polyester film, draw on the matte surface according to the manufacturer's directions. Use a cleaning solution to prepare the surface for ink, such as removing spots that might not take ink properly.

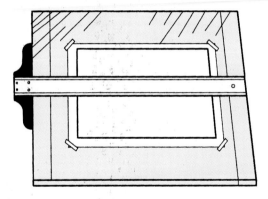

Figure 10.9 The T-square and drafting board are basic drafting tools.

10.3 Drafting Equipment

T-Square and Board
The T-square and drafting board are basic pieces of drafting equipment (**Fig. 10.9**). Always tape drawing paper to the board parallel to the blade of the T-square. By holding its head against the edge of the drawing board, you may then move the T-square and draw parallel horizontal lines. Most small drafting boards are made of basswood, which is lightweight and strong. Standard board sizes are 12 × 14 inches, 15 × 20 inches, and 21 × 26 inches.

Drafting Machine
The T-square will always be used to some extent in industry and the classroom. However, most professional drafters prefer the mechanical drafting machine (**Fig. 10.10**), which is attached to the drawing table top and has fingertip controls for drawing lines at any angle.

Figure 10.11 shows a modern, fully equipped drafting station. Today, most offices are equipped with computer graphics stations, which have almost replaced manual equipment (**Fig. 10.12**).

DRAFTING MACHINE

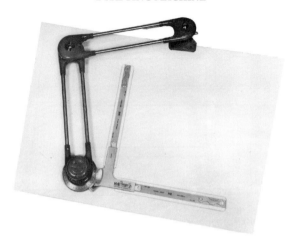

Figure 10.10 The drafting machine often is used instead of the T-square and drafting board. (Courtesy of Keuffel & Esser Company.)

Figure 10.11 The professional drafter or engineer may work in this type of environment. (Courtesy of Martin Instrument Company.)

Triangles

The two types of triangles used most often are the 45° triangle and the 30°–60° triangle. The 30°–60° triangle is specified by the longer of the two sides adjacent to the 90° angle (**Fig. 10.13**). Standard sizes of 30°–60° triangles range in 2-inch intervals from 4 to 24 inches.

Figure 10.12 Many professional workstations are equipped with computer graphics equipment.

THE 30°–60° TRIANGLE

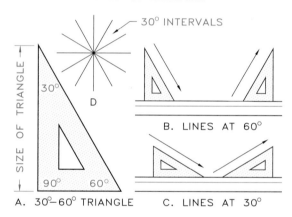

Figure 10.13 You may use a 30°–60° triangle to draw lines at 30° intervals throughout 360°.

The 45° triangle is specified by the length of the sides adjacent to the 90° angle. These range in 2-inch intervals from 4 to 24 inches, but the 6-inch and 10-inch sizes are adequate for most classroom applications. **Figure 10.14** shows the various angles that you may draw with this triangle. By using the 45° and 30°–60° triangles in combination, you may draw angles at 15° intervals throughout 360° (**Fig. 10.15**).

45° TRIANGLE

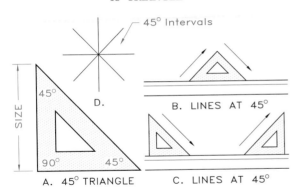

Figure 10.14 You may use a 45° triangle to draw lines at 45° intervals throughout 360°.

ANGLES DRAWN WITH TRIANGLES

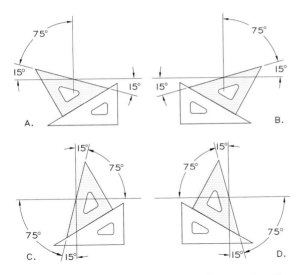

Figure 10.15 By using a 30°–60° triangle in combination with a 45° triangle, you may draw angles at 15° intervals.

Protractor

When drawing or measuring lines at angles other than multiples of 15°, use a protractor (**Fig. 10.16**). Protractors are available as semicircles (180°) or circles (360°). Adjustable triangles with movable edges that can be set at different angles with thumbscrews also are available.

THE PROTRACTOR

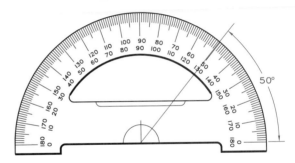

Figure 10.16 Use a semicircular protractor to measure angles.

DRAFTING INSTRUMENTS

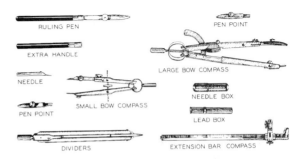

Figure 10.17 You may buy drafting instruments individually.

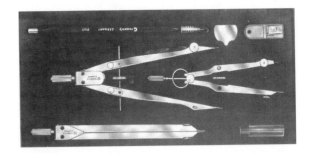

Figure 10.18 You also may buy instruments as a cased set. (Courtesy of Gramercy Guild.)

The Instrument Set

Figure 10.17 shows some of the basic drawing instruments. They are available individually or in cased sets, such as the one shown in **Fig. 10.18**.

USING THE COMPASS

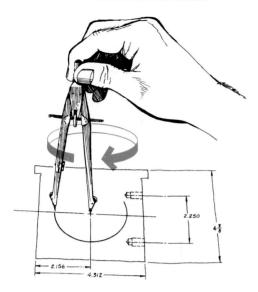

Figure 10.19 Use a compass for drawing circles.

SHARPENING THE COMPASS LEAD

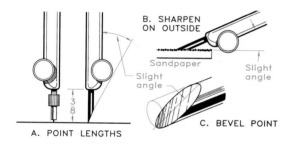

A. POINT LENGTHS

B. SHARPEN ON OUTSIDE

Sandpaper

Slight angle

Slight angle

C. BEVEL POINT

Figure 10.20
A Adjust the pencil point to be the same length as the compass point.
B Sharpen the lead from the outside with a sandpaper pad.
C Sharpen the lead to a beveled point.

Compass Use the compass to draw circles and arcs in pencil or ink **(Fig. 10.19)**. To draw circles well with a pencil compass, sharpen the lead on its outside with a sandpaper board **(Fig. 10.20)**. A bevel cut of this type gives the best point for drawing circles. You cannot draw a thick arc in pencil

SMALL BOW COMPASSES

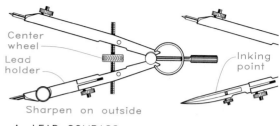

Center wheel
Lead holder
Sharpen on outside

Inking point

A. LEAD COMPASS   B. INK COMPASS

Figure 10.21 Use a small bow compass for drawing circles of up to 2-in. radius: (A) lead; (B) ink.

COMPASS EXTENSION

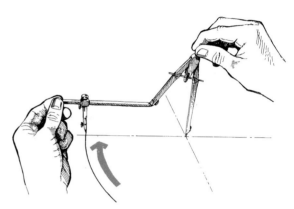

Figure 10.22 Use an extension bar with a bow compass to draw large arcs.

with a single sweep. **To thicken the pencil line of a circle, draw a series of thin concentric circles by adjusting the radius of the compass slightly.**

When setting the compass point in the drawing surface, insert it just enough for a firm set, not to the shoulder of the point. When the table top has a hard covering, place several sheets of paper under the drawing to provide a seat for the compass point.

Use a small bow compass **(Fig. 10.21)** to draw small circles of up to 2 inches in radius. For larger circles, use an extension bar to extend the range of the large bow compass **(Fig. 10.22)**. Use a beam

BEAM COMPASS

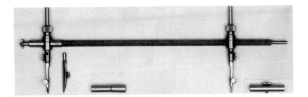

Figure 10.23 Use a beam compass to draw large circles. Ink attachments also are available.

CIRCLES WITH A TEMPLATE

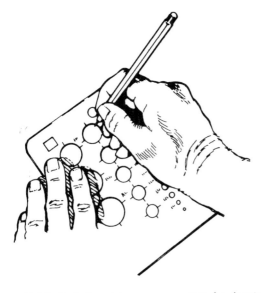

Figure 10.24 Circle templates are convenient for drawing small circles without a compass.

USING THE DIVIDERS

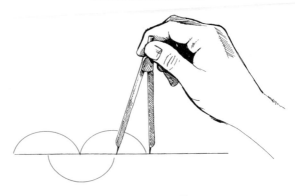

Figure 10.25 Use dividers to step off measurements.

TRANSFERRING DIMENSIONS

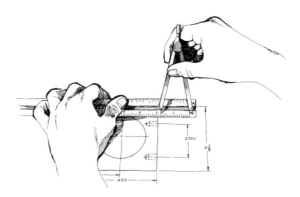

Figure 10.26 You may also use dividers to transfer dimensions from a scale to a drawing.

compass (**Fig. 10.23**) to produce still larger circles. You may draw very small circles conveniently with a circle template aligned with the centerlines of the circles (**Fig. 10.24**).

Dividers The dividers look like a compass but do not have pencil or pen points. Use them for laying off and transferring dimensions onto a drawing. For example, you can step off equal divisions rapidly and accurately along a line (**Fig. 10.25**). As you make each measurement, the dividers'

points make a slight impression mark in the drawing surface.

You may also use dividers to transfer dimensions from a scale to a drawing (**Fig. 10.26**) or to divide a line into a number of equal parts. Bow dividers (**Fig. 10.27**) are useful for transferring smaller dimensions, such as the spacing between lettering guidelines.

Proportional Dividers Proportional dividers allow transferral of dimensions from one scale to anoth-

SMALL BOW DIVIDERS

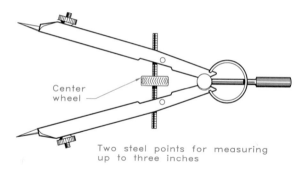

Center wheel

Two steel points for measuring up to three inches

Figure 10.27 Use a bow divider to transfer small dimensions, such as the spaces between guidelines for lettering.

PROPORTIONAL DIVIDERS

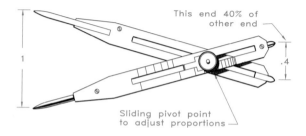

This end 40% of other end

Sliding pivot point to adjust proportions

Figure 10.28 Use a proportional divider to make measurements at one end that are proportional dimensions at the other end.

er. Moving the pivot point varies the ratio between the ends of the dividers (**Fig. 10.28**).

Templates

Various templates are available for use in drawing nuts and bolts, circles and ellipses, architectural symbols, and many other shapes. Templates work better when used with technical ink pens than with ruling pens.

10.4 Lines

The type of line produced by a pencil depends on the hardness of its lead, drawing surface, and your drawing technique. **Figure 10.29** shows examples

ALPHABET OF LINES

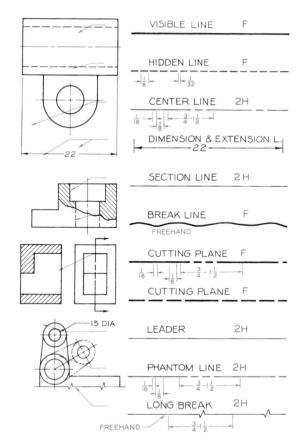

VISIBLE LINE   F

HIDDEN LINE   F

CENTER LINE   2H

DIMENSION & EXTENSION L.

SECTION LINE   2H

BREAK LINE   F
FREEHAND

CUTTING PLANE   F

CUTTING PLANE   F

LEADER   2H

PHANTOM LINE   2H

LONG BREAK   2H
FREEHAND

Figure 10.29 The alphabet of lines: Full-size lines and the pencil grades for drawing them appear in the right column.

of the standard lines, or **the alphabet of lines**, and the recommended pencils for drawing them, which vary somewhat with the drawing surface being used. Guidelines are light (just dark enough to be seen) and aid in lettering and laying out a drawing. A 2H or 4H pencil lead is recommended for drawing guidelines.

Horizontal Lines

To draw a horizontal line, use the upper edge of your horizontal straightedge and make strokes from left to right, if you are right-handed (**Fig. 10.30**), and from right to left if you are left-handed. Rotate the pencil about its axis so that its point will wear evenly

116 • **CHAPTER 10   DRAWING INSTRUMENTS**

HORIZONTAL LINES

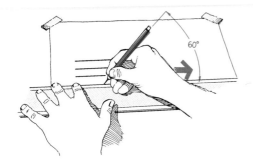

Figure 10.30 Draw horizontal lines along the upper edge of a straightedge while holding the pencil or pen in a plane perpendicular to the paper and at 60° to the surface.

PENCIL POSITION

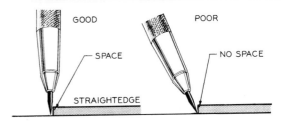

Figure 10.32 While drawing, hold the pencil or pen point in a plane perpendicular to the paper, leaving a space between the point and the straightedge.

ROTATION OF PENCIL

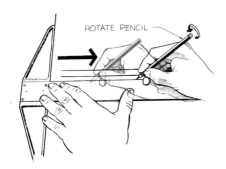

Figure 10.31 Rotate the pencil about its axis while drawing so that its point will wear evenly.

VERTICAL LINES

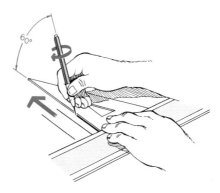

Figure 10.33 Draw vertical lines along the left side of a triangle (if you are right-handed) in an upward direction, holding the pencil or pen in a plane perpendicular to the paper and at 60° to the surface.

(Fig. 10.31). Darken pencil lines by drawing over them with multiple strokes. For drawing the best line, leave a small space between the straightedge and the pencil or pen point **(Fig. 10.32)**.

Vertical Lines

Use a triangle and a straightedge to draw vertical lines. Hold the straightedge firmly with one hand and position the triangle where needed and draw the vertical lines with the other hand **(Fig. 10.33)**. Draw vertical lines upward along the left side of the triangle if you are right-handed and upward along the right side of the triangle if you are left-handed.

Parallel Lines

You may draw a series of lines parallel to a given line by using a triangle and straightedge **(Fig. 10.34)**. Place the 45° triangle parallel to the given line and hold it against the straightedge (which may be another triangle). By holding the straightedge in one position, you may move the triangle to various positions along the straightedge and draw parallel lines.

DRAWING A PARALLEL LINE

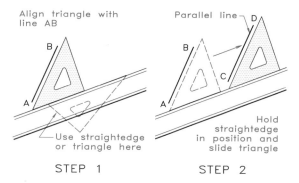

STEP 1   STEP 2

Figure 10.34 You may use a straightedge and a 45° triangle to draw a series of parallel lines. Hold the straightedge firmly in position and move the triangle into position to draw line CD.

DRAWING A PERPENDICULAR

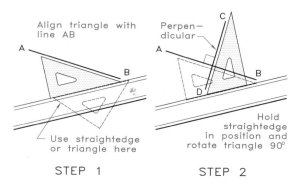

STEP 1   STEP 2

Figure 10.35 You may use a 30°–60° triangle and a straightedge to construct a line perpendicular to line AB. Align the triangle with line AB and then rotate the triangle to draw line CD.

Perpendicular Lines

You may construct perpendicular lines by using either of the standard triangles. Use a 30°–60° triangle and a straightedge, or another triangle, to draw line CD perpendicular to line AB (**Fig. 10.35**). Place one edge of the triangle parallel to line AB, with the straightedge in contact with the

DRAWING A 30° ANGLE

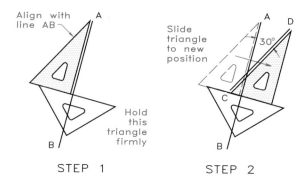

STEP 1   STEP 2

Figure 10.36

Step 1 Hold the 30°–60° triangle in contact with a straightedge and align the triangle with line AB.

Step 2 Hold the straightedge in position and slide the triangle to where you want to draw the 30° line.

USING THE IRREGULAR CURVE

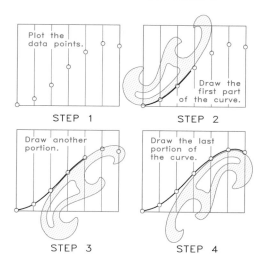

STEP 1   STEP 2

STEP 3   STEP 4

Figure 10.37

Step 1 Plot data points.

Step 2 Position the irregular curve to pass through as many points as possible and draw that portion of the curve.

Step 3 Reposition the irregular curve and draw another portion of the curve.

Step 4 Draw the last portion to complete the smooth curve.

Figure 10.38 Use an erasing shield for erasing in tight spots. Use a brush, not your hand, to brush away the erasure crumbs.

triangle (step 1). While holding the straightedge in place, rotate the triangle and move it to draw the perpendicular line CD (step 2).

Angular Lines

Figure 10.36 shows how to draw a 30° angle with a given line, AB, by using the 30°–60° triangle and another straightedge. You may use this same technique to draw angles of 60° and 45°.

Irregular Curves

You have to draw curves that are not arcs with an irregular curve (sometimes called French curves). These plastic curves come in a variety of sizes and shapes, but the one shown in **Fig. 10.37** is typical. Here, we used the irregular curve to connect a series of points to form a smooth curve.

Erasing Lines

Always use the softest eraser that will do a particular job. For example, do not use ink erasers to erase pencil lines because ink erasers are coarse and may damage the surface of the paper. When working in small areas, you should use an erasing shield to avoid accidentally erasing adjacent lines **(Fig. 10.38)**. Follow all erasing by brushing away the "crumbs" with a dusting brush. Wiping the crumbs away with your hands will smudge the drawing. A cordless electric eraser **(Fig. 10.39)** accommodates several grades of erasers for erasing both pencil and ink lines.

Figure 10.39 This cordless electric eraser is typical of those used by professional drafters.

TYPES OF SCALES

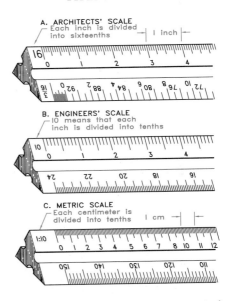

Figure 10.40 The architects' scale (A) measures in feet and inches. The engineers' scale (B) and the metric scale (C) are calibrated in decimal units.

10.5 Measurement

Scales

All engineering drawings require the use of scales for measuring lengths and sizes. Scales may be flat or triangular and are made of wood, plastic, or metal. **Figure 10.40** shows triangular architects',

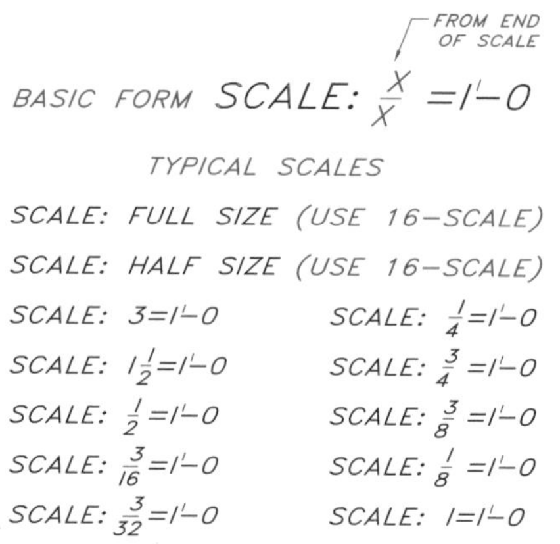

ARCHITECTS' SCALE

BASIC FORM SCALE: $\frac{X}{X}$ =1'-0   FROM END OF SCALE

TYPICAL SCALES

SCALE: FULL SIZE (USE 16-SCALE)
SCALE: HALF SIZE (USE 16-SCALE)
SCALE: 3=1'-0   SCALE: $\frac{1}{4}$=1'-0
SCALE: 1$\frac{1}{2}$=1'-0   SCALE: $\frac{3}{4}$=1'-0
SCALE: $\frac{1}{2}$=1'-0   SCALE: $\frac{3}{8}$=1'-0
SCALE: $\frac{3}{16}$=1'-0   SCALE: $\frac{1}{8}$=1'-0
SCALE: $\frac{3}{32}$=1'-0   SCALE: 1=1'-0

Figure 10.41 Use this basic form to indicate the scale on a drawing made with an architects' scale.

engineers', and metric scales. Most scales are either 6 or 12 inches long.

Architects' Scale Drafters use the architects' scale to dimension and scale architectural features such as room-size, cabinets, plumbing, and electrical layouts. Most indoor measurements are made in feet and inches with an architects' scale. **Figure 10.41** shows how to indicate on a drawing the scale you are using. Place this scale designation in the title block or in a prominent location on the drawing. Because dimensions measured with the architects' scale are in feet and inches, you should convert all dimensions to decimal equivalents (all feet or all inches) before making calculations.

Use the 16 scale for measuring full-size lines **(Fig. 10.42A)**. An inch on the 16 scale is divided into sixteenths to match the ruler used by carpenters. The measurement shown is 3 1/8". When the measurement is less than 1 ft, a zero may precede the inch measurements, with inch marks omitted, or 0'-3 1/8.

Figure 10.42B shows use of the 1 = 1'-0 scale to measure a line. Read the nearest whole foot (2 ft in this case) and then the remainder in inches from the end of the scale (3 1/2 in.) for a total of

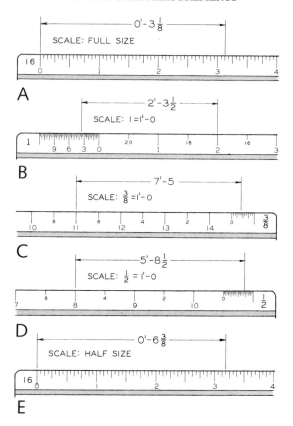

Figure 10.42 Lines measured with an architects' scale.

2'-3 1/2. Note that, at the end of each scale, a foot is divided into inches for use in measuring dimensions of less than a foot. The scale 1" = 1'-0 is the same as saying that 1 in. is equal to 12 in. or that the drawing is 1/12 the actual size of the object.

When you use the 3/8 = 1'-0 scale, 3/8 in. represents 12 in. on a drawing. Thus the line in **Fig. 10.42C** measures 7'-5. Similarly, at the 1/2 = 1'-0 scale, the line in **Fig. 10.42D** measures 5'-8 1/2.

To obtain a half-size measurement divide the full-size measurement by 2 and draw it with the 16 scale. This scale is sometimes specified as scale: 6 = 12 (inch marks omitted). The line in **Fig. 10.42E** measures to be 0'-6 3/8.

MARKING MEASUREMENTS

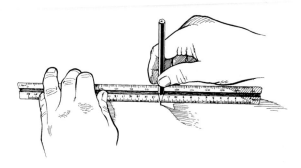

Figure 10.43 When marking off measurements along a scale, hold your pencil vertical for accuracy.

SPECIFYING FEET AND INCHES

Figure 10.44 Omit inch marks but show foot marks (according to current standards). When the inch measurement is less than a whole inch, use a leading zero.

When marking measurements, hold your pencil or pen vertical for the greatest accuracy (**Fig. 10.43**). Specify dimensions in feet and inches as shown in **Fig. 10.44**, with fractions twice as tall as whole numerals.

Engineers' Scale

On the engineers' scale, each inch is divided into multiples of 10. Because it is used for making drawings of outdoor projects—streets, structures, tracts of land, and other topographical features—it is sometimes called the civil engineers' scale.

With measurements already in decimal form, performing calculations is easy; there is no need to convert from one unit to another as when you use the architects' scale. **Figure 10.45** shows the form for specifying scales when using the engineers' scale. For example, scale: 1 = 10'.

ENGINEERS' SCALE

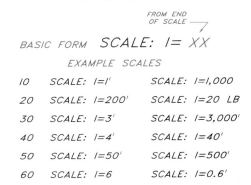

Figure 10.45 Use this basic form to indicate the scale on a drawing made with an engineers' scale.

ENGINEERS' SCALES

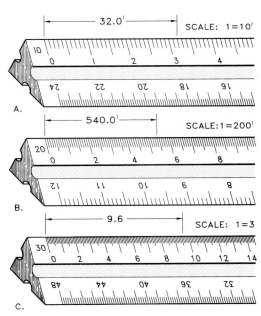

Figure 10.46 Lines measured with an engineers' scale.

Each end of the scale is labeled 10, 20, 30, and so on, which indicates the number of units per inch on the scale (Fig. 10.46). You may obtain many combinations simply by mentally moving the decimal places of a scale.

DECIMAL FRACTIONS

Figure 10.47 For decimal fractions in inches, omit leading zeros and inch marks. For feet, leave adequate space for decimal points between numbers and show foot marks.

METRIC PREFIXES AND ABBREVIATIONS

Value		Prefix	Symbol	Pronunciation
1 000 000 = 10^6	=	Mega	M	"Megah"
1 000 = 10^3	=	Kilo	k	"Keylow"
100 = 10^2	=	Hecto	h	"Heck tow"
10 = 10^1	=	Deka	da	"Dekah"
1	=			
0.1 = 10^{-1}	=	Deci	d	"Des sigh"
0.01 = 10^{-2}	=	Centi	c	"Cen'–ti"
0.001 = 10^{-3}	=	Milli	m	"Mill lee"
0.000 001 = 10^{-6}	=	Micro	μ	"Microw"

Figure 10.49 These prefixes and abbreviations indicate decimal placement for SI measurements.

ENGLISH UNITS

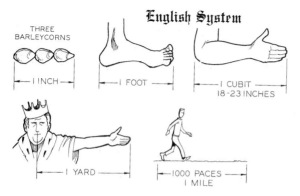

Figure 10.48 The units of the English system are based on arbitrary dimensions.

Figure 10.46A shows the use of the 10 scale to measure a line 32.0 ft long drawn at the scale of 1 = 10'. **Figure 10.46B** shows use of the 20 scale to measure a line 540.0 ft long drawn at a scale of 1 = 200'. **Figure 10.46C** shows use of the 30 scale to measure a line of 10.6 in. long at a scale of 1 = 3. **Figure 10.47** shows the proper format for indicating measurements in feet and inches.

English System of Units

The English (Imperial) system of units has been used in the United States, Great Britain (until recently), and Canada since it was established. This system is based on arbitrary units (of length) of the inch, foot, cubit, yard, and mile **(Fig. 10.48)**. Because there is no common relationship among these units, calculations are cumbersome. For example, finding the area of a rectangle that measures 25 in. × 6 3/4 yd first requires conversion of one unit to the other.

Metric System (SI) of Units

France proposed the metric system in the fifteenth century. In 1793 the French National Assembly agreed that the meter (m) would be one ten-millionth of the meridian quadrant of the earth and fractions of the meter would be expressed as decimal fractions. An international commission officially adopted the metric system in 1875. Scientists later found a slight error in the first measurement of the meter, so the meter was redefined as being equal to 1,650,763.73 wavelengths of the orange-red light given off by krypton-86.

The worldwide organization responsible for promoting the metric system is the International Standards Organization (ISO). It has endorsed the System International d'Unites (International System of Units), abbreviated SI.

Prefixes to SI units indicate placement of the decimal, as **Fig. 10.49** shows. **Figure 10.50** shows several comparisons of English and SI units.

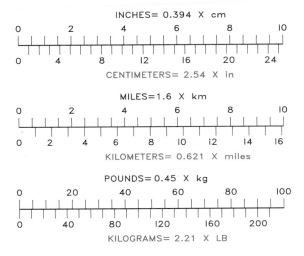

INCHES= 0.394 X cm

CENTIMETERS= 2.54 X in

MILES=1.6 X km

KILOMETERS= 0.621 X miles

POUNDS= 0.45 X kg

KILOGRAMS= 2.21 X LB

Figure 10.50 These scales show a comparison of the English system units with metric system units.

THE CENTIMETER

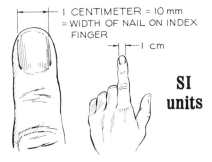

I CENTIMETER = 10 mm = WIDTH OF NAIL ON INDEX FINGER

I cm

SI units

Figure 10.51 The nail width of your index finger is approximately equal to 1 centimeter, or 10 millimeters.

Metric Scales

The basic metric unit of measurement for an engineering drawing is the millimeter (mm), which is one-thousandth of a meter, or one-tenth of a centimeter. Dimensions on a metric drawing are understood to be in millimeters unless otherwise specified.

The width of the fingernail of your index finger is a convenient way to approximate the dimen-

BASIC FORM *SCALE: I= XX* FROM END OF SCALE

EXAMPLE SCALES

SCALE: I:I (Imm=Imm; Icm=Icm)

SCALE: I:2 (Imm=2mm; Imm=20mm)

SCALE: I:3 (Imm=30mm; Imm=0.3mm)

SCALE: I:4 (Imm=4mm; Imm=40mm)

SCALE: I:5 (Imm=5mm; Imm=500mm)

SCALE: I:6 (Imm=6mm; Imm=60mm)

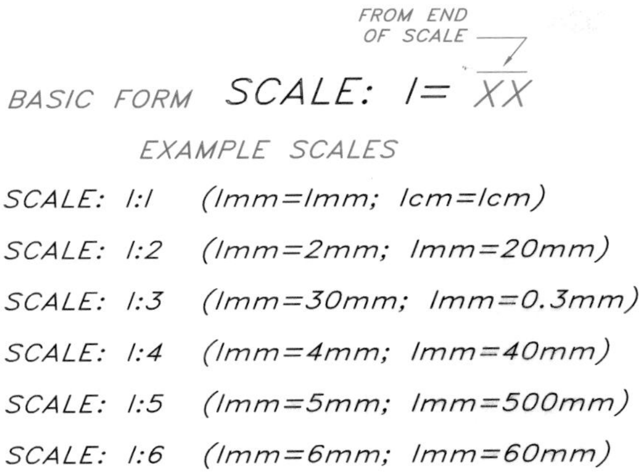

Figure 10.52 Use this basic form to indicate the scale of a drawing made with a metric scale.

METRIC FRACTIONS

| 22.0 | 0.15 | .15 |
| GOOD | GOOD | POOR |

Missing zero

Too little room for zero

| 46 | 14.6 | 14.6 |
| GOOD | GOOD | POOR |

Figure 10.53 In the metric system, a zero precedes the decimal. Allow adequate space for it.

sion of one centimeter, or ten millimeters (**Fig. 10.51**). Depicted in **Fig. 10.52** is the format for indicating metric scales on a drawing.

Decimal fractions are unnecessary on drawings dimensioned in millimeters. Thus dimensions usually are rounded off to whole numbers except for those measurements dimensioned with specified tolerances. For metric measurements of less than 1, a zero goes in front of the decimal. In the English system, the zero is omitted from measurements of less than an inch (**Fig. 10.53**).

METRIC UNITS

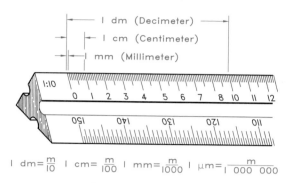

$$1 \ dm = \frac{m}{10} \quad 1 \ cm = \frac{m}{100} \quad 1 \ mm = \frac{m}{1000} \quad 1 \ \mu m = \frac{m}{1\ 000\ 000}$$

Figure 10.54 The decimeter is one tenth of a meter; the centimeter is one hundredth of a meter; a millimeter is one thousandth of a meter; and a micrometer is one millionth of a meter.

Metric scales are expressed as ratios: 1:20, 1:40, 1:100, 1:500, and so on. The scale ratios mean that one unit represents the number of units to the right of the colon. For example, 1:10 means that 1 mm equals 10 mm or 1 cm equals 10 cm or 1 m equals 10 m. The full-size metric scale (**Fig. 10.54**) shows the relationship between the metric units of the decimeter, centimeter, millimeter, and micrometer. The line shown in **Fig. 10.55A** measures 59 mm. Use the 1:2 scale when 1 mm represents 2 mm, 20 mm, 200 mm, and so on. The line shown in **Fig. 10.55B** measures 106 mm. **Figure 10.55C** shows a line measuring 165 mm, where 1 mm represents 3 mm.

Metric Symbols To indicate that drawings are in metric units, insert SI (**Fig. 10.56**) in or near the title block. The two views of the partial cone denote whether the orthographic views were drawn in accordance with the U.S. system (third-angle projection) or the European system (first-angle projection).

Expression of Metric Units **Figure 10.57** gives the general rules for expressing SI units. Do not use commas to separate digits in large numbers; instead, leave a space between them, as shown.

METRIC SCALES

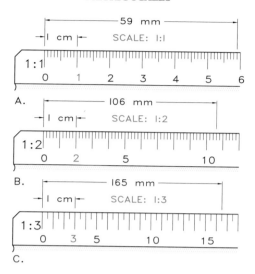

Figure 10.55 These lines are measured with a metric scale.

THE SI SYMBOL

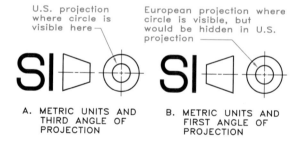

A. METRIC UNITS AND THIRD ANGLE OF PROJECTION

B. METRIC UNITS AND FIRST ANGLE OF PROJECTION

Figure 10.56 The large SI indicates that measurements are in metric units. The partial cones indicate whether the views are drawn in (A) the third-angle (U.S. system) or (B) the first-angle of projection (European system).

Scale Conversion

Appendix 2 gives factors for converting English to metric lengths and vice versa. For example, multiply decimal inches by 25.4 to obtain millimeters.

Multiply an architects' scale by 12 to convert it to an approximate metric scale. For example, Scale:

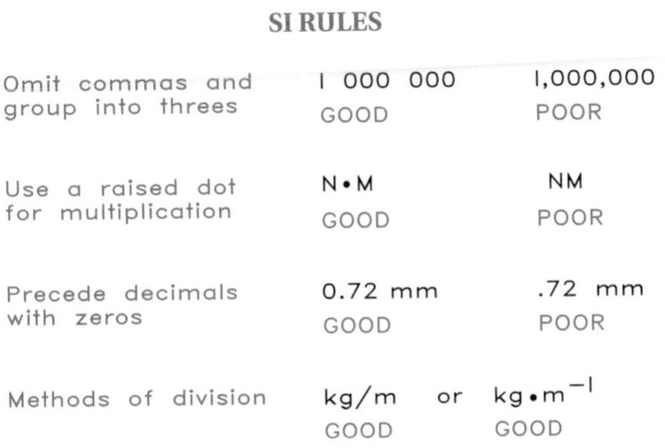

SI RULES
Omit commas and group into threes
I 000 000
GOOD
1,000,000
POOR
Use a raised dot for multiplication
N•M
GOOD
NM
POOR
Precede decimals with zeros
0.72 mm
GOOD
.72 mm
POOR
Methods of division
kg/m
GOOD
or
$kg \bullet m^{-1}$
GOOD

Figure 10.57 Follow the general rules for showing SI units.

SIZE AH SHEET (HORIZONTAL)
DIMENSIONS—INCHES
SIZE F G H I J
A 8.5 11 .50 .75 .38
B 11 17 .50 .50 .38
C 17 22 .75 .75 .38
D 22 34 .75 .75 .38
E 34 44 .75 .75 .38
I
F
H
H
G
J
H
ASSIGNMENT NAME
SEC 100 33 DON FOWLEY
JAN 1, 1994
22

Figure 10.59 This format is for size AH sheet (11″ × 8 1/2″, horizontal position) and other sheet sizes. The numbers in columns A–E are the various layout dimensions.

SIZE AV SHEET (VERTICAL)
8.50
11.00
.75
.25
.125" LETTERING
.625
ASSIGN. 33
LETTERING
GEORGE PATTON
NO 22 SEC 100 1-2-94
TIME
.25

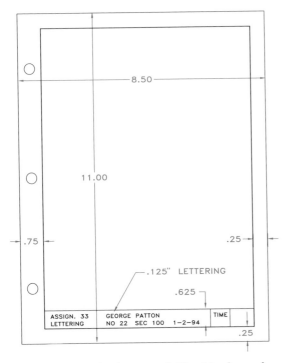

Figure 10.58 Use the format and title strip shown for a size AV sheet (8 1/2″ × 11″, vertical position) to present the solutions to problems at the end of each chapter.

1/8 = 1'-0 is the same as 1/8 in. = 12 in. or 1 in. = 96 in., which closely approximates the metric scale of 1:100. You cannot convert most metric scales exactly to English scales, but the metric scale of 1:60 does convert exactly to 1 = 5'.

10.6 Presentation of Drawings

The following formats are suggested for the presentation of drawings. The solutions to most problems may be drawn on 8 1/2-×-11-inch sheets with a title strip, as **Fig. 10.58** shows. When it is used vertically, we call this 8 1/2-×-11-inch sheet size AV throughout the rest of this textbook. When this sheet is used horizontally, as **Fig. 10.59** shows, we call it size AH.

Figure 10.59 shows the standard sizes of sheets, from size A through size E and an alternative title strip for sizes B, C, D, and E. Always use guidelines for lettering title strips.

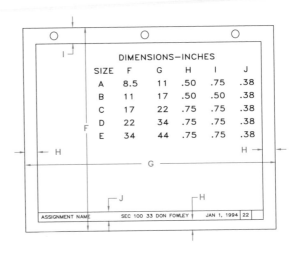

Figure 10.60 shows a smaller title block and parts list, which go in the lower right-hand corner of the sheet against the borders. When you use both on the same drawing, place the parts list directly above and in contact with the title block or title strip.

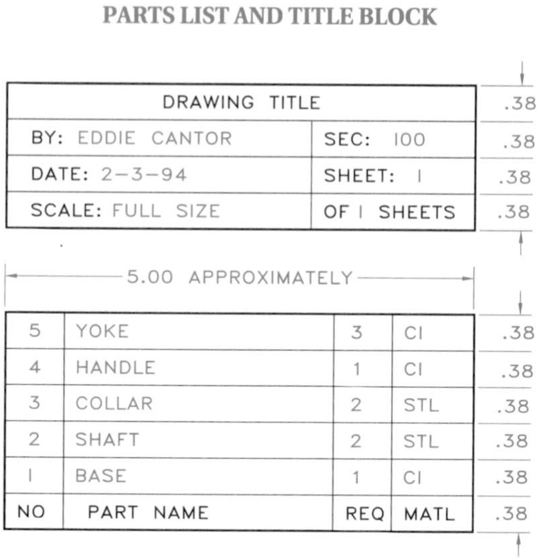

DRAWING TITLE			.38
BY: EDDIE CANTOR	SEC: 100		.38
DATE: 2-3-94	SHEET: 1		.38
SCALE: FULL SIZE	OF 1 SHEETS		.38

5.00 APPROXIMATELY

5	YOKE	3	CI	.38
4	HANDLE	1	CI	.38
3	COLLAR	2	STL	.38
2	SHAFT	2	STL	.38
1	BASE	1	CI	.38
NO	PART NAME	REQ	MATL	.38

Figure 10.60 You may use this title strip on sheet sizes B, C, D, and E instead of the one given in Fig. 10.58.

Problems

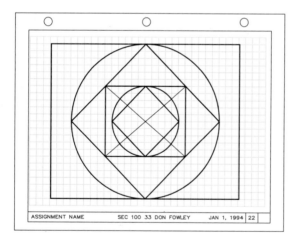

Figure 10.61 Problem 1.

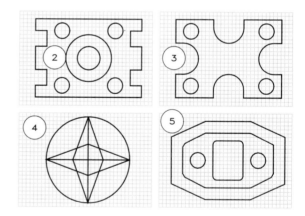

Figure 10.62 Problems 2–5.

Problems 1–9 (Figs. 10.61–10.63): Present your drawings on size AH paper, plain or with a printed grid, using the format shown in Fig. 10.61, in pencil or ink as assigned. Alternatively, you may present two half-size

drawings per sheet on size AV paper, using the side of each square of a grid to represent .25 in., or 6 mm. Problems 10–13 (Figs. 10.64–10.67): Draw full-size views on size AH sheets and omit the dimensions.

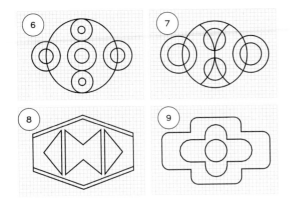

Figure 10.63 Problems 6–9.

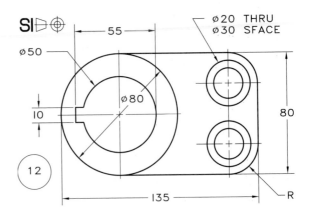

Figure 10.66 Problem 12.

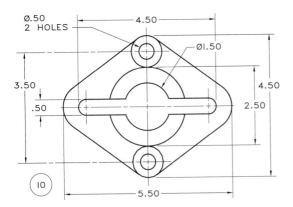

Figure 10.64 Problem 10.

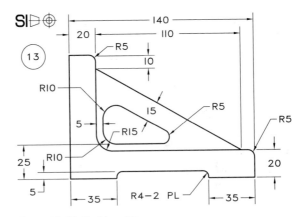

Figure 10.67 Problem 13.

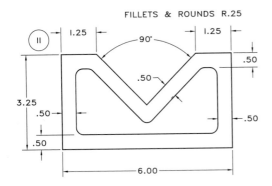

Figure 10.65 Problem 11.

11 Lettering

11.1 Introduction

Notes, dimensions, and specifications, which must be lettered, supplement all drawings. The ability to letter freehand is an important skill to develop because it affects the use and interpretation of drawings. It also displays an engineer's skill with graphics, and may be taken as an indication of professional competence.

11.2 Lettering Tools

The best pencils for lettering on most surfaces are the H, F, and HB grades, with an F grade pencil being the one most commonly used. Some papers and films are coarser than others and may require a harder pencil lead. To give the desired line width, round the point of the pencil slightly (**Fig. 11.1**), because a needle point will break off when you apply pressure.

While holding a pencil as shown in **Fig. 11.2**, revolve it slightly about its axis as you make

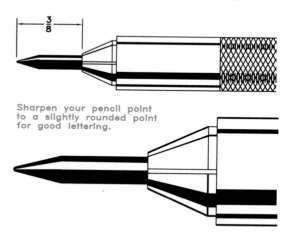

Figure 11.1 Good lettering begins with a properly sharpened pencil point. The F grade pencil is good for lettering.

strokes so that the lead will wear evenly. For good reproduction, bear down firmly to make letters black and bright with a single stroke. Prevent

Figure 11.2 Hold your pencil in a plane perpendicular to the paper and at 60° to the surface while lettering.

Figure 11.3 Use a protective sheet under your hand to prevent smudges and work from a comfortable position for natural strokes.

smudging while lettering by placing a sheet of paper under your hand to protect the drawing **(Fig. 11.3)**.

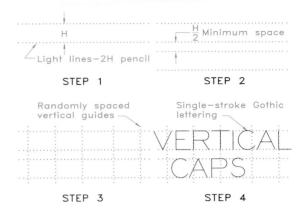

Figure 11.4 Use lettering guidelines in the following manner.

Step 1 Lay off letter heights, H, and draw light guidelines with a 2H pencil.

Step 2 Space lines no closer than H/2 apart.

Step 3 Draw vertical guidelines as light, thin, randomly spaced lines.

Step 4 Draw letters with single strokes using a medium-grade pencil: H, F, or HB. Do not erase the lightly drawn guidelines.

11.3 Guidelines

The most important rule of lettering is use guidelines at all times, whether you are lettering a paragraph or a single letter. Figure 11.4 shows how to draw and use guidelines. Use a sharp pencil in the 2H–4H grade range and draw light guidelines, just dark enough to be seen.

Most lettering on an engineering drawing is done with capital letters that are 1/8 inch (3 mm) high. The spacing between lines of lettering should be no closer than half the height of the capital letters, or 1/16 inch in this case.

Lettering Guides

Two instruments for drawing guidelines are the Braddock–Rowe lettering triangle and the Ames lettering instrument.

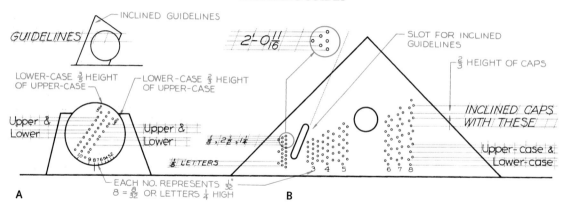

Figure 11.5

A The Ames lettering guide is used for drawing guidelines for lettering. Set the dial to the desired number of thirty-seconds of an inch for the height of uppercase letters.

B The Braddock–Rowe triangle also is used for drawing guidelines for lettering. The numbers near the guideline holes represent thirty-seconds of an inch.

The Braddock–Rowe triangle contains sets of holes for spacing guidelines (**Fig. 11.5B**). The numbers under each set of holes represent thirty-seconds of an inch. For example, the numeral 4 represents 4/32 inch or 1/8 inch for making uppercase (capital) letters. Some triangles have millimeter markings. Intermediate holes provide guidelines for lowercase letters, which are not as tall as capital letters (Fig. 11.6).

While holding a horizontal straightedge firmly in position, place the Braddock–Rowe triangle against its upper edge. Insert a sharp 2H pencil in the desired guideline hole to contact the drawing surface and guide the pencil point across the paper, drawing the guideline while the triangle slides along the straightedge. Repeat this procedure by moving the pencil point to each successive hole to draw other guidelines. Use the slanted slot in the triangle to draw guidelines, spaced randomly by eye, for inclined lettering.

The Ames lettering guide (**Fig. 11.5A**) is a similar device but has a circular dial for selecting proper guideline spacing. The numbers around the dial represent thirty-seconds of an inch. For example, the number 8 represents 8/32 inch, or

INCLINED AND VERTICAL GOTHIC

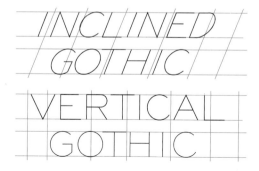

Figure 11.6 Use either vertical or inclined single-stroke Gothic lettering on engineering drawings.

guidelines for drawing capital letters that are 1/4 inch tall.

11.4 Freehand Gothic Lettering

The lettering recommended for engineering drawings is **single-stroke Gothic lettering**, so called because the letters are a variation of the Gothic

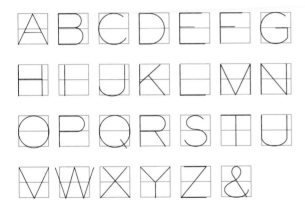

Figure 11.7 The alphabet is drawn here in single-stroke Gothic vertical uppercase letters. The letters are drawn inside a square to show their proportions.

COMMON ERRORS IN LETTERING

A. LETTERS POORLY DONE

B. STROKES TOO THIN

C. STROKES TOO THICK

D. BLACKER—BEAR DOWN WITH F PENCIL

Figure 11.8 Avoid these errors in lettering.

style made with a series of single strokes. Gothic lettering may be vertical or inclined **(Fig. 11.6)**, but only one style or the other should be used on a single drawing.

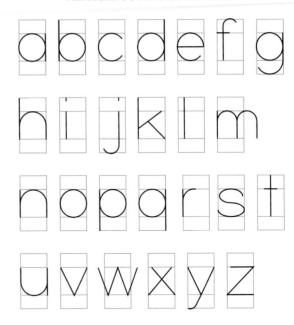

Figure 11.9 The alphabet is drawn here in single-stroke Gothic vertical lowercase letters.

Vertical Letters

Uppercase Figure 11.7 shows the alphabet in the single-stroke Gothic uppercase (capital) letters. Each letter is drawn inside a square box of guidelines to show their correct proportions. Draw each straight line with a single stroke; for example, draw the letter A with three single strokes. Letters composed of curves can best be drawn in segments; for example, draw the letter O by joining two semicircles.

The shape (form) of each letter is important. Small wiggles in strokes will not detract from your lettering if the letter forms are correct. **Figure 11.8** shows common errors in lettering that you should avoid.

Lowercase Figure 11.9 shows the alphabet in lowercase letters, which should be either two-thirds or three-fifths as tall as uppercase letters.

LETTER PROPORTIONS

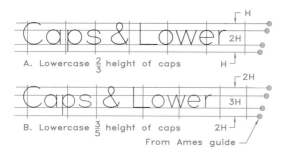

A. Lowercase $\frac{2}{3}$ height of caps

B. Lowercase $\frac{3}{5}$ height of caps

From Ames guide

Figure 11.10 The ratio of lowercase letters to the uppercase letters should be either two-thirds (A) or three-fifths (B). The Ames guide has both ratios, but the Braddock–Rowe triangle has only the two-thirds ratio.

VERTICAL NUMERALS

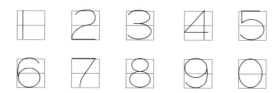

Figure 11.11 These numerals are used with single-stroke Gothic vertical letters.

Both lowercase ratios are labeled on the Ames guide, but only the two-thirds ratio is available on the Braddock–Rowe triangle.

Some lowercase letters, such as the letter b, have ascenders that extend above the body of the letter; some, such as the letter p, have descenders that extend below the body. Ascenders and descenders should be equal in length.

The guidelines in Fig. 11.9 that form squares about the body of each letter are used to illustrate their proportions. Letters that have circular bodies may extend slightly beyond the sides of the guideline squares. **Figure 11.10** gives examples of capital and lowercase letters used together.

INCLINED GOTHIC: UPPERCASE

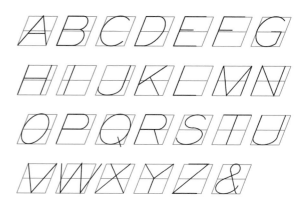

Figure 11.12 The alphabet is drawn here in single-stroke Gothic inclined uppercase letters.

Numerals **Figure 11.11** shows vertical numerals for use with single-stroke Gothic lettering, with each number enclosed in a square box of guidelines. Numbers should be the same height as the capital letters being used, usually 1/8 inch high. The numeral 0 (zero) is an oval, whereas the letter O is a circle in vertical lettering.

Inclined Letters
Uppercase Inclined uppercase (capital) letters have the same heights and proportions as vertical letters; the only difference is their 68° inclination **(Fig. 11.12)**. You may draw guidelines for inclined lettering with both the Braddock–Rowe triangle and the Ames guide.

Lowercase Draw inclined lowercase letters in the same manner as vertical lowercase letters **(Fig. 11.13)**, but draw circular features as ovals (ellipses). The angle of inclination is 68°, the same as for uppercase letters.

Numerals **Figure 11.14** shows the form of inclined numerals to be used with inclined lettering. **Figure 11.15** shows inclined letters and numbers in com-

INCLINED GOTHIC: LOWERCASE

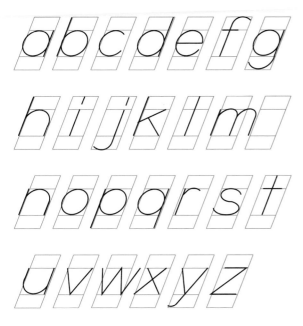

Figure 11.13 The alphabet is drawn here in single-stroke Gothic inclined lowercase letters.

INCLINED NUMERALS

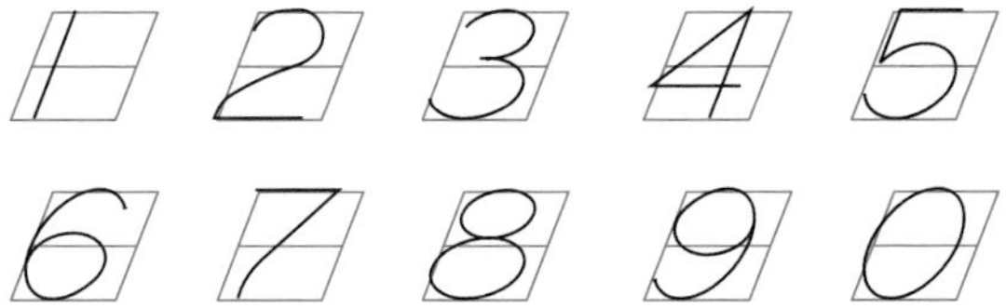

Figure 11.14 These numerals correspond to single-stroke Gothic inclined letters.

EXAMPLES OF NUMERALS

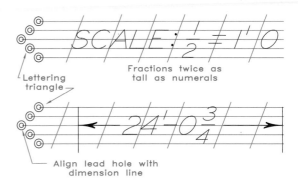

Figure 11.15 Make inclined common fractions twice as tall as single numerals. Omit inch marks.

GUIDELINES FOR FRACTIONS

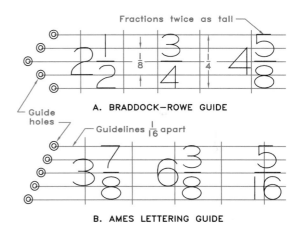

Figure 11.16 You may draw guidelines for common fractions, which should be twice as tall as single numerals, by using the (A) Braddock–Rowe triangle or the (B) Ames guide.

bination. **Figure 11.16** shows how to construct guidelines using the Braddock–Rowe triangle and the Ames lettering guide.

Spacing Numerals and Letters

Allow adequate space between numerals for the decimal point and fractions **(Fig. 11.17)**. Avoid the errors of lettering numerals shown.

Common fractions are twice as tall as single numerals (Fig. 11.17). Both the Braddock–Rowe triangle and the Ames guide have separate sets of holes, spaced 1/16 inch apart, for common fractions. The center guideline locates the fraction's crossbar.

When grouping letters to spell words, make the areas between the letters approximately equal

EXAMPLES OF FRACTIONS

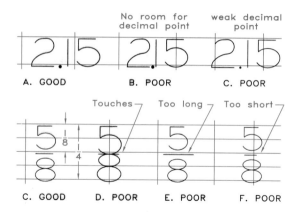

Figure 11.17 Avoid these errors in lettering.

SPACING OF LETTERS

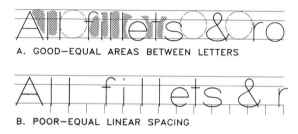

Figure 11.18 The areas between letters in a word should be approximately equal.

USING GUIDELINES

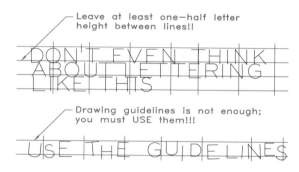

Figure 11.19 Always draw guidelines (vertical and horizontal) and use them whether you are lettering a paragraph or drawing a single letter.

INK PEN AND TEMPLATE

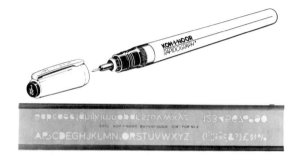

Figure 11.20 This technical ink pen and template may be used for lettering. (Courtesy of Koh-I-Noor Rapidograph, Incorporated.)

for the most visually pleasing result **(Fig. 11.18)**. **Figure 11.19** shows the incorrect use of guidelines and other violations of good lettering practice that you should avoid.

11.5 Mechanical Lettering

Drawings and illustrations that are to be reproduced by a printing process usually are drawn in India ink. Several mechanical aids are available for ink lettering.

The Rapidograph® lettering template **(Fig. 11.20)** is one type. Place it against a firmly held straightedge for aligning the letters. Move the template from position to position, and draw each letter (with a lettering pen) through the slots in the template.

Another system of mechanical lettering uses a grooved template and scriber that follows the grooves while an inkpoint draws the letters **(Fig. 11.21)**. A standard technical ink pen can be unscrewed from its barrel and attached to the scriber. Other templates of varying styles of letters and symbols are available.

SCRIBER AND TEMPLATE

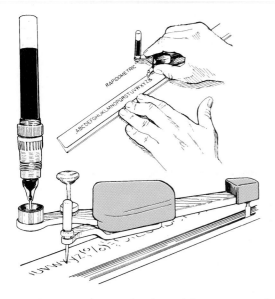

Figure 11.21 Templates and scribers of this type are useful for lettering.

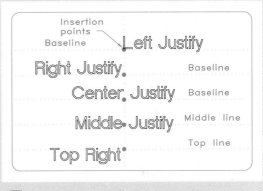

Figure 11.22 Text produced with the ROMANS font is best for engineering drawings.

Figure 11.23 Examples of text inserted in a drawing using various options under DTEXT/JUSTIFY.

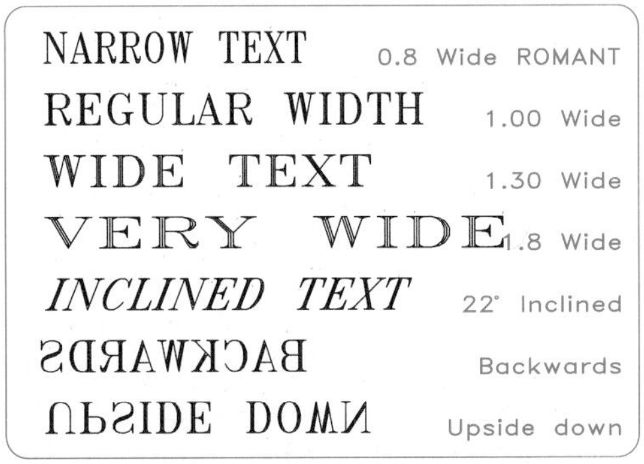

Figure 11.24 You may create different styles of lettering by changing the variables of the STYLE command.

11.6 Computer Lettering

Although AutoCAD offers a wide range of lettering (text) fonts, the ROMANS font shown in **Fig. 11.22** is the single-stroke Gothic text recommended for engineering drawings. DTEXT (dynamic text) is the command used to apply text to a drawing because it displays the text across the screen as it is typed:

```
Command: DTEXT

Justify/Style/<Start point>: J (CR)

Align/Fit/Center/Middle/Right/
TL/TC/TR/ML/MC/MR/BL/BC/BR:
```

The abbreviations represent the insertion points for a string of text as defined in **Fig. 11.23**, where various insertion points are given.

By using the STYLE command, you can modify a single font, ROMANT, to obtain a variety of letter forms, as shown in **Fig. 11.24**:

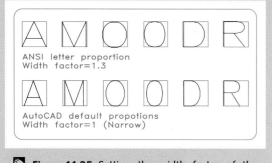

Figure 11.25 Setting the width factor of the ROMANS font at 1.30 matches the text to the ANSI standards for engineering lettering.

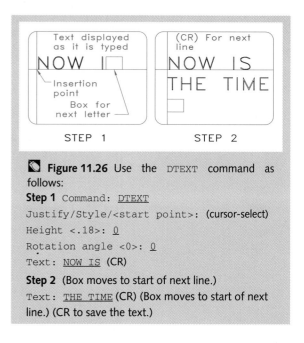

Figure 11.26 Use the DTEXT command as follows:

Step 1 Command: <u>DTEXT</u>
Justify/Style/<start point>: (cursor-select)
Height <.18>: <u>0</u>
Rotation angle <0>: <u>0</u>
Text: <u>NOW IS</u> (CR)

Step 2 (Box moves to start of next line.)
Text: <u>THE TIME</u> (CR) (Box moves to start of next line.) (CR to save the text.)

Command: <u>STYLE</u> (CR)

Text style name (or ?) <Standard>: <u>NEWFONT</u> (CR)

Existing style <TXT>: <u>ROMANT</u>

Height <0>: <u>0</u>

Width factor <1>: <u>1.3</u>

Obliquing angle <0>: (CR)

Backwards? <N>: (CR)

Upside-down? <N>: (CR)

Vertical? <N>: (CR)

NEWFONT is now the current style; it has a height of zero, and each letter has a width that is 130% of (1.3 times) the default width. The width of 1.3 matches the recommended letter proportions shown in **Fig. 11.25**. Assigning zero instead of a specific letter height permits you to change heights by responding when prompted by the DTEXT command. Had a height been specified, it would remain constant until you modified it by using the STYLE command.

Although you may respond to the prompts of the STYLE command in several ways to modify a text font, you must answer the prompt STYLE FILE with a name of an AutoCAD font (see Fig. 36.116). When using DTEXT, a box appears on the screen to represent the size of a letter (**Fig. 11.26**). You may use the select key of the mouse to move this box to any position on the screen and select a new starting point. When you press the enter button (carriage return) at the end of a line of text, the box will space down for the next line of text. The current text style will remain in use until you use the STYLE option of the DTEXT command to specify a new text style.

Present lettering problem solutions on size AH (11-×-8½-in.) paper, plain or with a grid, using the format shown in **Fig. 11.27**.

1. Practice drawing the alphabet in vertical uppercase letters, as shown in **Fig. 11.27**. Construct each letter three times: three A's, three B's, and so on. Use a medium-weight pencil—H, F, or HB.

2. Practice drawing vertical numerals and the alphabet in lowercase letters, as shown in Fig. **11.28**. Construct each letter and numeral two times: two 1's, two 2's, two a's, two b's, and so on. Use a medium-weight pencil—H, F, or HB.

3. Practice drawing the alphabet in inclined uppercase letters, as shown in Fig. 11.12. Construct each letter three times. Use a medium-weight pencil—H, F, or HB.

4. Practice lettering the vertical numerals and the alphabet in lowercase letters, as shown in Figs. 11.9 and 11.11. Construct each letter two times. Use a medium-weight pencil—H, F, or HB.

5. Construct guidelines for ⅛-in. capital letters starting ¼ in. from the top border. Each guideline should end ½ in. from the left and right borders. Using these guidelines, letter the first paragraph of the text of this chapter. Use all vertical capitals. Spacing between the lines should be ⅛ in.

6. Repeat Problem 5, but use all inclined capital letters. Use inclined guidelines to help you slant letters uniformly.

7. Repeat Problem 5, but use vertical capital and lowercase letters in combination. Capitalize only those words that are capitalized in the text.

8. Repeat Problem 5, but use inclined capital and lowercase letters in combination. Capitalize only those letters that are capitalized in the text.

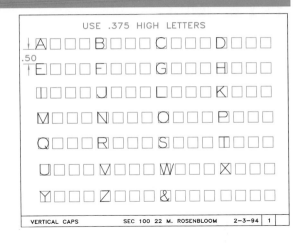

Figure 11.27 Problem 1.

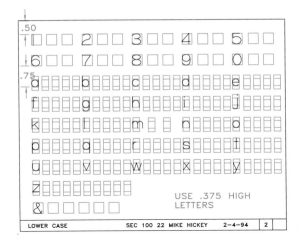

Figure 11.28 Problem 2.

Geometric Construction

12.1 Introduction

The solution of many graphical problems requires the use of geometry and geometric construction. Because mathematics was an outgrowth of graphical construction, the two areas are closely related. The proofs of many principles of plane geometry and trigonometry can be developed by using graphics. Moreover, graphical methods can be applied to solve some types of problems in algebra and arithmetic and virtually all types of problems in analytical geometry.

12.2 Angles

A fundamental application of geometric construction involves drawing lines at specified angles to each other. **Figure 12.1** gives names and definitions of various angles.

The unit of angular measurement is the degree, and a circle has 360 degrees. A degree (°) can be divided into 60 parts called minutes ('),

ANGLE DEFINITIONS

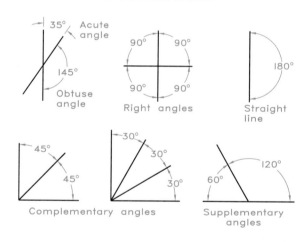

Figure 12.1 Standard angles and their names and definitions.

and a minute can be divided into 60 parts called seconds ("). An angle of 15°32'14" is an angle of 15 degrees, 32 minutes, and 14 seconds.

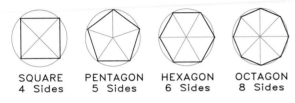

SQUARE PENTAGON HEXAGON OCTAGON
4 Sides 5 Sides 6 Sides 8 Sides

Figure 12.2 Regular polygons can be inscribed in circles.

TYPES OF TRIANGLES

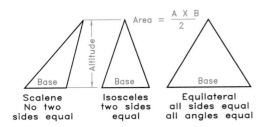

Area = $\dfrac{A \times B}{2}$

Base — Altitude — Base — Base

Scalene Isosceles Equilateral
No two two sides all sides equal
sides equal equal all angles equal

Figure 12.3 Types of triangles and their definitions.

12.3 Polygons

A **polygon** is a multisided plane figure of any number of sides. If the sides of a polygon are equal in length, the polygon is a **regular polygon**. A regular polygon can be inscribed in a circle and all its corner points will lie on the circle (**Figure 12.2**). Other regular polygons not pictured are the heptagon (7 sides), the nonagon (9 sides), the decagon (10 sides), and the dodecagon (12 sides).

The sum of the angles inside any polygon (interior angles) is $S = (n - 2) \times 180°$, where n is the number of sides of the polygon.

Triangles

A **triangle** is a three-sided polygon. The four types of triangles are **scalene, isosceles, equilateral,** and **right** (**Fig. 12.3**). The sum of the interior angles of a triangle is always $180°$ ($(3 - 2) \times 180°$).

Quadrilaterals

A **quadrilateral** is a four-sided polygon of any shape. The sum of the interior angles of a quadri-

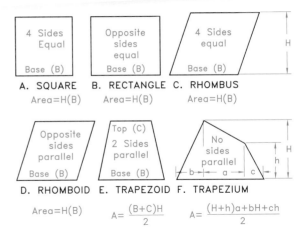

4 Sides Equal   Opposite sides equal   4 Sides equal

Base (B)   Base (B)   Base (B)   H

A. SQUARE B. RECTANGLE C. RHOMBUS
Area=H(B)   Area=H(B)   Area=H(B)

Opposite sides parallel   Top (C) 2 Sides parallel   No sides parallel

Base (B)   Base (B)   b — a — c   H h

D. RHOMBOID E. TRAPEZOID F. TRAPEZIUM

Area=H(B) $A = \dfrac{(B+C)H}{2}$ $A = \dfrac{(H+h)a + bH + ch}{2}$

Figure 12.4 Types of quadrilaterals, their definitions, and how to calculate their areas.

ELEMENTS OF CIRCLES

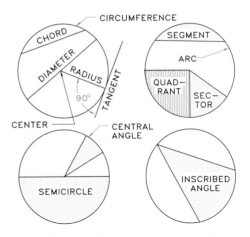

CIRCUMFERENCE
CHORD
DIAMETER
RADIUS
TANGENT
90°
CENTER
SEGMENT
ARC
QUAD-RANT
SEC-TOR
CENTRAL ANGLE
SEMICIRCLE
INSCRIBED ANGLE

Figure 12.5 Elements of a circle and their definitions.

lateral is $360°$ ($(4 - 2) \times 180°$). **Figure 12.4** shows the various types of quadrilaterals and the equations for their areas.

12.4 Circles

Figure 12.5 gives the names of the elements of a circle that are used throughout this textbook. A circle is constructed by swinging a radius from a

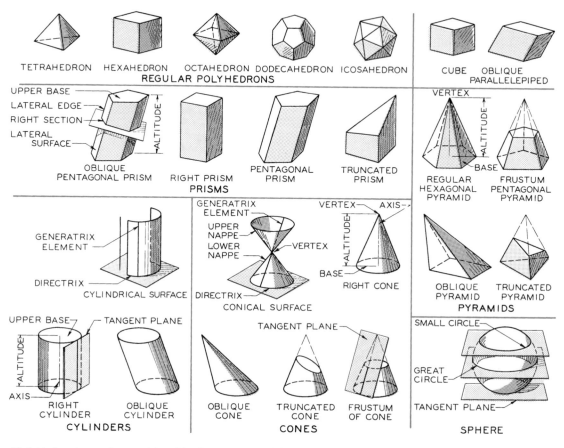

Figure 12.6 Various types of geometric solids, their names, and their elements.

fixed point through 360°. The area of a circle equals πR^2.

12.5 Geometric Solids

Figure 12.6 shows the various types of solid geometric shapes and their names and definitions.

Polyhedra

A **polyhedron** is a multisided solid formed by intersecting planes. If its faces are regular polygons, it is a regular polyhedron. Five regular polyhedra are

the **tetrahedron** (4 sides), the **hexahedron** (6 sides), the **octahedron** (8 sides), the **dodecahedron** (12 sides), and the **icosahedron** (20 sides).

Prisms

A **prism** has two parallel bases of equal shape connected by sides that are parallelograms. The line from the center of one base to the center of the other is the axis. If its axis is perpendicular to the bases, the prism is a right prism. If its axis is not perpendicular to the bases, the prism is an oblique prism. A prism that has been cut off to

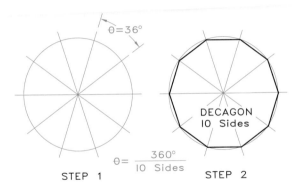

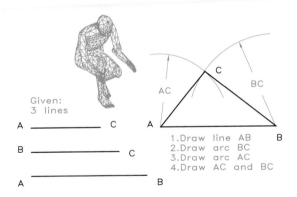

Figure 12.7 A regular polygon (sides of equal length):

Step 1 Divide circle into the required number of equal segments.

Step 2 Connect the segments where they intersect the circle.

Figure 12.8 A triangle can be constructed from three given sides.

form a base not parallel to the other is a truncated prism. A **parallelepiped** is a prism with bases that are either rectangles or parallelograms.

Pyramids

A **pyramid** is a solid with a polygon as a base and triangular faces that converge at a vertex. The line from the vertex to the center of the base is the axis. If its axis is perpendicular to the base, the pyramid is a right pyramid. If its axis is not perpendicular to the base, the pyramid is an oblique pyramid. A truncated pyramid is called a **frustum** of a pyramid.

Cylinders

A **cylinder** is formed by a line or an element (called a *generatrix*) that moves about the circle while remaining parallel to its axis. The axis of a cylinder connects the centers of each end of a cylinder. If the axis is perpendicular to the bases, it is the altitude of a right cylinder. If the axis does not make a 90° angle with the base, the cylinder is an oblique cylinder.

Cones

A **cone** is formed by a generatrix, one end of which moves about the circular base while the other end remains at a fixed vertex. The line from the center of the base to the vertex is the axis. If its axis is perpendicular to the base, the cone is a right cone. A truncated cone is called a frustum of a cone.

Spheres

A **sphere** is generated by revolving a circle about one of its diameters to form a solid. The ends of the axis of revolution of the sphere are *poles*.

12.6 Constructing Polygons

A regular polygon (having equal sides) can be inscribed in or circumscribed about a circle. When it is inscribed, all corner points will lie on the circle (**Fig. 12.7**). For example, constructing a 10-sided polygon involves dividing the circle into 10 sectors and connecting the points to form the polygon.

Triangles

When you know the lengths of all three sides of a triangle, you may construct it with a compass by *triangulation*, as **Fig. 12.8** shows. Any triangle

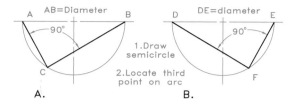

Figure 12.9 Any triangle inscribed in a semicircle is a right triangle.

HEXAGONS

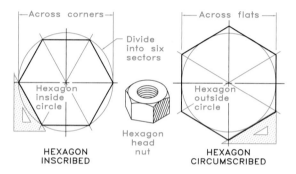

Figure 12.10 A hexagon can be inscribed in or circumscribed about a circle with a 30°–60° triangle.

OCTOGONS: CIRCLE METHOD

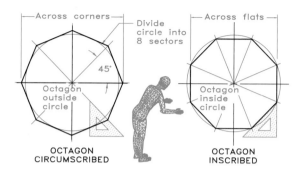

Figure 12.11 An octagon can be inscribed in or circumscribed about a circle with a 45° triangle.

OCTAGONS: SQUARE METHOD

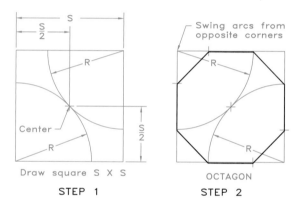

Figure 12.12 An octagon also can be constructed in a square.

inscribed in a semicircle, as **Fig. 12.9** shows, will be a right triangle.

Hexagons

The hexagon, a six-sided regular polygon, can be inscribed in or circumscribed about a circle (**Fig. 12.10**). Use a 30°–60° triangle to draw the hexagon. The circle represents the distance from corner to corner for an inscribed hexagon and from flat to flat for a circumscribed hexagon.

Octagons

The octagon, an eight-sided regular polygon, can be inscribed in or circumscribed about a circle (**Fig. 12.11**) or inscribed in a square (**Fig. 12.12**).

Use a 45° triangle to draw the octagon in the first case and a compass and straightedge in the second case.

Pentagons

The pentagon, a five-sided regular polygon, can be inscribed in or circumscribed about a circle. **Figure 12.13** shows another method of constructing a pentagon with a compass and straightedge.

PENTAGON: INSCRIBED

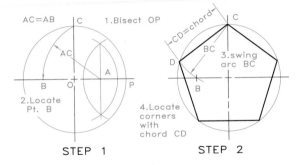

Figure 12.13 Constructing an inscribed pentagon:
Step 1 Bisect radius OP to locate point A. With A as the center and AC as the radius, locate point B on the diameter.
Step 2 With point C as the center and BC as the radius, locate point D. Use line CD as the chord to locate the other corners of the pentagon.

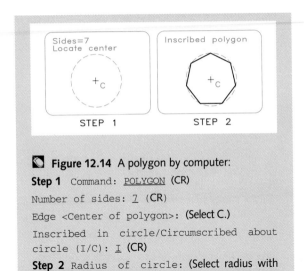

Figure 12.14 A polygon by computer:
Step 1 Command: POLYGON (CR)
Number of sides: 7 (CR)
Edge <Center of polygon>: (Select C.)
Inscribed in circle/Circumscribed about circle (I/C): I (CR)
Step 2 Radius of circle: (Select radius with cursor.) (Polygon is inscribed inside the imaginary circle.)

 Computer Method   You may use one of two POLYGON options from under the DRAW command to produce polygons. One option asks you to give the number of sides, select the center, specify the radius, and indicate whether the polygon is to be inscribed in or circumscribed about an imaginary circle (Fig. 12.14). The second option allows you to specify the number of sides and select the endpoints of one edge of the polygon before drawing it.

12.7 Bisecting Lines and Angles

Lines

Figure 12.15 shows two methods of finding the midpoint of a line. In the first method, a compass is used to construct a perpendicular bisector to a line. In the second method, a standard triangle and a straightedge are used.

BISECTING A LINE

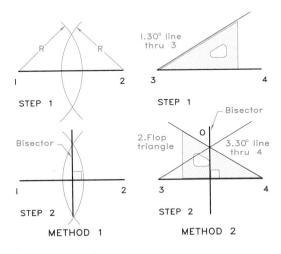

Figure 12.15 Bisecting a line:
Method 1 Use a compass and any radius.
Method 2 Use a standard triangle and a straightedge.

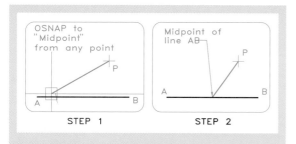

Figure 12.16 Midpoint of a line by computer:
Step 1 Command: <u>LINE</u> (CR) From point: <u>P</u>
(Locate P anywhere.)

To point: <u>OSNAP</u> (Select MIDPOINT mode.)

(Select any point on line AB.)

Step 2 The line from point P extends to the midpoint of the line AB.

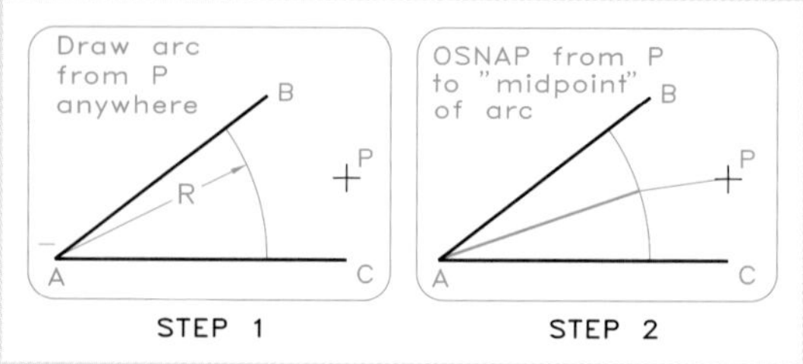

Figure 12.18 Bisecting an angle by computer:
Step 1 Using the ARC command, any radius, and center A, draw an arc that SNAPs to lines AC and AB.

Step 2 Command: <u>LINE</u> (CR) From point: (Select P anywhere.)

To point: <u>OSNAP</u>, Midpoint of (Select arc.)

(The line from P to the midpoint of the arc locates the bisector.)

BISECTING AN ANGLE

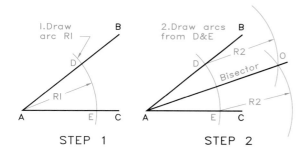

Figure 12.17 Bisecting an angle:
Step 1 Swing an arc of any radius R1 to locate points D and E.

Step 2 Draw equal arcs from D and E to locate point O. Line AO is the bisector of the angle.

Computer Method The midpoint of a line may be found **(Fig. 12.16)** by using the MIDPOINT mode of OSNAP while drawing a line from any point, P, to the line. The line will snap to the line's midpoint.

Angles

You may bisect angles by using a compass and drawing three arcs, as shown in **Fig. 12.17**.

Computer Method You may bisect an angle by first using the ARC command to draw an arc of a convenient radius between two lines, with its center at their intersection **(Fig. 12.18)**. Then use the DRAW command and the MIDPOINT mode of OSNAP to draw a line from any point, P, to the arc's midpoint and draw the bisector from point P to vertex A.

12.8 Revolution of Shapes

Figure 12.19 demonstrates how to rotate a triangle about point 1 of line 1–3. First rotate point 3 to its desired position (in this case vertically below point 1) with a compass. Then find point 2 by triangulation with arcs having radii 1–2 and 3–2. Use a straightedge to connect point 1 with new points 2 and 3 to complete the rotated view.

REVOLVING A FIGURE

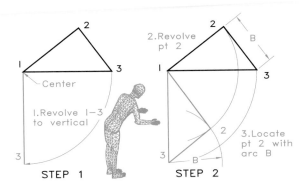

STEP 1   STEP 2

Figure 12.19 Revolving a figure:
Step 1 Revolve line 1–3 about point 1 with a compass.

Step 2 Locate point 2 by swinging arc 1–2 from point 1 and arc 3–2 from point 3.

ENLARGEMENT OF A FIGURE

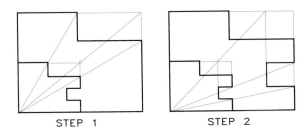

STEP 1   STEP 2

Figure 12.20 Enlarging a shape:
Step 1 To make a proportional enlargement, use a series of diagonals drawn through a single point, the lower left-hand corner in this case.

Step 2 Draw additional diagonals to locate the other features of the shape.

12.9 Enlargement and Reduction of Shapes

Figure 12.20 shows how to enlarge a shape by using a series of radial lines extending upward from its lower left corner. These lines pass through the corners of the shape and a rectangle

DIVIDING A LINE

STEP 1   STEP 2

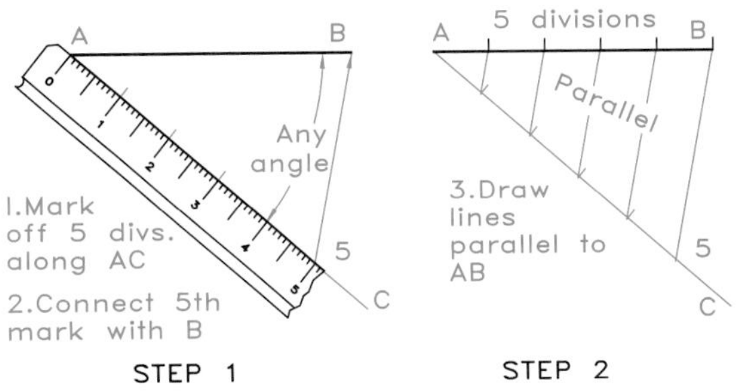

Figure 12.21 Dividing a line:
Step 1 To divide line AB into five equal lengths, lay off five equal divisions along line AC, and connect point 5 to end B with a construction line.

Step 2 Draw a series of five lines parallel to 5B to divide line AB.

drawn to enclose it. Use a straightedge to connect the points that form the larger shape and rectangle, including the notches. The larger shape is proportional to the smaller shape. This method may also be used to reduce a larger shape.

12.10 Division of Lines

Dividing a line into several equal parts often is necessary. **Figure 12.21** shows the method used to solve this type of problem—in this case dividing line AB into five equal lengths.

The same principle applies to locating equally spaced lines on a graph (**Fig. 12.22**). Lay scales with the desired number of units (0 to 3 and 0 to 5, respectively) across the graph up and down and then left to right. Make marks at each whole unit and draw vertical and horizontal index lines through these points. These index lines are used to show data in a graph.

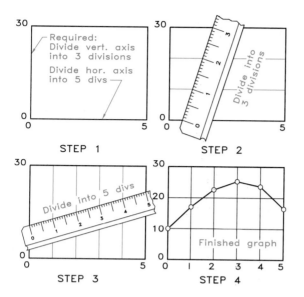

STEP 1 · STEP 2 · STEP 3 · STEP 4

Figure 12.22 Dividing a space:

Step 1 Draw the outline of the graph.

Step 2 To divide the y axis into three equal segments, lay a scale having three units of measurement spanning the graph with the 0 and 3 located on the top and bottom lines. Make a mark at points 1 and 2 and draw horizontal lines through them.

Step 3 To divide the x axis into five equal segments, lay a scale having at least five units of measurement across the graph with the 0 and 5 on the left- and right-hand vertical lines. Make marks at points 1, 2, 3, and 4 and draw vertical lines through them.

Step 4 Plot the data points and draw the curve of the graph.

12.11 Arcs

Through Three Points

You may draw an arc through three points by connecting the points with two lines and drawing perpendicular bisectors through each line to locate the center of the circle at C (**Fig. 12.23**). Draw the arc and lines AB and BD become chords of the arc.

To find the center of a circle or an arc, reverse this process by drawing two chords that intersect at a point on the circumference and bisecting

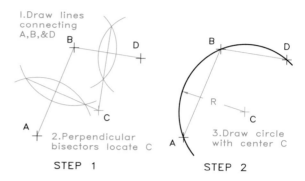

STEP 1 · STEP 2

Figure 12.23 An arc through three points:

Step 1 Connect points A, B, and D with two lines and draw their perpendicular bisectors to intersect at the center, C.

Step 2 Using the center C, and the distance to the points as the radius, R, draw the arc through the points.

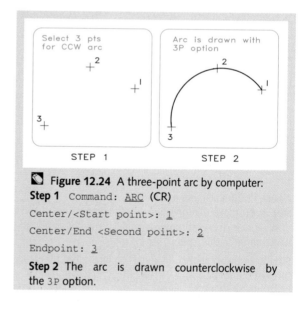

STEP 1 · STEP 2

Figure 12.24 A three-point arc by computer:

Step 1 Command: ARC (CR)

Center/<Start point>: 1

Center/End <Second point>: 2

Endpoint: 3

Step 2 The arc is drawn counterclockwise by the 3P option.

them. The perpendicular bisectors intersect at the center of the circle.

Computer Method You may use the ARC command and the 3P (3-point) option to produce a counterclockwise arc through any three points that you select on the screen (**Fig. 12.24**).

RECTIFYING AN ARC

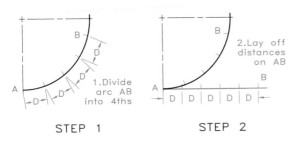

STEP 1   STEP 2

Figure 12.25 Rectifying an arc:

Step 1 Divide arc AB into a series of equal chords, D.

Step 2 Lay out the equal chords, D, along the tangent to the arc. The length of the arc is AB.

METHODS OF DRAWING PARALLEL LINES

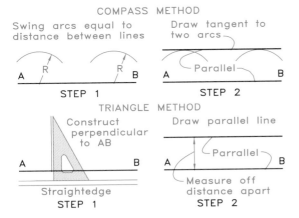

Figure 12.26 Drawing parallel lines:

Compass Method

Step 1 Swing two arcs from line AB.

Step 2 Draw the parallel line tangent to the arcs.

Triangle Method

Step 1 Draw a line perpendicular to AB.

Step 2 Measure the desired distance, R, along the perpendicular and draw the parallel line through it.

Rectifying Arcs

Rectifying an arc means laying out its length along a straight line, as **Fig. 12.25** shows. You may also rectify an arc mathematically to check the accuracy of

LOCATING A TANGENT POINT

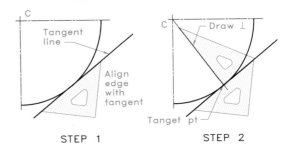

STEP 1   STEP 2

Figure 12.27 Locating a tangent point:

Step 1 Align a triangle with the tangent line while holding it against a firmly held straightedge.

Step 2 Hold the triangle in position, locate a second triangle perpendicular to it, and draw a line from the center to locate tangent point.

your construction. For example, because a circle has 360°, an arc of a 30° sector is one-twelfth of the full circumference. If the circumference of a circle is 12 inches, the rectified 30° arc would be 1 inch long.

12.12 Parallel Lines

You may draw one line parallel to another by using either method shown in **Fig. 12.26**. In the first method, use a compass and draw two arcs having radius R to locate a parallel line at the desired distance (R) from the first line. In the second method, measure the desired perpendicular distance R from the first line, mark it, and draw the parallel line through it with your T-square or drafting machine.

12.13 Tangents

Points of Tangency

A **point of tangency** is the theoretical point at which a line joins an arc or two arcs join without crossing. **Figure 12.27** shows how to find the point of tangency with triangles by constructing a perpendicular line to the tangent line from the arc's

MARKING TANGENT POINTS

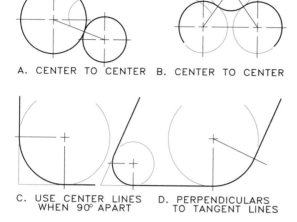

A. CENTER TO CENTER B. CENTER TO CENTER

C. USE CENTER LINES   D. PERPENDICULARS
   WHEN 90° APART   TO TANGENT LINES

Figure 12.28 Use thin (construction) lines that extend from the centers slightly beyond the arcs to mark points of tangency.

LINE TANGENT TO AN ARC

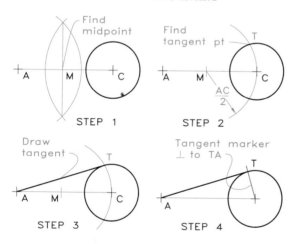

Find midpoint

Find tangent pt

STEP 1   STEP 2

Draw tangent

Tangent marker ⊥ to TA

STEP 3   STEP 4

Figure 12.29 A line tangent to an arc from a point:

Step 1 Connect point A with center C and locate point M by bisecting AC.

Step 2 Using point M as the center and MC as the radius, locate point T on the arc.

Step 3 Draw the line from A to T that is tangent to the arc of point T.

Step 4 Draw the tangent marker perpendicular to line TA from the center past the arc.

LINE TANGENT TO AN ARC

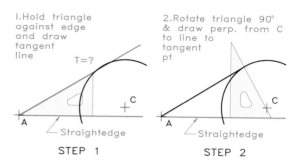

1.Hold triangle against edge and draw tangent line   T=?

2.Rotate triangle 90° & draw perp. from C to line to tangent pt

STEP 1   STEP 2

Figure 12.30 A tangent to an arc from a point:

Step 1 Hold a triangle against a straightedge and draw a line from point A that is tangent to the arc.

Step 2 Rotate your triangle 90° and locate the point of tangency by drawing a line through center C.

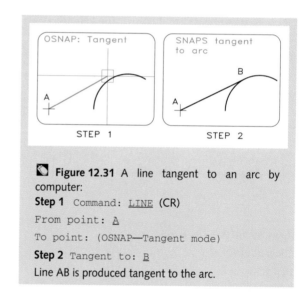

OSNAP: Tangent

SNAPS tangent to arc

STEP 1   STEP 2

◩ **Figure 12.31** A line tangent to an arc by computer:

Step 1 Command: <u>LINE</u> (CR)

From point: <u>A</u>

To point: (OSNAP—Tangent mode)

Step 2 Tangent to: <u>B</u>

Line AB is produced tangent to the arc.

center. **Figure 12.28** shows the conventional methods of marking points of tangency.

Line Tangent to an Arc

Figure 12.29 shows one way to find the point of tangency between a line drawn from point A and an arc. Connect point A to the arc's center and

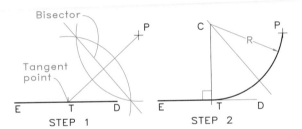

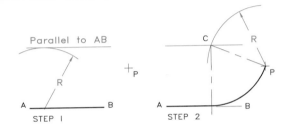

Figure 12.32 An arc through two points:
Step 1 To draw an arc through point P that is tangent to line DE at point T, draw the perpendicular bisector of TP.

Step 2 Construct a perpendicular to line DE at point T to intersect the bisector at point C and draw the arc from C with radius CT.

Figure 12.33 An arc tangent to a line through a point:
Step 1 When an arc is to be drawn tangent to line AB and through point P, first draw a line parallel to line AB and R distance from it.

Step 2 Draw an arc from point P with radius R to locate center C and draw the arc with radius R.

bisect line AC (step 1); swing an arc from point M through point C locating tangent point T (step 2); draw the tangent to T (step 3); and mark the tangent point (step 4). The point of tangency may be found by using a triangle, as **Fig. 12.30** shows.

◼ **Computer Method** A line can be drawn from a point tangent to an arc by using the TAN-GENT option of the OSNAP command **(Fig. 12.31)**. When prompted for the second point, select a point near the tangent point and the line will be drawn tangent to it.

Arc Tangent to a Line from a Point

To construct an arc that is tangent to line DE at T and that passes through point P **(Fig. 12.32)**, draw the perpendicular bisector of line TP. Draw a perpendicular to line DE at point T to locate the center at point C and swing an arc with radius OT. A similar problem **(Fig. 12.33)** requires drawing an arc of a given radius that is tangent to line AB and that passes through point P. An arc with its outer at C is tangent to the line.

◼ **Computer Method** An arc tangent to a line may be drawn from its endpoint by pressing (CR) to leave the LINE command and selecting the

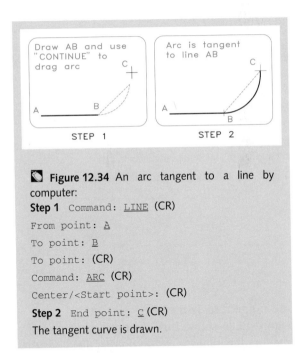

◼ **Figure 12.34** An arc tangent to a line by computer:
Step 1 Command: LINE (CR)

From point: A

To point: B

To point: (CR)

Command: ARC (CR)

Center/<Start point>: (CR)

Step 2 End point: C (CR)
The tangent curve is drawn.

ARC option and CONTINUE **(Fig. 12.34)**. The arc will begin at the end of the line, and you DRAG its other end to the location desired. This is an excellent method for drawing runouts.

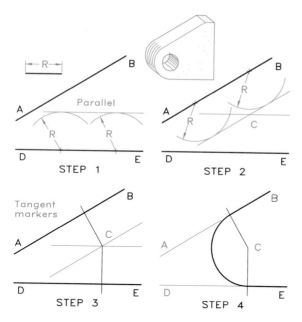

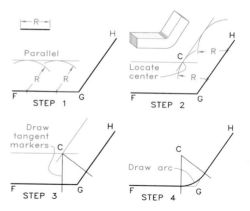

Figure 12.35 An arc tangent to lines making an acute angle:

Step 1 Construct a light line parallel to line DE with radius R.

Step 2 Draw a second light line parallel to and R distance from line AB to locate center C.

Step 3 Draw thin lines from center C perpendicular to lines AB and DE to locate the tangency points.

Step 4 Draw the tangent arc and darken your lines.

Figure 12.36 An arc tangent to lines making an obtuse angle:

Step 1 Using radius R, draw a light line parallel to line FG.

Step 2 Construct a light line parallel to line GH that is R distance from it to locate center C.

Step 3 Construct thin lines from center C perpendicular to lines FG and GH to locate the tangency points.

Step 4 Draw the tangent arc and darken your lines.

ARC TANGENT TO PERPENDICULAR LINES

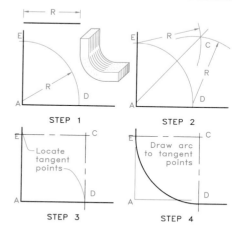

Figure 12.37 An arc tangent to perpendicular lines:

Step 1 Using radius R and center A, locate points D and E.

Step 2 Locate point C by swinging two arcs using radius R.

Step 3 Locate the tangent points with perpendiculars CE and CD.

Step 4 Draw the tangent arc and darken your lines.

Arc Tangent to Two Lines

Figure 12.35 shows how to construct an arc of a given radius tangent to two nonparallel lines that form an acute angle. The same steps apply to constructing an arc tangent to two lines that form an obtuse angle (**Fig. 12.36**). In both cases, the points of tangency are located with lines drawn from the centers perpendicular and past the original lines. **Figure 12.37** shows a technique for finding an arc tangent to perpendicular lines only.

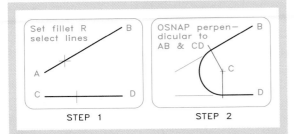

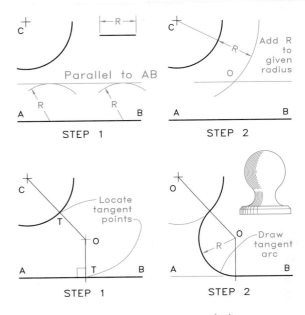

Figure 12.38 An arc tangent to two lines by computer:

Step 1 Command: <u>FILLET</u> (CR)

Polyline/Radius/<Select two objects>: <u>R</u> (CR)

Enter fillet radius <0.0000>: <u>.75</u> (CR)

Command: (**CR**)

FILLET Polyline/Radius/<Select two objects>: (Select points on AB and CD.)

Step 2 Command: <u>LINE</u> (CR)

From point. <u>CENTER</u> (CR) of (Select point on the arc.)

To point: <u>PERPEND</u> (CR) to (Select line AB and produce a perpendicular to the point of tangency. Extend the line beyond AB. Locate the tangent point on CD in the same manner.)

Figure 12.39 An arc tangent to an arc and a line:

Step 1 Draw a light line parallel to line AB that is R distance from it.

Step 2 Add radius R to the extended radius from center C. Swing the extended radius to locate the center, O.

Step 3 Draw lines OC and OT to locate the tangency points.

Step 4 Draw the tangent arc between the points of tangency with radius R and center O.

Computer Method To produce an arc tangent to two nonparallel lines, use the FILLET command (**Fig. 12.38**). When the radius length has been assigned and a point selected on each line, the arc is drawn and the lines are trimmed. To mark tangent points, snap to the center of the arc with the CENTER option of OSNAP and draw two lines perpendicular (use OSNAP's PERPEND option) to lines AB and CD from the center.

Arc Tangent to an Arc and a Line

Figure 12.39 shows the steps for constructing an arc tangent to an arc and a line. **Figure 12.40** shows a variation of this technique for an arc drawn tangent to a given arc and line with the arc reversed.

Arc Tangent to Two Arcs

Figure 12.41 shows how to draw an arc tangent to two arcs. Lines drawn between the centers locate the points of tangency. The resulting tangent arc is concave from the top. Drawing a convex arc tangent to the given arcs requires that its radius be greater than the radius of either of the given arcs, as shown in **Fig. 12.42**.

One variation of this problem (**Fig. 12.43**) is to draw an arc of a given radius tangent to the top of one arc and the bottom of the other. Another (**Fig. 12.44**) is to draw an arc tangent to a circle and a larger arc.

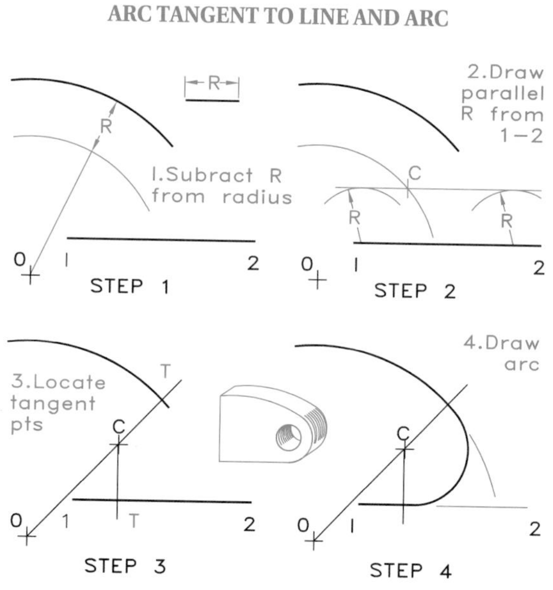

ARC TANGENT TO LINE AND ARC

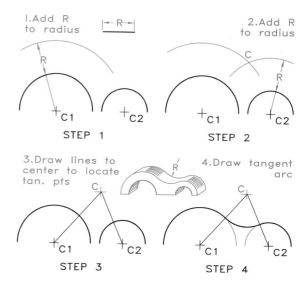

Figure 12.40 An arc tangent to an arc and a line:

Step 1 Subtract radius R from the radius through center O. Draw a concentric arc with this shortened radius.

Step 2 Draw a line parallel to line 1–2 and R distance from it to locate the center, C.

Step 3 Locate the tangency points with lines from O through C and from C perpendicular to line 1–2.

Step 4 Draw the tangent arc between the tangent points with radius R and center C.

ARC TANGENT TO TWO ARCS: CONCAVE

Figure 12.41 A concave arc tangent to two arcs:

Step 1 Extend the radius of one circle by adding the radius R to it. Use the extended radius to draw a concentric arc.

Step 2 Extend the radius of the other circle by adding radius R to it. Use this extended radius to construct an arc to locate center C.

Step 3 Connect center C with centers C1 and C2 with thin lines to locate the tangency points.

Step 4 Draw the tangent arc between the points of tangency using radius R and center C.

Computer Method Use the FILLET command **(Fig. 12.45)** to produce an arc tangent to two arcs. After entering the command, specify the radius when prompted. Press the carriage return (CR) twice to return to the COMMAND mode, and the FILLET command is ready for use.

Select the two arcs with your cursor; the tangent arc is drawn and the arcs are trimmed at the points of tangency. Mark the tangency points by drawing lines from centers C1 and C2.

Ogee Curves

The **ogee curve** is an S curve formed by tangent arcs. The ogee curve shown in **Fig. 12.46** is the

result of constructing two arcs tangent to three intersecting lines. **Figure 12.47** shows an unequal-arc ogee curve drawn to pass through points B, E, and C.

12.14 Conic Sections

Conic sections are plane figures that can be described both graphically and mathematically; they are formed by passing imaginary cutting planes through a right cone, as **Fig. 12.48** shows.

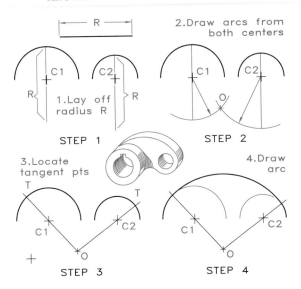

Figure 12.42 A convex arc tangent to two arcs:

Step 1 Extend the radius of each arc from the arc past its center and lay off radius R from the arcs along these lines.

Step 2 Use the distance from each center to the ends of the extended radii to swing arcs to locate center O.

Step 3 Draw thin lines from center O through centers C1 and C2 to locate the points of tangency.

Step 4 Draw the tangent arc between the tangent points using radius R and center O.

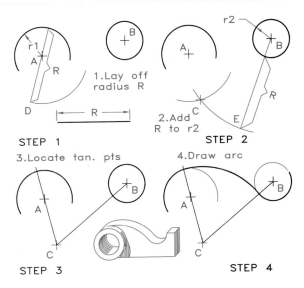

Figure 12.43 An arc tangent to two circles:

Step 1 Lay off radius R from the arc along an extended radius to locate point D.

Step 2 Extend the radius through center B and add radius R to it from point E. Use radius BE to locate center C.

Step 3 Draw thin lines from center C through centers A and B to locate the points of tangency.

Step 4 Draw the tangent arc between the tangent points using radius R and center C.

Ellipses

The **ellipse** is a conic section formed by passing a plane through a right cone at an angle (**Fig. 12.48B**). Mathematically, the ellipse is the path of a point that moves in such a way that the sum of the distances from two focal points is a constant. The largest diameter of an ellipse—the **major diameter**—is always the true length. The shortest diameter—the **minor diameter**—is perpendicular to the major diameter.

Revolving the edge view of a circle yields an ellipse (**Fig. 12.49**). The ellipse template shown in **Fig. 12.50** is used to draw the same ellipse. The angle between the line of sight and the edge of the circle is the angle of the ellipse template (or the one closest to this size) that should be used. Ellipse templates are available in 5° intervals and in major diameter sizes that vary in increments of about 1/8 inch (**Fig. 12.51**).

You may construct an ellipse inside a rectangle or parallelogram by plotting a series of points to form the ellipse (**Fig. 12.52**). Two circles can be used to construct an ellipse by making the diameter of the large circle equal to the major diameter and the diameter of the small circle equal to the minor diameter (**Fig. 12.53**).

ARC TANGENT TO TWO CIRCLES

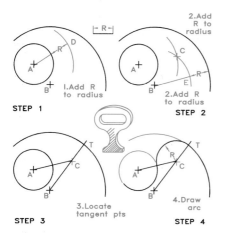

Figure 12.44 An arc tangent to two arcs:

Step 1 Add radius R to the radius from center A. Use radius AD to draw a concentric arc from center A.

Step 2 Subtract radius R from the radius through B. Use radius BE to draw an arc to locate center C.

Step 3 Draw thin lines to connect the centers and locate the points of tangency.

Step 4 Draw the tangent arc between the tangency points using radius R and center C.

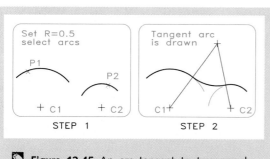

◆ Figure 12.45 An arc tangent to two arcs by computer:

Step 1 Command: FILLET (CR)

Polyline/Radius/<Select two lines>: R (CR)

Enter fillet radius <current>: .5 (CR)

Step 2 Command: FILLET (CR)

Polyline/Radius/<Select two objects>: P1, P2

(Points on each arc.) (The tangent arc is drawn, and the arcs are trimmed.) Tangent points are located by lines between centers.

AN OGEE CURVE

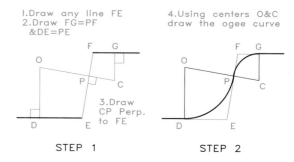

Figure 12.46 An ogee curve:

Step 1 To draw an ogee curve between two parallel lines, draw light line EF at any angle. Locate point P anywhere along EF. Find the tangent points by making FG equal to FP and DE equal to EP. Draw perpendiculars at G and D to intersect the perpendicular at O and C.

Step 2 Use radii CP and OP at centers O and C to draw two tangent arcs to complete the ogee curve.

OGEE CURVE: UNEQUAL ARCS

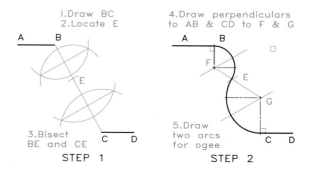

Figure 12.47 An unequal-arc ogee curve:

Step 1 Parallel lines are to be connected by an ogee curve passing through points B and C. Draw light line BC and select point E on it. Bisect BE and CE.

Step 2 Construct perpendiculars at points B and C to intersect the bisectors and locate centers F and G. Locate the points of tangency and draw the ogee curve using radii FB and GC.

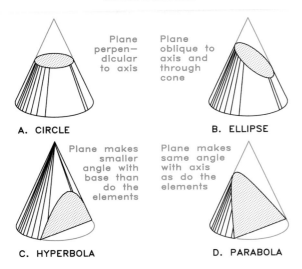

A. CIRCLE

B. ELLIPSE

Plane perpendicular to axis

Plane oblique to axis and through cone

Plane makes smaller angle with base than do the elements

Plane makes same angle with axis as do the elements

C. HYPERBOLA

D. PARABOLA

Figure 12.48 The conic sections are the (A) circle, (B) ellipse, (C) hyperbola, and (D) parabola. They are formed by passing cutting planes through a right cone.

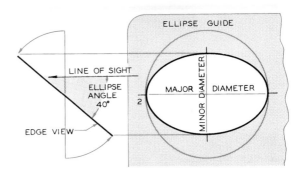

Figure 12.50 When the edge view of a circle is revolved so that the line of sight is not perpendicular to it, the circle appears as an ellipse. The angle between the line of sight and the edge view of the circle is the angle of the ellipse template.

ELLIPSE TEMPLATES

ELLIPSES BY REVOLUTION

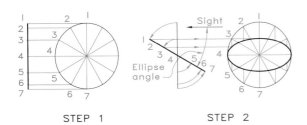

STEP 1   STEP 2

Figure 12.49 An ellipse by revolution:

Step 1 When the edge view of a circle is perpendicular to the projectors between its adjacent view, it appears as a circle. Mark equally spaced points around the circle's circumference and project them to the edge.

Step 2 Revolve the edge view of the circle and project the points to the circular view, which now appears as an ellipse. The points project vertically downward to their new positions.

Figure 12.51 Ellipse templates are calibrated at 5° intervals from 15° to 60°. (Courtesy of Timely Products, Incorporated.)

ELLIPSE: PARALLELOGRAM METHOD

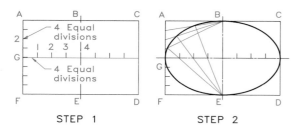

Figure 12.52 An ellipse by the parallelogram method:
Step 1 Draw an ellipse inside a rectangle or parallelogram by dividing the horizontal centerline into the same number of equal segments as the shorter sides, AF and CD.

Step 2 The method used to construct the curve is shown for one quadrant—sets of rays from E and B that intersect give the points on which to plot the curve.

ELLIPSE: CIRCLE METHOD

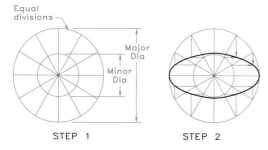

Figure 12.53 An ellipse by the circle method:
Step 1 Draw two concentric circles with the large one equal to the major diameter and the small one equal to the minor diameter. Divide them into equal sectors.

Step 2 Plot points on the ellipse by projecting downward from the large circle to intersect horizontal projectors drawn from the intersections on the small circle.

Computer Method Use the `ELLIPSE` command to produce an ellipse by selecting the endpoints of the major diameter and a third point that defines the length of minor radius **(Fig. 12.54)**. Another option of the `ELLIPSE` command allows you to select the center, the end of

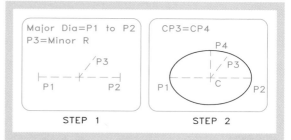

Figure 12.54 An ellipse by computer:
Step 1 `Command: ELLIPSE` (CR)

`<Axis endpoint 1> /Center:` (Select P1.)

`Axis endpoint 2:` (Select P2.)

Step 2 `<Other axis distance>/Rotatmon:` (Select radius P3 in any direction and the ellipse is produced.) (Alternatively, the option, Rotation, may be used to specify the orientation of the major diameter.)

the major radius, and the end of the minor radius to produce the ellipse. Isometric ellipses can be drawn in isometric pictorial drawings.

The mathematical equation of an ellipse is

$$\frac{x^2}{a^2}+\frac{y^2}{b^2}=1, \text{ where } a, b \neq 0.$$

Parabolas

The **parabola** is defined as a plane curve, each point of which is equidistant from a straight line (called a **directrix**) and a focal point. The parabola is the conic section formed when the cutting plane and an element on the cone's surface make the same angle with the cone's base (see Fig. 12.48D).

Figure 12.55 shows construction of a parabola from its mathematical definition. To draw a parabola geometrically, divide two perpendicular lines into the same number of equal segments,

PARABOLA: MATHEMATICAL METHOD

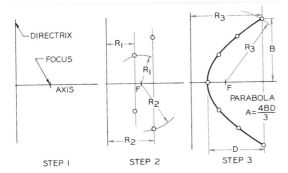

Figure 12.55 A parabola by the mathematical method:
Step 1 Draw an axis perpendicular to the directrix (a line). Choose a point for the focus, F.

Step 2 Use a series of selected radii to find points on the curve. For example, draw a line parallel to the directrix and R2 from it. Swing R2 from F to intersect the line and plot the point.

Step 3 Continue the process with a series of arcs of varying radii until you find an adequate number of points to complete the curve.

PARABOLA: TANGENT METHOD

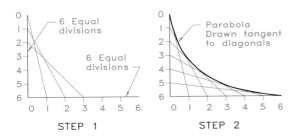

Figure 12.56 A parabola by the tangent method:
Step 1 Draw two thin lines at a convenient angle and divide each into the same number of segments. Connect the points with a series of diagonals.

Step 2 When finished, draw a smooth curve that is tangent to the diagonals.

PARABOLA: PARALLELOGRAM METHOD

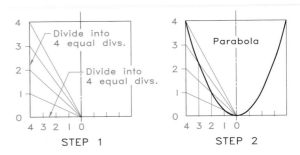

Figure 12.57 A parabola by the parallelogram method:
Step 1 Construct a parallelogram (shown here as a rectangle) to contain the parabola and locate its axis parallel to the sides through 0. Divide the sides into equal segments. Connect the segment division points to point 0.

Step 2 Construct lines parallel to the sides (vertical in this case) to locate points along the rays from 0 and draw a smooth curve through them.

connect them as shown in **Fig. 12.56**, and draw a smooth curve through the plotted points. **Figure 12.57** shows a third method of drawing a parabola, which involves the use of a rectangle or parallelogram. The mathematical equation of the parabola is

$$y = ax^2 + bx + c, \text{ where } a \neq 0.$$

Hyperbolas

The **hyperbola** is a two-part conic section defined as the path of a point that moves in such a way that the difference of its distances from two focal points is a constant (see Fig. 12.48C). **Figure 12.58** shows construction of a hyperbola according to this definition.

Figure 12.59 shows a method of constructing a hyperbola by drawing perpendicular lines through point B to serve as asymptotes. The hyperbolic curve comes closer and closer to the asymptotes as it is extended, but it never touches them.

HYPERBOLA CONSTRUCTION

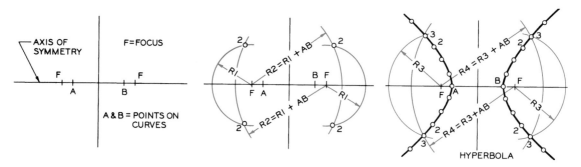

Figure 12.58 A hyperbola:
Step 1 Draw a perpendicular through the axis of symmetry. Locate focal points F equidistant from it on both sides. Locate points A and B equidistant from the perpendicular at a distance of your choosing, but between the focal points.

Step 2 Select radius R1 to draw arcs using focal points F as the centers. Add R1 to AB (the distance between the nearest points on the hyperbolas) to find R2. Draw arcs using radius R2 and the focal points as centers. The intersections of R1 and R2 establish points 2 on the hyperbola.

Step 3 Select other radii and add them to AB to locate additional points as in step 2. Draw a smooth curve through the points.

EQUILATERAL HYPERBOLA

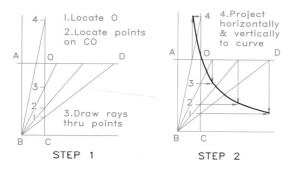

Figure 12.59 An equilateral hyperbola:
Step 1 Draw perpendiculars through B. Select any point as O and draw horizontal and vertical thin lines through O. Divide line CO into equal segments, and draw rays from B through them to horizontal line AD.

Step 2 Draw horizontal construction lines from the segment division points along line CO and project lines from AD vertically to points of intersection. Connect these points with a smooth curve.

SPIRAL

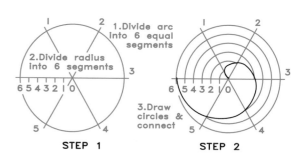

Figure 12.60 Constructing a spiral:
Step 1 Draw a circle and divide it into equal parts. Divide the radius into the same number of equal parts (six in this case).

Step 2 Begin inside and draw arc 0–1 to intersect radius 0–1. Then swing arc 0–2 to radius 0–2, and continue until you reach the last point, at 6, on the original circle, and connect the points.

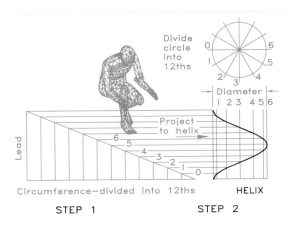

Divide circle into 12ths

Diameter
1 2 3 4 5 6

Project to helix

Lead

Circumference—divided into 12ths   HELIX

STEP 1   STEP 2

1.Divide into 12ths

3.Draw top v; project to front

2.Divide lead into 12ths

LEAD

STEP 1   STEP 2

Figure 12.61 A cylindrical helix:
Step 1 Divide the top view of the cylinder into equal parts and project them to the front view. Lay out the circumference and the height (lead: pronounced leed) of the cylinder. Divide the circumference into the same number of equal parts by taking the measurements from the top view.

Step 2 Project the points along the inclined rise to their respective points on the diameter and connect them with a smooth curve.

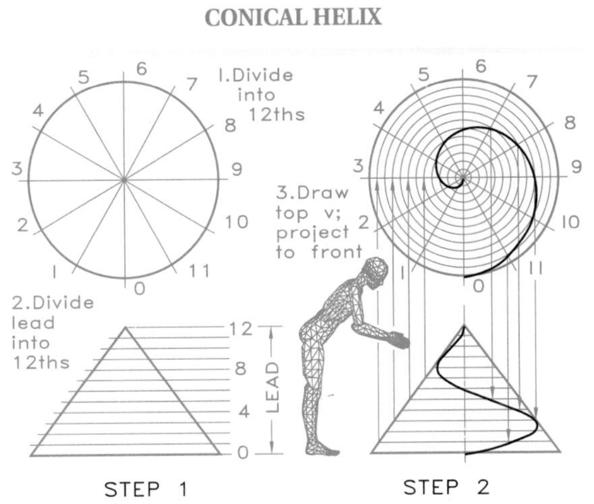

Figure 12.62 A conical helix:
Step 1 Divide the cone's base into equal parts. Pass a series of horizontal cutting planes through the front view of the cone. Use the same number as the number of divisions on the base (12 in this case).

Step 2 Project all the divisions along the front view of the cone to the line 3–9 and draw a series of arcs from the center to their respective radii in the top view and plot the points. Project the points to their respective cutting planes in the front view and connect them with a smooth curve.

12.15 Spirals

The **spiral** is a coil lying in a single plane that begins at a point and becomes larger as it travels around the origin. **Figure 12.60** shows the steps for constructing a spiral.

12.16 Helixes

The **helix** is a three-dimensional curve that coils around a cylinder or cone at a constant angle of inclination. Applications of helixes are corkscrews and the threads on a screw. **Figure 12.61** shows a helix constructed about a cylinder, and **Fig. 12.62** shows a helix constructed about a cone.

Problems

Present your solutions to these problems on size AV (8½-×-11-inch) paper similar in appearance to that shown in **Fig. 12.63**. The printed grid represents 0.20-in. intervals, so you can use your engineers' 10 scale to lay out the problems. By equating each grid interval to 5 mm, you also can use your full-sized metric scale to lay out and solve the problems. Show your construction and mark all points of tangency, as discussed in the chapter.

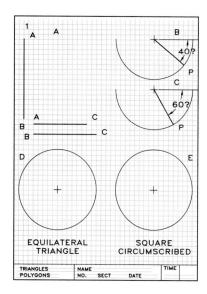

Figure 12.63 Problem 1(A–E). Basic constructions.

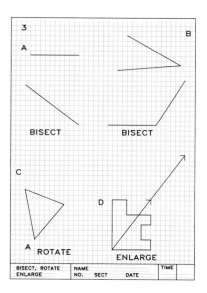

Figure 12.65 Problem 3(A–D). Basic constructions.

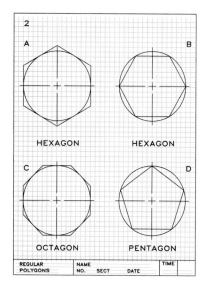

Figure 12.64 Problem 2(A–D). Constructions of regular polygons.

1. Basic constructions (**Fig. 12.63**)

    (**A**) Draw triangle ABC using the given sides.

    (**B–C**) Inscribe an angle in the semicircles with the vertexes at point P.

    (**D**) Inscribe a three-sided regular polygon inside the circle.

    (**E**) Circumscribe a four-sided regular polygon about the circle.

2. Construction of regular polygons (**Fig. 12.64**)

    (**A**) Circumscribe a hexagon about the circle.

    (**B**) Inscribe a hexagon in the circle.

    (**C**) Circumscribe an octagon about the circle.

    (**D**) Construct a pentagon inside the circle using the compass method.

3. Basic constructions (**Fig. 12.65**)

    (**A**) Bisect the lines.

    (**B**) Bisect the angles.

    (**C**) Rotate the triangle 60° clockwise about point A.

    (**D**) Enlarge the given shape to the size indicated by the diagonal.

4. Line division and tangency construction (**Fig. 12.66**)

    (**A**) Divide line AB into seven equal parts. Draw a construction line through point A.

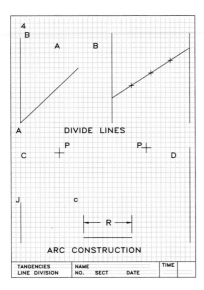

Figure 12.66 Problem 4(A–D). Tangency construction.

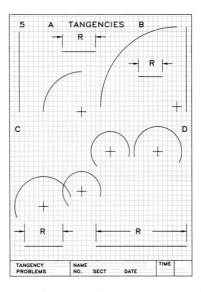

Figure 12.68 Problem 6(A–D). Tangency construction.

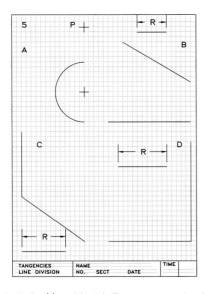

Figure 12.67 Problem 5(A–D). Tangency construction.

(**B**) Divide the space between the two vertical lines into four equal segments. Draw three vertical lines at the division points that are equal in length to the given lines.

(**C**) Construct an arc with radius R that is tangent to the line at J and that passes through point P.

(**D**) Construct an arc with radius R that is tangent to the line and passes through point P.

5. Tangency construction (**Fig. 12.67**)

(**A**) Construct a line from point P that is tangent to the semicircle. Locate the points of tangency. Use the compass method.

(**B–D**) Construct arcs with the given radii tangent to the lines.

6. Tangency construction (**Fig. 12.68**)

(**A–D**) Construct arcs that are tangent to the arcs or lines shown. The radii are given for each problem.

7. Ogee curve construction (**Fig. 12.69**)

(**A–D**) Construct ogee curves that connect the ends of the given lines and pass through point P.

8. Tangency construction (**Fig. 12.70**)

(**A–B**) Using the given radii, connect the circles with a tangent arc as indicated in the sketches.

9. Rectifying an arc and ellipse construction (**Fig. 12.71**)

(**A–B**) Rectify the arc along the given line by dividing the circumference into equal segments and laying them off with your dividers.

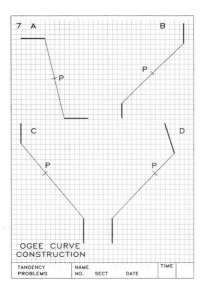

Figure 12.69 Problem 7(A–D). Ogee curve construction.

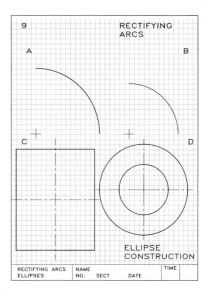

Figure 12.71 Problem 9(A–D). Rectifying an arc, ellipse construction.

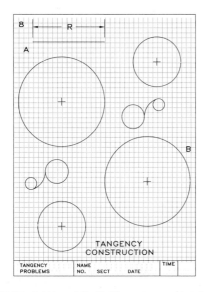

Figure 12.70 Problem 8(A–B). Tangency construction.

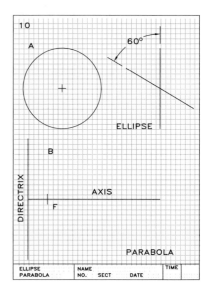

Figure 12.72 Problem 10(A–B). Ellipse and parabola construction.

(C) Construct an ellipse inside the rectangular layout.

(D) Construct an ellipse inside the large circle. The small circle represents the minor diameter.

10. Ellipse and parabola construction (Fig. 12.72)

(A) Construct an ellipse inside the circle when the edge view has been rotated as shown.

(B) Using the focal point F and the directrix, plot and draw the parabola formed by these elements.

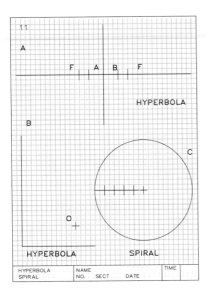

Figure 12.73 Problem 11(A–C). Hyperbola and spiral construction.

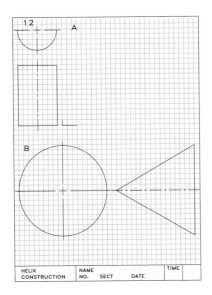

Figure 12.74 Problem 12(A–B). Helix construction.

11. Hyperbola and spiral construction (Fig. 12.73)

(A) Using the focal point F, points A and B on the curve, and the axis of symmetry, construct the hyperbola.

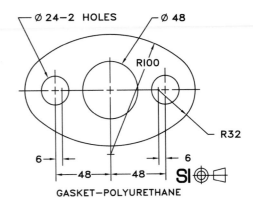

Figure 12.75 Problem 13. Gasket.

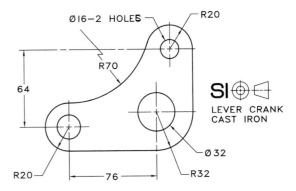

Figure 12.76 Problem 14. Lever crank.

(B) Construct a hyperbola that passes through O. The perpendicular lines are asymptotes.

(C) Construct a spiral by using the four divisions marked along the radius.

12. Helix construction (Fig. 12.74)

(A–B) Construct helixes that have a rise equal to the heights of the cylinder and cone. Show construction and the curve in all views.

13–22. Practical applications (Figs. 12.75–12.84). Construct the given shapes on size A sheets, one problem per sheet. Select the scale that will best fit the problem to the sheet. Mark all points of points of tangency and strive for good line quality.

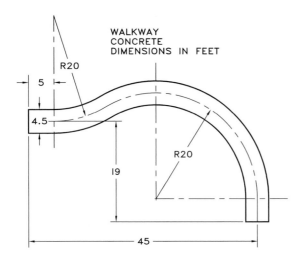

WALKWAY
CONCRETE
DIMENSIONS IN FEET

R20

5

4.5

R20

19

45

Figure 12.77 Problem 15. Road tangency.

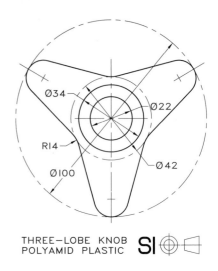

Ø34

Ø22

R14

Ø42

Ø100

THREE-LOBE KNOB
POLYAMID PLASTIC SI ⊕ ⊟

Figure 12.79 Problem 17. Three-lobe knob.

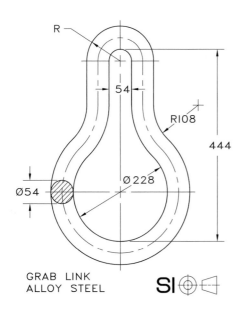

R

54

R108

444

Ø54

Ø228

GRAB LINK
ALLOY STEEL SI ⊕ ⊟

Figure 12.78 Problem 16. Grab link.

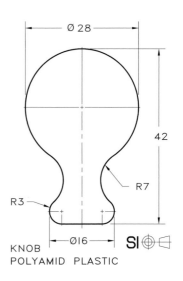

Ø 28

42

R7

R3

Ø16

KNOB
POLYAMID PLASTIC SI ⊕ ⊟

Figure 12.80 Problem 18. Knob.

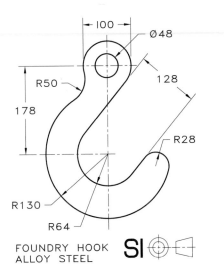

FOUNDRY HOOK
ALLOY STEEL   SI ⊕⊡

Figure 12.81 Problem 19. Foundry hook.

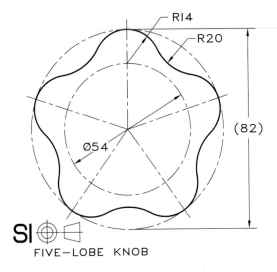

SI ⊕⊡
FIVE–LOBE KNOB

Figure 12.83 Problem 21. Five-lobe knob.

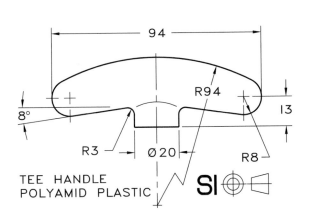

TEE HANDLE
POLYAMID PLASTIC   SI ⊕⊡

Figure 12.82 Problem 20. Tee handle.

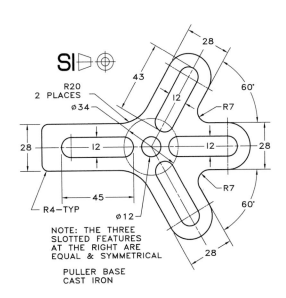

SI ⊡⊕

NOTE: THE THREE
SLOTTED FEATURES
AT THE RIGHT ARE
EQUAL & SYMMETRICAL

PULLER BASE
CAST IRON

Figure 12.84 Problem 22. Puller base.

Orthographic Sketching

13.1 Introduction

Sketching is a rapid, freehand method of drawing rather than drawing with instruments. **Moreover, sketching is a thinking process as much as a method of communication.** Designers usually develop their ideas by making many sketches before arriving at the final solution.

Designers with sketching skills can use their sketches to assign the drafting of finished drawings to assistants. Without sketching skills, designers are unable to utilize their helpers effectively. Many new products and projects have begun as sketches made on the back of an envelope or on a napkin at a restaurant table. Sketching also helps to communicate on the job when words are inadequate. The ability to communicate by any means is a great asset, and sketching is one of the best ways to transmit ideas.

13.2 Shape Description

Although the angle bracket in **Fig. 13.1** is a simple three-dimensional object, describing it with

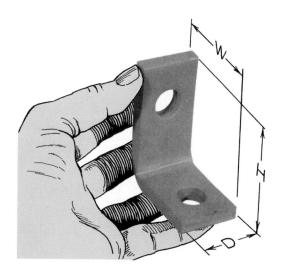

Figure 13.1 How can you sketch this angle bracket to convey its shape effectively?

words is difficult. Most untrained people would think that drawing it as a three-dimensional pictorial would be difficult. **To make drawing such**

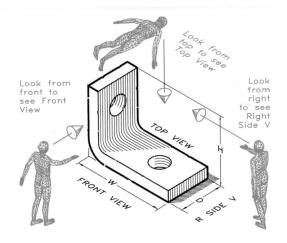

Figure 13.2 These positions give the viewpoints for three views of the angle bracket: top, front, and right side.

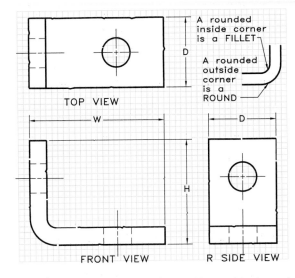

Figure 13.3 This sketch shows three orthographic views of the angle bracket.

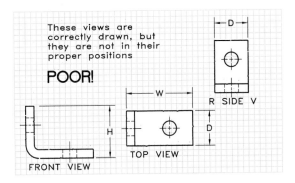

Figure 13.4 Views must be sketched in their standard orthographic positions. When they are incorrectly positioned as shown here, the object cannot be readily understood.

objects less difficult, engineers devised a standard system, called orthographic projection, for showing objects in different views.

In orthographic projection, separate views represent the object at 90° intervals as the viewer moves about it (**Fig. 13.2**). **Figure 13.3** shows two-dimensional views of the bracket from the front, top, and right side. The top view appears above the front view because both share the dimension of width. The side view appears to the right of the front view because both share the dimension of height.

The views of the bracket contain three types of lines: **visible lines, hidden lines,** and **centerlines.** Visible lines are the thickest. Thinner dashed lines, hidden lines, represent features that cannot be seen in a view. The thinnest lines are centerlines, or imaginary lines composed of long and short dashes to show the centers of arcs and axes of cylinders.

The space between views may vary, but the views must be positioned as shown here. This arrangement is logical, the views are easiest to interpret in this order, and the drawing process is most efficient because the views project from each other. **Figure 13.4** illustrates the lack of clarity when views are incorrectly positioned, even though each view is properly drawn.

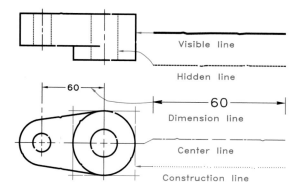

Figure 13.5 This series of lines comprises the alphabet of lines for sketching. The lines at the right are full size.

LINES FOR SKETCHING

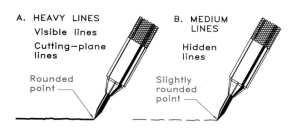

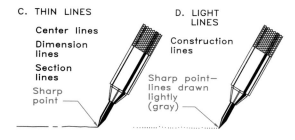

Figure 13.6 An F pencil is a good choice for all sketching lines if you sharpen it differently for varying line widths.

13.3 Sketching Techniques

You need to understand the application of line types used in sketching (freehand) orthographic views before continuing with the principles of projection. The so-called alphabet of lines for sketching is presented in **Fig. 13.5**. All lines,

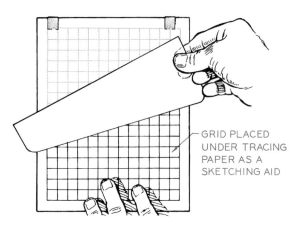

Figure 13.7 A grid placed under a sheet of tracing paper can aid you in freehand sketching.

except construction lines, should be black and dense. Recall that construction lines are drawn lightly so that they need not be erased. Using the proper widths of these lines is an important part of making good sketches.

Medium-weight pencils, such as H, F, or HB grades, are best for sketching the lines shown in Fig. 13.6. By sharpening the pencil point to match the desired line width, you may use the same grade of pencil for all these lines. Lines sketched freehand should have a freehand appearance; do not attempt to make them appear mechanical. Using a printed grid or laying translucent paper over a printed grid can aid your sketching technique (**Fig. 13.7**).

When you make a freehand sketch, lines will be vertical, horizontal, angular, and/or circular. By not taping your drawing to the table top, you can position the sheet for the most comfortable strokes, usually from left to right (**Fig. 13.8**). Examples of correctly sketched lines are contrasted with incorrectly sketched ones in **Fig. 13.9**.

HORIZONTAL LINES

VERTICAL LINES

ANGULAR LINES

CIRCLES AND ARCS

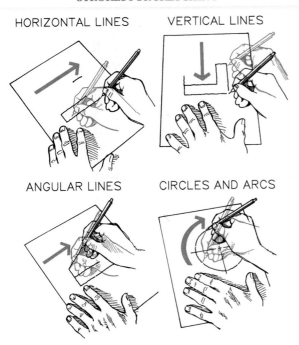

Figure 13.8 Sketch lines as shown here for the best results; rotate your sheet for comfortable sketching positions.

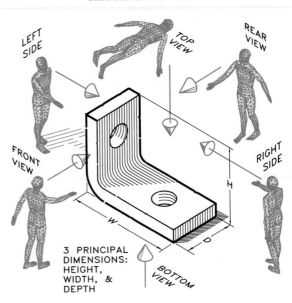

Figure 13.10 Six principal views of the angle bracket can be sketched from the viewpoints shown.

SKETCHING TECHNIQUE

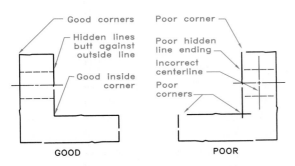

Figure 13.9 For good sketches, follow the examples of good technique and avoid the common errors of poor technique shown.

13.4 Six-View Sketching

The maximum number of principal views that may be drawn in orthographic projection is six, as the viewer changes position at 90° intervals (Fig. 13.10). In each view, two of the three dimensions of height, width, and depth are seen.

These views must be sketched in their standard positions **(Fig. 13.11)**. The width dimension is common to the top, front, and bottom views. The height dimension is common to the right-side, front, left-side, and rear views. Note the simple but effective dimensioning of each view with two dimensions. Seldom is an object so complex that it requires six orthographic views.

13.5 Three-View Sketching

You can adequately describe most objects with three orthographic views—usually the top, front,

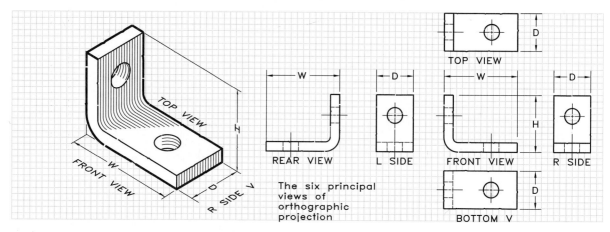

Figure 13.11 This six-view sketch of the angle bracket shows the six principal views of orthographic projection. Note the placement of dimensions on the views.

STANDARD ARRANGEMENT

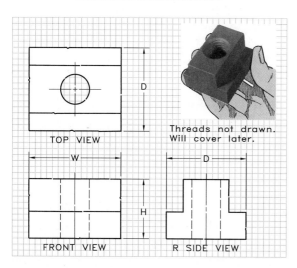

Figure 13.12 This sketch shows the standard orthographic arrangement for three views of a jaw nut, with dimensions and labels.

and right-side views. **Figure 13.12** shows a typical three-view sketch of a T-block with height, width, and depth dimensions and the front, top, and right-side views labeled.

A PART TO SKETCH

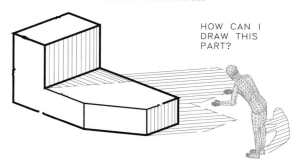

Figure 13.13 Sketches of three orthographic views describe this fixture block in Fig. 13.14.

The object shown in **Fig. 13.13** is represented by three orthographic views on a grid in **Fig. 13.14.** To obtain those views, first sketch the overall dimensions of the object, then sketch the slanted surface in the top view and project it to the other views. Finally, darken the lines; label the views; and letter the overall dimensions of height, width, and depth.

Slanted surfaces will appear as edges or foreshortened (not-true-size) planes in the principal

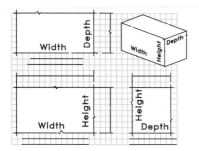

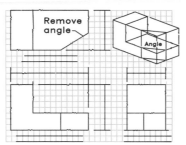

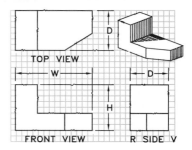

Figure 13.14 Three-view sketching:
Step 1 Block in the views with light construction lines that will not need to be erased. Allow proper spacing for labeling and dimensioning the views.

Step 2 Remove the notches and project from view to view.

Step 3 Check for correctness, darken the lines, and letter the labels and dimensions.

VIEW OF PLANS

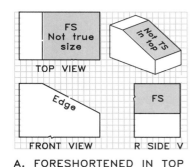

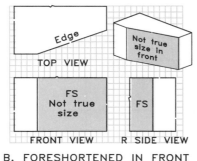

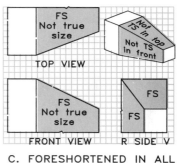

A. FORESHORTENED IN TOP   B. FORESHORTENED IN FRONT   C. FORESHORTENED IN ALL

Figure 13.15 Views of planes:
A The plane appearing as an angular edge in the front view is foreshortened in the top and side views.

B The plane appearing as an angular edge in the top view is foreshortened in the front and side views.

C Two sloping planes appear foreshortened in the side view and each appears as an edge in either the top or front views.

views of orthographic projection (**Fig. 13.15**). In **Fig. 13.16C**, two intersecting planes of the object slope in two directions; thus both appear foreshortened in the front, top, and right-side views.

A good way to learn orthographic projection is to construct a missing third view (the front view in **Fig. 13.16**) when two views are given. In **Fig. 13.17**, we construct the missing right-side view

from the given top and front views. To obtain the depth dimension for the right-side view, transfer it from the top view with dividers; to obtain the height dimension, project it from the front view.

Figure 13.18 shows a fixture pad sketched in three views. The pad has a finished surface, indicated by "✔" marks in the two views where the surface appears as edges, and four counterbored

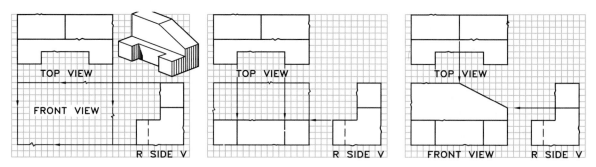

Figure 13.16 Sketching a missing front view:

Step 1 To sketch the front view, begin by blocking it in with light construction lines that will not need to be erased.

Step 2 Project the notch from the top view to the front view and sketch the lines as final lines.

Step 3 Project the ends of the angular notch from the top and right-side views, check the views, and darken the lines.

SKETCHING A MISSING SIDE VIEW

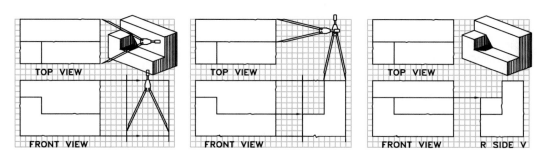

Figure 13.17 Sketching a missing side view:

Step 1 To find the right-side view, transfer the overall depth with dividers and project the height from the front view. Block in the view with light construction lines.

Step 2 Locate the notch in the side view with your dividers and project its base from the front view. Use light construction lines.

Step 3 Project the top of the notch from the front view, check for correctness, darken the lines, and label the views.

holes. Dimension lines for the height, width, and depth labels should be spaced at least three letter heights from the views. For example, when you use 1/8-inch letters, position them at least 3/8-inch from the views.

Apply the finish mark symbol to the edge views of any finished surfaces, visible or hidden, to specify that the surface is to receive machining to make it smoother. The surface in **Fig. 13.19** is being finished by grinding, which is one of many methods of smoothing a surface.

13.6 Circular Features

The pulley shaft depicted in **Fig. 13.20** in two views is composed of circular features. We enhanced

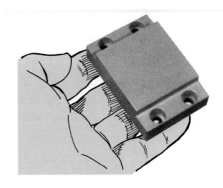

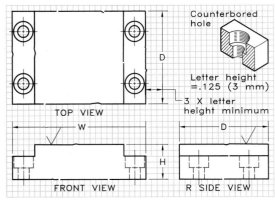

Figure 13.18 Three orthographic views adequately describe the rest pad. Space dimension lines at least three letter heights from the views. Finish marks (✔ marks) indicate that the top surface has been machined to a smooth finish. Counterbored holes allow bolt heads to be recessed.

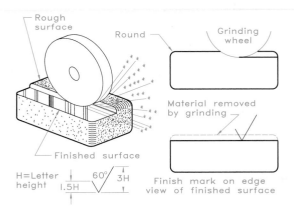

Figure 13.19 Place a finish mark on any edge view of a surface (visible or hidden) that has been (or is to be) smoothed by machining. Grinding is one of the methods used to finish a surface.

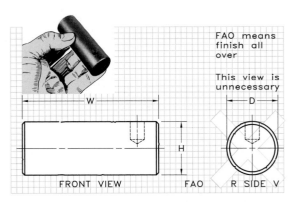

Figure 13.20 This pulley shaft is a typical cylindrical part that can be represented adequately by two views.

these features by adding centerlines. **Figure 13.21** shows how to apply centerlines to indicate the center of the circular ends of a cylinder and its vertical axis. Perpendicular centerlines cross in circular views to locate the center of the circle and extend beyond the arc by about 1/8 inch. Centerlines consist of alternating long and short dashes, about 1 inch and 1/8 inch in length, respectively.

When centerlines coincide with visible or hidden lines, the centerline should be omitted because object lines are more important and cen-terlines are imaginary lines. **Figure 13.22** shows the precedence of lines.

The centerlines shown in **Fig. 13.23** clarify whether the circles and arcs are concentric (share the same centers). **Figure 13.24** shows the correct manner of applying centerlines to orthographic views of an object composed of concentric cylinders.

BASICS OF CENTERLINES

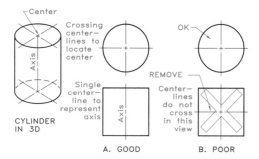

Figure 13.21 Centerlines identify the centers of circles and axes of cylinders. Centerlines cross only in the circular view and extend about ⅛ inch beyond the outside lines.

PRECEDENCE OF LINES

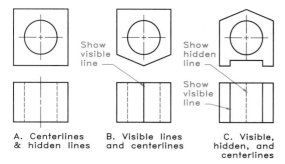

Figure 13.22 When visible lines coincide with hidden lines, show the visible lines. When hidden lines coincide with centerlines, show the hidden lines.

Sketching Circles

You may sketch circles by either of the methods shown in **Fig. 13.25**. Use light guidelines and dark centerlines to block in the circle. Drawing a freehand circle in one continuous arc is difficult, so draw short arcs with the help of the guidelines.

Figure 13.26 shows the steps involved in constructing three orthographic views of a part having circular features. **Figure 13.27** shows a typical

THE USE OF CENTERLINES

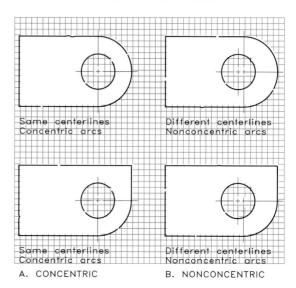

Figure 13.23
A Extend centerlines beyond the last arc that has the same center.

B Sketch separate centerlines when the arcs are not concentric.

RELATIVE LINE WEIGHTS

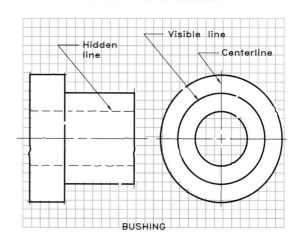

Figure 13.24 This orthographic sketch depicts the application of centerlines to concentric cylinders and the relative weights of various lines.

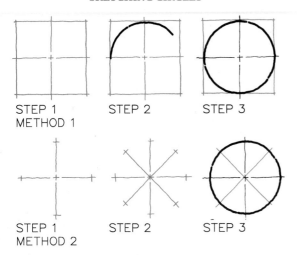

STEP 1   STEP 2   STEP 3
METHOD 1

STEP 1   STEP 2   STEP 3
METHOD 2

Figure 13.25 Sketching circles:

Method 1

Step 1 Using construction lines, block in the diameter of the circle about the centerlines.

Step 2 Sketch an arc tangent to the box through two tangent points.

Step 3 Complete the circle with other arcs.

Method 2

Step 1 Mark off radii on the centerlines.

Step 2 Mark off radii on two construction lines drawn at 45°.

Step 3 Sketch the circles with arcs passing through the marked points.

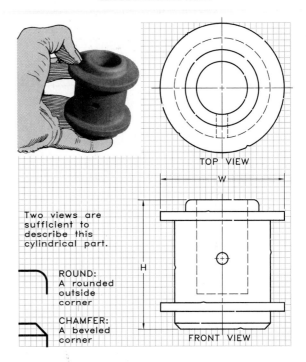

TOP VIEW

W

Two views are sufficient to describe this cylindrical part.

ROUND:
A rounded outside corner

CHAMFER:
A beveled corner

H

FRONT VIEW

Figure 13.27 Two orthographic views adequately describe this cylindrical pivot base.

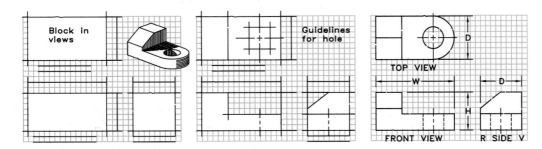

Block in views

Guidelines for hole

TOP VIEW
W   D
D

FRONT VIEW   R SIDE V
H

Figure 13.26 Sketching circular features in orthographic views:

Step 1 To sketch orthographic views of the part, begin by blocking in the overall dimensions with construction lines. Leave room for labels and dimensions.

Step 2 Construct the centerlines and the squares that block in the diameters of the circles. Find the slanted surface in the side view.

Step 3 Sketch the arcs, darken the lines, label the views, and show the dimensions W, D, and H.

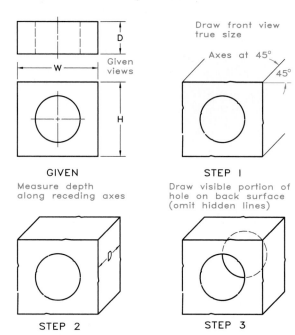

Figure 13.28 Sketching oblique pictorials:

Step 1 Sketch the front of the part as an orthographic front view. Sketch the receding lines at 45° to show the depth dimension.

Step 2 Measure the depth along the receding axes and sketch the back of the part.

Step 3 Locate the circle on the rear plane, show the visible portion of it, and omit the hidden lines.

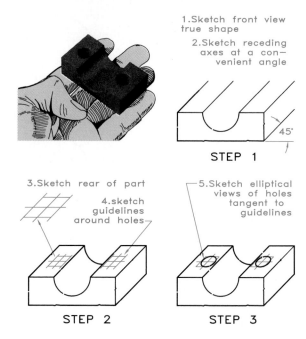

Figure 13.29 Sketching arcs in oblique pictorials:

Step 1 Sketch the front view of the mounting bracket saddle as a true front view. Sketch the receding axes from each corner.

Step 2 Sketch the rear of the part by measuring its depth along the receding axes. Sketch guidelines about the holes on the upper planes.

Step 3 Sketch the circular features as ellipses on the upper planes tangent to the guidelines.

part having circular features and two sketched views of it. Note the definitions of a **round** and a **chamfer**.

13.7 Oblique Pictorial Sketching

An **oblique pictorial** is a three-dimensional representation of an object's height, width, and depth. It approximates a photograph of an object, making the sketch easier to understand at a glance than orthographic views. Sketch the front of the object as a true-shape orthographic view (**Fig. 13.28**). Sketch the receding axes at an angle of between 20° and 60° oblique to the front view. Lay off the depth dimension as its true length along the receding axes. When the depth is true length, the oblique is a **cavalier oblique**.

The major advantage of an oblique pictorial is the ease of sketching circular features as circular arcs on the true-size front plane. **Figure 13.29** shows an oblique sketch of a shaft block. Circular features on the receding planes appear as ellipses, requiring slanted guidelines, as shown.

ISOMETRIC SKETCH

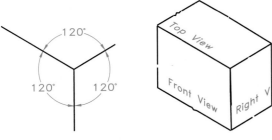

A. THE ISOMETRIC AXES   B. ISOMETRIC DRAWING

Figure 13.30 An isometric sketch:
A Begin an isometric pictorial by sketching three axes spaced 120° apart. One axis usually is vertical.

B Sketch the isometric shape parallel to the three axes and use its true measurements as the dimensions.

ISOMETRIC SKETCH

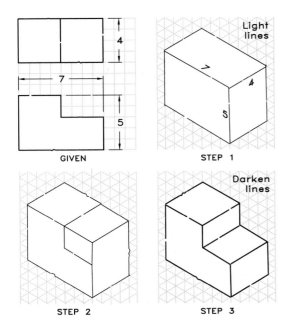

Figure 13.31 Sketching isometric pictorials:
Step 1 Use an isometric grid, transfer dimensions from the given views, and sketch a box having those dimensions.

Step 2 Locate the notch by measuring over four squares and down two squares, as shown in the orthographic views.

Step 3 Complete the pictorial by finishing the notch and darkening the lines.

ANGLES IN ISOMETRIC

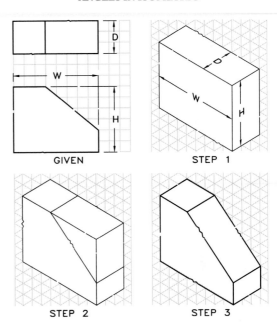

Figure 13.32 Sketching angles in isometric pictorials:
Step 1 Sketch a box from the overall dimensions given in the orthographic views.

Step 2 Angles cannot be measured with a protractor. Find each end of the angle with coordinates measured along the axes.

Step 3 Connect the ends of the angle and darken the lines.

13.8 Isometric Pictorial Sketching

Another type of three-dimensional representation is the **isometric pictorial**, in which the axes make 120° angles with each other (**Fig. 13.30**). Specially printed isometric grids with lines intersecting at 60° angles make isometric sketching easier (**Fig. 13.31**). Simply transfer the dimensions from the squares in the orthographic views to the isometric grid.

You cannot measure angles in isometric pictorials with a protractor; you must find them by connecting coordinates of the angle laid off along the isometric axes. In **Fig. 13.32**, locate the ends of the angular plane by using the coordinates for

13.8 ISOMETRIC PICTORIAL SKETCHING • 177

DOUBLE ANGLES IN ISOMETRIC

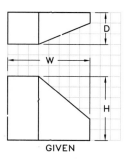
GIVEN

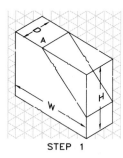
STEP 1

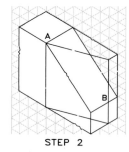
STEP 2

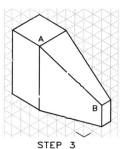
STEP 3

Figure 13.33 Sketching double angles in isometric pictorials:

Step 1 When part of an object has two sloping angles that intersect, begin by sketching the overall box and finding one of the angles.

Step 2 Find the second angle, which locates point B, the intersection line between the planes.

Step 3 Connect points A and B and darken the lines. Line AB is the line of intersection between the two sloping planes.

CIRCLES IN ISOMETRIC

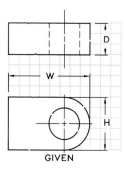
GIVEN

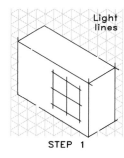
STEP 1

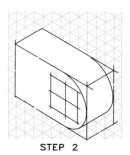
STEP 2

STEP 3

Figure 13.34 Sketching circles in isometric pictorials:

Step 1 Sketch a box using the overall dimensions given in the orthographic views. Sketch the centerlines and a rhombus of guidelines blocking in the circular hole.

Step 2 Sketch the isometric arcs tangent to the box. These arcs are elliptical rather than circular.

Step 3 Sketch the hole and darken the lines. Hidden lines usually are omitted in isometric pictorial sketches.

width and height. When a part has two sloping planes that intersect (**Fig. 13.33**), you have to sketch them one at a time to find point B. Line AB is found as the line of intersection between the planes.

Circles in Isometric Pictorials

Circles appear as ellipses in isometric pictorials. When you sketch them, begin with their centerlines and construction lines enclosing their diameters, as shown in **Fig. 13.34**. The end of the block

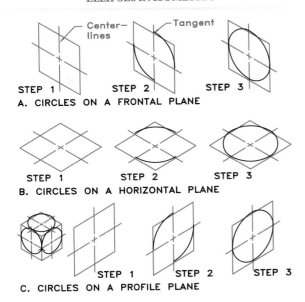

A. CIRCLES ON A FRONTAL PLANE

B. CIRCLES ON A HORIZONTAL PLANE

C. CIRCLES ON A PROFILE PLANE

Figure 13.35 Sketch circular features on the (A) frontal, (B) horizontal, and (C) profile plane of an isometric as ellipses:
Step 1 Lay out centerlines and guidelines.
Step 2 Sketch two arcs.
Step 3 Connect the ends of the arcs to complete the ellipses.

is semicircular in the front view, so its center must be equidistant from the top, bottom, and end of the front view. Circles and ellipses are easier to sketch if you use construction lines.

Figure 13.35 shows how to use centerlines and construction lines to draw ellipses in the three isometric planes: **frontal**, **horizontal**, and **profile** views. **Figure 13.36** shows an application of this technique to sketching a cylinder. Use this same technique to sketch an object having semicircular ends (**Fig. 13.37**). Hidden lines are visually omitted in isometric drawings.

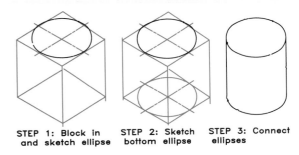

STEP 1: Block in and sketch ellipse STEP 2: Sketch bottom ellipse STEP 3: Connect ellipses

Figure 13.36 Sketching a cylinder as an isometric pictorial:
Step 1 Block in the cylinder and sketch the upper ellipse.
Step 2 Sketch the lower ellipse.
Step 3 Connect the ellipses with tangent lines and darken the lines.

ISOMETRIC SKETCH

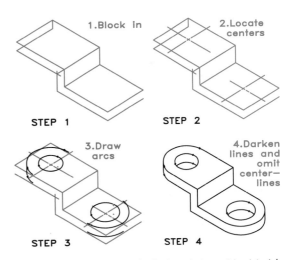

Figure 13.37 Sketching circular features in isometric pictorials:
Step 1 Block in the isometric shape of the object with light lines.
Step 2 Locate the centerlines of the holes and the rounded ends.
Step 3 Sketch the circular features of the ends of the part and the holes.
Step 4 Show the bottoms of the holes and darken the lines of all features.

Problems

Sketch your solutions to these problems on size A (8½-×-11-inch) paper, with or without a printed grid. **Figure 13.38** shows a format for this size sheet and a 0.20-in. grid (convertible to an approximate metric grid by equating each square to 5 mm). Execute all sketches and lettering by applying the principles covered in this chapter and Chapter 11. **Figures 13.39, 13.40,** and **13.41** contain the problems and instructions.

Figure 13.38 Use this layout of a size A sheet for sketching problems. You may sketch two problems on each sheet.

Figure 13.39 (A) Sketch top, front, and right-side views of the problems assigned, supplying lines that may be missing from all views. **(B)** Sketch obliques of the problems assigned. **(C)** Sketch isometrics of the problems assigned.

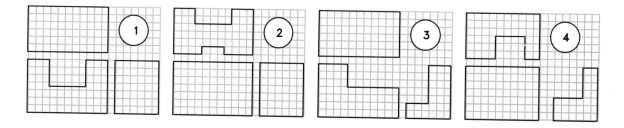

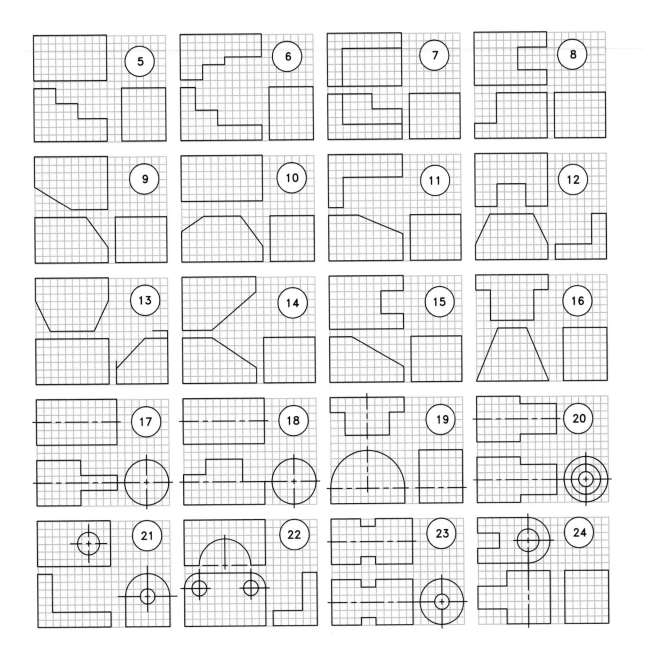

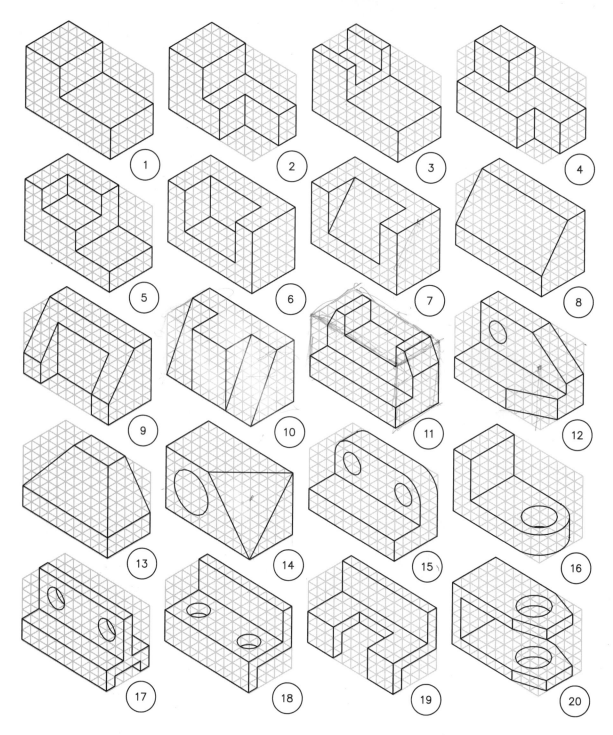

Figure 13.40 Sketch the top, front, and right-side views of the problems assigned, two problems per sheet.

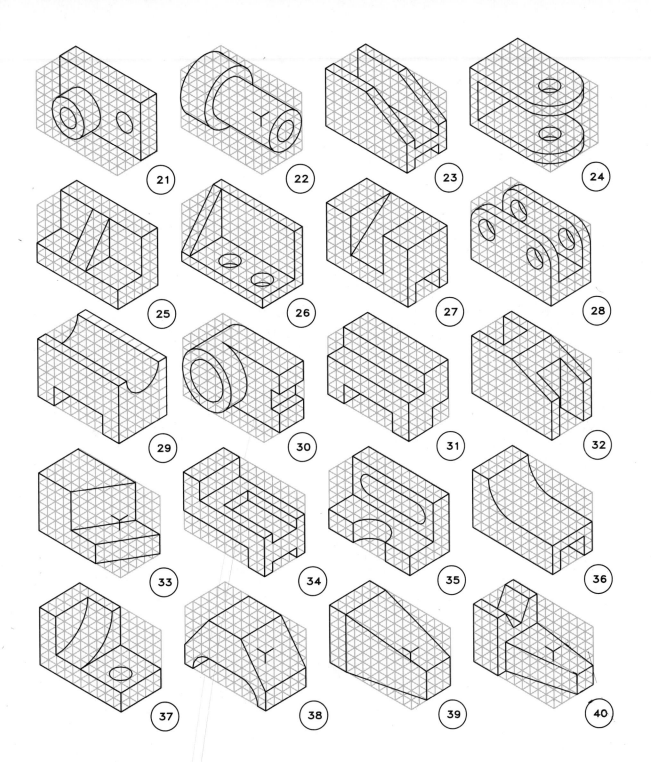

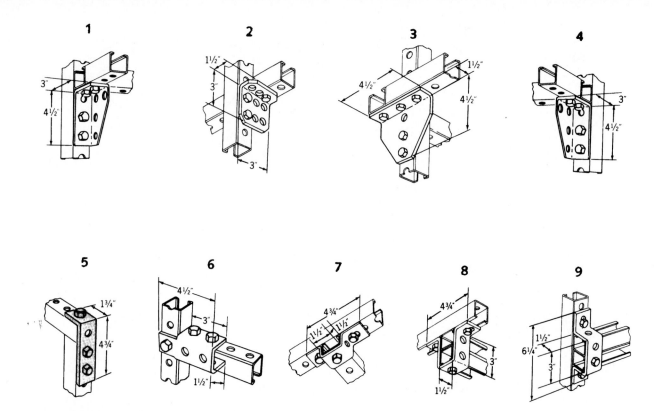

Figure 13.41 Sketch the top, front, and right-side views of the brackets assigned, one per size A sheet. (Courtesy Midland Ross Corporation.)

Orthographic Drawing with Instruments

14.1 Introduction

In Chapter 13, you were introduced to orthographic projection by freehand sketching, which is an excellent way to develop a design concept. Now, you must convert these sketches into orthographic views drawn to scale with instruments (or by computer) to more precisely define your design. Afterwards, you will add dimensions, notes, and specifications to convert these drawings into working drawings from which the design will become a reality.

Orthographic drawings are three-dimensional objects represented by separate views arranged in a standard manner that are readily understood by the technological team. Because multiview drawings usually are executed with instruments and drafting aids, they are often called **mechanical drawings.** They are called **working drawings**, or **detail drawings**, when sufficient dimensions, notes, and specifications are added to enable the product to be manufactured or built from the drawings.

14.2 Orthographic Projection

An artist is likely to represent objects impressionistically, but the engineer must represent them precisely. Orthographic projection is used to prepare precise, scaled, and clearly presented drawings from which the project depicted can be built.

Orthographic projection is the system of drawing views of an object by projecting them perpendicularly onto projection planes with parallel projectors. **Figure 14.1** illustrates this concept of projection by imagining that the object is inside a glass box and three of its views are projected to planes of the box.

Figure 14.2 illustrates the principle of orthographic projection where the front view is projected perpendicularly onto a vertical projection plane, called the frontal plane, with parallel projectors. The projected front view is two dimensional because it has only width and height and lies in a single plane. Similarly, the top view is projected onto a horizontal projection plane, and the

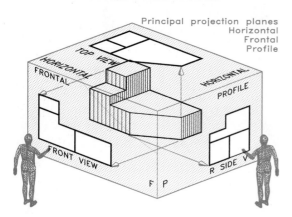

Principal projection planes
Horizontal
Frontal
Profile

Figure 14.1 Orthographic projection is the system of projecting views onto an imaginary glass box with parallel projectors to the three mutually perpendicular projection planes.

ORTHOGRAPHIC PROJECTION

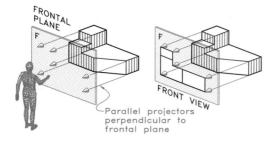

Figure 14.2 An orthographic view is found by projecting from the object to a projection plane with parallel projectors that are perpendicular to the projection plane.

side view is projected onto a second vertical projection plane.

Imagine that the box is opened into the plane of the drawing surface. **Figure 14.3A** illustrates how three planes of a glass box are opened into a single plane (**Fig. 14.3B**) to yield the standard positions for the three orthographic views. These views are the front, top, and right-side views.

The principal projection planes of orthographic projection are the horizontal (H), frontal

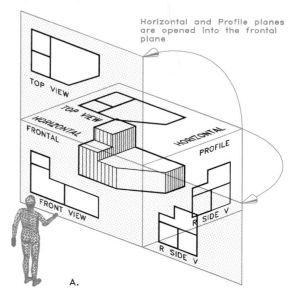

Horizontal and Profile planes
are opened into the frontal
plane

A.

The standard arrangement of three orthographic views:

Top View above the Front View
Right—Side View right of the Front View

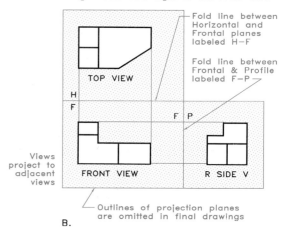

Fold line between Horizontal and Frontal planes labeled H—F

Fold line between Frontal & Profile labeled F—P

Views project to adjacent views

Outlines of projection planes are omitted in final drawings

B.

Figure 14.3 When the imaginary glass box is opened into the plane of the drawing surface, the resulting orthographic views and associated labeling will appear as shown here.

(F), and profile (P) planes. Views projected onto these principal planes are principal views. The dimensions used to give the sizes of principal views are height (H), width (W), and depth (D).

THE ALPHABET OF LINES

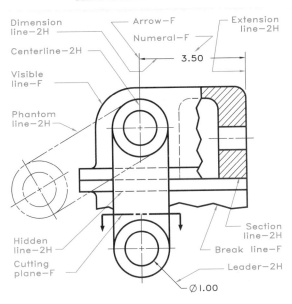

Figure 14.4 The alphabet of lines and recommended pencil grades for drawing orthographic views are shown here.

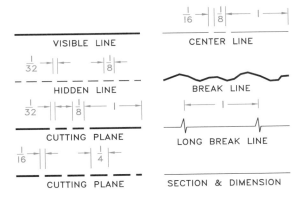

Figure 14.5 Full-size line weights recommended for drawing orthographic views are shown above.

14.3 Alphabet of Lines

Draw all orthographic views with dark and dense lines as if drawn with ink. Only the line widths should vary—except for guidelines and construction lines, which are drawn very lightly for layout

---

Visible: 0.7 mm	Visible: 0.7 mm
Hidden: 0.3 mm	Hidden: 0.3 mm
Center: 0.3 mm	Center: 0.3 mm
Cutting pl: 0.7 mm	Cutting pl: 0.7 mm
A. LTSCALE=0.4	B. LTSCALE=0.2

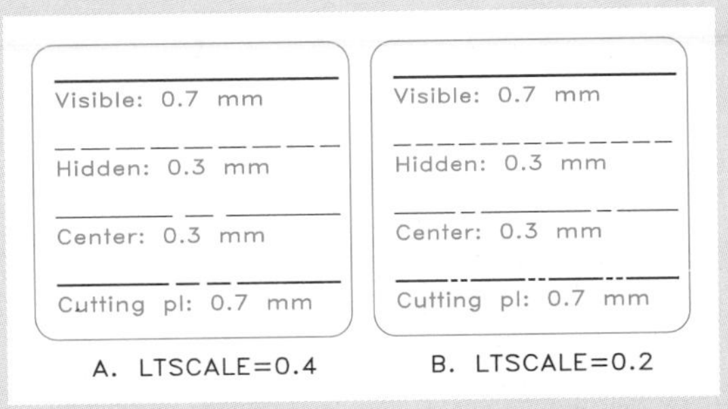

Figure 14.6 These lines were drawn using AutoCAD's standard LINETYPEs and two pens (P.3 and P.7 points). (A) The LTSCALE factor of 0.4 gives relatively long dashes and spaces between dashes. (B) By contrast, the LTSCALE factor of 0.2 produces dashes and spaces that are half as long as the factor of 0.4.

and lettering. **Figure 14.4** gives examples of lines used in orthographic projection and the recommended pencil grades for them. The lengths of dashes in hidden lines and centerlines are drawn longer as a drawing's size increases. **Figure 14.5** further describes these lines.

Computer Lines Computer graphics plotters use pens with points of varying widths, usually 0.7 mm and 0.3 mm wide, to draw lines of different thicknesses. To vary the lengths of dashes and the spaces between them in da shed lines, use LTSCALE **(Fig. 14.6)**. A large number produces long dashes and a smaller number produces shorter dashes. When LTSCALE is used, ALL noncontinuous (centerlines and hidden lines) on the drawing are changed at the same time.

A comparison of lines drawn with LTSCALEs of 0.2 and 0.4 and pen widths of 0.7 mm and 0.3 mm is shown in **Fig. 14.7**. Centerlines become solid lines and hidden lines become longer with LTSCALE=0.4. Try several LTSCALE settings to get a proper balance of the dashed lines in a drawing.

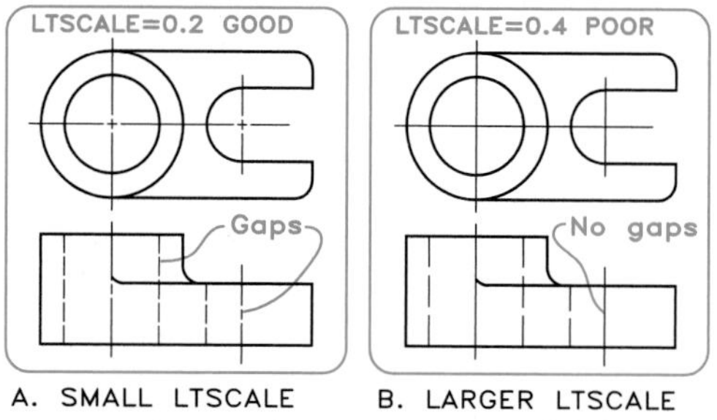

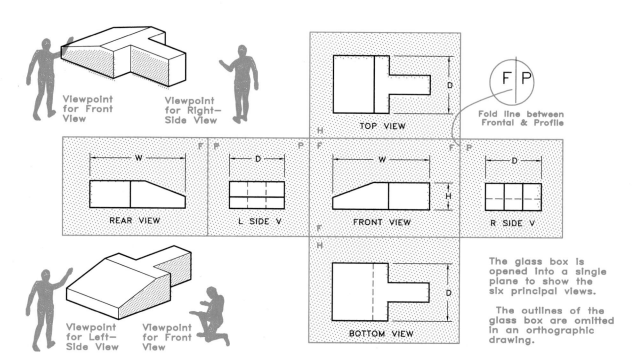

Figure 14.7
A This two-view drawing was made with an LTSCALE of 0.2, which gives properly spaced hidden lines and centerlines.
B The same drawing made with an LTSCALE of 0.4 does not produce the proper spacing between hidden lines and centerlines.

SIX PRINCIPAL VIEWS

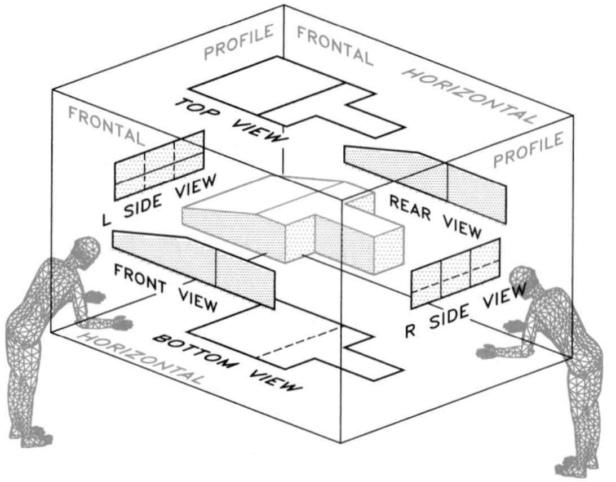

Figure 14.8 Six principal views of an object can be drawn in orthographic projection. Imagine that the object is in a glass box with the views projected onto its six planes.

SIX PRINCIPAL VIEWS

The glass box is opened into a single plane to show the six principal views.

The outlines of the glass box are omitted in an orthographic drawing.

Figure 14.9 Opening the box into a single plane positions the six views as shown to describe the object.

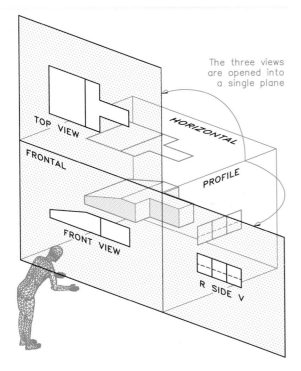

Figure 14.10 Three-view drawings are commonly used to describe small objects such as machine parts.

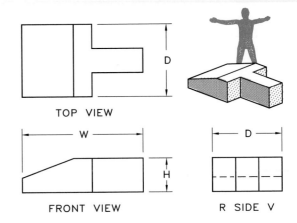

Figure 14.11 This three-view drawing depicts the object shown in Fig. 14.10.

14.4 Six-View Drawings

When you imagine that an object is inside a glass box you will see two horizontal planes, two frontal planes, and two profile planes (**Fig. 14.8**). **Therefore, the maximum number of principal views that can be used to represent an object is six.** The top and bottom views are projected onto horizontal planes, the front and rear views onto frontal planes, and the right- and left-side views onto profile planes.

To draw the six views on a sheet of paper, imagine the glass box is opened up into the plane of the drawing paper as shown in **Fig. 14.9**. Place the top view over and bottom view under the front view; place the right-side view to the right and the left-side view to the left of the front view; and place the rear view to the left of the left-side view.

Projectors align the views both horizontally and vertically about the front view. Each side of the fold lines of the glass box is labeled H, F, or P (horizontal, frontal, or profile) to identify the projection planes on each side of the imaginary fold lines (Fig. 14.9).

Height (H), width (W), and depth (D), the three dimensions necessary to dimension an object, are shown in their recommended positions in Fig. 14.9. The standard arrangement of the six views allows the views to share dimensions by projection. For example, the height dimension, which is shown only once between the front and right-side views, applies to the four horizontally aligned views. The width dimension is placed between the top and front views, but applies to the bottom view also.

14.5 Three-View Drawings

The most commonly used orthographic arrangement of views is the three-view drawing, consisting of front, top, and right-side views. Imagine that the views of the object are projected onto the planes of the glass box (**Fig. 14.10**) and the three planes are opened into a single plane, the frontal plane. **Figure 14.11** shows the resulting three-

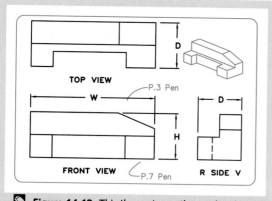

REVOLUTION TO RIGHT-SIDE VIEW

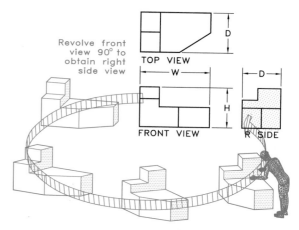

Figure 14.14 By rotating the front view 90° about a vertical axis, the side view is found.

REVOLUTION OF FRONT VIEW

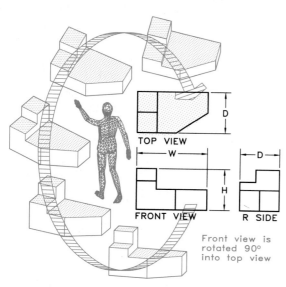

Figure 14.13 By rotating the front view 90° about a horizontal axis, the top view is obtained.

view drawing where the views are labeled and dimensioned with H, W, and D.

🚫 **Computer Views** The three-view drawing of the part shown in **Fig. 14.12** was drawn by using the LINE command, dimensioned by using the DIMension command, and plotted with pen widths of 0.7 mm and 0.3 mm.

14.6 Views by Revolution

In addition to using the glass-box approach for visualizing orthographic projection, imagine that the object is revolved until the desired view is obtained. For example, if the front view of the part in **Fig. 14.13** is revolved 90°, you will see its top view true shape (TS). Placed in its proper position, the top view aligns with the projectors from the front view. By revolving the front view 90°, you will see the right-side view true shape (**Fig. 14.14**). The side view is aligned with the projectors from the front view.

14.7 Arrangement of Views

Figure 14.15 shows the standard positions for a three-view drawing: The top and side views are projected from and aligned with the front view. Improperly arranged views that do not project from view to view are also shown. **Figure 14.16**

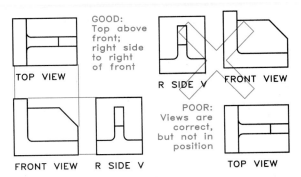

Figure 14.15 Orthographic views must be arranged in their proper positions in order for them to be interpreted correctly.

ALIGNMENT OF DIMENSIONS

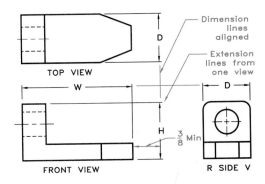

Figure 14.16 Dimension and extension lines used in three-view orthographic projection should be aligned. Draw extension lines from only one view when dimensions are placed between views.

illustrates the rules of projection and shows the proper alignment of dimensions. Orthographic projection shortens layout time, improves readability, and reduces the number of dimensions required because they are placed between and are shared by the views to which they apply.

14.8 Selection of Views

Select the sequence of orthographic views with the fewest hidden lines. **Figure 14.17A** shows that

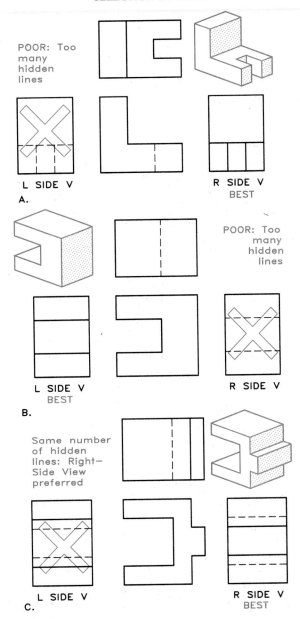

Figure 14.17 Selection of views:

A Select the sequence of views with the fewest hidden lines.

B Select the left-side view because it has fewer hidden lines than the right-side view.

C When both views have an equal number of hidden lines, select the right-side view.

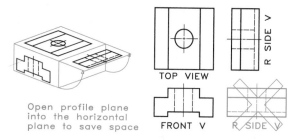

Open profile plane into the horizontal plane to save space

Figure 14.18 The side view can be projected from the top view instead of the front view. This alternate position saves space when the depth of an object is considerably greater than its height.

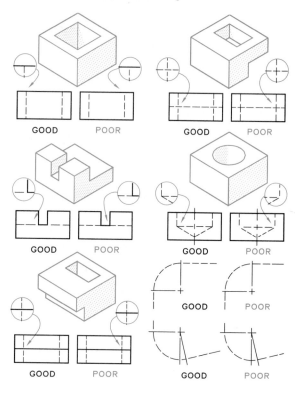

Figure 14.19 These drawings show proper intersections and other line techniques in orthographic views.

the right-side view is preferable to the left-side view because it has fewer hidden lines. Although the three-view arrangement of top, front, and right-side views is more commonly used, the top, front, and left-side view arrangement is acceptable **(Fig. 14.17B)** if the left-side view has fewer hidden lines than the right-side view.

The most descriptive view usually is selected as the front view. If an object, such as a chair, has predefined views that people generally recognize as the front and top views, you should label the accepted front view as the orthographic front view.

Although the right-side view usually is placed to the right of the front view, the side view can be projected from the top view **(Fig. 14.18)**. This alternative position is advisable when the object has a much larger depth than height.

14.9 Line Techniques

Figure 14.19 illustrates techniques for handling most types of intersecting lines, hidden lines, and arcs in combination. Proper application of these principles improves the readability of orthographic drawings.

Become familiar with the order of importance (precedence) of lines **(Fig. 14.20)**. **The most important line, the visible object line, is shown regardless of any other line lying behind it. Of**

next importance is the hidden line, which is more important than the centerline.

◐ **Centerlines by Computer** In **Fig. 14.21A**, the centerline dashes do not cross properly at the centers of the circle and concentric arc. **Figure 14.21B** shows how to make them cross properly. Centerlines must extend equal distances beyond the circle to center the short dashes, and line segments are added to these centerlines beyond the concentric arc.

14.10 Point Numbering

Some orthographic views are difficult to draw due to their complexity. By numbering the endpoints of the lines of the parts in each view as you con-

1. Visible over hidden lines
2. Hidden over centerlines

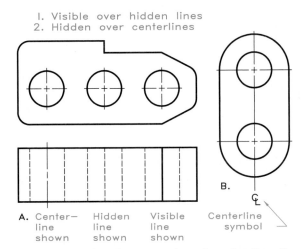

A. Center—
line
shown

Hidden
line
shown

Visible
line
shown

Centerline
symbol

B.

Figure 14.20 When lines coincide with each other, the more important lines take precedence (cover up) the other lines. The order of importance is: visible lines, hidden lines, and center lines.

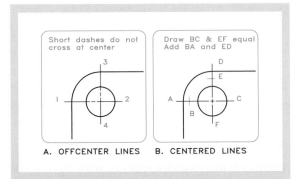

Short dashes do not cross at center

Draw BC & EF equal
Add BA and ED

A. OFFCENTER LINES

B. CENTERED LINES

Figure 14.21 Centerlines by computer:

A When centerlines for a rounded corner are drawn from 1 to 2 and from 3 to 4, the center dashes do not cross at the center of the hole.

B Centerline dashes can be made to cross at the center by drawing a line from A to B to C, where the BC segment extends equally beyond the center on both sides. Draw line DF from D to E to F in the same manner.

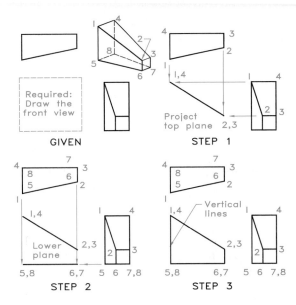

Required:
Draw the front view

GIVEN

Project
top plane 2,3

STEP 1

Lower plane

STEP 2

Vertical lines

STEP 3

Figure 14.22 Point numbering:

Required Find the front view.

Step 1 Number the corners of plane 1–2–3–4 in the top and side views and project these points to the front view.

Step 2 Number the corners of plane 5–6–7–8 in the top and side views and project these points to the front view.

Step 3 Connect the numbered lines to complete the missing front view.

struct it **(Fig. 14.22)**, the location of the object's features will be easier. For example, using numbers on the top and side views of this object aids in the construction of the missing front view. Projecting points from the top and side views to the intersections of the projectors locates the object's front view.

14.11 Lines and Planes

A line can appear true length (TL), foreshortened (FS), or as a point (PT) in an orthographic view (Fig. 14.23). A line that appears true length is parallel to the reference line in the adjacent view. The reference line in these examples represents the

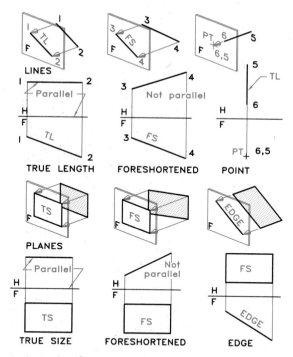

Figure 14.23 A line will appear in orthographic projection as true length, foreshortened, or as a point. A plane in orthographic projection appears as true size, foreshortened, or as an edge.

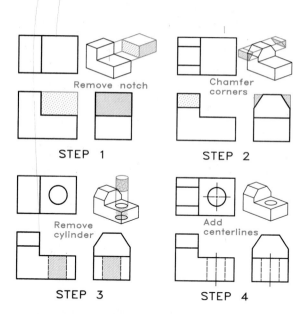

Figure 14.24 Views by subtraction:

Step 1 Block in the views of the object using overall dimensions of H, W, and D. Create the notch by removing the block.

Step 2 Bevel the corners of the object by removing the triangular volumes.

Step 3 Form the hole by removing a cylindrical volume.

Step 4 Add centerlines to complete the views.

fold line between the horizontal and frontal planes and therefore is labeled H-F. **A plane can appear true size (TS), foreshortened, or as an edge in orthographic projection (Fig. 14.23).**

Lines and planes that are true length or true size can be measured with a scale. Foreshortened lines and planes are not true length or true size and are less than full size.

14.12 Views by Subtraction

Figure 14.24 illustrates how three views of a part are drawn by beginning with a block having the overall height, width, and depth of the finished part and removing volumes from it. This drawing

procedure is similar to the steps of making the part in the shop.

14.13 Laying Out Three-View Drawings

The depth dimension applies to both the top and side views, but these views usually are positioned where depth does not project between them (**Fig. 14.25**). The depth dimension can be transferred between the top and side views with dividers or by using a 45° line.

Layout Rules From the material presented so far, we can summarize the rules to be followed in drawing orthographic views:

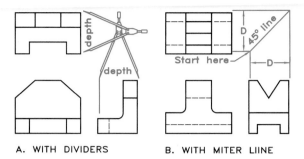

A. WITH DIVIDERS   **B. WITH MITER LIINE**

Figure 14.25

A Transfer the depth dimension to the side view with your dividers.

B Use a 45° miter line to transfer the depth dimension between the top and side views.

THREE-VIEW DRAWING

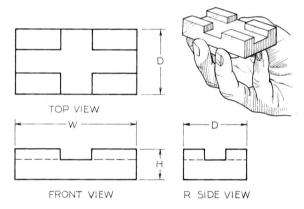

TOP VIEW

FRONT VIEW   R SIDE VIEW

Figure 14.26 This three-view drawing depicts an object that has only horizontal and vertical planes.

1. **Draw orthographic views in their proper positions.**
2. **Select the most descriptive view as the front view, if the object does not have a predefined front view.**
3. **Select the sequence of views with the fewest hidden lines.**
4. **Label the views; for example, top view, front view, and right-side view.**

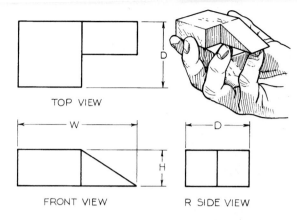

TOP VIEW

FRONT VIEW   R SIDE VIEW

Figure 14.27 This three-view drawing shows an object that has a sloping plane.

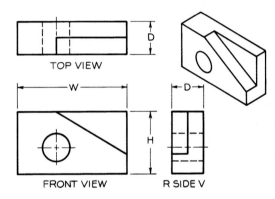

TOP VIEW

FRONT VIEW   R SIDE V

Figure 14.28 This three-view drawing shows an object that has a sloping plane with a cylindrical hole through it.

5. **Place dimensions between the views to which they apply.**
6. **Use the proper alphabet of lines.**
7. **Leave adequate room between the views for labels and dimensions.**
8. **Draw the views necessary to describe a part. Sometimes fewer and more views are required.**

   Figures 14.26–14.30 show examples of three-view orthographic drawings of objects drawn by applying these rules.

THREE-VIEW DRAWING

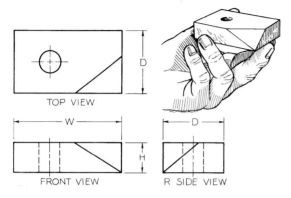

TOP VIEW

W

FRONT VIEW

R SIDE VIEW

D

H

D

Figure 14.29 This three-view drawing depicts an object that has a plane with a compound slope.

THREE-VIEW DRAWING

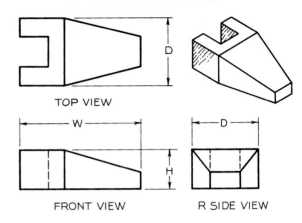

TOP VIEW

W

FRONT VIEW

R SIDE VIEW

D

H

D

Figure 14.30 This three-view drawing shows an object that has planes with compound slopes.

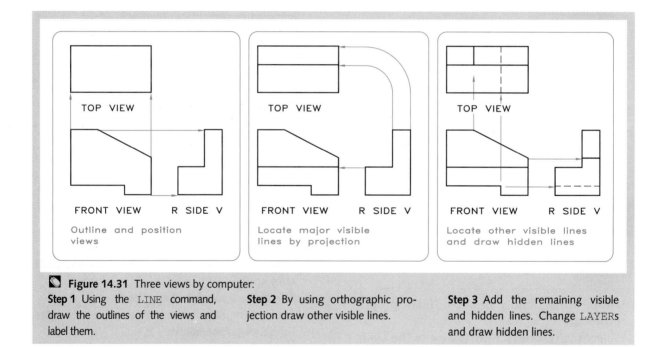

TOP VIEW

FRONT VIEW   R SIDE V

Outline and position views

TOP VIEW

FRONT VIEW   R SIDE V

Locate major visible lines by projection

TOP VIEW

FRONT VIEW   R SIDE V

Locate other visible lines and draw hidden lines

◩ **Figure 14.31** Three views by computer:

Step 1 Using the LINE command, draw the outlines of the views and label them.

Step 2 By using orthographic projection draw other visible lines.

Step 3 Add the remaining visible and hidden lines. Change LAYERs and draw hidden lines.

◩ **Three-View Drawings by Computer Figure 14.31** illustrates the steps for laying out three views of an object by computer, top, front, and side views. Use the LINE command to draw the overall outlines of the three views. Other lines are added in by projecting from view to view. Use the DTEXT command to label the views as shown in step 1.

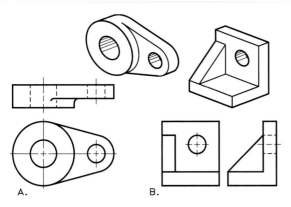

Figure 14.32 These objects can be adequately described with two orthographic views.

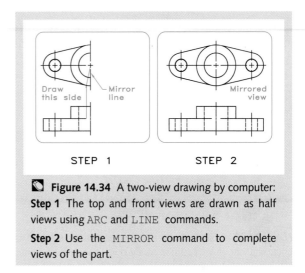

Figure 14.34 A two-view drawing by computer:
Step 1 The top and front views are drawn as half views using ARC and LINE commands.
Step 2 Use the MIRROR command to complete views of the part.

TWO-VIEW DRAWING

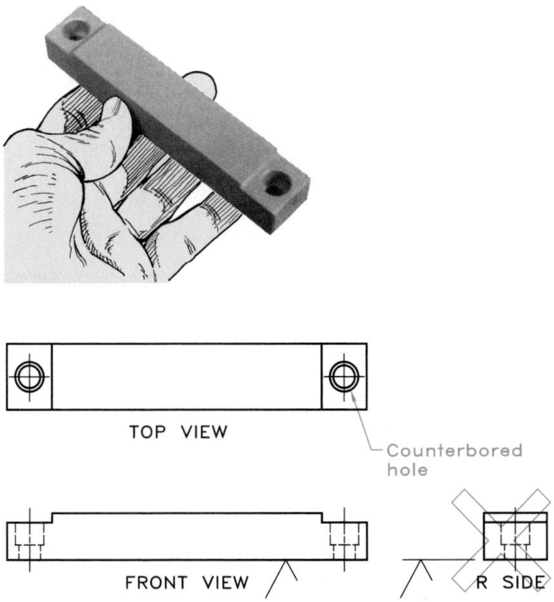

Figure 14.33 Two views adequately define this part.

14.14 Two-View Drawings

Economize on time and space by using only the views necessary to depict an object. **Figure 14.32** shows objects that require only two views. The fixture block in **Fig. 14.33** is another example of a part needing only two views to be adequately described.

 Two-View Drawings by Computer Figure **14.34** shows the steps in making a two-view drawing of a symmetrical object by computer. Use the ARC and LINE commands to draw the half views and MIRROR them to find their other halves.

14.15 One-View Drawings

Simple cylindrical parts and parts of a uniform thickness can be described by only one view as shown in **Fig. 14.35.** Supplementary notes clarify features that would have been shown in the omitted views. Diameters are labeled with diameter signs and thicknesses are noted.

14.16 Simplified and Removed Views

The right- and left-side views of the part in **Fig. 14.36** would be harder to interpret if all hidden lines were drawn by rigorously following the rules of orthographic projection. Simplified views in which confusing and unnecessary lines have been omitted are better and more readable.

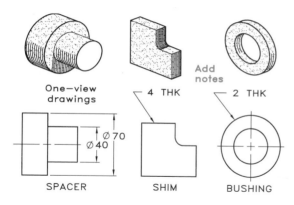

Figure 14.35 Objects that are cylindrical or of a uniform thickness can be described with only one orthographic view and supplementary notes.

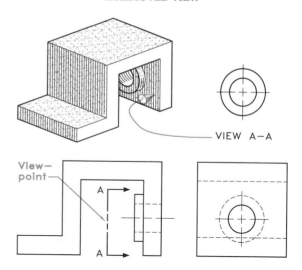

Figure 14.37 Use a removed view, indicated by the directional arrows, to show hard-to-see views in removed locations.

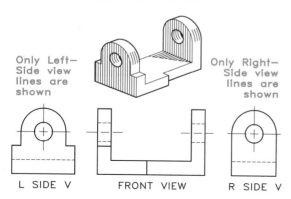

Figure 14.36 Use simplified views with unnecessary and confusing hidden lines omitted to improve clarity.

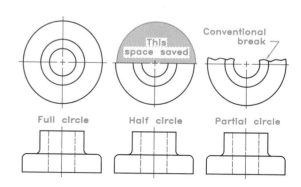

Figure 14.38 Save space and time by drawing the circular view of a cylindrical part as a partial view.

When it is difficult to show a feature with a standard orthographic view because of its location, a **removed view** can be drawn **(Fig. 14.37)**. The removed view, indicated by the directional arrows, is clearer when moved to an isolated position.

14.17 Partial Views

Partial views of symmetrical or cylindrical parts may be used to save time and space. Omitting the rear of the circular top view in **Fig. 14.38** saves space without sacrificing clarity. To clarify that a

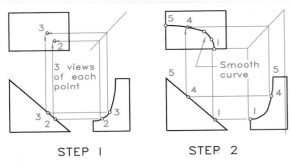

STEP 1   STEP 2

Figure 14.39 Curve plotting:
Step 1 Locate points 2 and 3 in the front and side views by projection. Project points 2 and 3 to the top view.

Step 2 Locate the remaining points in the three views and connect the points in the top view with a smooth curve.

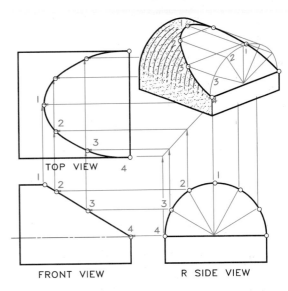

Figure 14.40 The ellipse in the top view was found by numbering points in the front and side views and then projecting them to the top view.

part of the view has been omitted, a conventional break is used in the top view.

14.18 Curve Plotting

You may draw an irregular curve by following the rules of orthographic projection, as **Fig. 14.39** shows. Begin plotting by locating and numbering points along the curve in the front and side views. Project these points to the top view where each point is located and connect them with a smooth curve drawn using your irregular curve.

Figure 14.40 shows an ellipse plotted in the top view from front and side-view projections. The points on the curve should be numbered as they are transferred.

14.19 Conventional Practices

The readability of an orthographic view may be improved if the rules of projection are violated. Violations of rules customarily made for the sake of clarity are called **conventional practices.**

Symmetrically spaced holes in a circular plate **(Fig. 14.41)** are drawn at their true radial distance from the center of the plate in the front view as a

conventional practice. Imagine that the holes are revolved to the centerline in the top view before projecting them to the front view.

This principle of revolution also applies to symmetrically positioned features such as ribs, webs, and the three lugs on the outside of the part shown in **Fig. 14.42. Figure 14.43** shows the applications of conventional practices to holes and ribs in combination.

Another conventional revolution is illustrated in **Fig. 14.44** where the front view of an inclined arm is revolved to a horizontal position so that it can be drawn true size in the top view. The revolved arm in the front view is not drawn because the revolution is imaginary.

Figure 14.45 shows how to improve views of parts by conventional revolution. By revolving the top views of these parts 45°, slots and holes no longer coincide with the centerlines and can be seen more clearly. Draw the front views of the

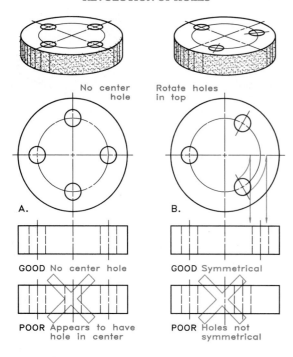

Figure 14.41

A Omit the center hole found by true projection that gives an impression that a hole passes through the center of the plate.

B Use a conventional view to show the holes located at their true radial distances from the center. They are imagined to be rotated to the centerline in the top view.

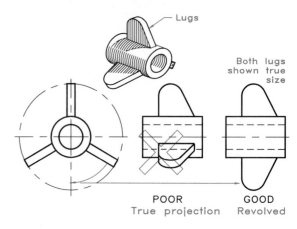

Figure 14.42 Symmetrically positioned external features, such as webs, ribs, and these lugs, are imagined to be revolved to their true-size positions for the best views.

RIBS AND HOLES
CONVENTIONAL PRACTICES

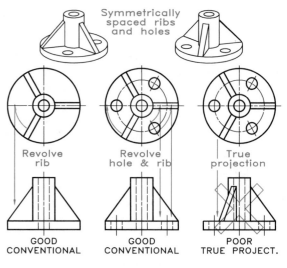

Figure 14.43 Conventional methods of revolving holes and ribs in combination improve clarity.

slots and holes true size by imagining that they have been revolved 45°.

Another type of conventional view is the true-size development of a curved sheet-metal part drawn as a flattened-out view (**Fig. 14.46**). The top view shows the part's curvature.

14.20 Conventional Intersections

In orthographic projection, lines are drawn to represent the intersections (joining lines) between planes of object. Wherever planes intersect, form-ing a sharp edge, this line of intersection is projected to its adjacent view. **Figure 14.47** gives examples of views where lines are required and not required.

CONVENTIONAL REVOLUTION

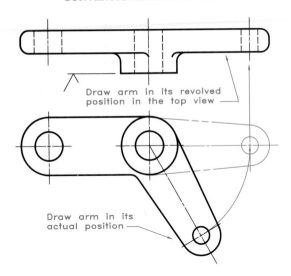

Draw arm in its revolved position in the top view

Draw arm in its actual position

Figure 14.44 The front view of the arm is imagined to be revolved so that its true length can be drawn in the top view. This is an accepted conventional practice.

SLOTS AND HOLES

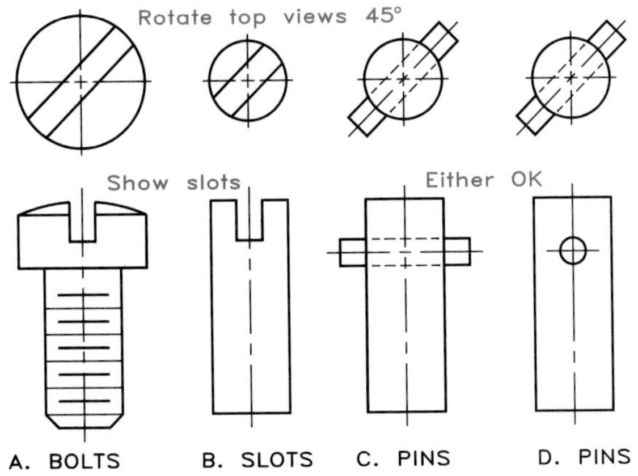

Rotate top views 45°

Show slots   Either OK

A. BOLTS   B. SLOTS   C. PINS   D. PINS

Figure 14.45 It is conventional practice to draw parts like these at 45° in the top view. Draw the front views with the slots and holes drawn true size.

Figure 14.48 shows how to draw intersections between cylinders rather than plotting more complex, orthographically correct lines of intersection. **Figures 14.48A** and **C** show conventional intersec-

DEVELOPED VIEWS

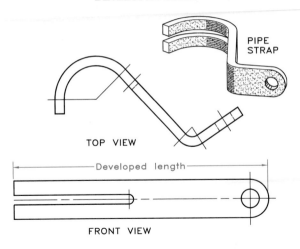

PIPE STRAP

TOP VIEW

Developed length

FRONT VIEW

Figure 14.46 It is conventional practice to use true-size developed (flattened-out) views of objects that have been made of bent sheet metal.

REPRESENTATION OF FEATURES

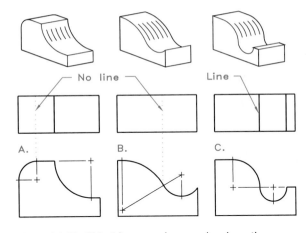

No line   Line

A.   B.   C.

Figure 14.47 Object lines are drawn only where there are sharp intersections or where arcs are tangent at their centerlines, as shown in (C).

tions, which means they are approximations drawn for ease of construction while being sufficiently representative of the object. **Figure 14.48B** shows an easy-to-draw intersection between

INTERSECTIONS: CYLINDERS

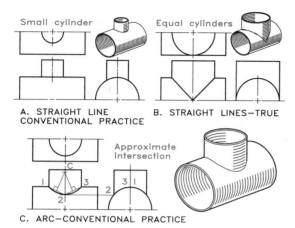

A. STRAIGHT LINE
CONVENTIONAL PRACTICE

B. STRAIGHT LINES—TRUE

Approximate intersection

C. ARC—CONVENTIONAL PRACTICE

Figure 14.48

A and **C** Use these conventional methods of showing intersections between cylinders for ease of construction.

B This intersection at B is easy to draw and also is a true projection.

INTERSECTIONS: CYLINDERS AND PRISMS

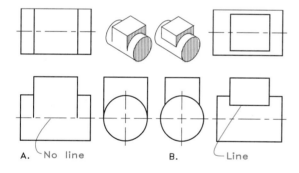

A. No line

B. Line

Figure 14.49 Conventional methods of drawing intersections between prisms and cylindrical shapes are easy to draw and are sufficiently clear.

cylinders of equal diameters, and this is a true intersection as well. **Figures 14.49** and **14.50** show other cylindrical intersections, and **Fig. 14.51** shows conventional practices for depicting intersections formed by holes in cylinders.

INTERSECTIONS: RECTANGULAR SHAPES

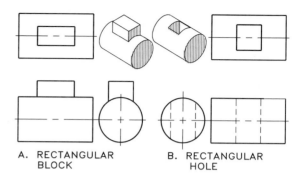

A. RECTANGULAR BLOCK

B. RECTANGULAR HOLE

Figure 14.50 Conventional intersections between prisms and cylindrical shapes also are easy to draw and are sufficiently clear.

14.21 Fillets and Rounds

Fillets and rounds are rounded intersections between the planes of a part that are used on castings, such as the body of the Collet Index Fixture in **Fig. 14.52**. **A fillet is an inside rounding and a round is an external rounding on a part.** The radii of fillets and rounds usually are small, about 1/4 inch. Fillets give added strength at inside corners and rounds improve appearance and remove sharp edges.

A casting will have square corners only when its surface has been **finished**, which is the process of machining away part of the surface to a smooth finish (**Fig. 14.53**). Indicate finished surfaces by placing a finish mark (V) on all edge views of finished surfaces whether the edges are visible or hidden. **Figure 14.54** shows four types of finish marks. A more detailed surface texture symbol is presented in Chapter 21.

Note in **Fig. 14.53C** that a boss is a raised cylindrical feature and that the curve formed by a fillet at its point of tangency is a **runout**. **Figure 14.55** illustrates several techniques for showing fillets and rounds on orthographic views with a circle template.

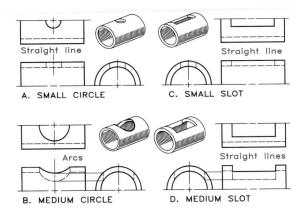

Figure 14.51 Conventional methods of illustrating holes in cylinders are easy to draw and are sufficiently clear.

Figure 14.52 The edges of this Collet Index Fixture are rounded to form fillets and rounds. The surface of the casting is rough except where it has been machined. (Courtesy of Hardinge Brothers, Inc.)

FILLETS AND ROUNDS

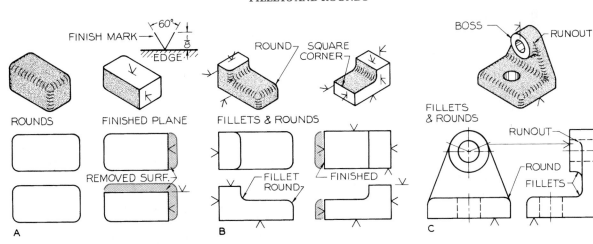

Figure 14.53 Fillets and rounds:
A Machining smooths (finishes) the surface of a casting and removes rounds, leaving square corners. A finish mark on an edge view of a surface indicates that it is to be finished.

B Fillets, rounded inside corners, and rounds, rounded outside corners, are removed when their surfaces are finished. The fillets can be seen only in the front view in this case.

C The views of an object with fillets and rounds must call attention to these features.

FINISH MARKS

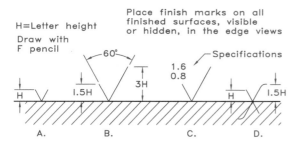

Figure 14.54 Any of these finish mark symbols are placed on all views of finished surfaces, whether hidden or visible.

FILLETS AND ROUNDS

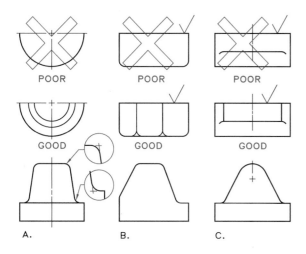

Figure 14.55 These examples show both poorly drawn and conventionally drawn fillets and rounds in orthographic views.

Computer Method A computer-drawn finish mark can be saved as a BLOCK and used repeatedly and rapidly **(Fig. 14.56)**. If the finish mark is drawn where the x-value (letter height) is 1 inch, it will be easy to scale it when INSERTed. For example, a scale factor of 0.125 will reduce it to ⅛ inch to correspond to ⅛-inch-high lettering. By saving the BLOCK as a WBLOCK, you can use it with different drawing files, not just the file it was created on.

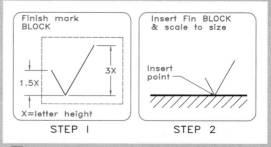

Figure 14.56 A finish mark by computer:
Step 1 Draw a finish mark where X, the letter height, is 1 inch high. Make a BLOCK of it named FIN. Select the insertion point to be the point of the vee.

Step 2 INSERT the block, FIN, by selecting the point on the edge view of the finished surface. Scale the block to 0.125, and its height will be ⅛ inch high.

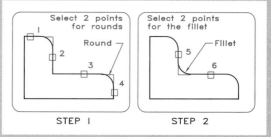

Figure 14.57 Fillets and rounds by computer:
Step 1 Command: FILLET (CR)
Polyline Radius <Select two objects>: R (CR)
Enter fillet radius <0.000>: 0.50 (CR)
Command: (CR) FILLET Polyline Radius/<Select two objects>: 1 and 2 (Repeat, selecting 3 and 4 to draw round.)

Step 2 Press Enter (CR), and select points 5 and 6 on the inside lines to construct a fillet. The lines are trimmed as the fillets and rounds applied to them.

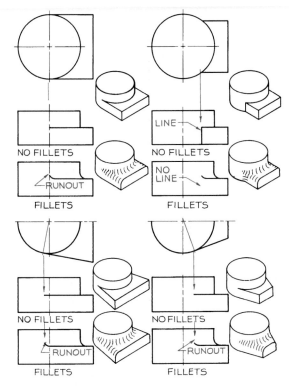

Figure 14.58 These examples show conventional ways of showing intersections between features of objects and runouts. Intersections with fillets have runouts drawn with a circle template.

■ **Computer Method** Fillets and rounds are drawn by computer as shown in **Fig. 14.57.** The object is first drawn with angular intersections, and the FILLET command is used to round the corners for both fillets and rounds.

Figure 14.58 gives a comparison of intersections and runouts of parts with and without fillets and rounds. Large runouts are constructed as an eighth of a circle with a compass as shown in **Fig. 14.59.** Small runouts are drawn with a circle template. Runouts on orthographic views reveal much about the details of an object. For example, the runout in the top view of **Fig. 14.60A** tells us that the rib has rounded corners, whereas the top view of **Fig. 14.60B** tells us the rib is completely round.

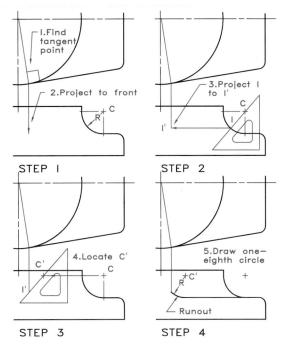

Figure 14.59 Plotting runouts:

Step 1 Find the point of tangency in the top view and project it to the front view.

Step 2 A 45° triangle is used to find point 1, which is projected to point 1'.

Step 3 Move the 45° triangle to locate point C', which is on the horizontal projector from center C.

Step 4 Use the radius of the fillet to draw the runout with C' as its center. The runout arc is equal to one-eighth of a circle.

Figures 14.61 and **14.62** illustrate other types of filleted and rounded intersections.

■ **Computer Method** **Figure 14.63** shows a drawing of a part with runouts at its tangent points. The runouts are plotted by using the LINE command, then using the ARC command, and finally pressing (CR) to obtain the first point of an arc tangent to and connected to the end of the line. Locate the other end of the arc to complete the runout.

RUNOUTS: RIBS

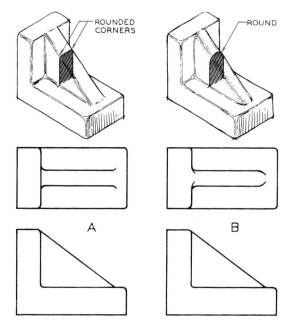

Figure 14.60 These drawings illustrate runouts for differently shaped ribs: (A) fillets and (B) rounded edges.

RUNOUTS: ROUNDED EDGES

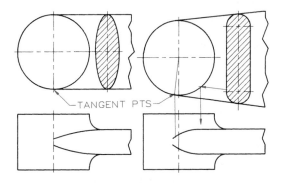

Figure 14.61 These drawings show conventional runouts for other differently shaped cross sections.

14.22 Left-Hand and Right-Hand Views

Two parts often are required that are "mirror images" of each other (**Fig. 14.64**). Drawing time is reduced by drawing views of one of the parts

RUNOUTS: ROUNDED FEATURES

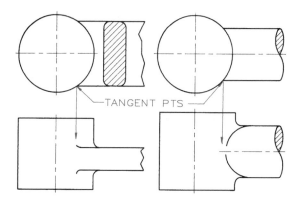

Figure 14.62 These drawings show conventional runouts for differently shaped cross-sections.

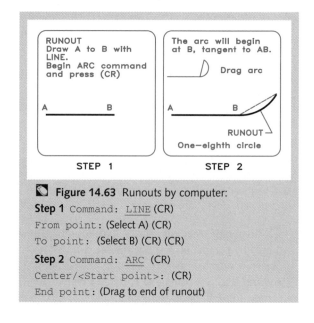

Figure 14.63 Runouts by computer:

Step 1 Command: LINE (CR)
From point: (Select A) (CR)
To point: (Select B) (CR) (CR)

Step 2 Command: ARC (CR)
Center/<Start point>: (CR)
End point: (Drag to end of runout)

and adding a note that the other matching part (mirrored part) has the same dimensions.

14.23 First-Angle Projection

The examples in this chapter are presented as third-angle projections, where the top view is

MIRROR-IMAGE PARTS

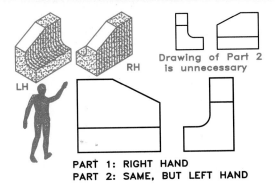

Figure 14.64 When left- and right-hand mirrored parts are needed, only one view is drawn and labeled. Indicate the other views by a note.

THIRD-ANGLE PROJECTION: U.S. SYSTEM

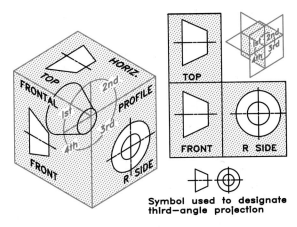

Figure 14.65 Third-angle projection is used for drawing orthographic views in the U.S., Great Britain, and Canada. The top view is placed over the front view and the right-side view is placed to the right of the front view. The truncated cone is the symbol that designates third-angle projection.

placed over the front view and the right-side view is placed to the right of the front view, as shown in **Fig. 14.65**. This method is used in the United States, Great Britain, and Canada. Most of the rest of the world uses first-angle projection.

FIRST-ANGLE PROJECTION: EUROPEAN SYSTEM

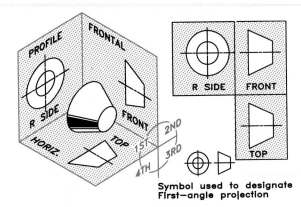

Figure 14.66 First-angle projection is used in most of the world. It shows the right-side view to the left of the front view and the top view under the front view. The truncated cone is the symbol that designates first-angle projection.

THE SI SYMBOL

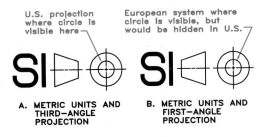

Figure 14.67 These symbols are placed on drawings to specify first-angle or third-angle projection and metric units of measurement.

The first-angle system is illustrated in **Fig. 14.66**, in which an object is placed above the horizontal plane and in front of the frontal plane. When these projection planes are opened onto the surface of the drawing paper, the front view projects over the top view, and the right-side view to the left of the front view.

The angle of projection used in making a drawing is indicated by placing the symbol showing a truncated cone in or near the title block (**Fig. 14.67**). **When metric units of measurement are used, the SI symbol is given in combination with the cone on the drawing.**

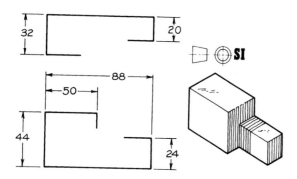

Figure 14.68 Problem 1. Guide block.

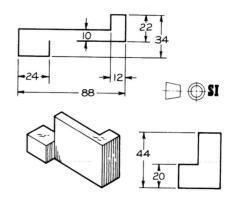

Figure 14.70 Problem 3. Adjustable stop.

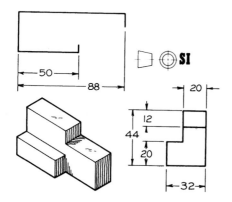

Figure 14.69 Problem 2. Double step.

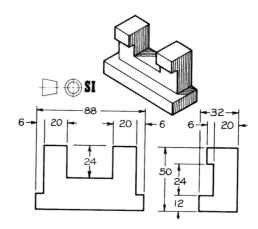

Figure 14.71 Problem 4. Lock catch.

Draw the following problems as orthographic views on size A or size B paper, one or two problems per sheet, as assigned by your instructor.

1–7. (Figs. 14.68–14.74) Draw the given views using the dimensions provided, and then construct the missing top, front, or right-side views.

8–17. (Figs. 14.75–14.84) Complete the three views of the objects. Each square grid is equal to 0.20 inches or 5 mm. Two problems can be placed on a size A sheet (AV format). Label the views and show the overall dimensions as W, D, and H.

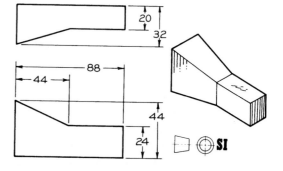

Figure 14.72 Problem 5. Two-way adjuster.

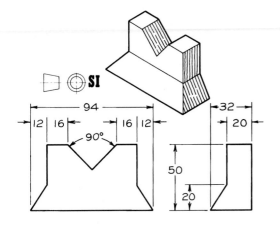

Figure 14.73 Problem 6. 90° Vee block.

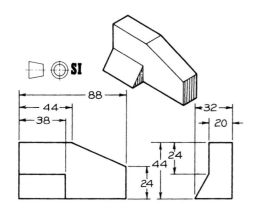

Figure 14.74 Problem 7. Filler.

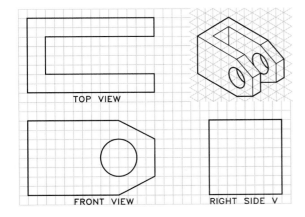

Figure 14.75 Problem 8. Clevis.

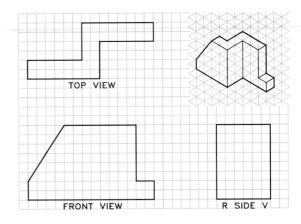

Figure 14.76 Problem 9. Corner box.

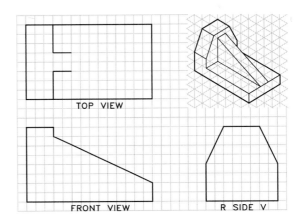

Figure 14.77 Problem 10. Angle block.

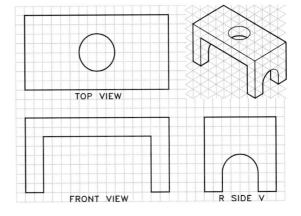

Figure 14.78 Problem 11. Fixture guide.

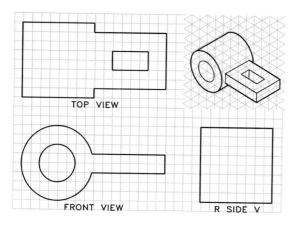

Figure 14.79 Problem 12. Lifting ring.

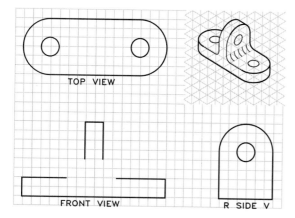

Figure 14.80 Problem 13. Pivot piece.

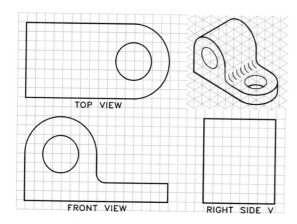

Figure 14.81 Problem 14. Bushing.

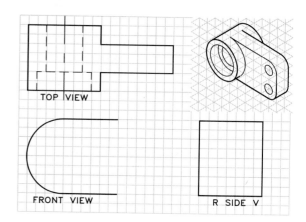

Figure 14.82 Problem 15. Journal.

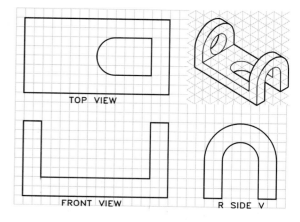

Figure 14.83 Problem 16. End guide.

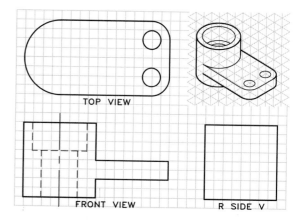

Figure 14.84 Problem 17. Left journal.

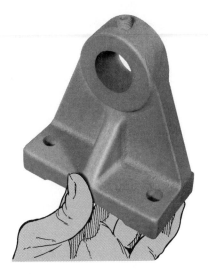

Figure 14.85 Problem 18. Tensioner base.

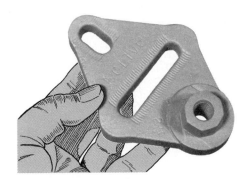

Figure 14.86 Problem 19. Base plate.

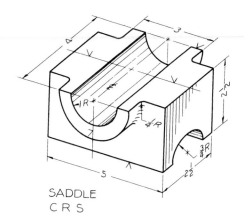

SADDLE
C R S

Figure 14.87 Problem 20. Saddle.

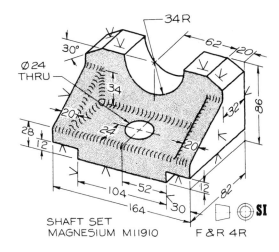

SHAFT SET
MAGNESIUM M11910

F & R 4R

SI

Figure 14.88 Problem 21. Shaft set.

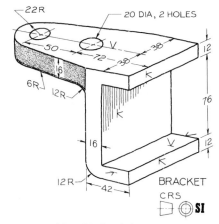

BRACKET
CRS

SI

Figure 14.89 Problem 22. Bracket.

18. (Fig. 14.85) Draw three full-size views in millimeters on a size A sheet. The overall dimensions are: height, 120 mm; width, 112 mm; and depth, 56 mm. Estimate the other dimensions.

19. (Fig. 14.86) Draw three full-size views in millimeters on a size A sheet. The overall dimensions are: height, 10 mm; width, 126 mm; and depth, 90 mm. Estimate the other dimensions.

20–45. (Figs. 14.87–14.112) Construct the necessary orthographic views to describe the objects. Label the views and show the overall dimensions of W, D, and H.

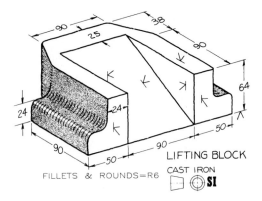

Figure 14.90 Problem 23. Lifting block.

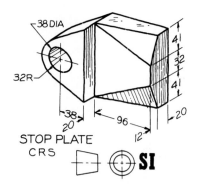

Figure 14.93 Problem 26. Stop plate.

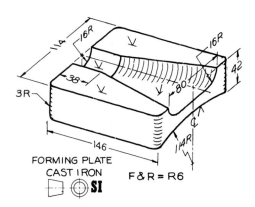

Figure 14.91 Problem 24. Forming plate.

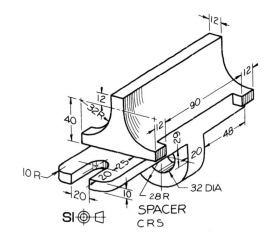

Figure 14.94 Problem 27. Spacer.

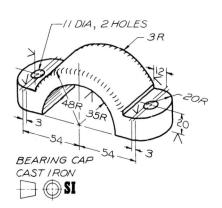

Figure 14.92 Problem 25. Bearing cap.

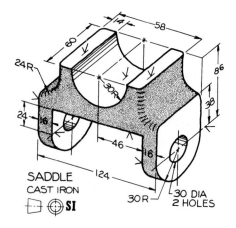

Figure 14.95 Problem 28. Saddle.

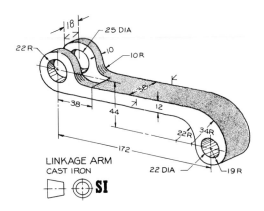

Figure 14.96 Problem 29. Linkage arm.

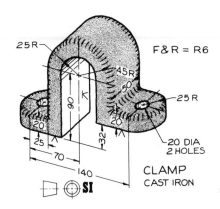

Figure 14.99 Problem 32. Clamp.

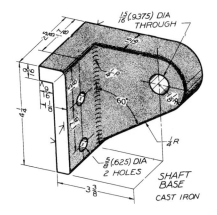

Figure 14.97 Problem 30. Shaft base.

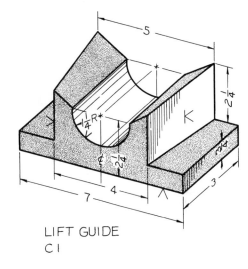

LIFT GUIDE
C I

Figure 14.100 Problem 33. Lift guide.

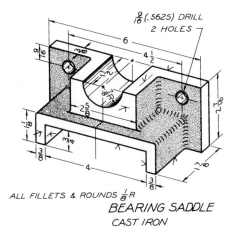

Figure 14.98 Problem 31. Bearing saddle.

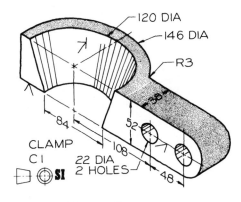

Figure 14.101 Problem 34. Clamp.

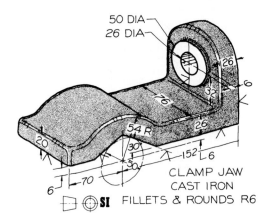

Figure 14.102 Problem 35. Clamp jaw.

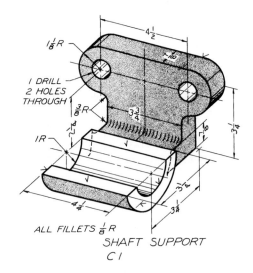

Figure 14.103 Problem 36. Shaft support.

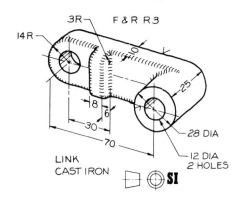

Figure 14.104 Problem 37. Link.

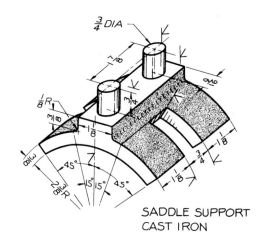

SADDLE SUPPORT
CAST IRON

Figure 14.105 Problem 38. Saddle support.

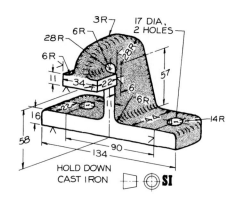

HOLD DOWN
CAST IRON

Figure 14.106 Problem 39. Hold down.

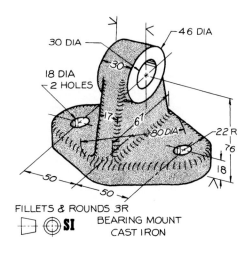

BEARING MOUNT
CAST IRON

Figure 14.107 Problem 40. Bearing mount.

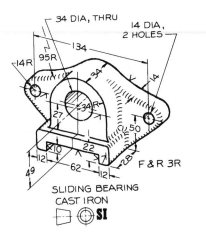

Figure 14.108 Problem 41. Sliding bearing.

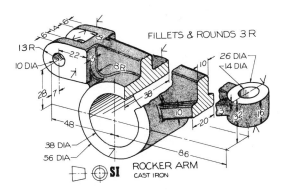

Figure 14.109 Problem 42. Rocker arm.

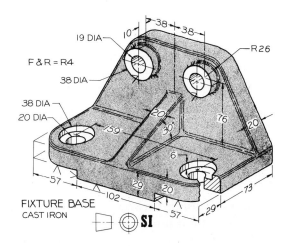

Figure 15.112 Problem 45. Fixture base.

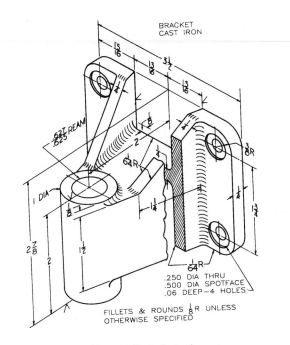

Figure 14.111 Problem 44. Swivel attachment.

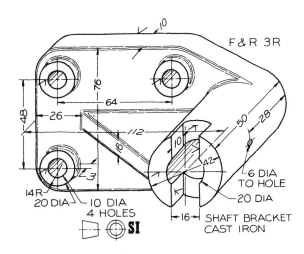

Figure 14.112 Problem 45. Shaft bracket.

Auxiliary Views

15.1 Introduction

Objects often are designed to have sloping or inclined surfaces that do not appear true size in principal orthographic views. A plane of this type is not parallel to a principal projection plane (horizontal, frontal, or profile) and is, therefore, a non-principal plane. Its true shape must be projected onto a plane that is parallel to it. This view is called an **auxiliary view**.

An auxiliary view projected from a primary view (principal view) is called a primary auxiliary view. An auxiliary view projected from a primary auxiliary view is a secondary auxiliary view. By the way, get out your dividers; you must use them all the time in drawing auxiliary views.

The inclined surface of the part shown in **Fig. 15.1A** does not appear true size in the top view because it is not parallel to the horizontal projection plane. However, the inclined surface will appear true size in an auxiliary view projected perpendicularly from its edge view in the front view **(Fig. 15.1B)**.

THE AUXILIARY VIEW

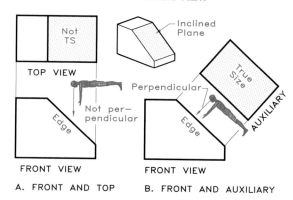

Figure 15.1 A surface that appears as an inclined edge in a principal view can be found true size by an auxiliary view. (A) The top view is foreshortened. (B) The inclined plane is true size in the auxiliary view.

The relationship between an auxiliary view and the view it was projected from is the same as that between any two adjacent orthographic views. **Figure 15.2A** shows an auxiliary view pro-

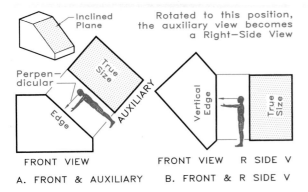

Figure 15.2 An auxiliary view has the same relationship with the view it is projected from as that of any two adjacent principal views.

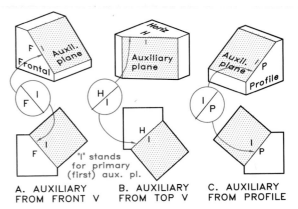

Figure 15.3 A primary auxiliary plane can be folded from the frontal, horizontal, or profile planes. The fold lines are labeled F–1, H–1, and P–1, with 1 on the auxiliary plane side and P on the principal-plane side.

jected perpendicularly from the edge view of the sloping surface. By rotating these views (the front and auxiliary views) so that the projectors are horizontal, the views have the same relationship as regular front and right-side views (**Fig. 15.2B**).

15.2 Folding-Line Theory

The three principal orthographic planes are the **frontal** (F), **horizontal** (H), and **profile** (P) planes. An auxiliary view is projected from a principal orthographic view (a top, front, or side view), and a primary auxiliary plane is perpendicular to one of the principal planes and oblique to the other two.

Think of auxiliary planes as planes that fold into principal planes along a folding line (**Fig. 15.3**). The plane in **Fig. 15.3A** folds at a 90° angle with the frontal plane and is labeled F–1. F is an abbreviation for frontal, and 1 represents first, or primary, auxiliary plane. **Figure 15.3B** and **15.3C** illustrate the positions for auxiliary planes that fold from the horizontal and profile planes, labeled H–1 and P–1, respectively.

It is important that reference lines be labeled as shown in Fig. 15.3, with the numeral 1 placed on the auxiliary side and the letter H, F, or P on the principal-plane side.

VIEWPIONTS FOR AUXILIARY VIEWS PROJECTED
FROM THE TOP VIEW

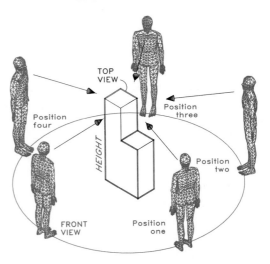

Figure 15.4 By moving your viewpoint around the top view of an object, you will see a series of auxiliary views in which the height dimension (H) is true length.

15.3 Auxiliary Views from the Top View

By moving your position about the top view of a part as shown in **Fig. 15.4**, each line of sight is perpendicular to the height dimension. One of the

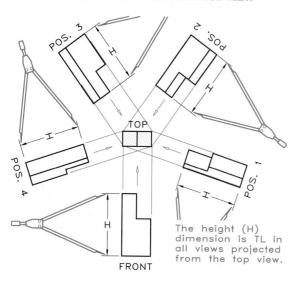

Figure 15.5 The views shown in Fig. 15.4 would be drawn as shown here with the same height dimensions common to each view.

views is a principal view, the front view, while the other positions see nonprincipal views called **auxiliary views**.

Figure 15.5 illustrates how these five views (one of which is a front view) are projected from the top view. The line of sight for each auxiliary view is parallel to the horizontal projection plane; therefore the height dimension is true length in each view projected from a top view.

Folding-Line Method

The inclined plane shown in **Fig. 15.6** is an edge in the top view and is perpendicular to the horizontal plane. If an auxiliary plane is drawn parallel to the inclined surface, the view projected onto it will be a true-size view of the inclined surface. **A surface must appear as an edge in a principal view before it can be found true size in a primary auxiliary view.**

The steps of drawing an auxiliary of a similar part are shown in **Fig. 15.7**. The true-size view of the inclined surface can be found by an auxiliary projected from the top view of the edge view of the plane. Draw the line H–1, the fold line, between the horizontal projection plane (H) and the primary auxiliary plane (1), at a convenient distance from the top view (step 1). Draw the fold line (labeled H–F) between the top and front views to represent the fold line between the horizontal and frontal projection planes.

Steps 2, 3, and 4 show how to transfer height distances, H, with dividers from the front view to the auxiliary view to locate the four corners of the inclined surface. Make each height measurement perpendicular to the reference line (fold line) in both the front and auxiliary views.

15.4 Rules of Auxiliary View Construction

The basic rules of drawing auxiliary views are shown in **Fig. 15.8**. **An auxiliary view must be projected perpendicularly from the edge view of the surface to be found true size.** Usually, the inclined surface, or a partial view, is all that is shown in the auxiliary view, but the entire object can be drawn in the auxiliary view if desired, as shown here.

Reference lines (fold lines) should be drawn as thin black lines with a 2H or 3H pencil. Draw the outlines of the auxiliary view as thick visible lines the same as visible lines in principal views, usually with an F or HB pencil. **Construction lines and projectors should be drawn as light gray lines, just dark enough to be seen, with a pencil in the 2H to 4H range.**

Always transfer dimensions from principal views (the front view in this case) perpendicularly from the fold lines to auxiliary views. Use your dividers to transfer these dimensions to the auxiliary view. Labeling points with numbers or letters in all views will be helpful. Do your lettering in a professional manner with guidelines.

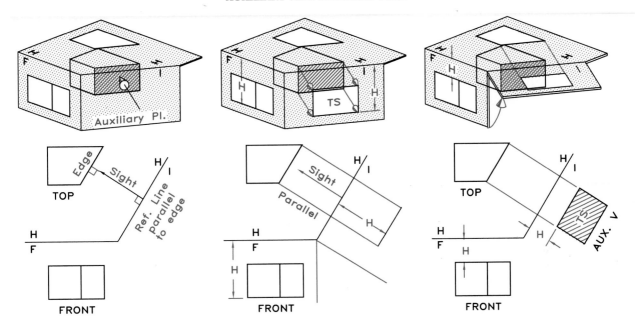

Figure 15.6 Auxiliary view from the top:
Required Find the true-size view of the inclined surface.

Step 1 Draw the line of sight perpendicular to the edge view of the inclined surface. Draw the H–1 line parallel to the edge. Draw the H–F reference line between the top and front views.

Step 2 Project from the edge view of the inclined surface parallel to the line of sight. Transfer the H dimensions from the front view to locate an auxiliary view of a line.

Step 3 Locate the other corners of the inclined surface by projecting to the auxiliary view. Locate the points by transferring the height dimensions (H) from the front view to the auxiliary view.

15.5 Auxiliary Views from the Top View: Application

Figure 15.9 illustrates how the folding-line method is used to find an auxiliary view of a part shown in an imaginary glass box. Because the inclined surface in the top view appears as an edge, it can be found true size in a primary auxiliary view. The height dimension (H) in the frontal plane will be the same as in the auxiliary plane, because both planes are perpendicular to the horizontal plane. In **Fig. 15.10**, the auxiliary plane is rotated about the H–1 fold line into the plane of the top view.

When drawn on a sheet of paper, the views of this object appear as shown in **Fig. 15.11**. The front view is a partial view because the omitted portion would have been hard to draw and would not have been true size. The auxiliary view also is drawn as a partial view because the front view shows the omitted features better.

■ **Computer Method** You may draw the true-size view of a plane by projecting from the top view, where the inclined plane appears as an edge **(Fig. 15.12)**. Use AutoCAD's ROTATE option of the SNAP command to rotate the grid to a position parallel to the edge view of the plane in the top view.

CONSTRUCTION OF AN AUXILIARY VIEW

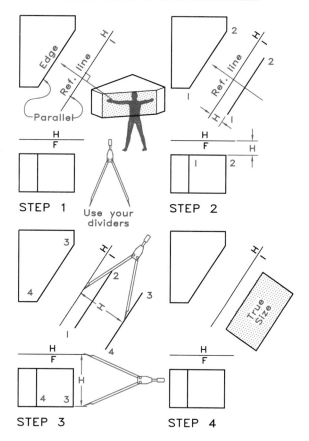

STEP 1 Use your dividers **STEP 2**

STEP 3   **STEP 4**

Figure 15.7 Auxiliary view construction:

Step 1 Draw a line of sight perpendicular to the edge view of the inclined surface. Draw the H–1 fold line parallel to the edge view of the inclined surface and insert an H–F fold line between the given views.

Step 2 Find points 1 and 2 by transferring the height (H) dimensions with your dividers from the front view to the auxiliary view.

Step 3 Find points 3 and 4 in the same manner by transferring the H dimensions.

Step 4 The corner points are connected to complete the true-size view of the inclined plane.

Reference-Plane Method

A second method of locating an auxiliary view uses **reference planes. Figure 15.13A** shows a horizontal reference plane (HRP) drawn through

RULES FOR DRAWING AUXILIARY VIEWS

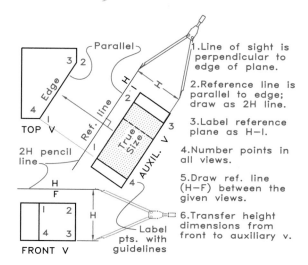

1. Line of sight is perpendicular to edge of plane.

2. Reference line is parallel to edge; draw as 2H line.

3. Label reference plane as H–I.

4. Number points in all views.

5. Draw ref. line (H–F) between the given views.

6. Transfer height dimensions from front to auxiliary v.

Figure 15.8 Apply these rules when drawing auxiliary views. Use thin black lines for reference lines and light gray construction lines (just dark enough to show) for projectors. Use guidelines, label points on the views, and label the fold lines.

VIEW PROJECTED ONTO AUXILIARY PLANE

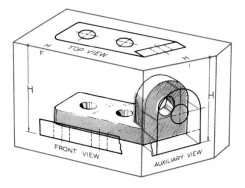

Figure 15.9 By imagining that the object is placed in a glass box, you can see the relationship of the auxiliary projection plane, on which the true-size view is projected, and the horizontal projection plane.

the center of the front view. Because this view is symmetrical, height dimensions can be conveniently transferred from the front view to the auxiliary view and laid off on both sides of the HRP.

AUXILIARY PLANE ROTATED INTO HORIZONTAL

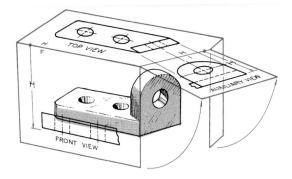

Figure 15.10 Fold the auxiliary plane into the horizontal projection plane by revolving it about the H–1 fold line.

DRAWING THE AUXILIARY VIEW

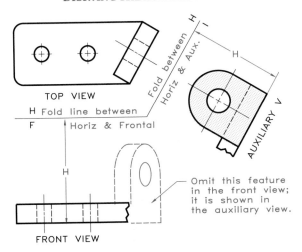

Figure 15.11 When the views of the object are drawn, they are laid out in this arrangement. The front view is drawn as a partial view because the omitted elliptical features are shown better as a true-size view in the auxiliary view.

The reference plane can be placed at the base of the front view as shown in **Fig. 15.13B**. In this case, the height dimensions are measured upward from the HRP in both the front and auxiliary views. You may draw a reference plane (the HRP in this example) in any convenient position in the front view: through the part, above it, or below it.

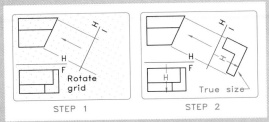

Figure 15.12 Auxiliary view by computer:
Step 1 To perpendicularly project an auxiliary from the edge of the inclined surface in the top view, use the SNAP command and the ROTATE option to rotate the grid parallel to the inclined surface. The H–1 reference line is drawn parallel to the edge, and projectors are drawn perpendicular to the H–1 line.

Step 2 Draw the auxiliary view by measuring height dimensions (H) in the front view and transferring them perpendicularly from the H–1 line. The auxiliary view is the true-size view of the inclined surface.

REFERENCE-PLANE METHOD

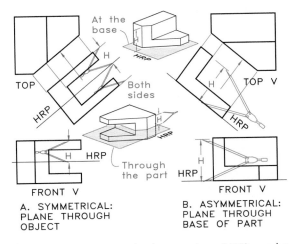

Figure 15.13 A horizontal reference plane (HRP) can be positioned through the part or in contact with it. The dimension of height (H) is measured from the HRP and transferred to the auxiliary view.

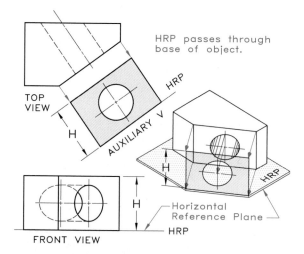

Figure 15.14 An auxiliary view projected from the top view is used to draw a true-size view of the inclined surface using a horizontal reference plane. The HRP is drawn through the bottom of the front view.

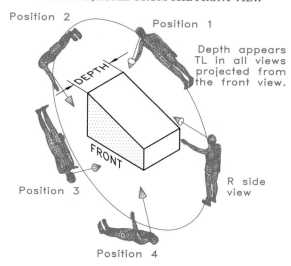

Figure 15.15 By moving your viewpoint around the frontal view of an object, you will see a series of auxiliary views in which the depth dimension (D) is true length.

A similar example of an auxiliary view drawn with a horizontal reference plane is shown in **Fig. 15.14**. In this example, the hole appears as circle instead of an ellipse.

15.6 Auxiliary Views from the Front View

By moving about the front view of the part as shown in **Fig. 15.15**, you will be looking parallel to the edge view of the frontal plane. Therefore the depth dimension (D) will appear true size in each auxiliary view projected from the front view. One of the positions gives a principal view, the right-side view, and position 1 gives a true-size view of the inclined plane. **Figure 15.16** illustrates the relationship between the auxiliary views projected from the front view.

Folding-Line Method

A plane of an object that appears as an edge in the front view (**Fig. 15.17**) is true size in an auxiliary

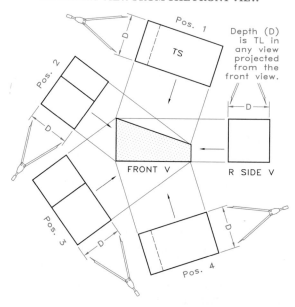

Figure 15.16 The auxiliary views shown in Fig. 15.15 would be seen in this arrangement when viewing the front view.

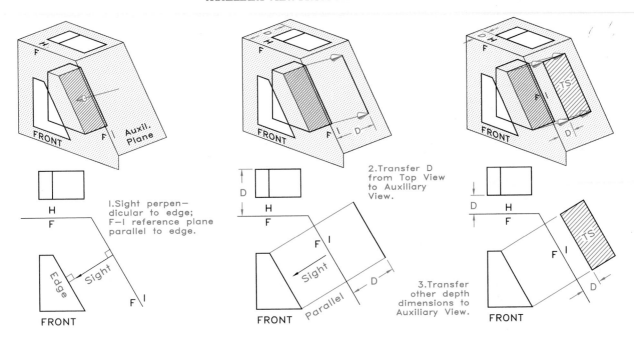

Figure 15.17 Auxiliary from the front—folding-line method:

Step 1 Draw the line of sight perpendicular to the edge of the plane and draw the F–1 reference parallel to the edge. Draw the H–F fold line between the top and front views.

Step 2 Project perpendicularly from the edge view of the inclined surface and parallel to the line of sight. Use the depth dimensions (D) transferred from the top view to locate a line in the auxiliary view.

Step 3 Locate the other corners of the inclined surface by projecting to the auxiliary view. Locate the points by transferring the depth dimensions (D) from the top to the auxiliary view.

view projected perpendicularly from it. Draw fold line F–1 parallel to the edge view of the inclined plane in the front view at a convenient location.

Draw the line of sight perpendicular to the edge view of the inclined plane in the front view. Observed from this direction, the frontal plane appears as an edge; therefore measurements perpendicular to the frontal plane—depth dimensions (D)—will be seen true length. Transfer depth dimensions from the top view to the auxiliary view with your dividers.

The object in **Fig. 15.18** is imagined to be enclosed in a glass box and an auxiliary plane is folded from the frontal plane to be parallel to the inclined surface. When drawn on a sheet of paper,

the views appear as shown in **Fig. 15.19**. The top and side views are drawn as partial views because the auxiliary view eliminates the need for complete views. The auxiliary view, located by transferring the depth dimension measured perpendicularly from the edge view of the frontal plane in the top view and transferred to the auxiliary view, shows the surface's true size.

◫ **Computer Method** Find the true-size view of the inclined surface that appears as an edge in the front view **(Fig. 15.20)** by using the LISP commands, PARALLEL and TRANSFER (see Section 27.2). While in AutoCAD's drafting mode, type PARALLEL to draw the reference line parallel to edge AB. Type TRANSFER to obtain

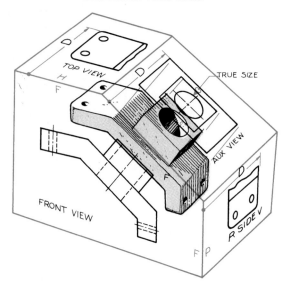

Figure 15.18 This object is shown in an imaginary glass box to illustrate the relationship of the auxiliary plane, on which the true-size view of the inclined surface is projected, with the principal planes.

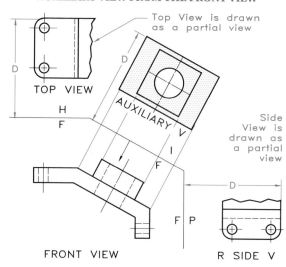

Figure 15.19 The layout and construction of an auxiliary view of the object shown in Fig. 15.18 is shown here.

prompts for transferring measurements from the top view as if you were using your dividers. Connect the circular points to complete the auxiliary view.

Reference-Plane Method

The object shown in **Fig. 15.21** has an inclined surface that appears as an edge in the front view; therefore this plane can be found true size in a primary auxiliary view. It is helpful to draw a reference plane through the center of the symmetrical top view because all depth dimensions can be located on each side of the frontal reference plane. Because the reference plane is a frontal plane, it is labeled FRP in the top and auxiliary views. In the auxiliary view, the FRP is drawn parallel to the edge view of the inclined plane at a convenient distance from it. By transferring depth dimensions from the FRP in the top view to the FRP in the auxiliary view, the symmetrical view of the part is drawn.

15.7 Auxiliary Views from the Profile View

By moving your position about the profile view (side view) of the part as shown in **Fig. 15.22**, you will be looking parallel to the edge view of the profile plane. Therefore, the width dimension will appear true size in each auxiliary view projected from the side view. One of the positions gives a principal view, the front view, and position 1 gives a true-size view of the inclined plane. **Figure 15.23** illustrates the arrangement of the auxiliary views projected from the side view if they were drawn on a sheet of paper.

Folding-Line Method

Because the inclined surface in **Fig. 15.24** appears as an edge in the profile plane, it can be found true size in a primary auxiliary view projected from the side view. The auxiliary fold line, P–1, is drawn parallel to the edge view of the inclined surface. A line of sight perpendicular to the auxiliary plane shows the profile plane as an edge.

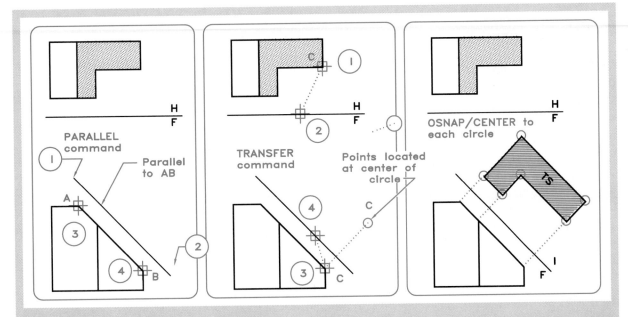

Figure 15.20 Auxiliary view by computer:

Step 1 While in AutoCAD, type PAR-ALLEL to receive the prompts for the first end of the reference line (1) and its approximate second endpoint (2). You are then prompted for the beginning and end of line A, to which the reference line is parallel.

Step 2 Type TRANSFER and you are prompted to select a point in the top view (1) and its distance from the reference line (2) to be transferred. You are then asked for the front view of the point to be projected from (3) and the reference plane (4). The point is projected to the auxiliary view and is located by a circle.

Step 3 Continue using TRANSFER to locate the other corner points of the inclined plane. Connect the points using the CENTER option of the OSNAP command to snap to the centers of the circles. ERASE the circles after connecting their centers.

Therefore width dimensions (W) transferred from the front view to the auxiliary view appear true length in the auxiliary view.

Reference-Plane Method

The object shown in **Fig. 15.25** has an inclined surface that appears as an edge in the right-side view, the profile view. This inclined surface may be drawn true size in an auxiliary view by using a profile reference plane (PRP) that is a vertical edge in the front view. Draw the PRP through the center of the front view because the view is symmetrical. Then find the true-size view of the inclined

plane by transferring equal width dimensions (W) with your dividers from the edge view of the PRP in the front view to both sides of the PRP in the auxiliary view.

15.8 Auxiliary Views of Curved Shapes

The cylinder shown in **Fig. 15.26** has an inclined surface that appears as an edge in the front view. The true-size view of this plane can be seen in an auxiliary view projected from the front view.

Because the cylinder is symmetrical, a frontal reference plane (FRP) is drawn through the center

AUXILIARY VIEW BY FRONTAL REFERENCE-PLANE METHOD

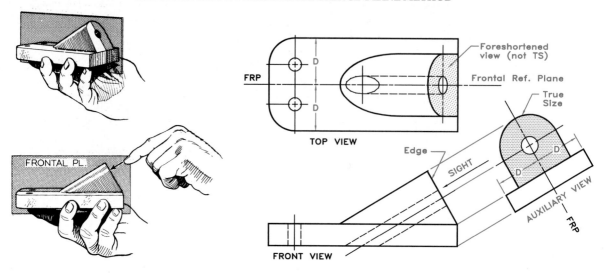

Figure 15.21 Because the inclined surface of this part is symmetrical, it is helpful to use a frontal reference plane (FRP) that passes through the object. Project the auxiliary view perpendicularly from the edge view of the plane in the front view. The FRP appears as an edge in the auxiliary view, and depth dimensions (D) are transferred from each side of it in the top view to locate points on the true-size view of the inclined surface.

VIEWPOINTS FOR VIEWS PROJECTED FROM THE PROFILE VIEW

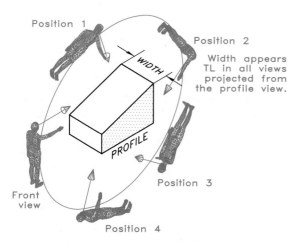

Figure 15.22 By moving your viewpoint around a profile (side) view of an object, you will obtain a series of auxiliary views in which the width dimension (W) is true length.

VIEWS PROJECTED FROM THE PROFILE VIEW

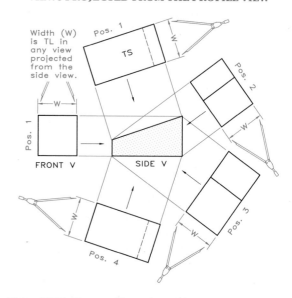

Figure 15.23 The auxiliary views shown in Fig. 15.22 would be seen in this arrangement when projected from the side view.

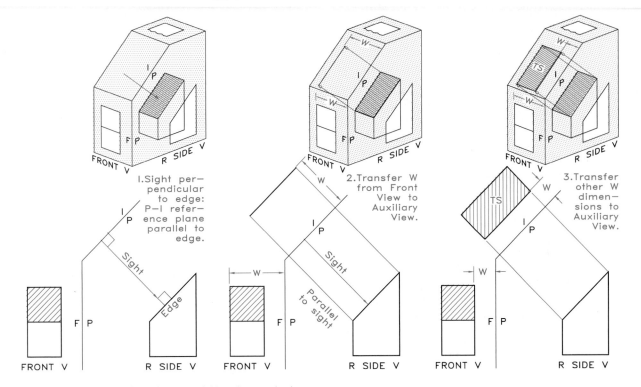

Figure 15.24 Auxiliary from the side—folding-line method:

Step 1 Draw a line of sight perpendicular to the edge view of the inclined surface. Draw the P–1 fold line parallel to the edge view, and draw the F–P fold line between the given views.

Step 2 Project the corners of the edge view parallel to the line of sight. Transfer the W dimensions from the front view to locate a line in the auxiliary view.

Step 3 Find the other corners of the inclined surface by projecting to the auxiliary view. Locate the points by transferring the width dimensions (W) from the front view to the auxiliary view.

of the side view so that equal dimensions can be laid off easily on both sides of it. Points located about the circular right-side view are projected to the edge view of the surface in the front view.

In the auxiliary view, the FRP is drawn parallel to the edge view of the plane in the front view, and the points are projected perpendicularly from the edge view of the plane. Dimensions A and B are shown as examples of depth dimensions used for locating points in the auxiliary view. To construct a smooth elliptical curve, more points than shown are needed.

A true-size auxiliary view of a surface bounded by an irregular curve is shown in **Fig. 15.27**. Project points from the curve in the top view to the front view. Locate these points in the auxiliary view by transferring depth dimensions (D) from the FRP in the top view to the auxiliary view.

15.9 Partial Views

Auxiliary views are used as supplementary views to clarify features that are difficult to depict with principal views alone. Consequently, portions of

PROFILE REFERENCE PLANE

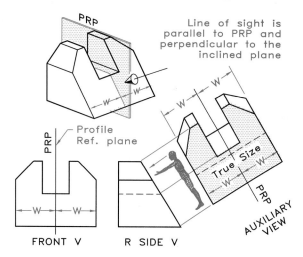

Figure 15.25 An auxiliary view is projected from the right-side view by using a profile reference plane (PRP) to show the true-size view of the inclined surface.

FRONTAL REFERENCE PLANE

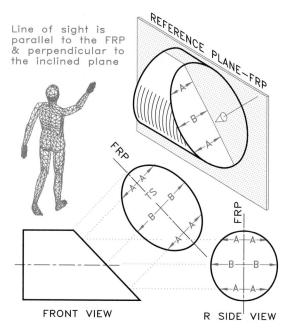

Figure 15.26 The auxiliary view of this elliptical surface was found by locating a series of points about its perimeter. The frontal reference plane (FRP) is drawn through its center in the side view since the object is symmetrical.

IRREGULAR SURFACE

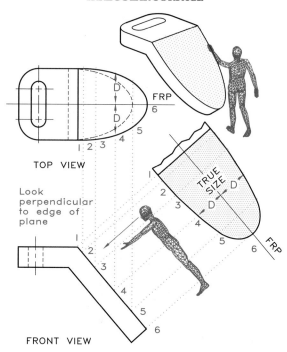

Figure 15.27 The auxiliary view of this curved surface required that a series of points be located in the top view, be projected to the front view, and then projected to the auxiliary view. The FRP was passed through the top view.

principal views and auxiliary views may be omitted, provided that the partial views adequately describe the part. The object shown in **Fig. 15.28** is composed of a complete front view, a partial auxiliary view, and a partial top view. These partial views are easier to draw and are more descriptive without sacrificing clarity.

15.10 Auxiliary Sections

In **Fig. 15.29**, a cutting plane labeled A–A is passed through the part to obtain the auxiliary section labeled section A–A. The auxiliary section provides a good and efficient way to describe features of the part that could not be as easily described by additional principal views.

PARTIAL VIEWS

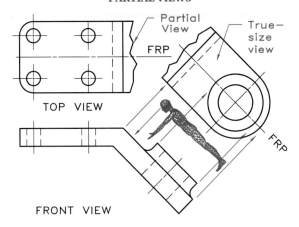

Figure 15.28 Partial views with foreshortened portions omitted can be used to represent objects. The FRP reference line is drawn through the center of the object in the top view because the object is symmetrical to make point location easier.

AUXILIARY SECTIONS

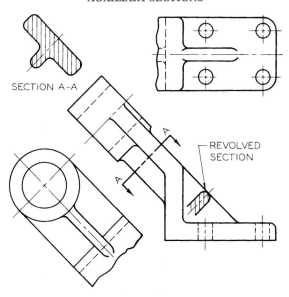

Figure 15.29 A cutting plane labeled A–A is passed through the object and the auxiliary section, section A–A, is drawn as a supplementary view to describe the part. The top and front views are drawn as partial views.

SECONDARY AUXILIARY VIEW

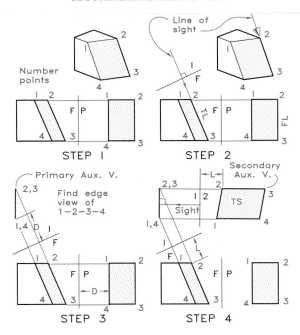

Figure 15.30 Secondary auxiliary views:

Step 1 Draw a fold line F–P between the front and side views. Label the corner points in both views.

Step 2 Line 2–3 is a true-length frontal line in the front view. Draw reference line F–1 perpendicular to line 2–3 at a convenient location with a line of sight parallel to line 2–3.

Step 3 Find the edge view of plane 1–2–3–4 by transferring depth dimensions (D) from the side view.

Step 4 Draw a line of sight perpendicular to the edge view of 1–2–3–4 and draw the 1–2 fold line parallel to the edge view. Find the true-size auxiliary view by transferring the dimensions (L) from the front view to the auxiliary view.

15.11 Secondary Auxiliary Views

Figure 15.30 shows how to project a secondary auxiliary view from a primary auxiliary view. An edge view of the oblique plane is found in the primary auxiliary view by finding the point view of a true-length line (2–3) that lies on the oblique surface. A line of sight perpendicular to the edge view

SECONDARY AUXILIARY AND PARTIAL VIEWS

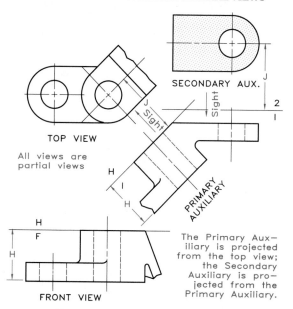

Figure 15.31 A secondary auxiliary view projected from a primary auxiliary view that was projected the top view is shown here. All views are drawn as partial views.

ELLIPSE GUIDE ANGLES

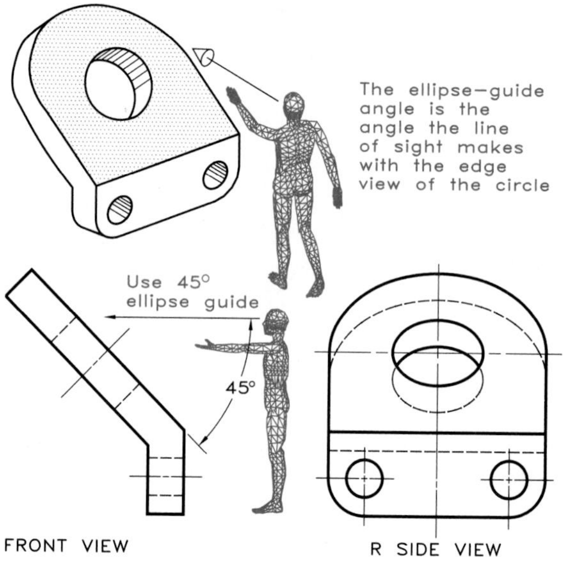

Figure 15.32 The ellipse guide angle is the angle that the line of sight makes with the edge view of the circular feature. The ellipse angle for the right-side view is 45°.

of the plane gives a secondary auxiliary view that shows the oblique plane as true size.

Note that the reference line between the primary auxiliary view and the secondary auxiliary view is labeled 1–2 to represent the fold line between the primary plane (1) and the secondary plane (2). The 1 label is placed on the primary side and the 2 label is placed on the secondary side.

Figure 15.31 illustrates the construction of a secondary auxiliary view that gives the true-size view of a surface on a part using these same principles and a combination of partial views. A secondary auxiliary view must be used in this case because the oblique plane does not appear as an edge in a principal view.

Find the point view of a line on the oblique plane to find the edge view of the plane in the primary auxiliary view. The secondary auxiliary view, projected perpendicularly from the edge view of the plane in the primary auxiliary view, gives a true-size view of the plane. In this example all of the views are drawn as partial views.

15.12 Elliptical Features

Occasionally, circular shapes will project as ellipses, which must be drawn with an irregular curve or an ellipse template. The ellipse template (guide) is by far the most convenient method of drawing ellipses. The angle of the ellipse template is the angle the line of sight makes with the edge view of the circular feature. In **Fig. 15.32** the angle is found to be 45° where the curve is an edge in the front view, so the right-side view of the curve is drawn as a 45° ellipse.

Problems

Solve the following problems on size A or size B sheets, as assigned by your instructor.

1–13. (Fig. 15.33) Using the example layout, change the top and front views by substituting the top views given at the right in place of the one given in the example. The angle of inclination in the front view is 45° for all problems, and the height is 38 mm (1.5 inches) in the front view. Construct auxiliary views that show the inclined surface true size. Draw two problems per size A sheet.

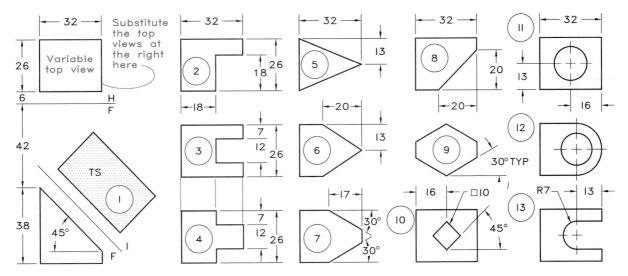

Figure 15.33 Problems 1–13. Primary auxiliary views.

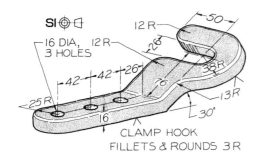

Figure 15.34 Problem 14. Clamp hook.

14–29. (Figs. 15.34–15.49) Draw the necessary primary and auxiliary views to describe the parts shown. Draw one per size A or size B sheet, as assigned. Adjust the scale of each to fit the space on the sheet.

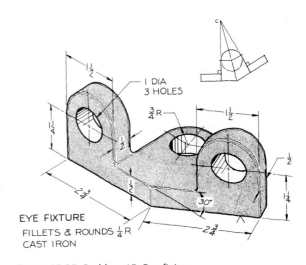

EYE FIXTURE
FILLETS & ROUNDS ¼ R
CAST IRON

Figure 15.35 Problem 15. Eye fixture.

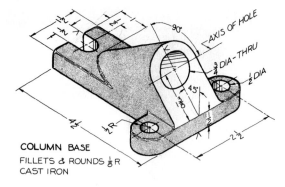

COLUMN BASE
FILLETS & ROUNDS $\frac{1}{8}$R
CAST IRON

Figure 15.36 Problem 16. Column base.

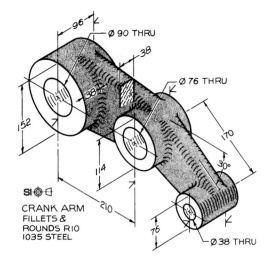

CRANK ARM
FILLETS &
ROUNDS R10
1035 STEEL

Figure 15.37 Problem 17. Crank arm.

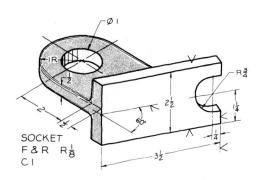

SOCKET
F & R R$\frac{1}{8}$
C I

Figure 15.38 Problem 18. Socket.

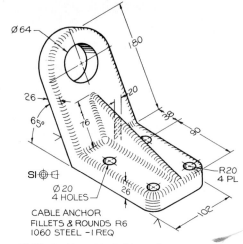

CABLE ANCHOR
FILLETS & ROUNDS R6
1060 STEEL –1 REQ

Figure 15.39 Problem 19. Cable anchor.

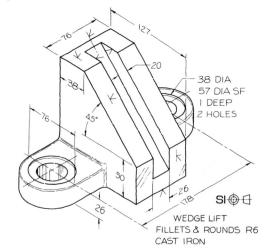

WEDGE LIFT
FILLETS & ROUNDS R6
CAST IRON

Figure 15.40 Problem 20. Wedge lift.

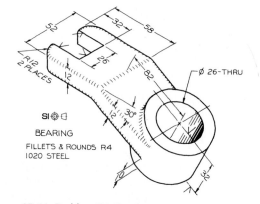

BEARING
FILLETS & ROUNDS R4
1020 STEEL

Figure 15.41 Problem 21. Bearing.

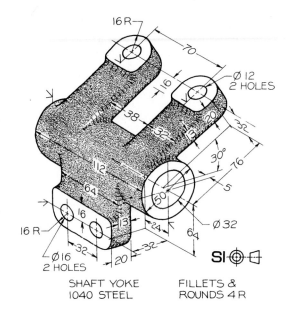

16 R
70
Ø 12
2 HOLES
16
38
32
20
112
64
16 R
16
30°
76
5
112
50
64
13
24
Ø 32
16 R
16
13
Ø16
2 HOLES
32
20
32
SI ⊕ ⊟

SHAFT YOKE
1040 STEEL

FILLETS &
ROUNDS 4 R

Figure 15.42 Problem 22. Shaft yoke.

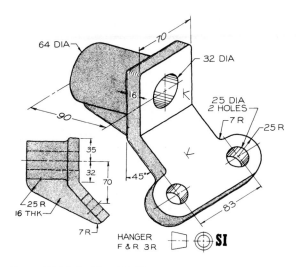

64 DIA
70
32 DIA
16
25 DIA
2 HOLES
90
7 R
25 R
35
32
70
45°
83
25 R
16 THK
7 R

HANGER
F & R 3 R
⊕ **SI**

Figure 15.44 Problem 24. Hanger.

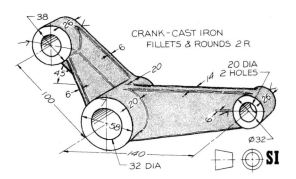

38
26
CRANK-CAST IRON
FILLETS & ROUNDS 2 R
6
45
20
20 DIA
2 HOLES
14
100
6
20
26
58
6
Ø32
140
32 DIA
⊟ ⊕ **SI**

Figure 15.45 Problem 25. Crank.

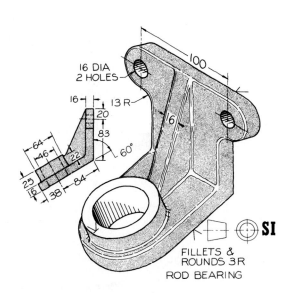

16 DIA
2 HOLES
100
16
13 R
16
20
83
64
46
22
60°
25
16
38
84
⊟ ⊕ **SI**

FILLETS &
ROUNDS 3 R
ROD BEARING

Figure 15.43 Problem 23. Rod bearing.

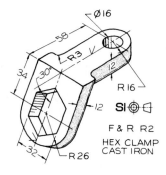

Ø16
58
R 3
12
34
R 16
30°
12
SI ⊕ ⊟
32
R 26
F & R R2
HEX CLAMP
CAST IRON

Figure 15.46 Problem 26. Hexagon angle.

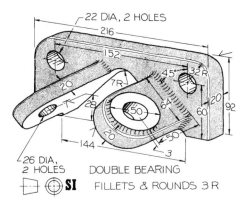

Figure 15.47 Problem 27. Double bearing.

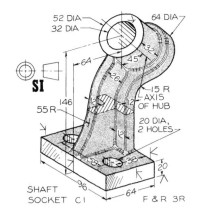

Figure 15.48 Problem 28. Shaft socket.

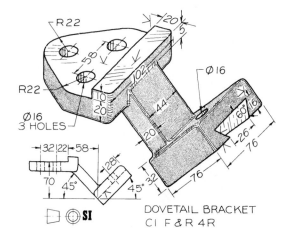

Figure 15.49 Problem 29. Dovetail bracket.

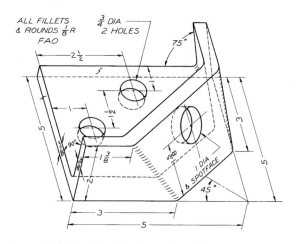

Figure 15.50 Problem 30. Corner bracket.

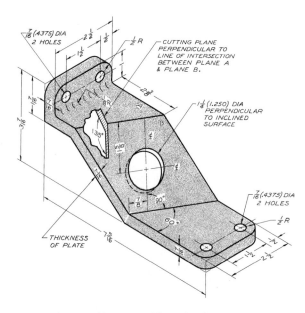

Figure 15.51 Problem 31. Oblique bracket.

30. (Fig. 15.50) Lay out these orthographic views on size B sheets, and complete the auxiliary and primary views.

31. (Fig. 15.51) Construct orthographic views of the given object, and using secondary auxiliary views, draw auxiliary views that give the true-size views of the inclined surfaces. Draw one per size B sheet.

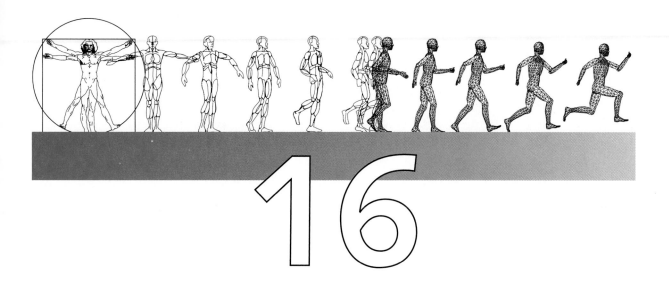

16

Sections

16.1 Introduction

Correctly drawn orthographic views that show all hidden lines may not clearly describe an object's internal details. This shortcoming can be overcome by imagining that part of the object has been cut away and shown in a cross-sectional view, called a **section**.

16.2 Basics of Sectioning

Figure 16.1 shows pictorially a section created by passing an imaginary cutting plane through the object to reveal its internal features. Think of the cutting plane as a knife-edge cutting through the object. **Figure 16.1A** shows the standard top and front views, and **Fig. 16.1B** shows the method of drawing a section. The front view is full section, with the portion cut by the imaginary plane cross-hatched. Hidden lines usually are omitted because they are not needed.

    Figure 16.2 shows two types of cutting planes. Either is acceptable although the one with pairs of

SECTIONS VS. VIEWS

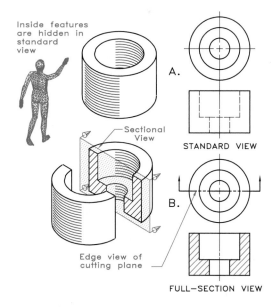

Inside features are hidden in standard view

Sectional View

Edge view of cutting plane

A.

STANDARD VIEW

B.

FULL—SECTION VIEW

Figure 16.1 This drawing compares a standard orthographic view with a full-section view that shows the internal features of the same object.

235

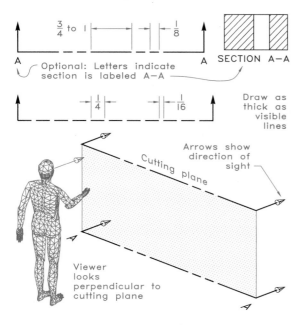

Figure 16.2 Use cutting-plane lines to represent sections (the cutting edge). The cutting plane marked A–A produces a section labeled A–A.

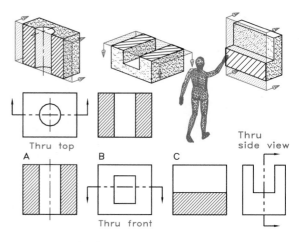

Figure 16.3 The three standard positions of cutting planes through orthographic (A) top, (B) front, and (C) side sectional views as sections. The arrows point in the direction of your line of sight for each section.

short dashes is most often used. The spacing and proportions of the dashes depend on the size of the drawing. The line thickness of the cutting plane is the same as the visible object line. Letters placed at each end of the cutting plane are used to label the sectional view, such as section A–A.

The sight arrows at the ends of the cutting plane are always perpendicular to the cutting plane. In the sectional view, the observer is looking in the direction of the sight arrows, perpendicular to the surface of the cutting plane.

Figure 16.3 shows the three basic positions of sections and their respective cutting planes. In each case perpendicular arrows point in the direction of the line of sight. For example, the cutting plane in **Fig. 16.3A** passes through and removes the front of the top view and the line of sight is perpendicular to the remainder of the top view.

The top view appears as a section when the cutting plane passes through the front view and the line of sight is downward (**Fig. 16.3B**). When the cutting plane passes vertically through the side view (**Fig. 16.3C**), the front view becomes a section.

16.3 Sectioning Symbols

Figure 16.4 shows the hatching symbols used to distinguish between different materials in sections. Although these symbols may be used to indicate the materials in a section, you should provide supplementary notes specifying the materials to ensure clarity.

The cast-iron symbol (evenly spaced section lines) may be used to represent any material and is the symbol used most often. Draw cast-iron symbols with a 2H pencil, slant the lines upward at 30°, 45°, or 60° angles, and space the lines about 1/16 inch apart (close together in small areas and farther apart in larger areas).

STANDARD HATCHING SYMBOLS

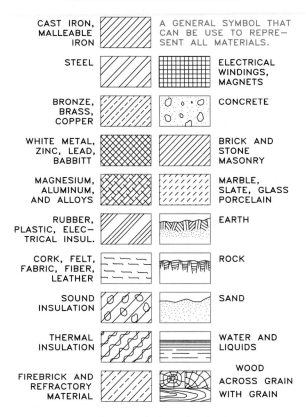

Figure 16.4 Use these symbols for hatching parts in section. The cast-iron symbol may be used for any material.

AUTOCAD HATCHING SYMBOLS

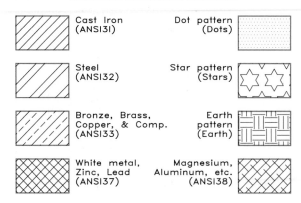

Figure 16.5 These are a few of the hatching symbols provided by AutoCAD. Use the pattern scale factor to vary the spacing between lines and dashes.

SECTION-LINE SPACING

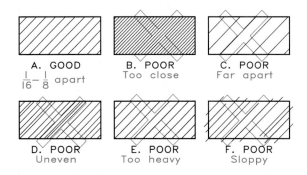

Figure 16.6 Techniques:

A Section lines are thin lines drawn $\frac{1}{16}$ to $\frac{1}{8}$ in. apart.

B–F Avoid these typical section lining errors.

Computer Method Figure 16.5 shows a few of the many cross-sectional symbols available with AutoCAD. You may vary the spacing between the lines and the dash lengths by changing the pattern scale factor.

Figure 16.6A shows properly drawn section lines: thin and evenly spaced. **Figure 16.6B–F** show common errors of section lining.

Section thin parts such as sheet metal, washers, and gaskets by completely blacking in the

areas (**Fig. 16.7**), because space does not permit the drawing of section lines. Show large parts with an outline section to save time and effort.

You should hatch sectioned areas with symbols that are neither parallel nor perpendicular to the outlines of the parts lest they be confused with serrations or other machining treatments of the surface. (**Fig. 16.8**).

LARGE AND THIN PARTS IN SECTION

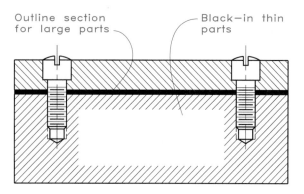

Figure 16.7 Black–in thin parts and hatch large areas around their outlines (outline sectioning) to save time and effort.

HATCH-LINE ANGLES

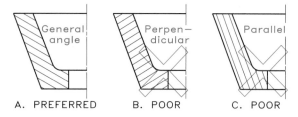

Figure 16.8 Draw section lines at angles that are neither parallel nor perpendicular to the outline of a part, so that they are not misunderstood as machining features.

Computer Method **Figure 16.9** shows the method of applying section symbols to an area with AutoCAD. After assigning the proper hatch symbol with the HATCH command, select the area to be sectioned with a window, and the section lines are drawn. To vary the spacing of the section lines, change the scale factor of the HATCH command. BHATCH, another hatching command, is covered in Chapter 36.

The lines used to depict sectioned areas must intersect perfectly at each corner point; no T-joints are permitted **(Fig. 16.10)**. Poor intersections may cause hatching symbols to fill the desired area improperly.

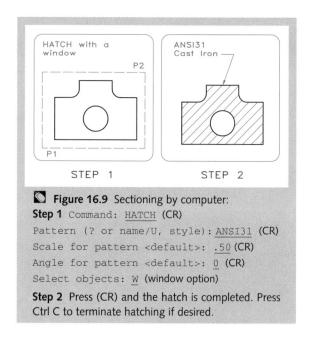

Figure 16.9 Sectioning by computer:
Step 1 Command: HATCH (CR)
Pattern (? or name/U, style): ANSI31 (CR)
Scale for pattern <default>: .50 (CR)
Angle for pattern <default>: 0 (CR)
Select objects: W (window option)
Step 2 Press (CR) and the hatch is completed. Press Ctrl C to terminate hatching if desired.

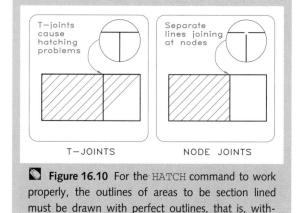

Figure 16.10 For the HATCH command to work properly, the outlines of areas to be section lined must be drawn with perfect outlines, that is, without T-joints, gaps, or overlaps.

16.4 Sectioning Assemblies of Parts

When sectioning an assembly of several parts, draw section lines at varying angles to distinguish the parts from each other **(Fig. 16.11A)**. Using dif-

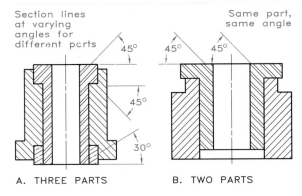

A. THREE PARTS   B. TWO PARTS

Figure 16.11 Hatching assemblies:
A Draw section lines of different parts in an assembly at varying angles to distinguish the parts.

B Draw section lines on separated portions of the same part (both sides of a hole here) in the same direction.

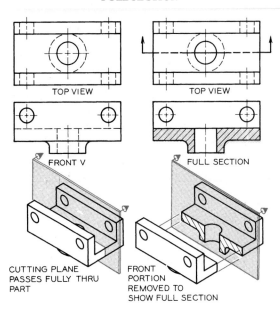

Figure 16.12 A full section is found by passing a cutting plane fully through the top view of this part, removing half of it. The arrows at each end of the cutting plane indicate the direction of your sight. The sectional view shows the part's internal features clearly.

ferent material symbols in an assembly also helps distinguish between the parts and their materials. Cross-hatch the same part at the same angle and with the same symbol even though portions of the part may be separated **(Fig. 16.11B)**.

16.5 Full Sections

A cutting plane passed fully through an object and removing half of it forms a full section view. **Figure 16.12** shows two orthographic views of an object with all its hidden lines. We can describe the part better by passing a cutting plane through the top view to remove half of it. The arrows on the cutting plane indicate the direction of sight. The front view becomes a full section, showing the surfaces cut by the cutting plane.

Figure 16.13 shows a full section through a cylindrical part, with half the object removed. **Figure 16.13A** shows the correctly drawn sectional view. A common mistake in constructing sections is omitting the visible lines behind the cutting plane **(Fig. 16.13B)**.

Omit hidden lines in sectional views unless you consider them necessary for a clear understanding of the view. Also, omit cutting planes if you consider them unnecessary. **Figure 16.14** shows a full section of a part from which the cutting plane was omitted because its path is obvious.

Parts Not Requiring Section Lining

Many standard parts, such as nuts and bolts, rivets, shafts, and set screws, do not require section lining even though the cutting plane passes through them **(Fig. 16.15)**. These parts have no internal features, so sections through them would be of no value. Other parts not requiring section lining are roller bearings, ball bearings, gear teeth, dowels, pins, and washers.

FULL SECTION: CYLINDRICAL PART

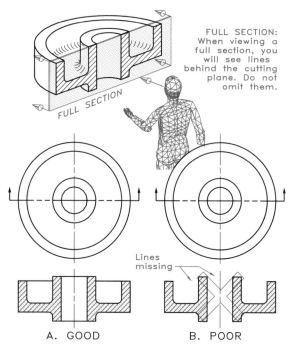

FULL SECTION:
When viewing a full section, you will see lines behind the cutting plane. Do not omit them.

FULL SECTION

Lines missing

A. GOOD   B. POOR

Figure 16.13

A When a cutting plane is passed through a cylinder to obtain a full section, you will see lines behind the plane, not just the cut surface.

B Showing only the lines at the cutting plane's surface yields an incomplete view.

Ribs

Ribs are not section lined when the cutting plane passes flatwise through them (**Fig. 16.16A**), because to do so would give a misleading impression of the rib. But ribs do require section lining when the cutting plane passes perpendicularly through them and shows their true thickness (**Fig. 16.16B**).

Figure 16.17 shows an alternative method of section lining webs and ribs. The outside ribs in **Fig. 16.17A** do not require section lining because the cutting plane passes flatwise through them

FULL SECTION: PULLEY ARM

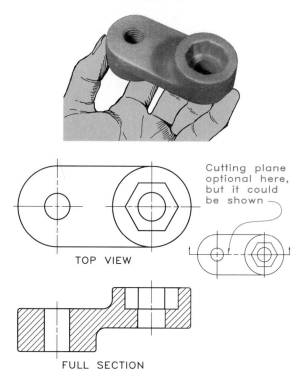

TOP VIEW

Cutting plane optional here, but it could be shown

FULL SECTION

Figure 16.14 The cutting plane of a section can be omitted if its location is obvious.

PARTS NOT HATCHED

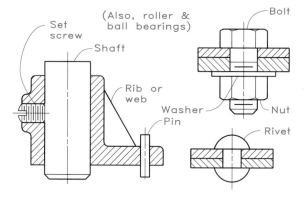

Set screw

Shaft

(Also, roller & ball bearings)

Bolt

Rib or web

Washer

Pin

Nut

Rivet

Figure 16.15 By conventional practice these parts are not section lined even though cutting planes pass through them.

RIBS IN SECTION

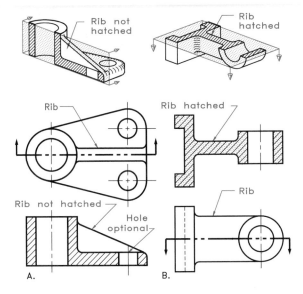

Figure 16.16

A Do not hatch a rib cut in a flatwise direction.

B Hatch ribs when cutting planes pass through them, showing their true thickness.

RIBS AND WEBS IN SECTION

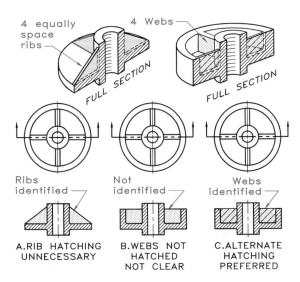

Figure 16.17

A You need not hatch well-defined outside ribs in section.

B Define poorly identified webs by (C) using alternative hatching to call attention to them.

RIBS IN SECTION

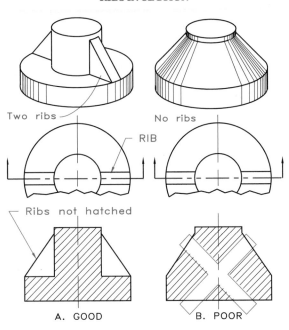

Figure 16.18 In this case (A) the ribs are not hatched to better describe the part than by (B) hatching them. When you use partial views to save space for drawing sections, remove the portion from the side adjacent to the section.

and they are well identified. As a rule, webs do not require cross-hatching, but the webs shown in **Fig. 16.17B** are not well identified in the front section and could go unnoticed. Therefore using alternate section lines as shown in **Fig. 16.17C** is better. Here, extending every-other section line through the webs ensures that they can be identified easily.

By not section lining the ribs in **Fig. 16.18A** we provided an effective section view of the part. If we had section lined the ribs, the section would give the impression that the part is solid and conical (**Fig. 16.18B**).

16.6 Half Sections

A half section is a view obtained by passing a cutting plane halfway through an object and removing a quarter of it to show both external and

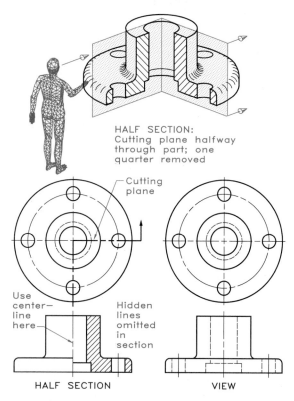

Figure 16.19 In a half section the cutting plane passes halfway through the object, removing a quarter of it, to show half the outside and half the inside of the object. Omit hidden lines unless you need them to clarify the view.

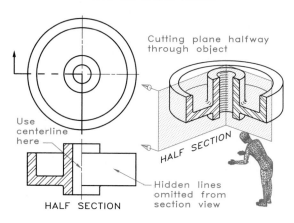

Figure 16.20 This half section describes the part that is shown orthographically and pictorially.

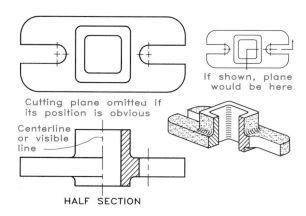

Figure 16.21 The cutting plane can be omitted when its location is obvious. The parting line between the section and the view may be a visible line or a centerline if the part is not cylindrical.

internal features. Half sections are used with symmetrical parts and with cylinders, in particular, as shown in **Fig. 16.19**. By comparing the half section with the standard front view, you can see that both internal and external features show more clearly in a half section than in a view. Hidden lines are unnecessary, and we've omitted them to simplify the section. **Figure 16.20** shows a half section of a pulley.

Note omission of the cutting plane from the half section shown in **Fig. 16.21** because the cutting plane's location is obvious. **Because the parting line of the half section is not at a centerline, you may use a solid line or a centerline to separate the sectional half from the half that appears as an external view.**

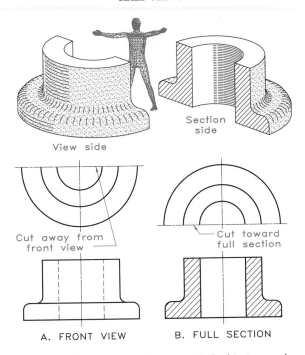

Section side

View side

Cut away from front view

Cut toward full section

A. FRONT VIEW

B. FULL SECTION

Figure 16.22 Half views of symmetrical objects can be used to conserve space and drawing time. (A) The omitted portion of the view is away from the front view. (B) The omitted portion of the top view is adjacent to the section. In half sections, the omitted half view may be either adjacent to or away from the section.

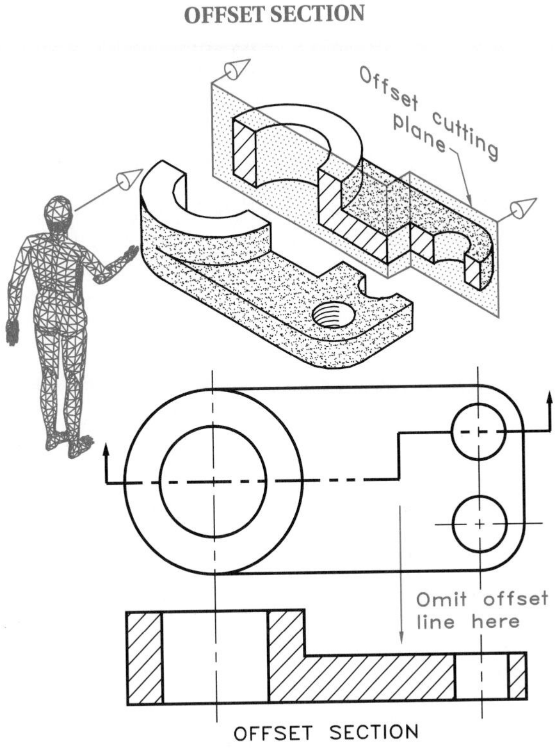

Offset cutting plane

Omit offset line here

OFFSET SECTION

Figure 16.23 An offset section is formed by a cutting plane that must be offset to pass through features not in a single plane. Here, the offset cutting plane is drawn in the top view and the front view is drawn as an offset section.

16.7 Half Views

Figure 16.22 shows **half views**, or conventional methods of representing symmetrical views that require less space and less time to draw than full views. A half top view is sufficient when drawn adjacent to the section view or front view. For half views (not sections), the removed half is the half away from the adjacent view **(Fig. 16.22A)**. For full sections, the removed half is the half nearest the section **(Fig. 16.22B)**. When drawing partial views with half sections, you may omit either the near or the far halves of the partial views.

16.8 Offset Sections

An **offset section** is a full section in which the cutting plane is offset to pass through important features that do not lie in a single plane. **Figure 16.23** shows an offset section in which the plane is offset to pass through the large hole and one of the small holes. The cut formed by the offset is not shown in the section because it is imaginary.

16.9 Broken-Out Sections

A **broken-out section** shows a partial view of a part's interior features. The broken-out section of the part shown in **Fig. 16.24** reveals details of the

BROKEN-OUT SECTION

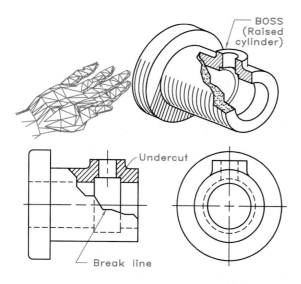

Figure 16.24 To find a broken-out section, imagine that part of the object has been broken away to reveal interior features.

BROKEN-OUT SECTION: PULLEY

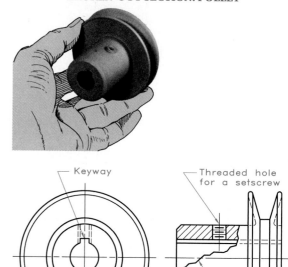

Figure 16.25 This broken-out section effectively shows the keyway and threaded hole for a setscrew in the pulley.

wall thickness to describe the part better. The irregular lines representing the break are conventional breaks (discussed later in this chapter).

The broken-out section of the pulley in **Fig. 16.25** clearly depicts the keyway and threaded hole for a setscrew. This method shows the part efficiently, with the minimum of views.

16.10 Revolved Sections

A **revolved section** describes a part when you revolve its cross section about an axis of revolution and place it on the view where the revolution occurred. Note the use of revolved sections to explain two cross sections of the shaft shown in **Fig. 16.26** (with and without conventional breaks). Conventional breaks are optional; you may draw a revolved section on the view without them.

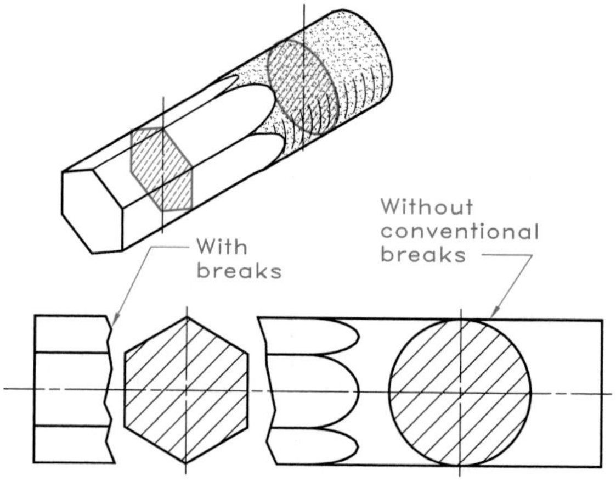

Figure 16.26 Revolved sections show cross-sectional features of a part to eliminate the need for supplementary orthographic views. You may superimpose revolved sections on the given views or use conventional breaks to separate them from the given views.

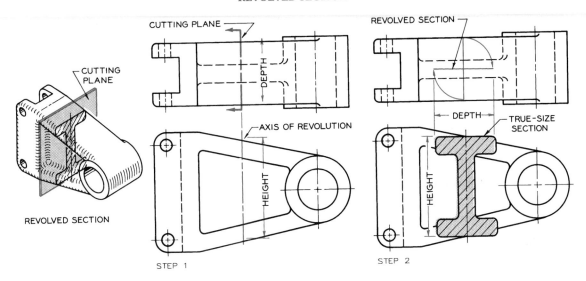

Figure 16.27 Drawing revolved sections:
Step 1 Show an axis of revolution in the front view. The cutting plane would appear as an edge in the top view if you were to show it.

Step 2 Revolve the vertical section in the top view to show the section at true size in the front view. Do not draw object lines through the revolved section.

REVOLVED SECTIONS

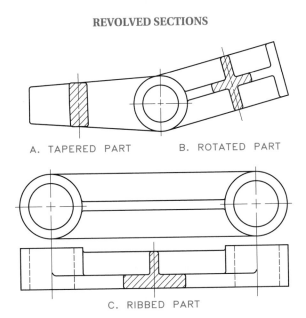

Figure 16.28 These revolved sections help describe the cross sections of the two parts and make complex orthographic views unnecessary.

A revolved section helps to describe the part shown in **Fig. 16.27**. Imagine passing a cutting plane through the top view of the part (step 1). Then imagine revolving the cutting plane in the top view to obtain a true-size revolved section in the front view (step 2). Conventional breaks could be used on each side of the revolved section.

Figure 16.28 demonstrates how to use typical revolved sections to show cross sections through parts without having to draw additional orthographic views.

16.11 Conventional Revolutions

In **Fig. 16.29A**, the middle hole is omitted because it does not pass through the center of the circular plate. However, in **Fig. 16.29B**, the hole does pass through the plate's center and is shown in the section. Although the cutting plane does not pass through one of the symmetrically spaced holes in

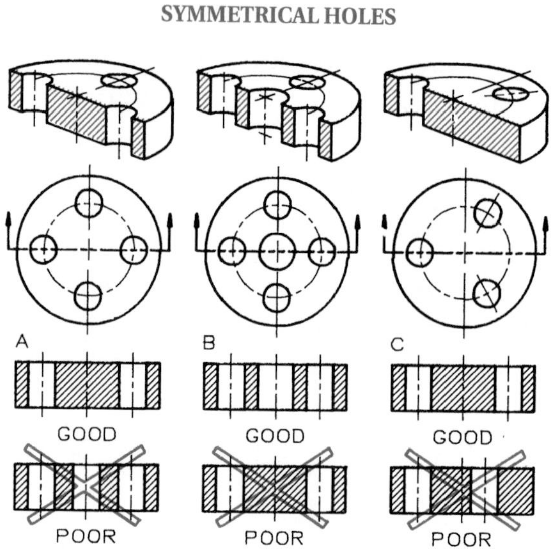

SYMMETRICAL HOLES

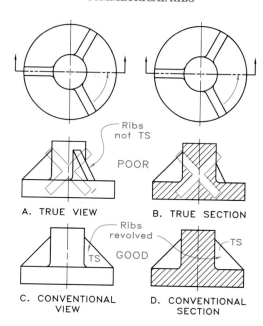

Figure 16.29 Revolve symmetrically spaced holes to show their true radial distances from their centers in sectional views. (A) Do not show the middle hole because it is not at the center of the plate. (B) Show the center hole because it is at the center of the plate. (C) Rotate one of the holes to the cutting plane to make the sectional view symmetrical and more descriptive.

SYMMETRICAL RIBS

A. TRUE VIEW

B. TRUE SECTION

C. CONVENTIONAL VIEW

D. CONVENTIONAL SECTION

Figure 16.30 Show symmetrically spaced ribs revolved in both orthographic and sectional views to show them true size as a conventional practice.

RIBS AND HOLES IN SECTION

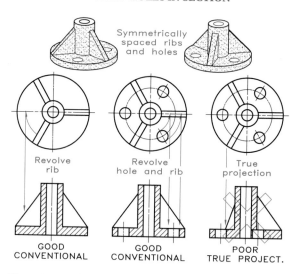

Figure 16.31 Show parts having symmetrically spaced ribs and holes in section with ribs rotated to show their true size and holes rotated to show them at their true radial distance from the center.

the top view (**Fig. 16.29C**), the hole is revolved to the cutting plane to show the full section.

When ribs are symmetrically spaced about a hub (**Fig. 16.30**), it is conventional practice to revolve them so that they appear true size in both views and sections. **Figure 16.31** illustrates the conventional practice of revolving both holes and ribs (or webs) of symmetrical parts. Revolution gives a better description of the parts in a manner that is easier to draw.

A cutting plane may be positioned in either of two ways shown in **Fig. 16.32**. Even though the cutting plane does not pass through the ribs and holes in **Fig. 16.32A**, they may appear in section as if the cutting plane passed through them. The path of the cutting plane also may be revolved, as shown in **Fig. 16.32B**. In this case the ribs are revolved to their true-size position in the section view, although the plane does not cut them.

CUTTING-PLANE POSITIONS

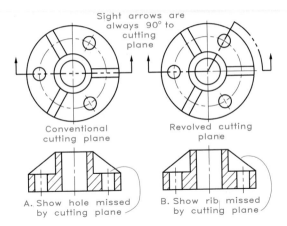

Figure 16.32 Show symmetrically located ribs true size in section whether the cutting plane passes through them or not. You may revolve the path of the cutting plane through certain features if you want, but the sight arrows are always perpendicular to the cutting plane.

SPOKES IN SECTION

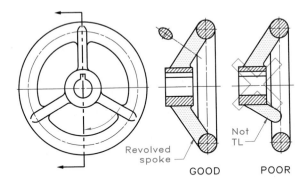

Figure 16.33 Revolve symmetrically spaced spokes to show them at true size in section. Do not section line spokes.

The same principles apply to symmetrically spaced spokes (**Fig. 16.33**). Draw only the revolved, true-size spokes and do not section line them. If the spokes shown in **Fig. 16.34A** were

WEBS AND SPOKES IN SECTION

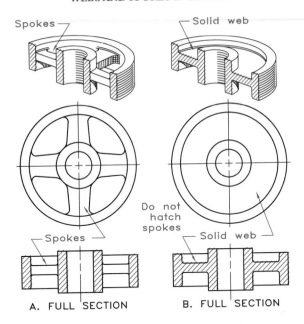

Figure 16.34
A Do not hatch spokes when the cutting plane passes through them.
B Hatch solid webs in sections of this type.

hatched, they could be misunderstood as a solid web, as shown in **Fig. 16.34B**.

Revolving the symmetrically positioned lugs shown in **Fig. 16.35** yields their true size in both the front view and section. The same principles of rotation apply to the part shown in **Fig. 16.36**, where the inclined arm appears in the section as if it had been revolved to the centerline in the top view and then projected to the sectional view.

16.12 Removed Sections

A **removed section** is a revolved section that is shown outside the view in which it was revolved (**Fig. 16.37**). Centerlines are used as axes of rotation to show the locations from which the sections are taken. Where space does not permit revolution on

LUGS IN SECTION

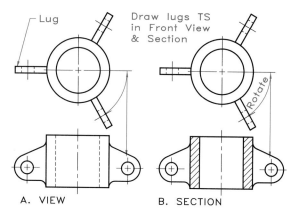

Figure 16.35 Revolve symmetrically spaced lugs (flanges) to show their true size in the (A) front view and (B) in sections.

REVOLVED FEATURES

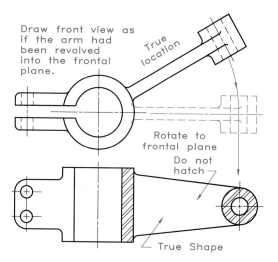

Figure 16.36 It is conventional practice to revolve a part with an inclined arm extending from a circular hub as if it were true shape in the sectional view.

the given view (**Fig. 16.38A**), removed sections must be used instead of revolved sections (**Fig. 16.38B**).

Removed sections do not have to position directly along an axis of revolution adjacent to the view from which they were revolved. Instead,

REMOVED SECTIONS

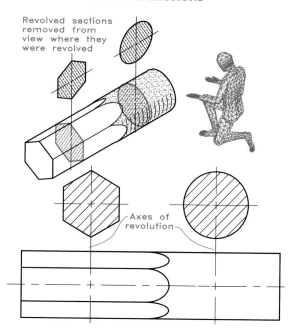

Figure 16.37 Removed sections are revolved sections that are drawn outside the object along their axes of revolution.

REVOLVED AND REMOVED SECTIONS

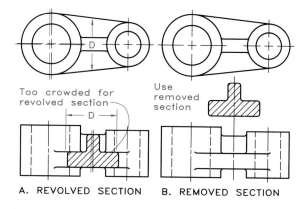

Figure 16.38 Removed sections are necessary when space does not permit the use of revolved sections.

REMOVED SECTIONS

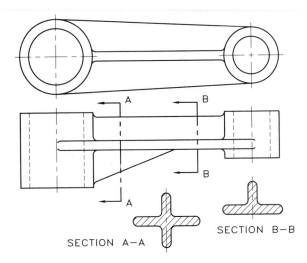

Figure 16.39 Lettering each end of a cutting plane (such as A–A) identifies the removed section labeled Section A–A shown elsewhere on the drawing.

CUTTING PLANE FOR REMOVED SECTIONS

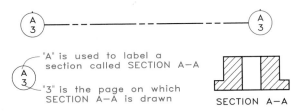

Figure 16.40 If placing a removed section on another page in a set of drawings is necessary, label each end of the cutting plane with a letter and a number. Here, the letters refer to Section A–A, and the numbers mean that Section A–A appears on page 3.

removed sections can be located elsewhere on a drawing if they are properly labeled (**Fig. 16.39**). For example, the plane labeled with an A at each end identifies the location of section A–A; the same applies to section B–B.

When a set of drawings consists of multiple sheets, removed sections and the views from

CONVENTIONAL BREAKS

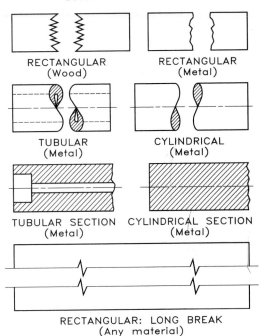

Figure 16.41 These conventional breaks indicate that a portion of an object is not shown.

which they are taken may appear on different sheets. When this method of layout is necessary, label the cutting plane in the view from which the section was taken and the sheet on which the section appears (**Fig. 16.40**).

16.13 Conventional Breaks

Figure 16.41 shows types of conventional breaks to use when you remove portions of an object. You may draw the "figure-eight" breaks used for cylindrical and tubular parts freehand (**Fig. 16.42**) or with a compass when they are larger (**Fig. 16.43**).

One use of conventional breaks is to shorten a long piece by removing the portion between the breaks so that it may be drawn at a larger size (**Fig. 16.44**). The dimension specifies the true length of

CYLINDRICAL BREAKS

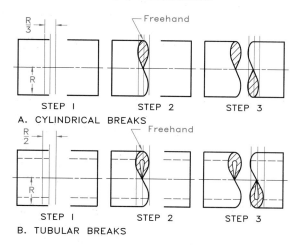

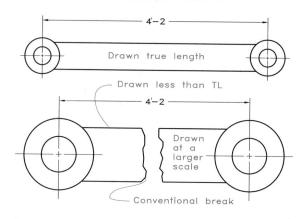

Figure 16.42 Guidelines aid in drawing conventional breaks in (A) cylindrical and (B) tubular sections freehand, as shown here. The radius, R, establishes the width of both "figure-eight" break symbols.

APPLICATION OF BREAKS

Figure 16.44 The use of conventional breaks allows this part to be drawn effectively at a larger scale. It is permissible to insert a revolved section between the breaks.

BREAKS WITH INSTRUMENTS

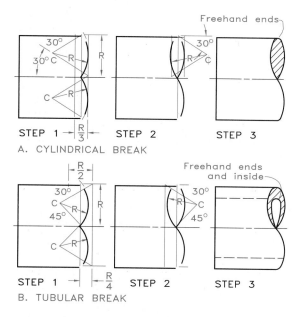

Figure 16.43 Instruments help in drawing conventional breaks in larger cylindrical and tubular parts.

the part, and the breaks indicate that a portion of the length has been removed.

16.14 Phantom (Ghost) Sections

A phantom or ghost section depicts parts as if they were being X-rayed. In **Fig. 16.45**, the cutting plane is used in the normal manner, but the section lines are drawn as dashed lines. If the object were shown as a regular full section, the circular hole through the front surface could not be shown in the same view.

16.15 Auxiliary Sections

You may use auxiliary sections to supplement the principal views of orthographic projections (**Fig. 16.46**). Pass auxiliary cutting plane A–A through the front view and project the auxiliary view from the cutting plane as indicated by the sight arrows. Section A–A gives a cross-sectional description of the part that would be difficult to depict by other principal orthographic views.

250 • **CHAPTER 16 SECTIONS**

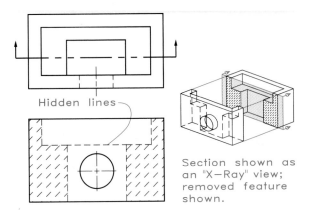

Hidden lines

Section shown as an "X—Ray" view; removed feature shown.

Figure 16.45 Phantom sections give an "X-ray" view of an object, allowing you to show the hole in the front of the cutting plane. Draw section lines as dashed lines.

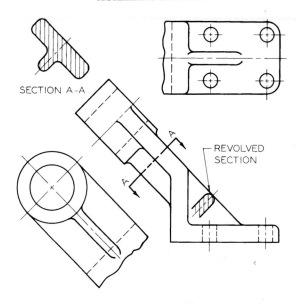

SECTION A-A

REVOLVED SECTION

Figure 16.46 Auxiliary sections can be used as supplementary views to add clarity to a drawing.

Problems

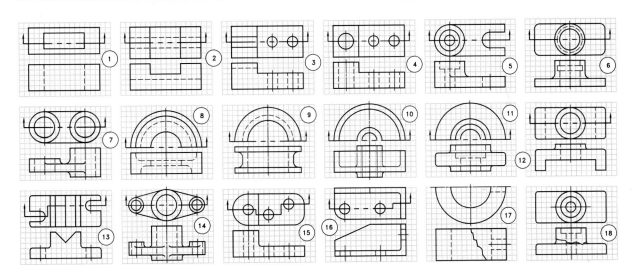

Figure 16.47 Problems 1–24. Introductory sections.

(continued)

Solve the problems shown in **Fig. 16.47** on size A sheets by drawing two solutions per sheet. Each grid space equals 0.20 in., or 5 mm.

1–16. Draw the sections indicated by the cutting planes.

17–20. Draw broken-out sections.

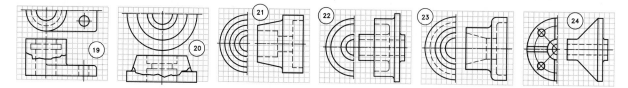

Fig. 16.47 continued

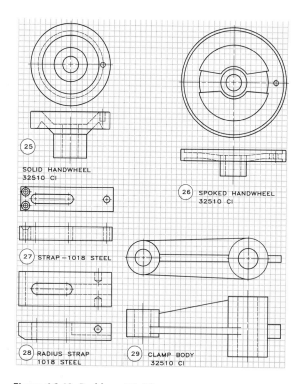

Figure 16.48 Problems 25–29.

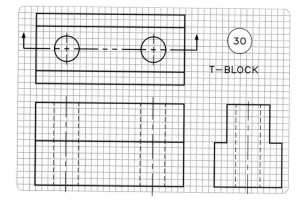

Figure 16.49 Problem 30. Full section.

21–24. Draw half sections.

25–29. (Fig. 16.48) Construct views of these fixtures on size A sheets and draw appropriate sections to describe them clearly. Exercise your design ability and add finish marks, fillets, and rounds where you believe they are needed but not shown. Use the grid—0.20 in., or 5 mm— to determine the full-size dimensions of the parts. Select the best scale and sheet size for each solution. (Courtesy of Jergens, Incorporated, Cleveland, Ohio.)

30–34. (Figs. 16.49–16.53) Complete these drawings as full sections. Draw one solution per size AH sheet. Each grid space equals 0.20 in., or 5 mm. Show the cutting planes in each solution.

35–36. (Figs. 16.54–16.55) Complete the drawings as half sections with one solution per size AH sheet. Each grid space equals 0.20 in., or 5 mm. Show the cutting planes in each solution.

37–38. (Figs. 16.56–16.57) Complete the drawings as offset sections. Draw one solution per size AH sheet. Each grid space equals 0.20 in., or 5 mm. Show the cutting planes in each solution.

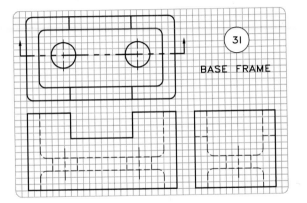

Figure 16.50 Problem 31. Full section.

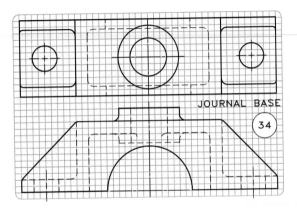

Figure 16.53 Problem 34. Full section.

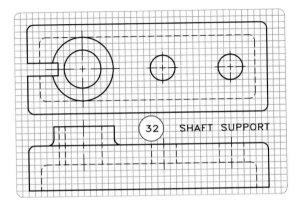

Figure 16.51 Problem 32. Full section.

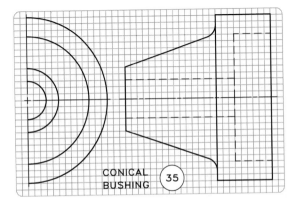

Figure 16.54 Problem 35. Half section.

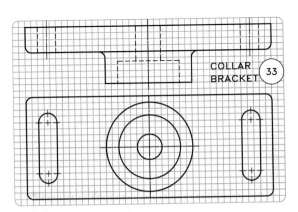

Figure 16.52 Problem 33. Full section.

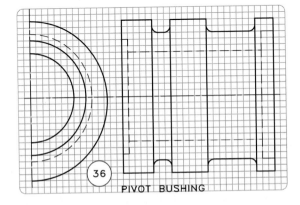

Figure 16.55 Problem 36. Half section.

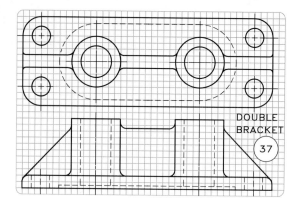

Figure 16.56 Problem 37. Offset section.

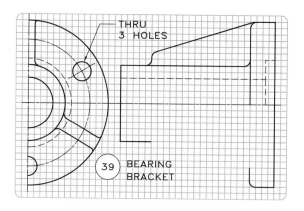

Figure 16.58 Problem 39. Full section.

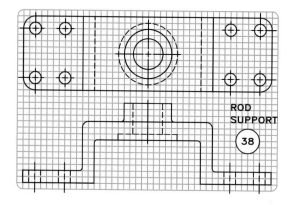

Figure 16.57 Problem 38. Offset section.

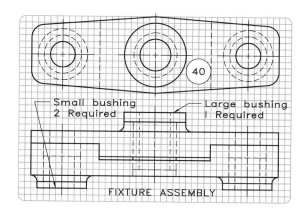

Figure 16.59 Problem 40. Full section in assembly.

39. (Fig. 16.58) Complete the partial view as a full section. Draw the views on a size AH sheet. Each grid space equals 0.20 in., or 5 mm. Show the cutting plane in your solution.

40. (Fig. 16.59) Complete the front view as a full section of the assembly. Draw the views on a size AH sheet. Each grid space equals 0.20 in., or 5 mm. Show the cutting plane in your solution.

Screws, Fasteners, and Springs

17.1 Introduction

Screws provide a fast and easy method of fastening parts together, adjusting the position of parts, and transmitting power. **Screws, sometimes called threaded fasteners, should be purchased rather than made as newly designed parts for each product.** Screws are available through commercial catalogs in countless forms and shapes for various specialized and general applications. Such screws are cheap, easy to replace, and interchangeable.

The types of threaded parts most often encountered in industry are covered by current ANSI Standards and include both Unified National (UN) and International Organization for Standardization (ISO) threads. Adoption of the UN thread in 1948 by the United States, Great Britain, and Canada (sometimes called the ABC Standards), a modification of the American Standard and the Whitworth thread, was a major step in standardizing threads. The ISO developed metric standards to unify thread specifications for even broader worldwide applications.

Other types of fasteners include **keys** and **rivets**. **Springs** resist and react to forces and have applications varying from pogo sticks to automobiles. Springs also are available in many forms and styles from specialty manufacturers who supply most of them to industry.

17.2 Threads

Terminology
Understanding threaded parts begins with learning their terminology, which we use throughout this chapter.

External thread: a thread on the outside of a cylinder, such as a bolt **(Fig. 17.1)**.

Internal thread: a thread cut on the inside of a part, such as a nut (Fig. 17.1).

Major diameter: the largest diameter on an internal or external thread **(Fig. 17.2)**.

Minor diameter: the smallest diameter on an internal or external thread (Fig. 17.2).

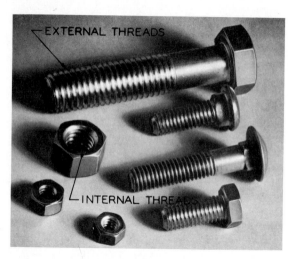

Figure 17.1 This photo shows examples of external threads (bolts) and internal threads (nuts). (Courtesy of Russell, Burdsall & Ward Bolt and Nut Company.)

THREAD TERMINOLOGY

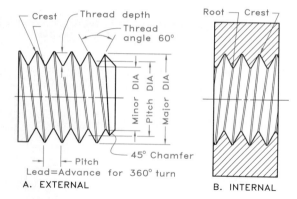

Figure 17.2 This drawing illustrates thread terminology for (A) external and (B) internal threads.

Pitch diameter: the diameter of an imaginary cylinder passing through the threads at the points where the thread width is equal to the space between the threads (Fig. 17.2 and Fig. 17.4).

Lead (pronounced leed): the distance a screw will advance when turned 360°.

THREAD FORMS

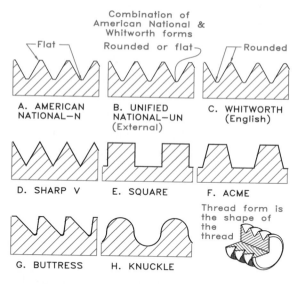

Figure 17.3 This drawing depicts standard thread forms for external threads.

Pitch (thread width): the distance between crests of threads, found by dividing 1 inch by the number of threads per inch of a particular thread (Fig. 17.2).

Crest: the peak edge of a screw thread (Fig. 17.2).

Thread angle: the angle between threads cut by the cutting tool, usually 60° (Fig. 17.2).

Root: the bottom of the thread cut into a cylinder to form the minor diameter (Fig. 17.2).

Thread form: the shape of the thread cut into a threaded part (Fig. 17.3).

Thread depth: the depth of the thread from the major diameter to the minor diameter; also measured as the root diameter (Fig. 17.2).

Thread series: the number of threads per inch for a particular diameter, grouped into coarse, fine, extra fine, and eight constant-pitch thread series.

Thread class: the closeness of fit between two mating parts. Class 1 represents a loose fit and Class 3 a tight fit.

Right-hand thread: one that will assemble when turned clockwise. A right-hand external thread slopes downward to the right when its axis is horizontal and in the opposite direction on internal threads.

Left-hand thread: one that will assemble when turned counterclockwise. A left-hand external thread slopes downward to the left when its axis is horizontal and in the opposite direction on internal threads.

Specifications (English System)

Form **Thread form** is the shape of the thread cut into a part (**Fig. 17.3**). The Unified National form is the most widely used form in the United States and is denoted UN in thread notes. The American National form is denoted by N, when it appears occasionally on old drawings.

Transmission of power is achieved with the Acme, square, and buttress threads, which are commonly used in gearing and other machinery applications (Fig. 17.3). The sharp V thread is used for set screws and in applications where friction in assembly is desired.

The Unified National Rolled form, denoted UNR, is specified only for external threads, never for internal threads, because internal threads cannot be formed by rolling. The standard UN form has a flat root (a rounded root is optional) (**Fig. 17.4A**), and the UNR form (**Fig. 17.4B**) has a rounded root that is formed by rolling a cylinder across a die. The UNR form is used instead of the UN form where precision of assembly is less critical.

Series The **thread series** designates the spacing of threads that vary with diameter. The American National and the Unified National (UN/UNR) forms include three graded series: coarse (C), fine (F), and extra fine (EF). Eight constant-pitch series (4, 6, 8, 12, 16, 20, 28, and 32 threads per inch) are also available.

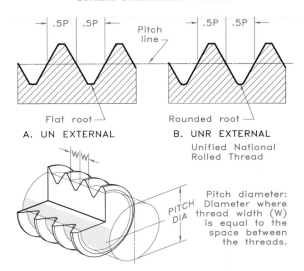

Figure 17.4
A The UN external thread has a flat root (rounded root is optional) and a flat crest.

B The UNR thread has a rounded root formed by rolling rather than cutting. The UNR form does not apply to internal threads.

A **coarse** Unified National form thread is denoted UNC or UNRC, which is a combination form and series designation. Thus an American National (N) form for a coarse thread is written NC. The coarse thread (UNC/UNRC or NC) has the largest pitch of any series and is suitable for bolts, screws, nuts, and general use with cast iron, soft metals, and plastics when rapid assembly is desired.

Fine threads (NF or UNF/UNRF) are used for bolts, nuts, and screws when a high degree of tightening is required. Fine threads are closer together than coarse threads, and their pitch is graduated to be smaller on smaller diameters.

Extra-fine threads (UNEF/UNREF or NEF) are suitable for sheet metal screws and bolts, thin nuts, ferrules, and couplings when the

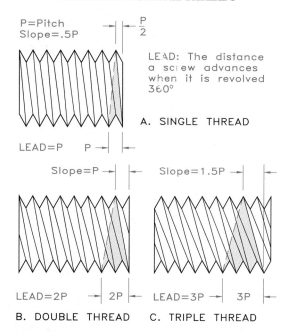

Figure 17.5 This drawing shows (A) single, (B) double, and (C) triple threads.

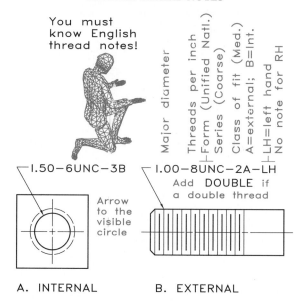

Figure 17.6 These English thread notes apply to (A) internal and (B) external threads.

length of engagement is limited and high stresses must be withstood.

Constant-pitch threads (4 UN, 6 UN, 8 UN, 12 UN, 16 UN, 20 UN, 28 UN, and 32 UN) are used on larger diameter threads (beginning near the 1/2-inch size) and have the same pitch size regardless of the diameter size. The most commonly used constant-pitch threads are 8 UN, 12 UN, and 16 UN members of the series, which are used on threads of about 1 inch in diameter and larger. Constant-pitch threads may be specified as UNR or N thread forms. The ANSI table in the Appendix shows constant-pitch threads that can be used on larger thread diameters instead of graded pitches of coarse, fine, and extra fine.

Class of Fit   The **class of fit** is the tightness of fit between two mating threads, as between a nut

and bolt, and is indicated in the thread note by the numbers 1, 2, or 3 followed by the letters A or B. **For UN forms, the letter A represents an external thread and the letter B represents an internal thread.** The letters A and B do not appear in notes for the now obsolete American National form (N).

Class 1A and 1B threads are used on parts that assemble with a minimum of binding and precision. Class 2A and 2B threads are general-purpose threads for bolts, nuts, and screws used in general and mass-production applications. Class 3A and 3B threads are used in precision assemblies where a close fit is required to withstand stresses and vibration.

Single and Multiple Threads   A **single thread (Fig. 17.5A)** is a thread that advances the distance of its pitch in a revolution of 360°; that is, its pitch is equal to its lead. The crest lines have a slope of 1/2 P since only 180° of the revolution is visible in the view.

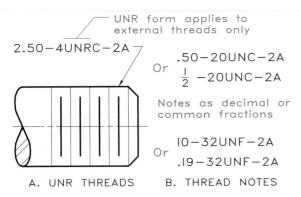

APPLICATION OF NOTES

UNR form applies to external threads only

2.50−4UNRC−2A

Or
.50−20UNC−2A

$\frac{1}{2}$ −20UNC−2A

Notes as decimal or common fractions

Or
10−32UNF−2A

.19−32UNF−2A

A. UNR THREADS   **B. THREAD NOTES**

Figure 17.7

A The UNR thread notes apply to external threads only.

B Thread notes may be given as decimal fractions or common fractions.

AMERICAN NATIONAL STANDARDS INSTITUTE (ANSI) UNIFIED INCH SCREW THREADS (UN AND UNR)

Nominal Diameter	Basic Diameter	Coarse NC & UNC		Fine NF & UNF		Extra Fine NEF/UNEF	
		Thds per In.	Tap Drill DIA	Thds per In.	Tap Drill DIA	Thds per In.	Tap Drill DIA
1	1.000	8	.875	12	.922	20	.953
1−1/16	1.063	...	...	...	...	18	1.000
1−1/8	1.125	7	.904	12	1.046	18	1.070
1−3/16	1.188	...	...	...	...	18	1.141
1−1/4	1.250	7	1.109	12	1,172	18	1.188
1−5/16	1.313	...	...	...	...	18	1.266
1−3/8	1.375	6	1.219	12	1.297	18	1.313
1−7/16	1.438	...	...	...	...	18	1.375
1−1/2	1.500	6	1.344	12	1.422	18	1.438

Figure 17.8 This portion of the ANSI tables for UN and UNR threads is from the Appendix.

Multiple threads are used wherever quick assembly is required. A **double thread** is composed of two threads that advance a distance of 2P when turned 360° (**Fig. 17.5B**); that is, its lead is equal to 2P. The crest lines have a slope of P because only 180° of the revolution is visible in the view. A **triple thread** advances 3P in 360° with a crest line slope of 1-1/2 P in the view where 180° of the revolution is visible (**Fig. 17.5C**).

Notes

Drawings of threads are only symbolic representations and are inadequate to give the details of a thread unless accompanied by notes (Fig. 17.6). In a thread note, first give the major diameter, then the number of threads per inch, the form and series, the class of fit, and the letter A or B to denote external or internal threads, respectively. For a double or triple thread, include the word **double** or **triple** in the note, and for left-hand threads add the letters LH.

Figure 17.7 shows a UNR thread note for the external thread. (UNR does not apply to internal threads.) When inches are the unit of measurement, write thread notes as decimal fractions,

although common fractions may be used. The information for thread notes comes from ANSI tables for English thread specifications.

Using Thread Tables

Figure 17.8 shows a portion of the Appendix, which gives the UN/UNR thread table from which specifications for standardized interchangeable threads can be selected. Note that a 1-1/2-inch-diameter bolt with fine thread (UNF) has 12 threads per inch. Therefore the thread note is written as

1.500-12 UNF-2A or 1-1/2-12 UNF-2A.

If the thread were internal (nut), the thread note would be the same, but you would use the letter B instead of the letter A.

For constant-pitch thread series selected for larger diameters, write the thread note as

1.750-12 UN-2A or 1-3/4-12 UN-2A.

For the UNR thread form (for external threads only) simply substitute UNR for UN in the last three columns, as, for example, UNREF for UNEF (extra fine). **Figure 17.9** shows the preferred

PLACEMENT OF THREAD NOTES

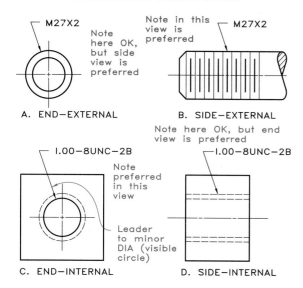

Figure 17.9 Place notes for external threads on the rectangular view of the threads. Place thread notes for internal threads in the circular view if space permits.

METRIC THREAD TABLE

COARSE		FINE	
MAJ. DIA & THD PITCH	TAP DRILL	MAJ. DIA & THD PITCH	TAP DRILL
M20 X 2.5	17.5	M20 X 1.5	18.5
M22 X 2.5	19.5	M22 X 1.5	20.5
M24 X 3	21.0	M24 X 2	22.0
M27 X 3	24.0	M27 X 2	25.0
M30 X 3.5	26.5	M30 X 2	28.0
M33 X 3.5	29.5	M33 X 2	31.0
M36 X 4	32.0	M36 X 2	33.0
M39 X 4	35.0	M39 X 2	36.0
M42 X 4.5	37.5	M42 X 2	39.0

Figure 17.10 This portion of the ISO tables, which present dimension specifications for metric threads, is from Appendix 10.

placement of thread notes (with leaders) for external and internal threads.

Metric Thread Notes

Metric thread notes usually given as a basic designation are suitable for general applications. However, for applications where the assembly of threaded parts is crucial, a complete thread designation should be noted. The ISO thread table in Appendix 10, a portion of which is shown in **Fig. 17.10**, contains specifications for metric thread notes.

Basic Designation **Figure 17.11** shows examples of metric screw thread notes. Each note begins with the letter M, designating the note as metric, followed by the major diameter size in millimeters and the pitch in millimeters separated by the multiplication sign, ×.

Complete Designation The first part of a complete designation note (**Fig. 17.12**) is identical to that for the basic designation, to which is added the toler-

METRIC THREAD NOTES

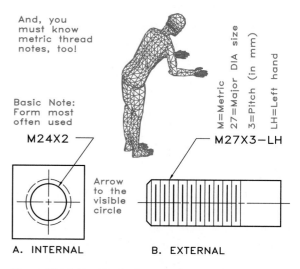

Figure 17.11 Use this basic type of metric thread note for (A) internal and (B) external threads.

ance class designation separated by a dash. (A **tolerance** is a specified maximum variation in size that ensures assembly of the threaded parts.) The 5g represents the pitch diameter tolerance, and the 6g represents the crest diameter tolerance.

COMPLETE METRIC DESIGNATION

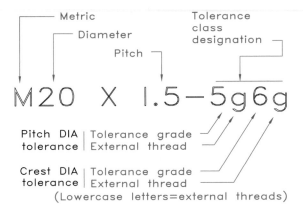

Figure 17.12 A complete designation note for metric threads adds tolerance specifications to the basic note.

TOLERANCE GRADES: METRIC THREADS

External Thread			Internal Thread	
Major Diameter	Pitch Diameter		Minor Diameter	Pitch Diameter
–	3		–	–
4	4	FINE	4	4
–	5		5	5
6	6	MEDIUM	6	6
–	7		7	7
8	8	COARSE	8	8
–	9		–	–

Figure 17.13 Tolerance grades for major and minor diameters and pitch diameters of each range from fine to coarse. Tolerance grades combined with position symbols (such as 6g) become tolerance classes.

The numbers 5 and 6 are thread tolerance grades (variations in size from the basic diameter) from **Fig. 17.13**. These grades are for the pitch diameter and the major and minor diameters of medium general-purpose thread similar to class 2A and 2B threads in the UN system. Grades of less than 6 are best suited to fine-series fits and

POSITION, ALLOWANCE, AND ENGAGEMENT

EXTERNAL THREADS (Lowercase letters)	INTERNAL THREADS (Uppercase letters)
e = Large allowance	G = Small allowance
g = Small allowance	H = No allowance
h = No allowance	

LENGTH OF ENGAGEMENT

S = Short   N = Normal   L = Long

EXAMPLE: Refer to Figure 17.13
Position   Lower-case = external threads
                Upper-case = internal threads
Allowance e, g, h, G, H = amount of allowance
Engagement S, N, & L Columns of Fig. 17.13

Figure 17.14 Use symbols to represent position, allowance, and engagement length. Position means either external or internal threads. A lowercase letter indicates an external thread and an uppercase letter signifies an internal thread.

short lengths of engagement. Grades greater than 6 are best suited to coarse-series fits and long lengths of engagement.

The letters following the grade numbers denote tolerance positions, that is, either external or internal threads. Lowercase letters designate external threads (bolts), as **Fig. 17.14** shows. The letters e, g, and h represent large allowance, small allowance, and no allowance, respectively. (**Allowance** is the permitted variation in size from the basic diameter.) Uppercase letters designate internal threads (nuts). The letters G and H, placed after the tolerance grade number, denote small allowance and no allowance, respectively. For example, 5g designates a medium tolerance with small allowance for the pitch diameter of an external thread, and 6H designates a medium tolerance with no allowance for the minor diameter of an internal thread.

Tolerance classes are fine, medium, and coarse (**Fig. 17.15**). They represent combinations of tolerance grades, tolerance positions, and lengths of engagement—short (S), normal (N), and long (L). Appendix 24 contains a table of lengths of engagement (the actual length of the assembled thread in

	External Threads (bolts)									Internal Threads (nuts)					
Quality	Tolerance position e (large allowance)			Tolerance position g (small allowance)			Tolerance position h (no allowance)			Tolerance position G (small allowance)			Tolerance position H (no allowance)		
	Length of engagement			Length of engagement			Length of engagement			Length of engagement			Length of engagement		
	Group S	Group N	Group L	Group S	Group N	Group L	Group S	Group N	Group L	Group S	Group N	Group L	Group S	Group N	Group L
Fine							3h4h	4h	5h4h				4H	5H	6H
Medium		**6e**	7e6e	5g6g	**6g**	7g6g	5h6h	6h	7h6h	5G	6G	7G	**5H**	**6H**	7H
Coarse					8g	9g8g					7G	8G		7H	8H

*In selecting tolerance class, select first from the commercial classes in boxes, second from the bold print, third from the medium-size print, and fourth from the small-size print.

Figure 17.15 Tolerance classes for large and medium allowances and for no allowance for internal and external threads are shown here. Select the most commonly used tolerance class, commercial threads (in boxes) first, then the classes in bold print, medium-size print, and small-size print, respectively.

mating parts). After deciding whether to use a fine, medium, or coarse class of fit, you should select the thread designation first from those in boxes (commercial threads), second from those in bold print, third from those in medium-size print, and fourth from those in small print. The 6H class is comparable to the 2B class of fit in the UN system for an interior thread, and the 6g class is similar to the 2A class of fit.

Figure 17.16 shows variations for complete designation thread notes. When the minor and pitch diameters have identical grades, the tolerance class symbol consists of one number and letter, such as 6H **(Fig. 17.16A)**. The uppercase H indicates that the position is internal, with no allowance added to the basic thread note. If necessary, add the length of engagement symbol (S, N, L) to the tolerance class designation **(Fig. 17.16B)**. For unknown lengths of thread engagement, use group N (normal).

TOLERANCE AND ENGAGEMENT SYMBOLS

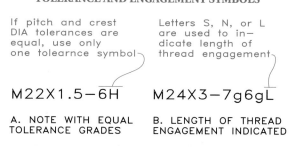

If pitch and crest DIA tolerances are equal, use only one tolearnce symbol

Letters S, N, or L are used to indicate length of thread engagement

M22X1.5—6H

M24X3—7g6gL

A. NOTE WITH EQUAL TOLERANCE GRADES

B. LENGTH OF THREAD ENGAGEMENT INDICATED

Figure 17.16 Complete notes:
A When both pitch and crest diameter tolerance grades are the same, show the tolerance class symbol only once.
B The letters S, N, and L indicate the length of the thread engagement.

Specify the fit between mating threads as shown in **Fig. 17.17**, with a slash separating the tolerance class designations of internal and external threads. Additional information about ISO

MATING THREADS

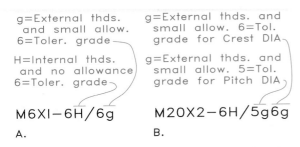

g=External thds.
and small allow.
6=Toler. grade

H=Internal thds.
and no allowance
6=Toler. grade

g=External thds. and
small allow. 6=Tol.
grade for Crest DIA

g=External thds. and
small allow. 5=Tol.
grade for Pitch DIA

M6X1—6H/6g

A.

M20X2—6H/5g6g

B.

Figure 17.17 A slash mark is used to separate the tolerance class designations of mating internal and external threads.

THREAD SYMBOLS

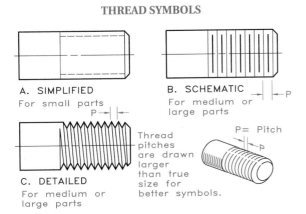

A. SIMPLIFIED
For small parts

B. SCHEMATIC
For medium or
large parts

C. DETAILED
For medium or
large parts

Thread
pitches
are drawn
larger
than true
size for
better symbols.

P= Pitch

Figure 17.18 The three types of thread symbols used to represent threads are (A) simplified, (B) schematic, and (C) detailed.

threads may be obtained from *ISO Metric Screw Threads,* a booklet of standards published by ANSI that was used as the basis for most of this section.

17.3 Drawing Threads

Threads may be represented by detailed, schematic, and simplified symbols (Fig. 17.18). Detailed symbols represent a thread most realistically, simplified symbols represent a thread least realistically, and schematic symbols are a compromise between the two.

DETAILED SYMBOLS: EXTERNAL

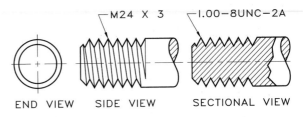

END VIEW   SIDE VIEW   SECTIONAL VIEW

Figure 17.19 These detailed symbols represent external threads in view and section.

DETAILED SYMBOLS: INTERNAL

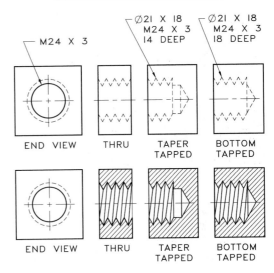

END VIEW   THRU   TAPER
TAPPED   BOTTOM
TAPPED

END VIEW   THRU   TAPER
TAPPED   BOTTOM
TAPPED

Figure 17.20 These detailed symbols represent internal threads. Approximate the minor diameter as 75% of the major diameter. Tap drill diameters (Ø21) are found in the Appendix.

Detailed Symbols

UN/UNR Threads **Figure 17.19** illustrates detailed thread symbols for external threads. Instead of helical curves, straight lines depict crest and root lines. **Figure 17.20** depicts detailed thread symbols for internal threads in views and sections.

DRAWING DETAILED THREADS

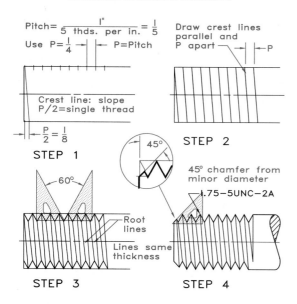

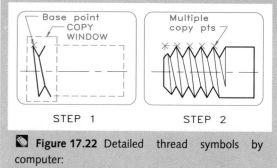

Figure 17.22 Detailed thread symbols by computer:

Step 1 Generate a typical detailed thread symbol at the end of the screw.

Step 2 Duplicate a typical set of threads with the COPY command and the MULTIPLE option along the predetermined snap points of the screw.

Figure 17.21 Detailed representation of threads:
Step 1 To draw a detailed representation of a 1.75-5 UNC-2A thread, determine the actual pitch by dividing 1 inch by the number of threads per inch, or 5 in this case. In this case, use a pitch of ¼ instead of ⅕ to avoid drawing the threads too close together. Lay off the pitch along the length of the thread and draw a crest line at a slope of P/2, or ⅛ inch in this case.

Step 2 Draw the other crest lines as dark, visible lines parallel to the first crest line.

Step 3 Find the root lines by constructing 60° vees between the crest lines. Draw the root lines from the bottom of the vees. Root lines are parallel to each other but not to crest lines.

Step 4 Construct a 45° chamfer at the end of the thread from the minor diameter. Darken all lines and add a thread note.

Figure 17.21 shows how to draw a detailed thread representation for both English and metric threads. When using the English tables, calculate the pitch by dividing 1 inch by the number of threads per inch. In the metric system, pitch is given in the tables. **You should draw the spacing between crest lines (the pitch) larger than the actual pitch size for a clear, uncrowded representation.**

Computer Method You can draw detailed thread symbols by computer **(Fig. 17.22)** and duplicate them with the COPY command's MULTIPLE option. The program produces a typical set of threads in step 1 and then copies it repetitively in step 2.

Square Threads **Figure 17.23** shows how to draw and notate a detailed representation of a square thread. Follow the same basic steps to draw views and sections of square internal threads **(Fig. 17.24)**. In section, draw both the internal crest and root lines, but in a view draw only the outline of the threads. Place thread notes for internal threads in the circular view, whenever possible, with the leader pointing toward the center and stopping at the visible circle.

When a square thread is long, you may represent it by using phantom lines, without drawing all the threads **(Fig. 17.25)**. This conventional practice saves time and effort without reducing the drawing's effectiveness.

Acme Threads A modified version of the square thread is the **Acme thread**, which has tapered (15°) sides for easier engagement than square

THE SQUARE THREAD

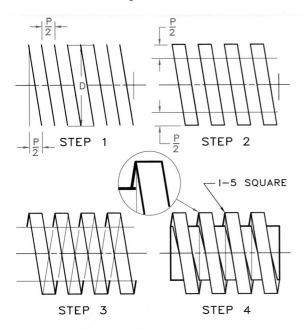

Figure 17.23 Drawing the square thread:
Step 1 Lay out the major diameter. Space the crest lines ½P apart and slope them downward to the right for right-hand threads.

Step 2 Connect every other pair of crest lines. Find the minor diameter by measuring ½P inward from the major diameter.

Step 3 Connect the opposite crest lines with light construction lines to establish the profile of the thread form.

Step 4 Connect the inside crest lines with light construction lines to locate the points on the minor diameter where the thread wraps around the minor diameter. Darken the final lines.

threads. **Figure 17.26** shows the steps involved in drawing detailed Acme threads. Like square threads, Acme threads are used for transmission of force and power, as in screw jacks and lathes. Appendix 11 contains the table for Acme thread specifications and dimensions.

Figure 17.27 shows internal Acme threads in view and section. Slope left-hand internal threads in section to appear the same as right-hand exter-

INTERNAL SQUARE THREADS

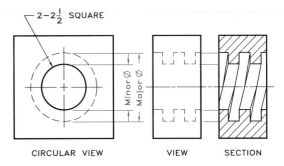

Figure 17.24 This drawing shows internal square threads in view and section.

SIMPLIFIED SQUARE THREADS

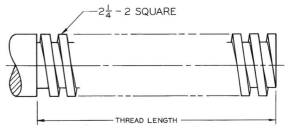

Figure 17.25 The conventional method of showing square threads is to draw sample threads at each end and to connect them with phantom lines.

nal threads. **Figure 17.28** shows a shaft being threaded with Acme threads on a lathe as the tool travels the length of the shaft.

Schematic Symbols
Figure 17.29 shows schematic representations of external threads, with metric notes. Draw schematic thread symbols by using thin parallel crest lines and thick root lines. Because schematic symbols are easy to draw and adequately represent threads, it is the thread symbol used most often for medium-size threads. Use the same schematic symbols for left-hand and right-hand threads and write LH in the thread note of left-

DRAWING ACME THREADS

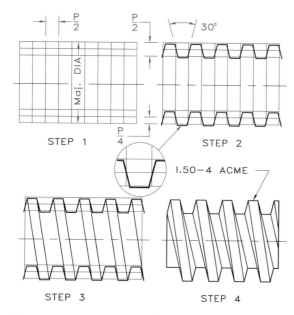

Figure 17.26 Drawing the Acme thread:

Step 1 Lay out the major diameter and thread length and divide the shaft into equal divisions ½P apart. Locate the minor and pitch diameters by using distances ½P and ¼P.

Step 2 Draw construction lines at 15° angles with the vertical along the pitch diameter to make a total angle of 30°.

Step 3 Draw the crest lines across the screw.

Step 4 Darken the lines, draw the root lines, and add the thread note to complete the drawing.

hand threads. Right-hand threads carry no designation in thread notes.

    Figure 17.30 shows schematic symbols representing internal threaded holes in view and section. The size of the tap drill diameter is approximately equal to the major diameter minus the pitch. However, the minor diameter usually is drawn a bit smaller to provide better separation between the circles representing the major and minor diameters.

    Figure 17.31 shows how to draw threads schematically for English specifications. Draw the

INTERNAL ACME THREADS

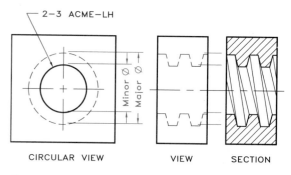

Figure 17.27 This drawing shows internal Acme threads in view and section.

Figure 17.28 This photo demonstrates cutting an Acme thread on a lathe. (Courtesy of Clausing Corporation.)

SCHEMATIC SYMBOLS: EXTERNAL

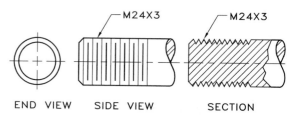

Figure 17.29 These schematic symbols represent external threads in view and section.

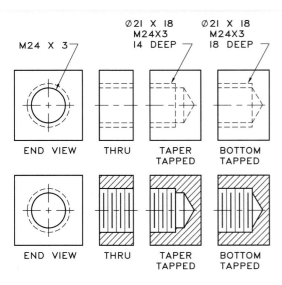

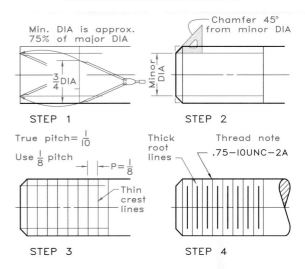

Figure 17.30 These schematic symbols represent internal threads in view and section. Tap drill diameters are found in the Appendix.

Figure 17.31 Schematic representation of threads:

Step 1 Lay out the major diameter and locate the minor diameter (about three-quarters of the major diameter). Draw the minor diameter as light construction lines.

Step 2 Chamfer the end of the threads with a 45° angle from the minor diameter.

Step 3 Find the true pitch of a .75-10UNC-2A thread (0.1) by dividing 1 inch by the number of threads per inch (10). Use a larger pitch, ⅛ inch in this case, to draw the thin crest lines ⅛ inch apart.

Step 4 Draw root lines as thick as the visible lines between the crest lines to the construction lines representing the minor diameter. Add a thread note.

minor diameter at approximately three-quarters of the major diameter and the chamfer (bevel) 45° from the minor diameter. Draw crest lines as thin lines and root lines as thick visible lines.

   Computer Method You may draw schematic thread symbols by computer **(Fig. 17.32)** and duplicate them with the COPY command's MULTI-PLE option. The program produces a typical set of threads in step 1 and copies it repetitively in step 2.

Simplified Symbols

Figure 17.33 illustrates the use of simplified symbols for external threads and a specifications note. **Figure 17.34** shows the use of simplified symbols for internal threads. **Simplified symbols are the easiest to draw and are the best suited for drawing small threads where schematic and detailed symbols would be crowded.** Draw the minor diameter as hidden lines spaced at about three-quarters the major diameter. **Figure 17.35** shows

the steps involved in drawing simplified threads. With experience, you will be able to approximate the location of the minor diameter of simplified threads by eye.

Drawing Small Threads

Instead of drawing small threads to actual size, draw minor diameters smaller to separate root and crest lines farther for a clear, uncrowded, easy to-draw representation (Fig. 17.36). Enlarge the thread's pitch to separate thread symbols when

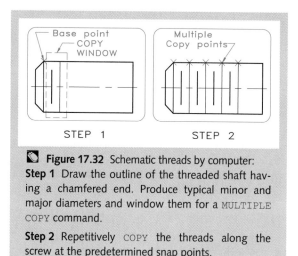

SIMPLIFIED SYMBOLS: INTERNAL

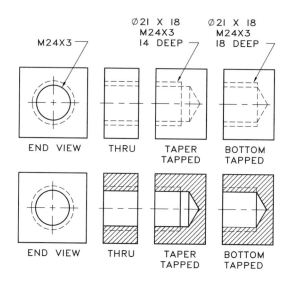

Figure 17.34 These simplified thread symbols represent internal threads in view and section. Draw minor diameters at about ¾ the major diameter.

SIMPLIFIED SYMBOLS: EXTERNAL

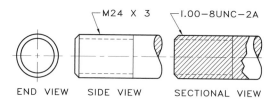

Figure 17.33 These simplified thread symbols represent external threads in view and section.

DRAWING SIMPLIFIED THREADS

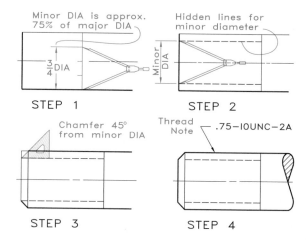

Figure 17.35 Simplified representation of threads:
Step 1 Lay out the major diameter. Locate the minor diameter (about three-quarters of the major diameter).

Step 2 Draw hidden lines to represent the minor diameter.

Step 3 Draw a 45° chamfer from the minor diameter to the major diameter.

Step 4 Darken the lines and add a thread note.

drawing all three types of symbols (simplified, schematic, and detailed). Add a thread note to the symbolic drawing to give the necessary specifications.

17.4 Nuts and Bolts

Nuts and bolts (**Fig. 17.37**) come in many forms and sizes and have many different applications. **Figure 17.38** depicts some common types of threaded fasteners. A **bolt** is a threaded cylinder with a head and is used with a **nut** to hold parts

CONVENTIONAL PRACTICES

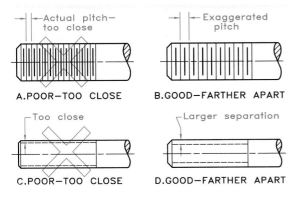

Figure 17.36 Most threads must be drawn using exaggerated dimensions instead of actual measurements to prevent the drawing from having lines drawn too closely together.

Figure 17.37 This photo shows a nut, a bolt, and washers in combination. (Courtesy of Lamson & Sessions.)

TYPES OF BOLTS AND SCREWS

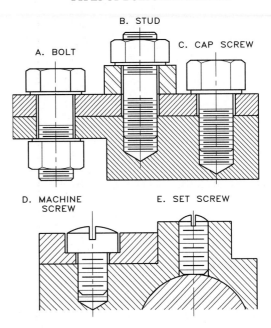

Figure 17.38 This drawing illustrates several types of threaded bolts and screws.

NUT AND BOLT HEAD TYPES

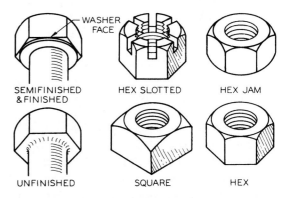

Figure 17.39 The types of finishes for nuts and bolt heads and several types of nuts.

together. A **stud** is a headless bolt, threaded at both ends, that is screwed into one part with a nut attached to the other end.

A **cap screw** usually does not have a nut, but passes through a hole in one part and screws into another threaded part. A **hexagon-head machine screw** is similar to, but smaller than, a cap screw. Machine screws also come with other types of heads. A **set screw** is used to hold one member

fixed in place with another, usually to prevent rotation, as with a pulley on a shaft.

Figure 17.39 shows the types of heads used on regular and heavy bolts and nuts. Heavy bolts

SCREW AND BOLT HEADS

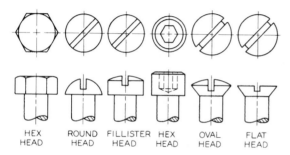

HEX   ROUND   FILLISTER HEX   OVAL   FLAT
HEAD   HEAD   HEAD   HEAD   HEAD   HEAD

Figure 17.40 These are common types of bolt and screw heads.

A DIMENSIONED BOLT

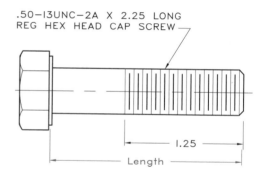

.50–13UNC–2A X 2.25 LONG
REG HEX HEAD CAP SCREW

1.25

Length

Figure 17.41 A properly dimensioned and noted hexagon-head bolt.

have thicker heads than regular bolts for heavier usage. A **finished head** (or nut) has a 1/64-inch-thick washer face (a circular boss) to provide a bearing surface for smooth contact. **Semifinished bolt heads** and nuts are the same as finished bolt heads and nuts. **Unfinished bolt heads** and nuts have no boss and no machined surfaces.

A **hexagon jam nut** does not have a washer face, but it is chamfered (beveled at its corners) on both sides. **Figure 17.40** shows other standard bolt and screw heads for cap screws and machine screws.

Dimensions

Figure 17.41 shows a properly dimensioned bolt. The ANSI tables in the Appendix give nut and bolt dimensions, but you may use the following guides for hexagon-head and square-head bolts.

Overall Lengths Hexagon-head bolts are available in 1/4-inch increments up to 8 inches long, in 1/2-inch increments from 8 to 20 inches long, and in 1-inch increments from 20 to 30 inches long. Square-head bolts are available in 1/8-inch increments from 1/2 to 3/4 inch long, in 1/4-inch increments from 3/4 inch to 5 inches long, in 1/2-inch increments from 5 to 12 inches long, and in 1-inch increments from 12 to 30 inches long.

Thread Lengths For both hexagon-head and square-head bolts up to 6 inches long,

Thread length = 2D + ¼ in.

where D is the diameter of the bolt. For bolts more than 6 inches long

Thread length = 2D + ½ inch.

Threads for bolts can be coarse, fine, or 8-pitch threads. The class of fit for bolts and nuts is understood to be 2A and 2B if no class is specified in the note.

Dimension Notes

Designate standard square-head and hexagon-head bolts by notes in one of three forms:

⅜-16 × 1½ SQUARE BOLT—STEEL;

½-13 × 3 HEX CAP SCREW—
SAE GRADE 8—STEEL;

.75 × 5.00 UNC-2A HEX HD LAG SCREW.

The numbers (left to right) represent bolt diameter, threads per inch (omit for lag screws), bolt length, screw name, and material (material designation is optional). When not specified in a note, each bolt is assumed to have a class 2 fit.

Three types of notes for designating nuts are

½-13 SQUARE NUT—STEEL;

¾-16 HEAVY HEX NUT;

1.00-8 UNC-2B HEX THICK SLOTTED
NUT—CORROSION-RESISTANT STEEL.

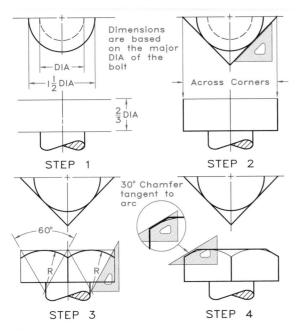

STEP 1 — STEP 2

STEP 3 — STEP 4

Figure 17.42 Drawing the square head:

Step 1 Draw the major diameter, DIA, of the bolt. Use 1.5 DIA to establish the hexagon-head's diameter and ⅔ DIA to establish its thickness.

Step 2 Draw the top view of the square head at a 45° angle to give an across-corners view.

Step 3 Show the chamfer in the front view by using a 30°–60° triangle to find the centers for the radii.

Step 4 Show a 30° chamfer tangent to the arcs in the front view. Darken the lines.

When nuts are not specified as heavy, they are assumed to be regular. When the class of fit is not specified in a note, it is assumed to be 2B for nuts.

Drawing Square Heads

The Appendix gives dimensions for square bolt heads and nuts. However, conventional practice is to draw nuts and bolts by using the general proportions shown in **Fig. 17.42**. The first step in drawing a bolt head or nut is to determine whether the view is to be across corners or across flats—

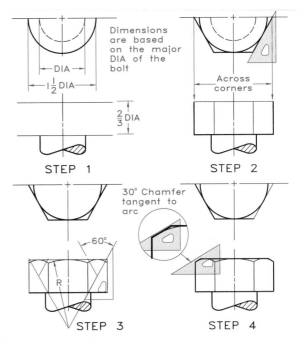

STEP 1 — STEP 2

STEP 3 — STEP 4

Figure 17.43 Drawing the hexagon head:

Step 1 Draw the major diameter, DIA, of the bolt and use it to establish the head diameter (as 1.5 DIA) and thickness (as ⅔ DIA).

Step 2 Construct a hexagon head with a 30°–60° triangle to give an across-corners view.

Step 3 Find arcs in the front view to show the chamfer of the head.

Step 4 Draw a 30° chamfer tangent to the arcs in the front view. Darken the lines.

that is, whether the lines at either side of the view represent the square's corners or flats. Drawing across corners represents nuts and bolts best, but ccasionally you must draw one across flats when the head or nut is truly in this orientation.

Drawing Hexagon Heads

Figure 17.43 shows how to draw the head of a hexagon bolt across corners by using the bolt's major diameter, D, as the basis for all other proportions. Begin by drawing the top view of the head as

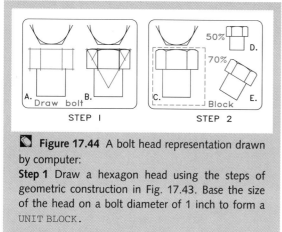

STEP 1   STEP 2

Figure 17.44 A bolt head representation drawn by computer:

Step 1 Draw a hexagon head using the steps of geometric construction in Fig. 17.43. Base the size of the head on a bolt diameter of 1 inch to form a UNIT BLOCK.

Step 2 Convert the drawing into a BLOCK by using a window. INSERT the block at any size and position as, for example, at D and E.

a circle of a diameter of 1-1/2 D. For a regular head the thickness is 2/3 D and for a heavy head it is 7/8 D. Circumscribe a hexagon about the circle. Then draw outside arcs in the rectangular view and tangent chamfers (bevels) to complete the drawing.

Computer Method You may draw a hexagon head for a bolt by computer, as shown in step 1 of **Fig. 17.44**. Use a bolt diameter of 1 inch for easy scaling. You may then scale and rotate the block as desired when you use the INSERT command. Step 2 shows the drawing BLOCKed and inserted at scales of 50% and 75%. To insert a thread with a diameter of 0.50 inch, assign a size factor of 0.50 when prompted by the INSERT and BLOCK commands.

Drawing Nuts

Use the same techniques to draw a square and a hexagon nut (shown across corners in **Fig. 17.45**) that you did to draw bolt heads. However, nuts are thicker than bolt heads: The thickness of a regular nut is 7/8 D, and the thickness of a heavy nut is 1 D, where D is the bolt diameter. You may insert hidden lines in the front view to indicate threads, or you

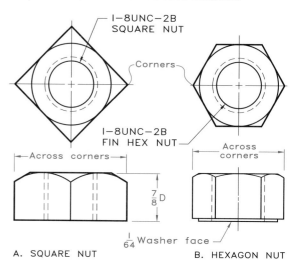

A. SQUARE NUT   B. HEXAGON NUT

Figure 17.45 Drawing square and hexagon nuts across corners involves the same steps used for drawing bolt heads. Add notes to give nut specifications.

DRAWING NUTS ACROSS FLATS

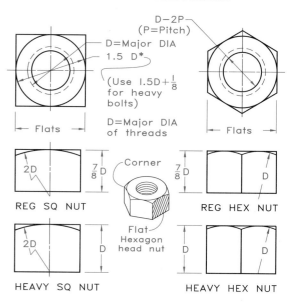

Figure 17.46 These square and hexagon nuts are drawn across flats, with notes added to give nut specifications. Square nuts always are unfinished.

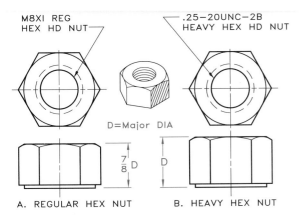

Figure 17.47 These regular and heavy hexagon nuts are drawn across corners, with specification notes added.

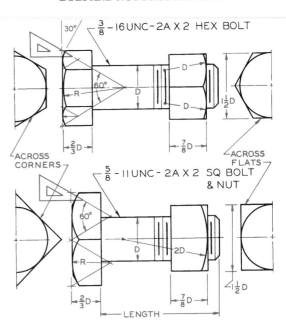

Figure 17.48 This drawing shows hexagon nuts and bolts and square nuts and bolts in assembly and their relative proportions.

may omit them because threading is understood. **Exaggerate the thickness of the 1/64-inch washer face on the finished and semifinished hexagon nuts to about 1/32 inch to make it more noticeable.** Place thread notes on circular views with leaders when space permits. Square nuts that are not labeled heavy are assumed to be regular nuts.

Figure 17.46 shows how to construct square and hexagon nuts across flats. For regular nuts, the distance across flats is 1-$1/2 \times D$ (D = major diameter of the thread), and 1-$5/8$ D for heavy nuts. Draw the top views in the same way you did across-corner top views, but rotate them to give across-flat front views. **Figure 17.47** depicts dimensioned and noted hexagon regular and heavy nuts drawn across corners.

Drawing Nut and Bolt Combinations

Apply the methods of drawing nuts and bolts to drawing nuts and bolts in assembly **(Fig. 17.48)**. Use the major diameter, D, of the bolt as the basis for other dimensions. Add a note to give the specifications of the nut and bolt. Here, the views of the bolt heads are across corners, and the views of the nuts are across flats, although both views could have been drawn across corners. Use the half end views to find the front views by projection.

17.5 Screws

Cap Screws

Cap screws are used to hold two parts together without a nut. The cap screw passes through a hole in one part and screws into a threaded hole in the other part. Cap screws are usually larger than machine screws and may also be used with nuts. **Figure 17.49** shows the standard types of cap screw heads drawn on a grid that can be used as a guide for drawing cap screws of other sizes. The Appendix gives cap screw dimensions, which can aid in drawing them.

Machine Screws

Smaller than most cap screws, **machine screws** usually are less than 1 inch in diameter. They screw into a threaded hole in a part or into a nut. Machine screws are fully threaded when their length is 2 inches or less. Longer screws have

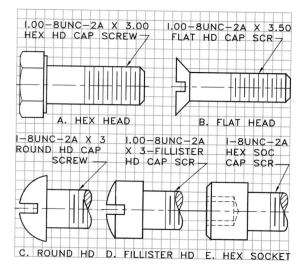

Figure 17.49 These cap screws are drawn on a grid to give the proportions for drawing them at different sizes. Notes give thread specifications, length, head type, and bolt name (cap screw).

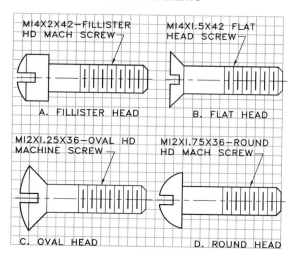

Figure 17.50 These are standard types of machine screws. The same proportions may be used to draw machine screws of all sizes.

thread lengths of 2D + 1/4 inch (D = major diameter of the thread). **Figure 17.50** shows four types of machine screws, along with notes, drawn on a grid that may be used as an aid in drawing them without dimensions from a table. Machine screws range in diameter from No. 0 (0.060 inch) to 3/4 inch, as shown in Appendix 22, which gives the dimensions of round-head machine screws.

Set Screws

Set screws are used to hold parts, such as pulleys and handles on a shaft, together and prevent rotation. **Figure 17.51** shows various types of set screws, with dimensions denoted by letters that correspond to the tables of dimensions in the Appendix.

Set screws are available in combinations of points and heads. The shaft against which the set screw is tightened may have a machined flat surface to provide a good bearing surface for a **dog** or **flat-point** set screw end to press against. The cup point

gives good friction when pressed against round shafts. The **Cone point** works best when inserted into holes drilled in the part being held. The **headless set screw** has no head to protrude above a rotating part. An **exterior square head** is good for applications in which greater force must be applied with a wrench to hold larger set screws in position.

Wood Screws

A **wood screw** is a p inted screw having sharp coarse threads that will screw into wood while making its own internal threads. **Figure 17.52** shows the three most common types of wood screws drawn on a grid to show their relative proportions.

Sizes of wood screws are specified by single numbers, such as 0, 6, or 16. From 0 to 10, each digit represents a different size. Beginning at 10, only even-numbered sizes are standard, that is, 10, 12, 14, 16, 18, 20, 22, and 24. Use the following formula to translate these numbers into the actual diameter sizes:

STANDARD SETSCREWS

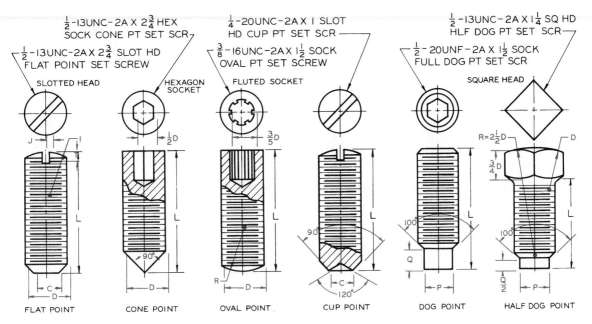

Figure 17.51 Setscrews are available with various combinations of heads and points. Notes give their measurements. (See the Appendix.)

WOOD SCREWS

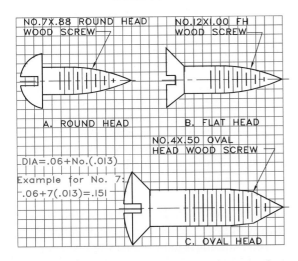

Figure 17.52 Standard types of wood screws are drawn on a grid that gives the proportions for drawing them at other sizes.

Actual DIA = 0.06 + (screw number × 0.013).

For example, the diameter for the No. 7 wood screw shown in Fig. 17.52 is calculated as follows:

$$DIA = 0.06 + 7(0.013) = 0.151.$$

17.6 Other Threaded Fasteners

We have space to cover only the more common types of nuts and bolts in this chapter. **Figure 17.53** shows a few of the many other types of threaded fasteners that have their own special applications. Three types of wing screws that are turned by hand are available in incremental lengths of 1/8 inch (**Fig. 17.54**). **Figure 17.55** shows two types of thumb screws, which serve the same purpose as wing screws, and **Fig. 17.56** shows wing nuts that can be screwed together by fingertip without wrenches or screwdrivers.

MISCELLANEOUS SCREWS

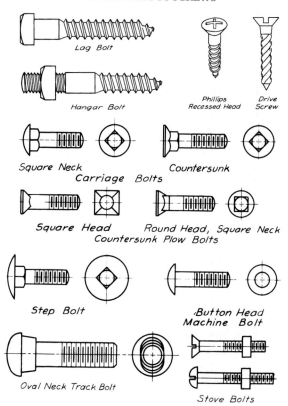

Figure 17.53 This drawing illustrates miscellaneous types of bolts and screws.

WING SCREWS

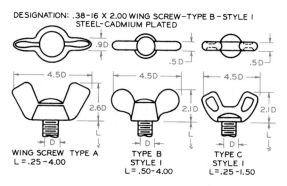

Figure 17.54 These wing screw proportions are for screw diameters of about 5/16 inch. The same proportions may be used to draw wing screws of any diameter. Type A screws are available in diameters of 4, 6, 8, 10, 12, 0.25", 0.313", 0.375", 0.438", 0.50", and 0.625". Type B screws are available in diameters of 10 to 0.625". Type C screws are available in diameters of 6 to 0.375".

THUMB SCREWS

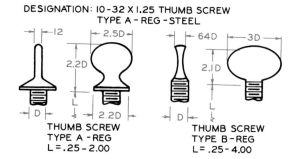

Figure 17.55 These thumb screw proportions are for screw diameters of about ¼ inch. The same proportions may be used to draw thumb screws of any diameter. Type A screws are available in diameters of 6, 8, 10, 12, 0.25", 0.313", and 0.375". Type B thumb screws are available in diameters of 6 to 0.50".

17.7 Tapping a Hole

An internal thread is made by drilling a hole with a tap drill with a 120° point **(Fig. 17.57)**. **The depth of the drilled hole is measured to the shoulder of the conical point, not to the point.** The diameter of the drilled hole is approximately equal to the root diameter, calculated as the major diameter of the screw thread minus its pitch. (See Appendix.) The hole is **tapped**, or threaded, with a tool called a **tap** of one of the types shown.

The taper, plug, and bottoming hand taps have identical measurements, except for the chamfered portion of their ends. The **taper tap** has a long chamfer (8 to 10 threads), the **plug tap** has a shorter

chamfer (3 to 5 threads), and the **bottoming tap** has the shortest chamfer (1 to 1-1/2 threads).

When tapping is to be done by hand in open or "through" holes, the taper tap should be used for coarse threads and in harder metals because it ensures straighter alignment and starting. The

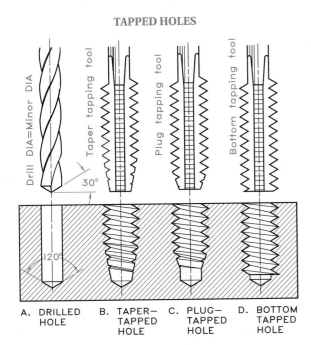

DESIGNATION: 10-32 TYPE A WING NUT-REG SERIES
STEEL-ZINC PLATED

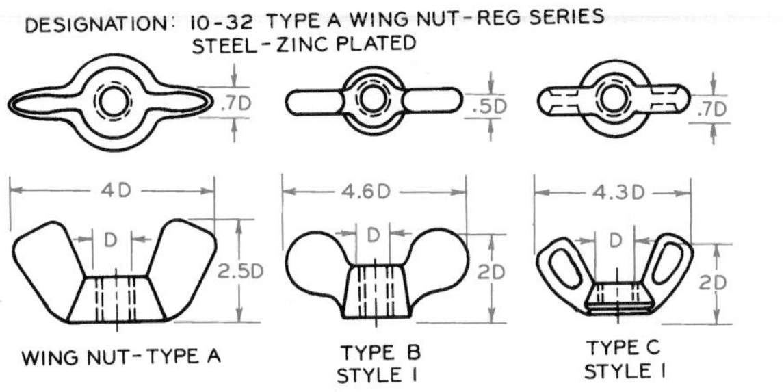

Figure 17.56 These wing nut proportions are for screw diameters of ⅜ inch. The same proportions may be used to draw thumb screws of any size. Type A wing nuts are available in screw diameters of 3, 4, 5, 6, 8, 10, 12, 0.25″, 0.313″, 0.375″, 0.438″, 0.50″, 0.583″, 0.625″, and 0.75″. Type B nuts are available in sizes from 5 to 0.75″. Type C nuts are available in sizes from 4 to 0.50″.

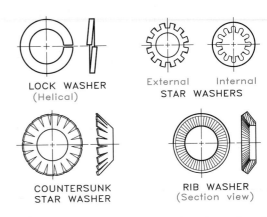

Figure 17.58 Lock washers are used to keep threaded parts from vibrating apart.

TAPPED HOLES

plug tap may be used in soft metals and for fine-pitch threads. **When a hole is tapped to its bottom, all three taps—taper, plug, and bottoming—are used in that sequence on the same internal threads.**

Notes may be added to specify the depth of a drilled hole and the depth of the threads within it. For example, a note reading 7/8 DIA-3 DEEP × 1–8 UNC-2A × 2 DEEP means that the hole is to be drilled deeper than it is threaded and that the last usable thread will be 2 inches deep in the hole.

17.8 Washers, Lock Washers, and Pins

Various types of **washers** are used with nuts and bolts to improve their assembly and increase their fastening strength. Plain washers are noted on a drawing as

.938 × 1.750 × 0.134 TYPE A PLAIN WASHER,

where the numbers (left to right) represent the washer's inside diameter, outside diameter, and thickness. (See Appendix 33.)

Lock washers reduce the likelihood that threaded parts will loosen because of vibration and movement. **Figure 17.58** shows several common types of lock washers. Appendix 33 contains

Figure 17.57 Three types of tapping tools are used to thread internal drilled holes: taper tap, plug tap, and bottom tap.

TYPES OF PINS

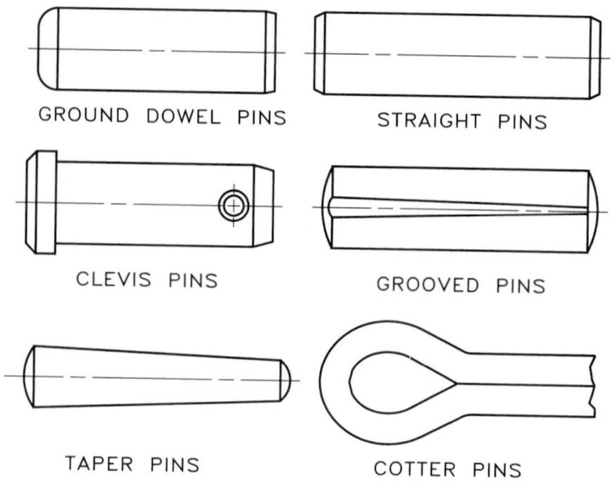

GROUND DOWEL PINS

STRAIGHT PINS

CLEVIS PINS

GROOVED PINS

TAPER PINS

COTTER PINS

Figure 17.59 Pins are used to hold parts together in assembly.

PIPE THREAD SYMBOLS

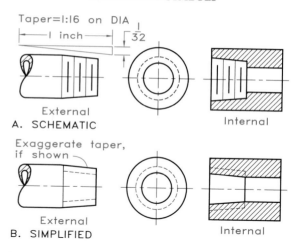

Taper=1:16 on DIA

1 inch

1/32

External

A. SCHEMATIC

Internal

Exaggerate taper, if shown

External

B. SIMPLIFIED

Internal

Figure 17.60 These schematic and simplified thread symbols represent pipe threads; NP stands for National Pipe.

a table of dimensions for regular and extra-heavy-duty helical-spring lock washers. Designate them with a note in the form:

HELICAL-SPRING LOCK WASHER
1/4 REGULAR—PHOSPHOR BRONZE,

where the 1/4 is the washer's inside diameter. Designate tooth lock washers with a note in one of two forms:

INTERNAL-TOOTH LOCK WASHER
1/4-TYPE A—STEEL;

EXTERNAL-TOOTH LOCK WASHER-562-TYPE B—STEEL.

Pins (Fig. 17.59) are used to hold parts together in a fixed position. Appendix 27 gives dimensions for taper pins. The **cotter pin** is another locking device that everyone who has had a toy wagon is familiar with. Appendix 28 contains a table of dimensions for cotter pins.

17.9 Pipe Threads and Fittings

Pipe threads are used for screwing together connecting pipes and tubing and for lubrication fittings. The most commonly used pipe thread is tapered at a ratio of 1 to 16 on its diameter, but straight pipe threads also are available **(Fig. 17.60).** Tapered pipe threads will engage only for an effective length of

$$L = (0.80D + 6.8)P,$$

where D is the outside diameter of the threaded pipe and P is the pitch of the thread.

The pipe threads shown in Fig. 17.60 have a taper exaggerated to 1:16 on radius (instead of on diameter) to emphasize it. Drawing them with no taper obviously is easier. You may use either schematic or simplified symbols to show the threaded features.

Use the following ANSI abbreviations in pipe thread notes. All begin with NP (for National Pipe thread).

NPT: national pipe taper

NPTF: national pipe thread (dryseal, for pressure-tight joints)

NPS: straight pipe thread

NPSC: straight pipe thread in couplings

NPSI: national pipe straight internal thread

PIPE THREAD NOTES

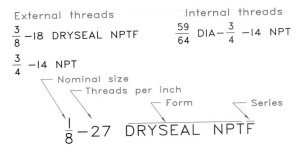

Figure 17.61 These are typical pipe thread notes.

NPSF: straight pipe thread (dryseal)

NPSM: straight pipe thread for mechanical joints

NPSL: straight pipe thread for locknuts and locknut pipe threads

NPSH: straight pipe thread for hose couplings and nipples

NPTR: taper pipe thread for railing fittings

To specify a pipe thread in note form, give the nominal pipe diameter (the common-fraction size of its internal diameter), the number of threads per inch, and the thread-type symbol:

1-1/4-11-1/2 NPT or 3-8 NPTR

Appendix 12 gives a table of dimensions for pipe threads. **Figure 17.61** shows how to present specifications for external and internal threads in note form. Dryseal threads, either straight or tapered, provide a pressure-tight joint without the use of a lubricant or sealer.

Grease Fittings

Grease fittings (Fig. 17.62) allow the application of grease to moving parts that must be lubricated. Threads of grease fittings are available as tapered and straight pipe threads. The ends where grease is inserted with a grease gun are available straight or at 90° and 45° angles. A one-way valve, formed

GREASE FITTINGS

Thread size	$\frac{1}{8}$ 3mm		$\frac{1}{4}$ 6mm		$\frac{3}{8}$ 10mm	
Overal length	L=in.	mm	L=in.	mm	L=in.	mm
Straight	.625	16	1.000	25	1.200	30
90° Elbow	.800	20	1.250	32	1.400	36
45° Angle	1.000	25	1.500	38	1.600	41

GREASE FITTINGS

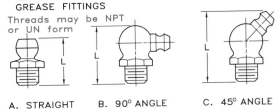

A. STRAIGHT   B. 90° ANGLE   C. 45° ANGLE

Figure 17.62 These three standard types of grease fittings permit lubrication of moving parts with a grease gun.

by a ball and spring, permits grease to enter the fitting (forced through by a grease gun) but prevents it from escaping.

17.10 Keys

Keys are used to attach pulleys, gears, or crank handles to shafts, allowing them to remain assembled while moving and transmitting power. The four types of keys shown in **Fig. 17.63** are the most commonly used. The Appendix contains tables of dimensions for keyways, keys, and keyseats.

17.11 Rivets

Rivets are fasteners that permanently join thin overlapping materials. The rivet is inserted in a hole slightly larger than the diameter of the rivet, and the application of pressure to the projecting end forms the headless end into shape. Forming may be done with either hot or cold rivets, depending on the application.

Figure 17.64 shows typical shapes and proportions of small rivets that vary in diameter from 1/16 to 1-3/4 inches. Rivets are used extensively in

STANDARD KEYS

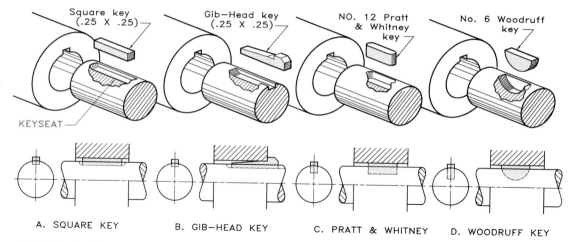

Figure 17.63 Standard keys are used to hold parts on a shaft.

TYPICAL RIVETS

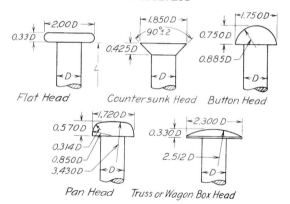

Figure 17.64 This drawing depicts types and proportions of small rivets, which have shank diameters of up to ½ inch.

pressure-vessel fabrication, heavy construction (such as bridges and buildings), and sheet-metal construction.

Figure 17.65 shows some of the standard ANSI symbols for representing rivets. Rivets that are driven in the shop are called **shop rivets**, and those assembled at the job site are called **field rivets**.

RIVET SYMBOLS

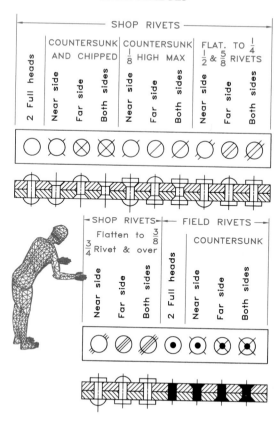

Figure 17.65 These symbols represent rivets in a drawing.

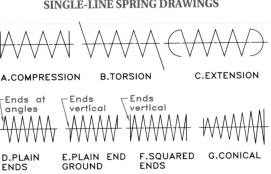

A.COMPRESSION   B.TORSION   C.EXTENSION

D.PLAIN ENDS   E.PLAIN END GROUND   F.SQUARED ENDS   G.CONICAL

H.SINGLE-LINE REPRESENTATIONS: SIMPLIFIED

Figure 17.66

A–C These are single-line representations of various types of springs.

D–G These single-line representations of springs show various types of ends.

H These are simplified single-line representations of the springs depicted in D–G.

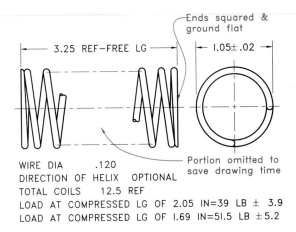

Figure 17.67 This conventional double-line drawing is of a compression spring and includes its specifications.

17.12 Springs

Springs are devices that absorb energy and react with an equal force. Most springs are **helical**, as in a bed, but they can also be **flat** (leaf), as in an automobile chassis. Some of the more common types of springs are **compression, torsion, extension, flat,** and **constant force springs**.

Figure 17.66A–C show single-line conventional representations of the first three types. **Figures 17.66D–F** represent the types of ends used on compression springs. Plain ends of springs simply end with no special modification of the coil. Ground plain ends are coils that have been machined by grinding to flatten the ends perpendicular to their axes. Squared ends are inactive coils that have been closed to form a circular flat coil at the end of a spring, which may also be ground.

Figure 17.66G represents a conical helical spring. **Figure 17.66H** shows schematic single-line representations of the springs depicted in Fig. 17.66D–G (conventional method of drawing springs), with phantom outlines instead of all the coils.

Figure 17.67 shows working drawing specifications of a compression spring drawn as a double-line representation. It shows both ends of the spring and the use of phantom lines to omit the central part of the spring's coils in order to save drawing time. Notate the diameter and free length of the spring on the drawing and give the remaining specifications a table near the drawing.

A working drawing of an extension spring (**Fig. 17.68**) is similar to that of a compression spring. An extending spring is designed to resist stretching, whereas a compression spring is designed to resist squeezing. In a drawing of a helical torsion spring, which resists and reacts to a twisting motion (**Fig. 17.69**), angular dimensions specify the initial and final positions of the spring as torsion is applied to it. Again, notate the dimensions on the drawing and add specifications to describe their details.

EXTENSION SPRING

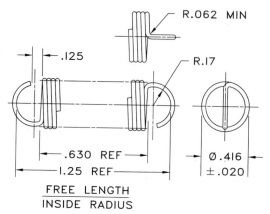

FREE LENGTH
INSIDE RADIUS

WIRE DIA	0.42
DIRECTION OF HELIX	OPTIONAL
TOTAL COILS	14 REF
RELATIVE POSITION OF ENDS	180° ±20°
EXTENDED LENGTH INSIDE ENDS WITHOUT PERMANENT SET	2.45 IN (MAX)
INITIAL TENSION	1.00 LB ±.10 LB
LOAD	4.0 LB ±.4 LB AT 1.56 IN
EXTENDED LG INSIDE ENDS LOAD	6.30 LB ±.63 LB AT 1.95

Figure 17.68 This conventional double-line drawing is of an extension spring and includes its specifications.

TORSION SPRING

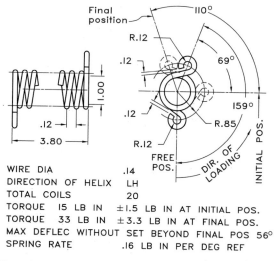

WIRE DIA	.14
DIRECTION OF HELIX	LH
TOTAL COILS	20
TORQUE 15 LB IN	±1.5 LB IN AT INITIAL POS.
TORQUE 33 LB IN	±3.3 LB IN AT FINAL POS.
MAX DEFLEC WITHOUT SET BEYOND FINAL POS	56°
SPRING RATE	.16 LB IN PER DEG REF

Figure 17.69 This conventional double-line drawing is of a helical torsion spring and includes its specifications.

DOUBLE-LINE DRAWINGS OF SPRINGS

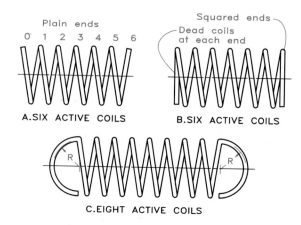

Figure 17.70

A This double-line drawing shows a spring with six active coils.

B This double-line drawing shows a spring with six coils and a "dead" coil (inactive coil) at each end.

C This double-line drawing shows an extension spring with eight active coils.

Drawing Springs

Springs may be represented with single-line drawings (see Fig. 17.66) or as more realistic double-line drawings (**Fig. 17.70**). Draw each type shown by first laying out the diameters of the coils and lengths of the springs and then dividing the lengths into the number of active coils (**Fig. 17.70A**). In **Fig. 17.70B**, both end coils are "dead" (inactive) coils, and only six coils are active. **Figure 17.70C** depicts an extension spring with eight active coils.

Figure 17.71 shows the steps involved in drawing a double-line representation of a compression spring. Here, the ends of the spring are to be squared and ground to give flat ends perpendicular to the axis of the spring.

DRAWING A SPRING: DETAILED

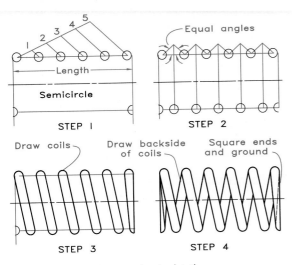

Figure 17.71 Drawing a spring in detail:

Step 1 Lay out the diameter and length of the spring and locate the five coils by the diagonal-line technique.

Step 2 Locate the coils on the lower side along the bisectors of the spaces between the coils on the upper side.

Step 3 Connect the coils on each side. This is a right-hand coil; a left-hand spring would slope in the opposite direction.

Step 4 Construct the back side of the spring and the end coils to complete the drawing. The spring has a square end that is to be ground.

Problems

Solve and draw these problems on size A sheets. Each grid space equals 0.20 inch, or 5 mm.

1. (Fig. 17.72) Draw detailed representations of Acme threads with major diameters of 2 in. Show both external and internal threads as views and sections. Provide a thread note by referring to Appendix 11.

2. Repeat Problem 1, but draw internal and external detailed representations of square threads.

3. Repeat Problem 1, but draw internal and external detailed representations of UN threads. Provide a thread note for a coarse thread with a class 2 fit.

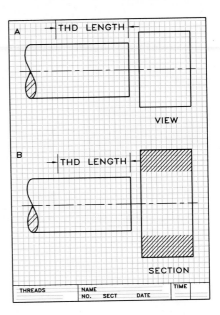

Figure 17.72 Problems 1–3.

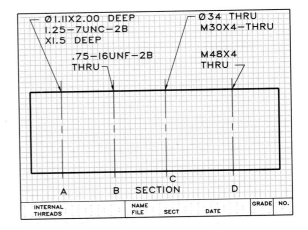

Figure 17.73 Problems 4–6.

4. Using the notes in **Fig. 17.73**, draw detailed representations of the internal threads and holes in section. Provide thread notes on each as specified.

5. Repeat Problem 4, but use schematic thread symbols.

6. Repeat Problem 4, but use simplified thread symbols.

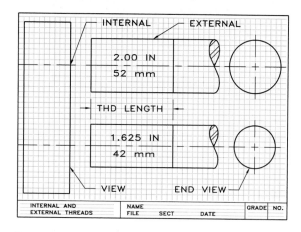

Figure 17.74 Problems 7–9.

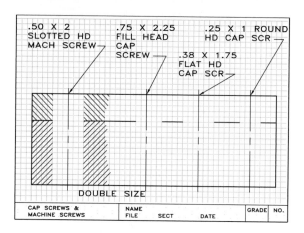

Figure 17.76 Problems 13–15.

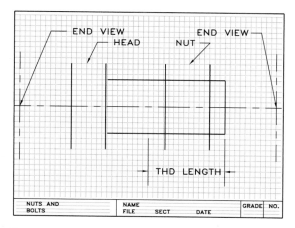

Figure 17.75 Problems 10–12.

Figure 17.77 Problem 16.

7. Using the partial views in **Fig. 17.74** and detailed thread symbols, draw external, internal, and end views of the full-size threaded parts. Provide thread notes for UNC threads with a class 2 fit.

8. Repeat Problem 7, but use schematic thread symbols.

9. Repeat Problem 7, but use simplified thread symbols.

10. (Fig. 17.75) Complete the drawing of the finished hexagon-head bolt and a heavy hexagon nut. Draw the bolt head and nut across corners using detailed thread symbols. Provide thread notes in either English or metric forms, as assigned.

11. Repeat Problem 10, but draw the nut and bolt as having unfinished square heads. Use schematic thread symbols.

12. Repeat Problem 10, but draw the bolt with a regular finished hexagon head across flats, using simplified thread symbols. Draw the nut across flats also and provide thread notes for both.

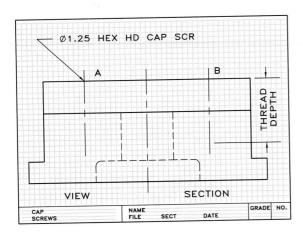

Figure 17.78 Problem 17.

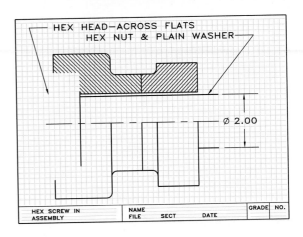

Figure 17.80 Problem 19.

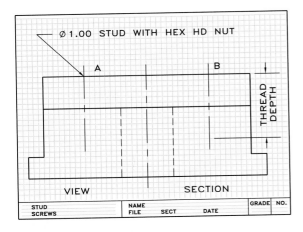

Figure 17.79 Problem 18.

and the method of using the set screw to hold the shaft and provide a thread note.

17. (Fig. 17.78) On axes A and B, construct hexagon-head cap screws (across flats), with UNC threads and a class 2 fit. The cap screws should not reach the bottoms of the threaded holes. Convert the view to a half section.

18. (Fig. 17.79) On axes A and B, draw studs having a hexagon-head nut (across flats) that hold the two parts together. The studs are to be fine series with a class 2 fit, and they should not reach the bottom of the threaded hole. Provide a thread note. Show the view as a half section.

19. (Fig. 17.80) Draw a 2.00-in. (50-mm) diameter hexagon-head bolt, with its head across flats, using schematic symbols. Draw a plain washer and regular nut (across corners) at the right end. Design the size of the opening in the part at the left end to hold the bolt head so that it will not turn. Use a UNC thread with a series 2 fit and provide a thread note.

20. (Fig. 17.81) Draw a 2.00-in. (50-mm) diameter hexagon-head cap screw that holds the two parts together. Determine the length of the bolt, show the threads with schematic thread symbols, and provide a thread note.

13. Use the notes in **Fig. 17.76** to draw the screws in section and complete the sectional view showing all cross-hatching. Use detailed thread symbols and provide thread notes to the parts.

14. Repeat Problem 13, but use schematic thread symbols.

15. Repeat Problem 13, but use simplified thread symbols.

16. (Fig. 17.77) The pencil pointer has a ¼-in. shaft that fits into a bracket designed to clamp onto a desk top. A set screw holds the shaft in position. Make a drawing of the bracket, estimating its dimensions. Show the details

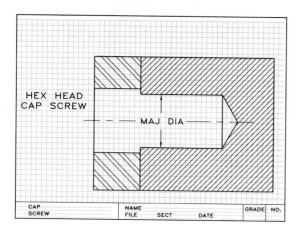

Figure 17.81 Problem 20.

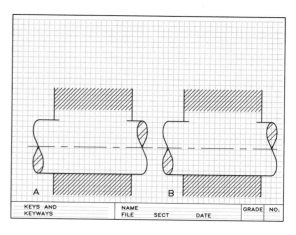

Figure 17.82 Problems 21–22.

21. (Fig. 17.82) The part at A is held on the shaft by a square key, and the part at B is held on the shaft by a gib-head key. Using Appendix 30, complete the drawings and provide the necessary notes.

22. Repeat Problem 21, but use Woodruff keys, one with a flat bottom and the other with a round bottom. Using Appendix 29, complete the drawings and provide the necessary notes.

23–26. (Fig. 17.83) Using Table 17.1, make a double-line drawing of the spring assigned.

27–30. Repeat Problem 23, but draw the springs using single-line representations.

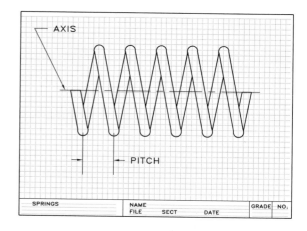

Figure 17.83 Problems 23–30.

Table 17.1
Problems 23-30

Problems	No. of Turns	Pitch	Size Wire	Inside Diameter	Outside Diameter	
23	4	1	No. 4 = 0.2253	3		RH
24	5	$\frac{3}{4}$	No. 6 = 0.1920		2	LH
25	6	$\frac{5}{8}$	No. 10 = 0.1350	2		RH
26	7	$\frac{3}{4}$	No. 7 = 0.1770		$1\frac{3}{4}$	LH

Gears and Cams

18.1 Introduction

Gears are toothed wheels whose circumferences mesh to transmit force and motion from one gear to the next. The three most common types are **spur gears, bevel gears**, and **worm gears (Fig. 18.1).**

Cams are irregularly shaped plates and cylinders that control the motion of a follower as they revolve to produce a type of reciprocating action. For example, cams make the needle of a sewing machine move up and down.

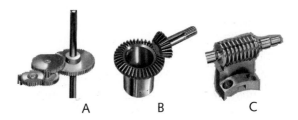

Figure 18.1 The three basic types of gears are (A) spur gears, (B) bevel gears, and (C) worm gears. (Courtesy of the Process Gear Company.)

18.2 Spur Gears

Terminology

The **spur gear** is a circular gear with teeth cut around its circumference. Two meshing spur gears transmit power from one shaft to a parallel shaft. When the two meshing gears are unequal in diameter, the smaller gear is called the **pinion** and the larger one the **spur**.

The following terms and corresponding formulas describe the parts of a spur gear, several of which are shown in **Fig. 18.2**.

Pitch circle (PC): the imaginary circle of a gear, as if it were a friction wheel without teeth that contacted another circular friction wheel.

Pitch diameter (PD): the diameter of the pitch circle; $PD = N/DP$, where N is the number of teeth and DP is the diametral pitch.

Diametral pitch (DP): the ratio between the number of teeth on a gear and its pitch diame-

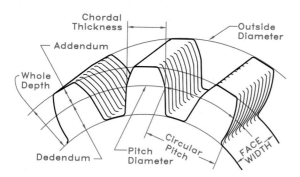

Figure 18.2 These terms apply to spur gears.

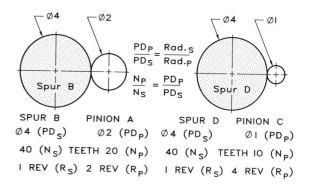

Figure 18.3 These are examples of ratios between meshing spur gears and pinion gears.

ter; DP = N/PD, where N is the number of teeth, and is expressed as teeth per inch of diameter.

Circular pitch (CP): the circular measurement from one point on a tooth to the corresponding point on the next tooth measured along the pitch circle; CP = 3.14/DP.

Center distance (CD): the distance from the center of a gear to its mating gear's center; CD = $(N_P + N_S)/(2DP)$, where N_P and N_S are the number of teeth in the pinion and spur, respectively.

Addendum (A): the height of a gear above its pitch circle; A = 1/DP.

Dedendum (D): the depth of a gear below the pitch circle; D = 1.157/DP.

Whole depth (WD): the total depth of a gear tooth; WD = A + D.

Working depth (WKD): the depth to which a tooth fits into a meshing gear; WKD = 2/DP, or WKD = 2A.

Circular thickness (CRT): the circular distance across a tooth measured along the pitch circle; CRT = 1.57/DP.

Chordal thickness (CT): is the straight-line distance across a tooth at the pitch circle; CT = PD (sin 90°/N), where N is the number of teeth.

Face width (FW): the width across a gear tooth parallel to its axis; a variable dimension, but usually three to four times the circular pitch; FW = 3CP to 4CP.

Outside diameter (OD): the maximum diameter of a gear across its teeth; OD = PD + 2A.

Root diameter (RD): the diameter of a gear measured from the bottom of its gear teeth; RD = PD – 2D.

Pressure angle (PA): the angle between the line of action and a line perpendicular to the centerline of two meshing gears; angles of 14.5° and 20° are standard for involute gears.

Base circle (BC): the circle from which an involute tooth curve is generated or developed; BC = PD cos PA.

Tooth Forms

The most common gear tooth is an involute tooth with a 14.5° pressure angle. The 14.5° angle is the angle of contact between two gears when the tangents of both gears are in contact. Gears with pressure angles of 20° and 25° also are used. Gear teeth with larger pressure angles are wider at the base and thus stronger than the standard 14.5° teeth.

Gear Ratios

The diameters of two meshing spur gears establish ratios that are important to their function **(Fig. 18.3)**. If the diameter of a gear is twice that of its pinion (the small gear), the gear has twice as many teeth as the pinion. The pinion then must make twice as many turns as the spur; therefore the revolutions per minute (RPM) of the pinion is twice that of the spur.

The relationship between two meshing gears may be determined by finding the velocity of a point on the pinion that is equal to $\pi PD \times RPM$. The velocity of a point on the large gear equals $\pi PD \times RPM$. The velocity of points on each gear must be equal, so

$$\pi\, PD_P\, (RPM_P) = \pi\, PD_S\, (RPM_S);$$

therefore

$$\frac{PD_P}{PD_S} = \frac{RPM_S}{RPM_P}.$$

If the radius of the pinion is 1 inch, the diameter of the spur is 4 inches, and the RPM of the pinion is 20, the RPM of the spur is

$$\frac{2(1)}{2(4)} = \frac{RPM_S}{20}.$$

or

$$RPM_S = \frac{2(20)}{2(4)} = 5\,RPM$$

Thus the RPM of the spur (5) is one-fourth that of the pinion (20).

The number of teeth on each gear is proportional to the diameters of a pair of meshing gears, or

$$\frac{N_P}{N_S} = \frac{PD_P}{PD_S}$$

where N_P and N_S are the number of teeth on the

pinion and spur, respectively, and PD_P and PD_S are their pitch diameters.

Calculations

Before starting a working drawing of a gear, you have to calculate the gear's dimensions.

Problem 1 Calculate the dimensions for a spur that has a pitch diameter of 5 in., a diametral pitch of 4, and a pressure angle of 14.5°.

Solution

Number of teeth: $PD(DP) = 5(4) = 20$.

Addendum: $1/4 = 0.25"$.

Dedendum: $1.157/4 = 0.2893"$.

Circular thickness: $1.5708/4 = 0.3927"$.

Outside diameter: $(20 + 2)/4 = 5.50"$.

Root diameter: $5 - 2(0.2893) = 4.421"$.

Chordal thickness:
  $5(\sin 90°/20) = 5(0.079) = 0.392"$.

Chordal addendum:
  $0.25 + [0.3927^2/(4 \times 5)] = 0.2577"$.

Face width: $3.5(0.79) = 2.75$.

Circular pitch: $3.14/4 = 0.785"$.

Working depth: $0.6366(3.14/4) = 0.4997"$.

Whole depth: $0.250 + 0.289 = 0.539"$.

Use these dimensions to draw the spur and to provide specifications necessary for its manufacture.

Problem 2 shows the method of determining design information for two meshing gears when you know their working ratios.

Problem 2 Find the number of teeth and other specifications for a pair of meshing gears with a driving gear that turns at 100 RPM and a driven gear that turns at 60 RPM. The diametral pitch for each is 10, and the center-to-center distance between the gears is 6 in.

Solution

Step 1 Find the sum of the teeth on both gears:

Total teeth = 2(center-to-center distance)(DP)

= 2(6)(10) = 120 teeth.

Step 2 Find the number of teeth for the driving gear:

$$\frac{\text{Driver RPM}}{\text{Driven RPM}} + 1 = \frac{100}{60} + 1 = 2.667,$$

so

$$\frac{\text{Total Teeth}}{\frac{100}{60} + 1} = \frac{120}{2.667} = 45 \text{ teeth.}$$

(The number of teeth must be a whole number since there cannot be fractional teeth on a gear.)

Step 3 Find the number of teeth for the driven gear:

Total teeth – teeth on driver = teeth on driven gear

120 – 45 = 75 teeth.

Step 4 Calculate the other dimensions for the gears as in Problem 1. Adjusting the center distance to yield a whole number of teeth may be necessary.

Drawing Spur Gears

Figure 18.4 shows a conventional drawing of a spur gear. Not having to draw a top view with gear teeth saves a lot of time. Showing only circular and sectional views of the gear and providing a table of dimensions called **cutting data** is acceptable. Circular phantom lines represent the root circle, pitch circle, and outside circle of the gear in the circular view.

A table of dimensions is a necessary part of a gear drawing (**Fig. 18.5**). You may calculate these data or get them from tables of standards in gear handbooks such as *Machinery's Handbook*.

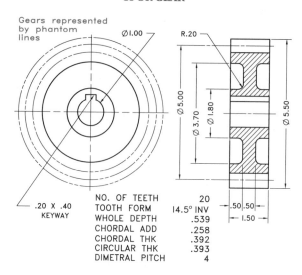

SPUR GEAR

NO. OF TEETH	20
TOOTH FORM	14.5° INV
WHOLE DEPTH	.539
CHORDAL ADD	.258
CHORDAL THK	.392
CIRCULAR THK	.393
DIMETRAL PITCH	4

Figure 18.4 This detail drawing of a spur gear contains a table of values that supplements the dimensions shown on the view and section.

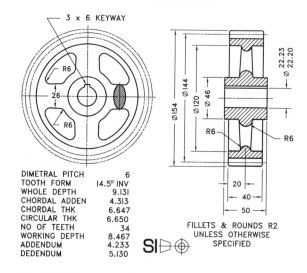

SPUR GEAR: METRIC

DIMETRAL PITCH	6
TOOTH FORM	14.5° INV
WHOLE DEPTH	9.131
CHORDAL ADDEN	4.313
CHORDAL THK	6.647
CIRCULAR THK	6.650
NO OF TEETH	34
WORKING DEPTH	8.467
ADDENDUM	4.233
DEDENDUM	5.130

FILLETS & ROUNDS R2
UNLESS OTHERWISE
SPECIFIED

Figure 18.5 This is a detail drawing of a spur gear that was produced on a computer.

BEVEL GEAR TERMINOLOGY

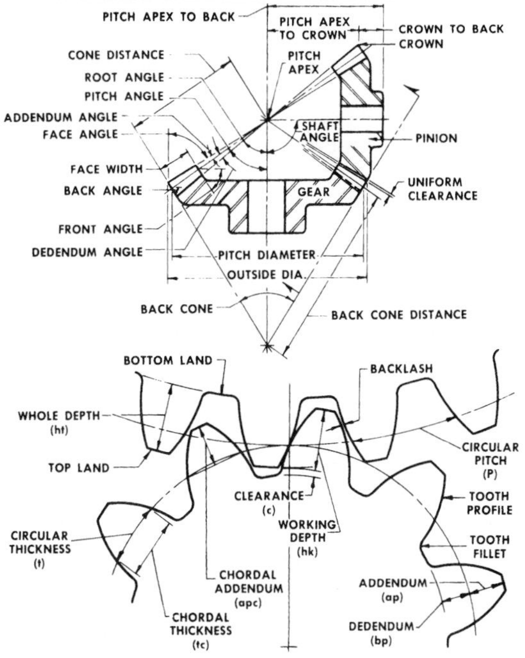

Figure 18.6 These terms apply to bevel gears. (Courtesy of Philadelphia Gear Corporation.)

18.3 Bevel Gears

Terminology

Bevel gears have axes that intersect at angles. The angle of intersection usually is 90°, but other angles also are used. The smaller of the two bevel gears is the pinion, as with spur gears; the larger is the gear.

Figure 18.6 illustrates the terminology of bevel gearing. A further explanation of and the corresponding formula for each feature follow. You may also use gear handbooks to find these dimensions.

Pitch angle of pinion (PA_p): $\tan PA_p = N_p/N_g$, where N_g and N_p are the number of teeth on the gear and pinion, respectively.

Pitch angle of gear (PA_g): $\tan PA_g = N_g/N_p$.

Pitch diameter (PD): the number of teeth, N, divided by the diametral pitch, DP; $PD = N/P$.

Addendum (A): measured at the large end of the tooth; $A = 1/DP$.

Dedendum (D): is measured at the large end of the tooth; $D = 1.157/DP$.

Whole tooth depth (WD): $WD = 2.157/DP$.

Thickness of tooth (TT): measured at the pitch circle; $TT = 1.571/DP$.

Diametral pitch: $DP = N/PD$, where N is the number of teeth.

Addendum angle (AA): the angle formed by the addendum and pitch cone distance; $\tan AA = A/PCD$.

Angular addendum: $AK = \cos PA \times A$.

Pitch cone distance:

(PCD): $PCD = PD/(2 \sin PA)$.

Dedendum angle (DA): the angle formed by the dedendum and the pitch cone distance; $\tan DA = D/PCD$.

Face angle (FA): the angle between the gear's centerline and the top of its teeth;

$$FA = 90° - (PCD + AA).$$

Cutting angle (or root angle) (CA): the angle between the gear's axis and the roots of the teeth; $CA = PCD - D$.

Outside diameter (OD): the greatest diameter of a gear across its teeth; $OD = PD + 2A$.

Apex to crown distance (AC): the distance from the crown of the gear to the apex of the cone measured parallel to the axis of the gear; $AC = OD/(2 \tan FA)$.

Chordal addendum (CA):

$$CA = A + [(TT^2 \cos PA)/4PD].$$

Chordal thickness (CT): measured at the large end of the tooth; $CT = PD (\sin 90°/N)$.

Face width (FW) can vary, but should be approximately equal to the pitch cone distance divided by 3; $FW = PCD/3$.

Calculations

Problem 3 demonstrates use of the preceding formulas. Some of the formulas result in specifications that apply to both gear and pinion.

Problem 3 Two bevel gears intersect at right angles and have a diametral pitch of 3. The gear has 60 teeth and the pinion has 45 teeth. Find the dimensions of the gear.

Solution

Pitch cone angle of gear:
tan PCA = 60/45 = 1.33; PCA = 53°7'.

Pitch cone angle of pinion:
tan PCA = 45/60; PCA = 36°52'.

Pitch diameter of gear: 60/3 = 20.00".

Pitch diameter of pinion: 45/3 = 15.00".

The following calculations yield the same dimensions for both gear and pinion:

Addendum: 1/3 = 0.333".

Dedendum: 1.157/3 = 0.3857".

Whole depth: 2.157/3 = 0.719".

Tooth thickness on pitch circle:
1.571/3 = 0.5237".

Pitch cone distance:
20/(2 sin 53°7') = 12.5015".

Addendum angle:
tan AA = 0.333/12.5015 = 1°32'

Dedendum angle:
DA = 0.3857/12.5015 = 0.0308 = 1°46'.

Face width: PCD/3 = 4.00".

The following dimensions must be calculated separately for gear and pinion:

Chordal addendum of gear:
0.333" + [(0.5237^2 cos 53°7')/(4 x 20)] = 0.336".

Chordal addendum of pinion:
0.333" + [(0.5237^2 cos 36°52')/(4 × 15)] = 0.338".

Chordal thickness of gear:
sin 90°/(60 × 20") = 0.524".

Chordal thickness of pinion:
sin 90°/(45 × 15") = 0.523".

Face angle of gear:
90° – (53°7' + 1°32') = 35°21'.

Face angle of pinion:
90° – (36°52' + 1°32') = 51°36'.

Cutting angle of gear: 53°7' – 1°46' = 51°21'.

Cutting angle of pinion: 36°52' – 1°46' = 35°6'.

Angular addendum of gear:
0.333" cos 53°7' = 0.1999".

Angular addendum of pinion:
0.333" cos 36°52' = 0.2667".

Outside diameter of gear:
20" + 2(0.1999") = 20.4000".

Outside diameter of pinion:
15" + 2(0.2667") = 15.533".

Apex-to-crown distance of gear:
(20.400"/2)(tan 35°7') = 7.173".

Apex-to-crown distance of pinion:
(15.533"/2)(tan 51°36') = 9.800".

Drawing Bevel Gears

Use the dimensions calculated to lay out bevel gears in a detail drawing. Many of these dimensions are difficult to measure with a high degree of accuracy on a drawing. Therefore, providing a table of cutting data for each gear is important.

Figure 18.7 shows the steps involved in drawing bevel gears. On the finished drawing, the views show certain dimensions, and the table of dimensions contains the rest.

18.4 Worm Gears

A worm gear consists of a threaded shaft called a worm and a circular gear called a spider (Fig. 18.8). When the worm is revolved, it causes the

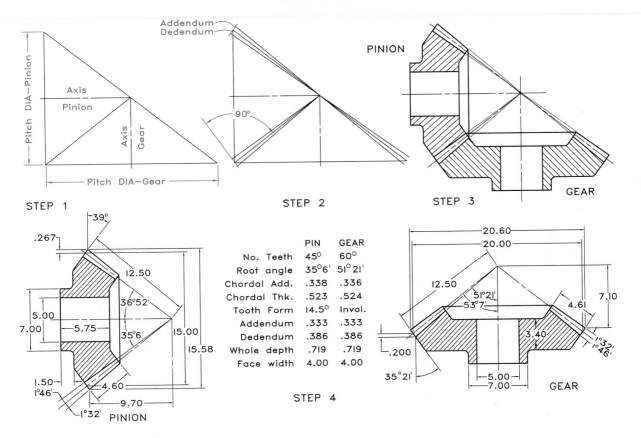

STEP 1

STEP 2

STEP 3

STEP 4

	PIN	GEAR
No. Teeth	45°	60°
Root angle	35°6'	51°21'
Chordal Add.	.338	.336
Chordal Thk.	.523	.524
Tooth Form	14.5°	Invol.
Addendum	.333	.333
Dedendum	.386	.386
Whole depth	.719	.719
Face width	4.00	4.00

Figure 18.7 Drawing bevel gears:

Step 1 Lay out the pitch diameters and axes of the two bevel gears.

Step 2 Draw construction lines to establish the limits of the teeth by using the addendum and dedendum dimensions.

Step 3 Draw the pinion and gear using the specified or calculated dimensions.

Step 4 Complete the detail drawings of both gears and provide a table of cutting data.

spider to revolve about its axis. **Figures 18.8** and **18.9** illustrate the terminology of worm gearing. The following lists further explain these terms and provide the formulas for calculating their dimensions for worm and spider.

Worm Terminology

Linear pitch (P): is the distance from one thread to the next, measured parallel to the worm's axis; P = L/N, where N is the number of threads (1 if a single thread, 2 if a double thread, and so on).

Lead (L): the distance a thread advances in a turn of 360°.

Addendum of tooth (AW): AW = 0.3183P.

Pitch diameter (PDW): PDW = OD – 2AW, where OD is the outside diameter.

WORM GEARS

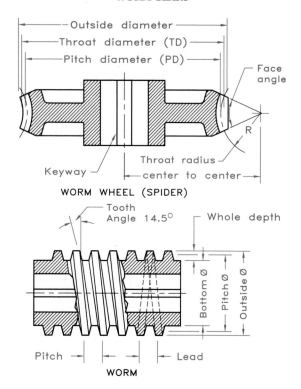

Figure 18.8 These terms apply to worm gears.

WORM GEAR: SPIDER

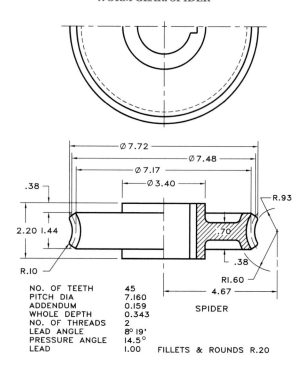

NO. OF TEETH	45
PITCH DIA	7.160
ADDENDUM	0.159
WHOLE DEPTH	0.343
NO. OF THREADS	2
LEAD ANGLE	8° 19'
PRESSURE ANGLE	14.5°
LEAD	1.00

FILLETS & ROUNDS R.20

Figure 18.9 This detail drawing is of a spider for a worm gear with a table of cutting data.

Whole depth of tooth (WDT): WDT = 0.6866P.

Bottom diameter of worm (BD):
$$BD = OD - 2WDT.$$

Width of thread at root (WT): WT = 0.31P.

Minimum length of worm (MLW):

MLW = $\sqrt{8PDS}$ (AW), where PDS is the pitch diameter of the spider.

Helix angle (HA): $\cot \beta = 3.14(PDW)/L$.

Outside diameter (OD): OD = PD + 2A.

Spider Terminology

Pitch diameter of spider (PDS):
PDS = N(P)/3.14, where N is the number of teeth on the spider.

Throat diameter of spider (TD):
$$TD = PDS + 2A.$$

Radius of spider throat (RST):
$$RST = OD \text{ of worm}/2 - 2A.$$

Face angle (FA): may be selected between 60° and 80° for the average application.

Center-to-center distance (CD):
measured between the worm and spider;
$$CD = PDW + PDS/2.$$

Outside diameter of spider (ODS):
$$ODS = TD + 0.4775P.$$

Face width of gear (FW): FW = 2.38P + 0.25.

Calculations
Problem 4 demonstrates use of the preceding formulas to find the dimensions for a worm gear.

Problem 4 Calculate the dimensions for a worm gear (worm and spider). The spider has 45 teeth, and the worm has an outside diameter of 2.50 in., a double thread, and a pitch of 0.5 in.

Solution

Lead: $L = 0.5"(2) = 1"$.

Worm addendum: $AW = 0.3183P = 0.1592"$.

Pitch diameter of worm:
$$PDW = 2.50" - 2(0.1592") = 2.1818".$$

Pitch diameter of spider:
$$PDS = (45" \times 0.5)/3.14 = 7.166".$$

Center distance between worm and spider:
$$CD = (2.182" + 7.166")/2 = 4.674".$$

Whole depth of worm tooth:
$$WDT = 0.687(0.5") = 0.3433".$$

Bottom diameter of worm:
$$BD = 2.50" - 2(0.3433") = 1.813".$$

Helix angle of worm:
$$\cot \beta = 3.14(2.1816)/1 = 8°19'.$$

Width of thread at root: $WT = 0.31(1) = 0.155"$.

Minimum length of worm:
$$MLW = \sqrt{8(0.1592)\ (7.1656)} = 3.02"$$

Throat diameter of spider:
$$TD = 7.1656" + 2(0.1592") = 7.484".$$

Radius of spider throat:
$$RST = (2.5/2) - (2 \times 0.1592") = 0.9318".$$

Face width: $FW = 2.38(0.5) + 0.25 = 1.44"$.

Outside diameter of spider:
$$ODS = 7.484 + 0.4775\ (0.5) = 7.723".$$

Drawing Worm Gears

Draw and dimension the worm and spider as shown in **Figs. 18.9** and **18.10**. The preceding calculations yield the dimensions needed for scaling and laying out the drawings and providing cutting data.

WORM GEAR

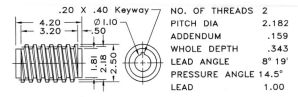

NO. OF THREADS	2
PITCH DIA	2.182
ADDENDUM	.159
WHOLE DEPTH	.343
LEAD ANGLE	8° 19'
PRESSURE ANGLE	14.5°
LEAD	1.00

.20 X .40 Keyway — Ø1.10
4.20 — .50
3.20
1.81 / 2.18 / 2.50

Figure 18.10 This detail drawing of a worm is based on calculated dimensions.

Figure 18.11 This photo shows three types of machined cams. (Courtesy of Ferguson Machine Company.)

18.5 Cams

Plate cams are irregularly shaped machine elements that produce motion in a single plane, usually up and down **(Fig. 18.11)**. As the cam revolves about its center, the cam's shape alternately raises and lowers the follower that is in contact with it. Cams utilize the principle of the inclined wedge, with the surface of the cam acting as the wedge, causing a change in the slope of the plane, and thereby producing the desired motion of the follower. **Cams are designed primarily to produce (1) uniform or linear motion (2) harmonic motion, (3) gravity motion (uniform acceleration), or (4) combinations of these motions.**

Uniform Motion

The uniform motion depicted in **Fig. 18.12A** represents the motion of the cam follower as the cam

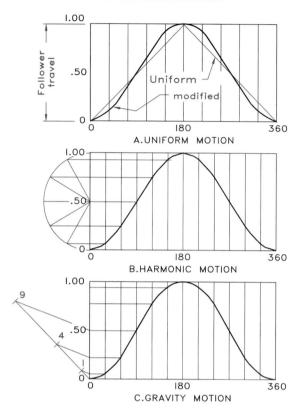

A.UNIFORM MOTION

B.HARMONIC MOTION

C.GRAVITY MOTION

Figure 18.12 These displacement diagrams show three standard motions: uniform, harmonic, and gravity.

rotates through 360°. This curve has sharp corners, indicating abrupt changes of velocity that cause the follower to bounce. Therefore uniform motion usually is modified to smooth the changes of velocity. The radius of the modifying arc varies up to a radius of one-half the total displacement, depending on the speed of operation.

Harmonic Motion

The harmonic motion plotted in **Fig. 18.12B** is a smooth, continuous motion based on the change of position of points on a circle. At moderate speeds this displacement gives a smooth operation.

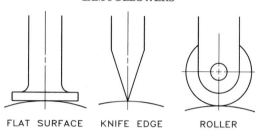

FLAT SURFACE   KNIFE EDGE   ROLLER

Figure 18.13 Three basic types of cam followers are the flat surface, roller, and knife edge.

Gravity Motion

The gravity motion (uniform acceleration) illustrated in **Fig. 18.12C** is used for high-speed operation. The variation of displacement is analogous to the force of gravity, with the difference in displacement being 1, 3, 5, 5, 3, 1, based on the square of the number. For instance, $1^2 = 1$; $2^2 = 4$; $3^2 = 9$ give a uniform acceleration. This motion is repeated in reverse order for the remaining half of the follower's motion. Intermediate points are obtained by squaring fractional increments, such as $(2.5)^2$.

Cam Followers

Three basic types of **cam followers** are the **flat surface**, **roller**, and **knife edge (Fig. 18.13)**. Use of flat-surface and knife-edge followers is limited to slow-moving cams, where minor force will be exerted during rotation. The roller follower is able to withstand higher speeds.

Designing Plate Cams

Harmonic Motion **Figure 18.14** shows the steps involved in designing a plate cam for harmonic motion. Before designing a cam, you must know the motion of the follower, rise of the follower, diameter of the base circle, and direction of rotation. The displacement diagram shown in step 1

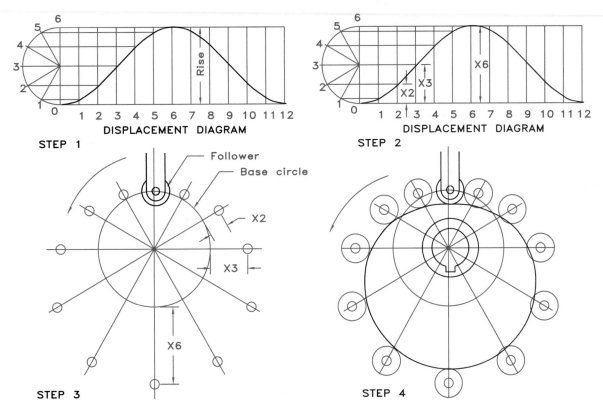

Figure 18.14 Drawing a plate cam for harmonic motion:
Step 1 Construct a semicircle whose diameter equals the rise of the follower. Divide the semicircle into the same number of segments as there are between 0° and 180° on the horizontal axis of the displacement diagram. Plot the displacement curve.

Step 2 Measure distances of rise and fall (X1, X2, X3, ..., X6) at each interval from the base circle.

Step 3 Construct the base circle and draw the follower. Divide the circle into the same number of sectors as there are divisions on the displacement diagram. Transfer distances from the displacement diagram to the respective radial lines of the circle, measuring outward from it.

Step 4 Draw circles to represent the positions of the roller as the cam revolves counterclockwise. Draw the cam profile tangent to all the rollers to complete the drawing.

of Fig. 18.14 gives the specifications graphically for the cam.

Gravity Motion **Figure 18.15** shows the steps involved in designing a cam for gravity motion. The same steps used in designing the cam for harmonic motion apply, but the displacement diagram and knife-edge follower are different.

Cam with an Offset Follower The cam shown in **Fig. 18.16** produces harmonic motion through 360°. In this case, plot the motion directly from the follower rather than from the usual displacement diagram.

Draw a semicircle with its diameter equal to the total motion of the follower. Draw the base circle to pass through the center of the roller of

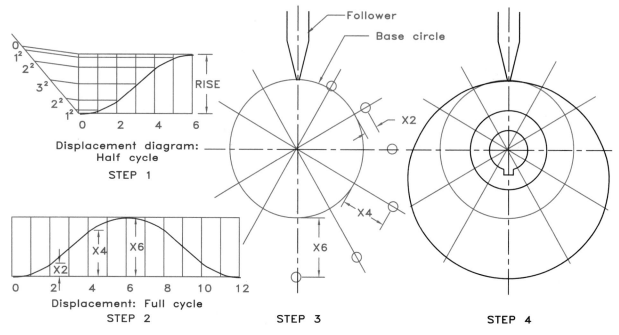

Displacement diagram:
Half cycle

STEP 1

Displacement: Full cycle

STEP 2

STEP 3

STEP 4

Figure 18.15 Drawing a plate cam for uniform acceleration:
Step 1 Construct a displacement diagram to represent the rise of the follower. Divide the horizontal axis into angular increments of 30°. Draw a construction line through point 0; locate the 1^2, 2^2, and 3^2 divisions and project them to the vertical axis to represent half the rise.

Step 2 Use the same construction to find the right half of the symmetrical curve.

Step 3 Construct the base circle and draw the knife-edge follower. Divide the circle into the same number of sectors as there are divisions in the displacement diagram. Transfer distances from the displacement diagram to their respective radial lines of the base circle, measuring outward from the base circle.

Step 4 Connect the points found in step 3 with a smooth curve to complete the cam profile. Also show the cam hub and keyway.

the follower. Extend the centerline of the follower downward and draw a circle tangent to the extension with its center at the center of the base circle. Divide the small circle into 30° intervals to establish points through which to draw construction lines tangent to the circle.

Lay out the distances from tangent points to the position points along the path of the follower

along the tangent lines drawn at 30° intervals. Locate these points by measuring from the base circle, as shown; for example, point 3 is located distance X from the base circle. Draw the circular roller in all views, and then draw the profile of the cam tangent to the rollers at all positions.

PLATE CAM: HARMONIC MOTION / OFFSET FOLLOWER

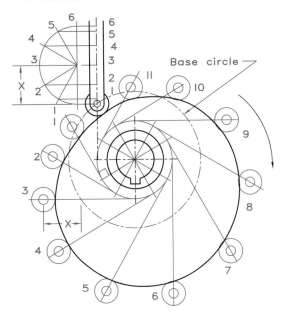

Figure 18.16 Drawing of a plate cam having an offset roller follower.

Problems

Gears

Use size A sheets for the following gear problems. Select appropriate scales so that the drawings will effectively use the available space.

1–5. Calculate the dimensions for the following spur gears, and make a detail drawing of each. Give the dimensions and cutting data for each gear. Provide any other dimensions needed.

Problem	Gear Teeth	Diametral Pitch	14.5° Involute
1	20	5	"
2	30	3	"
3	40	4	"
4	60	6	"
5	80	4	"

6–10. Calculate the gear sizes and number of teeth using the following ratios and data.

Problem	RPM Pinion	RPM Gear	Center to Center	Diametral Pitch
6	100 (driver)	60	6.0"	10
7	100 (driver)	50	8.0"	9
8	100 (driver)	40	10.0"	8
9	100 (driver)	35	12.0"	7
10	100 (driver)	25	14.0"	6

11–20. Make a detail drawing of each gear for which you made calculations in Problems 6–10. Provide a table of cutting data and other dimensions needed to complete the specifications.

21–25. Calculate the specifications for the bevel gears that intersect at 90°, and make detail drawings of each, including the necessary dimensions and cutting data.

Problem	Diametral Pitch	No. of Teeth on Pinion	No. of Teeth on Gear
21	3	60	15
22	4	100	40
23	5	100	60
24	6	100	50
25	7	100	30

26–30. Calculate the specifications for the worm gears and make a detail drawing of each, providing the necessary dimensions and cutting data.

Problem	No. of Teeth in Spider Gear	Outside DIA of Worm	Pitch of Worm	Thread of Worm
26	45	2.50	0.50	double
27	30	2.00	0.80	single
28	60	3.00	0.80	double
29	30	2.00	0.25	double
30	80	4.00	1.00	single

Cams

Use size B sheets for the following cam problems. The standard dimensions are base circle, 3.50 in.; roller follower, 0.60-in. diameter; shaft, 0.75-in. diameter; and hub, 1.25-in. diameter. The direction of rotation is clockwise. The follower is positioned vertically over the center of the base circle. Lay out the problems and displacement diagrams as shown in **Fig. 18.17**.

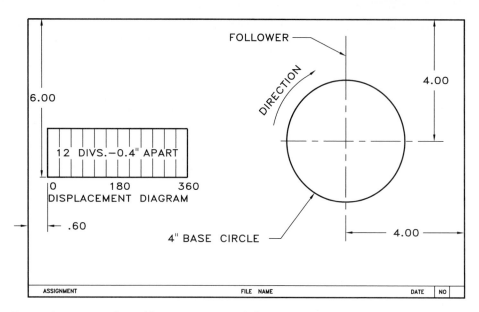

Figure 18.17 Layout for Problems 31–36 on size B sheets.

31. Draw a plate cam with a knife-edge follower for uniform motion and a rise of 1.00 in.

32. Draw a displacement diagram and a cam that will give a modified uniform motion to a knife-edge follower with a rise of 1.7 in. Modify the uniform motion with an arc of one-quarter the rise in the displacement diagram.

33. Draw a displacement diagram and a cam that will give a harmonic motion to a roller follower with a rise of 1.60 in.

34. Draw a displacement diagram and a cam that will give a harmonic motion to a knife-edge follower with a rise of 1.00 in.

35. Draw a displacement diagram and a cam that will give uniform acceleration to a knife-edge follower with a rise of 1.70 in.

36. Draw a displacement diagram and a cam that will give a uniform acceleration to a roller follower with a rise of 1.40 in.

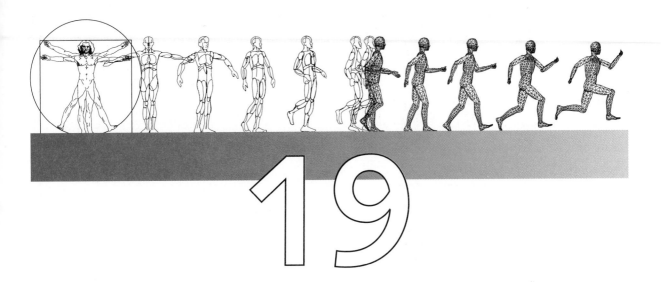

Materials and Processes

Figure 19.1 This furnace operator is pouring an aluminum alloy of manganese into ingots (shown at the right) that will be remelted and cast. (Courtesy of the Aluminum Company of America.)

19.1 Introduction

Various materials and manufacturing processes are commonly used to make parts similar to those discussed in this textbook. A large proportion of parts designed by engineers are made of metal, but other materials (such as plastics and ceramics) are available to the designer in increasingly useful applications.

Metallurgy, the study of metals, is a field that is constantly changing as new processes and alloys are developed **(Fig. 19.1)**. These developments affect the designer's specification of metals and their proper application for various purposes. Three associations have standardized and continually update guidelines for designating various types of metals: the American Iron and Steel Institute (AISI), the Society of Automotive Engineers (SAE), and the American Society for Testing Materials (ASTM).

19.2 Commonly Used Metals

Iron*

Metals that contain iron, even in small quantities, are called **ferrous metals**. Three common types of iron are **gray iron, white iron**, and **ductile iron**.

---

*This section on iron was developed by Dr. Tom Pollock, a metallurgist at Texas A&M University

DESIGNATIONS OF GRAY IRON (450 LB / CF)

ATSM Grade (1000 psi)	SAE Grade	Typical Uses
ASTM 25 CI	G 2500 CI	Small engine blocks, pump bodies, clutch plates, transmission cases
ASTM 30 CI	G 3000 CI	Auto engine blocks, heavy castings, flywheels
ASTM 35 CI	G 3500 CI	Diesel engine blocks, tractor transmission cases, heavy & high-strength parts
ASTM 40 CI	G 4000 CI	Diesel cylinders, pistons, camshafts

Figure 19.2 These are the numbering designations of gray iron and its typical uses.

Gray iron contains flakes of graphite, which result in low strength and low ductility and make it easy to machine. Gray iron resists vibration better than other types of iron. **Figure 19.2** shows designations of and typical applications for gray iron.

White iron contains carbide particles that are extremely hard and brittle, enabling it to withstand wear and abrasion. Although the composition of white iron differs from one supplier to another, there are no designated grades of white iron. It is used for parts on grinding and crushing machines, digging teeth on earthmovers and mining equipment, and wear plates on reciprocating machinery used in textile mills.

Ductile iron (also called nodular or spheroidized iron) contains tiny spheres of graphite, making it stronger and tougher than most types of gray iron and more expensive to produce. Three sets of numbers **(Fig. 19.3)** describe the most important features of ductile iron. **Figure 19.4** shows the designations of and typical applications for the commonly used alloys of ductile iron.

Malleable iron is made from white iron by a heat-treatment process that converts carbides into carbon nodules (similar to ductile iron).

DUCTILE IRON NOTE

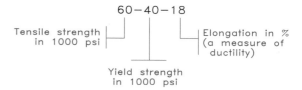

Figure 19.3 This note illustrates the numbering system for ductile iron.

DESIGNATIONS OF DUCTILE IRON (490 LB / CF)

Grade	Typical Uses
60-40-18 CI	Valves, steam fittings, chemical plant equipment, pump bodies
65-45-12 CI	Machine components that are shock loaded, disc brake calipers
80-55-6 CI	Auto crankshafts, gears, rollers
100-70-3 CI	High-strength gears and machine parts
120-90-2 CI	Very high-strength gears, rollers, and slides

Figure 19.4 These are the numbering designations of ductile iron and its typical uses.

Figure 19.5 shows the numbering system for designating grades of malleable iron. **Figure 19.6** shows some of the commonly used grades of malleable iron and their typical applications.

Cast iron is iron that is melted and poured into a mold to form it by casting, a commonly used process for producing machine parts. Although cheaper and easier to machine than steel, iron does not have steel's ability to withstand shock and force.

Steel

Steel is an alloy of iron and carbon, which often contains other constituents such as manganese, chromium, or nickel. Carbon (usually between 0.20% and 1.50%) is the ingredient having the greatest effect on the grade of steel. The three major types of steel are **plain carbon steels, free-cutting**

MALLEABLE IRON NOTES

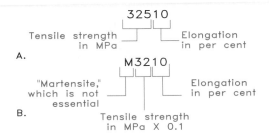

Figure 19.5 These notes illustrate the numbering designations for malleable iron.

DESIGNATIONS OF MALLEABLE IRON (490 LB / CF)

ASTM Grade	Typical Uses
35018 CI	Marine and railroad valves and fittings, "black−iron" pipe fittings (similar to 60−40−18 ductile CI)
45006 CI	Machine parts (similar to 80−55−6 ductile CI)
M3210 CI	Low−stress components, brackets
M4504 CI	Crankshafts, hubs
M7002 CI	High−strength parts, connecting rods, universal joints
M8501 CI	Wear−resistant gears and sliding parts

Figure 19.6 These are the numbering designations of malleable iron and its typical uses.

carbon steels, and **alloy steels**. **Figure 19.7** gives the types of steels and their designations by four-digit numbers. The first digit indicates the type of steel: 1 is carbon steel, 2 is nickel steel, and so on. The second digit gives content (as a percentage) of the material represented by the first digit. The last two or three digits give the percentage of carbon in the alloy: 100 equals 1%, and 50 equals 0.50%.

Steel weighs about 490 pounds per cubic foot. Some frequently used SAE steels are 1010, 1015, 1020, 1030, 1040, 1070, 1080, 1111, 1118, 1145, 1320, 2330, 2345, 2515, 3130, 3135, 3240, 3310, 4023, 4042, 4063, 4140, and 4320.

DESIGNATIONS OF STEEL (490 VBS / CF)

Type of steel	Number	Applications
Carbon steels		
Plain carbon	10XX	Tubing, wire, nails
Resulphurized	11XX	Nuts, bolts, screws
Manganese steel	13XX	Gears, shafts
Nickel steel	23XX	Keys, levers, bolts
	25XX	Carburized parts
	31XX	Axles, gears, pins
	32XX	Forgings
	33XX	Axles, gears
Molybdenum	40XX	Gears, springs
Chromium−moly.	41XX	Shafts, tubing
Nickel−chromium	43XX	Gears, pinions
Nickel−moly.	46XX	Cams, shafts
	48XX	Roller bearings, pins
Chromium steel	51XX	Springs, gears
	52XX	Ball bearings
Chrom. vanadium	61XX	Springs, forgings
Silicon manganese	92XX	Leaf springs

Figure 19.7 These are the numbering designations of steel and its applications.

Copper

One of the first metals discovered, **copper** is easily formed and bent without breaking. Because it is highly resistant to corrosion and is highly conductive, it is used for pipes, tubing, and electrical wiring. It is an excellent roofing and screening material because it withstands the weather well. Copper weighs about 555 pounds per cubic foot.

Copper has several alloys, including brasses, tin bronzes, nickel silvers, and copper nickels. **Brass** (about 530 pounds per cubic foot) is an alloy of copper and zinc, and **bronze** (about 548 pounds per cubic foot) is an alloy of copper and tin. Copper and copper alloys are easily finished by buffing or plating; joined by soldering, brazing, or welding; and machined.

Wrought copper has properties that permit it to be formed by hammering. A few of the numbered designations of wrought copper are C11000, C11100, C11300, C11400, C11500, C11600, C10200, C12000, and C12200.

ALUMINUM DESIGNATIONS (169 LB / CF)

Composition	Alloy Number	Application
Aluminum (99% pure)	1XXX	Tubing, tank cars
Aluminum alloys		
Copper	2XXX	Aircraft parts, screws, rivets
Manganese	3XXX	Tanks, siding, gutters
Silicon	4XXX	Forging, wire
Magnesium	5XXX	Tubes, welded vessels
Magnesium and silicon	6XXX	Auto body, pipes
Zinc	7XXX	Aircraft structures
Other elements	8XXX	

Figure 19.8 These are the numbering designations of aluminum and aluminum alloys and their applications.

ALUMINUM CASTINGS AND INGOT DESIGNATIONS

Composition	Alloy Number
Aluminum (99% pure)	1XX.X
Aluminum alloys	
Copper	2XX.X
Silicon with copper and/or magnesium	3XX.X
Silicon	4XX.X
Magnesium	5XX.X
Magnesium and silicon	6XX.X
Zinc	7XX.X
Tin	8XX.X
Other elements	9XX.X

Figure 19.9 These are the numbering designations of cast aluminum and aluminum alloys.

Aluminum

Aluminum is a corrosion-resistant, lightweight metal (approximately 169 pounds per cubic foot) that has numerous applications. Most materials called aluminum actually are aluminum alloys, which are stronger than pure aluminum.

The types of wrought aluminum alloys are designated by four digits (**Fig. 19.8**). The first digit (2 through 9) indicates the alloying element that is combined with aluminum. The second digit indicates modifications of the original alloy or impurity limits. The last two digits identify other alloying materials or indicate the aluminum's purity.

Figure 19.9 shows a four-digit numbering system used to designate types of cast aluminum and alloys. The first digit indicates the alloy group, and the next two digits identify the aluminum alloy or aluminum purity. The number to the right of the decimal point represents the aluminum form: XX.0 indicates castings, XX.1 indicates ingots with a specified chemical composition, and XX.2 indicates ingots with a specified chemical composition other than the XX.1 ingot. **Ingots** are blocks of cast metal to be remelted, and **billets** are castings of aluminum to be formed by forging.

Magnesium

Magnesium is a light metal (109 pounds per cubic foot) available in an inexhaustible supply because it is extracted from seawater and natural brines. Magnesium is an excellent material for aircraft parts, clutch housings, crankcases for air-cooled engines, and applications where lightness is desirable.

Magnesium is used for die and sand castings, extruded tubing, sheet metal, and forging. Magnesium and its alloys may be joined by bolting, riveting, or welding. Some numbered designations of magnesium alloys are M10100, M11630, M11810, M11910, M11912, M12390, M13320, M16410, and M16620.

19.3 Properties of Metals

All materials have properties that designers must utilize to the best advantage. The following terms describe these properties.

Ductility: a softness in some materials, such as copper and aluminum, which permits them to be formed by stretching (drawing) or hammering without breaking.

Brittleness: a characteristic that will not allow metals such as cast irons and hardened steels to stretch without breaking.

Malleability: the ability of a metal to be rolled or hammered without breaking.

Hardness: the ability of a metal to resist being dented when it receives a blow.

Toughness: the property of being resistant to cracking and breaking while remaining malleable.

Elasticity: the ability of a metal to return to its original shape after being bent or stretched.

Modifying Properties by Heat Treatment

The properties of metals can be changed by various types of heat treating. Although heat affects all metals, steels are affected to a greater extent than others.

Hardening: heating steel to a prescribed temperature and quenching it in oil or water.

Quenching: rapidly cooling heated metal by immersing it in liquids, gases, or solids (such as sand, limestone, or asbestos).

Tempering: reheating previously hardened steel and then cooling it, usually by air, to increase its toughness.

Annealing: heating and cooling metals to soften them, release their internal stresses, and make them easier to machine.

Normalizing: heating metals and letting them cool in air to relieve their internal stresses.

Case hardening: hardening a thin outside layer of a metal by placing the metal in contact with carbon or nitrogen compounds that it absorbs as it is heated; afterward, the metal is quenched.

Flame hardening: hardening by heating a metal to within a prescribed temperature range with a flame and then quenching the metal.

SAND MOLD

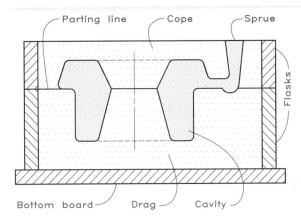

Figure 19.10 A two-section sand mold is used for casting a metal part.

19.4 Forming Metal Shapes

Casting

One of the two major methods of forming shapes is **casting**, which involves preparing a mold in the shape of the part desired, pouring molten metal into it, and cooling the metal to form the part. The types of casting, which differ in the way the molds are made, are **sand casting**, **permanent-mold casting**, **die casting**, and **investment casting**.

Sand Casting In the first step of sand casting, a wood or metal form or pattern is made in the shape of the part to be cast. The pattern is placed in a metal box called a **flask** and molding sand is packed around the pattern. When the pattern is withdrawn from the sand, it leaves a void forming the mold. Molten metal is poured into the mold through sprues or gates. After cooling, the casting is removed and cleaned (**Fig. 19.10**).

Cores formed from sand may be placed in a mold to create holes or hollows within a casting. After the casting has been formed, the cores are broken apart and removed, leaving behind the desired void within the casting.

Because the patterns are placed in and removed from the sand before the metal is

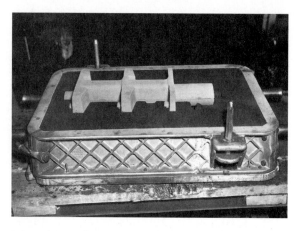

Figure 19.11 This pattern is held in the bottom half (the drag) of a sand mold to form a mold for a casting.

Figure 19.12 The tailstock casting for a lathe has raised bosses that were finished to improve the effectiveness of nuts and bolts. Fillets and rounds were added to the inside and outside corners. (Courtesy L. W. Chuck Company.)

poured, the sides of the patterns must be tapered, called **draft**, for ease of withdrawal from the sand. The angle of draft depends on the depth of the pattern in the sand and varies from 2° to 8° in most applications. **Figure 19.11** shows a pattern held in the sand by a lower flask. Patterns are made oversize to compensate for shrinkage that occurs when the casting cools.

Because sand castings have rough surfaces, features that come into contact with other parts must be machined by drilling, grinding, finishing, or shaping. The tailstock base of a lathe shown in **Fig. 19.12** illustrates raised bosses that have been finished. The casting must be made larger than finished size where metal is to be removed by machining.

Fillets and rounds are used at the inside and outside corners of castings to increase their strength by relieving the stresses in the cast metal (Fig. 19.13). Fillets and rounds also are used because forming square corners by the sand-casting process is difficult and because rounded edges make the finished product more attractive (Fig. 19.12).

Permanent-Mold Casting Permanent molds are made for the mass production of parts. They are generally made of cast iron and coated to prevent

APPLICATION OF FILLETS AND ROUNDS

A. SQUARE CORNERS

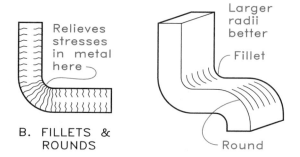

B. FILLETS & ROUNDS

Figure 19.13

A Square corners cause a failure line to form, causing a weakness at this point.

B Fillets and rounds make the corners of a casting stronger and more attractive.

C The larger the radii of fillets and rounds the stronger the casting will be.

PERMANENT-MOLD CASTING

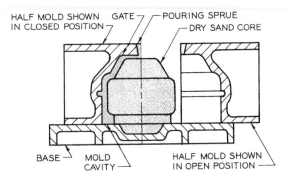

Figure 19.14 Permanent molds are made of metal for repetitive usage. Here, a sand core made from another mold is placed in the permanent mold to create a void within the casting.

DIE CASTING

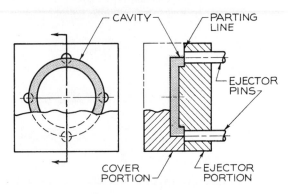

Figure 19.15 This die is used for casting a simple part. The metal is forced into the die to form the casting.

fusing with the molten metal poured into them (**Fig. 19.14**).

Die Casting Die castings are used for the mass production of parts made of aluminum, magnesium, zinc alloys, copper, and other materials. Die castings are made by forcing molten metal into dies (or molds) under pressure. They are inexpensive, meet close tolerances, and have good surface qualities. The same general principles of sand castings—using fillets and rounds, allowing for shrinkage, and specifying draft angles—apply to die castings (**Fig. 19.15**).

Investment Casting Investment casting is used to produce complicated parts or artistic sculptures that would be difficult to form by other methods (**Fig. 19.16**). A new pattern must be used for each investment casting, so a mold or die is made for casting a wax master pattern. The wax pattern, identical to the casting, is placed inside a container and plaster or sand is poured (invested) around it. Once the investment has cured, the wax pattern is melted, leaving a hollow cavity to serve as the mold for the molten metal. After the casting has set, the plaster or sand is broken away from it.

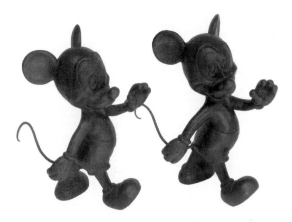

Figure 19.16 An investment casting (lost-wax process) is used to produce complex metal objects and art pieces.

Forgings

The second major method of forming shapes is **forging**, which is the process of shaping or forming heated metal by hammering or forcing it into a die. Drop forges and press forges are used to hammer metal billets into forging dies. Forgings have the high strength and resistance to loads and impacts required for applications such as aircraft landing gears (**Fig. 19.17**).

Figure 19.17 This aircraft landing-gear component was formed by forging. (Courtesy of Cameron Iron Works.)

FORGING DIES

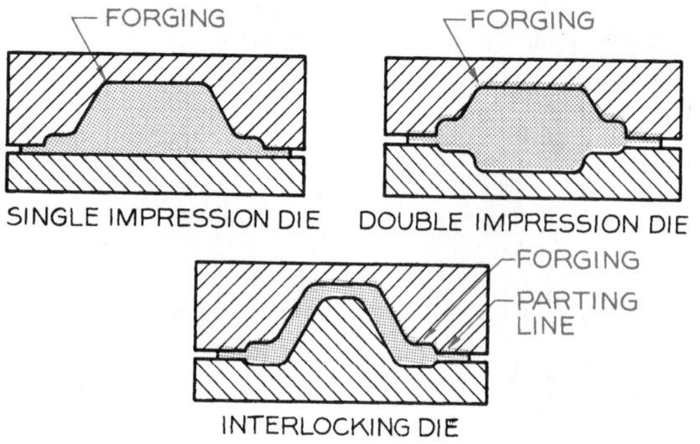

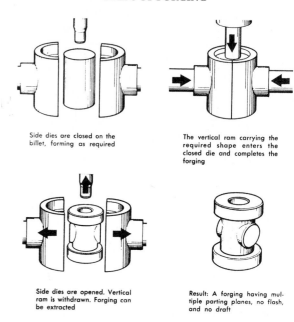

Figure 19.18 This drawing shows three types of forging dies.

Figure 19.18 shows three types of dies. A **single-impression die** gives an impression on one side of the parting line between the mating dies; a **double-impression die** gives an impression on both sides of the parting line; and the **interlocking dies** give an impression that may cross the parting line on either side. **Figure 19.19** shows how an object is forged with horizontal dies and a vertical ram to hollow the object.

STEPS OF FORGING

Side dies are closed on the billet, forming as required

The vertical ram carrying the required shape enters the closed die and completes the forging

Side dies are opened. Vertical ram is withdrawn. Forging can be extracted

Result: A forging having multiple parting planes, no flash, and no draft

Figure 19.19 These are the steps involved in forging a part with external dies and an internal ram. (Courtesy of Cameron Iron Works.)

Figure 19.20 illustrates the sequence of forging a part from a billet by hammering it into different dies. It is then machined to its proper size within specified tolerances.

Figure 19.21 shows a working drawing for making a forged part. When preparing forging drawings, you must consider (1) draft angles and parting lines, (2) fillets and rounds, (3) forging tolerances, (4) extra material for machining, and (5) heat treatment of the finished forging.

Draft, the angle of taper, is crucial to the forging process. The minimum radii for inside corners (fillets) are determined by the height of the feature (**Fig. 19.22**). Similarly, the minimum radii for the outside corners (rounds) are related to a feature's height (**Fig. 19.23**). **The larger the radius of a fillet or round, the better it is for the forging process.**

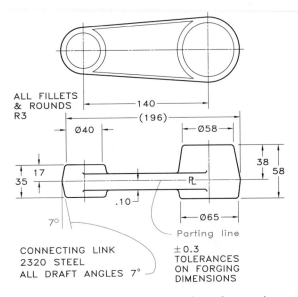

Figure 19.21 This working drawing for a forging shows draft angles and the parting line (PL) where the dies come together.

Figure 19.20 Steps A through G are required to forge a billet into the finished connecting rod. (Courtesy of the Drop Forging Association.)

Some of the standard steels used for forging are designated by the SAE numbers 1015, 1020, 1025, 1045, 1137, 1151, 1335, 1340, 4620, 5120, and 5140. Iron, copper, and aluminum also can be forged.

Rolling Rolling is a type of forging in which the stock is rolled between two or more rollers to shape it. Rolling can be done at right angles or parallel to the axis of the part **(Fig. 19.24)**. If a high degree of shaping is required, the stock usually is heated before rolling. If the forming requires only a slight change in shape, rolling can be done without heating the metal, which is called **cold rolling (CR)**; CRS means **cold-rolled steel. Figure 19.25** shows a cylindrical rod being rolled.

MINIMUM FILLET RADII FOR FORGINGS: INSIDE CORNERS

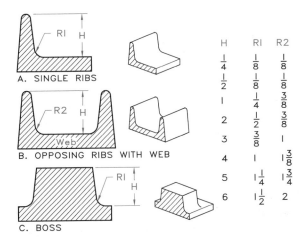

Figure 19.22 These guidelines are for determining the minimum radii for fillets (inside corners) on forged parts.

MINIMUM FILLET RADII ON FORGINGS: OUTSIDE CORNERS

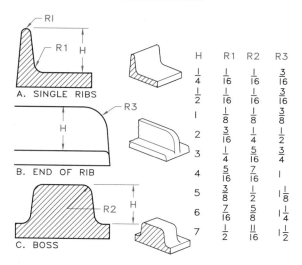

A. SINGLE RIBS

B. END OF RIB

C. BOSS

H	R1	R2	R3
$\frac{1}{4}$	$\frac{1}{16}$	$\frac{1}{16}$	$\frac{3}{16}$
$\frac{1}{2}$	$\frac{1}{16}$	$\frac{1}{16}$	$\frac{3}{16}$
1	$\frac{1}{8}$	$\frac{1}{8}$	$\frac{3}{8}$
2	$\frac{3}{16}$	$\frac{1}{4}$	$\frac{1}{2}$
3	$\frac{1}{4}$	$\frac{5}{16}$	$\frac{3}{4}$
4	$\frac{5}{16}$	$\frac{7}{16}$	1
5	$\frac{3}{8}$	$\frac{1}{2}$	$1\frac{1}{8}$
6	$\frac{7}{16}$	$\frac{5}{8}$	$1\frac{1}{4}$
7	$\frac{1}{2}$	$\frac{11}{16}$	$1\frac{1}{2}$

Figure 19.23 These guidelines are for determining the minimum radii of rounds (outside corners) on forged parts.

Figure 19.25 This cylindrical rod is being rolled to shape.

FORMING BY ROLLING

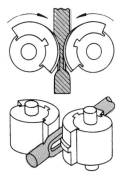

Figure 19.24 Features on parts may be formed by rolling. Here a part is being rolled parallel to its axes. (Courtesy of General Motors Corporation.)

CORNERS OF STAMPED PARTS

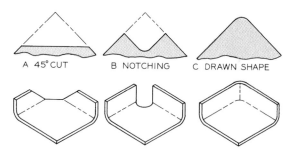

A 45° CUT B NOTCHING C DRAWN SHAPE

Figure 19.26 Box-shaped parts formed by stamping:

A A corner cut of 45° permits flanges to be folded with no further trimming.

B Notching has the same effect as the 45° cut and is often more attractive.

C A continuous corner flange requires that the blank be developed so that it can be drawn into shape.

Stamping

Stamping is a method of forming flat metal stock into three-dimensional shapes. The first step of stamping is to cut out the shapes, called **blanks**, which are formed by bending and pressing them

against forms. **Figure 19.26** shows three types of box-shaped parts formed by stamping, and **Fig. 19.27** shows a design for a flange to be formed by stamping. Holes in stampings are made by punching, extruding, or piercing (**Fig. 19.28**).

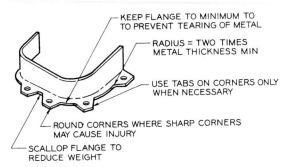

Figure 19.27 This drawing shows a sheet metal flange design with notes that explain design details.

HOLES IN SHEET METAL BY PUNCHING

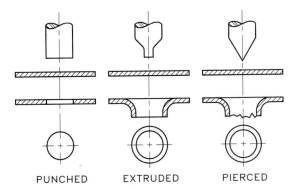

PUNCHED   EXTRUDED   PIERCED

Figure 19.28 These three methods are used to form holes in sheet metal.

19.5 Machining Operations

After metal parts have been formed, machining operations must be performed to complete them. The machines used most often are: **lathe, drill press, milling machine, shaper,** and **planer.** Some of these machines require manual operation; others are computer programmed and require minimal operator handling.

Lathe

The **lathe** shapes cylindrical parts while rotating the work piece between its centers **(Fig. 19.29).**

Figure 19.29 This typical metal lathe holds and rotates the work piece between its centers for machining. (Courtesy of the Clausing Corporation.)

LATHE OPERATIONS

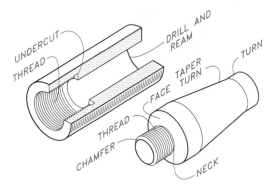

Figure 19.30 These are basic operations performed on a lathe.

The fundamental operations performed on the lathe are turning, facing, drilling, boring, reaming, threading, and undercutting (Fig. 19.30).

Turning forms a cylinder with a tool that advances against and moves parallel to the cylin-

TURNING ON A LATHE

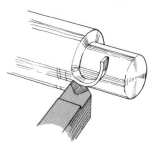

Figure 19.31 The most basic operation performed on the lathe is turning, whereby a continuous chip is removed by a cutting tool as the part rotates.

Figure 19.32 Facing on a lathe is the process of machining a surface that is perpendicular to the axis of revolution, as here with a wood cylinder. (Courtesy of Rockwell/Delta Corporation.)

der being turned between the centers of the lathe (**Fig. 19.31**). **Facing** forms flat surfaces perpendicular to the axis of rotation of the part being rotated (**Fig. 19.32**).

STEPS OF DRILLING

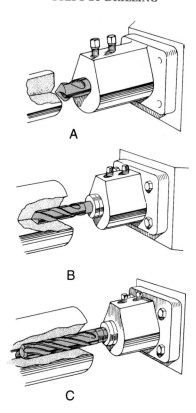

A

B

C

Figure 19.33 Three steps are involved in drilling a hole in the end of a cylinder: (A) start drilling, (B) twist drilling, and (C) core drilling, which enlarges the previously drilled hole to the required size.

BORING A HOLE

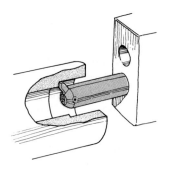

Figure 19.34 Boring is the method of enlarging holes that are larger than available drill bits with a cutting tool attached to a boring bar on the lathe.

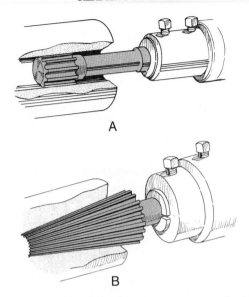

A

B

Figure 19.35 Fluted reamers can be used to finish inside (A) cylindrical and (B) conical holes within a few thousandths of an inch.

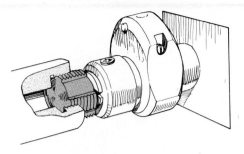

Figure 19.37 Internal threads can be cut on a lathe using a die called a tap. A recess, called a thread relief, was formed at the end of the threaded hole.

Figure 19.36 This hole is being reamed by honing. (Courtesy Barber-Coleman Company.)

Drilling is performed by mounting a drill in the tail stock of the lathe and rotating the work while the bit is advanced into the part (**Fig. 19.33**). **Boring** makes large holes that are too big to be drilled by enlarging smaller drilled holes (**Fig. 19.34**). **Reaming** removes only thousandths of an inch of material inside cylindrical and conical holes to enlarge them to their required tolerances (**Fig. 19.35**). **Figure 19.36** shows a close-up view of reaming by honing.

Threading of external shafts and internal holes can be done on the lathe. The die used for cutting internal holes is called a **tap** (**Fig. 19.37**). A tapping die being used to thread a hole held in the chuck of a lathe is illustrated in **Fig. 19.38**. **Undercutting** cuts a recess inside a cylindrical hole with a tool mounted on a boring bar. The groove is cut as the tool advances from the center of the axis of revolution into the part (**Fig. 19.39**).

The **turret lathe** is a programmable lathe that can perform sequential operations, such as drilling a series of holes, boring them, and then reaming them. The turret rotates each tool into position for its particular operation (**Fig. 19.40**).

Drill Press

The **drill press** is used to drill small- and medium-sized holes (**Fig. 19.41**). The stock being drilled is

Figure 19.38 This die is being used on a lathe to cut internal threads in a part. (Courtesy of the Landis Machine Company.)

CUTTING AN UNDERCUT

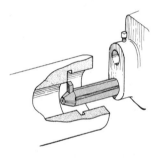

Figure 19.39 A recess (undercut) is formed by using the boring bar that is fed into the rotating stock.

Figure 19.40 A turret lathe performs a sequence of operations.

Figure 19.41 This small drill press is used to make holes in parts. (Courtesy of Clausing Corporation.)

held securely by fixtures or clamps **(Fig. 19.42). The drill press can be used for counterdrilling, countersinking, counterboring, spotfacing, and threading (Fig. 19.43).** Multiple-head drill presses can be programmed to perform a series of drilling operations for mass production applications **(Fig. 19.44).**

Measuring Cylinders The diameters of cylindrical features of parts made on a drill press (or lathe) are measured. Internal and external micrometer calipers **(Fig. 19.45)** can measure to within one ten-thousandth of an inch.

Figure 19.42 Work stock is held securely in place with fixtures and clamping devices during drilling.

Figure 19.44 This multiple-head drill press can be programmed to perform a series of operations sequentially.

HOLE-MACHINING OPERATIONS

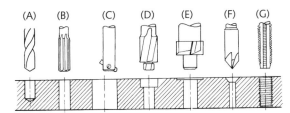

Figure 19.43 The basic operations performed on the drill press are (A) drilling, (B) reaming, (C) boring, (D) counterboring, (E) spotfacing, (F) countersinking, and (G) tapping (threading).

Broaching Machine

Cylindrical holes can be converted into square, rectangular, or hexagonal holes with a **broach** mounted on a special machine (**Fig. 19.46**). A broach has a series of teeth graduated in size along its axis, beginning with teeth that are nearly the size of the hole to be broached and tapering to the

MEASURING DIAMETERS

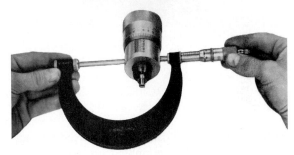

Figure 19.45 Internal and external micrometer calipers are used to measure the diameters of cylinders. (Courtesy of Scherr Tumico Company.)

BROACHING OPERATIONS

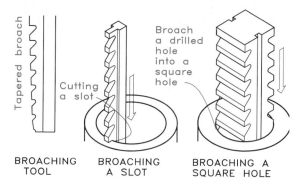

Figure 19.46 A broaching tool can be used to cut slots and holes with square corners on the interior and exterior of parts. Other shapes may be broached also.

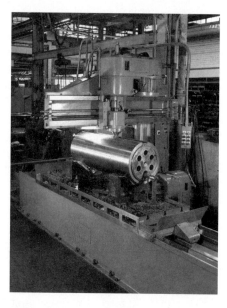

Figure 19.47 This milling machine is being used to cut a groove in a work piece. (Courtesy of the Brown & Sharpe Manufacturing Company.)

final size of the hole. The broach is forced through the hole by pushing or pulling in a single pass, with each tooth cutting more from the hole as it passes through. Broaches can be used to cut external grooves, such as keyways or slots, in a part.

SHAPER

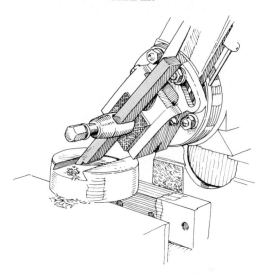

Figure 19.48 The shaper moves back and forth across the part, removing metal as it advances, to finish surfaces, cut slots, and perform other operations.

Milling Machine

The **milling machine** uses a variety of cutting tools, rotated about a shaft **(Fig. 19.47)**, to form different grooved slots, threads, and gear teeth. The milling machine can cut irregular grooves in cams and finish surfaces on a part within a high degree of tolerance.

Shaper

The **shaper** is a machine that holds a work piece stationary while the cutter passes back and forth across it to finish the surface or to cut a groove one stroke at a time **(Fig. 19.48)**. With each stroke of the cutting tool, the material is shifted slightly to align the part for the next overlapping stroke.

Planer

Unlike the shaper, which holds the work piece stationary, the **planer** passes the piece under the cutters **(Fig. 19.49)**. Like the shaper, the planer can cut grooves or slots and finish surfaces that must meet close tolerances.

Figure 19.49 This planer has stationary cutters and a 30-foot bed. Work is fed past the cutters to finish large surfaces. (Courtesy of Gray Corporation.)

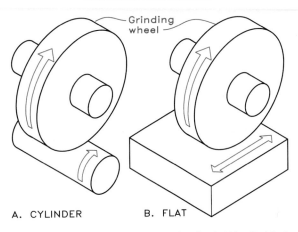

A. CYLINDER   B. FLAT

Figure 19.51 Grinding may be used to finish (A) cylindrical and (B) flat surfaces.

Figure 19.50 The operator is grinding the upper surface of this part to a smooth finish with a grinding wheel. (Courtesy of the Clausing Corporation.)

19.6 Surface Finishing

Surface finishing produces a smooth, uniform surface. It may be accomplished by grinding, polishing, lapping, buffing, and honing.

Grinding involves holding a flat surface against a rotating abrasive wheel (**Fig. 19.50**). Grinding is used to smooth surfaces, both cylin-

drical and flat, and to sharpen edges used for cutting, such as drill bits (**Fig. 19.51**). **Polishing** is done in the same way as grinding, except that the polishing wheel is flexible because it is made of felt, leather, canvas, or fabric.

Lapping produces very smooth surfaces. The surface to be finished is held against a lap, which is a large, flat surface coated with a fine abrasive powder that finishes a surface as the lap rotates. Lapping is done only after the surface has been previously finished by a less accurate technique, such as grinding or polishing. Cylindrical parts can be lapped by using a lathe with the lap.

Buffing removes scratches from a surface with a belt or rotating buffer wheel made of wool, cotton, felt, or other fabric. To enhance the buffing, an abrasive mixture is applied to the buffed surface during the process.

Honing finishes the outside or inside of holes within a high degree of tolerance (see Fig. 19.36). The honing tool is rotated as it is passed through the holes to produce the types of finishes found in gun barrels, engine cylinders, and other products requiring a high degree of smoothness.

E=Excellent
G=Good
F=Fair
P=Poor
A=Adhesives

	MACHINABILITY	FORMABILITY	CASTABILITY	WELDABILITY	CORROSION RES.	ABRASION RES.	LB/CU FT	YIELD: 1000 PSI	Typical Applications
THERMOPLASTICS									
ACRYLIC	G	G	E	A	E	F	74	9	Aircraft windows, TV parts, Lenses, skylights
ABS	G	G	G	a	E	G	66	66	Luggage, boat hulls, tool handles, pipe fittings
POLYMIDES (NYLON)	E	G	G	—	G	E	73	15	Helmets, gears, drawer slides, hinges, bearings
POLYETHYLENE	G	F	G	A	F	F	58	2	Chemical tubing, containers, ice trays, bottles
POLYPROPYLENE	G	G	G	A	E	G	56	5.3	Card files, cosmetic cases, auto pedals, luggage
POLYSTYRENE	G	E	G	A	P	G	67	7	Jugs, containers, furniture, lighted signs
POLYINYL CHLORIDE	E	E	G	A	G	G	78	4.8	Rigid pipe & tubing, house siding, packaging
THERMOSETS									
EPOXY	F	G	G	—	E	G	69	17	Circuit boards, boat bodies, coatings for tanks
SILICONE	F	G	G	—	G	G	109	28	Flexible hoses, heart valves, gaskets
ELASTOMERS									
POLYURETHANE	G	G	G	A	G	E	74	6	Rigid: Solid tires, bumpers; Flexible: Foam, sponges
SBR RUBBER	—	—	E	—	F	E	39	3	Belts, handles, hoses, cable coverings
GLASSES									
GLASS	F	G	—	—	F	F	160	10+	Bottles, windows, tumblers, containers
FIBERGLASS	G	—	E	A	G	G	109	20+	Boats, shower stalls, auto bodies, chairs, signs

Figure 19.52 These are the characteristics of and typical applications for commonly used plastics and other materials.

19.7 Plastics and Other Materials*

Plastics (polymers) are widely used in numerous applications ranging from clothing, containers, and electronics to automobile bodies and components. Plastics are easily formed into irregular shapes, have a high resistance to weather and chemicals, and are available in limitless colors. The three basic types of plastics are **thermoplastics, thermosetting plastics**, and **elastomers**.

Thermoplastics may be softened by heating and formed to the desired shape. If a polymer returns to its original hardness and strength after being heated, it is classified as a thermoplastic. In contrast, **thermosetting plastics** cannot be changed in shape by reheating after they have permanently set. **Elastomers** are rubberlike polymers that are soft, expandable, and elastic, which permits them to be deformed greatly and then return to their original size.

Figure 19.52 shows commonly used plastics and other materials, including glass and fiber-

Figure 19.53 The use of Dow plastic in this motorized wheelchair (made by Amigo, Inc.) reduced the number of parts by 97% and weight by 10%. It is also safe and easy to clean. (Courtesy of Dow Chemical Corporation.)

glass. The weights and yields of the materials are give, along with examples of their applications.

The motorized wheelchair shown in **Fig. 19.53** is made of plastic. It has fewer parts and weighs less than motorized wheelchairs made of metal. Its rounded corners make it safe, eliminate joints, and make it easy to fabricate and clean.

Ceramics is a material that is being more widely used in products. More innovative materials are expected to come into broad use in the future.

---

*Parts of this section are based on *Manufacturing Engineering and Technology, 2nd ed.*, Serope Kalpakjian, Addison-Wesley, 1992.

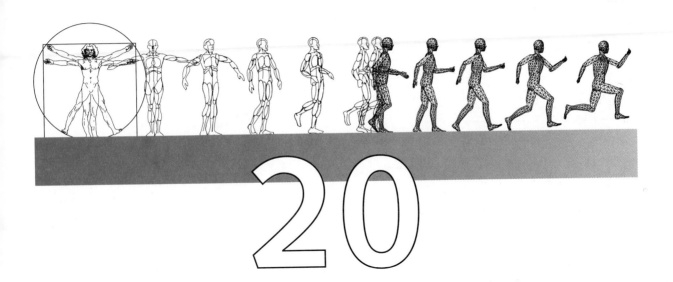

20

Dimensioning

20.1 Introduction

Working drawings must show dimensions and contain notes conveying sizes, specifications, and other information. With the addition of dimensions and notes, drawings will serve as construction documents and legal contracts.

The techniques of dimensioning presented are based primarily on the standards of the American National Standards Institute (ANSI), especially Y14.5M, *Dimensioning and Tolerancing for Engineering Drawings*. Standards of companies such as the General Motors Corporation also are used.

20.2 Terminology

The strap shown in **Fig. 20.1** is described in **Fig. 20.2** with orthographic views to which dimensions were added. Refer to this drawing as we introduce dimensioning terms.

Dimension lines: thin lines (2H–4H pencil) with arrows at each end and numbers placed near their midpoints to specify size.

Figure 20.1 This tapered strap is a part of a clamping device.

Extension lines: thin lines (2H–4H pencil) extending from the part and between which dimension lines are placed.

Centerlines: thin lines (2H–4H pencil) used to locate the centers of cylindrical parts such as holes.

319

A DIMENSIONED PART

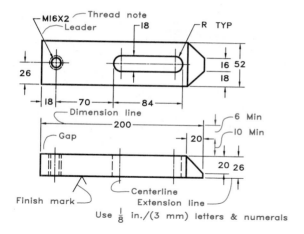

Figure 20.2 This typical dimensioned drawing of the tapered strap shown in Fig. 20.1 introduces the terminology of dimensioning.

Leaders: thin lines (2H–4H pencil) drawn from a note to the feature to which it applies.

Arrowheads: drawn at the ends of dimension lines and leaders and the same length as the height of the letters or numerals, usually 1/8 inch, as shown in **Fig. 20.3**.

Dimension numbers: placed near the middle of the dimension line and usually 1/8-inch high, with no units of measurement (", in., or mm) shown.

20.3 Units of Measurement

The two commonly used units of measurement are the decimal inch in the English (imperial) system, and the millimeter in the metric system (SI) **(Fig. 20.4)**. Giving fractional inches as decimals rather than common fractions makes arithmetic easier.

Figure 20.5 demonstrates the proper and improper dimensioning techniques with millimeters, decimal inches, and fractional inches. **In general, round off dimensions in millimeters to whole numbers without fractions.** However, when you

ARROWHEADS

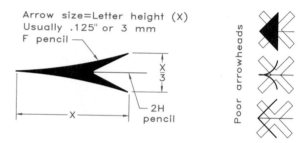

Figure 20.3 Draw arrowheads as long as the height of the letters used on the drawing and one third as wide as they are long.

MILLIMETERS VS. INCHES

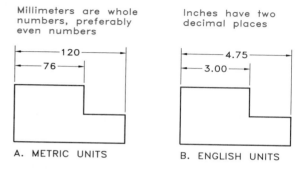

Figure 20.4 For the metric system, round millimeters to the nearest whole number. For the English system, show inches with two decimal places, even for whole numbers such as 3.00.

must show a metric dimension of less than a millimeter, use a zero before the decimal point. Do not use a zero before the decimal point when inches are the unit.

Show decimal inch dimensions with two-place decimal fractions, even if the last numbers are zeros. Omit units of measurement from the dimension because they are understood to be in millimeters or inches. For example, use 112 (not 112 mm) and 67 (not 67" or 5'-7").

Architects use combinations of feet and inches in dimensioning. They show foot marks but omit inch marks, as in 7'-2. Engineers use feet and dec-

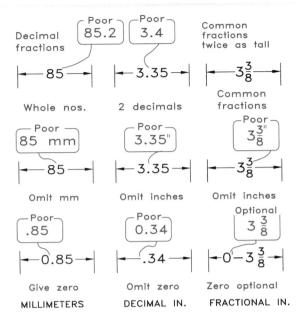

Figure 20.5 This drawing compares the proper and improper ways of notating units in SI and the English system.

imal fractions of feet to dimension large-scale projects such as road designs, as in 252.7'.

20.4 English/Metric Conversions

To convert dimensions in inches to millimeters, multiply by 25.4. Similarly, to convert dimensions in millimeters to inches, divide by 25.4.

When millimeter fractions are required as a result of conversion from inches, one-place fractions usually are sufficient, but two-place fractions are used in some cases. Find the decimal digit by applying the following rules.

- Retain the last digit unchanged if it is followed by a number less than 5; for example, round 34.43 to 34.4.

- Increase the last digit retained by 1 if it is followed by a number greater than 5; for example, round 34.46 to 34.5.

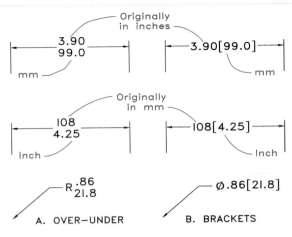

Figure 20.6 In dual dimensioning, place size equivalents in millimeters under or to the right of the inches (in brackets). Place the equivalent measurement in inches under or to the right of millimeters (in brackets). You may need to show millimeters converted from inches as decimal fractions.

- Retain the last digit unchanged if it is even and is followed by the digit 5; for example, round 34.45 to 34.4.

- Increase the last digit retained by 1 if it is odd and is followed by the digit 5; for example, round 34.75 to 34.8.

20.5 Dual Dimensioning

On some drawings you may have to give both metric and English units, called **dual dimensioning (Fig. 20.6)**. Place the millimeter equivalent either under or over the inch units, or place the converted dimension in brackets to the right of the original dimension. Be consistent in the arrangement you use on any set of drawings.

◨ **Computer Method** As **Fig. 20.7** shows, the dimensioning variable DIMALT must be set to ON to obtain alternative (dual) dimensions in brackets following the units originally used. Set the variable DIMALTF (scale factor) to the value of the multiplier to be used to change the first

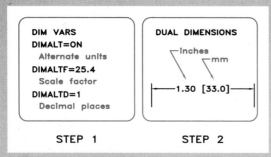

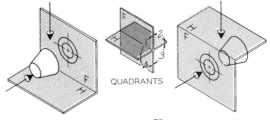

Figure 20.7 Producing alternate (dual) dimensions by computer:

Step 1 Set the Dim Vars to DIMALT to set dual dimensions to ON, assign the scale factor (DIMALTF), and specify the number of decimal places (DIMALTD).

Step 2 Find the linear dimensions by using the same steps as shown in Fig. 20.34. The dimension in brackets is the metric equivalent of the inch dimensions.

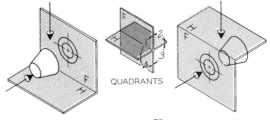

Figure 20.8

A The SI system uses the first angle of orthographic projection, which places the top view under the front view.

B The American system uses the third angle of projection, which places the top view over the front view.

Figure 20.9

A The SI symbol indicates that the millimeter is the unit of measurement, and the truncated cone specifies that third-angle projection was used to position the orthographic views.

B Again, the SI symbol denotes use of the millimeter, but the truncated cone designates that the first-angle projection was used.

dimension. Use DIMALTD to assign the desired number of decimal places for the second dimension. Then select dimensions by using the DIM command the same way you do to find single-value dimensions (see Fig. 20.34).

20.6 Metric Designation

Recall that in the metric system (SI) the first-angle of projection positions the front view over the top view and the right-side view to the left of the front view **(Fig. 20.8)**. You should label metric drawings with one of the symbols shown in **Fig. 20.9** to designate the angle of projection. Display either the letters *SI* or the word *metric* prominently in or near the title block to indicate that the measurements are metric.

20.7 Numeric and Symbolic Dimensioning

Vertical Dimensions

Vertical numeric dimensions on a drawing may be aligned or unidirectional. In the **unidirectional method** all dimensions appear in the standard

horizontal position **(Fig. 20.10A). In the aligned method numerals are parallel with vertical and angular dimension lines and read from the right-hand side of the drawing, never from the left-hand side (Fig. 20.10B).** Aligned dimensions are used almost entirely in architectural drawings where dimensions composed of feet, inches, and fractions are too long to fit well unidirectionally (such as 22'-10 1/2).

UNIDIRECTIONAL AND ALIGNED DIMENSIONS

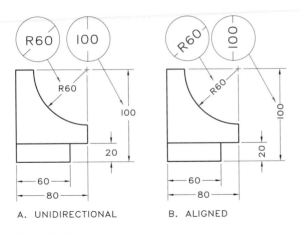

A. UNIDIRECTIONAL   B. ALIGNED

Figure 20.10

A Letter all unidirectional dimensions horizontally on a drawing.

B Aligned dimensions are lettered parallel to angular and vertical dimension lines to read from the right-hand side of the drawing (never from the left).

Computer Method The variable `DIMTIH` (dimensioning text inside dimension lines is horizontal), a `Dim Vars` of the `DIM` command of AutoCAD, must be set to `OFF` for aligned dimensions and to `ON` for unidirectional dimensions. The `DIMTOH` mode controls the positioning of text lying outside dimension lines where the numerals do not fit within a short dimension line. When `DIMTOH` is `ON`, the dimensions will be horizontal; when `OFF`, the dimensions will align with the dimension line.

Placement

Dimensions should be placed on the most descriptive views of the part being dimensioned. The first row of dimensions should be at least three times the letter height (3H) from the object **(Fig. 20.11)**. Successive rows of dimensions should be spaced equally at least two times the letter height apart (0.25 inch, or 6 mm, when 1/8-inch letters are used). Use the Braddock–Rowe lettering guide triangle to space the dimension lines **(Fig. 20.12)**.

PLACEMENT OF DIMENSIONS

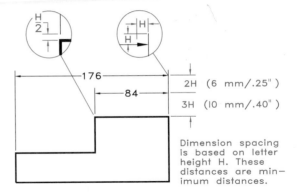

Figure 20.11 Place dimensions on a view as shown here, where all dimensioning geometry is based on the letter height (H) used.

COMMON FRACTION DIMENSIONS

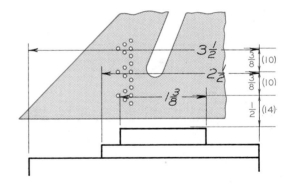

Figure 20.12 For common-fraction dimensions, draw guidelines for them by aligning the center holes in the Braddock–Rowe triangle with the dimension line.

Figure 20.13 illustrates how to place dimensions in limited spaces. **Regardless of space limitations, do not make numerals smaller than they appear elsewhere on the drawing.**

Computer Method **Figure 20.14** shows dimensioning variables and their minimum settings. You may change variables set at these proportions at the same time by using `DIMSCALE`. For example,

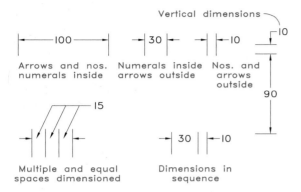

Figure 20.13 When space permits, place numerals and arrows inside extension lines. For smaller spaces use other placements, as shown.

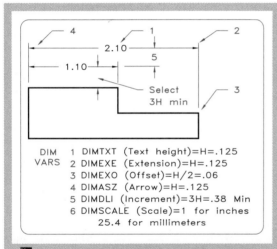

Figure 20.14 The assignment of dimensioning variables (Dim Vars) to be applied by AutoCAD is shown here. Once set, these variables remain active with the file in use. You may use DIMSCALE to enlarge or reduce all of these variables.

DIMSCALE = 25.4 would convert inch proportions to millimeter proportions. **Figure 20.15** shows other AutoCAD dimensioning variables (see Chapter 36 also).

Dim Vars	Default	Description
DIMALT	OFF	Alternate units selected
DIMALTD	2	Alternate unit decimal places
DIMALTF	25.4	Alternate unit scale factor
DIMAPOST	—	Default suffix for alternate text
DIMASO	ON	Create associative dimensions
DIMASZ	.125	Arrow length
DIMBLK	—	Arrow block name
DIMBLK1	—	First arrow block name
DIMBLK2	—	Second arrow block name
DIMCEN	.09	Center mark size
DIMCLRD	BYBLOCK	Dimension line color
DIMCLRE	BYBLOCK	Extension line & leader color
DIMCLRT	BYBLOCK	Dimension & extension color
DIMDLE	0	Dimension line extension
DIMDLI	.38	Dim. increment for cintinuation
DIMEXE	.125	Extension beyond dimension line
DIMEXO	.06	Extension line offset
DIMGAP	.06	Gap from dimension line to text
DIMLFAC	1	Length factor
DIMLIM	OFF	Gives tolerances in limit form
DIMPOST	—	Character suffix after dimensions
DIMRND	0	Rounding value for distances
DIMSAH	OFF	Separate arrowheads at each end
DIMSCALE	1	Scale factor for all dim. vars.
DIMSE1	OFF	Suppress first extension line
DIMSE2	OFF	Suppress second extension line
DIMSHO	ON	Changes dimens. while dragging
DIMSOXD	OFF	Suppress outside dimension lines
DIMSTYLE	*UNNAMED	Current dimensioning stype
DIMTAD	OFF	Text placed above dimension line
DIMTFAC	1	Tolerance text scale factor
DIMTIH	ON	Text inside extension lines horiz.
DIMTIX	OFF	Text forced inside extension lines
DIMTM	0	Minus tolerance value
DIMTP	0	Plus tolerance value
DIMTOFL	OFF	Forces dim. line inside, text out
DIMTOH	ON	Text outside ext. lines is horiz
DIMTOL	OFF	Applies tolerances to dimensions
DIMTSZ	0	Assigns size to ticks
DIMTVP	0	Text over or under dimen. line
DIMTXT	.125	Text height in dimension line
DIMZIN	0	Suppress 0 inches in ft & inch

Figure 20.15 The dimensioning variables (Dim Vars) shown here are introduced in this chapter and covered more thoroughly in Chapter 36.

Symbols

Figure 20.16 shows standard dimensioning symbols and their sizes based on the letter height, usually 1/8 inch. Use of these symbols instead of words saves drawing time.

20.8 Dimensioning by Computer

The AutoCAD program offers many features that are helpful in dimensioning. For example, **Fig. 20.17** shows several combinations of Dim Vars

DIMENSIONING SYMBOLS

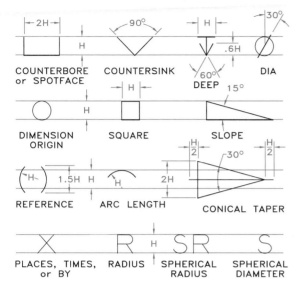

Figure 20.16 These symbols can be used instead of words to dimension parts. Their proportions are based on the letter height, H, which usually is ⅛ inch.

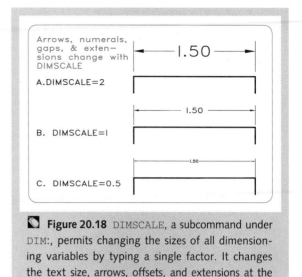

Figure 20.18 DIMSCALE, a subcommand under DIM:, permits changing the sizes of all dimensioning variables by typing a single factor. It changes the text size, arrows, offsets, and extensions at the same time.

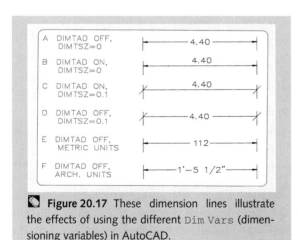

Figure 20.17 These dimension lines illustrate the effects of using the different Dim Vars (dimensioning variables) in AutoCAD.

(dimensioning variables) that are available with the DIM command.

You may place text inside the dimension line (DIMTAD: OFF) or above it (DIMTAD: ON). You may place arrowheads at the ends of dimension lines DIMASZ>0 or use tick marks (slashes) instead when DIMTSZ is set to a value greater than 0, usually about half the letter height. You may select UNITS as architectural (feet and inches), metric (no decimal fractions), decimal inches (two or more decimal fractions), or engineering units (feet and decimal inches).

You may use DIMSCALE (an option with the DIM command) to change all variables of dimensioning by typing a new DIMSCALE value **(Fig. 20.18)**. When invoked, the command changes the lengths of arrows, text size, and extension line offsets, without affecting other previously applied dimensions.

The current text STYLE is the dimensioning text. Had you set text to a specified height under the STYLE command, you could not change it

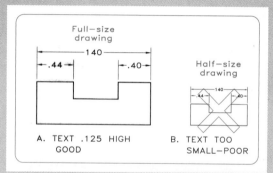

Figure 20.19 When dimensioning a drawing, be aware of its final plotted size so that you can size properly the dimensioning variables for reduction or enlargement. DIMSCALE is the most efficient command for assigning the proper scale to dimensioning variables.

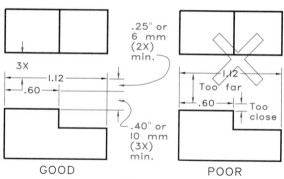

Figure 20.21 Place the first row of dimensions at least three times the letter height from the object. Successive rows should be at least two times the letter height apart.

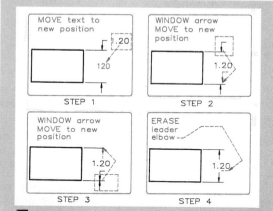

Figure 20.20 Editing dimensions by computer:

Step 1 You may want to MOVE dimension text inside the extension lines.

Step 2 You may MOVE arrows inside extension lines also.

Step 3 WINDOW a second arrow and MOVE it inside the extension lines.

Step 4 ERASE the unneeded leader extension to complete the modified dimension.

with a different DIMSCALE setting. For text height to be changeable, the text height set by the STYLE command must be 0 (zero). Then DIMSCALE will enlarge or reduce text height along with the other dimensioning variables (**Fig. 20.19**).

AutoCAD automatically performs many dimensioning operations, but occasionally you may want to modify the placement of arrows and text. **Figure 20.20** shows how to edit a dimension with the MOVE and ERASE commands.

20.9 Dimensioning Rules

Prisms

There are many rules of dimensioning, and you should become familiar with them in order to place dimensions and notes on drawings most effectively. **From time to time rules of dimensioning must be violated due to the complexity of the part or shortage of space. Figures 20.21–20.33** illustrate the fundamental rules of dimensioning prisms. We simplified these examples to focus on specific rules and to emphasize the logic behind them.

RULE 2: Place dimensions between the views

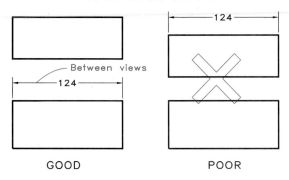

GOOD   POOR

Figure 20.22 Place dimensions between the views sharing these dimensions.

RULE 3: Dimension the most descriptive views

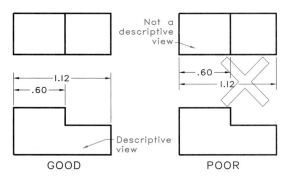

GOOD   POOR

Figure 20.23 Place dimensions on the most descriptive views of an object.

RULE 4: Dimension from visible lines, not hidden lines

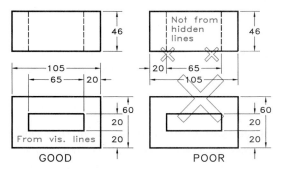

GOOD   POOR

Figure 20.24 Dimension visible features, not hidden features.

RULE 5: Give an overall dimension and omit one of the chain dimensions

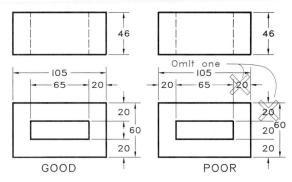

GOOD   POOR

Figure 20.25 Leave the last dimension blank in a chain of dimensions when you also give an overall dimension.

RULE 5 (DEVIATION): When one chain dimension is not omitted, mark one dimension as a reference dimension

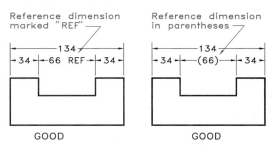

GOOD   GOOD

Figure 20.26 If you give all dimensions in a chain, mark the reference dimension (the one that would be omitted) with REF or place it in parentheses. Giving a reference dimension is a way of eliminating mathematical calculations in the shop.

RULE 6: Organize and align dimensions for ease of reading

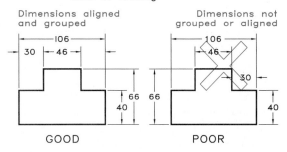

GOOD   POOR

Figure 20.27 Place dimensions in well-organized lines for uncluttered drawings.

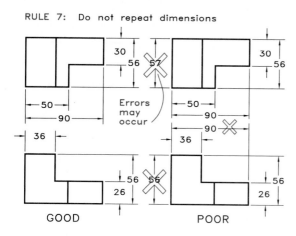

Figure 20.28 Do not duplicate dimensions on a drawing to avoid errors or confusion.

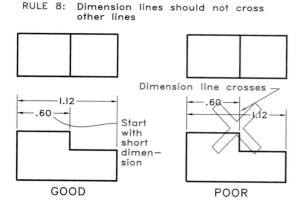

Figure 20.29 Dimension lines should not cross any other lines unless absolutely necessary.

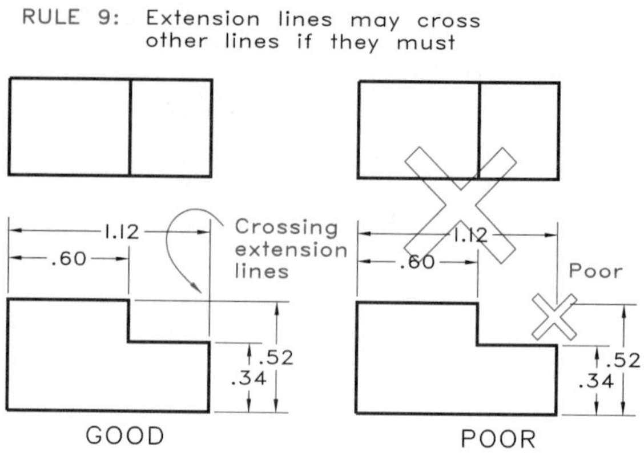

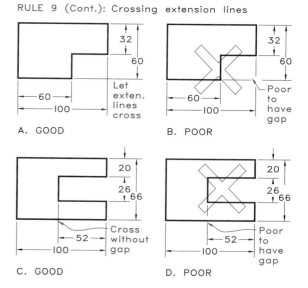

Figure 20.30 Extension lines may cross other extension lines or object lines if necessary.

Figure 20.31 Leave a small gap from the edges of an object to extension lines that extend from them. Do not leave gaps where extension lines cross object lines or other extension lines.

○ **Computer Method** Dimension the part shown in **Fig. 20.34** by entering the DIM command and selecting the LINEAR and HORIZONTAL options. Obtain extension lines automatically by responding to the First extension line origin prompt with a (CR) to get the prompt Select line, arc, or circle. Select the line to be measured and locate the dimension line when prompted.

If you set the dimensioning variable DIMASO to ON, dimensions will be associative dimensions.

That is, use of the STRETCH command updates the dimensioning numerals as you increase the part's size (**Fig. 20.35**). Associative dimensions will erase as a single unit (extension lines, dimension lines, arrows, and number). When DIMSHO

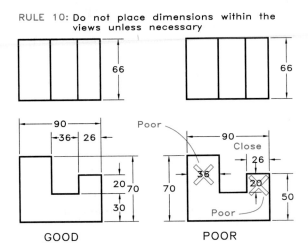

RULE 10: Do not place dimensions within the views unless necessary

66

66

90
├─36─┤ 26 ├─

90
Poor
Close

36
26

20 70
30

70 70

20

50

Poor

GOOD

POOR

Figure 20.32 Whenever possible place dimensions outside objects rather than inside their outlines.

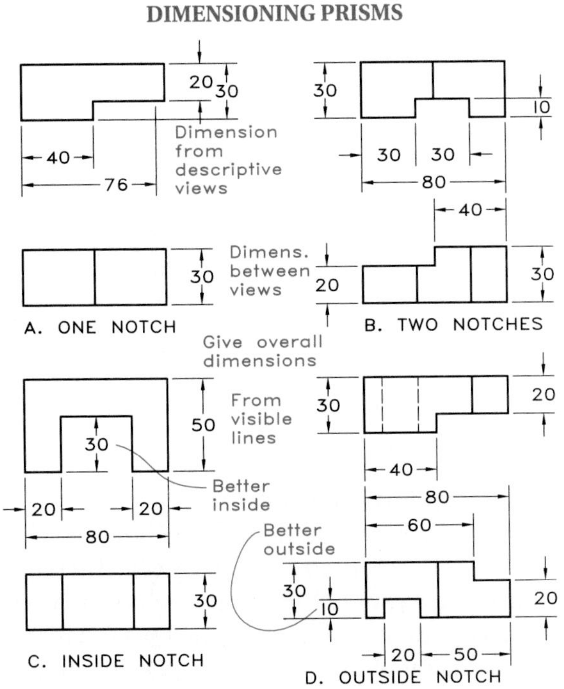

DIMENSIONING PRISMS

20 30
Dimension from descriptive views
├─40─┤
├─76─┤

30
10

├─30─┤30├─
├─80─┤

Dimens. between views
30
20

├─40─┤

30

A. ONE NOTCH

B. TWO NOTCHES

Give overall dimensions

From visible lines

50
30

30
20 20
├─80─┤

Better inside

20
30

├─40─┤
├─80─┤
├─60─┤

Better outside

30

30
10

20

├─20─┤─50─┤

C. INSIDE NOTCH

D. OUTSIDE NOTCH

Figure 20.33

A and **B** Dimension prisms from descriptive views and between views.

C You may dimension a notch inside the object if doing so improves clarity.

D Dimension visible lines, not hidden lines.

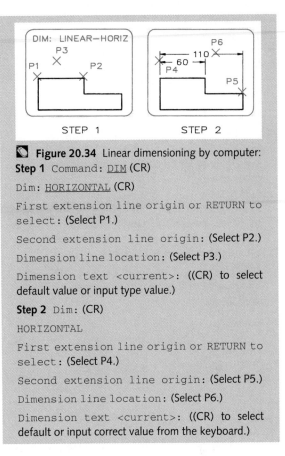

DIM: LINEAR—HORIZ
P3
P1 P2

P6
110
60
P4
P5

STEP 1   STEP 2

Figure 20.34 Linear dimensioning by computer:
Step 1 `Command:` `DIM` (CR)

`Dim:` `HORIZONTAL` (CR)

`First extension line origin or RETURN to select:` (Select P1.)

`Second extension line origin:` (Select P2.)

`Dimension line location:` (Select P3.)

`Dimension text <current>:` ((CR) to select default value or input type value.)

Step 2 `Dim:` (CR)

`HORIZONTAL`

`First extension line origin or RETURN to select:` (Select P4.)

`Second extension line origin:` (Select P5.)

`Dimension line location:` (Select P6.)

`Dimension text <current>:` ((CR) to select default or input correct value from the keyboard.)

is ON, you will be able to see the dimensioning numerals changing dynamically on the screen as you STRETCH the part to a new size.

Angles

You may dimension angles either by coordinates or angular measurements in degrees (**Fig. 20.36**). Recall that the units for angular measurements are degrees, minutes, and seconds; there are 60 minutes in a degree and 60 seconds in a minute. Seldom will you need to measure angles to the nearest second. **Figures 20.37** and **20.38** illustrate rules for dimensioning angles.

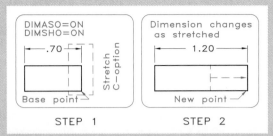

STEP 1   STEP 2

📎 **Figure 20.35** Producing associative dimensions by computer:

Step 1 Set `Dim Vars`, `DIMASO`, and `DIMSHO` to `ON` to associate the dimensions with the size of the part to which they apply. Use the `C` option of the `STRETCH` command to window the ends of an extension line and dimension lines.

Step 2 Select a base point and a new point. Dragging the size of the part recalculates the dimensions dynamically.

ANGULAR DIMENSIONS

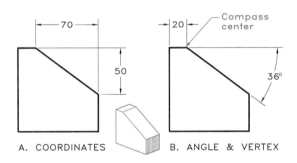

A. COORDINATES   B. ANGLE & VERTEX

Figure 20.36

A Dimension angular planes by using coordinates.

B Measure angles by locating the vertex and measuring the angle in degrees. When accuracy is essential, specify angles in degrees, minutes, and seconds.

📎 **Computer Method**   Select the `ANGULAR` option under the `DIM` command to dimension angles **(Fig. 20.39)**. If room is not available for the arrows between the extension lines, AutoCAD will generate them outside the extension lines.

RULE 11: Place angular dimensions outside the angle

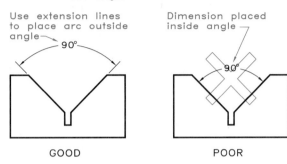

GOOD   POOR

Figure 20.37 Place angular dimensions outside angular notches by using extension lines.

RULE 12: Dimension rounded corners to the theoretical intersection

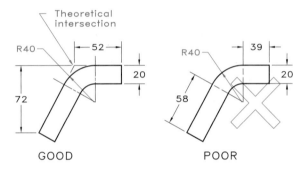

GOOD   POOR

Figure 20.38 Dimension a bent surface rounded corner by locating its theoretical point of intersection with extension lines.

Cylindrical Parts and Holes

The diameters of cylinders are measured with a micrometer **(Figs. 20.40** and **20.41)**. **Therefore dimension cylinders in their rectangular views with a diameter (Figs. 20.42 and 20.43).** Precede all diametral dimensions with the symbol Ø, which indicates that the dimension is a diameter. In the English system you also may use the abbreviation DIA after the diametral dimension.

Stagger dimensions for concentric cylinders to avoid crowding **(Fig. 20.44)**. Dimension cylindrical holes in their circular view with leaders **(Fig.**

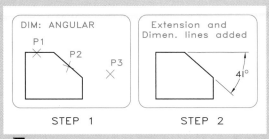

Figure 20.39 Producing angular dimensions by computer:

Step 1 Command: DIM (CR)

Dim: ANGULAR (CR)

Select first line: (Select P1.)

Second line: (Select P2.)

Enter dimension line arc location: (Select P3.)

Dimension text <41>: ((CR) to accept default value 41.)

Step 2 Enter text location: (CR)

Dim: ((CR) to continue angular dimensioning or Ctrl-C to return to a Command prompt.)

MEASURING EXTERNAL DIAMETER

Figure 20.41 This external micrometer caliper measures the diameter of a cylinder.

RULE 13: Dimension cylinders in their rectangular views with diameters

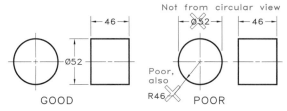

Figure 20.42 Dimension the diameter (not the radius) of a cylinder in the rectangular view.

MEASURING INTERNAL DIAMETER

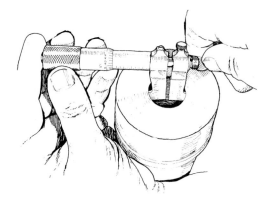

Figure 20.40 This internal micrometer caliper measures internal cylindrical diameters (radii cannot be measured).

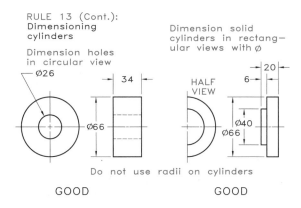

Figure 20.43 Dimension holes in their circular views with leaders. Dimension concentric cylinders with a series of diameters. Use half views to save drawing time.

RULE 14: Stagger dimension numerals to prevent crowding

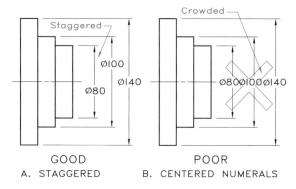

GOOD
A. STAGGERED

POOR
B. CENTERED NUMERALS

Figure 20.44 Dimensions on concentric cylinders are easier to read if they are staggered within their dimension lines.

RULE 15: Hole sizes are best given as dia— meters with leaders in circular views

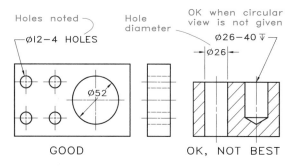

GOOD

OK, NOT BEST

Figure 20.45 Dimension holes in their circular view with leaders whenever possible, but dimension them in their rectangular views if necessary.

RULE 16: Leaders should have horizontal elbows and point toward the hole centers

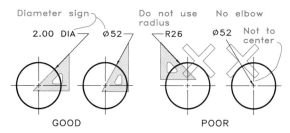

GOOD

POOR

Figure 20.46 Draw leaders pointing toward the centers of holes.

NOTES ON HOLES

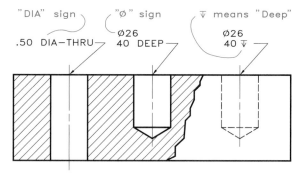

Figure 20.47 These are examples of holes noted in their rectangular views.

20.45). Draw leaders specifying hole sizes as shown in (**Fig. 20.46**). When you must place diameter notes for holes in the rectangular view instead of the circular view, draw them as shown in **Fig. 20.47**. **Figures 20.48** and **20.49** show examples of correctly dimensioned parts having cylindrical features.

◼ **Computer Method** Dimension circles in the circular view, as shown in **Fig. 20.50**. Dimension lines begin with the point selected on the circle and pass through the circle's center. The diameter symbol appears in front of the dimension numerals. Small circles are dimensioned with the arrows inside the circle and the dimension numerals outside, connected by a leader. Even smaller circles have both the arrows and dimension outside the circle.

Pyramids, Cones, and Spheres

Figure 20.51A–C show three methods of dimensioning pyramids. **Figure 20.51D** and **E** show two acceptable methods of dimensioning cones. Dimension a complete sphere by giving its diameter (**Fig. 20.51F**); give the radius if it is less than a hemisphere (**Fig. 20.51G**). Only one view is needed to describe a sphere.

CYLINDRICAL FEATURES

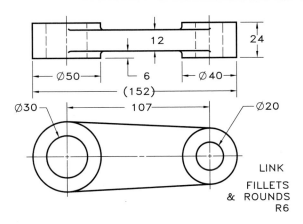

Figure 20.48 This drawing illustrates the application of dimensions to cylindrical features. (F&R R6 means that fillets and rounds have a 6 mm radius.)

DIMENSIONING A CYLINDRICAL PART

A

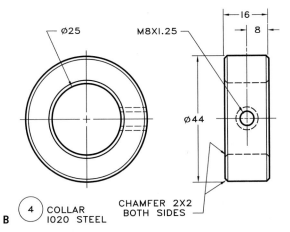

B

Figure 20.49 (A) This cylindrical part, a collar, is shown drawn and dimensioned in (B).

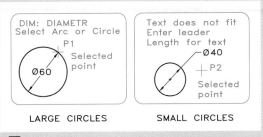

Figure 20.50 Dimensioning circles by computer: **Step 1** Select the `Diametr` option under the `DIM` command to dimension a circle. Select P1 on the arc (an endpoint of the dimension) to produce the dimension. You have the option of replacing the dimension measured by the computer with a different value.

Step 2 When text does not fit, you receive the prompt `Text does not fit, Enter leader length for text`. Select a point, P2, and the program generates the leader and dimension.

PYRAMIDS, CONES, AND SPHERES

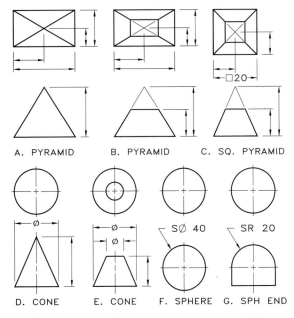

Figure 20.51 This drawing shows the proper way to dimension pyramids, cones, and spheres.

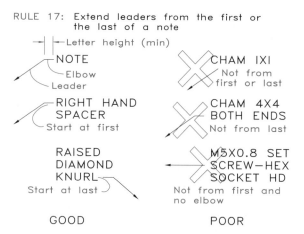

RULE 17: Extend leaders from the first or the last of a note

Letter height (min)

NOTE
Elbow
Leader

CHAM IXI
Not from
first or last

RIGHT HAND
SPACER
Start at first

CHAM 4X4
BOTH ENDS
Not from last

RAISED
DIAMOND
KNURL
Start at last

M5X0.8 SET
SCREW—HEX
SOCKET HD
Not from first and
no elbow

GOOD

POOR

Figure 20.52 Extend leaders from the first or the last word of a note with a horizontal elbow.

Leaders

Used to reference notes and dimensions to features, **leaders** most often are drawn at standard angles of triangles (see Fig. 20.46). As **Fig. 20.52** shows, leaders should begin at either the first or last word of a note, with a short horizontal line (elbow) from the note, and extend to the feature being described.

Computer Method To produce a leader that begins with an arrow, select the LEADER option under the DIM command **(Fig. 20.53)**. To ensure that the arrow touches the circle, use OSNAP and NEAREST to snap the point of the arrow to the circumference. The program will prompt you To point until you press (CR). Then the diameter of the circle will appear on the screen. You must type in the desired diameter because this command does not measure.

If the previous option was DIAMETR, the diameter symbol will precede the dimension. If the previous option was RADIUS, an R, the radius symbol, will precede the dimension.

Arcs and Radii

Dimension arcs with radii **(Fig. 20.54)**. Current standards specify that radii be dimensioned with

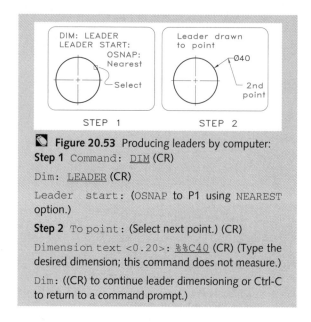

STEP 1   STEP 2

Figure 20.53 Producing leaders by computer:
Step 1 Command: DIM (CR)

Dim: LEADER (CR)

Leader start: (OSNAP to P1 using NEAREST option.)

Step 2 To point: (Select next point.) (CR)

Dimension text <0.20>: %%C40 (CR) (Type the desired dimension; this command does not measure.)

Dim: ((CR) to continue leader dimensioning or Ctrl-C to return to a command prompt.)

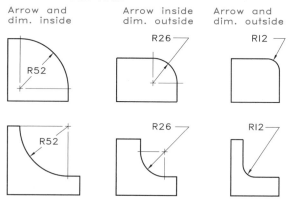

RULE 18: Dimension arcs (less than 180°) with radii

Arrow and dim. inside

Arrow inside dim. outside

Arrow and dim. outside

R52

R26

RI2

R52

R26

RI2

Figure 20.54 When space permits, place dimensions and arrows between the center and the arc. When space is not available for the number, place the arrow between the center and the arc number outside. If there is no space for the arrow inside, place both the dimension and arrow outside the arc with a leader.

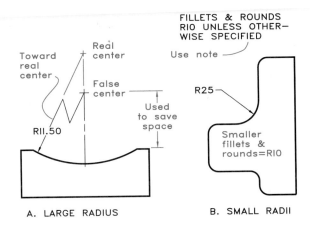

Figure 20.55 Dimensioning radii:

A Show a long radius with a false radius (a line with a zigzag) to indicate that it is not true length. Show its false center on the centerline of the true center.

B Specify fillets and rounds with a note to reduce repetitive dimensions of small arcs.

an R preceding the dimension (for example, R10). Previously, the standard was for R to follow the dimension (such as 10R). Thus both methods are seen on drawings.

You may dimension large arcs with a false radius (**Fig. 20.55**) by drawing a zigzag to indicate that the line does not represent the true radius. Where space is not available for radii, dimension small arcs with leaders.

■ Computer Method Use the RADIUS option of the DIM command to dimension arcs by selecting a point on the arc as the starting point of the arrow (**Fig. 20.56**). If space permits, the dimension will appear between the center and the arrow. For smaller arcs, the arrows will appear inside and the dimensions outside the arc.

When space is not available for arrows inside, both the arrow and dimension will appear outside the arc. Decimal values are preceded with a zero unless you override them by typing in R and the value without a preceding zero. Leading zeros will be omitted if the variable DIMZIN = 4.

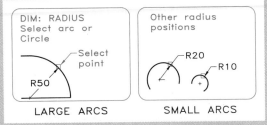

■ Figure 20.56 Use the RADIUS option of the DIM command to select a point on the arc. The program generates the dimension with its arrow at this point and precedes the dimension with an R. The program dimensions smaller arcs by positioning the dimension outside the arc or, because of even more limited space, by placing both the arrow and dimension outside the arc.

FILLETS AND ROUNDS

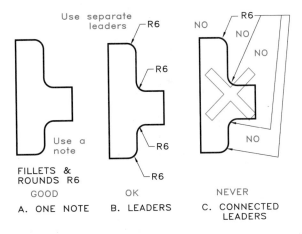

Figure 20.57 Indicate fillets and rounds by (A) notes or (B) separate leaders and dimensions. Never use confusing leaders (C).

Fillets and Rounds

When all fillets and rounds are equal in size, you may place a note on the drawing stating that condition or use separate notes (**Fig. 20.57**). If most, but not all, of the fillets and rounds have equal radii, the note may read **ALL FILLETS AND**

NOTES ON PARTS

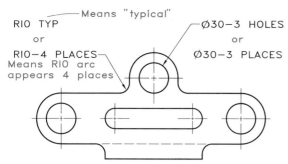

Figure 20.58 Use notes to indicate that identical features and dimensions are repeated on drawings to avoid having to dimension them individually.

ARCS AND CYLINDERS

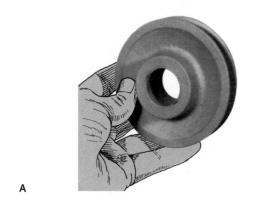

A

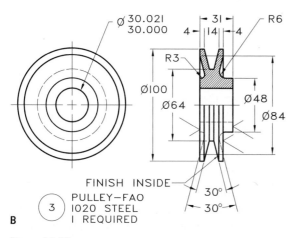

B

Figure 20.59

A This pulley comprises arcs and circles.

B This drawing illustrates how to dimension the part.

PARTS WITH ARCS

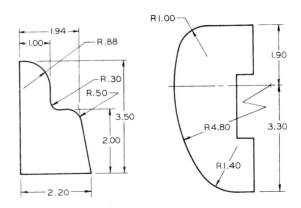

Figure 20.60 Dimension parts comprising a combination of tangent arcs and lines by using radii.

ROUNDS R6 UNLESS OTHERWISE SPECIFIED (or abbreviated as **F&R R6**), with the fillets and rounds of different radii dimensioned separately.

You may note repetitive features as shown in **Fig. 20.58** by using the notes **TYPICAL**, or **TYP**, which means that the dimensioned feature is typical of those not dimensioned. You may use the note **PLACES**, or **PL**, to specify the number of places that identical features appear, although only one is dimensioned.

Figure 20.59A shows a pulley that is drawn and dimensioned in **Fig. 20.59B**. The drawing demonstrates proper application of many of the rules discussed in this section.

20.10 Dimensioning Other Features

Curved Surfaces

You may dimension an irregular shape comprised of tangent arcs of varying sizes by using a series of radii (**Fig. 20.60**). Irregular curves may be dimensioned by using coordinates from two datum lines to locate a series of points along the curve (**Fig. 20.61**). Placing extension lines at an angle provides additional space for showing dimensions.

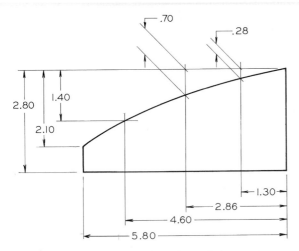

Figure 20.61 Dimensioning points located along an irregular curve requires use of coordinates.

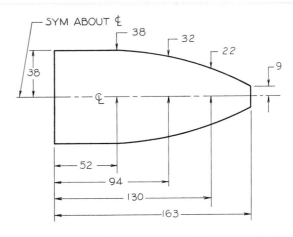

Figure 20.62 Dimensioning symmetrical parts bounded by irregular curves also requires the use of coordinates.

Symmetrical Objects

Dimension an irregular symmetrical curve with coordinates **(Fig. 20.62)**. Note the use of dimension lines as extension lines, a permissible violation of dimensioning rules.

Dimension other symmetrical objects by using coordinates to imply that the dimensions are symmetrical about the centerline (abbreviated **C**), as shown in **Fig. 20.63A**. **Figure 20.63B** shows a better method of dimensioning this type of object, where symmetry is explicit, not implicit.

20.11 Finished Surfaces

Parts formed in molds, called castings, have rough exterior surfaces. If these parts are to assemble with and move against other parts, they will not function well unless their contact surfaces are machined to a smooth finish by grinding, shaping, lapping, or a similar process.

To indicate that a surface is to be finished, finish marks are drawn on edge views of surfaces to be finished (Fig. 20.64). Finish marks should be shown in every view where finished surfaces appear as edges even if they are hidden lines.

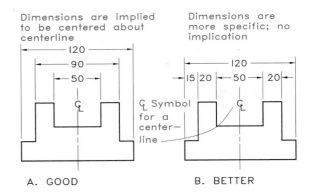

Figure 20.63

A You may dimension symmetrical parts implicitly about their centerlines.

B The better way to dimension symmetrical parts is explicitly about their centerlines.

The preferred finish mark for the general cases is the uneven mark shown in **Fig. 20.64B**. **When an object is finished on all surfaces, the note FINISHED ALL OVER (abbreviated FAO) is placed on the drawing.**

FINISH MARKS

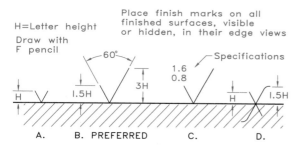

Figure 20.64 Finish marks indicate that a surface is to be machined to a smooth surface.

A The traditional V can be used for general applications.

B The unequal finish mark is the best one for general applications.

C Where surface texture must be specified, this finish mark is used with texture values.

D The f-mark is the least used symbol.

LOCATION DIMENSIONS

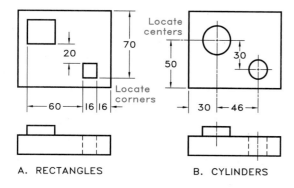

Figure 20.65 Location dimensions give the positions of geometric features with respect to other geometric features, but not their sizes.

20.12 Location Dimensions

Location dimensions give the positions, not the sizes, of geometric shapes (**Fig. 20.65**). Locate rectangular shapes by using coordinates of their corners and cylindrical shapes by using coordinates of their centerlines. In each case dimension the view that shows both measurements. Always

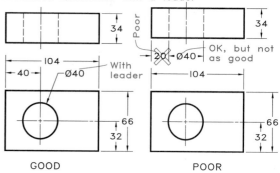

Figure 20.66 Locate cylindrical holes in their circular views by coordinates to their centers.

LOCATING CYLINDERS

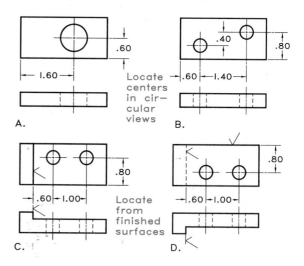

Figure 20.67

A Locate cylindrical holes in their circular views from two surfaces of the object.

B Locate multiple holes from center to center.

C Locate holes from finished surfaces, even if the finished surfaces are hidden as in (D).

extend coordinates from any finished surfaces (even if a finished surface is a hidden line) because smooth machined surfaces allow the most accurate measurements. Locate and dimension single holes as shown in **Fig. 20.66** and multi-

LOCATION DIMENSIONS

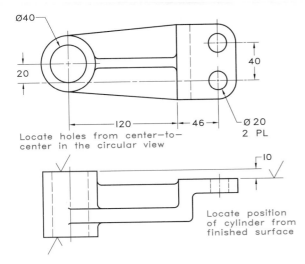

Figure 20.68 This example of a dimensioned shaft arm shows the application of location dimensions, with sizes omitted.

BASELINE DIMENSIONING

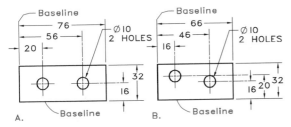

Figure 20.69 Measuring holes from two datum planes is the most accurate way to locate them and reduces the accumulation of errors possible in chain dimensioning.

BOLT CIRCLE

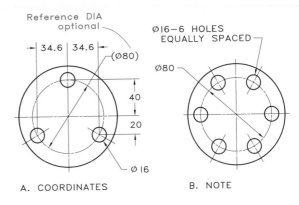

Figure 20.70 Locate holes on a bolt circle by using (A) coordinates or (B) notes.

CIRCLE OF CENTERS

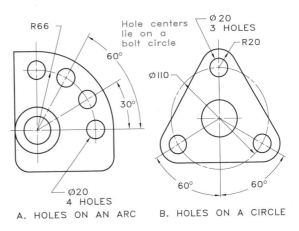

Figure 20.71 Locate centers of holes by using (A) a combination of radii and degrees or (B) a circle of centers.

ple holes as shown in **Fig. 20.67**. **Figure 20.68** shows the application of location dimensions to a typical part, with size dimensions omitted.

Baseline dimensions extend from two baselines in a single view (**Fig. 20.69**). The use of baselines eliminates the possible accumulation of errors in size from chain dimensioning.

You may locate holes through circular plates by using coordinates or a note (**Fig. 20.70**).

Dimension the diameter of the imaginary circle passing through the centers of the holes in the circular view as a reference dimension and locate the holes by using coordinates (**Fig. 20.70A**) or a note (**Fig. 20.70B**). This imaginary circle is called the **bolt circle** or **circle of centers**.

You may also locate holes with radial dimensions and their angular positions in degrees (**Fig. 20.71**). Holes may be located on their bolt circle even if the shape of the object is not circular.

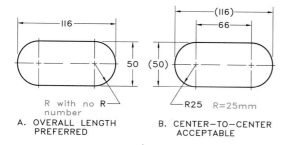

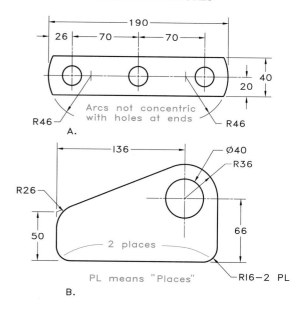

Figure 20.72

A Dimension objects having rounded ends from end-to-end and give the height.

B A less desirable choice is to dimension the rounded ends from center-to-center and give the radius.

Figure 20.73 These drawings show how to dimension parts having (A) rounded ends that are not concentric with the holes and (B) rounds and cylinders.

Objects with Rounded Ends

Dimension objects with rounded ends from one rounded end to the other (**Fig. 20.72A**) and show their radii as R without dimensions to specify that the ends are arcs. Obviously, the end radius is half the height of the part.

If you dimension the object from center to center (**Fig. 20.72B**), you must give the radius size. You may specify the overall width as a reference dimension (116) to eliminate the need for calculations.

Dimension parts with partially rounded ends as shown in **Fig. 20.73A**. Dimension objects with rounded ends that are smaller than a semicircle with a radius and locate the arc's center (**Fig. 20.73B**).

Dimension a single slot with its overall width and height (**Fig. 20.74A**). When there are two or more slots, dimension one slot and use a note to indicate that there are other identical slots (**Fig. 20.74B**).

The tool holder table shown in **Fig. 20.75** illustrates dimensioning of arcs and slots. To prevent dimension lines from crossing, several dimensions are placed on a less descriptive view.

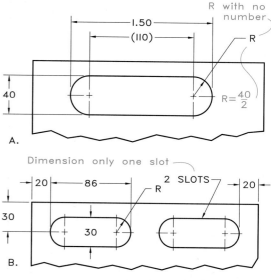

Figure 20.74 These drawings illustrate methods of dimensioning parts that have (A) one slot and (B) more than one slot.

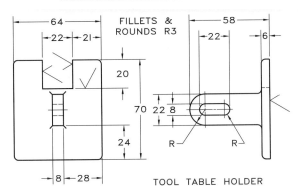

Figure 20.75 This dimensioned part has both slots and arcs.

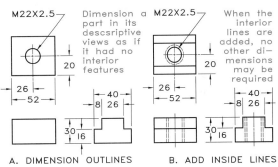

Figure 20.76 Outline dimensioning is the placement of dimensions on views as if they had no internal lines. Practice in applying this concept will help you place dimensions on their most descriptive views.

20.13 Outline Dimensioning

Now that you are familiar with most of the rules of dimensioning, you can better understand outline dimensioning, which is a way of applying dimensions to a part's outline (silhouette). By taking this approach, you have little choice but to place dimensions in the most descriptive views of the part. For example, imagine that the T-block shown in **Fig. 20.76** has no lines inside its outlines. It is dimensioned beginning with its location dimensions. When the inside lines are added, additional dimensions are seldom needed.

Figure 20.77 shows an example of outline dimensioning. Note the extension of all dimensions from the outlines of well-defined features.

20.14 Machined Holes

Machined holes are formed by machine operations, such as drilling or boring (**Fig. 20.78**). Give the diameter of the hole with the symbol Ø in front of its dimension (for example, Ø32) with a leader extending from the circular view. You may also note hole diameters with DIA after their size (for example, 2.00 DIA). Occasionally a machining operation is included in the note, such as 32 DRILL, but the omission of machining operations is preferable.

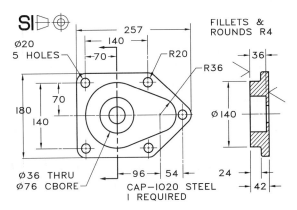

Figure 20.77 This drawing illustrates use of the outline method to dimension the cap in its most descriptive views.

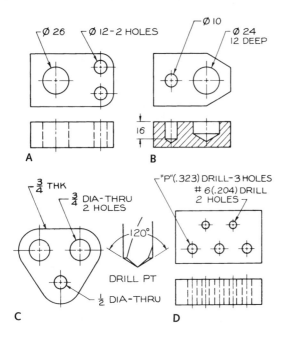

Figure 20.78

A and **B** You may dimension cylindrical holes by either of these methods.

C and **D** When you use only one view, you have to note the THRU holes or specify their depths.

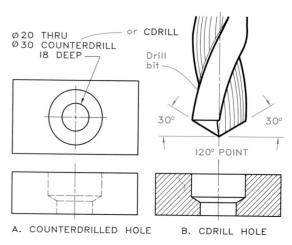

Figure 20.79 Counterdrilling notes give the specifications for drilling a larger hole inside a smaller hole. Do not dimension the 120° angle because it is a byproduct of the drill point. Noting the counterdrill with a leader from the circular view is preferable.

Drilling **Drilling** is the basic method of making holes. Dimension the size of a drilled hole with a leader extending from its circular view. You may give its depth in the note or dimension it in the rectangular view (**Fig. 20.78B**). **Dimension the depth of a drilled hole to the usable part of the hole, to the shoulder of the conical point.**

Counterdrilling involves drilling a large hole inside a smaller hole to enlarge it (**Fig. 20.79**). The drill point leaves a 120° angle as a byproduct of counterdrilling.

Countersinking is the process of forming conical holes for receiving screw heads (**Fig. 20.80**). Give the diameter of a countersunk hole (the maximum diameter on the surface) and the angle of the countersink in a note. Countersunk holes also are used as guides in shafts, spindles, and other cylindrical parts held between the centers of a lathe.

Spotfacing is the process of finishing the surface around holes to provide bearing surfaces for washers or bolt heads (**Fig. 20.81A**). **Figure 20.81B** shows the method of spotfacing a boss (a raised cylindrical element).

Boring **Boring** is used to make large holes and it is usually done on a lathe with a bore or a boring bar (**Fig. 20.82**). **Counterboring** is the process of enlarging the diameter of a drilled hole (**Fig. 20.83**) to give a flat hole bottom without the tapers as in counterdrilled holes.

Reaming is the operation of finishing or slightly enlarging drilled or bored holes within their prescribed tolerances. A ream is similar to a drill bit.

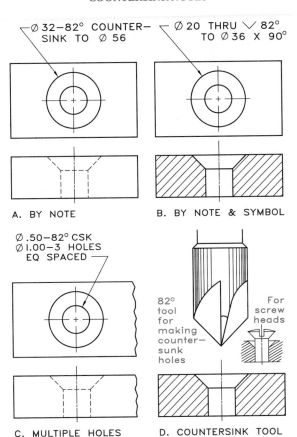

A. BY NOTE   B. BY NOTE & SYMBOL

C. MULTIPLE HOLES   D. COUNTERSINK TOOL

Figure 20.80 These illustrations show methods of noting and specifying countersunk holes for receiving screw heads.

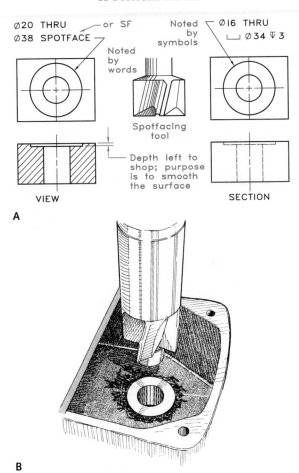

A

B

Figure 20.81

A Spotfacing is the smoothing of the surface for contact with a washer or a nut or bolt; dimension for spotfacing as shown.

B This spotfacing tool is used to finish the cylindrical boss to provide a smooth seat for a bolt head.

20.15 Chamfers

Chamfers are beveled edges cut on cylindrical parts, such as shafts and threaded fasteners, to eliminate sharp edges and to make them easier to assemble. When the chamfer angle is 45°, use a note in either of the forms shown in **Fig. 20.84A**. Dimension chamfers of other angles as shown in **Fig. 20.84B**. When inside openings of holes are chamfered, dimension them as shown in **Fig. 20.85**.

20.16 Keyseats

A **keyseat** is a slot cut into a shaft for aligning and holding a pulley or a collar on a shaft. **Figure 20.86** shows how to dimension keyways and keyseats with dimensions taken from the tables in the Appendix. The double dimensions on the diameter are tolerances (discussed in Chapter 12).

Figure 20.82 This photo shows the use of a lathe to bore a large hole with a boring bar. (Courtesy of Clausing Corporation.)

CHAMFERS

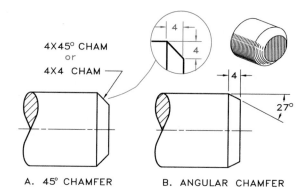

A. 45° CHAMFER   B. ANGULAR CHAMFER

Figure 20.84
A Dimension 45° chamfers by one of the methods shown.
B Dimension chamfers of all other angles as shown.

INSIDE CHAMFERS

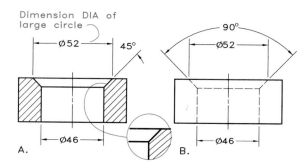

Figure 20.85 Dimension chamfers on the insides of cylinders as shown.

COUNTERBORED HOLES

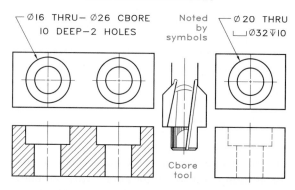

Figure 20.83 Counterbored holes are similar to counterdrilled holes but have flat bottoms instead of tapered sides. Dimension them as shown.

20.17 Knurling

Knurling is the operation of cutting **diamond-shaped** or **parallel** patterns on cylindrical surfaces for gripping, decoration, or press fits between mating parts that are permanently assembled. Draw and dimension diamond knurls and straight knurls as shown in **Fig. 20.87**, with notes that specify type, pitch, and diameter.

The abbreviation DP means diametral pitch, or the ratio of the number of grooves on the cir-

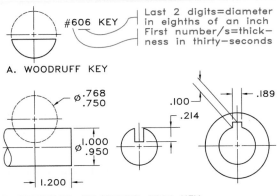

A. WOODRUFF KEY

B. KEYSEATS: WOODRUFF #606 KEY

Figure 20.86 These drawings show methods of dimensioning (A) Woodruff keys and (B) keyways used to hold a part on a shaft. Appendix 29 gives their tables of sizes.

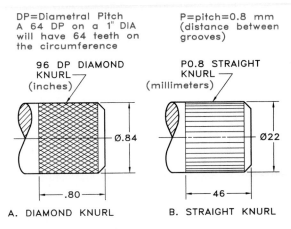

A. DIAMOND KNURL   B. STRAIGHT KNURL

Figure 20.87 The diamond knurl has a diametral pitch, DP, of 96 and the straight knurl has a linear pitch, P, of 0.8 mm. Pitch is the distance between the grooves on the circumference.

NECKS

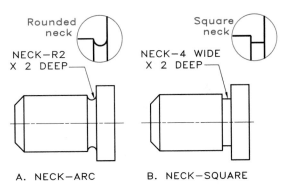

A. NECK—ARC   B. NECK—SQUARE

Figure 20.88 Necks are recesses cut in cylinders, with rounded or square bottoms, usually at the intersections of concentric cylinders. Dimension necks as shown.

cumference (N) to the diameter (D) expressed as DP = N/D. The preferred diametral pitches for knurling are 64 DP, 96 DP, 128 DP, and 160 DP.

For diameters of 1 inch, knurling of 64 DP, 96 DP, 128 DP, and 160 DP will have 64, 96, 128, and 160 teeth, respectively, on the circumference. The note P0.8 means that the knurling grooves are 0.8 mm apart. Make knurling calculations in inches and then convert them to millimeters. Specify knurls for press fits with the diameter size before knurling and with the minimum diameter size after knurling.

20.18 Necks and Undercuts

A **neck** is a groove cut around the circumference of a cylindrical part. If cut where cylinders of different diameters join (**Fig. 20.88**), a neck ensures that the assembled parts fit flush at the shoulder of the larger cylinder and allows trash that would cause binding to drop out of the way.

An **undercut** is a recessed neck inside a cylindrical hole (**Fig. 20.89A**). A **thread relief** is a neck that has been cut at the end of a thread to ensure that the head of the threaded part will fit flush against the part it screws into (**Fig. 20.89B**).

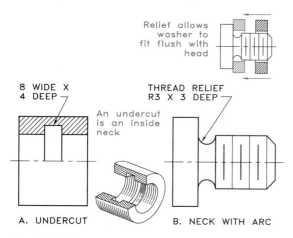

Figure 20.89

A An undercut is a groove cut inside a cylinder.

B A thread relief is a groove cut at the end of a thread to improve the screw's assembly. Dimension both types of necks as shown.

TAPER NOTES

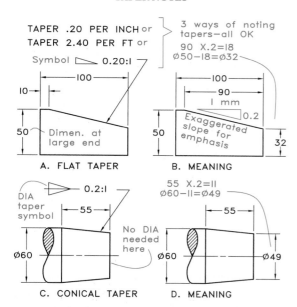

Figure 20.90 Tapers may be specified for either flat or conical surfaces: dimensioning and interpretation for (A and B) a flat taper and (C and D) for a conical taper.

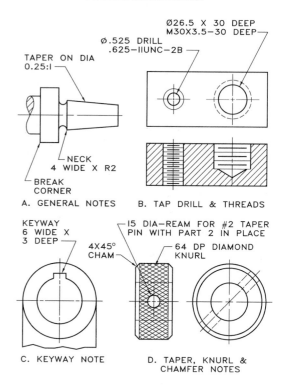

Figure 20.91

A The notes for this part indicate a neck, a taper, and a break corner, which is a slight round to remove sharpness from a corner.

B Threaded holes are sometimes dimensioned by giving the tap drill size in addition to the thread specifications, but selection of the tap drill size usually is left to the shop.

C This note is for dimensioning a keyway.

D The notes for this collar call for knurling, chamfering, and drilling for a #2 taper pin.

20.19 Tapers

Tapers for both flat planes and conical surfaces may be specified with either notes or symbols. **Flat taper** is the ratio of the difference in the heights at each end of a surface to its length (**Fig. 20.90A** and **B**). Tapers on flat surfaces may be

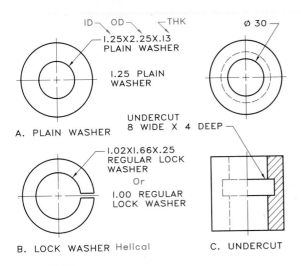

Figure 20.92

A and **B** Dimension washers and lock washers as shown by taking sizes from the tables in the Appendix.

C Dimension an undercut with a note.

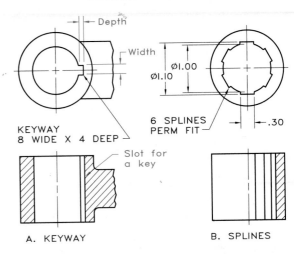

Figure 20.93 These drawings illustrate how to dimension (A) keyways and (B) splines.

expressed as inches per inch (.20 per inch), inches per foot (2.40 per foot), or millimeters per millimeter (0.20:1).

Conical taper is the ratio of the difference in the diameters at each end of a cone to its length (**Fig. 20.90C** and **D**). Tapers on conical surfaces may be expressed as inches per inch (.25 per inch), inches per foot (3.00 per foot), or millimeters per millimeter (0.25:1).

20.20 Miscellaneous Notes

Notes on detail drawings provide information and specifications that would be difficult to represent by drawings alone (**Figs. 20.91–20.93**). Place notes horizontally on the sheet whenever possible, because they are easier to letter and read in that position. Several notes in sequence on the same line should be separated with short dashes between them (for example, 15 DIA-30 DIA SPOTFACE). Use standard abbreviations (see Appendix 1) in notes to save space and time.

Problems

1–24 (a–b). (**Figs. 20.94** and **20.95**) Solve these problems on size A paper, one per sheet, if you draw them full size. If you draw them double size, use size B paper. The views are drawn on a 0.20-in. (5-mm) grid.

You will need to vary the spacing between views to provide adequate room for the dimensions. Sketching the views and dimensions to determine the required spacing before laying out the solutions with instruments would be helpful. Supply lines that may be missing in all views.

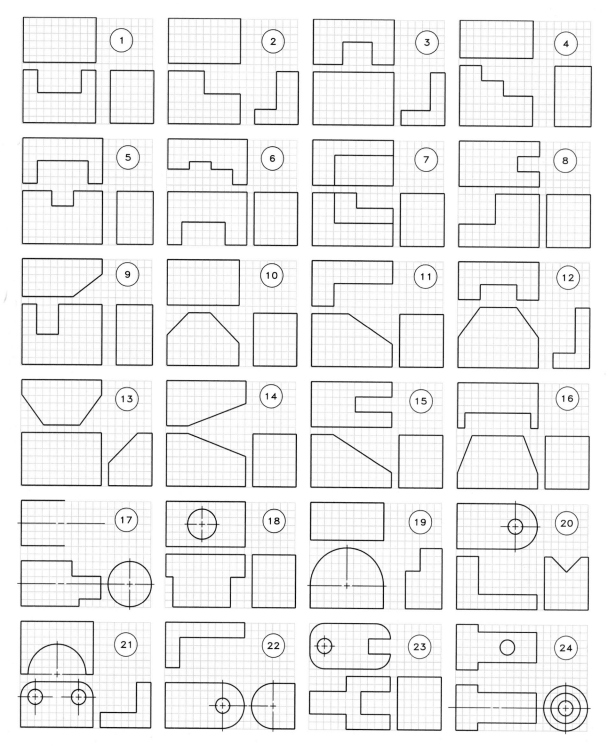

Figure 20.94 Problems 1–24(a). Lay out the views, supply missing lines, and dimension them.

Figure 20.95 Problems 1–24(b). Lay out the views, supply missing lines, and dimension them.

Tolerances

21.1 Introduction

Today's technology requires that parts be specified with increasingly exact dimensions. Many parts made by different companies at widely separated locations must be interchangeable, which requires precise size specifications and production.

The technique of dimensioning parts within a required range of variation to ensure interchangeability is called **tolerancing**. Each dimension is allowed a certain degree of variation within a specified zone, or **tolerance**. For example, a part's dimension might be expressed as 20 ± 0.50, which allows a tolerance (variation in size) of 1.00 mm.

A tolerance should be as large as possible without interfering with the function of the part to minimize production costs. Manufacturing costs increase as tolerances become smaller.

The cut-away view of the shaft journal in **Fig 21.1** illustrates parts that must rotate smoothly on ball bearings. If these parts were not toleranced and manufactured to a high degree of accuracy, they would not function properly.

Figure 21.1 This shaft journal would not rotate smoothly on the ball bearings if adjacent parts had not been manufactured within extremely close tolerances.

21.2 Tolerance Dimensions

Figure 21.2 shows three methods of specifying tolerances on dimensions: Unilateral, bilateral, and limit forms. When plus-or-minus tolerancing

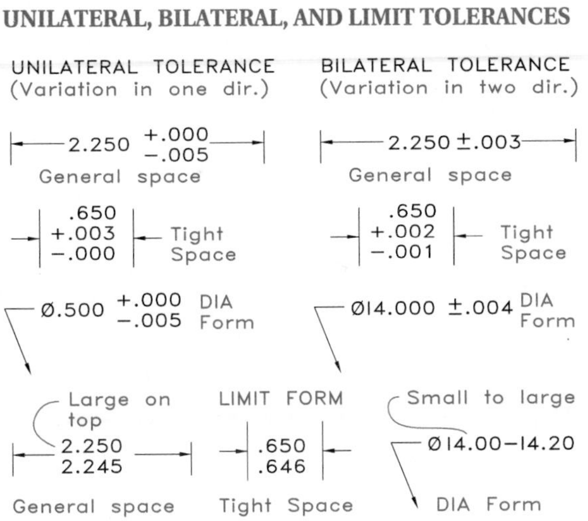

UNILATERAL, BILATERAL, AND LIMIT TOLERANCES
UNILATERAL TOLERANCE
(Variation in one dir.)
BILATERAL TOLERANCE
(Variation in two dir.)
2.250 +.000 −.005
General space
2.250 ±.003
General space
.650 +.003 −.000
Tight Space
.650 +.002 −.001
Tight Space
Ø.500 +.000 −.005
DIA Form
Ø14.000 ±.004
DIA Form
Large on top
2.250
2.245
General space
LIMIT FORM
.650
.646
Tight Space
Small to large
Ø14.00−14.20
DIA Form

Figure 21.2 These methods properly position and indicate tolerances in unilateral, bilateral, and limit forms for both general and tight spaces.

ORDER OF NUMBERS
Large limit on top
22.20
22.00
A. LIMITS
Plus tolerance on top
46.00 +0.40 −0.20
B. PLUS−MINUS
Small limit first
Ø26.00−26.40
C. LIMITS
Ø76.0 ±0.2
D. PLUS−MINUS

Figure 21.3 Place upper limits either above or to the right of lower limits. In plus-and-minus tolerancing, place the plus limits above the minus limits.

POSITIONING OF NUMBERS
H=1/8
Same no. of decimal places
H/2=1/16
H MAX
2.0000 +.0040 −.0020
A. PLUS−MINUS TOLERANCES
H/2=1/16
2.0400
1.9980
B. LIMIT−FORM TOLERANCES

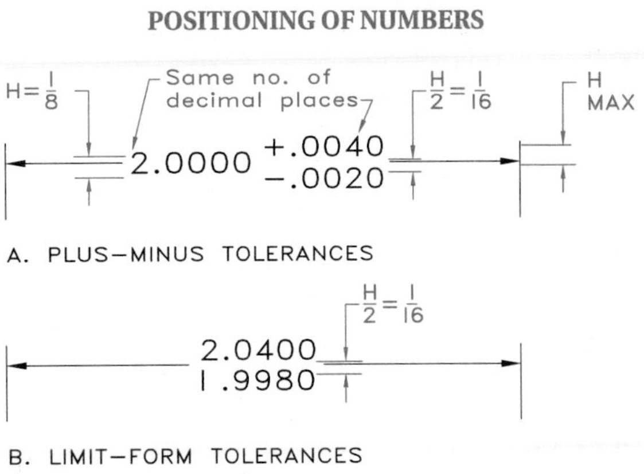

Figure 21.4 This drawing shows the spacing and ratios of numerals used to specify tolerances on dimensions.

is used, it is applied to a theoretical dimension called the **basic dimension**. When dimensions can vary in only one direction from the basic dimension (either larger or smaller) tolerancing is **unilateral**. Tolerancing that permits variation in both directions from the basic dimension (larger and smaller) is **bilateral**.

Tolerances may also be given in **limit form**, with dimensions representing the largest and smallest sizes for a feature. When tolerances are shown in limit form, the basic dimension will be unknown.

Figure 21.3 shows the customary methods of applying tolerance values on dimension lines. **Figure 21.4** shows the ratios of tolerance numerals placed in dimension lines.

Tolerances by Computer

AutoCAD allows toleranced dimensions to be shown in limit form or plus-and-minus form **(Fig. 21.5)**. `DIMLIM` must be turned `ON` to obtain dimensions in limit form. Assign tolerances to the `DIMTM` and `DIMTP` modes under the `DIM`: command.

In addition to linear dimensions, diametral and radial dimensions are automatically given with either a circle symbol or an R preceding the dimensions **(Figs. 21.6** and **21.7)**. You may tolerance angular measurements by using the plus-and-minus form or the limit form **(Fig. 21.8)**. The program measures the angle and automatically computes the upper and lower limits. You may edit

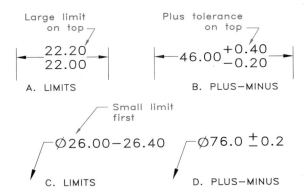

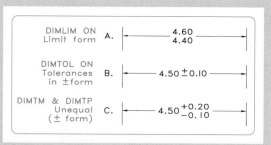

Figure 21.5 AutoCAD gives toleranced dimensions in the forms shown. When DIMLIM is ON, tolerances will be applied in limit form; when DIMTOL is ON, tolerances will be in plus-or-minus form.

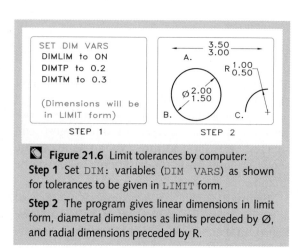

Figure 21.6 Limit tolerances by computer:
Step 1 Set DIM: variables (DIM VARS) as shown for tolerances to be given in LIMIT form.

Step 2 The program gives linear dimensions in limit form, diametral dimensions as limits preceded by Ø, and radial dimensions preceded by R.

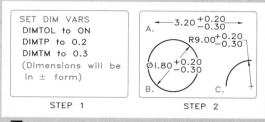

Figure 21.7 Plus-or-minus tolerances by computer:
Step 1 Set DIM: variables (DIM VARS) as shown with DIMTOL set to ON.

Step 2 The program will give linear dimensions as a basic diameter followed by plus-and-minus tolerances, diametral dimensions as linear dimensions preceded by Ø, and radial dimensions preceded by R.

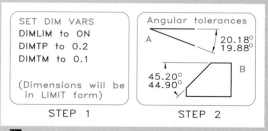

Figure 21.8 Angular tolerances by computer:
Step 1 Set DIM VARS as shown above. Use the UNITS command to preassign the desired number of decimal places for angular units.

Step 2 Select the lines forming the angles and the location of the arc, and the toleranced measurements will be given in limit form. Using the UNITS command, you may obtain angular measurements as decimal degrees, minutes and seconds, grads, radians, or surveyor's units.

a toleranced dimension by erasing and moving its parts to make the desired changes (**Fig. 21.9**).

21.3 Mating Parts

Mating parts must be toleranced to fit within a prescribed degree of accuracy (**Fig. 21.10**). The upper part is dimensioned with limits indicating its maximum and minimum sizes. The notch in the lower part is toleranced to be slightly larger, allowing the parts to assemble with a clearance fit.

Mating parts also may be cylindrical forms, such as a pulley, bushing, and shaft (**Fig. 21.11**).

The bushing should force fit inside the pulley to provide a good bearing surface for the rotating shaft. At the same time, the shaft and the bushing should mate so that the pulley and bushing will rotate on the shaft with a free running fit.

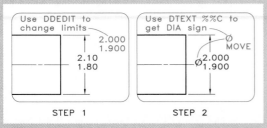

Figure 21.9 Editing tolerances by computer:
Step 1 When you want to change a toleranced dimension, use DDEDIT, select the given tolerances, and type new values.

Step 2 To add a diameter symbol, Ø, type % % C, and move it in front of the new limits.

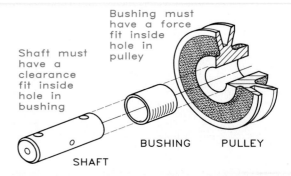

Figure 21.11 These parts must be assembled with cylindrical fits that give a clearance and an interference fit.

MATING PARTS

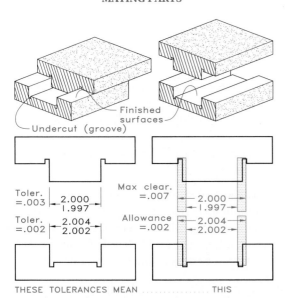

Figure 21.10 These mating parts have tolerances (variations in size) of 0.003″ and 0.002″ respectively. The allowance (tightest fit) between the assembled parts is 0.002″.

ANSI tables (see Appendixes 34–38) prescribe cylindrical-fit tolerances for different applications. Familiarity with the terminology of cylindrical tolerancing is essential to use these tables.

21.4 Tolerancing: English Units

Terminology

Figure 21.12, showing mating of cylindrical parts, illustrates the following tolerancing terminology and definitions.

Tolerance: the difference between the limits prescribed for a single feature, or 0.0025 in. for the shaft and 0.0040 for the hole **(Fig. 21.12A)**.

Limits of tolerance: the extreme measurements permitted by the maximum and minimum sizes of a feature, or 1.4925 and 1.4950 for the shaft and 1.5000 and 1.5040 for the hole **(Fig. 21.12B)**.

Allowance: the tightest fit between the two mating parts, or +0.0050 **(Fig. 21.12C)**, allowance is negative for an interference fit.

Nominal size: an approximate size of shaft and hole, usually expressed with common fractions, or 1.50 in. (1 1/2 in.) **(Fig. 21.12)**.

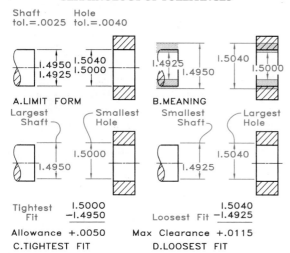

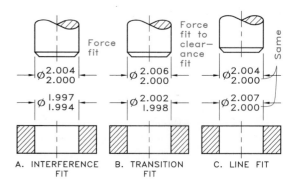

Figure 21.12 The allowance (tightest fit) between these assembled parts is +0.005″. The maximum clearance is +0.0115″.

Figure 21.13 This drawing shows three types of fits between mating parts in addition to the clearance fit shown in the Fig. 21.12.

Basic size: the exact theoretical size from which limits are derived by the application of plus-and-minus tolerances, or 1.5000 **(Fig. 21.12)**. The basic diameter cannot be determined if the tolerances are expressed in limit form.

Actual size: the measured size of the finished part.

Fit: the tightness between two assembled parts. The four types of fit are: clearance, interference, transition, and line.

Clearance fit: the clearance between two assembled mating parts—the fit between the shaft and the hole that permits a minimum clearance of 0.0050 in. and a maximum clearance of 0.0115 in. **(Fig. 21.12C** and **D)**.

Interference fit: results in an interference between the two assembled parts—the shaft is larger than the hole, requiring a **force** or **press fit**, an effect similar to welding the two parts **(Fig. 21.13A)**.

Transition fit: may result in either an interference or a clearance between the assembled parts—the shaft may be either smaller or larger than the hole and still be within the prescribed tolerances **(Fig. 21.13B)**.

Line fit: may result in surface contact or clearance when the limits are approached **(Fig. 21.13C)**.

Selective assembly: a method of selecting and assembling parts by trial and error and by hand, allowing parts to be made with greater tolerances at less cost as a compromise between a high manufacturing accuracy and ease of assembly.

Single limits: dimensions designated by either minimum (MIN) or maximum (MAX), but not by both **(Fig. 21.14)**; depths of holes, lengths, threads, corner radii, chamfers, and so on are sometimes dimensioned in this manner.

21.5 Basic Hole System

The basic hole system utilizes the smallest hole size as the basic diameter for calculating tolerances and allowances. The basic hole system is

MAXIMUM AND MINIMUM TOLERANCES

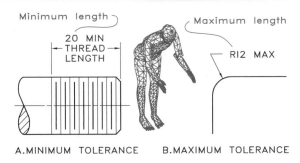

Figure 21.14 Single tolerances can be given in applications of this type in maximum (MAX) or minimum (MIN) form.

efficient when standard drills, reamers, and machine tools are available to give precise hole sizes. The smallest hole size is the basic diameter because a hole can be enlarged by machining but not reduced in size.

For example, in Fig. 21.12 the smallest diameter of the hole is 1.500 in. Subtract the allowance, 0.0050, from it to find the diameter of the largest shaft, 1.4950 in. To find the smallest limit for the shaft diameter, subtract the tolerance from 1.4950 in.

21.6 Basic Shaft System

The basic shaft system is applicable when shafts are available in highly precise standard sizes. The largest diameter of the shaft is the basic diameter for applying tolerances and allowances. The largest shaft size is used as the basic diameter because shafts can be machined to smaller size but not enlarged.

For example, if the largest permissible shaft size is 1.500 in., add the allowance to this dimension to obtain the smallest hole diameter into which the shaft fits. If the parts are to have an allowance of 0.0040 in., the smallest hole would have a diameter of 1.5040 in.

21.7 Cylindrical Fits

The ANSI B4.1 standard gives a series of fits between cylindrical features in inches for the basic hole system. The types of fit covered in this standard are:

RC: running or sliding clearance fits

LC: clearance locational fits

LT: transition locational fits

LN: interference locational fits

FN: force and shrink fits

Appendixes 34–38 list these five types of fit, each of which has several classes.

Running or sliding clearance fits (RC) provide a similar running performance, with suitable lubrication allowance, throughout the range of sizes. The clearance for the first two classes (RC 1 and RC 2), used chiefly as slide fits, increases more slowly with diameter size than other classes to maintain an accurate location even at the expense of free relative motion.

Locational fits (LC, LT, LN) determine the location of mating parts and may provide rigid or accurate location (interference fits) or some freedom of location (clearance fits). Locational fits are divided into three groups: **clearance fits (LC), transition fits (LT),** and **interference fits (LN).**

Force fits (FN) are interference fits characterized by the maintenance of constant bore pressures throughout the range of sizes. The interference varies almost directly with diameter, and the difference between its minimum and maximum values is small enough to maintain the resulting pressures within reasonable limits.

Figure 21.15 illustrates how to apply values from the tables in Appendix 34 for an RC 9 fit. The basic diameter of 2.5000 in. falls between 1.97 and 3.15 in. in the size column of the table. Limits are in thousandths, so convert the values by moving the decimal point three places to the left; for example, +7 is +0.0070 in.

Add the limits (+0.007 and 0.000 in.) to the basic diameter to find the upper and lower limits of the hole (2.5070 and 2.5000 in.). Determine the upper

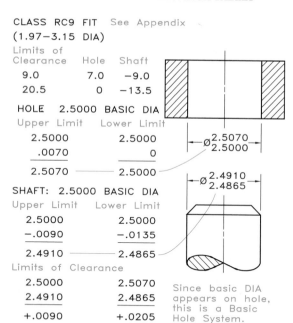

CLASS RC9 FIT See Appendix
(1.97–3.15 DIA)

Limits of
Clearance   Hole   Shaft
9.0   7.0   −9.0
20.5   0   −13.5

HOLE   2.5000 BASIC DIA
Upper Limit   Lower Limit
2.5000   2.5000
.0070   0

2.5070   2.5000

Ø 2.5070 / 2.5000

SHAFT: 2.5000 BASIC DIA
Upper Limit   Lower Limit
2.5000   2.5000
−.0090   −.0135

2.4910   2.4865

Ø 2.4910 / 2.4865

Limits of Clearance
2.5000   2.5070
2.4910   2.4865

+.0090   +.0205

Since basic DIA
appears on hole,
this is a Basic
Hole System.

Figure 21.15 This example shows how to calculate limits and allowances for an RC9 fit between a shaft and hole with a basic diameter of 2.5000 inches. Refer to Appendix 34.

CLASS RC9 Running & Clearance Fit (From Appendix)	Nominal Size Range Inches	Limits of Clearance	Limits	
			Hole	Shaft
	1.97–3.15	9.0	+7.0	−9.0
		20.5	0	−13.5

Complete chart below by using table above:
Class of fit: RC9 Basic Hole System
Basic Diameter: 2.5000

Hole Limits: Shaft Limits:
+7.0 = +.0070   −9.0 = −.0090
  0 = .0000   −13.5 = −.0135

Max Hole   Min Hole   Max Shaft   Min Shaft
2.5000   2.5000   2.5000   2.5000
+.0070   .0000   −.0090   −.0135

2.5070   2.5000   2.4910   2.4865

Hole Tolerance: .0070   Shaft Tolerance: .0045

Max Clearance   Min Clear. (Allowance)

Largest Hole   2.5070   Smallest Hole 2.5000
Smallest Shaft 2.4865   Largest Shaft 2.4910

              .0205   .0090

Figure 21.16 Cylindrical fit information may be calculated in an organized manner as this chart demonstrates. (Thanks to Steve Horton.)

and lower limits of the shaft (2.4910 and 2.4865 in.) by subtracting the two limits (−0.0090 and −0.0135 in.) from the basic diameter (2.5000 in.).

To get the tightest fit between the assembled parts (+0.0090 in.) and the loosest fit (+0.0205 in.), subtract the minimum sizes from the maximum sizes of the holes and shafts. These values appear in the Limit column of the table (Appendix 34).

This same method of using tables of fits applies to other types of fits and their respective tables: force fit, interference fit, transition fit, and locational fit. Subtract negativg limits from the basic diameter and add positive limits to it. **A minus sign preceding a limits of clearance in the tables indicates an interference fit between the assembled features, and a positive limit of clearance indicates a clearance fit.**

Figure 21.16 presents a calculation chart for the tolerances shown in Fig. 21.15. The possibility

of errors is less if you take the values from the tables and write them down in this manner.

21.8 Tolerancing: Metric System

Terminology

The system recommended by the International Standards Organization (ISO) in ANSI B4.2 for metric measurements are fits that usually apply to cylinders—hole and shaft—but you may also use these tables to specify fits between parallel contact surfaces, such as a key in a slot. **Figures 21.17–21.19** illustrate most of the definitions of metric limits and fits.

Basic size: the theoretical size, usually a diameter from which limits or deviations are calculated (**Fig. 21.17**); select it from the table shown in **Fig. 21.18** under the First Choice column.

METRIC SYSTEM TERMINOLOGY

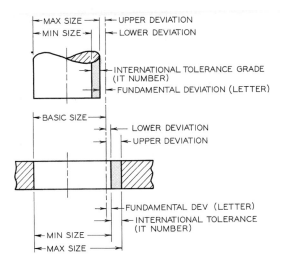

Figure 21.17 The terminology shown here relates to metric fits and limits.

PREFERRED BASIC SIZES (MILLIMETERS)

First Choice	Second Choice	First Choice	Second Choice	First Choice	Second Choice
I		10		100	
	1.1		11		110
1.2		12		120	
	1.4		14		140
1.6		16		160	
	1.8		18		180
2		20		200	
	2.2		22			220
2.5		25		250	
	2.8		28		280
3		30		300	
	3.5		35			350
4		40		400	
	4.5		45		450
5		50		500	
	5.5		55		550
6		60		600	
	7		70		700
8		80		800	
	9		90		900
				1000	

Figure 21.18 Basic sizes for metric fits should be selected first from the first-choice column and then from the second-choice column.

TOLERANCE SYMBOLS

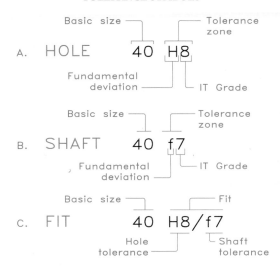

Figure 21.19 These tolerance symbols and their definitions apply to holes and shafts.

Deviation: the difference between the hole or shaft size and the basic size.

Upper deviation: the difference between the maximum permissible size of a part and its basic size (**Fig. 21.17**).

Lower deviation: the difference between the minimum permissible size of a part and its basic size (**Fig. 21.17**).

Fundamental deviation: the deviation closest to the basic size (**Fig. 21.17**); in the note 40 H8 in **Fig. 21.19**, the H represents the fundamental deviation for a hole, and in the note 40 f7, the f represents the fundamental deviation for a shaft.

Tolerance: the difference between the maximum and minimum allowable sizes of a single part.

International tolerance (IT) grade: a series of tolerances that vary with basic size to provide a uniform level of accuracy within a given grade (**Fig. 21.17**); in the note 40 H8 in **Fig. 21.19**, the 8 represents the IT grade; there are 18 IT grades: IT01, IT0, IT1, . . . , IT16.

ISO SYMBOLS FOR PREFERRED FITS

Hole Basis	Shaft Basis	Description	
H11/c11	C11/h11	**Loose Running Fit** for wide commerical tolerances on external members	
H9/d9	D9/h9	**Free Running Fit** for large temperature variations, high running speeds, or high journal pressures	Clearance Fits
H8/f7	F8/h7	**Close Running Fit** for accurate location and moderate speeds and journal pressures	
H7/g6	G7/h6	**Sliding Fit** for accurate fit and location and free moving and turning, not free running	
H7/h6	H7/h6	**Locational Clearance** for snug fits for parts that can be freely assembled	
H7/k6	K7/h6	**Locational Transition Fit** for accurate locations	Transition Fits
H7/n6	N7/h6	**Locational Transition Fit** for more accurate locations and greater interference	
H7/p6	P7/h6	**Locational Interference Fit** for rigidity and alignment without special bore pressures	
H7/s6	S7/h6	**Medium Drive Fit** for shrink fits on light sections; tightest fit usable for cast iron	Interference Fits
H7/u6	U7/h6	**Force Fit** for parts that can be highly stressed and for shrink fits.	

Figure 21.20 This list gives the preferred hole-basis and shaft-basis fits for the metric system.

Tolerance zone: a combination of the fundamental deviation and the tolerance grade; the H8 portion of the 40 H8 note in **Fig. 21.19**, is the tolerance zone.

Hole basis: a system of fits based on the minimum hole size as the basic diameter, with fundamental deviations; Appendixes 40 and 41 give hole-basis data for tolerances.

Shaft basis: a system of fits based on the maximum shaft size as the basic diameter, with fundamental deviations; Appendixes 42 and 43 give shaft-basis data for tolerances.

Clearance fit: a fit resulting in a clearance between two assembled parts under all tolerance conditions.

Interference fit: a force fit between two parts, requiring that they be driven together.

Transition fit: may result in either a clearance or an interference fit between assembled parts.

Tolerance symbols: notes giving the specifications of tolerances and fits (**Fig. 21.19**); the basic size is a number, followed by the fundamental deviation letter and the IT number, which combined give the tolerance zone; **uppercase letters indicate the fundamental deviations for holes, and lowercase letters indicate fundamental deviations for shafts.**

Preferred Sizes and Fits

The table in **Fig. 21.18** shows the preferred basic sizes for computing tolerances. Under the First Choice heading, each number increases by about 25% from the preceding value. Each number in the Second Choice column increases by about 12%. **To minimize cost, select basic diameters from the first column because they correspond to standard stock sizes for round, square, and hexagonal metal products.**

Figure 21.20 shows preferred clearance, transition, and interference fits for the hole-basis and shaft-basis systems. Appendixes 40–43 contain the complete tables.

Preferred Fits: Hole-Basis System   Figure 21.21

shows the types of fits for the hole-basis system, in which the **smallest hole is the basic diameter.** Clearance, transition, or interference fits are possible when toleranced with the options of the hole-basis-system. **Figure 21.22** compares the preferred fits for a hole-basis system. The lower deviation of the hole is zero, which means that the smallest hole is the basic size. Variations in fit between parts range from a clearance fit of H11/c11 to an interference fit of U7/u6 (see Fig. 21.20).

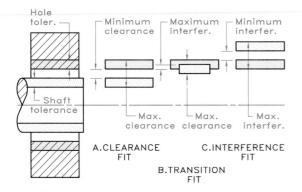

Figure 21.21 Types of fits: (A) clearance fit, (B) transition fit, where there can be either interference or clearance, and (C) interference fit, where the parts must be forced together.

PREFERRED FITS: HOLE-BASIS SYSTEM

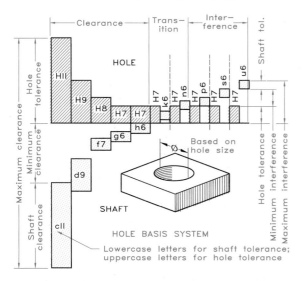

Figure 21.22 This diagram illustrates the preferred fits for the hole-basis system listed in Fig. 21.20. Appendixes 40 and 41 give values for these fits.

Preferred Fits: Shaft-Basis System **Figure 21.23** compares the preferred fits of the shaft-basis system, in which **the largest shaft is the basic diameter.** Variations in fit between parts range from a

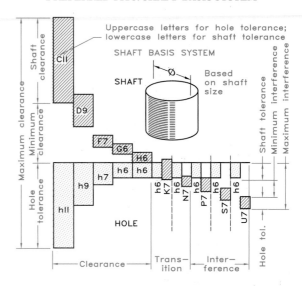

Figure 21.23 This diagram illustrates the preferred fits for a shaft-basis system listed in Fig. 21.20. Appendixes 42 and 43 give values for these fits.

clearance fit of C11/h11 to an interference fit of U7/h6 (see Fig. 21.20).

Standard Cylindrical Fits

The following examples demonstrate how to calculate and apply tolerances to cylindrical parts. The solutions involve the use of Appendix 40, Fig. 21.18, and Fig. 21.20.

Example 1 (Fig. 21.24)

Required: Use the hole-basis system, a close running fit, and a basic diameter of 49 mm.

Solution: Use a preferred basic diameter of 50 mm **(Fig. 21.18)** and fit of H8/f7 **(Fig. 21.20)**.

Hole: Find the upper and lower limits of the hole in Appendix 40 under H8 and across from 50 mm. These limits are 50.000 and 50.039 mm.

Shaft: Find the upper and lower limits of the shaft under f7 and across from 50 mm in Appendix 40. These limits are 49.950 and 49.975 mm.

CALCULATION OF METRIC FITS

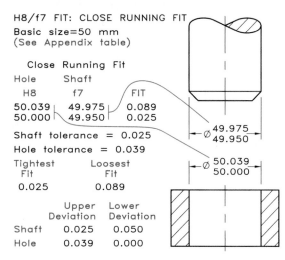

H8/f7 FIT: CLOSE RUNNING FIT

Basic size=50 mm
(See Appendix table)

Close Running Fit

Hole	Shaft	
H8	f7	FIT
50.039	49.975	0.089
50.000	49.950	0.025

Shaft tolerance = 0.025

Hole tolerance = 0.039

Tightest Fit	Loosest Fit
0.025	0.089

	Upper Deviation	Lower Deviation
Shaft	0.025	0.050
Hole	0.039	0.000

Figure 21.24 This drawing shows how to calculate and apply metric limits and fits to a shaft and hole (Appendix 40).

LOCATIONAL TRANSITION FIT—H7/K6

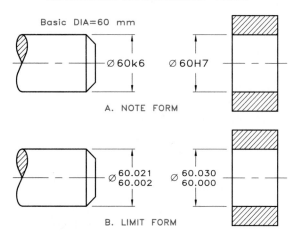

Basic DIA=60 mm

A. NOTE FORM

B. LIMIT FORM

Figure 21.25 These methods are for applying metric tolerances to a hole and shaft with a transition fit (Appendix 41).

Symbols: **Figure 21.24** shows how to apply toleranced dimensions to the hole and shaft.

Example 2 (Fig. 21.25)

Required: Use the hole-basis system, a location transition fit, and a basic diameter of 57 mm.

MEDIUM DRIVE FIT—H7/S6

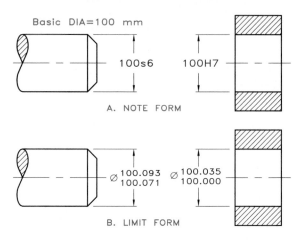

Basic DIA=100 mm

A. NOTE FORM

B. LIMIT FORM

Figure 21.26 These methods are for applying metric tolerances to a hole and shaft with an interference fit (Appendix 41).

Solution: Use a preferred basic diameter of 60 mm (Fig. 21.18) and a fit of H7/k6 (Fig. 21.20).

Hole: Find the upper and lower limits of the hole in Appendix 41 under H7 and across from 60 mm. These limits are 60.000 and 60.030 mm.

Shaft: Find the upper and lower limits of the shaft under k6 and across from 60 mm in Appendix 41. These limits are 60.021 and 60.002 mm.

Symbols: **Figure 21.25** shows two methods of applying the tolerance symbols to a drawing.

Example 3 (Fig. 21.26)

Required: Use the hole-basis system, a medium drive fit, and a basic diameter of 96 mm.

Solution: Use a preferred basic diameter of 100 mm (Fig. 21.18) and a fit of H7/s6 (Fig. 21.20).

Hole: Find the upper and lower limits of the hole in Appendix 41 under H7 and across from 100 mm. These limits are 100.035 and 100.000 mm.

FIT: H8/f7 Ø45 BASIC DIA

From Appendix

Hole H8	Shaft f7	HOLE LIMITS	45.039 45.000
0.039 0.000	−0.025 −0.050	SHAFT LIMITS	44.975 44.950

Figure 21.27 This calculation is for an H8/f7 fit for a nonstandard diameter of 45 mm (Appendixes 44 and 45).

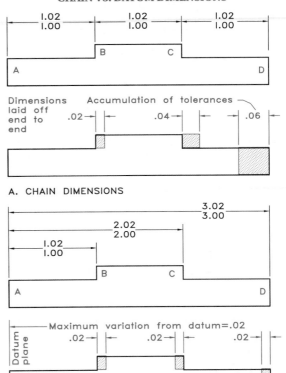

A. CHAIN DIMENSIONS

B. DATUM PLANE (BASELINE) DIMENSIONS

Figure 21.28
A Dimensions given end to end in a chain fashion may result in an accumulation of tolerances of up to 0.06″ at D instead of the specified 0.02″.

B When dimensioned from a single datum, the variations of B, C, and D cannot deviate more than the specified 0.02″ from the datum.

Shaft: Find the upper and lower limits of the shaft under s6 and across from 100 mm in Appendix 41. These limits are 100.093 and 100.071 mm. Appendix 41 gives the tightest fit as an interference of −0.093 mm, and the loosest fit as an interference of −0.036 mm. Minus signs in front of these numbers indicate an interference fit.

Symbols: **Figure 21.26** shows how to apply toleranced dimensions to the hole and shaft.

Nonstandard Fits: Nonpreferred Sizes

You may calculate limits of tolerances for any of the preferred fits shown in Fig. 21.20 for nonstandard sizes that do not appear in Appendixes 40–43. Limits of tolerances for nonstandard hole sizes are in Appendix 44, and limits of tolerances for nonstandard shaft sizes are in Appendix 45.

 Figure 21.27 shows the hole and shaft limits for an H8/f7 fit and a 45-mm DIA. The tolerance limits of 0.000 and 0.039 mm for an H8 hole are from Appendix 44, across from the size range of 40–50 mm. The tolerance limits of −0.025 and −0.050 mm for the shaft are from Appendix 45. Calculate the hole limits by adding the positive tolerances to the 45-mm basic diameter and the shaft limits by subtracting the negative tolerances from the 45-mm basic diameter.

21.9 Chain versus Datum-Plane Dimensions

When parts are dimensioned to locate surfaces or geometric features by a chain of dimensions laid end to end (**Fig. 21.28A**), variations may accumulate in excess of the specified tolerance. For example, the tolerance between surfaces A and B is

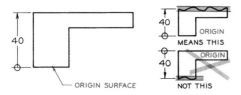

Figure 21.29 Selection of the shorter surface as the origin surface for locating a longer parallel surface gives the greatest accuracy.

0.02, between A and C it is 0.04, and between A and D it is 0.06.

You may eliminate an accumulation of tolerances by measuring from a single plane called a **datum plane** or **baseline**. A datum plane is usually on the object, but it can also be on the machine used to make the part. Because each plane in **Fig. 21.28B** is located with respect to a datum plane, the tolerances between the intermediate planes do not exceed the maximum tolerance of 0.02. Always base the application of tolerances on the function of a part in relationship to its mating parts.

Origin Selection
When you need to specify a surface as the origin (datum plane) for locating a parallel surface, selection of the shorter one gives more accurate results (**Fig. 21.29**). However, the angular variation permitted is less for the longer surface in this case than it would be if the longer surface had been selected as the datum plane.

21.10 Conical Tapers
Recall that taper is a ratio of the difference in the diameters of two circular sections of a cone to the distance between the sections. **Figure 21.30** shows a method of specifying a conical taper by giving a basic diameter and basic taper. The basic diameter of 20 mm is located midway in the length of the cone with a toleranced dimension.

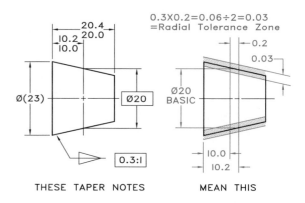

Figure 21.30 Indicate taper with a combination of tolerances and taper symbols. Here, the variation in diameter at any point is 0.06 mm, or 0.03 mm in radius.

Figure 21.30B shows how to calculate the radial tolerance zone.

21.11 Tolerance Notes
You should tolerance all dimensions on a drawing either by using the rules previously discussed or by placing a note in or near the title block. For example, the note

$$\text{TOLERANCE } \pm \frac{1}{64}$$

might be given on a drawing for less critical dimensions.

Some industries give dimensions in inches with two, three, and four decimal place fractions. A note for dimensions with two and three decimal places might be given on the drawing as

TOLERANCES XX.XX ±0.10; XX.XXX ±0.005.

Tolerances of four places would be given directly on the dimension lines.

The most common method of noting tolerances is to give as large a tolerance as feasible in a note, such as

TOLERANCES ± 0.05

TOLERANCED ANGLES

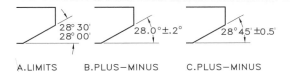

28° 30'
28° 00'
28.0°±.2°
28°45'±0.5'
A.LIMITS
B.PLUS–MINUS
C.PLUS–MINUS

Figure 21.31 You may tolerance angles by using any of these techniques.

INTERNATIONAL TOLERANCE GRADES

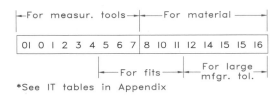

For measur. tools
For material
01 0 1 2 3 4 5 6 7 8 10 11 12 14 15 15 16
For fits
For large mfgr. tol.
*See IT tables in Appendix

Figure 21.32 This diagram shows the international tolerance (IT) grades and their applications (Appendix 39).

TOLERANCE OF MACHINING

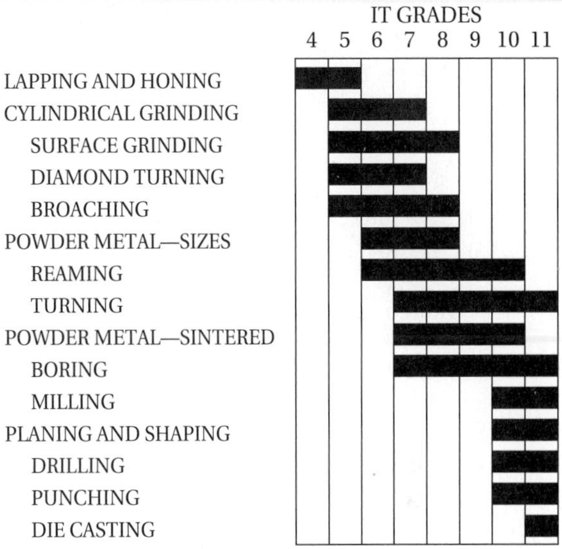

IT GRADES
4 5 6 7 8 9 10 11
LAPPING AND HONING
CYLINDRICAL GRINDING
SURFACE GRINDING
DIAMOND TURNING
BROACHING
POWDER METAL—SIZES
REAMING
TURNING
POWDER METAL—SINTERED
BORING
MILLING
PLANING AND SHAPING
DRILLING
PUNCHING
DIE CASTING

Figure 21.33 Tolerance values may be selected from the international tolerance grades applicable to various machining processes.

and to give tolerances on the dimension lines for dimensions requiring smaller tolerances. Give angular tolerances in a general note in or near the title block, such as

ANGULAR TOLERANCES ±0.5° or ±30'.

Use one of the techniques shown in **Fig. 21.31** to give specific angular tolerances directly on angular dimensions.

21.12 General Tolerances—Metric Units

All dimensions on a drawing must fall within certain tolerance ranges, even though the tolerances are not shown on dimension lines. Tolerances for dimensions not shown on dimension lines are given by a **general tolerance note** on the drawing.

Linear Dimensions Tolerance linear dimensions by indicating plus and minus (±) one half of an international tolerance (IT) grade as given in Appendix 39. You may select the IT grade from the chart in **Fig. 21.32**, where IT grades for mass-produced items range from IT12 through IT16. You may also select IT grades from **Fig. 21.33** for the particular machining process being used.

General tolerances using IT grades may be expressed in a note as follows:

UNLESS OTHERWISE SPECIFIED ALL

UNTOLERANCED DIMENSIONS ARE $\frac{\pm IT14}{2}$.

This note means that a tolerance of ±0.700 mm is allowed for a dimension between 315 and 400 mm. The value of the tolerance, 1.400 mm, is taken from Appendix 39.

Figure 21.34 shows recommended tolerances for fine, medium, and coarse series for graduated-size dimensions. A medium tolerance, for example, may be specified by the following note:

GENERAL TOLERANCES SPECIFIED IN ANSI B4.3
MEDIUM SERIES APPLY.

Equivalent tolerances may be given in table form **(Fig. 21.35)** on the drawing, the grade—medium in

GENERAL TOLERANCES: LINEAR DIMENSIONS (MM)

Basic Dimensions	Fine series	Medium series	Coarse series
0.5 to 3	± 0.05	± 0.1	--
Over 3 to 6	± 0.05	± 0.1	± 0.2
Over 6 to 30	± 0.1	± 0.2	± 0.5
Over 30 to 120	± 0.15	± 0.3	± 0.8
Over 120 to 315	± 0.2	± 0.5	± 1.2
Over 315 to 1000	± 0.3	± 0.8	± 2
Over 1000 to 2000	± 0.5	± 1.2	± 3

Figure 21.34 You may select general tolerance values from this table for fine, medium, and coarse series. Tolerances vary with dimensions.

TABLE OF GENERAL TOLERANCES

Get values from previous table

Specifies a Medium Series

DIMENSIONS (mm) GENERAL TOLRANCES UNLESS OTHERWISE SPECIFIED, THE FOLLOWING TOLERANCES ARE APPLICABLE							
LINEAR	Over to	0.5 6	6 30	30 120	120 315	315 1000	1000 2000
TOL.	+ −	0.1	0.2	0.3	0.5	0.8	1.2

Figure 21.35 This table for a medium series of values was extracted from Fig. 21.34 for insertion on a working drawing to provide the tolerances for a medium series of sizes.

this example—selected from Fig. 21.34. General tolerances may be given in a table for dimensions expressed with one or no decimal places (**Fig. 21.36**).

General tolerances may also be notated in the following form:

UNLESS OTHERWISE SPECIFIED ALL
UNTOLERANCED DIMENSIONS ARE ±0.8 mm.

Use this method only when the dimensions on a drawing are similar in size.

TOLERANCES FOR ONE AND NO DECIMAL PLACES

Medium series for numbers with one decimal place

Coarse series for numbers with no decimal places

DIMENSIONS (mm) GENERAL TOLRANCES UNLESS OTHERWISE SPECIFIED, THE FOLLOWING TOLERANCES ARE APPLICABLE					
LINEAR	OVER TO	− 120	120 315	315 1000	1000 −
TOL.	ONE DECIMAL ±	0.3	0.5	0.8	1.2
	NO DECIMALS ±	0.8	1.2	2	3

Figure 21.36 Placed on a drawing, this table of tolerances would indicate the tolerances for dimensions having one or no decimal places, such as 24.0 and 24, denoting medium and coarse series.

ANGULAR AND TAPER TOLERANCES

Length of shorter leg (mm)	Up to 10	Over 10 to 50	Over 50 to 120	Over 120 to 400
Degrees	± 1°	± 0° 30'	± 0° 20'	± 0° 10'
mm per 100	± 1.8	± 0.9	± 0.6	± 0.3
Millimeters	± 18	± 9	± 6	± 3

Figure 21.37 General tolerances for angular and taper dimensions may be taken from this table of values.

Angular Tolerances Express angular tolerances as (1) an angle in decimal degrees or in degrees and minutes, (2) a taper expressed in percentage (mm per 100 mm), or (3) milliradians. (To find milliradian, multiply the degrees of an angle by 17.45.) **Figure 21.37** shows the suggested tolerances for each of these units, based on the length of the shorter leg of the angle.

General angular tolerances may be notated on the drawing as follows:

UNLESS OTHERWISE SPECIFIED THE GENERAL
TOLERANCES IN ANSI B4.3 APPLY.

TABLE OF ANGULAR TOLERANCES

ANGULAR TOLERANCES				
LENGTH OF SHORTER LEG (mm)	UP TO 10	OVER 10 TO 50	OVER 50 TO 120	OVER 120 TO 400
TOLERANCE	±1°	± 0°30'	± 0° 20'	± 0°10'

Values in degrees and minutes taken from previous table

Figure 21.38 Extracted from Fig. 21.37, this table of values inserted on a drawing would indicate general tolerances for angles in degrees and minutes.

GEOMETRIC SYMBOLS

	TOLERANCE	CHARACTERISTIC	SYMBOL
INDIVIDUAL FEATURES	FORM	STRAIGHTNESS	—
		FLATNESS	▱
		CIRCULARITY	○
		CYLINDRICITY	⌭
INDIVIDUAL OR RELATED FEATURES	PROFILE	PROFILE OF A LINE	⌒
		PROFILE OF A SURFACE	⌓
RELATED FEATURES	ORIENTATION	ANGULARITY	∠
		PERPENDICULARITY	⊥
		PARALLELISM	//
	LOCATION	POSITION	⊕
		CONCENTRICITY	◎
	RUNOUT	CIRCULAR RUNOUT	↗
		TOTAL RUNOUT	↗↗

Figure 21.39 These symbols specify the geometric characteristics of a part's features.

A second method involves showing a portion of the table from Fig. 21.37 as a table of tolerances on the drawing (**Fig. 21.38**). A third method is a note with a single tolerance such as:

UNLESS OTHERWISE SPECIFIED ANGULAR TOLERANCES ARE ±0°30' (or ±0.5°).

PROPORTIONS OF SYMBOLS

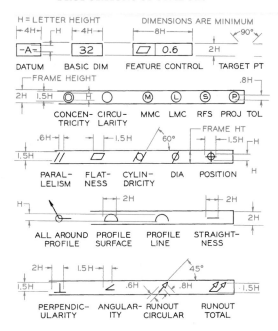

Figure 21.40 Use these general proportions (based on the letter height used) for drawing feature control symbols and frames.

21.13 Geometric Tolerances

Geometric tolerancing specifies tolerances that control location form, profile, orientation, and runout on a dimensioned part as covered by the ANSI *Y14.5M-1982 Standards* and the *Military Standards (Mil-Std)* of the U.S. Department of Defense. Before discussing those types of tolerancing, however, we need to introduce you to symbols, size limits, rules, three-datum-plane concepts, and applications.

Symbols

Figure 21.39 shows various symbols used to represent geometric characteristics of dimensioned drawings. **Figure 21.40** shows additional symbols, feature control frames, and their proportions, which are based on the letter height, H. On most drawings, a 1/8-in. or 3-mm letter height is rec-

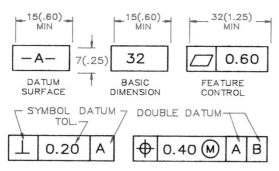

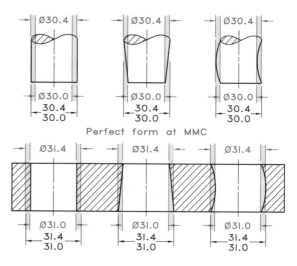

Perfect form at MMC

Figure 21.41 These are examples of geometric tolerancing frames used to indicate datum planes, basic dimensions, and feature control symbols.

Figure 21.43 When only a tolerance of size is specified on a feature, the limits prescribe the form of the features, as shown for these shafts and holes having identical limits.

MAXIMUM MATERIAL CONDITION (MMC)

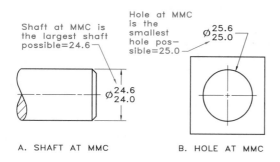

Figure 21.42 A shaft is at maximum material condition (MMC) when it is at the largest size permitted by its tolerance. A hole is at MMC when it is at its smallest size.

ommended. **Figure 21.41** depicts some feature control frames and their proportions.

Size Limits

Three conditions of size are used when geometric tolerances are applied: **maximum material condition, least material condition, and regardless of feature size.**

In the **maximum material condition (MMC)**, a feature contains the maximum amount of material.

For example, the shaft shown in **Fig. 21.42** is at MMC when it has the largest permitted diameter of 24.6 mm. The hole is at MMC when it has the most material, or the smallest diameter of 25.0 mm.

The **least material condition (LMC)** indicates that a feature contains the least amount of material. The shaft in Fig. 21.42 is at LMC when it has the smallest diameter of 24.0 mm. The hole is at LMC when it has the least material, or the largest diameter of 25.6 mm.

The **regardless of feature size (RFS)** condition indicates that tolerances apply to a geometric feature regardless of its size. These sizes range from MMC to LMC.

21.14 Rules for Tolerancing

Three general rules of tolerancing geometric features should be followed.

Rule 1 (Individual Feature Size) When only a tolerance of size is specified on a feature, the limits of size control the variation in its geometric form.

TOLERANCES OF POSITION

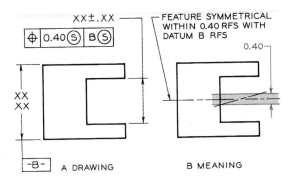

Figure 21.44 Tolerances of position should include the note of M, S, or L to indicate maximum material condition, regardless of feature size, or least material condition.

THE THREE DATUM PLANES

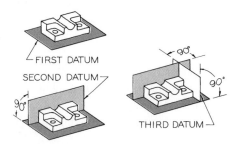

Figure 21.45 When an object is referenced to a primary datum plane, it comes into contact with the datum plane at at least three points. The vertical surface contacts the secondary datum plane at at least two points. The third datum plane comes into contact with at least one point on the object. The datum planes are listed order of priority in the feature control frame.

The forms of the shaft and hole shown in **Fig. 21.43** are permitted to vary within the tolerance ranges of the dimensions.

Rule 2 (Tolerances of Position) When a tolerance of position is specified on a drawing, MMC, LMC, or RFS must be specified with respect to the tolerance, datum, or both. The specification of symmetry of the part in **Fig. 21.44** is based on a tolerance at RFS from a datum at RFS.

THREE-PLANE REFERENCE SYSTEM

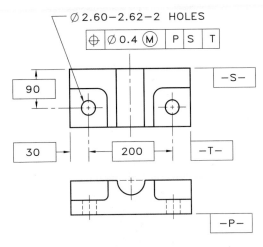

Figure 21.46 Label the three planes of the reference system where they appear as edges. The primary datum plane (P) is given first in the feature control frame; the secondary plane (S), second; and the tertiary plane (T), third. Numbers in frames are exact basic dimensions.

Rule 3 (All Other Geometric Tolerances) The RFS condition applies to all other geometric tolerances for individual tolerances and datum references if no modifying symbol is given in the feature control frame. If a feature is to be at MMC, it must be specified.

Three-Datum-Plane Concept

A datum plane is used as the origin of a part's features that have been toleranced. Datum planes usually relate to manufacturing equipment, such as machine tables or locating pins.

Three mutually perpendicular datum planes are required to dimension a part accurately. For example, the part shown in **Fig. 21.45** sits on the primary datum plane, with at least three points of its base in contact with the datum. The part is related to the secondary plane by at least two contact points. The third (tertiary) datum is in contact with at least one point on the object.

The priority of datum planes is presented in sequence in feature control frames. For example in **Fig. 21.46**, the primary datum is surface P, the

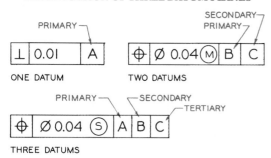

Figure 21.47 Use feature control frames to indicate from one to three datum planes in order of priority.

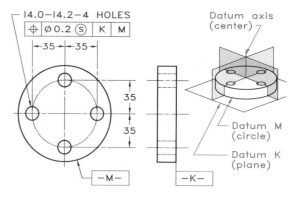

Figure 21.48 These true-position holes are located with respect to primary datum K and secondary datum M. Because datum M is a circle, the implication is that the holes are located about two intersecting datum planes formed by the crossing centerlines in the circular view, satisfying the three-plane concept.

secondary datum is surface S, and the tertiary datum is surface T. **Figure 21.47** lists the order of priority of datum planes A–C sequentially in the feature control frames.

21.15 Cylindrical Datum Features

Figure 21.48 illustrates a part with a cylindrical datum feature that is the axis of a true cylinder. Datum K is the primary datum. Datum M is associated with two theoretical planes—the second and third in a three-plane relationship.

The two theoretical planes are represented in the circular view by perpendicular centerlines that intersect at the point view of the datum axis. All dimensions originate from the datum axis perpendicular to datum K; the other two intersecting datum planes are used for measurements in the x and y directions.

The priority of the datum planes in the feature control frame is significant in the manufacturing and inspection processes. The part shown in **Fig. 21.49** is dimensioned in three ways to show the effects of datum-plane selection and material condition on the location of the hole pattern.

Figure 21.49B illustrates the effect of specifying diameter A at RFS as the primary datum plane and surface B as the secondary datum plane. During production the part is centered on cylinder A. The

part is mounted in a chuck, mandrel, or centering device on the processing equipment, which centers the part at RFS. Any variation from perpendicular in surfaces A and B will affect the degree of contact of surface B with its datum plane.

If surface B were specified as the primary datum feature, it would contact datum plane B at no fewer than three points (**Fig. 21.49C**). The axis of datum cylinder A will be gauged by the smallest cylinder that is perpendicular to the first datum that will contact cylinder A at RFS. This cylinder identifies variation from perpendicular between planes A and B and size variations.

In **Fig. 21.49D**, plane B is specified as the primary datum feature and cylinder A as the secondary datum feature at MMC. The part is mounted on the processing equipment so that at least three points on feature B come into contact with datum B. The datum axis is the axis of a circumscribed cylinder of a fixed size that is perpendicular to datum B. Using the modifier to specify MMC gives a more liberal tolerance zone than when RFS is specified.

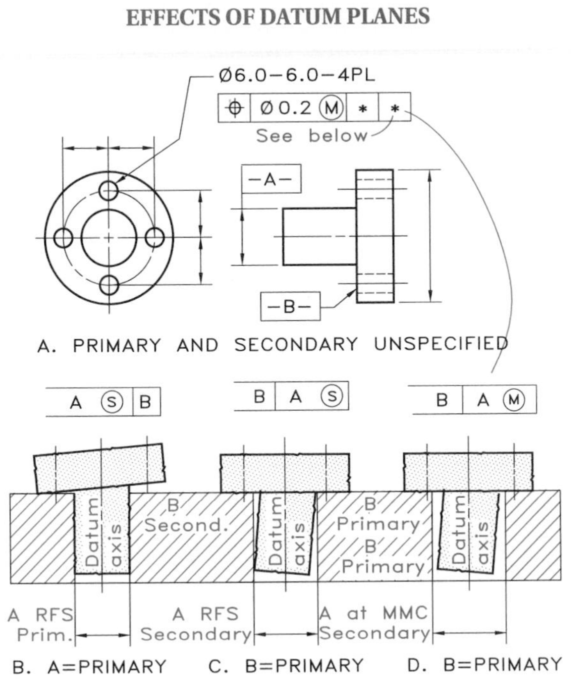

Ø6.0−6.0−4PL

⊕ | Ø 0.2 Ⓜ | * | *

See below

−A−

−B−

A. PRIMARY AND SECONDARY UNSPECIFIED

| A Ⓢ B | | B | A Ⓢ | | B | A Ⓜ |

B. A=PRIMARY   C. B=PRIMARY   D. B=PRIMARY

Figure 21.49 For the unspecified datum planes in (A), examples (B)–(D) illustrate the effects of selecting the datum planes in order of priority and of RFS and MMC.

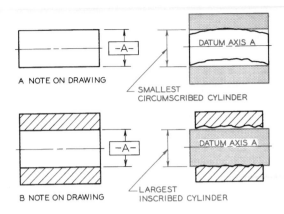

A NOTE ON DRAWING

−A−

SMALLEST CIRCUMSCRIBED CYLINDER

DATUM AXIS A

B NOTE ON DRAWING

−A−

LARGEST INSCRIBED CYLINDER

DATUM AXIS A

Figure 21.50 The datum axis of a shaft is the smallest circumscribed cylinder in contact with the shaft. The datum axis of a hole is the centerline of the largest inscribed cylinder in contact with the hole.

Datum Features at RFS

When size dimensions are applied to a feature at RFS, the processing equipment that comes into contact with surfaces of the part establishes the datum. Variable machine elements, such as chucks or center devices, are adjusted to fit the external or internal features and establish datums.

Primary Diameter Datums For an external cylinder (shaft) at RFS, the datum axis is the axis of the smallest circumscribed cylinder that contacts the cylindrical feature (**Fig. 21.50A**). That is, the largest diameter of the part making contact with the smallest cylinder of the machine element holding the part is the datum axis.

    For an internal cylinder (hole) at RFS, the datum axis is the axis of the largest inscribed cylinder making contact with the hole. That is, the smallest diam-

eter of the hole making contact with the largest cylinder of the machine element inserted in the hole is the datum axis (**Fig. 21.50B**).

Primary External Parallel Datums The datum for external features at RFS is the center plane between two parallel planes—at minimum separation—that contact the planes of the object (**Fig. 21.51A**). These are planes of a viselike device at minimum separation that holds the part.

Primary Internal Parallel Datums The datum for internal features is the center plane between two parallel planes—at their maximum separation—that contact the inside planes of the object (**Fig. 21.51B**).

Secondary Datums The secondary datum (axis or center plane) for both external and internal diameters (or distances between parallel planes) has the additional requirement that the cylinder in contact with the parallel elements of the hole be perpendicular to the primary datum (**Fig. 21.52**). Datum axis B is the axis of cylinder B.

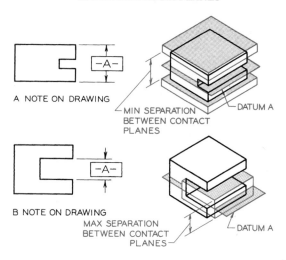

A NOTE ON DRAWING

MIN SEPARATION
BETWEEN CONTACT
PLANES

DATUM A

B NOTE ON DRAWING

MAX SEPARATION
BETWEEN CONTACT
PLANES

DATUM A

Figure 21.51 The datum plane for external parallel surfaces is the center plane between two contact parallel
planes at their minimum separation. The datum plane for
internal parallel surfaces is the center plane between two
contact parallel surfaces at their maximum separation.

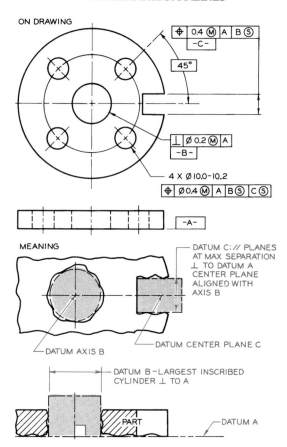

ON DRAWING

MEANING

DATUM C: // PLANES
AT MAX SEPARATION
⊥ TO DATUM A
CENTER PLANE
ALIGNED WITH
AXIS B

DATUM AXIS B

DATUM CENTER PLANE C

DATUM B — LARGEST INSCRIBED
CYLINDER ⊥ TO A

PART

DATUM A

Figure 21.52 A part located with respect to primary, secondary, and tertiary datum planes.

Tertiary Datums The third datum (axis or center
plane) for both external and internal features has
the further requirement that either the cylinder or
parallel planes be oriented angularly to the secondary datum. Datum C in Fig. 21.52 is the tertiary datum plane.

21.16 Location Tolerancing

**Tolerances of location deal with position, con-
centricity, and symmetry.**

Position Toleranced location dimensions yield a
square (or rectangular) coordinate tolerance zone
for the center of a hole (**Fig. 21.53A**). **In contrast,
true-position dimensions, called basic dimensions, locate the exact position of a hole's center,
about which a circular tolerance zone is specified (Fig. 21.53B).**

In both the coordinate and true-position methods, the hole diameter is toleranced by identical

notes. In the true-position method, a feature control frame specifies the diameter of the circular tolerance zone inside which the hole's center must
lie. **A circular position zone gives a more precise
tolerance of the hole's true position than a square.**

Figure 21.54 shows an enlargement of the
square tolerance zone resulting from the use of
coordinates to locate a hole's center. The diagonal
across the square zone is greater than the specified tolerance by a factor of 1.4. Therefore the
true-position method, shown enlarged in **Fig.
21.55** can have a larger circular tolerance zone by

SQUARE VS. CIRCULAR TOLERANCE ZONES

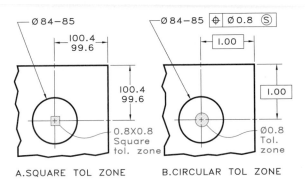

A.SQUARE TOL ZONE   B.CIRCULAR TOL ZONE

Figure 21.53

A These dimensions give a square tolerance zone for the axis of the hole.

B Basic dimensions (in frames) locate the true center about which a circular tolerance zone of 0.8 mm is specified.

THE SQUARE TOLERANCE ZONE

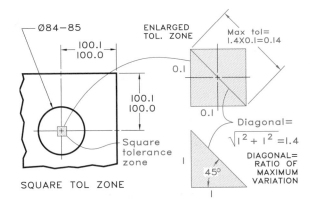

SQUARE TOL ZONE

Figure 21.54 The coordinate method of tolerancing gives a square tolerance zone with a diagonal that exceeds the specified tolerance by a factor of 1.4.

a factor of 1.4 and still have the same degree of accuracy specified by the 0.1 square zone. If a variation of 0.14 across the diagonal of the square tolerance zone is acceptable in the coordinate method, a circular tolerance zone of 0.14, which is

THE CIRCULAR TOLERANCE ZONE (TRUE POSITION TOLERANCING)

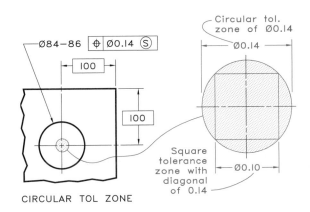

CIRCULAR TOL ZONE

Figure 21.55 The true-position method of tolerancing gives a circular tolerance zone with its center at the true position of the hole. The circular tolerance zone can be 1.4 times greater than the square tolerance zone and still be as accurate.

greater than the 0.1 tolerance permitted by the square zone, should be acceptable in the true-position tolerance method.

The circular tolerance zone specified in the circular view of a hole extends the full depth of the hole. **Therefore the tolerance zone for the centerline of the hole is a cylindrical zone inside which the axis must lie.** Because both the size of the hole and its position are toleranced, these two tolerances establish the diameter of a gauge cylinder for checking conformance of hole sizes and their locations against specifications (**Fig. 21.56**).

Subtracting the true-position tolerance from the hole at MMC (the smallest permissible hole) yields the circle that represents the least favorable condition when the part is gauged or assembled with a mating part. When the hole is not at MMC, it is larger and permits greater tolerance and easier assembly.

Gauging a Two-Hole Pattern **Gauging** is a technique of checking dimensions to determine

CYLINDRICAL TOLERANCE ZONE

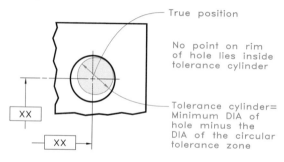

Figure 21.56 When a hole at MMC is located at true position, no element of the hole will be inside the imaginary cylinder obtained by subtracting the circular tolerance zone from the minimum diameter of the hole.

HOLES AT MMC (SMALLEST SIZE)

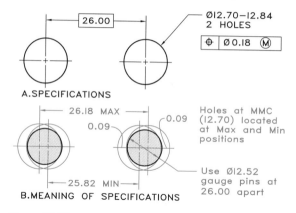

Figure 21.57

A These two holes at MMC are to be located at true position as specified.

B The two holes may be gauged with pins 12.52 mm in diameter located 26.00 mm apart.

HOLES NOT AT MMC (AT LARGEST SIZE)

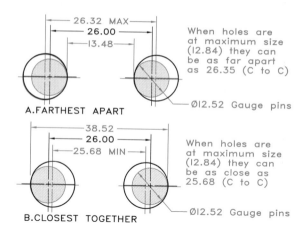

Figure 21.58

A These two holes at MMC, may have their centers spaced as far apart as 26.32 mm apart and still be acceptable.

B The holes may be placed as close as 25.68 mm apart when they are at maximum size.

tolerance, 0.18), as **Fig. 21.57B** shows. Thus two pins with diameters of 12.52 mm spaced exactly 26.00 mm apart could be used to check the diameters and positions of the holes at MMC, the most critical size. If the pins can be inserted into the holes, the holes are properly sized and located.

When the holes are not at MMC, or larger than the minimum size, these gauge pins permit a greater range of variation (**Fig. 21.58**). When the holes are at their maximum size of 12.84 mm, they can be located as close as 25.68 mm from center to center or as far apart as 26.32 mm from center to center.

Concentricity Concentricity is a feature of location because it specifies the relationship of two cylinders that share the same axis. In **Fig. 21.59**, the large cylinder is labeled as datum A, which means the large diameter is used as the datum for locating the small cylinder's axis.

whether they meet specified tolerances (**Fig. 21.57**). The two holes, with diametral size limits of 12.70–12.84, are located at true position 26.00 mm apart within a diameter of 0.18 at MMC. The gauge pin diameter is calculated to be 12.52 mm (the smallest hole's size, 12.70, minus the true-position

CONCENTRICITY

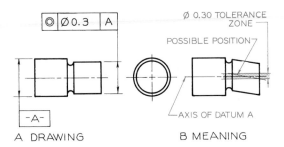

Figure 21.59 Concentricity is a tolerance of location. Here, the feature control frame specifies that the axis of the small cylinder be concentric to datum cylinder A, within a tolerance of 0.3 mm diameter.

FEATURE CONTROL FRAME

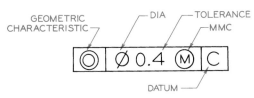

FEATURE CONTROL FRAME

Figure 21.60 This typical feature control frame indicates that a surface is concentric to datum C within a cylindrical diameter of 0.4 mm at MMC.

SYMMETRY

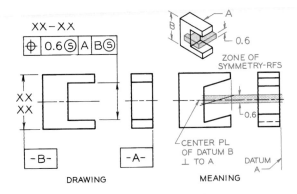

DRAWING   MEANING

Figure 21.61 Symmetry is a tolerance of location. It specifies that a part's features be symmetrical about the center plane between parallel surfaces of the part.

FLATNESS

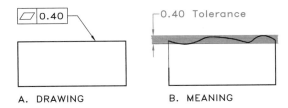

A. DRAWING   B. MEANING

Figure 21.62 Flatness is a tolerance of form. It specifies a tolerance zone within which an object's surface must lie.

We use feature control frames of the type shown in **Fig. 21.60** to specify concentricity and other geometric characteristics throughout the remainder of this chapter.

Symmetry Symmetry also is a feature of location in which a feature is symmetrical with the same contour and size on opposite sides of a central plane. **Figure 21.61A** shows how to apply a symmetry feature symbol to the notch that is symmetrical about the part's central datum plane B for a zone of 0.6 mm (**Fig. 21.61B**).

21.17 Form Tolerancing

Flatness A surface is flat when all its elements are in one plane. A feature control frame specifies flatness within a 0.4 mm tolerance zone in **Fig. 21.62** where no point on the surface may vary more than 0.40 from the highest to the lowest point.

Straightness A surface is straight if all its elements are straight lines within a specified tolerance zone. The feature control frame shown in **Fig. 21.63** specifies that the elements of a cylinder must be straight

STRAIGHTNESS

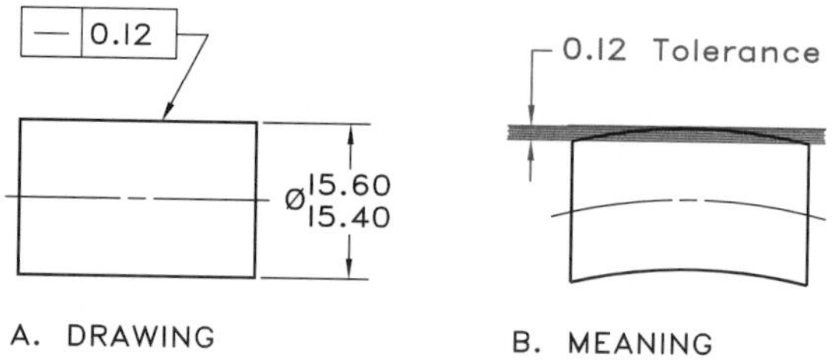

A. DRAWING   B. MEANING

Figure 21.63 Straightness is a tolerance of form. It indicates that elements of a surface are straight lines. The tolerance frame is applied to the views in which elements appear as straight lines.

ROUNDNESS: CYLINDER

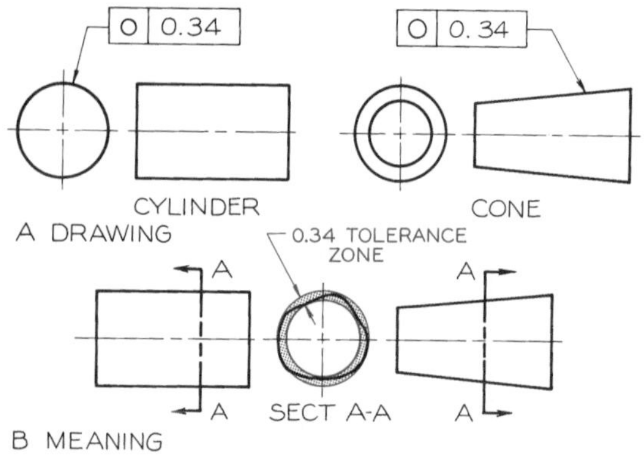

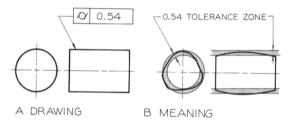

A DRAWING

B MEANING

Figure 21.64 Roundness is a tolerance of form. It indicates that a cross section through a surface of revolution is round and lies within two concentric circles.

within 0.12 mm. On flat surfaces, straightness is measured in a plane passing through control-line elements, and it may be specified in two directions (usually perpendicular) if desired.

Roundness A surface of revolution (a cylinder, cone, or sphere) is round when all points on the surface intersected by a plane perpendicular to its axis are equidistant from the axis. In **Fig. 21.64** the feature control frame specifies roundness of a cone and cylinder, permitting a tolerance of 0.34

ROUNDNESS: SPHERE

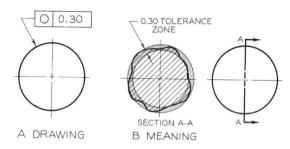

A DRAWING   B MEANING

Figure 21.65 Roundness of a sphere means that any cross section through it is round within the specified tolerance.

CYLINDRICITY

A DRAWING   B MEANING

Figure 21.66 Cylindricity is a tolerance of form that is a combination of roundness and straightness. It indicates that the surface of a cylinder lies within a tolerance zone formed by two concentric cylinders.

mm on the radius. **Figure 21.65** specifies a 0.30 mm tolerance zone for the roundness of a sphere.

Cylindricity A surface of revolution is cylindrical when all its elements lie within a cylindrical tolerance zone, which is a combination of tolerances of roundness and straightness (**Fig. 21.66**). Here, a cylindricity tolerance zone of 0.54 mm on the radius of the cylinder is specified.

21.18 Profile Tolerancing

Profile tolerancing involves specifying tolerances for a contoured shape formed by arcs or irregular curves and can apply to a surface or a single line.

PROFILE: PLANE

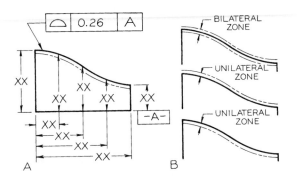

Figure 21.67 Profile is a tolerance of form for irregular curves of planes. (A) The curving plane is located by coordinates and is toleranced unidirectionally. (B) The tolerance may be applied by any of these methods.

PROFILE: LINE

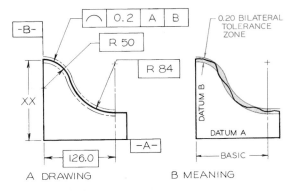

Figure 21.68 The profile of a line is a tolerance of form that specifies the variation allowed from the path of a line. Here, the line is formed by tangent arcs. The tolerance zone may be either bilateral or unilateral, as shown in Fig. 21.70.

The surface with the unilateral profile tolerance shown in **Fig. 21.67A** is defined by coordinates. **Figure 21.67B** shows how to specify bilateral and unilateral tolerance zones.

A profile tolerance for a single line is specified as shown in **Fig. 21.68**. The curve is formed by

PARALLELISM: PLANE

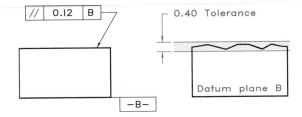

Figure 21.69 Parallelism is a tolerance of form. It indicates that a plane is parallel to a datum plane within specified limits. Here, plane B is the datum plane.

tangent arcs whose radii are given as basic dimensions. The radii are permitted to vary ±0.10 mm from the basic radii.

21.19 Orientation Tolerancing

Tolerances of orientation include **parallelism, perpendicularity, and angularity.**

Parallelism A surface or line is parallel when all its points are equidistant from a datum plane or axis. Two types of parallelism tolerance zones are:

1. A **planar tolerance** zone parallel to a datum plane within which the axis or surface of the feature must lie (**Fig. 21.69**). This tolerance also controls flatness.

2. A **cylindrical tolerance** zone parallel to a datum feature within which the axis of a feature must lie (**Fig. 21.70**).

Figure 21.71 shows the effect of specifying parallelism at MMC, where the modifier M is given in the feature control frame. Tolerances of form apply at RFS when not specified. Specifying parallelism at MMC means that the axis of the cylindrical hole must vary no more than 0.20 mm when the holes are at their smallest permissible size.

As the hole approaches its upper limit of 30.30, the tolerance zone increases to a maximum of

PARALLELISM: CYLINDER

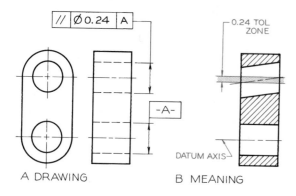

A DRAWING   B MEANING

Figure 21.70 You may specify parallelism of one centerline to another by using the diameter of one of the holes as the datum.

PARALLELISM AT MMC

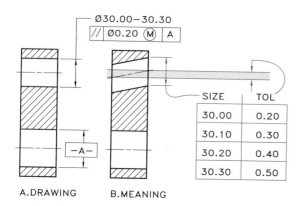

A.DRAWING   B.MEANING

Figure 21.71 The critical tolerance exists when features are at MMC. (A) The upper hole must be parallel to the hole used as datum A within a 0.20 DIA. (B) As the hole approaches its maximum size of 30.30 mm, the tolerance zone approaches 0.50 mm.

0.50 DIA. Therefore a greater variation is given at MMC than at RFS.

Perpendicularity **Figure 21.72** specifies perpendicularity of a plane to a datum plane. The feature control frame shows that the surface perpendicu-

PERPENDICULARITY: PLANE

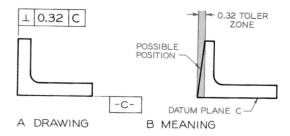

A DRAWING   B MEANING

Figure 21.72 Perpendicularity is a tolerance form that gives a tolerance zone for a plane perpendicular to a specified datum plane.

PERPENDICULARITY: CYLINDER

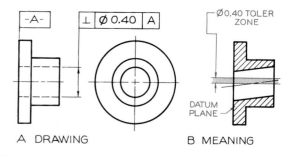

A DRAWING   B MEANING

Figure 21.73 Perpendicularity can apply to the axis of a feature, such as the centerline of a cylinder.

lar to datum plane C has a tolerance of 0.32 in. In **Fig. 21.73** a hole is specified as perpendicular to datum plane A.

Angularity A surface or line is angular when it is at an angle (other than 90°) from a datum or an axis. The angularity of the surface shown in **Fig. 21.74** is dimensioned with a basic angle (exact angle) of 30° and an angularity tolerance zone of 0.25 mm inside of which the plane must lie.

21.20 Runout Tolerancing

Runout tolerancing is a way of controlling multiple features by relating them to a common datum

ANGULARITY

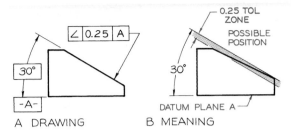

A DRAWING   B MEANING

Figure 21.74 Angularity is a tolerance of form specifying the tolerance zone for an angular surface with respect to a datum plane. Here, the 30° angle is a true, or basic, angle to which a tolerance of 0.25 mm is applied.

RUNOUT: CIRCULAR

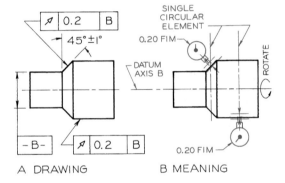

A DRAWING   B MEANING

Figure 21.75 Runout tolerance, a composite of several tolerance of form characteristics, is used to specify concentric cylindrical parts. The part is mounted on the datum axis and is gauged as it is rotated.

axis. Features so controlled are surfaces of revolution about an axis and surfaces perpendicular to the axis.

The datum axis, such as diameter B in **Fig. 21.75**, is established by a circular feature that rotates about the axis. When the part is rotated about this axis, the features of rotation must fall within the prescribed tolerance at **full indicator movement (FIM)**.

The two types of runout are **circular runout** and **total runout**. One arrow in the feature control

RUNOUT: TOTAL

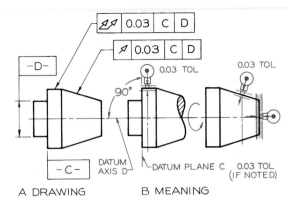

A DRAWING   B MEANING

Figure 21.76 Here, runout tolerance is measured by mounting the object on the primary datum plane C and the secondary datum cylinder D. The cylinder and conical surface are gauged to check their conformity to a tolerance zone of 0.03 mm. The runout at the end of the cone could have been noted.

frame indicates circular runout; two arrows indicate total runout.

Circular Runout Rotating an object about its axis 360° determines whether a circular cross section exceeds the permissible runout tolerance at any point (**Fig. 21.76**). This same technique is used to measure the amount of wobble in surfaces perpendicular to the axis of rotation.

Total Runout Used to specify cumulative variations of circularity, straightness, coaxiality, angularity, taper, and profile of a surface (Fig. 21.76), total runout tolerances are measured for all circular and profile positions as the part is rotated 360°. When applied to surfaces perpendicular to the axis, total runout tolerances control variations in perpendicularity and flatness.

Conclusion
The dimensioned part shown in **Fig. 21.77** illustrates several of the techniques of geometric tolerancing described in this and previous sections.

APPLICATION OF GEOMETRIC TOLERANCES

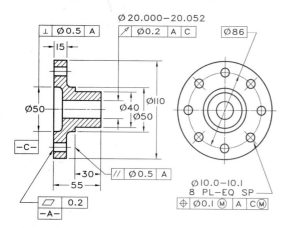

Figure 21.77 A combination of notes and symbols describe this part's geometric features.

SURFACE TEXTURE

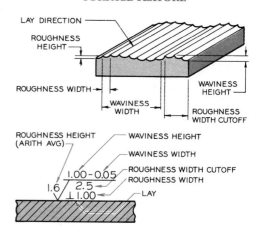

Figure 21.78 These are the definitions of surface texture for a finished surface.

21.21 Surface Texture

Because the surface texture of a part affects its function, it must be precisely specified instead of giving an unspecified finished mark such as a ✔. **Figure 21.78** illustrates most of the terms that apply to surface texture (surface control).

Surface texture: the variation in a surface, including roughness, waviness, lay, and flaws.

Roughness: the finest of the irregularities in the surface caused by the manufacturing process used to smooth the surface.

Roughness height: the average deviation from the mean plane of the surface measured in microinches (μin.) or micrometers (μm), or millionths of an inch and a meter, respectively.

Roughness width: the width between successive peaks and valleys forming the roughness measured in microinches or micrometers.

Roughness width cutoff: the largest spacing of repetitive irregularities that includes average roughness height (measured in inches or millimeters); when not specified, a value of 0.8 mm (0.030 in.) is assumed.

SURFACE TEXTURE SYMBOLS

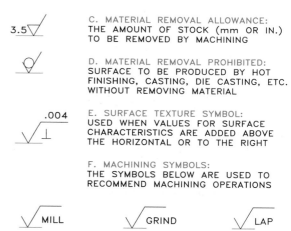

Figure 21.79 Use surface texture symbols to specify surface finish on the edge views of finished surfaces.

Waviness: a widely spaced variation that exceeds the roughness width cutoff measured in inches or millimeters; roughness may be

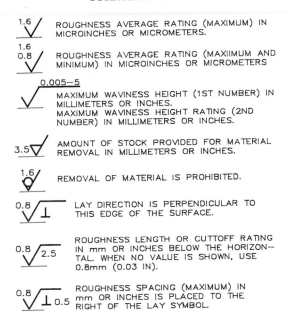

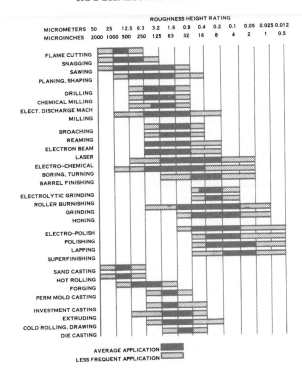

Figure 21.80 Values may be added to surface control symbols for more precise specifications.

Figure 21.81 Various types of production methods result in the surface roughness heights shown in micrometers and microinches (millionths of a meter or an inch).

regarded as a surface variation superimposed on a wavy surface.

Waviness height: the peak-to-valley distance between waves measured in inches or millimeters.

Waviness width: the spacing between wave peaks or wave valleys measured in inches or millimeters.

Lay: the direction of the surface pattern caused by the production method used.

Flaws: irregularities or defects occurring infrequently or at widely varying intervals on a surface, including cracks, blow holes, checks, ridges, scratches, and the like; the effect of flaws is usually omitted in roughness height measurements.

Contact area: the surface that will make contact with a mating surface.

Figure 21.79 shows symbols for specifying surface texture. The point of the ✔ must touch the edge view of the surface, an extension line from the surface, or a leader pointing to the surface. **Figure 21.80** shows how to specify values as a part of surface texture symbols.

Roughness height values are related to the processes used to finish surfaces and may be taken from the table in **Fig. 21.81**. The preferred values of roughness height are listed in **Fig. 21.82**.

The preferred roughness width cutoff values in **Fig. 21.83** are for specifying the sampling width used to measure roughness height. A value of 0.80 mm is assumed if no value is given. When required, maximum waviness height values may be selected from the recommended values shown in **Fig. 21.84**.

PREFERRED ROUGHNESS AVERAGE VALUES

Micro— meters μm	Micro— inches μin.	Micro— meters μm	Micro— inches μin.
0.025	1	1.6	63
0.050	2	3.2	125
0.10	4	6.3	250
0.20	8	12.5	500
0.40	16	25	1000
0.80	32	Mircometers=0.001 mm	

Figure 21.82 This range of roughness heights is recommended in ANSI Y14.36 standards.

ROUGHNESS WIDTH CUTOFF VALUES

MILLIMETERS	0.08	0.25	0.80	2.5	8.0	25
INCHES	.003	.010	.030	.1	.3	1

Figure 21.83 This range of roughness width cutoff values is recommended in ANSI Y14.36 standards.

MAXIMUM WAVINESS HEIGHT VALUES

mm	in.	mm	in.
0.0005	.00002	0.025	.001
0.0008	.00003	0.05	.002
0.0012	.00005	0.08	.003
0.0020	.00008	0.12	.005
0.0025	.0001	0.20	.008
0.005	.0002	0.25	.010
0.008	.0003	0.38	.015
0.012	.0005	0.50	.020
0.020	.0008	0.80	.030

Figure 21.84 This range of maximum waviness height values is recommended in ANSI Y14.36 standards.

Lay symbols indicating the direction of texture (markings made by the machining operation) on a surface (Fig. 21.85) may be added to surface texture symbols as shown in Fig. 21.86. The perpendicular sign indicates that lay is perpendicular to the edge view of the surface in this view (where the surface control symbol appears). **Figure 21.87** illustrates how to apply a variety of surface texture symbols to a part.

LAY SYMBOLS

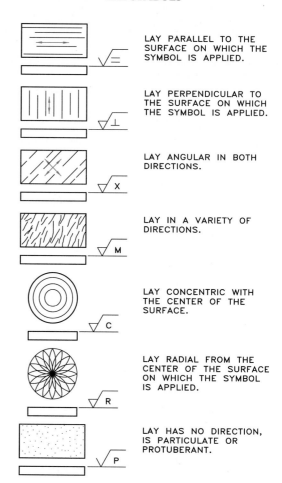

Figure 21.85 These symbols are used to indicate the direction of lay with respect to the surface where the control symbol is placed.

SURFACE TEXTURE SYMBOLS

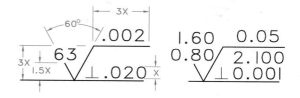

Figure 21.86 These are examples and sizes of typical fully specified surface texture symbols.

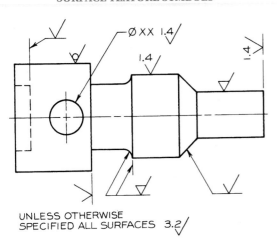

THE APPLICATION OF
SURFACE TEXTURE SYMBOLS

UNLESS OTHERWISE
SPECIFIED ALL SURFACES 3.2

Figure 21.87 Techniques of applying surface texture symbols to a part are illustrated here.

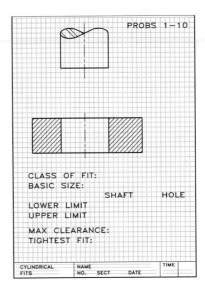

Figure 21.88 Problems 1–10.

Problems

Solve the following problems on size A sheets laid out on a grid of 0.20 in. or 5 mm.

Cylindrical Fits

1. (Fig. 21.88) Draw the shaft and hole shown (it need not be to scale), give the limits for each diameter, and complete the table of values. Use a basic diameter of 1.00 in. (25 mm) and a class RC 1 fit or a metric fit of H8/f7.

2. Repeat Problem 1, but use a basic diameter of 1.75 in. (45 mm) and a class RC 9 fit or a metric fit of H11/c11.

3. Repeat Problem 1, but use a basic diameter of 2.00 in. (51 mm) and a class RC 5 fit or a metric fit of H9/d9.

4. Repeat Problem 1, but use a basic diameter of 12.00 in. (305 mm) and a class LC 11 fit or a metric fit of H7/h6.

5. Repeat Problem 1, but use a basic diameter of 3.00 in. (76 mm) and a class LC 1 fit or a metric fit of H7/h6.

6. Repeat Problem 1, but use a basic diameter of 8.00 in. (203 mm) and a class LC 1 fit or a metric fit of H7/k6.

7. Repeat Problem 1, but use a basic diameter of 102 in. (2591 mm) and a class LN 3 fit or a metric fit of H7/n6.

8. Repeat Problem 1, but use a basic diameter of 11.00 in. (279 mm) and a class LN 2 fit or a metric fit of H7/p6.

9. Repeat Problem 1, but use a basic diameter of 6.00 in. (152 mm) and a class FN 5 fit or a metric fit of H7/s6.

10. Repeat Problem 1, but use a basic diameter of 2.60 in. (66 mm) and a class FN 1 fit or a metric fit of H7/u6.

Position Tolerancing

11. (Fig. 21.89) Make an instrument drawing of the part shown. Locate the two holes with a size tolerance of 1.00 mm and a position tolerance of 0.50 DIA. Insert the proper symbols and dimensions.

12. Repeat Problem 11, but locate three holes using the same tolerances for size and position.

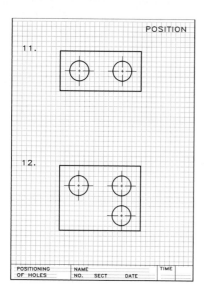

Figure 21.89 Problems 11–13.

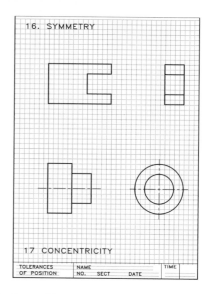

Figure 21.91 Problems 16 and 17.

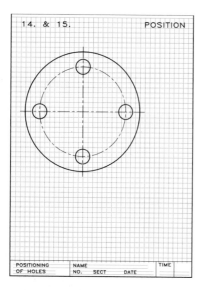

Figure 21.90 Problems 14 and 15.

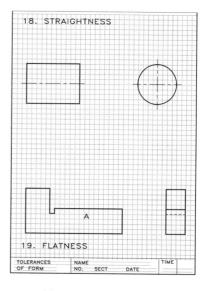

Figure 21.92 Problems 18 and 19.

13. Give the specifications for a two-pin gauge that can be used to measure the correctness of the two holes specified in Problem 11. Make a sketch of the gauge and show the proper dimensions on it.

14. (Fig. 21.90) Using positioning tolerances, locate the holes and properly note them to provide a size

tolerance of 1.50 mm and a locational tolerance of 0.60 DIA.

15. Repeat Problem 14, but locate six equally spaced, equally sized holes using the same tolerances of position.

16. (Fig. 21.91) Using a feature control symbol and the necessary dimensions, indicate that the notch is

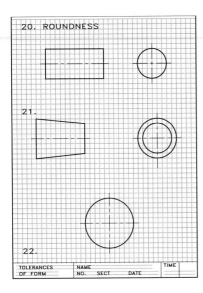

Figure 21.93 Problems 20–22.

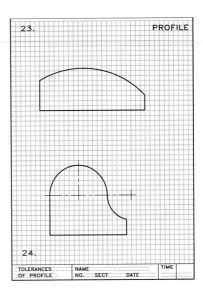

Figure 21.94 Problems 23 and 24.

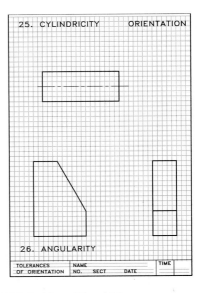

Figure 21.95 Problems 25 and 26.

symmetrical to the left-hand end of the part within 0.60 mm.

17. (Fig. 21.91) Using a feature control symbol and the necessary dimensions, indicate that the small cylinder is concentric with the large one (the datum cylinder) within a tolerance of 0.80.

18. (Fig. 21.92) Using a feature control symbol and the necessary dimensions, indicate that the elements of the cylinder are straight within a tolerance of 0.20 mm.

19. (Fig. 21.92) Using a feature control symbol and the necessary dimensions, indicate that surface A of the object is flat within a tolerance of 0.08 mm.

20–22. (Fig. 21.93) Using feature control symbols and the necessary dimensions, indicate that the cross sections of the cylinder, cone, and sphere are round within a tolerance of 0.40 mm.

23. (Fig. 21.94) Using a feature control symbol and the necessary dimensions, indicate that the profile of the irregular surface of the object lies within a bilateral or unilateral tolerance zone of 0.40 mm.

24. (Fig. 21.94) Using a feature control symbol and the necessary dimensions, indicate that the profile of

the line formed by tangent arcs lies within a bilateral or unilateral tolerance zone of 0.40 mm.

25. (Fig. 21.95) Using a feature control symbol and the necessary dimensions, indicate that the cylindricity of the cylinder is 0.90 mm.

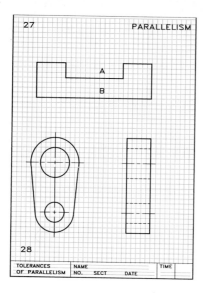

Figure 21.96 Problems 27 and 28.

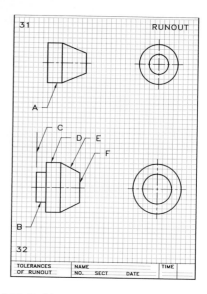

Figure 21.98 Problems 31 and 32.

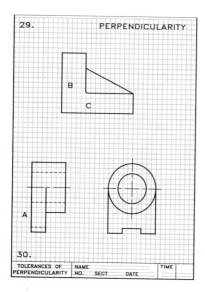

Figure 21.97 Problems 29 and 30.

26. (Fig. 21.95) Using a feature control symbol and the necessary dimensions, indicate that the angularity tolerance of the inclined plane is 0.7 mm from the bottom of the object, the datum plane.

27. (Fig. 21.96) Using a feature control symbol and the necessary dimensions, indicate that surface A of the object is parallel to datum B within 0.30 mm.

28. (Fig. 21.96) Using a feature control symbol and the necessary dimensions, indicate that the small hole is parallel to the large hole, the datum, within a tolerance of 0.80 mm.

29. (Fig. 21.97) Using a feature control symbol and the necessary dimensions, indicate that the vertical surface B is perpendicular to the bottom of the object, the datum C, within a tolerance of 0.20 mm.

30. (Fig. 21.97) Using a feature control symbol and the necessary dimensions, indicate that the hole is perpendicular to datum A within a tolerance of 0.08 mm.

31. (Fig. 21.98) Using a feature control symbol and cylinder A as the datum, indicate that the conical feature has a runout of 0.80 mm.

32. (Fig. 21.98) Using a feature control symbol with cylinder B as the primary datum and surface C as the secondary datum, indicate that surfaces D, E, and F have a runout of 0.60 mm.

22

Welding

22.1 Introduction

Welding is the process of permanently joining metal by heating a joint to a suitable temperature with or without applying pressure and with or without using filler material. The welding practices described comply with the standards developed by the American Welding Society and the American National Standards Institute (ANSI).

Welding is done in shops, on assembly lines, or in the field, as shown in **Fig. 22.1**, where a welder is joining pipes. Welding is a widely used method of fabrication that you must become familiar with in order to make and read drawings containing welding notes and specifications.

Advantages of welding over other methods of fastening include (1) simplified fabrication, (2) economy, (3) increased strength and rigidity, (4) ease of repair, (5) creation of gas- and liquid-tight joints, and (6) reduction in weight and size.

Figure 22.1 This welder is joining two pipes in accordance with specifications on a set of drawings. (Courtesy of Texas Eastern; *TE Today*; photo by Bob Thigpen.)

22.2 Welding Processes

Figure 22.2 shows various types of welding processes. The three main types are **gas welding**, **arc welding**, and **resistance welding**.

TYPES OF WELDING PROCESSES

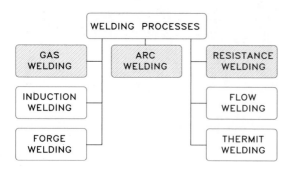

Figure 22.2 The three main types of welding processes are gas welding, arc welding, and resistance welding.

ARC WELDING

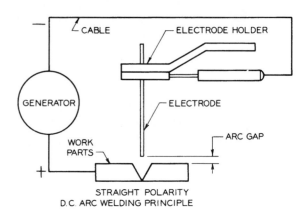

Figure 22.4 In arc welding, either AC or DC current is passed through an electrode to heat the joint.

GAS WELDING PROCESS

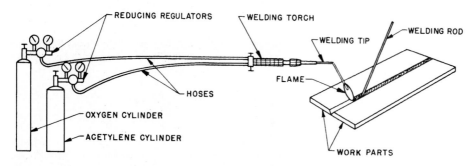

Figure 22.3 The gas welding process burns gases such as oxygen and acetylene in a torch to apply heat to a joint. The welding rod supplies the filler material. (Courtesy of General Motors Corporation.)

Gas welding involves the use of gas flames to melt and fuse metal joints. Gases such as acetylene or hydrogen are mixed in a welding torch and burned with air or oxygen (**Fig. 22.3**). The oxyacetylene method is widely used for repair work and field construction.

Most oxyacetylene welding is done manually with a minimum of equipment. Filler material in the form of welding rods is used to deposit metal at the joint as it is heated. Most metals, except for low- and medium-carbon steels, require fluxes to aid the process of melting and fusing the metals.

Arc welding involves the use of an electric arc to heat and fuse joints, with pressure sometimes required in addition to heat (**Fig. 22.4**). The filler material is supplied by a consumable or nonconsumable electrode through which the electric arc is transmitted. Metals well-suited to arc welding are wrought iron, low- and medium-carbon steels, stainless steel, copper, brass, bronze, aluminum,

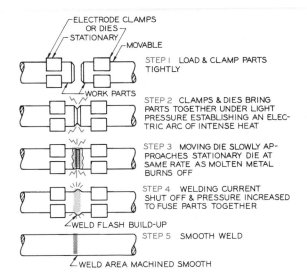

STEP I LOAD & CLAMP PARTS TIGHTLY

STEP 2 CLAMPS & DIES BRING PARTS TOGETHER UNDER LIGHT PRESSURE ESTABLISHING AN ELECTRIC ARC OF INTENSE HEAT

STEP 3 MOVING DIE SLOWLY APPROACHES STATIONARY DIE AT SAME RATE AS MOLTEN METAL BURNS OFF

STEP 4 WELDING CURRENT SHUT OFF & PRESSURE INCREASED TO FUSE PARTS TOGETHER

STEP 5 SMOOTH WELD

Figure 22.5 Flash welding, a type of arc welding, uses a combination of electric current and pressure to fuse two parts.

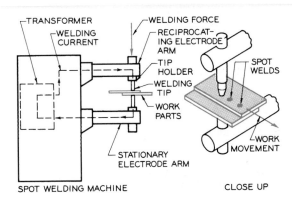

Figure 22.6 Resistance spot welding may be used to join lap and butt joints.

and some nickel alloys. In electric-arc welding, **the flux is a material coated on the electrodes that forms a coating on the metal being welded.** This coating protects the metal from oxidation so that the joint will not be weakened by overheating.

Flash welding is a form of arc welding, but it is similar to resistance welding because both pressure and electric current **(Fig. 22.5)** are applied. The pieces to be welded are brought together, and an electric current is passed through them, causing heat to build up between them. As the metal burns the current is turned off, and the pressure between the pieces is increased to fuse them.

Resistance welding comprises several processes by which metals are fused both by the heat produced from the resistance of the parts to an electric current and by pressure. Fluxes and filler materials normally are not used. All resistance welds are either lap- or butt-type welds.

Resistance spot welding is performed by pressing the parts together, and an electric current fuses them as illustrated in the lap joint weld in

Fig. 22.6. A series of small welds spaced at intervals, called **spot welds**, secure the parts. **Figure 22.7** lists welding processes that may be used for different materials.

22.3 Weld Joints and Welds

Figure 22.8 shows the five standard weld joints. The **butt joint** can be joined with the square groove, V-groove, bevel groove, U-groove, and J-groove welds. The **corner joint** can be joined with these welds and with the fillet weld. The **lap joint** can be joined with the bevel groove, J-groove, fillet, slot, plug, spot, projection, and seam welds. The **edge joint** uses the same welds as the lap joint along with the square groove, V-groove, U-groove, and seam welds. The **tee joint** can be joined by the bevel groove, J-groove, and fillet welds.

Figure 22.9 depicts commonly used welds and their corresponding ideographs (symbols). The **fillet weld** is a built-up weld at the intersection (usually 90°) of two surfaces. The **square, bevel, V-groove, J-groove,** and **U-groove welds** all have grooves, and the weld is made in these grooves. **Slot** and **plug**

RESISTANCE WELDING

Material	Spot Welding	Flash Welding
Low—carbon mild steel		
SAE 1010	Rec.	Rec.
SAE 1020	Rec.	Rec.
Medium—carbon steel		
SAE 1030	Rec.	Rec.
SAE 1050	Rec.	Rec.
Wrought alloy steel		
SAE 4130	Rec.	Rec.
SAE 4340	Rec.	Rec.
High—alloy austenitic stainless steel		
SAE 30301—30302	Rec.	Rec.
SAE 30309—30316	Rec.	Rec.
Ferritic and martensistic stainless steel		
SAE 51410—51430	Satis.	Satis.
Wrought heat—resisting alloys		
19—9—DL	Satis.	Satis.
16—25—6	Satis.	Satis.
Cast iron	NA	Not Rec.
Gray iron	NA	Not Rec.
Aluminum & alum. alloys	Rec.	Satis.
Nickel & nickel alloys	Rec.	Satis.

Rec.—Recommended
Not Rec.— Not recommended

Satis.—Satisfactory
NA—Not applicable

Figure 22.7 Resistance welding processes for various materials are shown here.

STANDARD WELD JOINTS

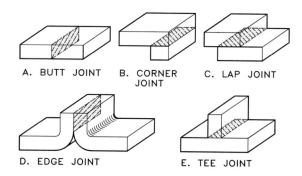

A. BUTT JOINT   B. CORNER JOINT   C. LAP JOINT

D. EDGE JOINT   E. TEE JOINT

Figure 22.8 These diagrams depict the five standard weld joints.

STANDARD WELDS AND IDEOGRAPHS

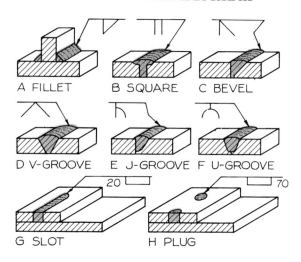

A FILLET   B SQUARE   C BEVEL

D V-GROOVE   E J-GROOVE   F U-GROOVE

G SLOT   H PLUG

Figure 22.9 These views illustrate standard welds and their corresponding ideographs.

WELDING SYMBOL

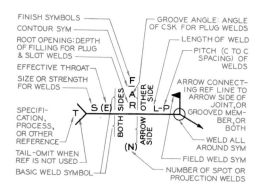

FINISH SYMBOLS
CONTOUR SYM
ROOT OPENING: DEPTH OF FILLING FOR PLUG & SLOT WELDS
EFFECTIVE THROAT
SIZE OR STRENGTH FOR WELDS
SPECIFI-CATION, PROCESS, OR OTHER REFERENCE
TAIL-OMIT WHEN REF IS NOT USED
BASIC WELD SYMBOL

GROOVE ANGLE: ANGLE OF CSK FOR PLUG WELDS
LENGTH OF WELD
PITCH (C TO C SPACING) OF WELDS
ARROW CONNECT-ING REF LINE TO ARROW SIDE OF JOINT, OR GROOVED MEM-BER, OR BOTH
WELD ALL AROUND SYM
FIELD WELD SYM
NUMBER OF SPOT OR PROJECTION WELDS

Figure 22.10 The welding symbol. Usually it is modified to a simpler form for use on drawings.

welds have intermittent holes or openings where the parts are welded. Holes are unnecessary when resistance welding is used.

22.4 Welding Symbols

If a drawing has a general welding note such as **ALL JOINTS ARE WELDED THROUGHOUT**, the

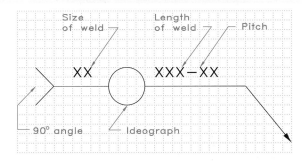

Figure 22.11 Welding symbol proportions are based on the letter height used on a drawing, usually ⅛ in. or 3 mm. This grid is equal to the letter height.

WELDING IDEOGRAPHS

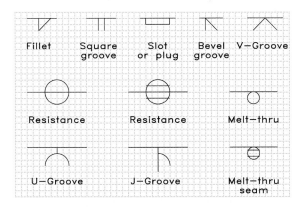

Figure 22.12 The sizes of the ideographs shown on the ⅛-in. (3 mm) grid (the letter height) are proportional to the size of the welding symbol (Fig. 22.11).

designer has transferred responsibility to the welder. Welding is too important to be left to chance and should be specified more precisely.

Symbols convey welding specifications on a drawing. **Figure 22.10** shows the symbol in its complete form, but it usually appears on a drawing in modified form (less detail). The scale of the welding symbol is based on the letter height used on the drawing, or the size of the grid on which the symbol is drawn, as shown in **Fig. 22.11**. The

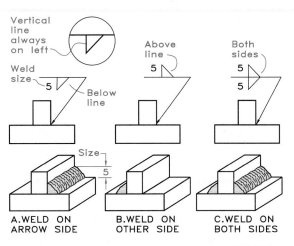

Figure 22.13 Fillet welds may be noted with abbreviated symbols. (A) When the ideograph appears below the horizontal line, it specifies a weld on the arrow side. (B) When it is above the line, it specifies a weld on the opposite side. (C) When it is on both sides of the line, it specifies a weld on each side.

standard height of lettering on a drawing is usually 1/8 in. or 3 mm.

The **ideograph** is the symbol that denotes the type of weld desired, and it generally depicts the cross section representation of the weld. **Figure 22.12** shows the ideographs used most often. They are drawn to scale on the 1/8-in. (3-mm) grid (equal to the letter height), which represents their full size when added to the welding symbol.

22.5 Application of Symbols

Fillet Welds

In **Fig. 22.13A**, placement of the fillet weld ideograph below the horizontal line of the symbol indicates that the weld is at the joint on the arrow side—the right side in this case. The vertical leg of the ideograph is always on the left side.

A numeral (either a common fraction or a decimal value) to the right of the ideograph indicates the size of the weld. You may omit this number

FILLET WELDS: ALL AROUND

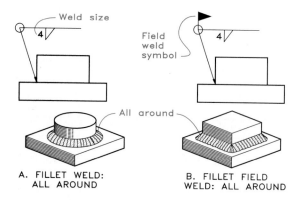

Figure 22.14 These symbols indicate fillet welds all around two types of parts.

FILLET-WELD SYMBOLS

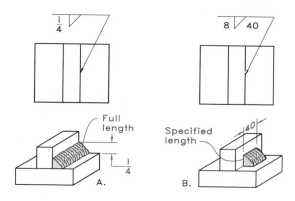

Figure 22.15

A This symbol indicates full-length fillet welds.

B This symbol indicates fillet welds of specified, but less than, full length.

from the symbol if you insert a general note elsewhere on the drawing to specify the fillet size, such as:

ALL FILLET WELDS 1/4 IN. UNLESS
OTHERWISE NOTED.

Placing the ideograph above the horizontal line **Fig. 22.13B** indicates that the weld is to be on the other side, that is, the joint on the other side of the part away from the arrow. When the part is to be welded on both sides, use the ideograph shown in **Fig. 22.13C**. You may omit the tail and other specifications from the symbol when you provide detailed specifications elsewhere.

A single arrow often is used to specify a weld that is to be made all around two joining parts **(Fig. 22.14A)**; a circle, 6 mm (twice the letter height) in diameter, drawn at the bend in the leader of the symbol denotes this type of weld. If the welding is to be done in the field rather than in the shop, a solid black triangular "flag" is used **(Fig. 22.14B)**.

You may specify a fillet weld that is to run the full length of the two parts as in **Fig. 22.15A**. The ideograph is on the lower side of the horizontal line, so the weld is on the arrow side. You may specify a fillet weld that is to run shorter than full

INTERMITTENT WELDING

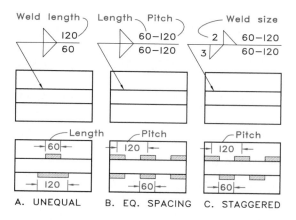

Figure 22.16 These symbols specify intermittent welds of varying lengths and alignments.

length as in **Fig. 22.15B**, where 40 represents the weld's length in millimeters.

You may specify fillet welds to run different lengths and be positioned on both sides of a part as in **Fig. 22.16A**. The dimension on the lower side of the horizontal gives the length of the weld on the arrow side, and the dimension on the upper

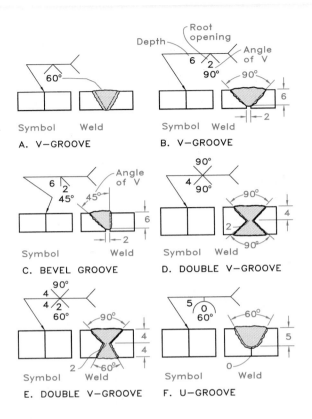

Figure 22.17 This drawing shows the various types of groove welds and their general specifications.

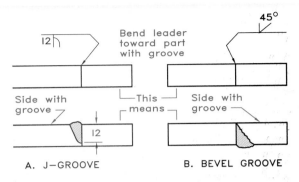

Figure 22.18 J-groove welds and bevel welds are specified by bent arrows pointing to the side of the joint to be grooved or beveled.

side of the horizontal gives the length on the opposite side.

Intermittent welds have a specified length and are spaced uniformly, center to center, at an interval called the **pitch**. In **Fig. 22.16B**, the welds are equally spaced on both sides, are 60 mm long, and have pitches of 120 mm, as indicated by the symbol shown. The symbol shown in **Fig. 22.16C** specifies intermittent welds that are staggered in alternate positions on opposite sides.

Groove Welds
Figure 22.17 shows standard types of groove welds. When you do not give the depth of the grooves,

angle of the chamfer, and root openings on a symbol, you must specify them elsewhere on the drawing or in supporting documents. In **Fig. 22.17A and B**, the angles of the V-joints are labeled 60° and 90° under the ideographs. In **Fig. 22.17B**, the depths of the weld (6) and the root opening (2)—the gap between the two parts—are given.

In a **bevel groove** weld, only one of the parts is beveled. The symbol's leader is bent and pointed toward the beveled part to call attention to it (**Figs. 22.17C** and **22.18B**). This practice also applies to J-groove welds, where one side is grooved and the other is not (**Fig. 22.18A**).

Notate double V-groove welds by weld size, bevel angle, and root opening (**Fig. 22.17D and E**). Omit root opening sizes or show a zero on the symbol when parts fit flush. Give the angle and depth of the groove in the symbol for a **U-groove** weld (**Fig. 22.17F**).

Seam Welds
A seam weld joins two lapping parts with either a continuous weld or a series of closely spaced spot welds. The seam weld process to be used is identified by abbreviations in the tail of the weld symbol (**Fig. 22.19**). The circular ideograph for a resistance weld is about 12 mm (four times the letter height) in diameter and is centered over the hori-

WELDING PROCESS SYMBOLS

CAW	Carbon-arc w.	IB	Induction brazing
CW	Cold welding	IRB	Infrared brazing
DB	Dip brazing	OAW	Oxyacetylene w.
DFW	Diffusion welding	OHW	Oxyhydrogen w.
EBW	Electric beam w.	PGW	Pressure gas w.
ESW	Electroslag welding	RB	Resist. brazing
EXW	Explosion welding	RPW	Projection weld.
FB	Furnace brazing	RSEW	Resist. seam w.
FOW	Forge welding	RSW	Resist. spot w.
FRW	Friction welding	RW	Resist. welding
FW	Flash welding	TB	Torch brazing
GMAW	Gas metal arc w.	UW	Upset welding
GTAW	Gas tungsten w.		w.=welding

Figure 22.19 These abbreviations represent the various types of welding processes and are used in welding symbols.

SPOT WELDS

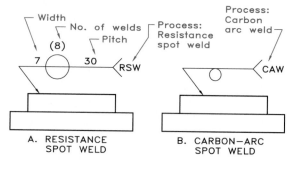

Figure 22.21 The process to be used for (A) resistance spot welds and (B) arc spot welds is indicated in the tail of the symbol. For the arc weld the symbol must specify the arrow side or the other side of the piece.

SEAM WELDS

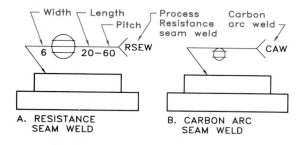

Figure 22.20 The process used for (A) resistance seam welds and (B) arc-seam welds is indicated in the tail of the symbol. For the arc weld the symbol must specify the arrow side or the other side of the piece.

SURFACE BUILT-UP WELD

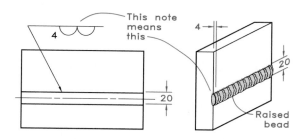

Figure 22.22 Use this method to apply a symbol to a built-up weld on a surface.

zontal line of the symbol **(Fig. 22.20A)**. The weld's width, length, and pitch are given.

When the seam weld is to be made by arc welding (CAW), the diameter of the ideograph is about 6 mm (twice the letter height) and goes on the upper or lower side of the symbol's horizontal line to indicate whether the seam is to be applied to the arrow side or opposite side **(Fig. 22.20B)**. When the length of the weld is not shown, the seam weld is understood to extend between abrupt changes in the direction of the seam.

Spot welds are similarly specified with ideographs and specifications by diameter, number of welds, and pitch between the welds. The process of resistance spot welding (RSW) is noted in the tail of the symbol **(Fig. 22.21A)**. For arc welding, the arrow side or other side must be indicated by a symbol **(Fig. 22.21B)**.

Built-Up Welds

When the surface of a part is to be enlarged, or built-up, by welding, indicate this process with a

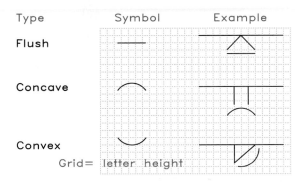

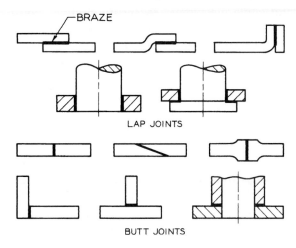

Figure 22.23 These contour symbols specify the desired surface finish of a weld.

USING CONTOUR SYMBOLS

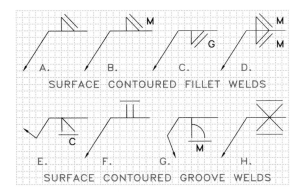

Figure 22.24 These examples of contoured weld symbols with letters added indicate the type of finishing to be applied to the weld (M, machining; G, grinding; C, chipping).

symbol, as shown in **Fig. 22.22**. Dimension the width of the built-up weld in the view. Specify the height of the weld above the surface in the symbol to the left of the ideograph. The radius of the circular segment is 6 mm (twice the letter height).

22.6 Surface Contouring

Contour symbols are used to indicate which of the three types of contours, flush, concave, or convex, is desired on the surface of the weld. **Flush** con-

Figure 22.25 The two basic types of brazing joints are lap joints and butt joints.

tours are smooth with the surface or flat across the hypotenuse of a fillet weld. **Concave** contours bulge inward with a curve, and **convex** contours bulge outward with a curve (**Fig. 22.23**).

Finishing the weld by an additional process to obtain the desired contour often is necessary. These processes, which may be indicated by their abbreviations, are **chipping** (C), **grinding** (G), **hammering** (H), **machining** (M), **rolling** (R), and **peening** (P), as shown in **Fig. 22.24**.

22.7 Brazing

Brazing is a method much like welding for joining pieces of metal. Brazing entails heating joints to more than 800°F and distributing by capillary action a nonferrous filler material, with a melting point below that of the base materials, between the closely fitting parts.

Before brazing the parts must be cleaned and the joints fluxed. The brazing filler is added before or just as the joints are heated beyond the filler's melting point. After the filler material has melted, it is allowed to flow between the parts to form the joint. As **Fig. 22.25** shows, there are two basic brazing joints: **lap joints** and **butt joints**.

Brazing is used to join parts, to provide gas- and liquid-tight joints, to ensure electrical conductivity, and to aid in repair and salvage. Brazed joints withstand more stress, higher temperature, and more vibration than soft-soldered joints.

22.8 Soft Soldering

Soldering is the process of joining two metal parts with a third metal that melts below the temperature of the metals being joined. Solders are alloys of nonferrous metals that melt below 800°F. Widely used in the automotive and electrical industries, soldering is one of the basic techniques of welding and often is done by hand with a soldering iron like the one depicted in **Fig. 22.26**. The iron is placed on the joint to heat it and to melt the solder. Figure 22.26 also shows how to notate a soldered joint.

SOLDERING IRON

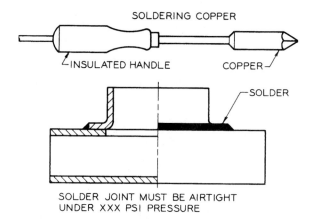

Figure 22.26 This typical hand-held soldering iron is used to soft-solder two parts together. The method of notating a drawing for soldering also is shown.

Problems

Solve these problems on size A sheets laid out on a grid of 0.20 inches (5 mm).

1–8. (Fig. 22.27) Give welding notes to include the information specified for each problem. You may omit instructional information from the solution.

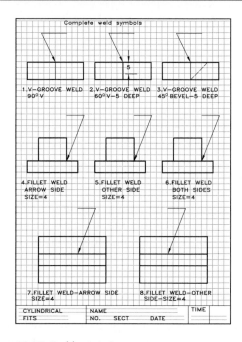

Figure 22.27 Problems 1–8.

23

Working Drawings

23.1 Introduction

Working drawings are the drawings from which a design is implemented. All principles of orthographic projection and techniques of graphics can be used to communicate the details of a project in working drawings. A **detail drawing** is a working drawing of a single part (or detail) within the set of working drawings.

Specifications are the written instructions that accompany working drawings. When the design can be represented on a few sheets, the specifications are usually written on the drawings to consolidate the information into a single format.

All parts must interact with other parts to some degree to yield the desired function from a design. Before detail drawings of individual parts are made the designer must thoroughly analyze the working drawing to ensure that the parts fit properly with mating parts, that the correct tolerances are applied, that the contact surfaces are properly finished, and that the proper motion is possible between the parts.

Much of the work in preparing working drawings is done by the drafter, but the designer, who is usually an engineer, is responsible for their correctness. It is working drawings that bring products and systems into being.

23.2 Working Drawings as Legal Documents

Working drawings are legal contracts that document the design details and specifications as directed by the engineer. Therefore drawings must be as clear, precise, and thorough as possible. Revisions and modifications of a project at the time of production or construction are much more expensive than when done in the preliminary design stages.

Poorly executed working drawings result in wasted time and resources and increase implementation costs. To be economically competitive drawings must be as error-free as possible.

Working drawings specify all aspects of the design, reflecting the soundness of engineering and function of the finished product and economy of fabrication. The working drawing is the instrument that is most likely to establish the responsibility for any failure to meet specifications during implementation.

23.3 Dimensions and Units

English System

The inch is the basic unit of the English system, and virtually all shop drawings are dimensioned in inches. This practice is followed even when dimensions are several feet in length.

The clamp illustrated in **Fig. 23.1** is detailed in working drawings (**Figs. 23.2–23.4**), which are dimensioned in inches. Decimal fractions are

Figure 23.1 This revolving clamp assembly holds parts while they are being machined. (Courtesy of Jergens, Incorporated.)

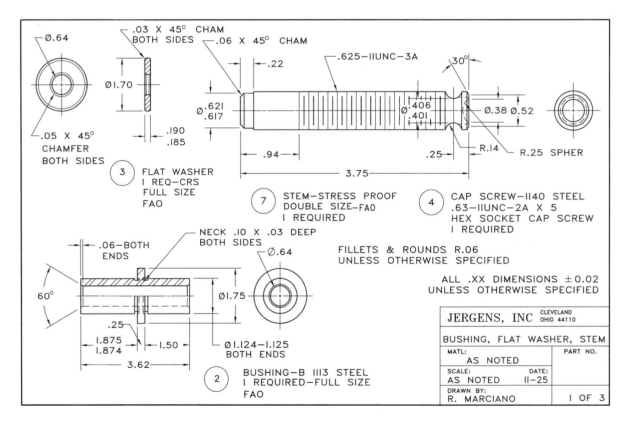

Figure 23.2 Sheet 1 of 3: This computer-produced detail drawing depicts parts of the clamp assembly shown in Fig. 23.1. (Figures 23.2–23.4 courtesy of Jergens, Incorporated.)

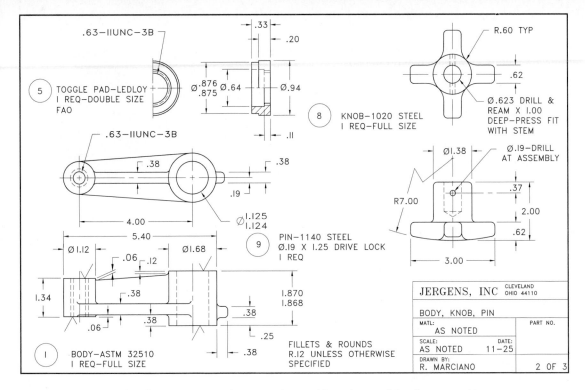

Figure 23.3 Sheet 2 of 3: This continuation of Fig. 23.2 shows additional parts of the clamp assembly.

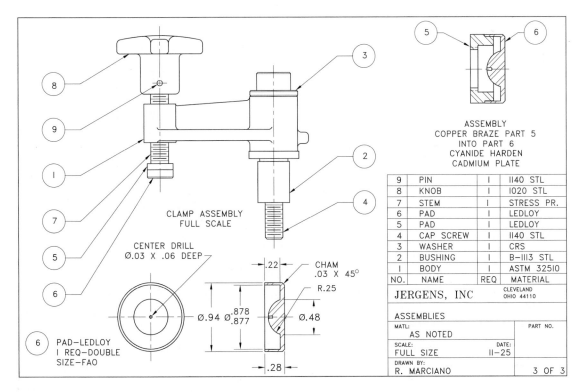

Figure 23.4 Sheet 3 of 3: This continuation of Fig. 23.2 shows the clamp assembly at full scale, an orthographic assembly drawing of the clamp assembly, and a parts list.

preferable to common fractions, although common fractions are still used (mostly by architects). Arithmetic can be done with greater ease with decimal fractions than with common fractions.

Inch marks (") are omitted from dimensions on working drawings because the units are understood to be in inches, and their omission saves drafting time. Usually, several dimensioned orthographic views of parts may be shown on each sheet. However, some companies have policies that views of only one part be drawn on a sheet, even if the part is extremely simple, such as a threaded fastener.

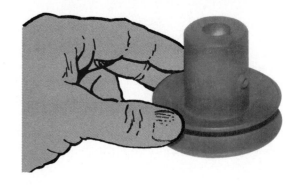

Figure 23.5 This photo is of a pulley and setscrew.

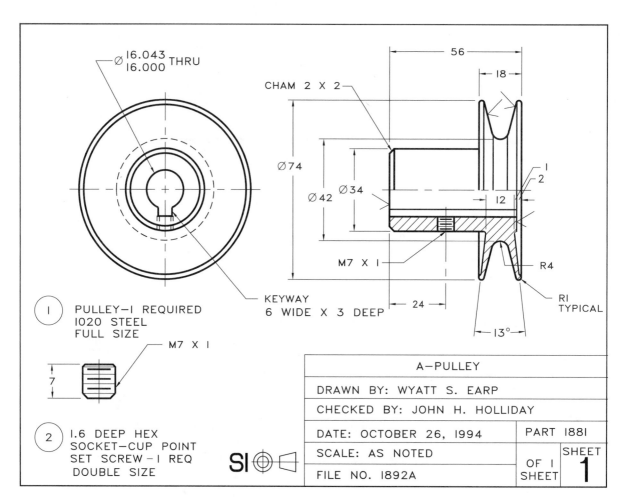

Figure 23.6 This computer-produced working drawing (dimensions in mm) gives details of the pulley and setscrew shown in Fig. 23.5.

The arrangement of views of parts on the sheet has no relationship to how the parts fit together; the views are simply positioned to best fit the available space on the sheet. **The views of each part are labeled with a part number, a name for identification, the material it is made of, and any other notes necessary to explain manufacturing procedures.**

The purpose of the orthographic **assembly drawing** shown on sheet 3 (**Fig. 23.4**) is to illustrate how the parts are to fit together. Each part is numbered and cross-referenced with the numbers in the **parts list**, which serves as a bill of materials.

Metric System

The millimeter is the basic unit of the metric system, and dimensions usually are given to the nearest whole millimeter without decimal fractions (except to specify tolerances). Metric abbreviations (mm) after the numerals are omitted from dimensions because the SI symbol near the title block indicates that all units are metric. If you have trouble relating to the length of a millimeter, recall that the fingernail of your index finger is about 10 mm wide.

The pulley (**Fig. 23.5**) is depicted and dimensioned in millimeters in the working drawings shown in **Fig. 23.6**. Dimensions and notes along with the descriptive views give the information needed to construct the pieces.

The left-end handcrank (**Fig. 23.7**) is detailed on two size B sheets in **Figs. 23.8** and **23.9**. The orthographic, sectioned assembly drawing of the left-end handcrank shown in Fig. 23.9 demonstrates how the parts are to be put together. The numbers in the balloons provide a cross-reference in the parts list, which goes just above the title block.

Figure 23.10 is a photograph of a lifting device used to level equipment such as lathes and milling machines; **Figs. 23.11** and **23.12** show working drawings that give details of its parts. The SI symbol indicates that the dimensions are in millimeters, and the truncated cone indicates third-angle

Figure 23.7 This photo is of a left-end handcrank.

projection views. The assembly drawing in Fig. 23.12 illustrates how the parts are to be assembled after they have been made.

Dual Dimensions

Some working drawings carry both inch and millimeter dimensions as shown in **Fig. 23.13** where the dimensions in parentheses or brackets are millimeters. The units may also appear as millimeters first and then be converted and shown in brackets as inches. Converting from one unit to the other results in fractional round-off errors. An explanation of the primary unit system for each drawing should be noted in the title block.

23.4 Laying Out a Detail Drawing

By Computer

You may lay out a simple working drawing of a single part by computer by beginning with the border (**Fig. 23.14**). Although this example is elementary, it is useful in describing the procedures for laying out more complex drawings. Allow a margin of at least 0.25 in. (7 mm) between the edge of the sheet and the borders. Outline space for the title block in the lower right-hand corner of the sheet to ensure that the drawing or notes do not occupy this area.

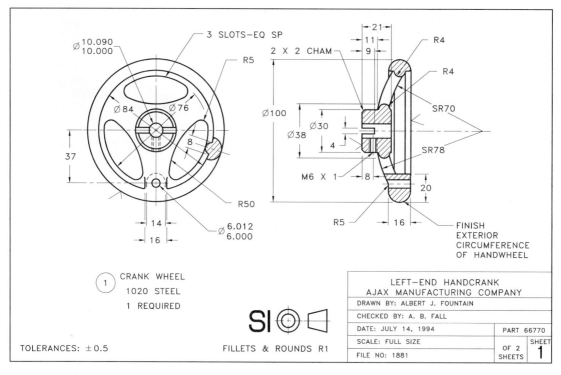

Figure 23.8 Sheet 1 of 2: This set of working drawings (dimensions in mm) depicts the crank wheel of the left-end handcrank shown in Fig. 23.7.

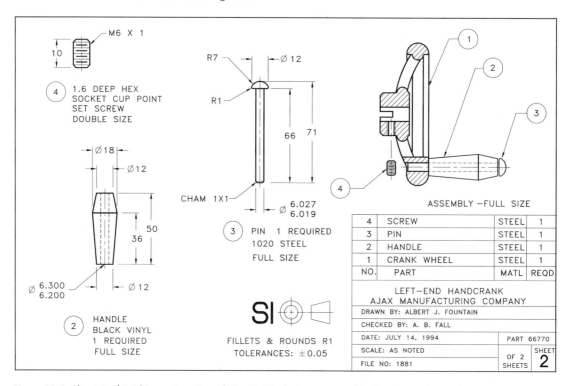

Figure 23.9 Sheet 2 of 2: This continuation of Fig. 23.8 includes an assembly drawing and parts list.

Figure 23.10 This Lev-L-Line lifting device is used to level heavy machinery. (Courtesy of Unisorb Machinery Installation Systems.)

First, make a freehand sketch to determine the necessary views, the number of dimensions, and their placement so that you can select the proper scale. Then use the computer to produce the views close together to make projection from view to view easier.

Next, MOVE the views apart to make room for notes and dimensions. Finally, add dimensions, notes, SI symbol, and title block to complete the drawing.

By Hand

When making a drawing by hand with instruments on paper or film, first lay out the views and dimensions on a different sheet of paper. Then

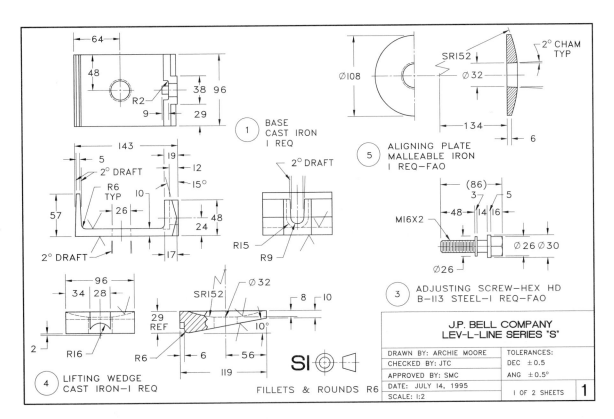

Figure 23.11 Sheet 1 of 2: This working drawing (dimensioned in SI units) is of the lifting device shown in Fig. 23.10. (Courtesy of Unisorb Machinery Installation Systems.)

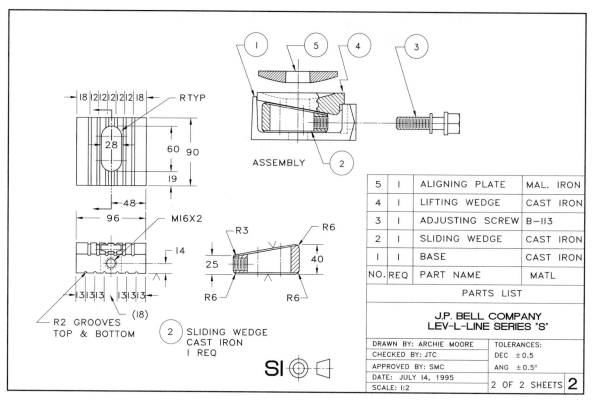

Figure 23.12 Sheet 2 of 2: This continuation of Fig. 23.11 is a further working drawing and assembly drawing of the lifting device. (Courtesy of Unisorb Machinery Installation Systems.)

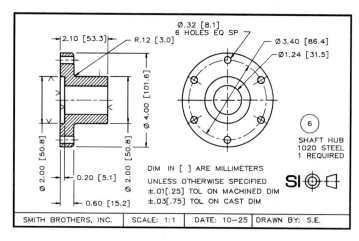

Figure 23.13 In this dual-dimensioned drawing, dimensions are shown in millimeters with their equivalents in inches given in brackets.

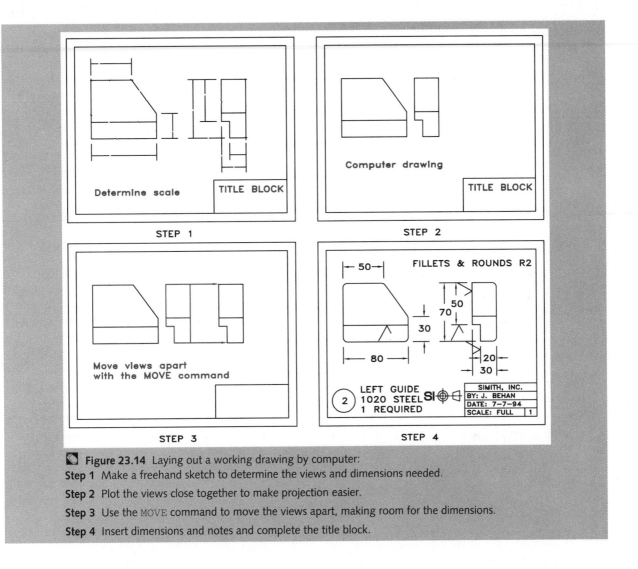

Figure 23.14 Laying out a working drawing by computer:

Step 1 Make a freehand sketch to determine the views and dimensions needed.

Step 2 Plot the views close together to make projection easier.

Step 3 Use the MOVE command to move the views apart, making room for the dimensions.

Step 4 Insert dimensions and notes and complete the title block.

overlay the drawing with vellum or film and trace it to obtain the final drawing. You must use guidelines for lettering for each dimension and note. Lightly draw them or underlay the drawing with a sheet containing guidelines.

Figure 23.15 shows the standard sheet sizes for working drawings. Paper, film, cloth, and reproductior materials are available in these modular sizes, so good practice requires that you make drawings in one of these standard sizes.

23.5 Notes and Other Information

Title Blocks and Parts Lists
Figure 23.16 shows a **title block** and **parts list** suitable for student assignments. Title blocks usu-

DRAWING-SHEET SIZES

ENGLISH SIZES			METRIC SIZES		
A	11 x 8.5		A4	297 X 210
B	17 X 11		A3	420 X 297
C	22 X 17		A2	594 X 420
D	34 X 22		A1	841 X 594
E	44 X 34		A0	1189 X 841

Figure 23.15 These are the standard sheet sizes for working drawings dimensioned in inches and millimeters.

PARTS LIST AND TITLE BLOCK

2	SHAFT	2	1020 STL	.38
1	BASE	1	CAST IRON	
NO	PART NAME	REQ	MATERIAL	
	PARTS LIST			

5" (Approximately)

TITLE		
BY: RED GRANGE	SECT: 500	.38
DATE: MAY 2, 1994	SHEET 1	
SCALE: FULL SIZE	OF 1 SHEETS	

$\frac{1}{8}$" LETTERS

Figure 23.16 This typical title block and parts list is suitable for most student assignments.

ally are placed in the lower right-hand corner of the drawing sheet against the borders. The parts list (Fig. 23.16) should be placed directly over the title block (see also Figs. 23.9 and 23.12).

Title Blocks In practice, title blocks usually contain the title or part name, drafter, date, scale, company, and sheet number. Other information, such as tolerances, checkers, and materials, also may be given. **Figure 23.17** shows another example of a title block, which is typical of those used by various industries. Any modifications or changes added after the first version to improve the design is shown in the revision blocks.

A TYPICAL TITLE BLOCK

REVISIONS	COMPANY NAME COMPANY ADDRESS		
CHG. HEIGHT	TITLE: LEFT—END BEARING		
FAO	DRAWN BY: JOHNNY RINGO		
	CHECKED BY: FRED DODGE		
	DATE: JULY 14, 1995		
	SCALE HALF SIZE	SHEET OF 3 SHEETS 1	

Figure 23.17 This title block, which includes a revision block, is typical of those used in industry.

Depending on the complexity of the project, a set of working drawings may contain from one to more than a hundred sheets. Therefore, giving the number of each sheet and the total number of sheets in the set on each sheet is important (for example, sheet 2 of 6, sheet 3 of 6, and so on).

Computer Method Figure 23.18 shows a computer shortcut for filling in a title block that will be used on several sheets. You may produce the title block only once, filling it in with dummy values to establish the positions of the text. Use the DDEDIT command to replace the dummy entries with updated values when the block is inserted or copied into a new drawing.

A similar shortcut in the design of a title block involves the use of ATTRIBUTES (see Chapter 36). In that case, the program will prompt you to INSERT entries into the title block.

Parts List The part numbers and part names in the parts list correspond to those given to each part depicted on the working drawings. In addition, the number of identical parts required are given along with the material used to make each part. Because the exact material (for example, 1020 STEEL) is designated for each part on the drawing, the material in the parts list may be shortened to STEEL, which requires less space.

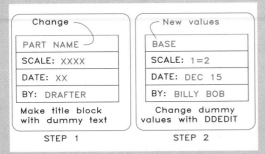

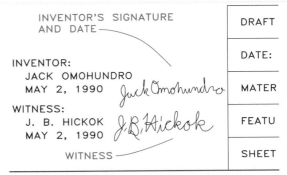

Figure 23.18 Producing a title block by computer:

Step 1 Draw the title block and add dummy text values, using the desired text style and size. Make a BLOCK of the title block.

Step 2 Position the title block against the lower bottom and right borders. EXPLODE the BLOCK and use the DDEDIT command to convert the dummy values to actual values.

Figure 23.19 This note next to the title block names the inventor and is witnessed by an associate to establish ownership of a design for patent purposes.

INDICATION OF SCALES

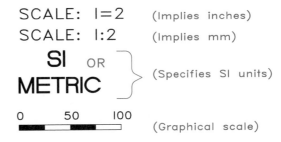

Figure 23.20 Specify scales in English and SI units on working drawings with these methods.

Patent Rights Note

Figure 23.19 shows the use of a note near the title block that names Jack Omohundro as the inventor of the part or process. An associate, J. B. Hickok, signs and dates the drawing as a witness to the designer's work. This type of note establishes ownership of the ideas and dates of their development to help the inventor obtain a patent. An even better case for design ownership is made if a second witness signs and dates the drawing.

Scale Specification

If all working drawings in a set are the same scale, you need to indicate it only once in the title block on each sheet. If several detail drawings on a working drawing are different scales, indicate them on the drawing under each set of views. In this case, indicate "as shown" in the title block opposite *scale*. When a drawing is not to scale, place the abbreviation NTS (not to scale) in the title block.

Figure 23.20 shows several methods of indicating scales. Use of the colon (for example, 1:2)

implies the metric system; use of the equal sign (for example, 1=2) implies the English system—but these are not absolute rules. The SI symbol or *metric* designation on a drawing specifies that millimeters are the units of measurement.

In some cases, you may want to show a graphical scale with calibrations on a drawing to permit the interpretation of linear measurements by transferring them with dividers from the drawing to the scale.

Tolerances

Recall from Chapter 21 that you may use general notes on working drawings to specify the dimension tolerances. **Figure 23.21** shows a table of val-

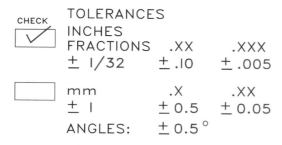

Figure 23.21 General tolerance notes on working drawings specify the dimension tolerances permitted.

Figure 23.22 Name and number each part on a working drawing for use in the parts list.

ues with boxes in which you can make a check mark to indicate whether the units are in inches or millimeters. Position plus-and-minus tolerances under each common or decimal fraction. For example, this table specifies that each dimension with two-place decimals will have a tolerance of ±0.10 in. You may also give angular tolerances in general notes (±0.5°, for example).

Part Names and Numbers

Give each part a name and number, using letters and numbers 1/8-in. (3 mm) high **(Fig. 23.22)**. **Place part numbers inside circles, called** *balloons*, **having diameters approximately four times the height of the numbers.**

Place part numbers near the views to which they apply, so their association will be clear. On assembly drawings, balloons are especially important because the same parts numbers are used in the parts list.

23.6 Checking a Drawing

People who check drawings must have special qualifications that enable them to identify errors and to suggest revisions and modifications that will result in a better product at a lower cost. A checker may be a chief drafter experienced in drafting and manufacturing processes or the engineer or designer who originated the project. In large companies, personnel in the various shops

involved in production review the drawings to ensure that the most efficient production methods are specified for each part.

Checkers never check the original drawing; instead, they mark corrections with a colored pencil on a diazo (blue-line) print. They return the marked-up print to the drafter who revises the original and makes another print for final approval.

In **Fig. 23.23**, the various modifications made by checkers are labeled with letters that are circled and placed near the revisions. The drafter lists and dates changes in the revision record, which lists the revisions made.

Checkers inspect a working or detail drawing for correctness and soundness of design. In addition, they are responsible for the drawing's completeness, quality, readability, and clarity, which reflect the lettering and drafting techniques used. **Lettering and text quality is especially important** because the shop person is guided by lettered notes and dimensions.

Checking Students' Work

The best way for you to check your drawings for adequate dimensions is to rapidly make a scale drawing of the part using the dimensions from the working drawings. You can identify missing dimensions more easily in this way than by reading a drawing by eye. **Figure 23.24** shows a grading scale for checking working drawings prepared by students, with hypothetical grading shown.

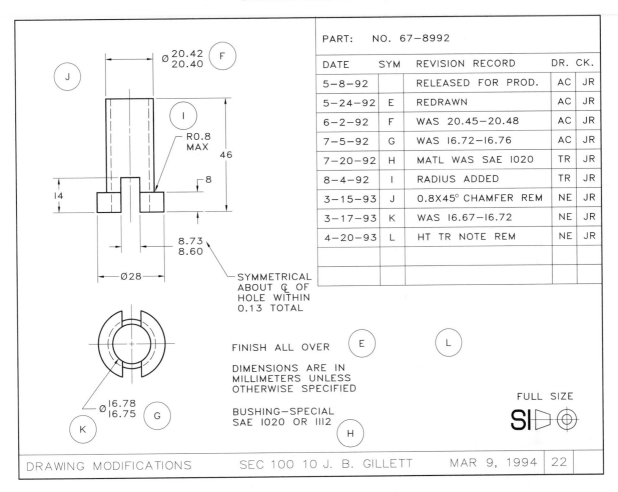

Figure 23.23 The modifications to this working drawing are noted near the details revised. The letters in balloons correspond to those in the revision table.

Use this list as an outline for reviewing working drawings to ensure that you have met the main requirements.

23.7 Drafter's Log

In addition to the individual revision records, drafters should keep a log of all changes made during a project. As the project progresses, the drafter should record the changes, dates, and people involved. Such a log allows anyone reviewing the project in the future to understand easily and clearly the process used to arrive at the final design.

Calculations often are made during a drawing's preparation. If they are lost or poorly done, they may have to be redone; therefore they should be a permanent part of the log.

TITLE BLOCK (5 points)		
Student's name	1	1
Checker's name	1	1
Date	1	1
Scale	1	1
Sheet number	1	1

DRAWING DETAILS (19 pts)		
Properly drawn views	10	9
Spacing of views	5	4.5
Part names and numbers	2	1.5
Sections & conventions	2	1.5

DESIGN INFORMATION (12 pts)		
Proper tolerances	5	4
Surface texture symbols	2	2
Thread notes & symbols	5	4.5

DIMEN. PRACTICES (20 pts)		
Proper arrowheads	3	2.5
Positions of dimensions	3	2.5
Completeness of dimens	4	3
Fillet & round notes	2	2
Machined-hole notes	2	1.5
Inch/mm marks omitted	2	2
Dimensions from best views	4	3.5

DRAFTSMANSHIP (19 points)		
Thicknesses of lines	6	5.5
Lettering	6	5
Neatness	3	1.5
Reproduction quality	4	3

GENERAL NOTES (5 points)		
SI symbol	2	2
Third angle symbol	2	2
General tolerance note	1	1

ASSEMBLY DRAWING (15 points)		
Descriptive views	6	5.5
Clarity of assembly	3	2.8
Parts list completeness	4	3.6
Part numbers in balloons	2	2

PRESENTATION (5 points)		
Properly stapled	2	2
Properly trimmed	1	1
Properly folded	1	1
Grade sheet attached	1	1
Total 100		**88**

Figure 23.24 This checklist may be used to evaluate a student's working-drawing assignment.

Figure 23.25 An assembly drawing explains how the parts of a product, such as this Ford tractor, are to be assembled. (Courtesy of Ford Motor Company.)

23.8 Assembly Drawings

After parts have been made according to the specifications of the working drawings, they will be assembled **(Fig. 23.25)** in accordance with the directions of an **assembly drawing**. Two general types of assembly drawings are **orthographic assemblies** and **pictorial assemblies**. Dimensions usually are omitted from assembly drawings.

The lifting device shown in Fig. 23.10 is depicted in an isometric assembly in **Fig. 23.26**. Each part is numbered with a balloon and leader to cross-reference them to the parts list, where more information about each part is given.

Figure 23.27 shows an orthographic exploded assembly drawing. In many applications, the arrangement of parts may be easier to understand when the parts are shown *exploded* along their centerlines. These views are shown as regular orthographic views, with some lines shown as hidden lines and others omitted.

PICTORIAL ASSEMBLY: ASSEMBLED

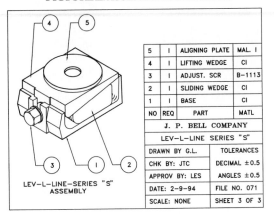

5	I	ALIGNING PLATE	MAL. I
4	I	LIFTING WEDGE	CI
3	I	ADJUST. SCR	B-1113
2	I	SLIDING WEDGE	CI
1	I	BASE	CI
NO	REQ	PART	MATL

J. P. BELL COMPANY	
LEV-L-LINE SERIES "S"	

DRAWN BY G.L.	TOLERANCES
CHK BY: JTC	DECIMAL ±0.5
APPROV BY: LES	ANGLES ±0.5
DATE: 2-9-94	FILE NO. 071
SCALE: NONE	SHEET 3 OF 3

LEV-L-LINE-SERIES "S"
ASSEMBLY

Figure 23.26 This isometric assembly drawing depicts the parts of the lifting device shown in Fig. 23.10 fully assembled. Dimensions usually are omitted from assembly drawings and a parts list is given.

ORTHOGRAPHIC ASSEMBLY: EXPLODED

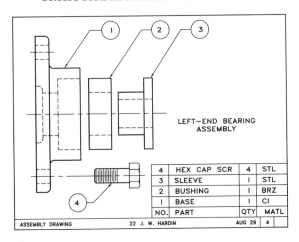

LEFT-END BEARING
ASSEMBLY

4	HEX CAP SCR	4	STL
3	SLEEVE	I	STL
2	BUSHING	I	BRZ
I	BASE	I	CI
NO.	PART	QTY	MATL

ASSEMBLY DRAWING   22 J. W. HARDIN   AUG 29   4

Figure 23.27 This exploded orthographic assembly illustrates how the parts shown are to be put together.

Assembly of the same part is shown in **Fig. 23.28** in an orthographic assembly drawing, in which the parts are depicted in their assembled positions. The views are sectioned to make them easier to understand.

ORTHOGRAPHIC ASSEMBLY: SECTIONED AND ASSEMBLED

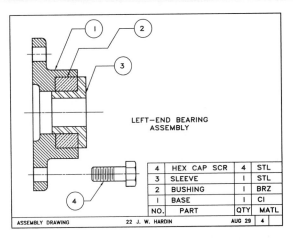

LEFT-END BEARING
ASSEMBLY

4	HEX CAP SCR	4	STL
3	SLEEVE	I	STL
2	BUSHING	I	BRZ
I	BASE	I	CI
NO.	PART	QTY	MATL

ASSEMBLY DRAWING   22 J. W. HARDIN   AUG 29   4

Figure 23.28 This sectioned orthographic assembly shows the parts from Fig. 23.27 in their assembled positions, except for the exploded bolt.

EXPLODED ASSEMBLY

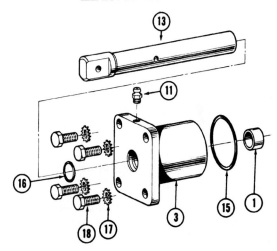

Figure 23.29 This exploded pictorial assembly drawing is of a shaft bearing assembly. (Courtesy of Cameron Iron Works, Incorporated.)

Figure 23.29 shows a brake-pedal assembly in an exploded pictorial assembly drawing, illustrating how the parts fit together. Special balloons refer to a variety of information given in notes on the complete drawing.

A FREEHAND WORKING DRAWING

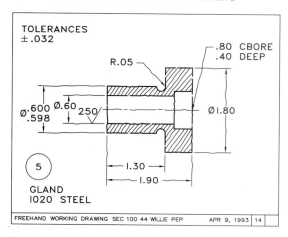

Figure 23.30 A freehand working drawing with the essential dimensions can be as adequate as an instrument-drawn detail drawing for simple parts.

Figure 23.31 The left part is a blank that has been forged. It will look like the part on the right after it has been machined.

TWO-PART SAND MOLD

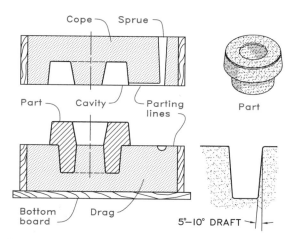

Figure 23.32 A two-part sand mold is used to produce a casting. A draft of from 5° to 10° is needed to permit withdrawal of the pattern from the sand. Some machining is usually required to finish various features of the casting within specified tolerances.

23.9 Freehand Working Drawings

A freehand sketch can serve the same purpose as an instrument drawing, provided that the part is sufficiently simple and that the essential dimensions are shown (**Fig. 23.30**). Use the same principles of making working drawings with instruments when making working drawings freehand.

23.10 Working Drawings for Forged Parts and Castings

The two parts shown in **Fig. 23.31** illustrate the difference between a forged part and a machined part. Recall from Chapter 19 that a forging is a rough form made by hammering (forging) the metal into shape or pressing it between two forms (called dies). The forging is then machined to its finished dimensions and tolerances.

A casting, like a forging, must be machined so that it will fit and function with other parts when assembled; therefore additional material is added to the areas where metal will be removed by machining. Recall (Chapter 19) that a casting is formed by pouring molten metal into a mold

formed by a pattern that is slightly larger than the finished part to compensate for metal shrinkage (**Fig. 23.32**). For the pattern to be removable from the sand that forms the mold, its sides must be tapered. This taper of from 5° to 10° is called **draft**.

In some industries, working drawings for forged and cast parts are prepared separately (**Fig. 23.33A**). More often, however, they are shown on the regular working drawing, with the understanding that the features to be machined by operations such as grinding or shaping are made oversize.

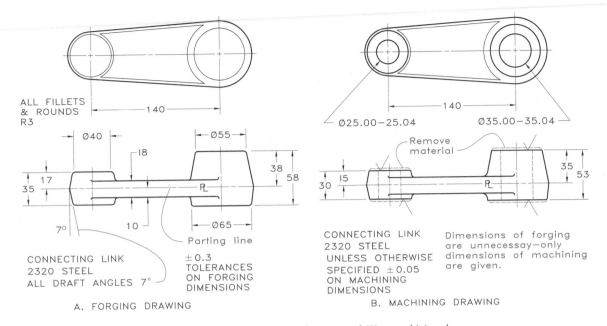

A. FORGING DRAWING

B. MACHINING DRAWING

Figure 23.33 These separate working drawings, (A) a forging drawing and (B) a machining drawing, give the details of the same part. Often, this information is combined into a single drawing.

Problems

General

The dimensions given in the problems do not always represent good dimensioning practices because of space limitations and the nature of three-dimensional pictorial drawings, but they are adequate to provide dimensions for making the working drawings. In cases where dimensions may be missing, use your own judgment to approximate them. Provide all the necessary information, notes, and dimensions to describe the views completely. Use any of the previously covered principles, conventions, and techniques to present the views with the maximum clarity and simplicity.

Working Drawing Practice

Reproduce the drawings shown in **Figs. 23.34–23.41**, as directed. Some have only one sheet and others have more. You may use the dimensions given or convert them to the other system (from millimeters to inches, for example). The purpose of these assignments is to give you experience in laying out a working drawing and improving your draftsmanship on the board or at the computer.

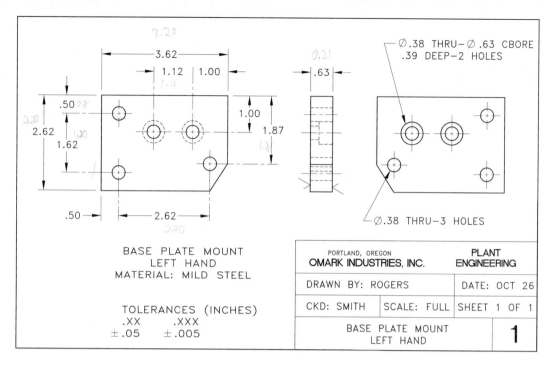

Figure 23.34 Duplicate the working drawing of the base plate mount by computer or drafting instruments on a size B sheet. (Courtesy of Omark Industries, Incorporated.)

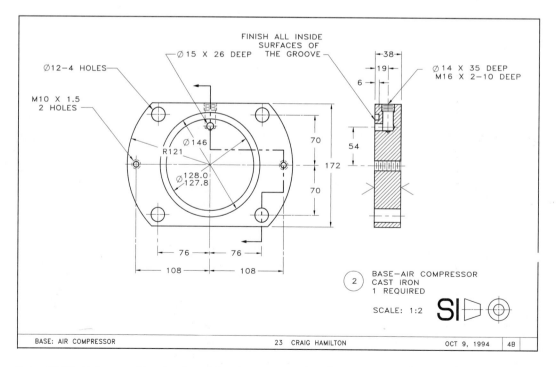

Figure 23.35 Make a working drawing of the air compressor base by computer or drafting instruments on a size B sheet.

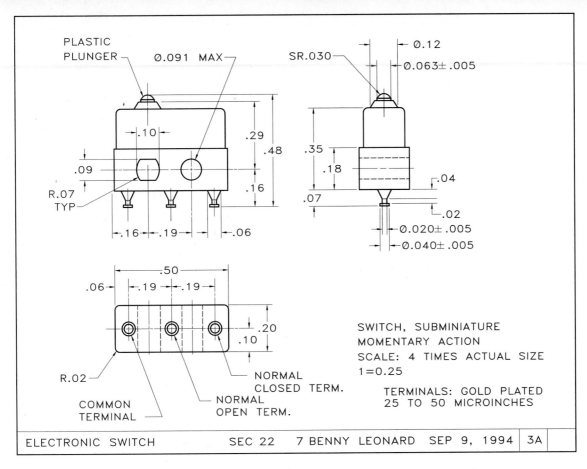

Figure 23.36 Make a drawing of the switch by computer or drafting instruments on a size A sheet. Note that the views are enlarged by a factor of 4.

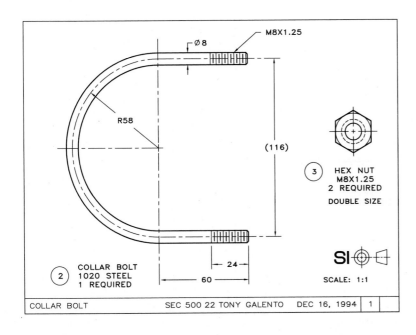

Figure 23.37 Sheet 1 of 3: Make full-size working drawings, with an assembly drawing (in mm), of a pipe hanger on size A sheets.

413

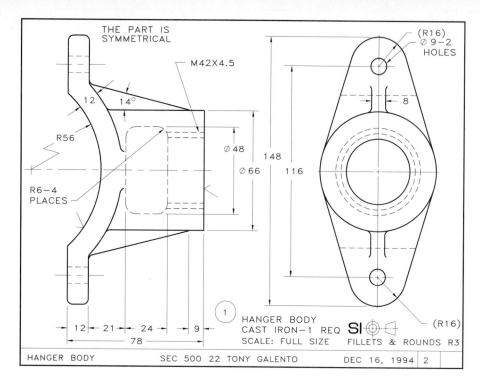

Figure 23.38 Sheet 2 of 3: The pipe hanger working drawings.

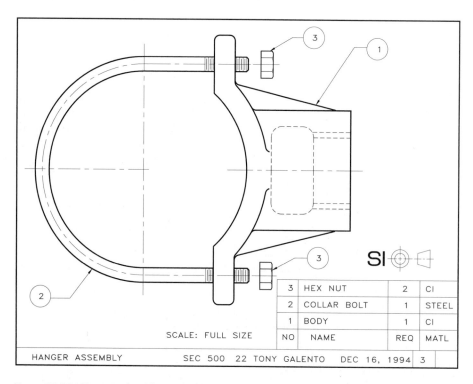

Figure 23.39 Sheet 3 of 3: The assembly of the pipe hanger.

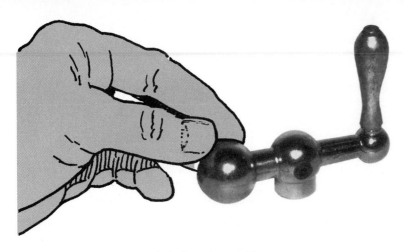

Figure 23.40 This photo is of a ball crank assembly.

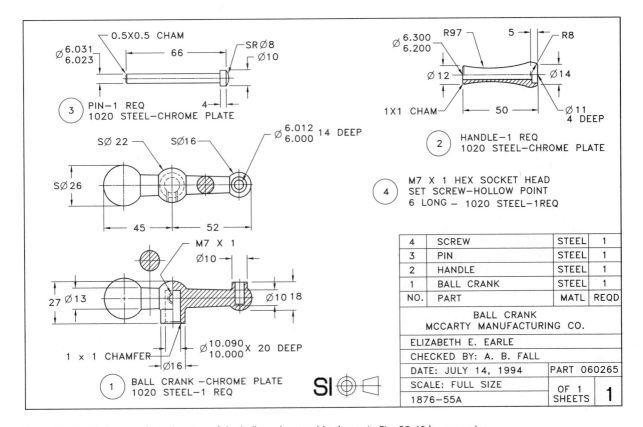

0.5X0.5 CHAM

Ø 6.031 / 6.023

66

SR Ø8
Ø10

4

③ PIN—1 REQ
1020 STEEL—CHROME PLATE

SØ 22   SØ16   Ø 6.012 / 6.000 14 DEEP

SØ 26

45   52

M7 X 1
Ø10

27 Ø13   Ø10 18

1 x 1 CHAMFER   Ø 10.090 / 10.000 X 20 DEEP
Ø16

① BALL CRANK —CHROME PLATE
1020 STEEL—1 REQ

SI

Ø 6.300 / 6.200   R97   5   R8

Ø 12   Ø14

1X1 CHAM   50   Ø 11
4 DEEP

② HANDLE—1 REQ
1020 STEEL—CHROME PLATE

④ M7 X 1 HEX SOCKET HEAD
SET SCREW—HOLLOW POINT
6 LONG — 1020 STEEL—1REQ

4	SCREW	STEEL	1
3	PIN	STEEL	1
2	HANDLE	STEEL	1
1	BALL CRANK	STEEL	1
NO.	PART	MATL	REQD

BALL CRANK
MCCARTY MANUFACTURING CO.

ELIZABETH E. EARLE	
CHECKED BY: A. B. FALL	
DATE: JULY 14, 1994	PART 060265
SCALE: FULL SIZE	OF 1 SHEETS 1
1876—55A	

Figure 23.41 Make a working drawing of the ball crank assembly shown in Fig. 23.40 by computer or drafting instruments on a size B sheet. On a second size B sheet make an assembly drawing of the parts of the ball crank.

Figure 23.42 Make a half-size detail drawing of this radial link on a size B sheet.

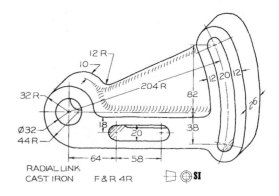

RADIAL LINK
CAST IRON   F & R 4R   SI

Figure 23.43 Make a full-size detail drawing of this shaft bearing on a size B sheet.

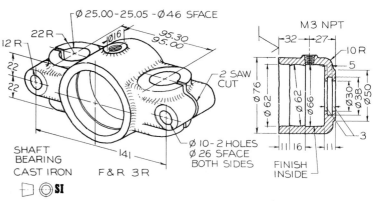

SHAFT
BEARING
CAST IRON   F & R 3R   SI

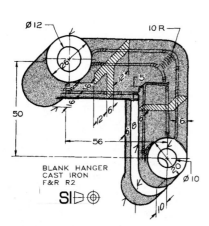

BLANK HANGER
CAST IRON
F&R R2
SI

Figure 23.44 Make a full-size detail drawing of this blank hanger on a size B sheet.

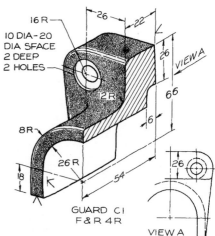

GUARD CI
F & R 4R

VIEW A

Figure 23.45 Make a full-size detail drawing of this guard on a size B sheet.

Working Drawings of Single Parts

Make working drawings by hand or by computer, as assigned, of the single parts shown in **Figs. 23.42–23.58** on the sheet sizes specified for each. Include a title block and the dimensions and notes necessary for constructing the part.

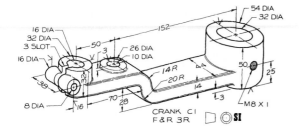

Figure 23.46 Make a half-size detail drawing of this crank on a size B sheet.

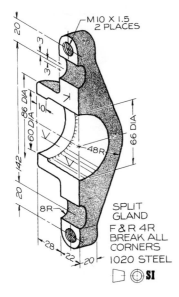

Figure 23.47 Make a half-size detail drawing of this split gland on a size B sheet.

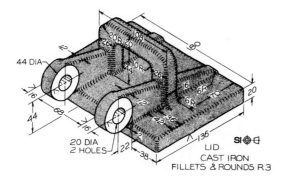

Figure 23.48 Make a half-size detail drawing of this lid on a size B sheet.

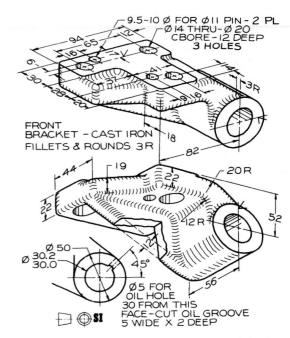

Figure 23.49 Make a half-size detail drawing of this front bracket drawing on a size B sheet.

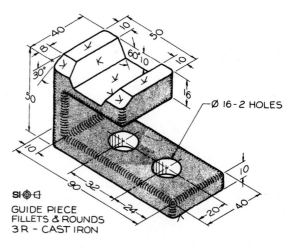

Figure 23.50 Make a full-size detail drawing of this guide piece on a size B sheet.

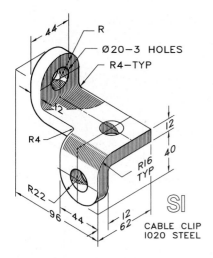

Figure 23.51 Make a half-size detail drawing of this spindle head on a size B sheet.

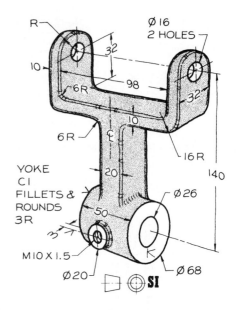

Figure 23.52 Make a double-size detail drawing of this yoke on a size B sheet.

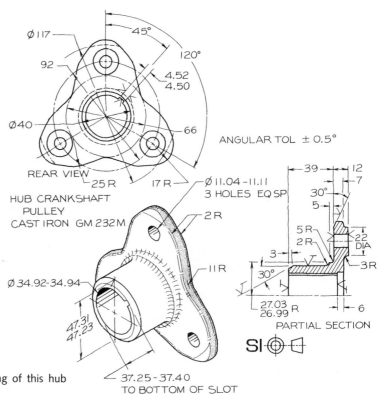

Figure 23.53 Make a full-size detail drawing of this hub crank shaft on a size B sheet.

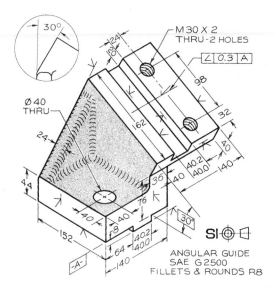

Figure 23.54 Make a half-size detail drawing of this angular guide on a size B sheet.

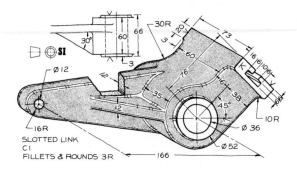

Figure 23.56 Make a half-size detail drawing of this slotted link on a size B sheet.

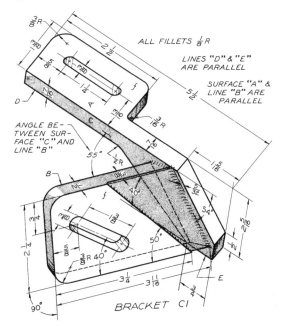

Figure 23.55 Make a full-size detail drawing of this bracket brace on a size B sheet.

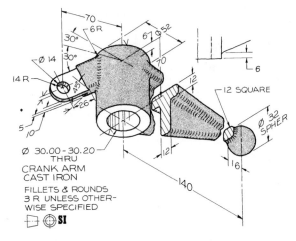

Figure 23.57 Make a full-size detail drawing of this crank arm on a size B sheet.

Figure 23.58 Make a full-size detail drawing of this corner brace on a size B sheet.

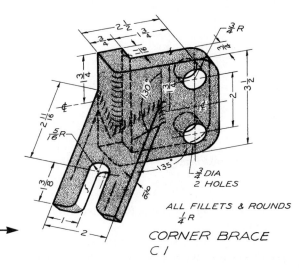

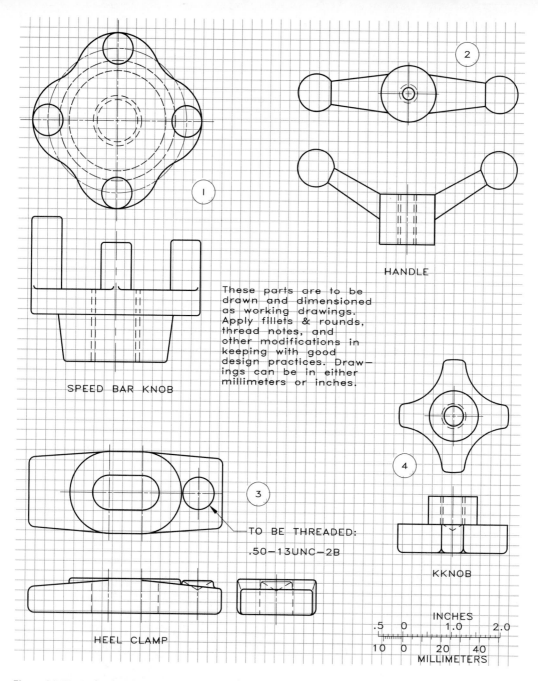

These parts are to be drawn and dimensioned as working drawings. Apply fillets & rounds, thread notes, and other modifications in keeping with good design practices. Drawings can be in either millimeters or inches.

HANDLE

SPEED BAR KNOB

TO BE THREADED:

.50—13UNC—2B

KKNOB

HEEL CLAMP

INCHES
.5   0   1.0   2.0
10   0   20   40
MILLIMETERS

Figure 23.59 Make detail drawings of the parts shown, with one per size A sheet.

Working Drawings of Single Parts Involving Design Features

Make working drawings of the single parts shown in **Figs. 23.59–23.66** on size A sheets. These problems require that you use your judgment and creativity to depict design features. Include a title block and the dimensions and notes necessary for making the part.

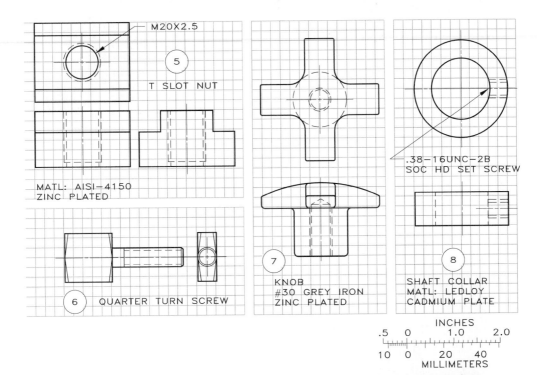

INCHES
.5 0   1.0   2.0
10 0   20   40
MILLIMETERS

Figure 23.60 Make detail drawings of the parts shown, with one per size A sheet.

Figure 23.61 Make detail drawings of this 7-in. diameter pulley on a size A sheet.

Figure 23.62 Make detail drawings of this 4-in. high angle on a size A sheet.

Figure 23.63 Make detail drawings of this shaft arm that fits shafts of 1.00 and 2.00 in. in diameter, with 6.00 in. between centers, on a size A sheet.

Figure 23.64 Make detail drawings of this 7.5-in. diameter driving cup on a size A sheet.

Figure 23.65 Make detail drawings of this 3-in.-ID shaft socket on a size A sheet.

Figure 23.66 Make detail drawings of this hanger bracket that extends 8 in. from the wall and supports a 0.50-in.-diameter bolt on a size A sheet.

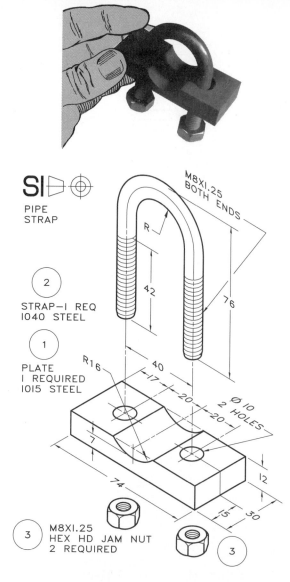

Figure 23.67 Make full-size working drawings of this U-bolt pipe strap on size A sheets.

Working Drawings of Products with Multiple Parts

Make working drawings by hand or by computer, as assigned, of the products consisting of multiple parts shown in **Figs. 23.67–23.89** on the sheet sizes specified. Include a title block and the dimensions and notes necessary for manufacturing the part. Draw an assembly that shows how the parts fit together. More than one sheet may be required for a solution.

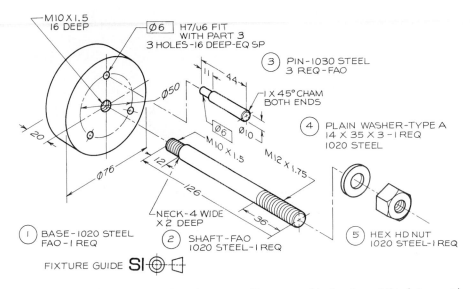

Figure 23.68 Make full-size working drawings, with an assembly drawing, of this fixture guide on size A sheets.

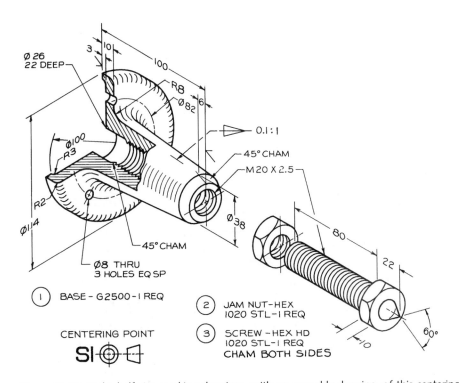

Figure 23.69 Make half-size working drawings, with an assembly drawing, of this centering point on size A sheets.

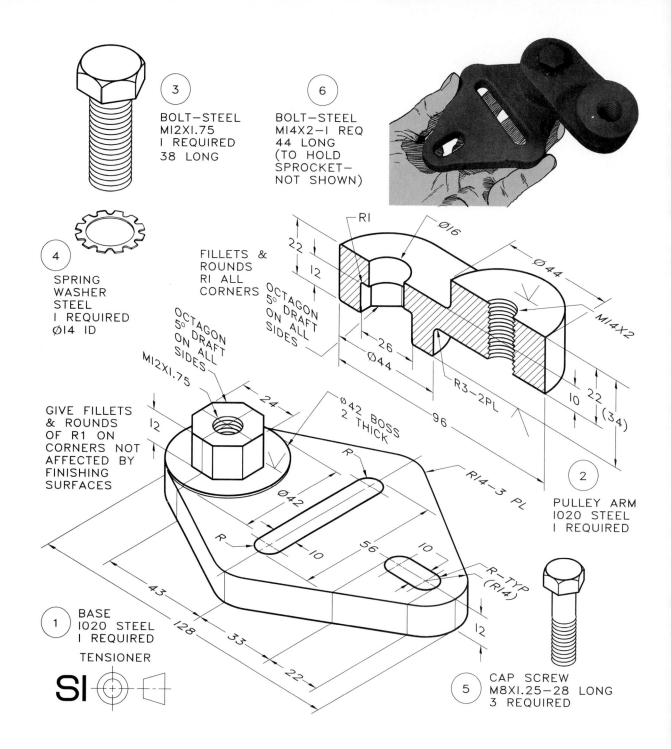

③ BOLT—STEEL
M12X1.75
1 REQUIRED
38 LONG

⑥ BOLT—STEEL
M14X2—1 REQ
44 LONG
(TO HOLD
SPROCKET—
NOT SHOWN)

④ SPRING
WASHER
STEEL
1 REQUIRED
Ø14 ID

FILLETS &
ROUNDS
R1 ALL
CORNERS

OCTAGON
5° DRAFT
ON ALL
SIDES

R1

Ø16

Ø44

M14X2

22

12

26

Ø44

R3—2PL

96

22
10
(34)

② PULLEY ARM
1020 STEEL
1 REQUIRED

OCTAGON
5° DRAFT
ON ALL
SIDES

M12X1.75

GIVE FILLETS
& ROUNDS
OF R1 ON
CORNERS NOT
AFFECTED BY
FINISHING
SURFACES

24

12

Ø42

Ø42 BOSS
2 THICK

R

R14—3 PL

R

56

10

10

R—TYP
(R14)

12

43

128

33

22

① BASE
1020 STEEL
1 REQUIRED

TENSIONER

SI

⑤ CAP SCREW
M8X1.25—28 LONG
3 REQUIRED

Figure 23.70 Make full-size working drawings, with an assembly drawing, of this tensioner on size B sheets.

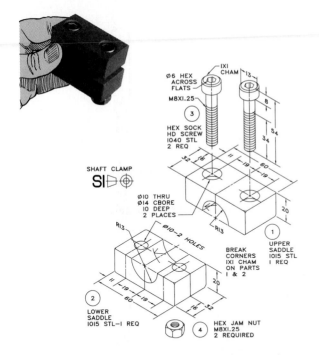

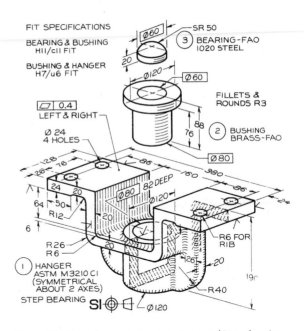

Figure 23.71 Make full-size working drawings, with an assembly drawing, of this shaft clamp on size A sheets.

Figure 23.72 Make one-quarter size working drawings, with an assembly drawing, of this step bearing on size B sheets.

Figure 23.73 Make full-size working drawings, with an assembly drawing, of this drill press vise on size B sheets.

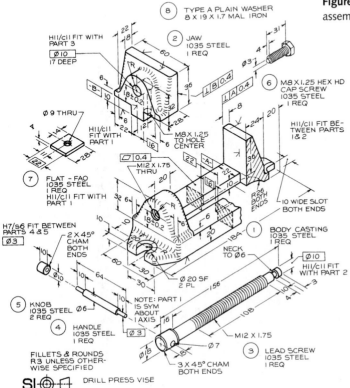

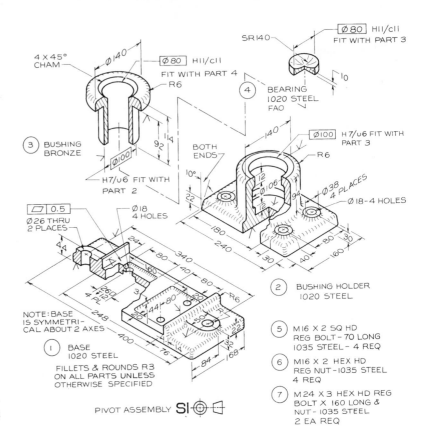

Figure 23.74 Make a working drawing, with an assembly drawing, of this pivot assembly on size B sheets.

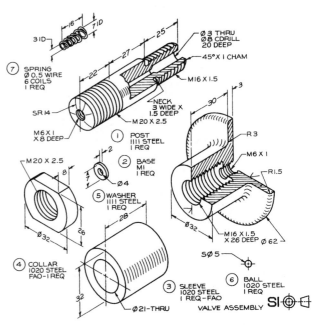

Figure 23.75 Make full-size working drawings, with an assembly drawing, of this valve assembly on size B sheets.

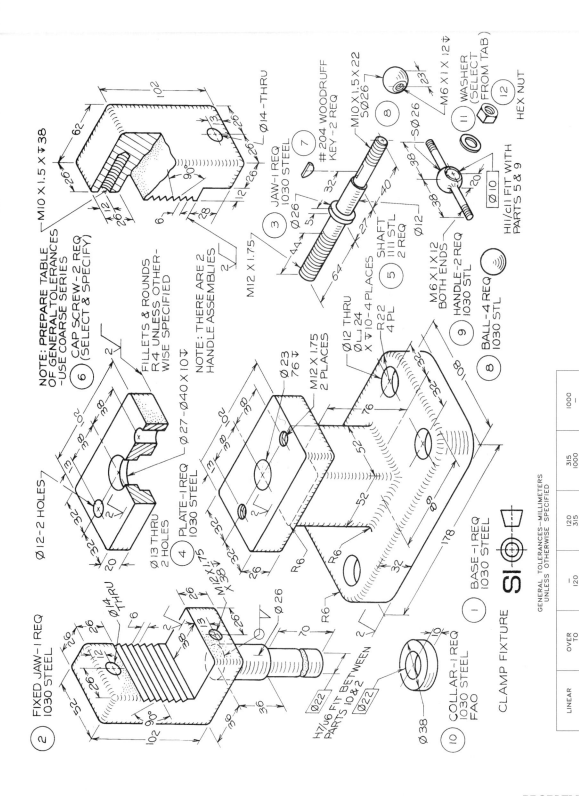

Figure 23.76 Make half-size working drawings, with an assembly drawing, of this clamp fixture on size B sheets.

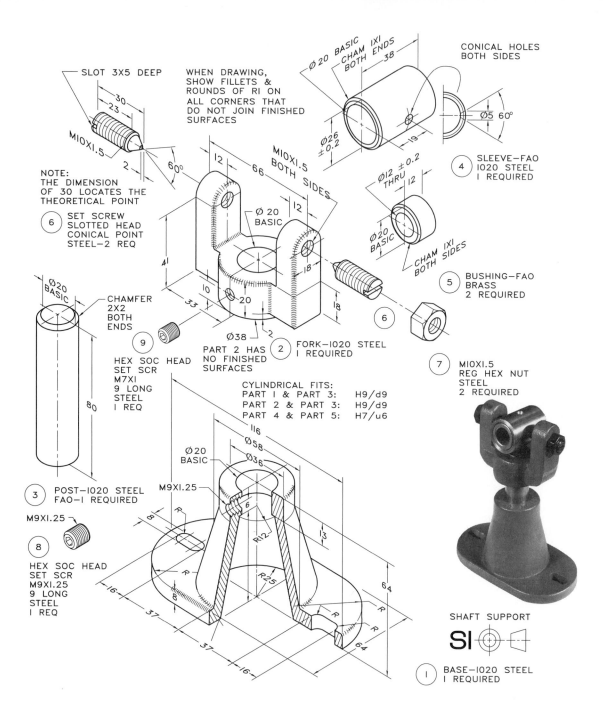

SLOT 3X5 DEEP

30
23

M10X1.5

2

60°

NOTE:
THE DIMENSION
OF 30 LOCATES THE
THEORETICAL POINT

⑥ SET SCREW
SLOTTED HEAD
CONICAL POINT
STEEL-2 REQ

WHEN DRAWING,
SHOW FILLETS &
ROUNDS OF R1 ON
ALL CORNERS THAT
DO NOT JOIN FINISHED
SURFACES

Ø 20 BASIC
CHAM 1X1
BOTH ENDS
38

CONICAL HOLES
BOTH SIDES

Ø26
±0.2

19

Ø5 60°

④ SLEEVE-FAO
1020 STEEL
1 REQUIRED

M10X1.5
BOTH SIDES

12
66

12

Ø 20
BASIC

18

Ø12 ±0.2
THRU 12

Ø20
BASIC

CHAM 1X1
BOTH SIDES

⑤ BUSHING-FAO
BRASS
2 REQUIRED

Ø20
BASIC

CHAMFER
2X2
BOTH
ENDS

⑨

HEX SOC HEAD
SET SCR
M7X1
9 LONG
STEEL
1 REQ

41

10

33

20

Ø38

2

PART 2 HAS
NO FINISHED
SURFACES

18

② FORK-1020 STEEL
1 REQUIRED

⑥

⑦ M10X1.5
REG HEX NUT
STEEL
2 REQUIRED

80

CYLINDRICAL FITS:
PART 1 & PART 3:   H9/d9
PART 2 & PART 3:   H9/d9
PART 4 & PART 5:   H7/u6

③ POST-1020 STEEL
FAO-1 REQUIRED

M9X1.25

⑧

HEX SOC HEAD
SET SCR
M9X1.25
9 LONG
STEEL
1 REQ

116
Ø58
Ø36

Ø20
BASIC

M9X1.25

8

6

R

R12

13

R25

64

16

8

37

R

R

37

R

64

16

SHAFT SUPPORT

SI

① BASE-1020 STEEL
1 REQUIRED

Figure 23.77 Make full-size working drawings, with an assembly drawing, of this shaft support on size B sheets.

428 • **CHAPTER 23   WORKING DRAWINGS**

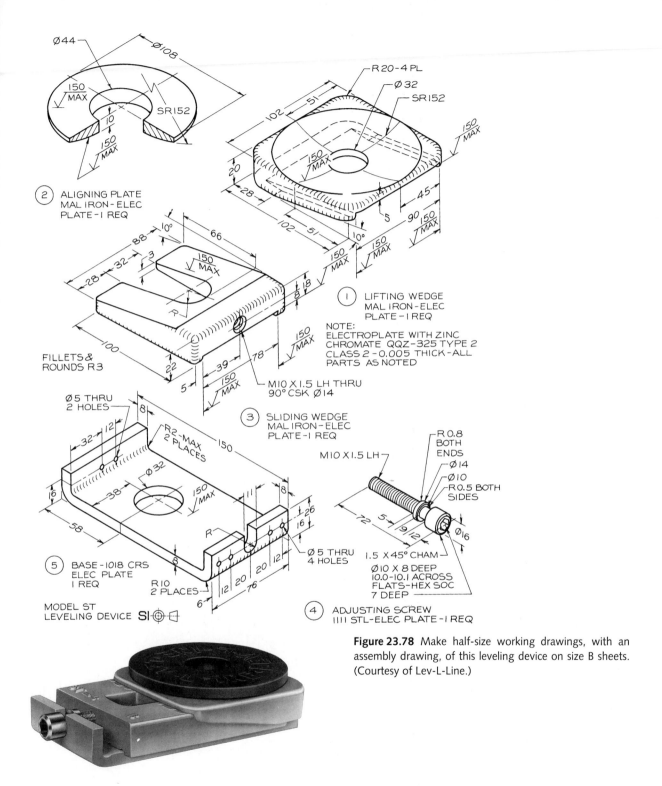

Ø44
Ø108
150 MAX
10
150 MAX
150 MAX
SR152

② ALIGNING PLATE
MAL IRON – ELEC
PLATE – 1 REQ

R 20 – 4 PL
Ø 32
SR152
102
51
20
28
150 MAX
45
5
90
150 MAX
102
51
10°
150 MAX
150 MAX
150 MAX

① LIFTING WEDGE
MAL IRON – ELEC
PLATE – 1 REQ

88
10°
66
28 32 3
150 MAX
18
8
100
R
22
39
78
5
150 MAX
150 MAX

FILLETS &
ROUNDS R3

NOTE:
ELECTROPLATE WITH ZINC
CHROMATE QQZ-325 TYPE 2
CLASS 2 - 0.005 THICK - ALL
PARTS AS NOTED

M10 X 1.5 LH THRU
90° CSK Ø14

③ SLIDING WEDGE
MAL IRON – ELEC
PLATE – 1 REQ

Ø5 THRU
2 HOLES
8
32 12
R2 MAX
2 PLACES
150
Ø32
38
150 MAX
11
8
16
58
26
16
R
8
12 20 20 12
Ø 5 THRU
4 HOLES
R10
2 PLACES
76
6

⑤ BASE – 1018 CRS
ELEC PLATE
1 REQ

MODEL ST
LEVELING DEVICE SI ⊕ ⊟

R 0.8
BOTH
ENDS
Ø14
Ø10
R 0.5 BOTH
SIDES

M10 X 1.5 LH
72
5
9 12
Ø16

1.5 X 45° CHAM
Ø10 X 8 DEEP
10.0-10.1 ACROSS
FLATS-HEX SOC
7 DEEP

④ ADJUSTING SCREW
1111 STL-ELEC PLATE-1 REQ

Figure 23.78 Make half-size working drawings, with an assembly drawing, of this leveling device on size B sheets. (Courtesy of Lev-L-Line.)

Figure 23.79 Make half-size working drawings, with an assembly drawing, of this C-clamp on size B sheets.

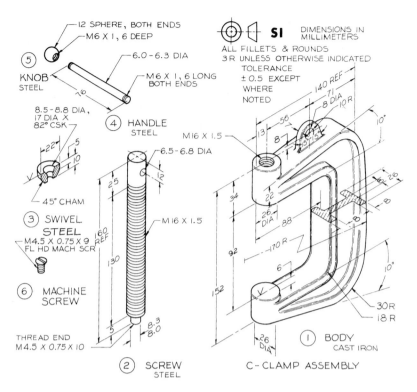

Figure 23.80 Make half-size working drawings, with an assembly drawing, of this indicating lever on size B sheets.

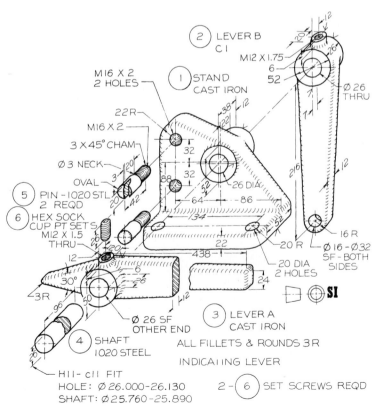

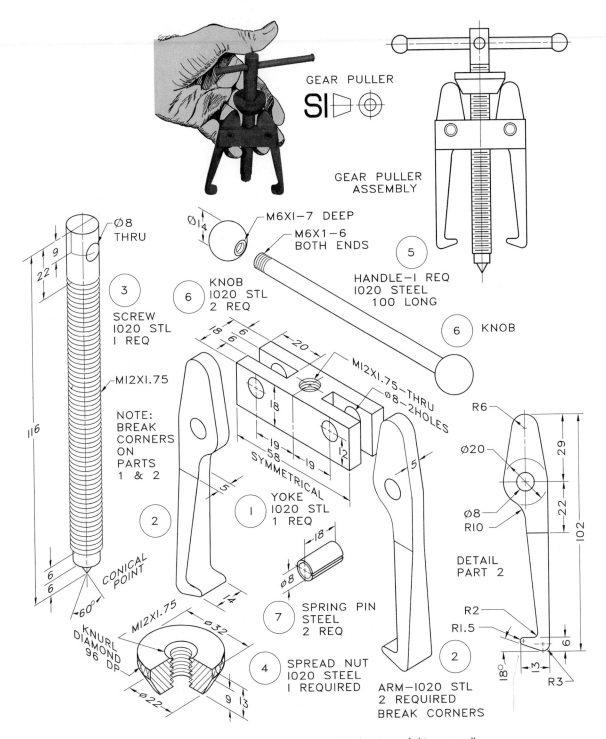

GEAR PULLER

SI ▷ ⊕

GEAR PULLER
ASSEMBLY

Ø8
THRU

22 9

③

SCREW
1020 STL
1 REQ

M12X1.75

116

NOTE:
BREAK
CORNERS
ON
PARTS
1 & 2

6
6

60°

CONICAL
POINT

KNURL
DIAMOND
96 DP

M12X1.75

Ø32

Ø22

9 13

④ SPREAD NUT
1020 STEEL
1 REQUIRED

Ø14

M6X1-7 DEEP

M6X1-6
BOTH ENDS

⑤

HANDLE-1 REQ
1020 STEEL
100 LONG

⑥ KNOB
1020 STL
2 REQ

⑥ KNOB

18
6

20

M12X1.75-THRU

Ø8-2HOLES

18

19
58

19

12

SYMMETRICAL

① YOKE
1020 STL
1 REQ

5

5

18

Ø8

⑦ SPRING PIN
STEEL
2 REQ

4

② ARM-1020 STL
2 REQUIRED
BREAK CORNERS

R6

Ø20

Ø8
R10

29

22

102

DETAIL
PART 2

R2

R1.5

18°

13

6

R3

②

Figure 23.81 Make double-size working drawings, with an assembly drawing, of this gear puller on size B sheets.

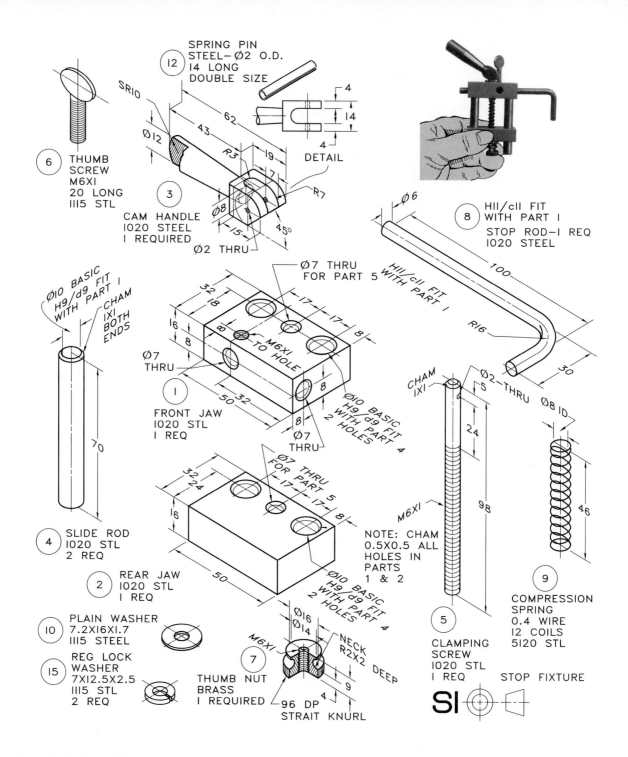

SPRING PIN
STEEL—Ø2 O.D.
14 LONG
DOUBLE SIZE ⑫

4
14
4
DETAIL

⑥ **THUMB
SCREW
M6XI
20 LONG
1115 STL**

SR10
Ø12
43
62
R3
19
7
R7
Ø8
15
45°
Ø2 THRU

③ **CAM HANDLE
1020 STEEL
I REQUIRED**

Ø6
⑧ **HII/cII FIT
WITH PART I
STOP ROD–I REQ
1020 STEEL**
100
R16
30
Ø8 ID

Ø10 BASIC
H9/d9 FIT
WITH PART I
1XI CHAM
BOTH
ENDS
70

④ **SLIDE ROD
1020 STL
2 REQ**

Ø7 THRU
FOR PART 5
HII/cII FIT
WITH PART I

32
18
16
8
8
M6XI
TO HOLE
Ø7
THRU
17
17
8
50
32
8
Ø7
THRU
Ø10 BASIC
H9/d9 FIT
WITH PART 4
2 HOLES

① **FRONT JAW
1020 STL
I REQ**

CHAM
1XI
Ø2–THRU
5
24
98
M6XI

NOTE: CHAM
0.5X0.5 ALL
HOLES IN
PARTS
1 & 2

⑤ **CLAMPING
SCREW
1020 STL
I REQ**

32
24
16
Ø7 THRU
FOR PART 5
17
17
8
50
Ø10 BASIC
H9/d9 FIT
WITH PART 4
2 HOLES

② **REAR JAW
1020 STL
I REQ**

⑨ **COMPRESSION
SPRING
0.4 WIRE
12 COILS
5120 STL**
46

⑩ **PLAIN WASHER
7.2X16X1.7
1115 STEEL**

⑮ **REG LOCK
WASHER
7X12.5X2.5
1115 STL
2 REQ**

Ø16
Ø14
M6XI
NECK
R2X2
DEEP
⑦
9
4
96 DP
STRAIT KNURL

**THUMB NUT
BRASS
I REQUIRED**

STOP FIXTURE

SI

Figure 23.82 Make full-size working drawings, with an assembly drawing, of this stop fixture on size B sheets.

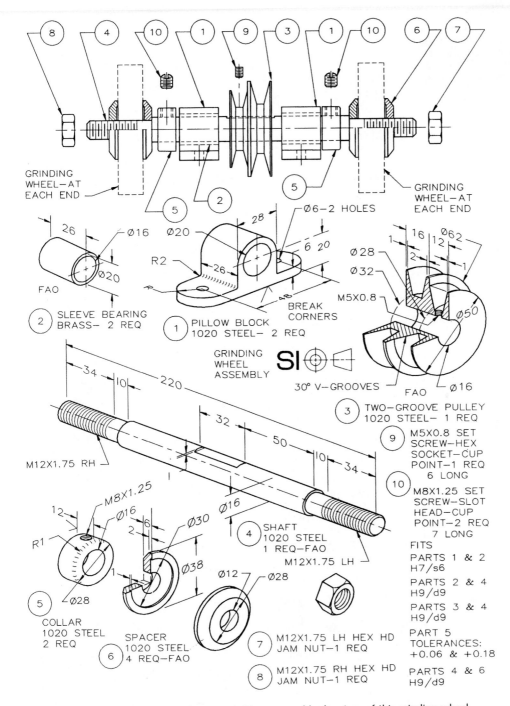

8   4   10   1   9   3   1   10   6   7

GRINDING
WHEEL—AT
EACH END

5   2

GRINDING
WHEEL—AT
EACH END

5

26   Ø16   Ø20

Ø20

FAO

2   SLEEVE BEARING
BRASS— 2 REQ

28

Ø6—2 HOLES

6   20

R2   26

BREAK
CORNERS

1   PILLOW BLOCK
1020 STEEL— 2 REQ

GRINDING
WHEEL
ASSEMBLY

SI

30° V—GROOVES

16   Ø62
1   12
Ø28   2
Ø32   1

M5X0.8   Ø50

FAO   Ø16

3   TWO—GROOVE PULLEY
1020 STEEL— 1 REQ

9   M5X0.8 SET
SCREW—HEX
SOCKET—CUP
POINT—1 REQ
6 LONG

10   M8X1.25 SET
SCREW—SLOT
HEAD—CUP
POINT—2 REQ
7 LONG

34   10   220

32

50

10   34

M12X1.75 RH

Ø16

Ø16

4   SHAFT
1020 STEEL
1 REQ—FAO

M12X1.75 LH

M8X1.25

12   Ø16
2   6
R1   Ø30

Ø38

Ø12   Ø28

5   Ø28
1

COLLAR
1020 STEEL
2 REQ

SPACER
1020 STEEL
6   4 REQ—FAO

7   M12X1.75 LH HEX HD
JAM NUT—1 REQ

8   M12X1.75 RH HEX HD
JAM NUT—1 REQ

FITS

PARTS 1 & 2
H7/s6

PARTS 2 & 4
H9/d9

PARTS 3 & 4
H9/d9

PART 5
TOLERANCES:
+0.06 & +0.18

PARTS 4 & 6
H9/d9

Figure 23.83 Make full-size working drawings, with an assembly drawing, of this grinding wheel assembly on size B sheets.

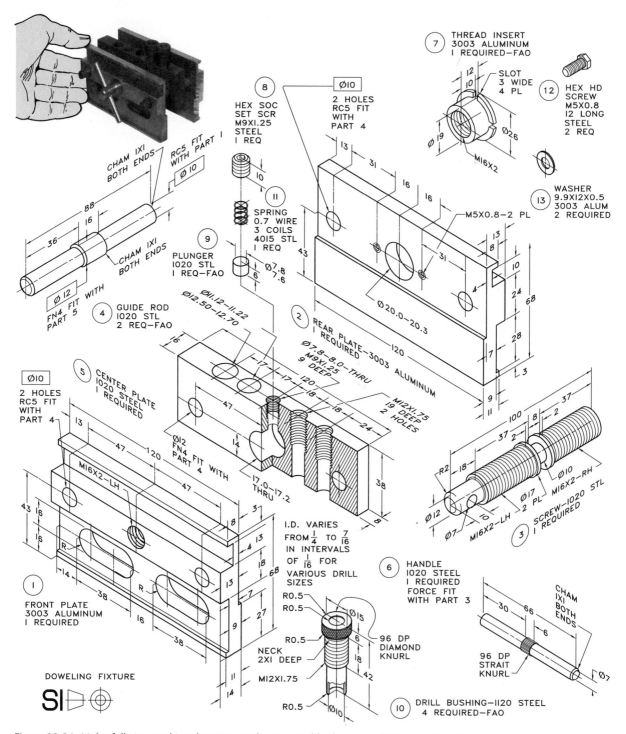

Figure 23.84 Make full-size working drawings, with an assembly drawing, of this dowelling fixture on size B sheets.

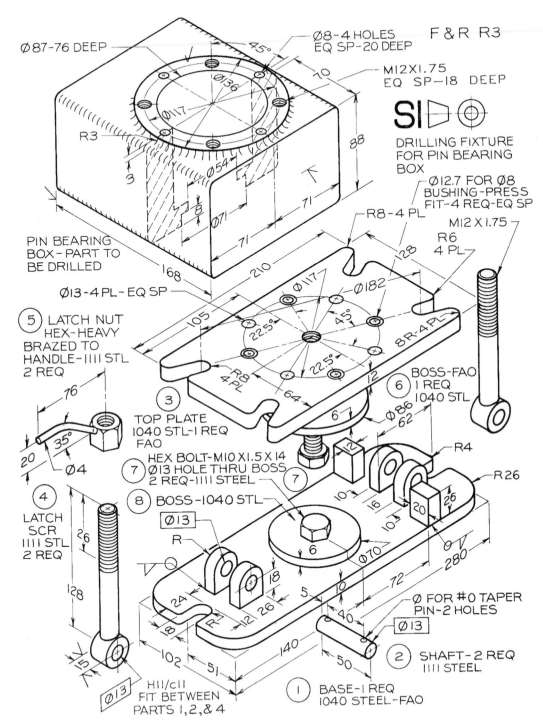

Ø87-76 DEEP

Ø8-4 HOLES
EQ SP-20 DEEP

F & R R3

45°

Ø136

70

Ø117

M12X1.75
EQ SP-18 DEEP

R3

88

DRILLING FIXTURE
FOR PIN BEARING
BOX

3

Ø54

Ø12.7 FOR Ø8
BUSHING-PRESS
FIT-4 REQ-EQ SP

8

Ø71

71

R8-4 PL

M12 X 1.75

71

R6
4 PL

PIN BEARING
BOX-PART TO
BE DRILLED

168

210

128

Ø13-4 PL-EQ SP

Ø117

Ø182

45°

8R-4 PL

⑤ LATCH NUT
HEX-HEAVY
BRAZED TO
HANDLE-1111 STL
2 REQ

105

22.5°

22.5°

12

BOSS-FAO
⑥ 1 REQ
1040 STL

R8
4 PL

64

6

Ø86

62

76

③

TOP PLATE
1040 STL-1 REQ
FAO

R4

35°

HEX BOLT-M10 X1.5 X 14

12

R26

20

26

④

Ø4

⑦

Ø13 HOLE THRU BOSS
2 REQ-1111 STEEL

⑦

10

16

LATCH
SCR
1111 STL
2 REQ

26

⑧ BOSS-1040 STL

Ø13

10

Ø70

R

6

128

R

18

Ø FOR #0 TAPER
PIN-2 HOLES

24

12

26

5

10

72

280

8

R

40

Ø13

15

Ø13

102

51

140

50

② SHAFT-2 REQ
1111 STEEL

H11/c11
FIT BETWEEN
PARTS 1, 2, & 4

① BASE-1 REQ
1040 STEEL-FAO

Figure 23.85 Make working drawings, with an assembly drawing, of this drilling fixture on size B
sheets.

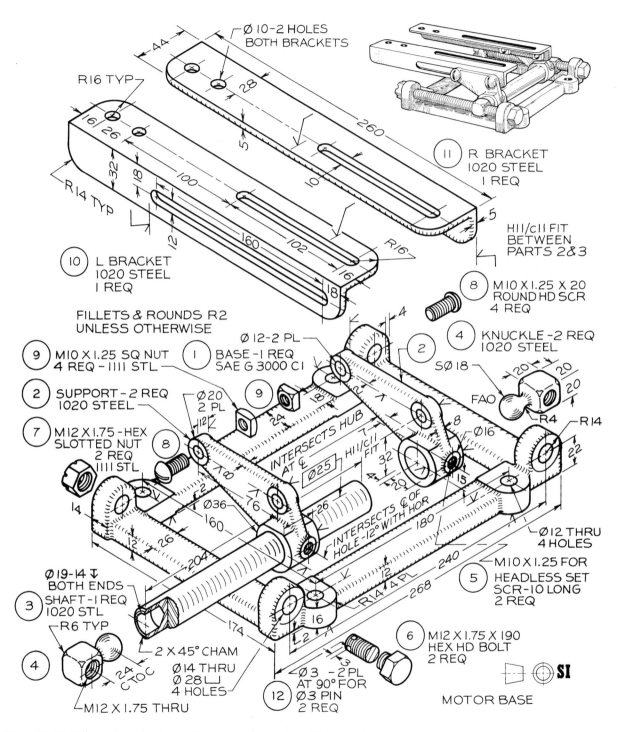

Ø10-2 HOLES
BOTH BRACKETS

R16 TYP

44

28

260

11 R BRACKET
    1020 STEEL
    1 REQ

16 26

32

18

R14 TYP

100

5

10

R16

5

HII/CII FIT
BETWEEN
PARTS 2&3

10 L BRACKET
    1020 STEEL
    1 REQ

12

160

102

16

18

8   M10 X1.25 X 20
    ROUND HD SCR
    4 REQ

FILLETS & ROUNDS R2
UNLESS OTHERWISE

Ø12-2 PL

2

4   KNUCKLE -2 REQ
    1020 STEEL

SØ18

9   M10 X1.25 SQ NUT
    4 REQ - 1111 STL

1   BASE -1 REQ
    SAE G 3000 CI

9

Ø20
2 PL

18

FAO

20
20
20

R4   R14

2   SUPPORT - 2 REQ
    1020 STEEL

12

24

INTERSECTS HUB
AT ℄

Ø25 HII/CII
     FIT

32

8

Ø16

22

7   M12 X1.75 -HEX
    SLOTTED NUT
    2 REQ
    1111 STL

8

80

2

76

Ø36

26

180

15

14

160

12

Ø12 THRU
4 HOLES

INTERSECTS ℄ OF
HOLE -12° WITH HOR

26

12

240

M10 X1.25 FOR
5   HEADLESS SET
    SCR-10 LONG
    2 REQ

Ø19-14
BOTH ENDS

3   SHAFT -1 REQ
    1020 STL

R6 TYP

204

R14 4 PL

268

16

4

2 X 45° CHAM

24
C TO C

Ø14 THRU
Ø 28
4 HOLES

174

12

Ø3 -2 PL
AT 90° FOR

6   M12 X1.75 X 190
    HEX HD BOLT
    2 REQ

12 Ø3 PIN
    2 REQ

M12 X1.75 THRU

⊏ ⊙ **SI**

MOTOR BASE

Figure 23.86 Make working drawings, with an assembly drawing, of this motor base on size B sheets.

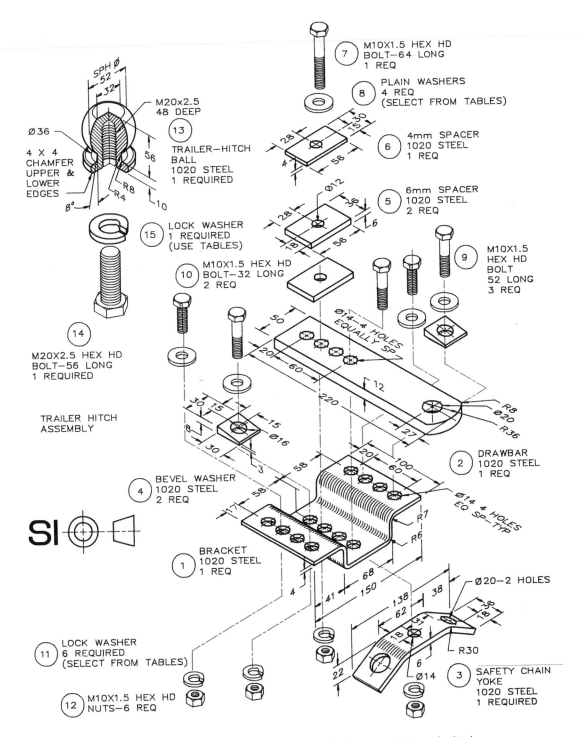

SPH Ø
52
32

M20x2.5
48 DEEP

13

TRAILER—HITCH
BALL
1020 STEEL
1 REQUIRED

Ø36

4 X 4
CHAMFER
UPPER &
LOWER
EDGES

R8
R4
8°

56

10

7 M10X1.5 HEX HD
   BOLT—64 LONG
   1 REQ

8 PLAIN WASHERS
   4 REQ
   (SELECT FROM TABLES)

28   30
15
4   58

6 4mm SPACER
   1020 STEEL
   1 REQ

Ø12
36
28   6
18   56

5 6mm SPACER
   1020 STEEL
   2 REQ

9 M10X1.5
   HEX HD
   BOLT
   52 LONG
   3 REQ

15 LOCK WASHER
    1 REQUIRED
    (USE TABLES)

10 M10X1.5 HEX HD
    BOLT—32 LONG
    2 REQ

14

M20X2.5 HEX HD
BOLT—56 LONG
1 REQUIRED

50
20
60
220
Ø14—4 HOLES
EQUALLY SP
12

TRAILER HITCH
ASSEMBLY

30   15
8
30   15
Ø16
3

R8
Ø20
R36
27

2 DRAWBAR
   1020 STEEL
   1 REQ

4 BEVEL WASHER
   1020 STEEL
   2 REQ

58   20   100
58   60

Ø14 4 HOLES
EQ SP TYP

R7
R6

SI

1 BRACKET
   1020 STEEL
   1 REQ

17

4   41   68
     150

Ø20—2 HOLES

138   38
62
18

R30

11 LOCK WASHER
    6 REQUIRED
    (SELECT FROM TABLES)

22   6
Ø14

3 SAFETY CHAIN
   YOKE
   1020 STEEL
   1 REQUIRED

12 M10X1.5 HEX HD
    NUTS—6 REQ

Figure 23.87 Make full-size working drawings, with an assembly drawing, of this trailer hitch on size B sheets.

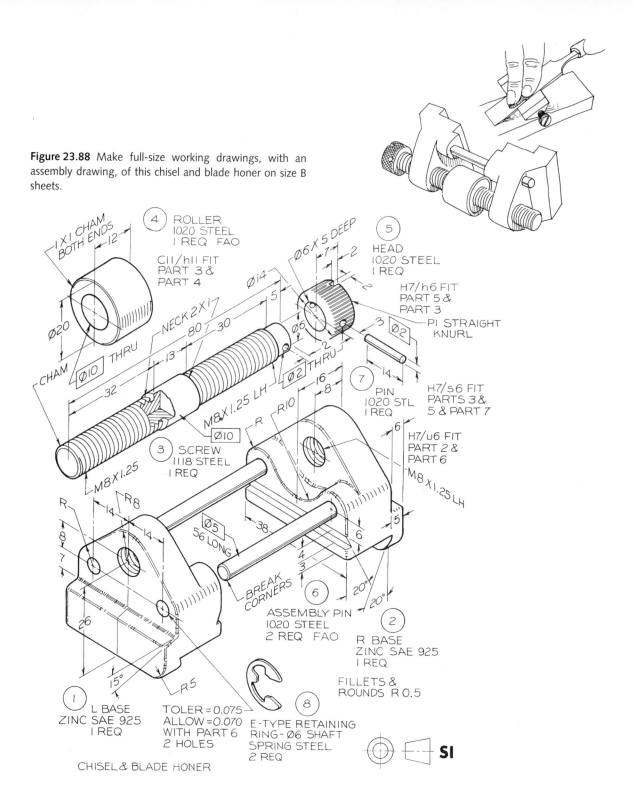

Figure 23.88 Make full-size working drawings, with an assembly drawing, of this chisel and blade honer on size B sheets.

④ ROLLER
1020 STEEL
1 REQ   FAO

C11/h11 FIT
PART 3 &
PART 4

1 X 1 CHAM
BOTH ENDS
12

⌀20

⌀10 THRU
CHAM

32

NECK 2 X 1
80   30
13

Ø6 X 5 DEEP
⌀14
5
2

⑤ HEAD
1020 STEEL
1 REQ

H7/h6 FIT
PART 5 &
PART 3

P1 STRAIGHT
KNURL

Ø6
Ø6

2 THRU
Ø2 THRU
M8 X 1.25 LH

⌀10

③ SCREW
1118 STEEL
1 REQ

M8 X 1.25

3
⌀2
14

⑦ PIN
1020 STL
1 REQ

H7/s6 FIT
PARTS 3 &
5 & PART 7

R   R10
6   8

H7/u6 FIT
PART 2 &
PART 6

M8 X 1.25 LH

R8
14
14

⌀5
56 LONG

38
6
5

R
8
7

26

BREAK
CORNERS

4
3

⑥ ASSEMBLY PIN
1020 STEEL
2 REQ   FAO

20°   20°

② R BASE
ZINC SAE 925
1 REQ

FILLETS &
ROUNDS R 0.5

15°
R5

① L BASE
ZINC SAE 925
1 REQ

TOLER = 0.075
ALLOW = 0.070
WITH PART 6
2 HOLES

⑧ E-TYPE RETAINING
RING - Ø6 SHAFT
SPRING STEEL
2 REQ

SI

CHISEL & BLADE HONER

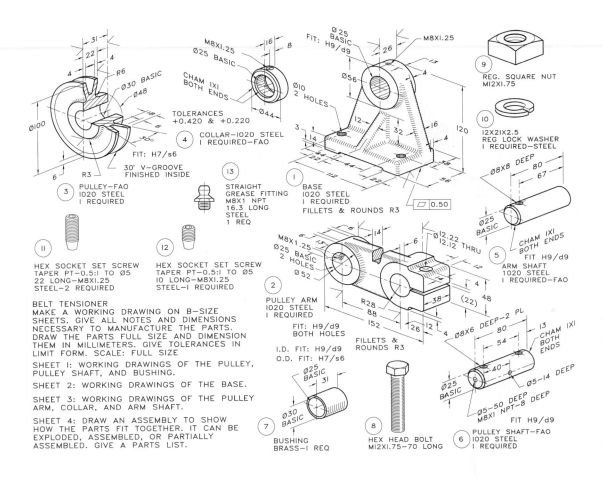

31
22
4
4
R6
Ø30 BASIC
Ø48
Ø100
6
R3
FIT: H7/s6
30° V-GROOVE
FINISHED INSIDE

3 PULLEY-FAO
   1020 STEEL
   1 REQUIRED

M8X1.25
Ø25 BASIC
CHAM 1X1
BOTH ENDS
Ø10
2 HOLES
Ø44
TOLERANCES
+0.420 & +0.220

4 COLLAR-1020 STEEL
   1 REQUIRED-FAO

13 STRAIGHT
   GREASE FITTING
   M8X1 NPT
   16.3 LONG
   STEEL
   1 REQ

11 HEX SOCKET SET SCREW
   TAPER PT-0.5:1 TO Ø5
   22 LONG-M8X1.25
   STEEL-2 REQUIRED

12 HEX SOCKET SET SCREW
   TAPER PT-0.5:1 TO Ø5
   10 LONG-M8X1.25
   STEEL-1 REQUIRED

BELT TENSIONER
MAKE A WORKING DRAWING ON B-SIZE
SHEETS. GIVE ALL NOTES AND DIMENSIONS
NECESSARY TO MANUFACTURE THE PARTS.
DRAW THE PARTS FULL SIZE AND DIMENSION
THEM IN MILLIMETERS. GIVE TOLERANCES IN
LIMIT FORM. SCALE: FULL SIZE

SHEET 1: WORKING DRAWINGS OF THE PULLEY,
PULLEY SHAFT, AND BUSHING.

SHEET 2: WORKING DRAWINGS OF THE BASE.

SHEET 3: WORKING DRAWINGS OF THE PULLEY
ARM, COLLAR, AND ARM SHAFT.

SHEET 4: DRAW AN ASSEMBLY TO SHOW
HOW THE PARTS FIT TOGETHER. IT CAN BE
EXPLODED, ASSEMBLED, OR PARTIALLY
ASSEMBLED. GIVE A PARTS LIST.

FIT: H9/d9
Ø25 BASIC
26
M8X1.25
Ø56
13
12
32
16
120
4
3
14
44
112
22
44
56

1 BASE
   1020 STEEL
   1 REQUIRED
   FILLETS & ROUNDS R3

□ 0.50

9 REG. SQUARE NUT
   M12X1.75

10 12X21X2.5
   REG LOCK WASHER
   1 REQUIRED-STEEL

Ø8X8 DEEP
80
67
Ø25 BASIC
CHAM 1X1
BOTH ENDS
FIT H9/d9

5 ARM SHAFT
   1020 STEEL
   1 REQUIRED-FAO

M8X1.25
Ø25 BASIC
2 HOLES
Ø52
14
6
Ø12.22 THRU
12.12
6
12
4
48
R28
88
152
26
12
4
(22)
38

2 PULLEY ARM
   1020 STEEL
   1 REQUIRED
   FIT: H9/d9
   BOTH HOLES

FILLETS &
ROUNDS R3

I.D. FIT: H9/d9
O.D. FIT: H7/s6

Ø25 BASIC
31
Ø30 BASIC

7 BUSHING
   BRASS-1 REQ

8 HEX HEAD BOLT
   M12X1.75-70 LONG

Ø8X6 DEEP-2 PL
80
54
13
CHAM 1X1
BOTH ENDS
Ø25 BASIC
40
Ø5-14 DEEP
Ø5-50 DEEP
M8X1 NPT-8 DEEP
FIT H9/d9

6 PULLEY SHAFT-FAO
   1020 STEEL
   1 REQUIRED

Figure 23.89 Make full-size working drawings, with an assembly drawing, of this belt tensioner on size B sheets.

Working Drawings of Multiple Parts Involving Design Features

Make dimensioned working drawings of the multiple parts shown in **Figs. 23.90–23.101** on a sheet size of your choice with the necessary dimensions and notes to fabricate the parts. Each part is given in a general format, which requires some design on your part. You must consider the addition of fillets and rounds, the application of finish marks, and the modification of features of the parts to make them functional and practical. Apply the tolerance to the parts in limit form by using the tables of cylindrical fits in the Appendix. Make an assembly drawing and parts list to show how the parts are to be put together.

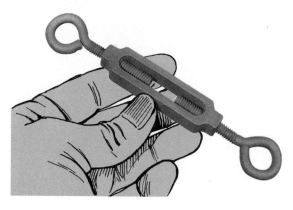

Figure 23.90 Design: Make working drawings, with an assembly drawing, of this turn buckle on size B sheets.

Figure 23.91 Design: Make working drawings, with an assembly drawing, of this I-beam clamp with a ³/₄-in.-diameter screw size B sheets. (Courtesy of Grinnel Corporation.)

Figure 23.92 Design: Make working drawings, with an assembly drawing, of this lathe dog with a ³/₄-in.-diameter screw on size B sheets.

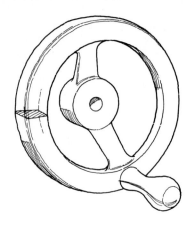

Figure 23.93 Design: Make working drawings, with an assembly drawing, of this hand wheel having a 10-in.-OD on size B sheets.

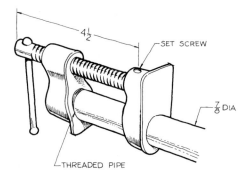

Figure 23.94 Design: Make working drawings, with an assembly drawing, of this clamping assembly on size B sheets.

Figure 23.95 Design: Make working drawings, with an assembly drawing, of this roller chain puller (for stretching a chain for assembly) on size B sheets. The prongs should join when closed.

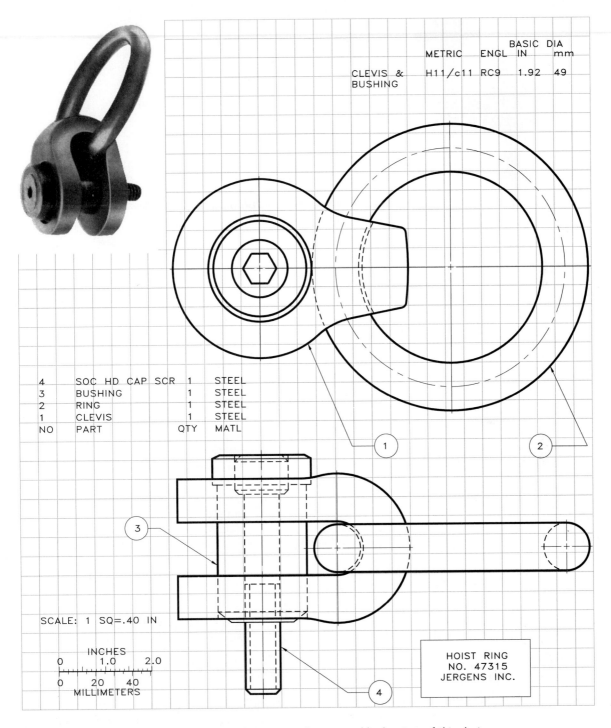

		BASIC DIA		
	METRIC	ENGL	IN	mm
CLEVIS & BUSHING	H11/c11	RC9	1.92	49

4	SOC HD CAP SCR	1	STEEL
3	BUSHING	1	STEEL
2	RING	1	STEEL
1	CLEVIS	1	STEEL
NO	PART	QTY	MATL

SCALE: 1 SQ=.40 IN

INCHES
0   1.0   2.0

0   20   40
MILLIMETERS

HOIST RING
NO. 47315
JERGENS INC.

Figure 23.96 Design: Make full-size working drawings, with an assembly drawing, of this clevis and bushing on size B sheets. (Courtesy of Jergens, Incorporated.)

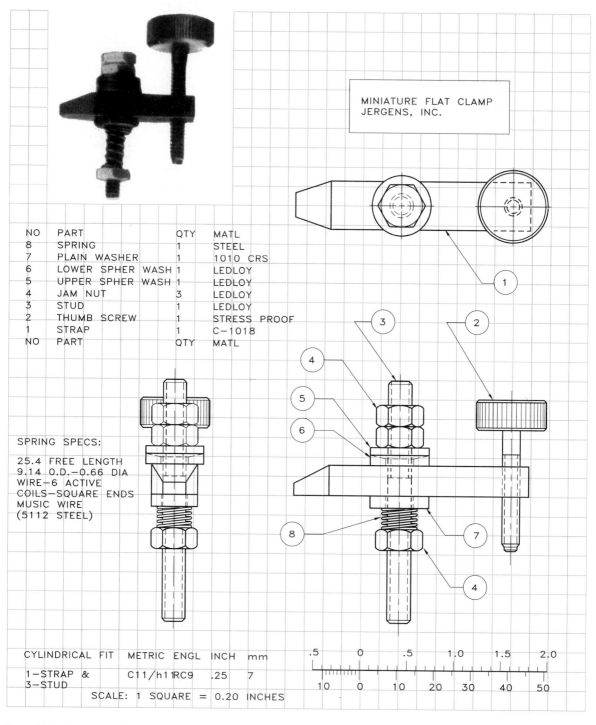

MINIATURE FLAT CLAMP
JERGENS, INC.

NO	PART	QTY	MATL
8	SPRING	1	STEEL
7	PLAIN WASHER	1	1010 CRS
6	LOWER SPHER WASH	1	LEDLOY
5	UPPER SPHER WASH	1	LEDLOY
4	JAM NUT	3	LEDLOY
3	STUD	1	LEDLOY
2	THUMB SCREW	1	STRESS PROOF
1	STRAP	1	C−1018
NO	PART	QTY	MATL

SPRING SPECS:

25.4 FREE LENGTH
9.14 O.D.−0.66 DIA
WIRE−6 ACTIVE
COILS−SQUARE ENDS
MUSIC WIRE
(5112 STEEL)

CYLINDRICAL FIT	METRIC	ENGL	INCH	mm
1−STRAP & 3−STUD	C11/h11	RC9	.25	7
	SCALE: 1 SQUARE = 0.20 INCHES			

Figure 23.97 Design: Make double-size working drawings, with an assembly drawing, of this miniature flat clamp on size B sheets. (Courtesy of Jergens, Incorporated.)

442 • CHAPTER 23 WORKING DRAWINGS

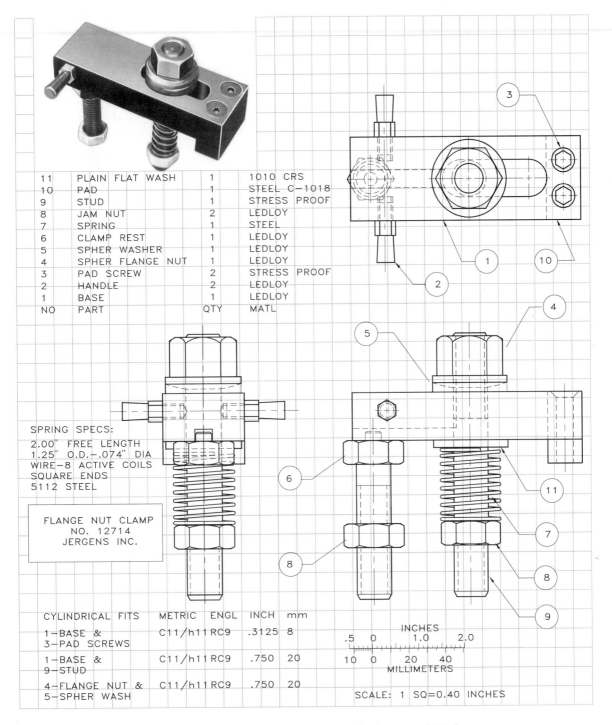

NO	PART	QTY	MATL
11	PLAIN FLAT WASH	1	1010 CRS
10	PAD	1	STEEL C-1018
9	STUD	1	STRESS PROOF
8	JAM NUT	2	LEDLOY
7	SPRING	1	STEEL
6	CLAMP REST	1	LEDLOY
5	SPHER WASHER	1	LEDLOY
4	SPHER FLANGE NUT	1	LEDLOY
3	PAD SCREW	2	STRESS PROOF
2	HANDLE	2	LEDLOY
1	BASE	1	LEDLOY

SPRING SPECS:
2.00" FREE LENGTH
1.25" O.D.-.074" DIA
WIRE-8 ACTIVE COILS
SQUARE ENDS
5112 STEEL

FLANGE NUT CLAMP
NO. 12714
JERGENS INC.

CYLINDRICAL FITS	METRIC	ENGL	INCH	mm
1-BASE & 3-PAD SCREWS	C11/h11	RC9	.3125	8
1-BASE & 9-STUD	C11/h11	RC9	.750	20
4-FLANGE NUT & 5-SPHER WASH	C11/h11	RC9	.750	20

INCHES
.5 0   1.0   2.0
10 0   20   40
MILLIMETERS

SCALE: 1 SQ=0.40 INCHES

Figure 23.98 Design: Make full-size working drawings, with an assembly drawing, of this flange nut clamp on size B sheets. (Courtesy of Jergens, Incorporated.)

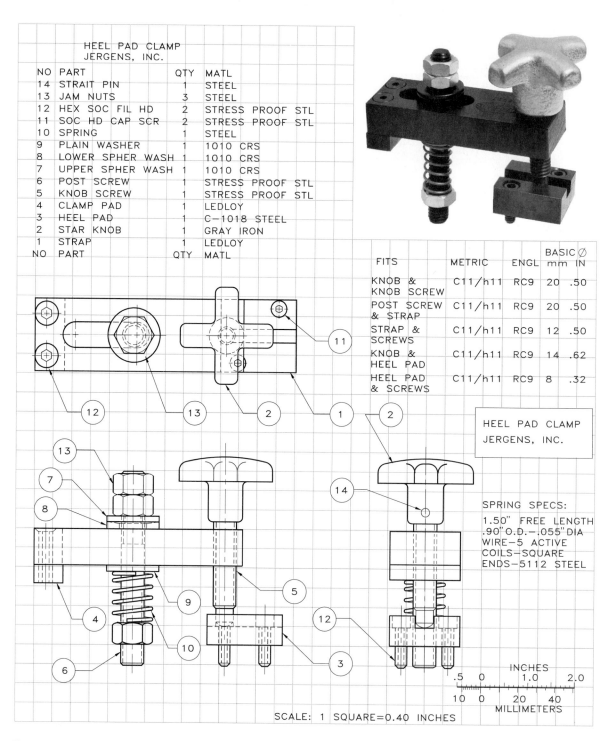

HEEL PAD CLAMP
JERGENS, INC.

NO	PART	QTY	MATL
14	STRAIT PIN	1	STEEL
13	JAM NUTS	3	STEEL
12	HEX SOC FIL HD	2	STRESS PROOF STL
11	SOC HD CAP SCR	2	STRESS PROOF STL
10	SPRING	1	STEEL
9	PLAIN WASHER	1	1010 CRS
8	LOWER SPHER WASH	1	1010 CRS
7	UPPER SPHER WASH	1	1010 CRS
6	POST SCREW	1	STRESS PROOF STL
5	KNOB SCREW	1	STRESS PROOF STL
4	CLAMP PAD	1	LEDLOY
3	HEEL PAD	1	C-1018 STEEL
2	STAR KNOB	1	GRAY IRON
1	STRAP	1	LEDLOY
NO	PART	QTY	MATL

FITS	METRIC	ENGL	BASIC Ø mm	IN
KNOB & KNOB SCREW	C11/h11	RC9	20	.50
POST SCREW & STRAP	C11/h11	RC9	20	.50
STRAP & SCREWS	C11/h11	RC9	12	.50
KNOB & HEEL PAD	C11/h11	RC9	14	.62
HEEL PAD & SCREWS	C11/h11	RC9	8	.32

HEEL PAD CLAMP
JERGENS, INC.

SPRING SPECS:
1.50" FREE LENGTH
.90" O.D.-.055" DIA
WIRE-5 ACTIVE
COILS-SQUARE
ENDS-5112 STEEL

INCHES
.5   0   1.0   2.0
10   0   20   40
MILLIMETERS

SCALE: 1 SQUARE=0.40 INCHES

Figure 23.99 Design: Make full-size working drawings, with an assembly drawing, of this heel pad clamp on size B sheets. (Courtesy of Jergens, Incorporated.)

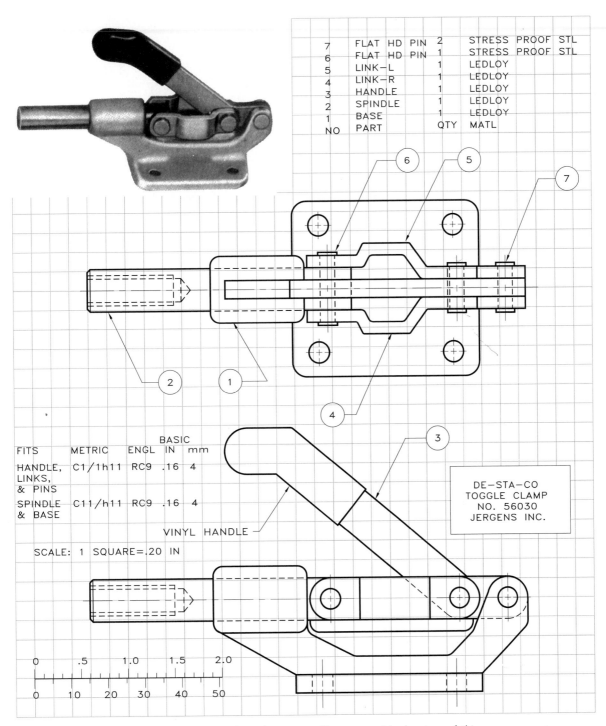

7	FLAT HD PIN	2	STRESS PROOF STL
6	FLAT HD PIN	1	STRESS PROOF STL
5	LINK—L	1	LEDLOY
4	LINK—R	1	LEDLOY
3	HANDLE	1	LEDLOY
2	SPINDLE	1	LEDLOY
1	BASE	1	LEDLOY
NO	PART	QTY	MATL

FITS	METRIC	ENGL	BASIC IN	mm
HANDLE, LINKS, & PINS	C1/1h11	RC9	.16	4
SPINDLE & BASE	C11/h11	RC9	.16	4

VINYL HANDLE

SCALE: 1 SQUARE=.20 IN

DE—STA—CO
TOGGLE CLAMP
NO. 56030
JERGENS INC.

0   .5   1.0   1.5   2.0

0   10   20   30   40   50

Figure 23.100 Design: Make double-size working drawings, with an assembly drawing, of this toggle clamp on size B sheets. (Courtesy of Jergens, Incorporated.)

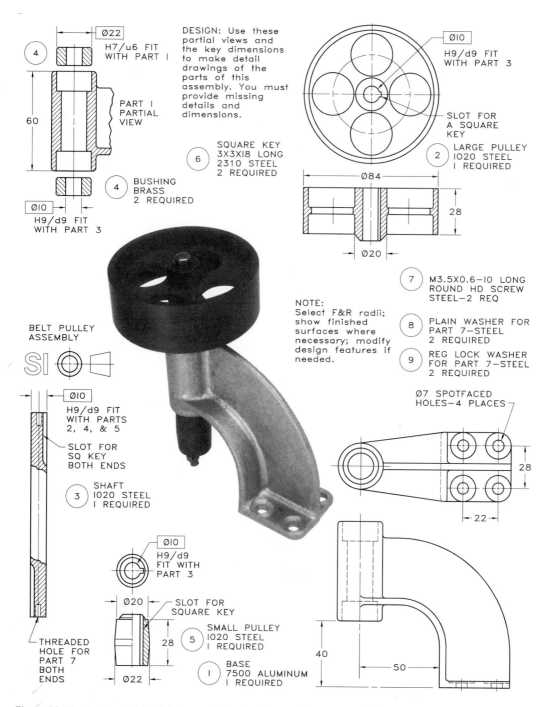

Ø22

4

H7/u6 FIT
WITH PART I

DESIGN: Use these
partial views and
the key dimensions
to make detail
drawings of the
parts of this
assembly. You must
provide missing
details and
dimensions.

Ø10

H9/d9 FIT
WITH PART 3

PART I
PARTIAL
VIEW

60

SQUARE KEY
3X3X18 LONG
2310 STEEL
2 REQUIRED

6

SLOT FOR
A SQUARE
KEY

Ø84

LARGE PULLEY
1020 STEEL
I REQUIRED

2

4

BUSHING
BRASS
2 REQUIRED

Ø10

H9/d9 FIT
WITH PART 3

28

Ø20

7

M3.5X0.6-10 LONG
ROUND HD SCREW
STEEL-2 REQ

NOTE:
Select F&R radii;
show finished
surfaces where
necessary; modify
design features if
needed.

8

PLAIN WASHER FOR
PART 7-STEEL
2 REQUIRED

BELT PULLEY
ASSEMBLY

9

REG LOCK WASHER
FOR PART 7-STEEL
2 REQUIRED

SI

Ø10

H9/d9 FIT
WITH PARTS
2, 4, & 5

Ø7 SPOTFACED
HOLES-4 PLACES

SLOT FOR
SQ KEY
BOTH ENDS

28

3

SHAFT
1020 STEEL
I REQUIRED

22

Ø10

H9/d9
FIT WITH
PART 3

Ø20

SLOT FOR
SQUARE KEY

THREADED
HOLE FOR
PART 7
BOTH
ENDS

28

5

SMALL PULLEY
1020 STEEL
I REQUIRED

40

50

Ø22

I

BASE
7500 ALUMINUM
I REQUIRED

Figure 23.101 Design: Make full-size working drawings, with an assembly drawing, of this belt
pulley assembly on size B sheets. Use your design ability to supply missing details, specifications,
and notes.

Reproduction of Drawings

24.1 Introduction

So far we have discussed the preparation of drawings and specifications through the working-drawing stage where detailed drawings are completed on tracing film or paper. Now the drawings must be reproduced, folded, and prepared for transmittal to those who will use them to prepare bids or to fabricate the parts. Several methods of reproduction are available to engineers and technologists for making copies of their drawings. However, most reproduction methods require strong, well-executed line work on the originals in order to produce good copies.

24.2 Types of Reproduction

Drawings made by a drafter are of little use in their original form. If original drawings were handled by checkers and by workers in the field or shop, they would quickly be soiled and damaged, and no copy would be available as a permanent record of the job. Therefore the reproduction of drawings is necessary for making inexpensive expendable copies for use by the people who need to use them.

The most often used processes of reproducing engineering drawings are (1) diazo printing, (2) microfilming, (3) xerography, and (4) photostating.

Diazo Printing

The **diazo print** more correctly is called a **whiteprint** or **blue-line print** than a blueprint because it has a white background and blue lines. Other colors of lines are available, depending on the type of diazo paper used. (Blueprinting, which creates a print with white lines and a blue background, is a wet process that is almost obsolete at the present.) **Figure 24.1** shows a typical diazo printer.

Diazo printing requires that original drawings be made on semitransparent tracing paper, cloth, or film that allows light to pass through the drawing. The diazo paper on which the blue-line print is made is chemically treated, giving it a yellow

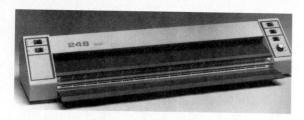

Figure 24.1 This typical whiteprinter operates on the diazo process. (Courtesy of Blu-Ray, Incorporated, Essex, CT.)

tint on one side. Diazo paper must be stored away from heat and light to prevent spoilage.

Making a diazo print first requires that the tracing-paper or film drawing be placed face up on the yellow side of the diazo paper **(Fig. 24.2A)** and then run through the diazo-process machine, exposing the drawing to a built-in light **(Fig. 24.2B)**. Light rays pass through the tracing paper and burn away the yellow tint on the diazo paper except where the drawing lines have shielded the paper from the light, similar to how a photographic negative works. The exposed diazo paper becomes a duplicate of the original drawing except that the lines are light yellow and not permanent. When the diazo paper is passed through the developing unit of the diazo machine **(Fig. 24.2C and D)**, ammonia fumes develop the yellow lines on it into permanent blue lines **(Fig. 24.2E)**.

The speed at which the drawing passes under the light determines the darkness of the copy; **the faster the speed, the darker the print is.** A slow speed burns out more of the yellow and produces a clear white background, but some of the lighter lines of the drawing may be lost. Most diazo copies are made at a speed fast enough to give a light tint of blue in the background and dark lines on the copy. Ink drawings give the best reproductions.

Microfilming

Microfilming is a photographic process that converts large drawings into film copies—either aperture cards or roll film. Drawings are photographed on either 16-mm or 35-mm film **(Fig. 24.3)**.

The roll film or aperture cards are placed in a microfilm enlarger–printer, where the individual drawings can be viewed on a built-in screen **(Fig. 24.4)**. The selected drawings can be printed from the film in standard sizes. Microfilm copies are usually made smaller than the original drawings to save paper and make the drawings easier to use.

Microfilming eliminates the need for large, bulky files of drawings because hundreds of drawings can be stored in miniature on a small amount of film. The aperture cards shown in **Fig. 24.4** are data processing cards that can be cataloged and recalled by a computer to make them accessible with a minimum of effort.

Xerography

Xerography is an electrostatic process of duplicating drawings on ordinary, unsensitized paper. Originally developed for business and clerical uses, xerography more recently has been used for the reproduction of engineering drawings. The xerographic process is used to reduce the sizes of the drawings being copied **(Fig. 24.5)**. The Xerox 2080 can reduce a 24-×-36-inch drawing to 8 × 10 inches.

Photostating

Photostating is a method of enlarging or reducing drawings photographically. **Figure 24.6** shows a combination camera and processor used for photographing drawings and producing high-contrast copies.

The drawing is placed under the glass of the exposure table, which is lit by built-in lamps. The image appears on a glass plate inside the darkroom where it is exposed on photographically sensitive paper. The exposed negative paper is placed in contact with receiver paper, and the two are fed through the developing solution to obtain a photostatic copy. Photostating also can be used to make reproductions on transparent films and for reproducing halftones (photographs with tones of gray).

24.3 Assembling Drawing Sets

After the original drawings have been copied, they should be stored flat without folding in a file for

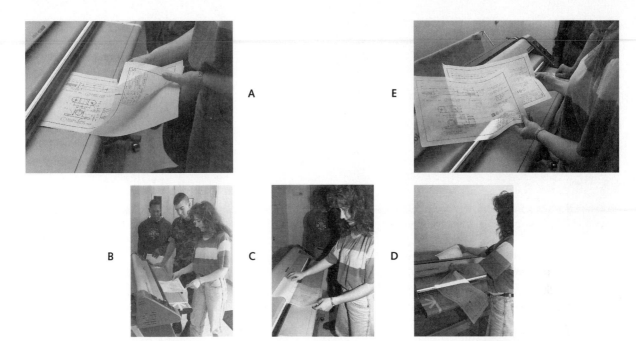

Figure 24.2 Blue-line prints are made by (A) placing the original readable side up and on top of the yellow side of the diazo paper, (B) feeding them under the light of the diazo machine, and (C) feeding the exposed diazo sheet through the ammonia chamber for developing the yellow lines into permanent blue lines. When the final print emerges (D), it is a full-size copy of the original (E).

Figure 24.3 The Micro-Master 35-mm camera and copy table are used for microfilming engineering drawings. (Courtesy of Keuffel & Esser Company, Morristown, NJ.)

Figure 24.4 The Bruning 1200 microfilm enlarger–printer makes drawings up to 18″ × 24″ from aperture cards and roll film. (Courtesy of Bruning Company.)

future use and updating. Prints made from the originals, however, usually are folded or rolled for ease of transmittal from office to office. **Figure 24.7** shows how to fold size B, C, D, and E sheets so that the image will appear on the outside of the fold. Drawings should be folded so that the title block always shows on the outside at the right, usually in the lower right-hand corner of the page **(Fig. 24.8)**. The final size after folding is 8 1/2 × 11 inches (or 9 × 12 inches).

An alternative method of folding and stapling size B sheets often is used for student assignments

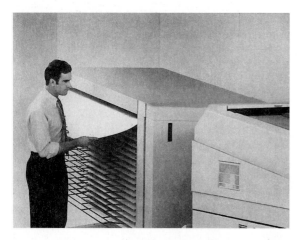

Figure 24.5 This Xerox Flat Sorter-36 and the Xerox 5080 copier offer plain-paper copying and sorting of drawings larger than 11″ × 17″. (Courtesy of Xerox Corporation.)

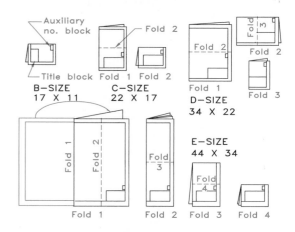

Figure 24.7 All standard drawing sheets can be folded to 8½″ × 11″ size for filing and storage.

Figure 24.6 This camera–processor enlarges and reduces drawings to be reproduced as photostats. (Courtesy of the Duostat Corporation.)

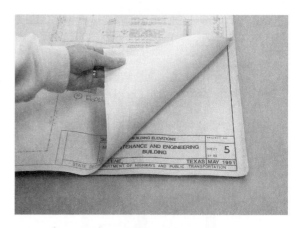

Figure 24.8 The title block should appear at the right, usually in the lower right-hand corner of the sheet.

so that they can be kept in a three-ring notebook **(Fig. 24.9)**. The basic rules of assembling drawings are listed in **Fig. 24.10**.

24.4 Transmittal of Drawings

Prints of drawings must be delivered to contractors, manufacturers, fabricators, and others who must use the drawings for implementing the project. Prints usually are placed in standard 9-×-12-inch envelopes and are delivered by hand or mail. Sets of large drawings, which may be 30 × 40 inches in size and contain four or more sheets, usually are rolled and mailed in a mailing tube when folding becomes impractical.

An advanced method of transmitting drawings is by use of the Xerox 7124 engineering fax

FOLDING B-SIZE SETS

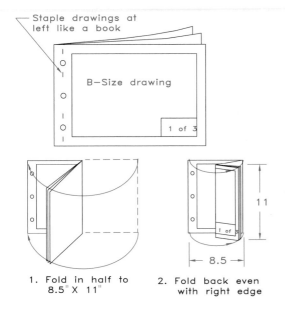

1. Fold in half to
   8.5" X 11"

2. Fold back even
   with right edge

Figure 24.9 A set of size B drawings can be assembled by stapling, punching, and folding, as shown here for safe-keeping in a three-ring notebook with the title block visible on top.

WORKING DRAWING CHECKLIST

1. Staple along left edge, like a book. Use several staples, never just one.

2. Fold with drawing on outside.

3. Fold drawings as a set, not one at a time separately.

4. Fold to and 8.5" X 11" modular size.

5. The title block must be visible after folding.

6. Sheets of a set should be uniform in size.

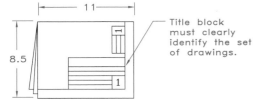

Title block must clearly identify the set of drawings.

Figure 24.10 Follow these basic rules for assembling sets of working drawing prints.

Figure 24.11 Fax machines, such as this Xerox 7124, can transmit large drawings throughout the country within three minutes over the Kinko's Copy Center network. (Courtesy of Xerox Corporation.)

machines at Kinko's Copy Centers **(Fig. 24.11)**. Within three minutes, large documents can be scanned and transmitted to more than 150 sites throughout the country.

Computer drawings can be transmitted on disk by mailing them to their destination, where hard copies can be plotted and reproduced. This procedure offers substantial savings in shipping charges.

In the future, more drawings will be sent electronically as data and as scanned images over telephone wires, making them available instantaneously at the desired location. What was once a fantasy is now a reality.

Three-Dimensional Pictorials

25.1 Introduction

A **three-dimensional pictorial** is a drawing that shows an object's three principal planes, much as they would be captured by a camera. This type of pictorial is an effective means of illustrating a part that is difficult to visualize when only orthographic views are given. Pictorials are especially helpful when a design is complex and when the reader of the drawings is unfamiliar with orthographic drawings.

Sometimes called **technical illustrations**, pictorials are widely used to describe products in catalogs, parts manuals, and maintenance publications. The ability to sketch pictorials rapidly to explain a detail to an associate in the field is an important communication skill.

The four commonly used types of pictorials are (1) obliques, (2) isometrics, (3) axonometrics, and (4) perspectives **(Fig. 25.1)**.

Oblique pictorials: three-dimensional drawings made by projecting from the object with

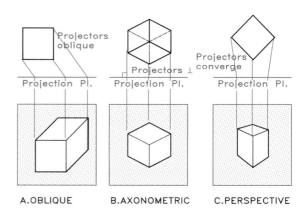

Figure 25.1 There are three pictorial projection systems: (A) oblique pictorials, with parallel projectors oblique to the projection plane; (B) axonometric (including isometric) pictorials, with parallel projectors perpendicular to the projection plane; and (C) perspectives, with converging projectors that make varying angles with the projection plane.

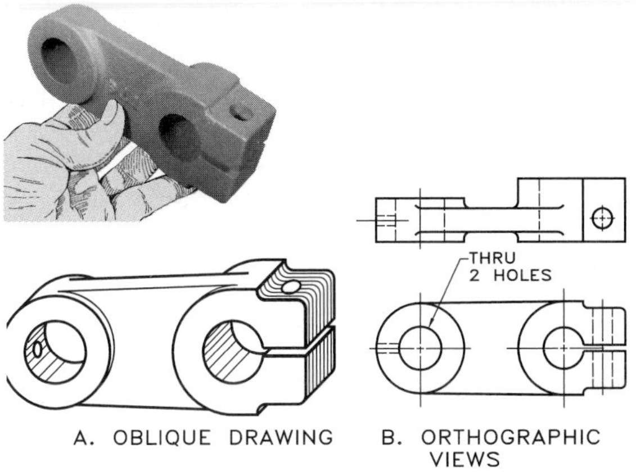

Figure 25.2 The oblique drawing of this part makes it easier to visualize than its representation by orthographic views.

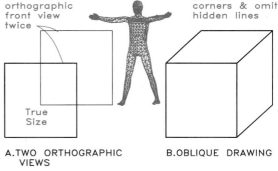

Figure 25.3 Draw two true-size surfaces of a box, connect them at the corners, and you have an oblique.

parallel projectors that are oblique to the picture plane (**Fig. 25.1A**).

Isometric and axonometric pictorials: three-dimensional drawings made by projecting from the object with parallel projectors that are perpendicular to the picture plane (**Fig. 25.1B**).

Perspective pictorials: three-dimensional drawings made with projectors that converge at the viewer's eye and make varying angles with the picture plane (**Fig. 25.1C**).

25.2 Oblique Drawings

The pulley arm shown in **Fig. 25.2** is illustrated by orthographic views and an oblique pictorial. Because most parts are drawn before they actually exist, photographs cannot be taken; therefore the next best option is to draw a three-dimensional pictorial of the part. Details can usually be drawn with more clarity than can be shown in a photograph.

Oblique pictorials are easy to draw. If you can drawn an orthographic view of a part, you are but one step away from drawing an oblique. For example, **Fig. 25.3**, shows that drawing a front view of a box twice and connecting its corners yields an oblique drawing.

Thus an oblique is an orthographic view with a receding axis, drawn at an angle to show the depth of the object. An oblique is a pictorial that does not exist in reality (a camera cannot give an oblique). This type of pictorial is called an **oblique** because its parallel projectors from the object are oblique to the picture plane. These projection principles are covered in Section 25.3.

Types of Obliques

The three basic types of oblique drawings are: (1) cavalier, (2) cabinet, and (3) general (**Fig. 25.4**). For each type, the angle of the receding axis with the horizontal can be at any angle between 0° and 90°. **Measurements along the receding axes of the cavalier oblique are laid off true length, and measurements along the receding axes of the cabinet oblique are laid off half size. The general oblique has measurements along the receding axes that are greater than half size and less than full size.**

Figure 25.5 shows three examples of cavalier obliques of a cube. The receding axes for each is drawn at a different angle, but the receding axes are drawn true length. **Figure 25.6** compares cavalier with cabinet obliques.

TYPES OF OBLIQUES

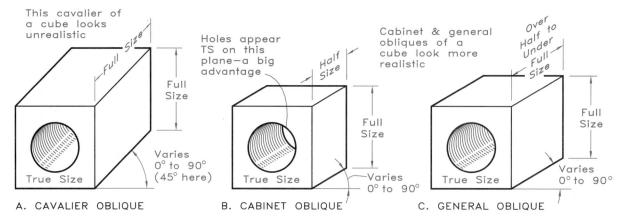

A. CAVALIER OBLIQUE

B. CABINET OBLIQUE

C. GENERAL OBLIQUE

Figure 25.4 There are three types of obliques:

A The cavalier oblique has a receding axis at any angle and true-length measurements along the receding axis.

B The cabinet oblique has a receding axis at any angle and half-size measurements along the receding axis.

C The general oblique has a receding axis at any angle and measurements along the receding axis larger than half size and less than full size.

CAVALIER OBLIQUES

Drawn with standard angles

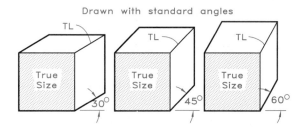

Figure 25.5 A cavalier oblique usually has its receding axis as one of the standard angles of drafting triangles. Each gives a different view of a cube.

Constructing Obliques

You can easily begin a cavalier oblique by drawing a box using the overall dimensions of height, width, and depth with light construction lines. As demonstrated in **Fig. 25.7**, first draw the front view as a true-size orthographic view. True mea-

CAVALIER VS. CABINET OBLIQUES

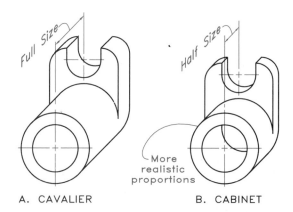

A. CAVALIER

B. CABINET

Figure 25.6 Measurements along the receding axis of a cavalier oblique are full size, and those in a cabinet oblique are half size.

surements must be made parallel to the three axes and transferred from the orthographic views with your dividers. Then remove the notches

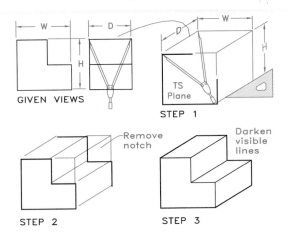

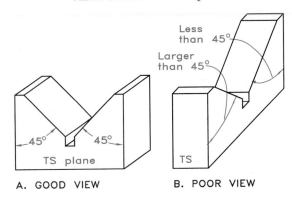

A. GOOD VIEW

B. POOR VIEW

Figure 25.7 Constructing a cavalier oblique:

Step 1 Draw the front surface of the object as a true-size plane. Draw the receding axis at a convenient angle and measure its depth as the true distance D transferred from the side view with dividers.

Step 2 Draw the notch in the front plane and project it to the rear plane.

Step 3 Darken the lines to complete the drawing.

Figure 25.8 Objects with angular features should be drawn in oblique so that the angles appear true size, the good view is as descriptive as possible, and its construction is simple.

LOCATING ANGLES WITH COORDINATES

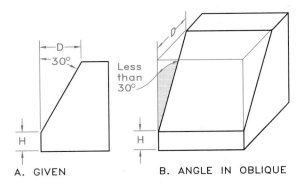

A. GIVEN

B. ANGLE IN OBLIQUE

from the blocked-in construction box to complete the oblique.

Angles

Angular measurements can be made on the true-size plane of an oblique, but not on the other two planes. Note in **Fig. 25.8** that a true angle can be measured on a true-size surface, but in **Fig. 25.9** angles along receding planes are either smaller or larger than their true sizes. **A better, easier-to-draw oblique is obtained when angles are drawn to appear true size.**

To construct an angle in an oblique on one of the receding planes, you must use coordinates, as shown in Fig. 25.9. To find the surface that slopes 30° from the front surface, locate the vertex of the angle, H distance from the bottom. To find the upper end of the sloping plane, measure the distance D along the receding axis. Transfer H and D

Figure 25.9 Angles that do not lie in a true-size plane of an oblique must be located with coordinates.

to the oblique with your dividers. The angle in the oblique is not equal to the 30° angle in the orthographic view.

Cylinders

The major advantage of an oblique is that circular features can be drawn as true circles on its frontal plane (Fig. 25.10). Draw the centerlines of the circular end at A and construct the receding axis at

CAVALIER OBLIQUE CYLINDER

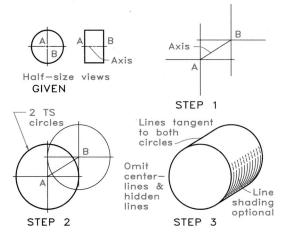

Figure 25.10 Drawing an oblique of a cylinder:
Step 1 Draw axis AB and locate the centers of the circular ends of the cylinder at A and B. Because the axis is true length this is a cavalier oblique.

Step 2 Draw a true-size circle with its center at A by using a compass or computer-graphics techniques.

Step 3 Draw the other circular end with its center at B and connect the circles with tangent lines parallel to axis AB.

SEMICIRCULAR FEATURES

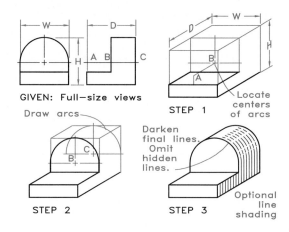

Figure 25.11 Drawing semicircular features in oblique:
Step 1 Block in the overall dimensions of the cavalier oblique with light construction lines, ignoring the semicircular features.

Step 2 Locate centers B and C and draw arcs with a compass or by computer tangent to the sides of the construction boxes.

Step 3 Connect the arcs with lines tangent to each arc and parallel to axis BC and darken the lines.

the desired angle. Locate the end at B by measuring along the axis, draw circles at each end at centers A and B, and draw tangents to both circles.

These same principles apply to construction of the object having semicircular features shown in **Fig. 25.11**. Position the oblique so that the semicircular features are true size. Locate centers A, B, and C and the two semicircles. Then complete the cavalier oblique.

Circles

Circular features drawn as true circles on a true-size plane of an oblique pictorial appear on the receding planes as ellipses.

The four-center ellipse method is a technique of constructing an approximate ellipse with a compass and four centers (**Fig. 25.12**). The ellipse is tangent to the inside of a rhombus drawn with

sides equal to the circle's diameter. Drawing the four arcs produces the ellipse.

The four-center ellipse method will not work for the cabinet or general oblique, but coordinates must be used. **Figure 25.13** illustrates the method of locating coordinates on the planes of cavalier and cabinet obliques. For the cabinet oblique, the coordinates along the receding axis are half size, and the coordinates along the horizontal axis (true-size axis) are full size. Draw the ellipse with an irregular curve or an ellipse template that approximates the plotted points.

Whenever possible, oblique drawings of objects with circular features should be positioned so circles can be drawn as true circles instead of ellipses. The view in **Fig. 25.14A** is better than the one in **Fig. 25.14B** because it gives a more descriptive view of the part and is easier to draw.

FOUR-CENTER ELLIPSE

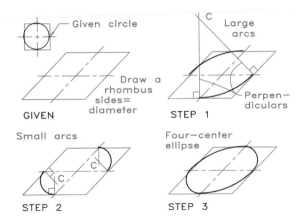

GIVEN — Given circle — Draw a rhombus sides= diameter

STEP 1 — C — Large arcs — Perpendiculars

Small arcs — STEP 2 — C — C

Four–center ellipse — STEP 3

Figure 25.12 Constructing a four-center ellipse in oblique: **Given** Block in the circle to be drawn in oblique with a square tangent to the circle. This square becomes a rhombus on the oblique plane.

Step 1 Draw construction lines perpendicular to the points of tangency to locate the centers for drawing two segments of the ellipse.

Step 2 Locate the centers for the two remaining arcs with perpendiculars drawn from adjacent tangent points.

Step 3 Draw the four arcs, which yield an approximate ellipse.

CIRCLES ON FACES OF CUBES

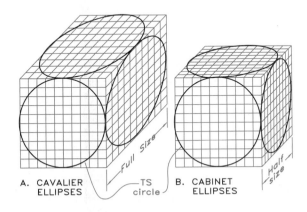

A. CAVALIER ELLIPSES — Full Size — TS circle — B. CABINET ELLIPSES — Half size

Figure 25.13 Circular features on the faces of cavalier and cabinet obliques are compared here. Ellipses on the receding planes of cabinet obliques must be plotted by coordinates. The spacing of the coordinates along the receding axis of cabinet obliques is half size.

VIEWPOINTS FOR OBLIQUES

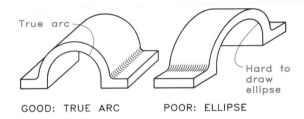

True arc — Hard to draw ellipse

GOOD: TRUE ARC — POOR: ELLIPSE

Figure 25.14 An oblique should be positioned so that circular and curving features may be drawn most easily.

CURVES BY COORDINATES

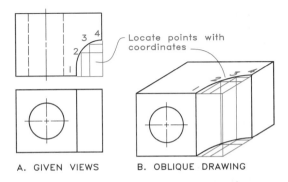

Locate points with coordinates

A. GIVEN VIEWS — B. OBLIQUE DRAWING

Figure 25.15 Coordinates are used to find points along irregular curves in oblique. Projecting the points downward a distance equal to the height of the object yields the lower curve.

Curves

Irregular curves in oblique pictorials must be plotted point by point with coordinates (**Fig. 25.15**). Transfer the coordinates from the orthographic to the oblique view and draw the curve through the plotted points with an irregular curve. If the object has a uniform thickness, plot the points for the lower curve by projecting vertically downward from the upper points a distance equal to the object's height.

To obtain the elliptical feature on the inclined surface shown in **Fig 25.16**, use a series of coordinates to locate points along its curve. Connect the

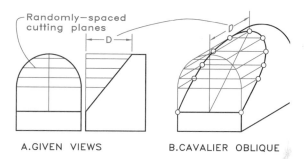

A.GIVEN VIEWS   B.CAVALIER OBLIQUE

Figure 25.16 Construction of an elliptical feature on an inclined surface in oblique requires the use of three-dimensional coordinates to locate points on the curve.

plotted points by using an irregular curve or ellipse template.

Sketching

Understanding the principles of oblique construction is essential for sketching obliques freehand. The sketch of the part shown in **Fig. 25.17** is based on the principles discussed, but its proportions were determined by eye instead of with scales and dividers.

Lightly drawn guidelines need not be erased when you darken the final lines. When sketching on tracing vellum, you can place a printed grid under the sheet to provide guidelines. Refer to Chapter 13 to review sketching techniques if needed.

Dimensioned Obliques

Dimensioned sectional views of obliques provide excellent, easily understood descriptions of objects (**Fig. 25.18**). Apply numerals and lettering in oblique pictorials by using either the aligned method (with numerals aligned with the dimension lines) or the unidirectional method (with numerals positioned horizontally regardless of the direction of the dimension lines), as shown in **Fig. 25.19**. Notes connected with leaders are positioned horizontally in both methods.

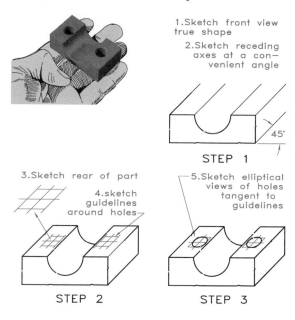

1.Sketch front view true shape
2.Sketch receding axes at a convenient angle

STEP 1

3.Sketch rear of part
4.sketch guidelines around holes
5.Sketch elliptical views of holes tangent to guidelines

STEP 2   STEP 3

Figure 25.17 Sketching obliques:

Step 1 Sketch the front of the object as true-size surface and draw a receding axis from each corner.

Step 2 Lay off the depth, D, along the receding axes to locate the rear of the part. Lightly sketch pictorial boxes as guidelines for drawing the holes.

Step 3 Sketch the holes inside the boxes and darken all lines.

25.3 Oblique Projection Theory

Now that you have a general understanding of oblique pictorials, you should know the theory on which this system is based. **Oblique projection** (**Fig. 25.20**) is the basis of oblique drawings. Receding axis 1–2 is perpendicular to the frontal projection plane. Projectors drawn from point 2 at 45° to the projection plane yield lengths on the front surface that are the same length as 1–2 (true length, in other words). Infinitely many 45° projectors form a cone of projectors with its apex at 2.

The true-length projections of lines 1–2' represent receding axes that can be used for cavalier obliques, which by definition, have true-length

DIMENSIONING AN OBLIQUE

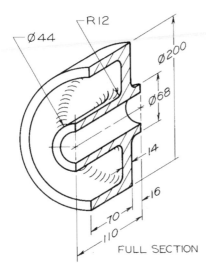

Figure 25.18 Oblique pictorials can be drawn as sections and dimensioned to serve as working drawings.

ALIGNED AND UNIDIRECTIONAL DIMENSIONS

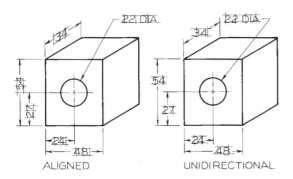

ALIGNED   UNIDIRECTIONAL

Figure 25.19 Either of these methods—aligned or unidirectional—is acceptable for dimensioning obliques.

dimensions along their receding axes. Do not use a vertical or a horizontal receding axis, but one between those limits.

To distinguish oblique projection from oblique drawing, as described in this chapter so far, observe the top and side views of a part and the picture planes shown in **Fig. 25.21**. In an oblique projection, projectors from the top and

PRINCIPLES OF CAVALIER PROJECTION

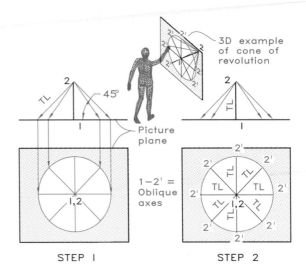

STEP 1   STEP 2

Figure 25.20 This drawing demonstrates the underlying principle of the cavalier oblique by using a series of projectors to form a cone:

Step 1 Each element from point 2 makes a 45° angle with the picture plane.

Step 2 The projected lengths of 1–2′ are equal in length to line 1–2, which is perpendicular to the picture plane. Thus the receding axis of a cavalier oblique is true length and at any angle.

PRINCIPLES OF OBLIQUE PROJECTION

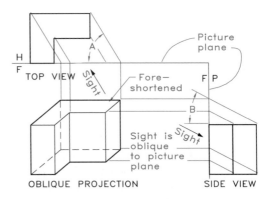

OBLIQUE PROJECTION   SIDE VIEW

Figure 25.21 An oblique projection may be drawn at varying angles of sight. However, a line of sight making an angle of less than 45° with the picture plane would result in a receding axis longer than its true length, thereby distorting the pictorial.

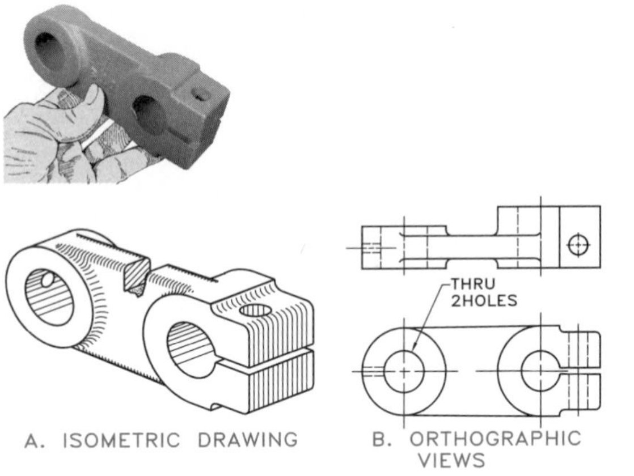

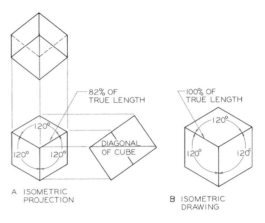

Figure 25.22 An isometric drawing gives a more realistic view than an oblique, and it is easier to visualize than are orthographic views.

Figure 25.23

A A true isometric projection is found by constructing a view in which the diagonal of a cube appears as a point and the axes are foreshortened.

B An isometric drawing is not a true projection because the dimensions are true size rather than foreshortened.

side views are oblique to the edge views of the projection planes, hence the name oblique.

Your line of sight can yield obliques with receding axes longer than true length (which should be avoided). Because of this shortcoming and the complexity of construction, **oblique pictorials usually are oblique drawings rather than oblique projections.**

25.4 Isometric Pictorials

In **Fig. 25.22** the pulley arm is drawn in orthographic views and in a three-dimensional pictorial drawing. The pictorial is an isometric drawing in which the three planes of the object are equally foreshortened, representing the object more realistically than an oblique drawing can.

With more realism comes more difficulty of construction. In particular, circles and curves do not appear true shape on any of the three isometric planes.

Isometric Projection versus Drawing
In isometric projection, parallel projectors are perpendicular to the imaginary projection (picture)

plane in which the diagonal of a cube appears as a point (Fig. 25.23). An isometric pictorial constructed by projection is called an **isometric projection**, with the three axes foreshortened to 82% of their true lengths and 120° apart. The name *isometric*, which means equal measurement, aptly describes this type of projection because the planes are equally foreshortened.

An **isometric drawing** is a convenient approximate isometric pictorial in which the measurements are shown full size along the three axes rather than at 82% as in isometric projection (**Fig. 25.24**). Thus the isometric drawing method allows you to measure true dimensions with standard scales and lay them off with dividers along the three axes. The only difference between the two is the larger size of the drawing. Consequently, **isometric drawings are used much more often than isometric projections.**

The axes of isometric drawings are separated by 120° (**Fig. 25.25**), but more often than not, one of the axes selected is vertical, since most objects

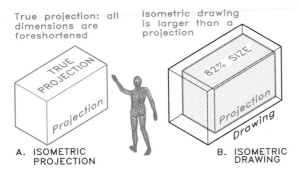

True projection: all dimensions are foreshortened

Isometric drawing is larger than a projection

A. ISOMETRIC PROJECTION

B. ISOMETRIC DRAWING

Figure 25.24 The true isometric projection is foreshortened to 82% of full size. The isometric drawing is drawn full size for convenience.

POSITIONING OF AXES

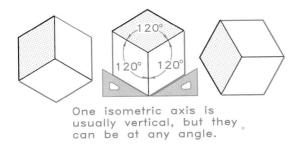

One isometric axis is usually vertical, but they can be at any angle.

Figure 25.25 Isometric axes are spaced 120° apart, but they can be revolved into any position. Usually, one axis is vertical, but it can be at any angle with axis spacing remaining the same.

have vertical lines. Isometrics without a vertical axis are still isometrics.

25.5 Isometric Drawings

An isometric drawing is begun by drawing three axes 120° apart. Lines parallel to these axes are called **isometric lines (Fig. 25.26A)**. You can make true measurements along isometric lines but not along nonisometric lines. The three surfaces of a cube in an isometric drawing are called **isometric**

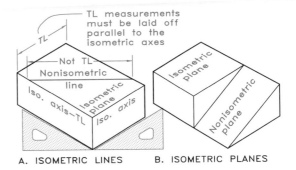

TL measurements must be laid off parallel to the isometric axes

Not TL Nonisometric line

A. ISOMETRIC LINES

B. ISOMETRIC PLANES

Figure 25.26

A Isometric lines (parallel to the three axes) give true measurements, but nonisometric lines do not.

B Here, the three isometric planes are equally foreshortened, and the nonisometric plane is inclined at an angle to one of the isometric planes.

planes (Fig. 25.26B). Planes parallel to those planes also are isometric planes.

To draw an isometric pictorial, you need a scale, dividers, and a 30°–60° triangle **(Fig. 25.27)**. Begin selecting the three axes and then constructing a plane of the isometric from the dimensions of height, H, and depth, D. Add the third dimension of width, W, and complete the isometric drawing.

Use light construction lines to block in all isometric drawings **(Fig. 25.28)** and the overall dimensions W, D, and H. Take other dimensions from the given views with dividers and measure along their isometric lines to locate notches in the blocked-in drawing.

Figure 25.29 shows an isometric drawing of a slightly more complex object, with two notches. The object was blocked in by using the H, W, and D dimensions. The notches in the block are removed to complete the drawing.

Angles

You cannot measure an angle's true size in an isometric drawing because the surfaces of an isometric are not true size. Instead, you must locate

ISOMETRIC OF A BOX

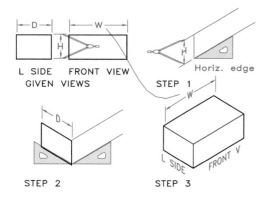

Figure 25.27 Drawing an isometric of a box:
Step 1 Use a 30°–60° triangle and a horizontal straight edge to construct a vertical line equal to the height, H, and draw two isometric lines through each end.

Step 2 Draw two 30° lines and locate the depth, D, by transferring this dimension from the given views with dividers.

Step 3 Locate the width, W, of the object, complete the surfaces of the isometric box, and darken the lines.

ISOMETRIC DRAWING

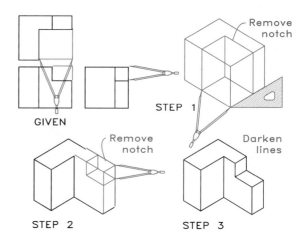

Figure 25.29 Laying out an isometric drawing:
Step 1 Use the overall dimensions given to block in the object with light lines and remove the large notch.

Step 2 Remove the small notch.

Step 3 Darken the lines to complete the drawing.

ISOMETRIC OF A SIMPLE OBJECT

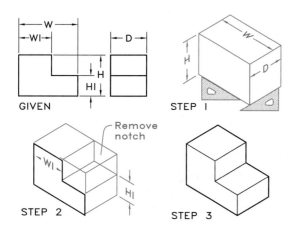

Figure 25.28 Constructing an isometric of a simple part:
Step 1 Construct an isometric drawing of a box by using the overall dimensions W, D, and H from the given views.

Step 2 Locate the notch in the box by transferring dimensions W1 and H1 from the given views with dividers.

Step 3 Darken the lines to complete the drawing.

ISOMETRIC: INCLINED SURFACES

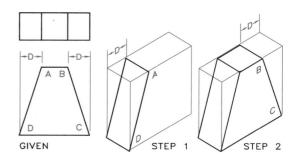

Figure 25.30 Use coordinates measured along the isometric axes to obtain inclined surfaces. Angular lines are not true length in isometric.

angles with isometric coordinates measured parallel to the axes **(Fig. 25.30)**. Lines AD and BC are equal in length in the orthographic view, but they are shorter and longer than true length in the isometric drawing. **Figure 25.31** shows a similar situation, where two angles drawn in isometric are

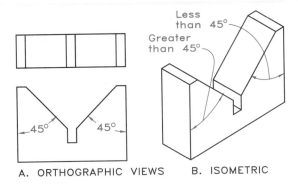

Figure 25.31 Angles in isometric may appear larger or smaller than they actually are.

ISOMETRIC: INCLINED PLANES

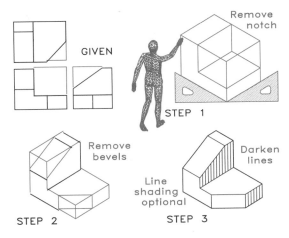

Figure 25.32 Drawing inclined planes in isometric:
Step 1 Block the object with light lines, using the overall dimensions, and remove the notch.

Step 2 Locate the ends of the inclined planes by using measurements parallel to the isometric axes.

Step 3 Darken the lines to complete the drawing.

less than and greater than their true dimensions in the orthographic view.

Figure 25.32 shows how to construct an isometric drawing of an object with inclined surfaces. Blocking in the object with its overall

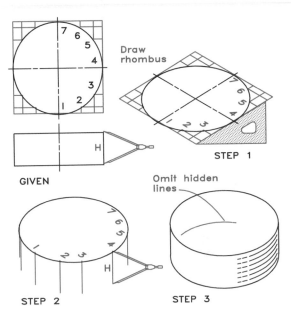

Figure 25.33 Plotting circles in isometric:
Step 1 Block in the circle by using its overall dimensions. Transfer the coordinates used to locate points on the circle to the isometric plane and connect them with a smooth curve.

Step 2 Drop each point a distance equal to the height of the cylinder to obtain the lower ellipse.

Step 3 Connect the two ellipses with tangent lines and darken the lines.

dimensions with light construction lines is followed by removal of the inclined portions.

Circles

Three methods of constructing circles in isometric drawings are (1) **point plotting**, (2) **four-center ellipse construction**, and (3) **ellipse template usage**.

Point plotting is a method of using a series of x and y coordinates to locate points on a circle in the given orthographic views. The coordinates are then transferred with dividers to the isometric drawing to locate the points on the ellipse one at a time (**Fig. 25.33**).

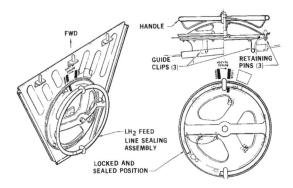

Figure 25.34 This handwheel assembly proposed for use in an orbital workshop is an example of parts with circular features drawn as ellipses in isometric. (Courtesy of NASA.)

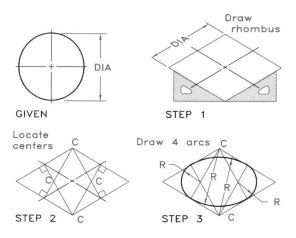

Figure 25.35 The four-center ellipse method:

Step 1 Use the diameter of the given circle to draw an isometric rhombus and the centerlines.

Step 2 Draw light construction lines perpendicularly from the midpoints of each side to locate four centers.

Step 3 Draw four arcs from the centers to represent an ellipse tangent to the rhombus.

Block in the cylinder with light construction lines and show the centerlines. Draw coordinates on the upper plane and use the height dimension to locate the points on the lower plane. Draw the ellipses with an irregular curve or an ellipse template.

A plotted ellipse is a true ellipse and is equivalent to a 35° ellipse drawn on an isometric plane. An example of a design composed of circular features drawn in isometric is the handwheel shown in **Fig. 25.34**.

Four-center ellipse construction is the method of producing an approximate ellipse by using four arcs drawn with a compass (**Fig. 25.35**). Draw an isometric rhombus with its sides equal to the diameter of the circle to be represented. Find the four centers by constructing perpendiculars to the sides of the rhombus at the midpoints of each side, and draw the four arcs to complete the ellipse. You may draw four-center ellipses on all three isometric planes because each plane is equally foreshortened (**Fig. 25.36**). Although it is only an approximate ellipse, the four-center ellipse technique is acceptable for drawing large ellipses and as a way to draw ellipses when an ellipse template is unavailable.

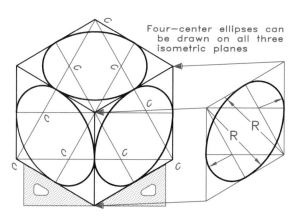

Figure 25.36 Four-center ellipses may be drawn on all three surfaces of an isometric drawing.

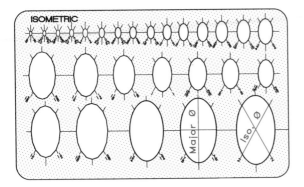

Figure 25.37 The isometric template (a 35° ellipse) is designed for drawing elliptical features in isometric. The isometric diameters of the ellipses are not their major diameters but are diameters that are parallel to the isometric axes.

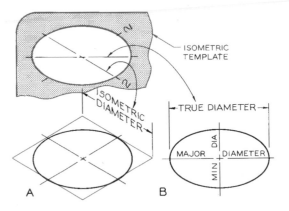

Figure 25.38
A Measure the diameter of a circle along the isometric axes. The major diameter of an isometric ellipse thus is larger than the measured diameter.

B The minor diameter is perpendicular to the major diameter.

Isometric ellipse templates are specially designed for drawing ellipses in isometric **(Fig. 25.37)**. The numerals on the templates represent the isometric diameters of the ellipses because diameters are measured parallel to the isometric axes of an isometric drawing **(Fig. 25.38)**. Recall that the **maximum diameter across the ellipse is its major diameter, which is a true diameter.** Thus the size of the diameter marked on the template is less than the ellipse's major diameter. You may use the isometric ellipse template to draw an ellipse by constructing centerlines of the ellipse in isometric and aligning the ellipse template with those isometric lines (Fig. 25.38).

Cylinders

A cylinder may be drawn in isometric by using the four-center ellipse method **(Fig. 25.39)**. Use the isometric axes and centerline axis to construct a rhombus at each end of the cylinder. Then draw the ellipses at each end, connect them with tangent lines, and darken the lines to complete the drawing.

An easier way to draw a cylinder is to use an isometric ellipse template **(Fig. 25.40)**. **Draw the**

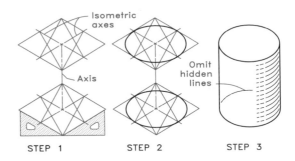

Figure 25.39 A cylinder using the four-center method:
Step 1 Draw an isometric rhombus at each end of the cylinder's axis.

Step 2 Draw a four-center ellipse within each rhombus.

Step 3 Draw lines tangent to each rhombus to complete the drawing.

axis of the cylinder and construct perpendiculars at each end. Because the axis of a right cylinder is perpendicular to the major diameter of its elliptical ends, position the ellipse template with its major diameter perpendicular to the axis.

CYLINDER: ELLIPSE TEMPLATE

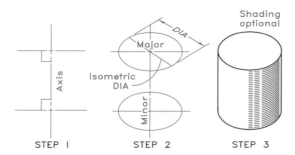

Figure 25.40 A cylinder using ellipse template method:
Step 1 Establish the length of the axis of the cylinder and draw perpendiculars at each end.

Step 2 Draw the elliptical ends by aligning the major diameter of the ellipse template with the perpendiculars at the ends of the axis. The isometric diameters of the isometric ellipse template will align with two isometric axes.

Step 3 Connect the ellipses with tangent lines to complete the drawing and omit hidden lines.

CYLINDRICAL HOLES IN ISOMETRIC

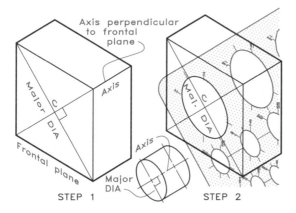

Figure 25.41 Constructing cylindrical holes through a block:
Step 1 Locate the center of the hole on a face of the isometric drawing. Draw the axis of the cylinder from the center parallel to the isometric axis perpendicular to the plane of the circle. The major diameter is perpendicular to this axis.

Step 2 Use the 2-in. ellipse template to draw the ellipse by aligning guidelines on the template with the major and minor diameters drawn on the front surface.

USING THE ELLIPSE TEMPLATE

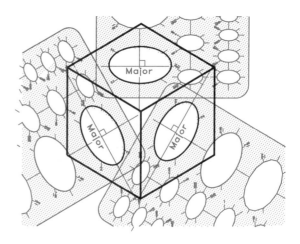

Figure 25.42 Position the isometric ellipse template as shown for drawing ellipses of various sizes on the three isometric planes.

Draw the ellipses at each end, connect them with tangent lines, and darken the visible lines to complete the drawing.

To construct a cylindrical hole in a block (**Fig. 25.41**), begin by locating the center of the hole on the isometric plane. Draw the axis of the cylinder parallel to the isometric axis that is perpendicular to the plane of the ellipse through its center. Align the ellipse template with the major diameter, which makes a 90° angle with the cylindrical axis and complete the elliptical view of the cylindrical hole.

The isometric ellipse template can be used to draw ellipses on all three planes of an isometric drawing. On each plane, the major diameter is perpendicular to the isometric axis of the adjacent perpendicular plane. The isometric diameters marked on the template align with the isometric axes. All ellipses drawn on isometric planes must align in the directions shown in **Fig. 25.42**.

Rounded Corners

The rounded corners of an object may be drawn with an ellipse template (**Fig. 25.43**). Block in each corner with light construction lines, draw center-

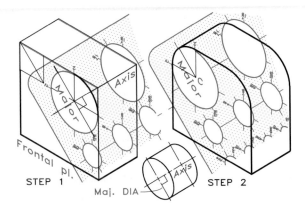

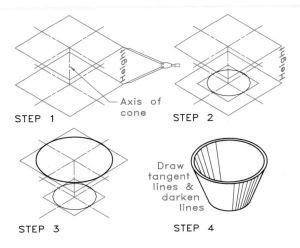

Figure 25.43 Drawing rounded corners:
Step 1 Draw the centerlines and isometric axes at the corners. Align the ellipse template with these guidelines and draw one quarter of the ellipse.

Step 2 Draw the other elliptical corner in the same manner with the same size ellipse.

Figure 25.44 Constructing a cone in isometric:
Step 1 Draw the axis of the cone and block in the larger end at both ends.

Step 2 Block in the smaller end of the cone.

Step 3 Connect the ellipses with tangents, draw the cone's wall thickness, and darken the lines to complete the drawing.

lines, draw the major diameter, and construct ellipses at each corner by positioning the template as shown. The rounded corners may also be constructed by using the four-center ellipse method (see Fig. 25.39) or by plotting points with coordinates (see Fig. 25.33).

A similar drawing involving the construction of ellipses is the conical shape shown in **Fig. 25.44**. Block in the ellipses on the upper and lower surfaces. Then draw the circular features by using a template or the four-center method, and draw lines tangent to each ellipse.

Inclined Planes
Inclined planes in isometric may be located by coordinates, but they cannot be measured with a protractor because they do not appear true size. **Figure 25.45** illustrates the coordinate method. Use horizontal and vertical coordinates (in the x and y directions) to locate key points on the orthographic views. Transfer these coordinates to the isometric drawing with dividers to show the features of the inclined surface.

INCLINED SURFACE

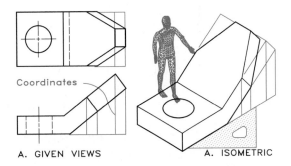

Figure 25.45 Inclined surfaces in isometric must be located with three-dimensional coordinates parallel to the isometric axes. True angles cannot be measured in isometric drawings.

Curves
Irregular curves in isometric must be plotted point by point, with coordinates locating each point. Locate points A through F in the ortho-

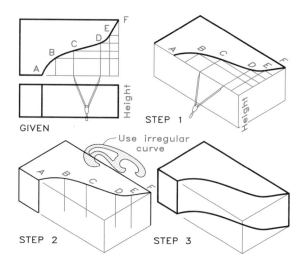

Figure 25.46 Plotting irregular curves:

Given Locate a series of points on the irregular curve with coordinates parallel to the W, D, and H dimensions.

Step 1 Block in the shape by using the overall dimensions. Locate points on the irregular curve with coordinates transferred from the orthographic views.

Step 2 Project these points downward the distance H from the upper points to obtain the lower curve.

Step 3 Connect the points and darken the lines.

graphic view with coordinates of width and depth **(Fig. 25.46)**. Then transfer them to the isometric view of the blocked-in part and connect them with an irregular curve.

Project points on the upper curve downward a distance of H, the height of the part, to locate points on the lower ellipse. Connect these points with an irregular curve and darken the lines to complete the isometric.

Ellipses on Nonisometric Planes

Ellipses on nonisometric planes in an isometric drawing, such as the one shown in **Fig. 25.47**, must be found by locating a series of points on the curve. Locate three-dimensional coordinates in the orthographic views and then transfer them to

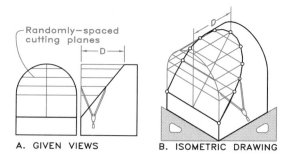

A. GIVEN VIEWS   B. ISOMETRIC DRAWING

Figure 25.47 To construct ellipses on inclined planes, draw coordinates to locate points in the orthographic views. Then transfer the three-dimensional coordinates to the isometric drawing and connect them with a smooth curve.

CIRCULAR FEATURES IN ISOMETRIC

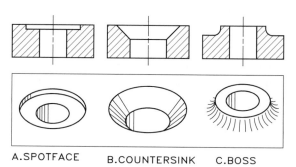

A.SPOTFACE   B.COUNTERSINK   C.BOSS

Figure 25.48 These examples of circular features in isometric may be drawn by using ellipse templates.

the isometric with dividers. Connect the plotted points with an irregular curve or an ellipse template selected to approximate the plotted points. It will not be an isometric ellipse template, but one that fits the plotted points.

25.6 Technical Illustration

Machine Parts

Figure 25.48 shows orthographic and isometric views of a **spotface**, **countersink**, and **boss**. These

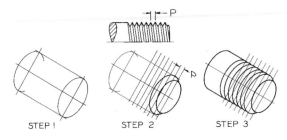

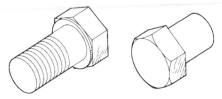

Figure 25.51 Isometric drawings of the lower and upper sides of a hexagon-head bolt.

Figure 25.49 Threads in isometric:
Step 1 Using an ellipse template, draw the cylinder to be threaded.

Step 2 Lay off perpendiculars, spacing them apart at a distance equal to the pitch of the thread, P.

Step 3 Draw a series of ellipses to represent the threads. Draw the chamfered end by using an ellipse whose major diameter is equal to the root diameter of the threads.

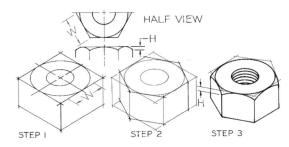

Figure 25.50 Constructing a nut:
Step 1 Use the overall dimensions of the nut to block in the nut.

Step 2 Construct the hexagonal sides at the top and bottom.

Step 3 Draw the chamfer with an irregular curve. Draw the threads to complete the drawing.

features may be drawn in isometric by point-by-point plotting of the circular features, the four-center or ellipse template method (the easiest method).

A threaded shaft may be drawn in isometric as shown in **Fig. 25.49**. First draw the cylinder in isometric. Draw the major diameters of the crest lines equally separated by distance P, the pitch of the thread. Then draw ellipses by aligning the major diameter of the ellipse template with the perpendiculars to the cylinder's axis. Use a smaller ellipse at the end for the 45° chamfered end.

Figure 25.50 shows how to draw a hexagon-head nut with an ellipse template. Block in the nut and draw an ellipse tangent to the rhombus. Construct the hexagon by locating distance W across a flat parallel to the isometric axes. To find the other sides of the hexagon, draw lines tangent to the ellipse. Lay off distance H at each corner to establish the chamfers.

Figure 25.51 depicts a hexagon-head bolt in two positions. The washer face is on the lower side of the head, and the chamfer is on the upper side.

Figure 25.52 shows how to use a portion of a sphere to draw a round-head screw. Construct a hemisphere and locate the centerline of the slot along one of the isometric planes. Measure the head's thickness, E, from the highest point on the sphere.

Sections

A full section drawn in isometric can clarify internal details that might otherwise be overlooked (**Fig. 25.53**). Half sections also may be used advantageously.

Dimensioned Isometrics

When you dimension isometric drawings, place numerals on the dimension lines, using either

BOLTHEAD IN ISOMETRIC

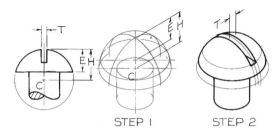

STEP 1   STEP 2

Figure 25.52 Drawing spherical features:
Step 1 Use an isometric ellipse template to draw the elliptical features of a round-head screw.

Step 2 Draw the slot in the head and darken the lines to complete the drawing.

ALIGNED VS. UNIDIRECTIONAL DIMENSIONS

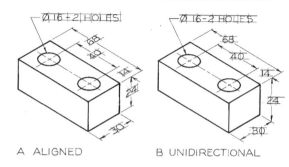

A ALIGNED   B UNIDIRECTIONAL

Figure 25.54 Either of the techniques shown—aligned or unidirectional—is acceptable for placing dimensions on isometric drawings. Guidelines should always be used for lettering.

SECTION IN ISOMETRIC

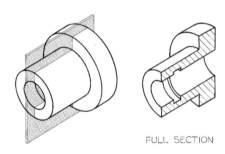

FULL SECTION

Figure 25.53 Isometric sections can be used to clarify the internal features of a part.

FILLETS & ROUNDS IN ISOMETRIC

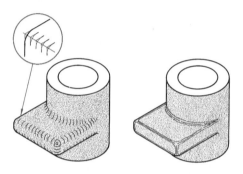

Figure 25.55 These are two methods of representing fillets and rounds on a part.

aligned or unidirectional numerals (**Fig. 25.54**). In both cases, notes connected with leaders usually are positioned horizontally, but drawing them to lie in an isometric plane is permissible. Always use guidelines for your lettering and numerals.

Fillets and Rounds

Fillets and rounds in isometric may be represented by either of the techniques shown in **Fig. 25.55** for added realism. The enlarged detail in the balloon shows how to draw intersecting guidelines equal in length to the radii of the fillets and rounds with arcs drawn tangent to them. These

arcs may be drawn either freehand or with an ellipse template. The stipple shading was applied by using an adhesive overlay film.

When fillets and rounds of a dimensional part are shown, it is much easier to understand its features than it is when the part is represented by orthographic views (**Fig. 25.56**).

Assemblies

Assembly drawings illustrate how to put parts together. **Figure 25.57A** shows common mistakes in applying leaders and balloons to an assembly, and **Fig. 25.57B** shows the correct method of

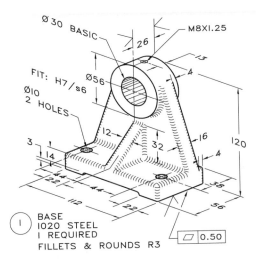

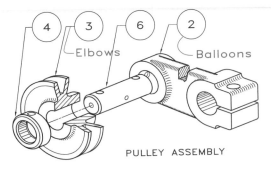

PULLEY ASSEMBLY

Figure 25.58 This exploded isometric assembly shows how parts are to be put together.

Figure 25.56 This three-dimensional drawing has been drawn to show fillets and rounds, and dimensions and notes are given so it can be used as a working drawing.

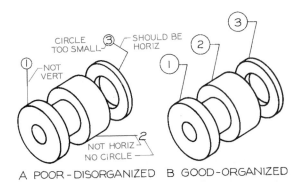

A POOR-DISORGANIZED   B GOOD-ORGANIZED

Figure 25.57 This drawing shows (A) common mistakes in applying leaders and part numbers in balloons of an assembly, and (B) acceptable techniques of applying leaders and part numbers to an assembly.

applying them. The numbers in the balloons correspond to the part numbers in the parts list. **Figure 25.58** shows an exploded assembly that illustrates the relationship of four mating parts. Illustrations of this type are excellent for inclusion in parts catalogs and maintenance manuals.

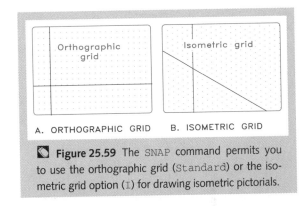

A. ORTHOGRAPHIC GRID   B. ISOMETRIC GRID

Figure 25.59 The SNAP command permits you to use the orthographic grid (Standard) or the isometric grid option (I) for drawing isometric pictorials.

Isometrics by Computer

AutoCAD provides an ISOMETRIC grid for drawing isometrics. The STYLE option of the SNAP command allows changing the rectangular GRID, called STANDARD (S), to ISOMETRIC (I) with dots shown vertically and at 30° to the horizontal (**Fig. 25.59**). In this mode, you can make the cursor's cross hairs SNAP to the grid points and align with the axes of isometric drawings.

Isometric drawings made with this system (**Fig. 25.60**) are not a true three-dimensional drawings system. Instead, they are two-dimensional isometrics that cannot be rotated to show other views.

The ISOCIRCLE option of the ELLIPSE command draws isometric ellipses automatically.

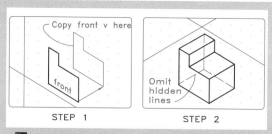

Figure 25.60 Producing isometrics by computer:
Step 1 Set the isometric grid on the screen (SNAP and I), and set SNAP to the grid. Draw the front view as an isometric and copy it to the backside with the COPY command.

Step 2 Connect the visible corner points and ERASE hidden lines to complete the drawing.

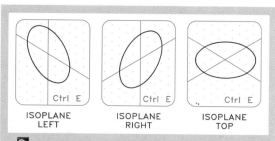

Figure 25.61 Use the ELLIPSE command and the ISOCIRCLE option to draw circles in isometric. By pressing Ctrl-E, you may alternatively rotate the isometric ellipses 120° to fit the three isometric planes. You may use the ISOPLANE command instead of Ctrl-E for the same purpose.

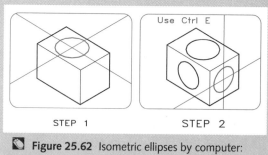

Figure 25.62 Isometric ellipses by computer:
Step 1 Use SNAP's Isometric-grid mode to draw isometric ellipses.

Command: ELLIPSE (CR)

<Axis endpoint 1>:/Center/
Isocircle: I (CR)

Center of Circle: (Select with cursor.)

<Circle radius>/Diameter: (Select radius with cursor.)

Step 2 Change the orientation of the cursor for drawing isometric ellipses on the other two planes by pressing Ctrl-E. Repeat the process in Step 1.

When using this command, the cursor is aligned with each of the three isometric planes by pressing Ctrl-E on the keyboard (**Fig. 25.61**). When the cursor is aligned with the proper axes of an isometric plane, you may select the center of the isometric ellipse or its diameter's endpoints (**Fig. 25.62**).

The ISOPLANE command changes the position of the cursor in the same way Ctrl-E does. ISOPLANE will prompt you to select from

Left/Top/Right/<Toggle>: options. To use the Toggle option, press (CR) to successively move the cursor position from plane to plane.

25.7 Axonometric Projection

An axonometric projection is a type of orthographic projection in which the **pictorial view is projected perpendicularly onto the picture plane with parallel projectors**. The object is positioned at an angle to the picture plane so that its pictorial projection will be a three-dimensional view. The three types of axonometric projections are: (1) **isometric**, (2) **dimetric**, or (3) **trimetric (Fig. 25.63)**.

Recall that the isometric projection is the type of pictorial in which the diagonal of a cube is seen as a point, the three axes and planes of the cube are equally foreshortened, and the axes are equally spaced 120° apart. Measurements along the three

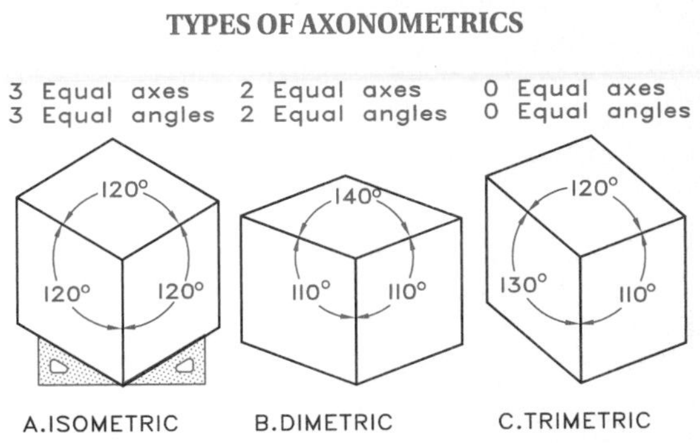

3 Equal axes   2 Equal axes   0 Equal axes
3 Equal angles   2 Equal angles   0 Equal angles

120°   140°   120°

120° 120°   110° 110°   130° 110°

A.ISOMETRIC   B.DIMETRIC   C.TRIMETRIC

Figure 25.63 This drawing illustrates the three types of axonometric projection.

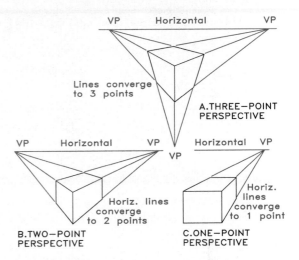

Figure 25.64 This drawing compares three-point, two-point, and one-point perspectives.

axes will be equal but less than true length because the isometric projection is true projection.

A **dimetric projection** is a pictorial in which two planes are equally foreshortened and two of the axes are separated by equal angles. Measurements along two axes of the cube are equal.

A **trimetric projection** is a pictorial in which all three planes are unequally foreshortened. The lengths of the axes are unequal, and the angles between them are different.

25.8 Perspective Pictorials

A perspective pictorial most closely resembles the view seen by the eye or camera and is the most realistic form of pictorial. In a perspective, parallel lines converge at **vanishing points** (VPs) as the lines recede from the observer. The three basic types of perspectives are (1) **one point**, (2) **two point**, and (3) **three point**, depending on the number of vanishing points used in their construction **(Fig. 25.64)**.

One-point perspectives have one surface of the objective that is parallel to the picture plane, making it a true shape. The other sides vanish to a single vanishing point on the horizon.

Two-point perspectives are positioned with two sides at an angle to the picture plane, requiring two vanishing points. All horizontal lines con-

verge at the vanishing points on the horizon, but vertical lines remain vertical and have no vanishing point.

Three-point perspectives have three vanishing points because the object is positioned so that all of its sides make an angle with the picture plane. Three-point perspectives are used in drawing large objects, such as buildings. They are the most realistic perspectives and the most complex to draw. Because of their complexity, we do not show how to construct them in this section.

Construction of One-Point Perspectives

Figure 25.65 shows how to draw a one-point perspective. It shows the top and side views of the object, picture plane, station point, horizon, and ground line. The **picture plane** (PP) appears as an edge in the top view and is the plane onto which the perspective is projected.

The **station point** (SP) is the location of the observer's eye in the top view and lies on the horizon in the front view. The **horizon** is a horizontal line in the front view that represents an infinite horizontal, such as the surface of the ocean, and is

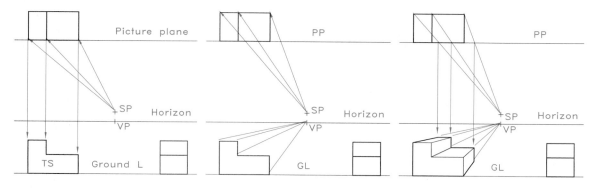

Figure 25.65 Constructing a one-point perspective:

Step 1 The object is parallel to the picture plane, so there will be only one vanishing point, located on the horizon below the station point. Projections from the top and side views establish the front plane, which is true size because it lies in the picture plane.

Step 2 Draw projectors from the station point to the rear points of the object in the top view and from the front view to the vanishing point on the horizon. In a one-point perspective the vanishing point is the front view of the station point.

Step 3 Construct vertical projectors from the top view to the front view from the points where the projectors cross the picture plane. These projectors intersect the lines leading to the single vanishing point.

aligned with the viewer's eye. The **ground line** (GL) is an infinite horizontal line parallel to the horizon from which vertical measurements are made.

When drawing any perspective, you should position the station point far enough away from the object that the perspective can be contained in a cone of vision of 30° or less (**Fig. 25.66**). A larger cone of vision will distort the perspective.

Constructing Two-Point Perspectives

If two surfaces of an object are positioned at angles to the picture plane, two vanishing points are required to draw it as a perspective. Placing the horizon above the ground line and the height of the object in the front view yields an **aerial view** (**Fig. 25.67**). Placing the ground line and horizon on top of each other in the front view gives a **ground-level view** (worm's-eye view). Placing the horizon above the ground line and through the

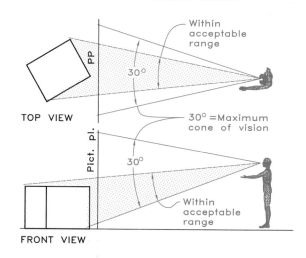

ACCEPTABLE CONE OF VISION

Figure 25.66 The station point should be far enough away from the object to permit the cone of vision to be less than 30° to reduce distortion.

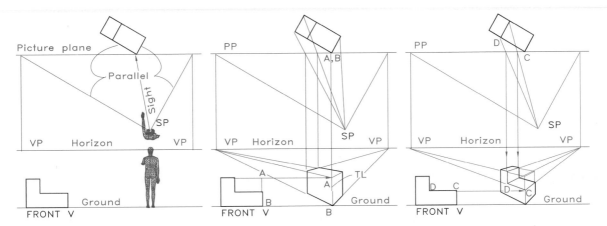

Figure 25.67 Constructing a two-point perspective:

Step 1 Extend projectors from the top view of the station point to the picture plane parallel to the forward edges of the object. Project these points vertically to the horizon in the front view to locate vanishing points. Draw the ground line below the horizon and construct the side view on the ground line.

Step 2 Lines in the picture plane are true length, so AB is true length. Project AB from the side view to determine its height. Project each end of AB to the vanishing points. Draw projectors from the station point to the exterior edges of the top view and project the intersections of these projectors with the picture plane to the front view.

Step 3 Find point C in the front view by projecting from the side view to line AB. Draw a projector from point C to the left vanishing point. Point D lies on this projector beneath the point where a projector from the station point to the top view of point D crosses the picture plane. Complete the notch by projecting to the respective vanishing points.

object, usually at a person's height for large objects such as buildings, results in a **general view**.

Figure 25.67 shows how to construct a two-point perspective. Because line AB lies in the picture plane, it will be true length in the perspective. All height dimensions originate at this vertical line because it is the only true-length line of the object.

The object shown in **Fig. 25.68** does not come into contact with the picture plane in the top view as it does in Fig. 25.67. To draw a perspective of this object, the planes of the object are extended to the picture plane. Measure the height on this line and draw an infinite plane to the right vanishing point. Locate the corner of the object on this

infinite plane by projecting the object's right corner to the picture plane in the top view with a projector from the station point and then projecting this point downward to the infinite plane.

Arcs

Draw arcs in perspective by using coordinates to locate points along the curves (**Fig. 25.69**). Transfer points 1–7 from the semicircular arc in the orthographic view to the perspective by projecting coordinates from the top and side views. All heights are projected to the TL line from the orthographic view. These points do not form a true ellipse but an egg-shaped oval. Connect the points with an irregular curve.

TWO-POINT PERSPECTIVE

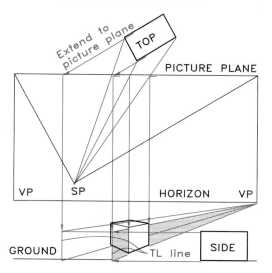

Figure 25.68 This two-point perspective is of an object that does not come into contact with the picture plane.

ARC IN PERSPECTIVE

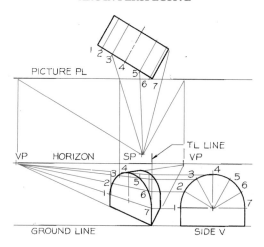

Figure 25.69 This two-point perspective shows an object having a semicircular feature.

25.9 Computer Software for Three-Dimensional Graphics

The number of software programs available for producing true three-dimensional (3D) pictorials on the microcomputer is increasing rapidly. Here, we

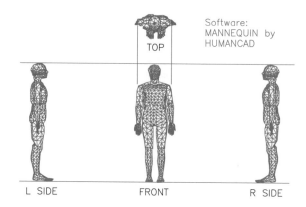

Figure 25.70 MANNEQUIN by HUMANCAD® can be used to draw human figures in three dimensions. (Courtesy of HUMANCAD.)

cover briefly three of these software programs: (1) MANNEQUIN, (2) Design Board Professional®, and (3) DynaPerspective®. AutoCAD's 3D options and solid modeling are covered further in Chapter 36.

MANNEQUIN

MANNEQUIN is a software package produced by HUMANCAD that may be used by itself or in conjunction with other software, including AutoCAD. MANNEQUIN provides a data base of three-dimensional males and females—large and small, young and old—that can be placed in a drawing to determine how well a design fits the users it has been developed for.

Figure 25.70 shows an example of a heavy man of medium height in top, front, left-side, and right-side views. These figures are true 3D figures that can be manipulated one finger at a time or one limb at a time from the keyboard. They can be moved to different positions and made to bend, reach, walk, and even see. For example, the mannequin can show you what a person would see when placed in a particular position, and this view will appear on the computer screen.

You can walk the mannequin across the screen in orthographic, isometric, or perspective. You can make it walk in a straight line or on a curved

Figure 25.71 Human figures drawn by MANNEQUIN can be made to move and walk in orthographic, isometric, and perspective. (Courtesy of HUMANCAD.)

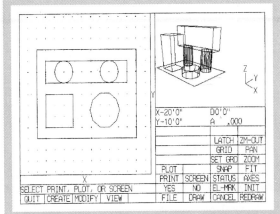

Figure 25.72 The CREATE screen of MegaCADD's Design Board Professional® enables you to create the orthographic views of a model that you want to convert to perspective views.

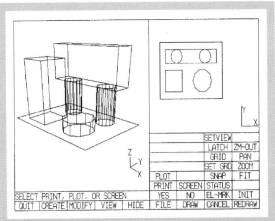

Figure 25.73 The VIEW screen of DBPro® shows the plan view of the model and its corresponding perspective view. You may change the viewpoint by selecting different observers' positions and target positions in the plan view.

path. **Figure 25.71** shows an example of movements of a 3D mannequin.

The ultimate challenge of graphics is to draw the human form in a natural and believable way. With the development of computer software packages such as MANNEQUIN, the goal of drawing the human figure easily and effectively has been largely attained.

Design Board Professional

Design Board Professional®, developed by MegaCADD, Inc., provides three modes for developing orthographic and perspective drawings of objects: CREATE, VIEW, and MODIFY. The CREATE mode (**Fig. 25.72**) provides a screen on which you can create a perspective by drawing 3D shapes in plan view.

The VIEW mode provides a screen on which you may select a viewpoint and look at the orthographic views (**Fig. 25.73**). The resulting perspective view appears in the large area at the left of the screen. The MODIFY option gives three ortho-

graphic views of the model along with its perspective view. You may delete, change, move, enlarge, and edit parts of the drawing in several other ways. The DRAW command produces much of the finished drawing.

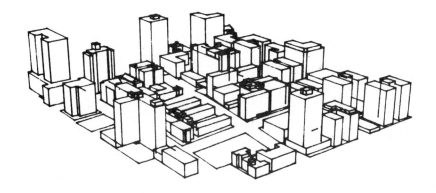

Figure 25.74 This perspective drawing (with hidden lines removed) of an urban area was plotted with MegaCADD's Design Board 3D. (Courtesy of MegaCADD, Incorporated.)

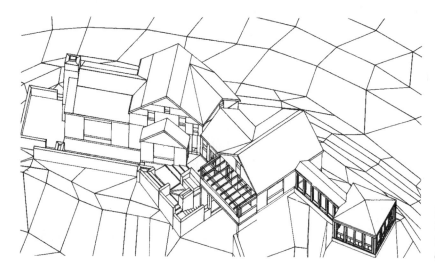

Figure 25.75 This perspective of a house and its site was produced with DynaPerspective® software as a 3D drawing.

Figure 25.74 shows an example of a plotted drawing produced with MegaCADD's Prohouse—a 3D representation of a city's downtown area. You may select viewpoints for different views of the city and generate a series of views to represent a "walk-through" by changing viewpoints and target points.

DynaPerspective

Developed by Dynaware Corporation, Dyna-Perspective is a powerful program that produces 3D orthographic and perspective drawings. You may use this solid-modeling package to draw, color, and rotate objects about three axes.

A portion of a complex example is the 3D drawing of a house positioned on an irregular site shown in **Fig. 25.75**. Because the drawing is truly

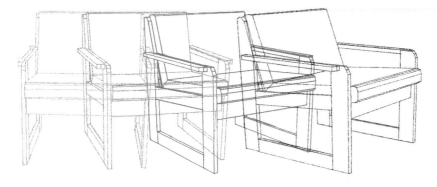

Figure 25.76 DynaPerspective software enables you to rotate objects about three axes. In this case, the chair is being rotated about its vertical axis.

3D, you may view the house both from the outside and from within.

Furniture pictorials, such as that of the chair shown in **Fig. 25.76**, are part of the software provided for architects and designers. They can be inserted in the drawing or viewed as orthographics or perspectives on the screen. After inserting it in the drawing, you may rotate the chair (as well as other furniture) about any of the three axes.

The Future
The future of 3D graphics is truly exciting. What is available today for the microcomputer was not possible even on much larger and more expensive computers just a few years ago. No doubt the capabilities of 3D programs will rapidly become more powerful and easier to use. Future graphics will include solid modeling, animation, and sound effects. Get ready for an exciting trip!

Problems

Draw your solutions to the following problems (**Fig. 25.77**) on size A or B sheets, as assigned. Select an appropriate scale to take advantage of the space available on each sheet. By letting each square represent 0.20 in. (5 mm), you can draw two solutions on each size A sheet. By setting each square to 0.40 inch (10 mm), you can draw one solution on each size B sheet.

Oblique Pictorials
1–24. Construct cavalier, cabinet, or general obliques of the parts assigned.

Isometric Pictorials
1–24. Construct isometrics of the parts assigned.

Perspective Pictorials
1–24. On size B sheets lay out perspective views of the parts assigned.

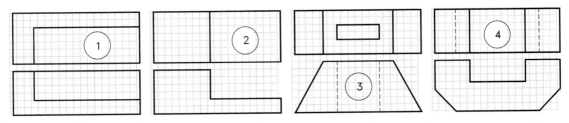

Figure 25.77 Problems 1–24.

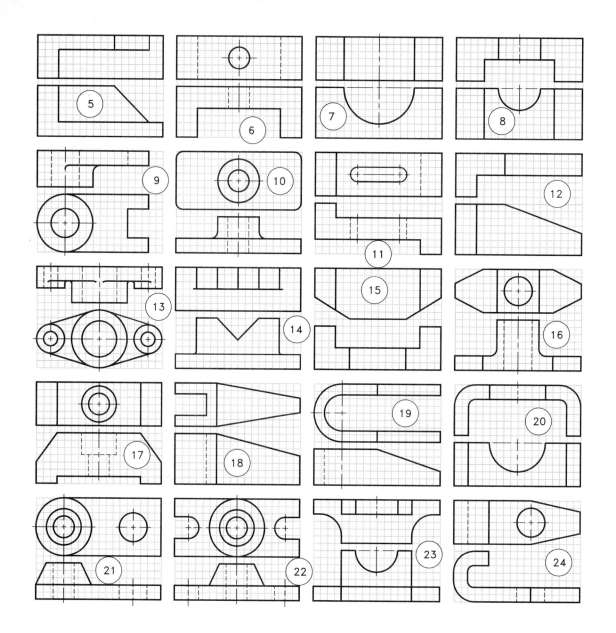

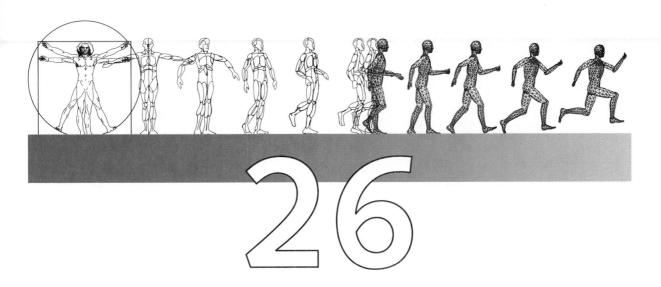

Points, Lines, and Planes

26.1 Introduction

Points, lines, and planes are the basic geometric elements used in three-dimensional (3D) spatial geometry, called descriptive geometry. You need to understand how to locate and manipulate these elements in their simplest form because they will be applied to 3D spatial problems in Chapters 26 through 31.

The unmanned spacecraft designed to explore Mars comprises many points, lines, and planes. Its geometry had to be established with great precision in order for it to function properly **(Fig. 26.1)**.

The labeling of points, lines, and planes is an essential part of 3D projection because it is your means of analyzing their spatial relationships. **Figure 26.2** illustrates the fundamental requirements for properly labeling these elements in a drawing:

> **Lettering:** use 1/8-inch letters with guidelines for labels; label lines at each end and planes at each corner with either letters or numbers.

Figure 26.1 The geometry of points, lines, and planes must be established with great precision in the design of this unmanned Mars orbiter spacecraft. (Courtesy of The Boeing Company.)

> **Points:** mark with two short perpendicular dashes forming a cross, not a dot; each dash should be approximately 1/8 inch long.

481

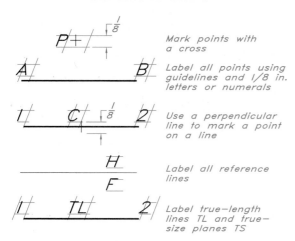

Mark points with a cross

Label all points using guidelines and 1/8 in. letters or numerals

Use a perpendicular line to mark a point on a line

Label all reference lines

Label true-length lines TL and true-size planes TS

Figure 26.2 These are standard practices for labeling points, lines, and planes.

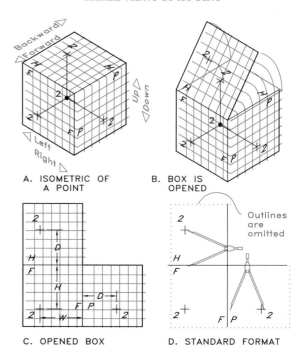

A. ISOMETRIC OF A POINT

B. BOX IS OPENED

C. OPENED BOX

D. STANDARD FORMAT

Figure 26.3

A This drawing shows three projections of point 2 pictorially.

B The projection planes are opened into a single plane.

C In the opened box, point 2 is 5 units to the left of the profile, 5 units below the horizontal, and 4 units behind the frontal view.

D The outlines of the projection are omitted in orthographic projection.

Points on lines: mark with a short perpendicular dash crossing the line, not a dot.

Reference lines: label these thin dark lines as described in Chapter 14.

Object lines: draw these lines used to represent points, lines, and planes heavier than reference lines with an H or F pencil; draw hidden lines thinner than visible lines.

True-length lines: label TRUE LENGTH or TL.

True-size planes: label TRUE SIZE or TS.

Projection lines: draw precisely with a 2H or 4H pencil as thin lines, just dark enough to be visible so they need not be erased.

26.2 Projection of Points

A **point** is a theoretical location in space having no dimensions other than its location. However, a series of points establishes lengths, areas, and volumes of complex shapes.

A point must be located in at least two adjacent orthographic views to establish its position in 3D space (**Fig. 26.3**). When the planes of the projection box (**Fig. 26.3A**) are opened onto the plane of the drawing surface (**Fig. 26.3C**), the projectors from each view of point 2 are perpendicular to the reference lines between the views. Letters H, F, and P represent the horizontal, frontal, and profile planes, the three principal projection planes.

A point may be located from verbal descriptions with respect to the principal planes. For

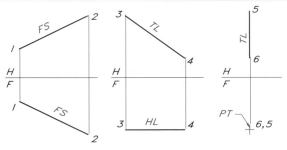

A. FORESHORTENED   B. TRUE LENGTH   C. POINT

Figure 26.4 A line in orthographic projection can appear as foreshortened (FS), true length (TL), or a point (PT).

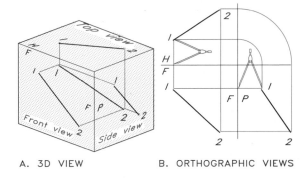

A. 3D VIEW   B. ORTHOGRAPHIC VIEWS

Figure 26.5

A This pictorial shows three views of a line in 3D space.

B These are the standard three orthographic views of a line.

example, point 2 in Fig. 26.3 may be described as being (1) 5 units left of the profile plane, (2) 5 units below the horizontal plane, and (3) 4 units behind the frontal plane.

When you look at the front view of the box, the horizontal and profile planes appear as edges. In the top view, the frontal and profile planes appear as edges. In the side view, the frontal and horizontal planes appear as edges.

26.3 Lines

A **line** is the straight path between two points in 3D space. A line may appear as (1) foreshortened, (2) true-length, or (3) a point **(Fig. 26.4).**

Oblique lines are neither parallel nor perpendicular to a principal projection plane **(Fig. 26.5).** When line 1–2 is projected onto the horizontal, frontal, and profile planes, it appears foreshortened in each view.

Principal lines are parallel to at least one of the principal projection planes. A principal line is true length in the view where the principal plane to which it is parallel appears true size. The three types of principal lines are **horizontal, frontal,** and **profile** lines.

Figure 26.6A shows a **horizontal line** (HL) which appears true length in the horizontal (top) view. Any line shown in the top view will appear true length as long as it is parallel to the horizontal plane.

When looking at the top view, you cannot tell whether the line is horizontal. You must look at the front or side views to do so. In those views, an HL will be parallel to the edge view of the horizontal, the HF fold line **(Fig 26.7).** A line that projects as a point in the front view is a combination horizontal and profile line.

A **frontal line** (FL) is parallel to the frontal projection plane. It appears true length in the front view because your line of sight is perpendicular to it in this view. In **Fig. 26.6B** line 3–4 is an FL because it is parallel to the edge of the frontal plane in the top and side views.

A **profile line** (PL) is parallel to the profile projection planes and appears true length in the side (profile) views. To tell whether a line is a PL, you must look at a view adjacent to the profile view, or the top or front view. In **Fig. 26.6C**, line 5–6 is parallel to the edge view of the profile plane in both the top and side views.

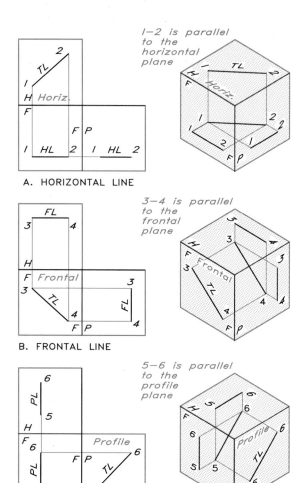

A. HORIZONTAL LINE

B. FRONTAL LINE

C. PROFILE LINE

1-2 is parallel to the horizontal plane

3-4 is parallel to the frontal plane

5-6 is parallel to the profile plane

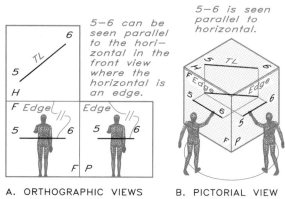

5-6 can be seen parallel to the horizontal in the front view where the horizontal is an edge.

5-6 is seen parallel to horizontal.

A. ORTHOGRAPHIC VIEWS   B. PICTORIAL VIEW

Figure 26.7 In order to determine that a line is horizontal, you must use the front or side views in which the horizontal projection plane is an edge. Line 5–6 is seen parallel to the horizontal and is therefore a horizontal line, too.

LOCATE A POINT ON A LINE

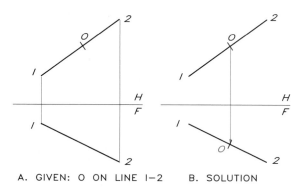

A. GIVEN: O ON LINE 1–2   B. SOLUTION

Figure 26.8 Point O on the top view of line 1–2 may be found in the front view by projection. The projector is perpendicular to the HF reference line between the views.

Figure 26.6 Principal lines:

A The horizontal line is true length in the horizontal (top) view. It is parallel to the edge view of the horizontal plane in the front and side views.

B The frontal line is true length in the front view. It is parallel to the edge view of the frontal plane in the top and side views.

C The profile line is true length in the profile (side) view. It is parallel to the edge view of the profile plane in the top and front views.

Locating a Point on a Line
Figure 26.8 shows the top and front views of a line 1–2 with point O located at its midpoint. To find the front view of the point, recall that, in orthographic projection, the projector between the views is perpendicular to the HF fold line. Use

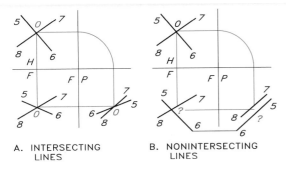

Figure 26.9

A These lines intersect because O, the point of intersection, projects as a common point of intersection in all views.

B The lines cross in the top and front views, but they do not intersect because there is no common point of intersection in all views.

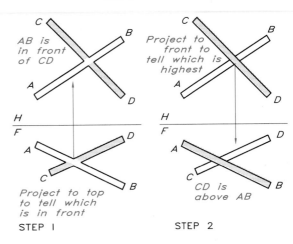

Figure 26.10 Determining visibility of lines:
Required Find the visibility of the lines in both views.

Step 1 Project the crossing point from the front to the top view. This projector strikes line AB before it strikes line CD, indicating that line AB is in front and thus is visible in the front view.

Step 2 Project the crossing point from the top view to the front view. This projector strikes line CD before it strikes line AB, indicating that line CD is above line AB and thus is visible in the top view.

that projector to project point O to line 1–2 in the front view. A point located at a line's midpoint will be at the line's midpoint in all orthographic views of the line.

Intersecting and Nonintersecting Lines

Lines that intersect have a common point of intersection lying on both lines. Point O in **Fig. 26.9A** is a point of intersection because it projects to a common crossing point in all three views.

However, the crossing point of the lines in **Fig. 26.9B** in the top and front views is not a point of intersection. Point O does not project to a common crossing point in the top and front views, so the lines do not intersect; they simply cross, as shown in the profile view.

26.4 Visibility

Crossing Lines

In **Fig. 26.10** nonintersecting lines AB and CD cross in certain views. Therefore portions of the

lines are visible or hidden at the crossing points (here, line thickness is exaggerated for purposes of illustration). Determining which line is above or in front of the other is referred to as finding a line's *visibility,* a requirement of many 3D problems.

You have to determine line visibility by analysis. For example, select a crossing point in the front view and project it to the top view to determine which line is in front of the other. Because the projector contacts line AB first, you know that line AB is in front of CD and is visible in the front view.

Repeat this process by projecting downward from the intersection in the top view to find that line CD is above line AB and is visible in the top view. If only one view were available, visibility would be impossible to determine.

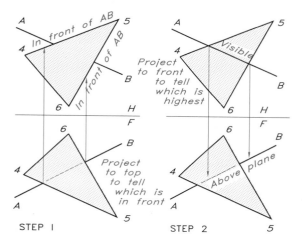

STEP 1   STEP 2

Figure 26.11 Determining visibility of a line and a plane:
Required Find the visibility of the plane and the line in both views.

Step 1 Project the points where line AB crosses the plane from the front view to the top view. These projectors intersect lines 4–6 and 5–6 of the plane first, indicating that the plane is in front of the line and making line AB hidden in the front view.

Step 2 Project the points where line AB crosses the plane in the top view to the front view. These projectors encounter line AB first, indicating that line AB is higher than the plane and making the line visible in the top view.

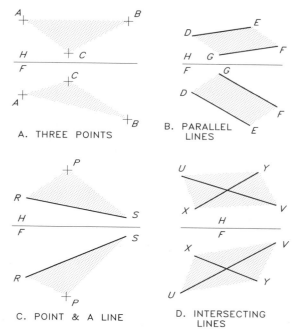

A. THREE POINTS   B. PARALLEL LINES

C. POINT & A LINE   D. INTERSECTING LINES

Figure 26.12 A plane may be represented as (A) three points not on a straight line, (B) two parallel lines, (C) a line and a point not on the line or its extension, and (D) two intersecting lines.

A Line and a Plane

The principles of visibility analysis also apply to determining visibility for a line and a plane (**Fig. 26.11**). First, project the intersections of line AB with lines 4–5 and 5–6 to the top view to determine that the lines of the plane (4–5 and 5–6) lie in front of line AB in the front view. Therefore line AB is a hidden line in the front view.

Similarly, project the two intersections of line AB in the top view to the front view, where line AB is found to lie above lines 4–5 and 5–6 of the plane. Because line AB is above the plane, it is a visible line in the top view.

26.5 Planes

A **plane** may be represented in orthographic projection by any of the four combinations shown in **Fig. 26.12**. In orthographic projection, a plane may appear as (1) **an edge**, (2) **a true-size plane**, or (3) **a foreshortened plane** (**Fig. 26.13**).

Oblique planes (the general case) are not parallel to principal projection planes in any view (**Fig. 26.14**). **Principal planes** are parallel to principal projection planes (**Fig. 26.15**). The three types of principal planes are **horizontal, frontal,** and **profile** planes.

A **horizontal plane** is parallel to the horizontal projection plane and is true size in the top view

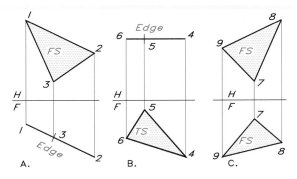

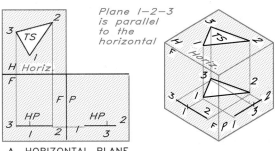

A. HORIZONTAL PLANE

Figure 26.13 A plane in orthographic projection can appear as (A) an edge, (B) true size (TS), or (C) foreshortened (FS). A plane that is foreshortened in all principal views is an oblique plane.

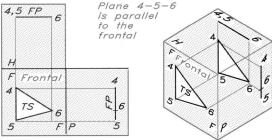

B. FRONTAL PLANE

OBLIQUE PLANE—GENERAL CASE

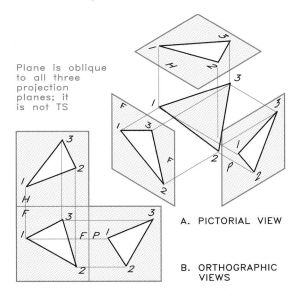

A. PICTORIAL VIEW

B. ORTHOGRAPHIC VIEWS

Figure 26.14 An oblique plane is neither parallel nor perpendicular to a projection plane. It is the general-case plane.

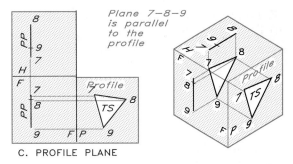

C. PROFILE PLANE

Figure 26.15 Principal planes:
A The horizontal plane is true size in the horizontal (top) view. It is parallel to the edge view of the horizontal plane in the front and profile views.

B The frontal plane is true size in the front view. It is parallel to the edge view of the frontal plane in the top and profile views.

C The profile plane is true size in the profile view. It is parallel to the edge view of the profile plane in the top and front views.

(Fig. 26.15A). To determine that the plane is horizontal, you must observe the front or profile views, where you can see its parallelism to the edge view of the horizontal plane.

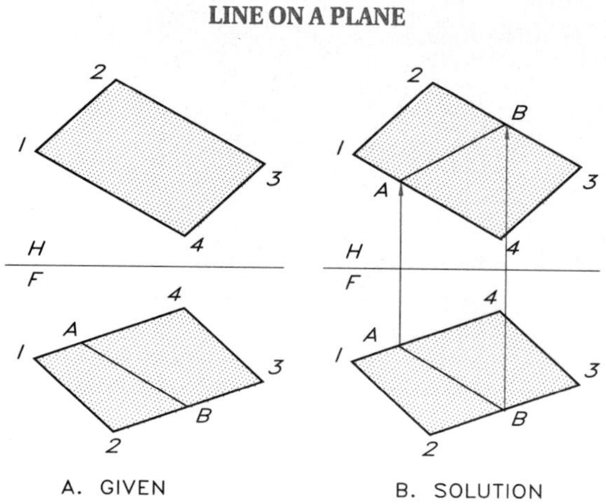

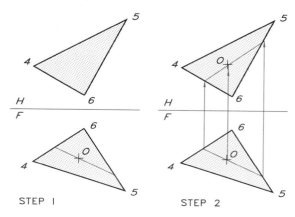

A. GIVEN   B. SOLUTION

Figure 26.16 If line AB lying on the plane is given, the line's top view can be found. Project points A and B to lines 1–4 and 2–3 and connect them to get the top view of line AB.

STEP 1   STEP 2

Figure 26.17 Locating a point on a plane:
Required Find the top view of point O on the plane.

Step 1 In the front view, draw a line through point O in any convenient direction except vertical.

Step 2 Project the ends of the line to the top view and draw the line. Project point O to this line.

A **frontal plane** is parallel to the frontal projection plane and appears true size in the front view (**Fig. 26.15B**). To determine that the plane is frontal, you must look at the top or profile views, where you can see its parallelism to the edge view of the frontal plane.

A **profile plane** is parallel to the profile projection plane and is true size in the side view (**Fig. 26.15C**). To determine that the plane is profile, you must observe the top or front views, where you can see its parallelism to the edge view of the profile plane.

A Line on a Plane

Line AB on the plane in the front view is given in **Fig. 26.16**, and it is to be found in the top view. Project points A and B, lying on lines 1–4 and 2–3, to the top view of lines 1–4 and 2–3. Connecting points A and B gives the top view of line AB.

A Point on a Plane

Point O on the front view of plane 4–5–6 in **Fig. 26.17** is to be located on the plane in the top view. First, draw a line in any direction (except vertical)

through the point to establish a line on the plane. Then project this line to the top view and project point O from the front view to the top view of the line.

Principal Lines on a Plane

Principal lines may be found in any view of a plane when at least two orthographic views of the plane are given. Any number of principal lines can be drawn on any plane.

Figure 26.18A shows a **horizontal line** parallel to the edge view of the horizontal projection plane in the front view. When projected to the top view, this line is true length.

Figure 26.18B shows a **frontal line** parallel to the edge view of the frontal projection plane in the top view. When projected to the front view, this line is true length.

Figure 26.18C shows a **profile line** parallel to the edge view of the profile projection plane in the front. When projected to the profile view, this line is true length.

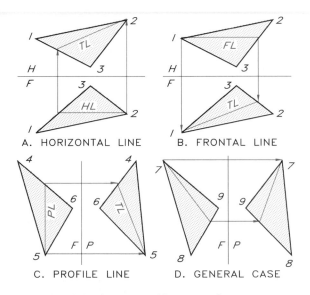

Figure 26.18 Finding principal lines on a plane:

A First, draw a horizontal line in the front view parallel to the edge view of the horizontal plane. Then, project it to the top view, where it is true length.

B First, draw a frontal line in the top view parallel to the edge view of the frontal plane. Then, project it to the front view, where it is true length.

C First, draw a profile line in the front view parallel to the edge view of the profile plane. Then project it to the profile view, where it is true length.

D A general-case line is not parallel to the frontal, horizontal, or profile planes and is not true length in any principal view.

In the **general case (oblique)**, a line is not parallel to the edge view of any principal projection plane **(Fig. 26.18D)**. Therefore it is not true length in any principal view.

26.6 Parallelism

Lines

Two parallel lines appear parallel in all views, except in views where both appear as points. Parallelism of lines in 3D space cannot be deter-

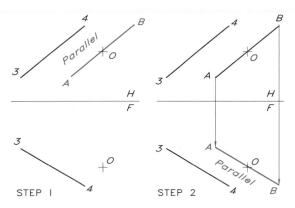

Figure 26.19 Constructing a line parallel to a line:
Required Draw a line through point O parallel to line 3–4.

Step 1 Draw line AB parallel to the top view of line 3–4, with its midpoint at O.

Step 2 Draw the front view of line AB parallel to the front view of 3–4 through point O.

mined without at least two adjacent orthographic views. In **Fig. 26.19**, line AB was drawn parallel to the horizontal view of line 3–4 and through point O, which is the midpoint. Projecting points A and B to the front view with projectors perpendicular to the HF reference plane yields the length of line AB, which is parallel to line 3–4 through point O.

A Line and a Plane

A line is parallel to a plane when it is parallel to any line in the plane. In **Fig. 26.20**, a line with its midpoint at point O is to be drawn parallel to plane 1–2–3. In this case line AB was drawn parallel to a line in the plane, or line 1–3, in the top and front views. The line could have been drawn parallel to any line in the plane, making infinite solutions possible.

Figure 26.21 shows a similar example. Here, a line parallel to the plane, with its midpoint at O, was drawn. In this case, the plane is represented by two intersecting lines instead of an outlined area.

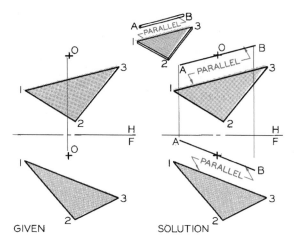

Figure 26.20 A line may be drawn through point O parallel to plane 1–2–3 if the line is parallel to any line in the plane. Draw line AB parallel to line 1–3 of the plane in the front and top views, making it parallel to the plane.

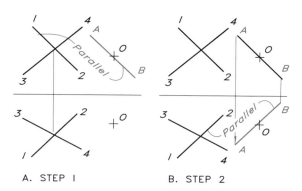

A. STEP I   B. STEP 2

Figure 26.21 Constructing a line parallel to plane: **Required** Draw a line through point O that is parallel to plane 1–2–3–4 represented by intersecting lines.

Step 1 Draw line AB parallel to line 1–2 through point O.

Step 2 Draw line AB parallel to the same line, line 1–2, in the front view, which makes line AB parallel to the plane.

Planes

Two planes are parallel when intersecting lines in one plane are parallel to intersecting lines in the other (Fig. 26.22). Determining whether planes are parallel is easy when both appear as edges in a view.

In **Fig. 26.23**, a plane is to be drawn through point O parallel to plane 1–2–3. First, draw line EF through point O parallel to line 1–2 in the top and front views. Then draw a second line through point O parallel to line 2–3 of the plane in the front and top views. These two intersecting lines form a plane parallel to plane 1–2–3, as intersecting lines on one plane are parallel to intersecting lines on the other.

26.7 Perpendicularity

Lines

When two lines are perpendicular, draw them with a true 90° angle of intersection in views where one or both of them appears true length

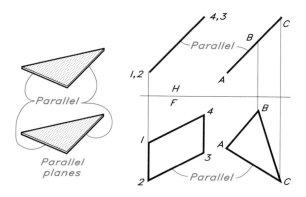

Figure 26.22 Two planes are parallel when intersecting lines in one are parallel to intersecting lines in the other. When parallel planes appear as edges, their edges are parallel.

(Fig. 26.24). In a view where neither of two perpendicular lines is true length, the angle between is not a true 90° angle.

In Fig. 26.24, the axis is true length in the front view; therefore any spoke of the circular wheel is perpendicular to the axis in the front view. Spokes

PLANE PARALLEL TO A PLANE

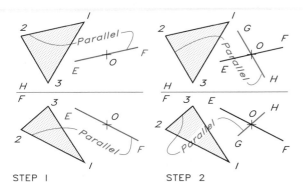

STEP 1   STEP 2

Figure 26.23 Constructing a plane through a point parallel to a plane:

Required Draw a plane through point O parallel to the given plane.

Step 1 Draw line EF parallel to any line in the plane (line 1–2 in this case). Show the line in both views.

Step 2 Draw a second line parallel to line 2–3 in the top and front views. These intersecting lines passing through O represent a plane parallel to 1–2–3.

PERPENDICULARITY: TWO LINES

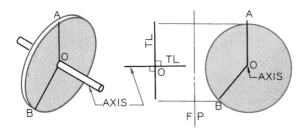

Figure 26.24 Perpendicular lines have a true angle of 90° between them in a view where one or both of them appear true length.

OA and OB are examples of true length foreshortened axes, respectively, in the front view.

A Line Perpendicular to a Principal Line In **Fig. 26.25**, a line is to be constructed through point O perpendicular to frontal line 5–6, which is true length

LINE PERPENDICULAR TO A LINE

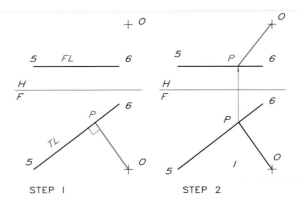

STEP 1   STEP 2

Figure 26.25 Constructing a line perpendicular to a principal line:

Required Draw a line from point O perpendicular to line 5–6.

Step 1 Line 5–6 is a frontal line and is true length in the front view, so a perpendicular from point O makes a true 90° angle with it in the front view.

Step 2 Project point P to the top view and connect it to point O. As neither line is true length in the top view, they do not intersect at 90° in this view.

in the front view. First, draw OP perpendicular to line 5–6 because it is true length. Then, project point P to the top view of line 5–6. In the top view, line OP is not perpendicular to line 5–6 because neither of the lines is true length in this view.

A Line Perpendicular to an Oblique Line In **Fig. 26.26**, a line is to be constructed from point O perpendicular to oblique line 1–2. First, draw a horizontal line from O to some convenient length in the front view, say, to E.

Locate point O in the top view by projection and draw line OE to make a 90° angle with the top view of 1–2. Line OE is true length in the top view, so it makes a true 90° angle with line 1–2.

Planes

A line is perpendicular to a plane when it is perpendicular to any two intersecting lines in the

LINE PERPENDICULAR TO AN OBLIQUE LINE

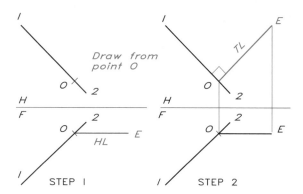

Figure 26.26 Constructing a line perpendicular to an oblique line:

Required Draw a line from point O on line 1–2 perpendicular to the line.

Step 1 Draw a horizontal line (OE) from point O in the front view.

Step 2 Horizontal line OE is true length in the top view, so draw it perpendicular to line 1–2 in this view.

PERPENDICULARITY: LINES & PLANES

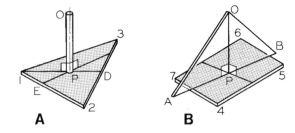

Figure 26.27

A A line is perpendicular to a plane when it is perpendicular to two intersecting lines on the plane.

B A plane is perpendicular to another plane if it contains a line that is perpendicular to the other plane.

plane (Fig. 26.27A). A plane is perpendicular to another plane when a line in one plane is perpendicular to the other plane (Fig. 26.27B).

A Line Perpendicular to a Plane

In **Fig. 26.28**, a line is to be drawn perpendicular to the plane from point O on the plane. First, draw a frontal line on the plane in the top view through O. Project the line to the front view, where it is true length. Draw line P at a convenient length perpendicular to the true-length line.

Then, draw a horizontal line through point O in the front view and project it to the top view of the plane, perpendicular to the true-length line. This construction results in a line perpendicular to the plane because the line is perpendicular to two intersecting lines, a horizontal and a frontal line, in the plane.

LINE PERPENDICULAR TO A PLANE

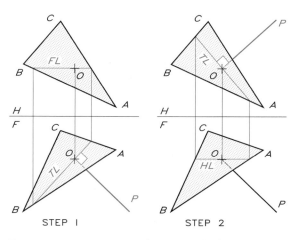

Figure 26.28 Constructing a line perpendicular to a plane:

Required From point O on the plane, draw a line perpendicular to the plane.

Step 1 Construct a frontal line on the plane through O in the top view. This line is true length in the front view, so draw line OP perpendicular to this true-length line.

Step 2 Construct a horizontal line through point O in the front view. This line is true length in the top view, so draw line OP perpendicular to it.

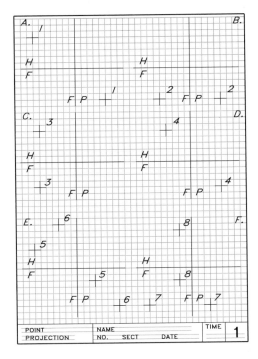

Figure 26.29 Problem 1 (A–F).

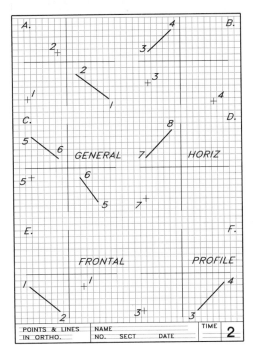

Figure 26.30 Problem 2 (A–F).

Use size A sheets for the following problems and lay out your solutions with instruments. Each square on the grid is equal to 0.20 in. or 5 mm. You can use either grid paper or plain paper. Label all reference planes and points in each problem with $^1/_8$-in. letters and numbers, using guidelines.

1. (Fig. 26.29)

(A–D) Draw three views (top, front, and right-side views) of the given point.

(E–F) Draw the three views of the points and connect them to form lines.

2. (Fig. 26.30)

(A–C) Draw three views (top, front, and right-side views) of the partially drawn lines.

(D–F) Draw the missing views of the lines so that 7–8 is a horizontal line, 1–2 is a frontal line, and 3–4 is a profile line.

3. (Fig. 26.31)

(A–B) Draw the right-side view of line 1–2 and plane 3–4–5.

(C–E) Draw the missing views of the planes so that 6–7–8 is a frontal plane, 1–2–3 is a horizontal plane, and 4–5–6 is a profile plane.

(F) Complete the top and side views of the plane that appears as an edge in the front view.

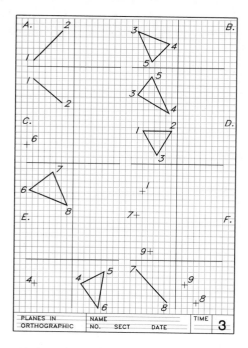

Figure 26.31 Problem 3 (A–F).

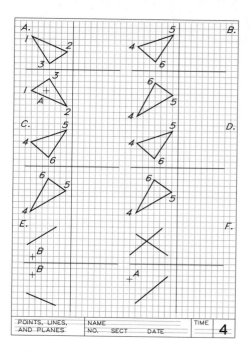

Figure 26.32 Problem 4 (A–F).

4. (Fig. 26.32)

   (A) Draw the side view of the plane and locate point A on the plane in all views.

   (B) Draw the side view and two horizontal lines on each view.

   (C) Draw the side view and two frontal lines on each view.

   (D) Draw the side view and two profile lines on each view.

   (E) Draw a line through B that is parallel to and equal in length to the given line.

   (F) Complete the three views of the intersecting lines.

5. (Fig. 26.33)

(A–B) Draw 1.50-in. lines that pass through point O and are parallel to their respective planes.

(C–D) Through point O draw the top and front views of lines that are perpendicular to their respective lines.

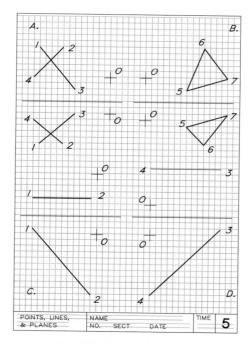

Figure 26.33 Problem 5 (A–D).

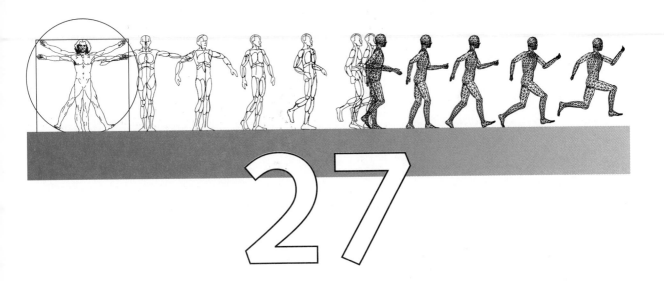

27

Primary Auxiliary Views in Descriptive Geometry

27.1 Introduction

Descriptive geometry is the projection of three-dimensional (3D) orthographic views onto a two-dimensional (2D) plane of paper to allow graphical determination of lengths, angles, shapes, and other geometric information. Orthographic projection is the basis for laying out and solving problems by descriptive geometry.

The **primary auxiliary view**, which permits analysis of 3D geometry, is essential to descriptive geometry. For example, the design of the spacecraft shown in **Fig. 27.1** contains many complex geometric elements (lines, angles, and surfaces) that were analyzed by descriptive geometry prior to its fabrication.

27.2 Descriptive Geometry by Computer

Five useful computer routines for solving descriptive geometry problems are covered in this section and documented in Appendix 47. The commands are `PERPLINE`, `PARALLEL`, `TRANSFER`, `COPY-`

Figure 27.1 This lunar vehicle could not have been designed without the use of descriptive geometry to determine lengths, angles, and areas of its component parts. (Courtesy of Ryan Aircraft Corporation.)

`DIST`, and `BISECT` and each is executed by typing in its name.[*]

---

[*]These LISP commands were written by Professor Leendert Kersten of the University of Nebraska at Lincoln.

495

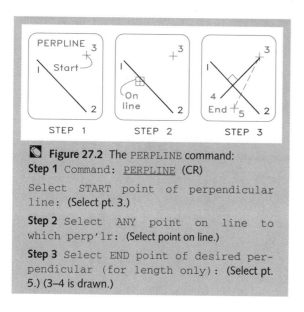

Figure 27.2 The PERPLINE command:
Step 1 Command: <u>PERPLINE</u> (CR)

Select START point of perpendicular line: (Select pt. 3.)

Step 2 Select ANY point on line to which perp'lr: (Select point on line.)

Step 3 Select END point of desired perpendicular (for length only): (Select pt. 5.) (3–4 is drawn.)

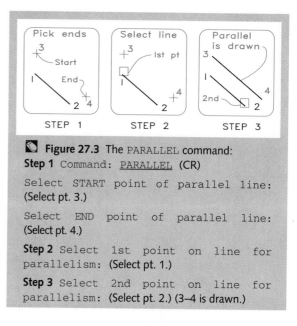

Figure 27.3 The PARALLEL command:
Step 1 Command: <u>PARALLEL</u> (CR)

Select START point of parallel line: (Select pt. 3.)

Select END point of parallel line: (Select pt. 4.)

Step 2 Select 1st point on line for parallelism: (Select pt. 1.)

Step 3 Select 2nd point on line for parallelism: (Select pt. 2.) (3–4 is drawn.)

Figure 27.2 shows how to draw line 3–4 perpendicular to line 1–2 with PERPLINE. Locate the starting point, select the line to which the constructed line is to be perpendicular, and pick a third point in the general area of the line's endpoint.

In **Fig. 27.3**, line 3–4 is drawn parallel to line 1–2 with PARALLEL. Locate the starting point of the parallel (point 3) and the general location of its endpoint (point 4). Select the endpoints of line 1–2 in the same order (from 1 to 2), the line is drawn.

The TRANSFER command (**Fig. 27.4**) transfers distances from reference lines in the same way in which dividers are used. Select endpoint 2 in the front view and the reference line. Select endpoint 2 in the top view and the auxiliary reference line. A circle locates the endpoint in the auxiliary view.

A distance may be copied from one position to another with COPYDIST, as **Fig. 27.5** shows. Select the endpoints of the line to be copied; locate its beginning point in the new position, and locate the direction of the copied line. A circle locates the endpoint of the line at the exact distance being copied.

The BISECT command can bisect an angle (**Fig. 27.6**) after selecting the vertex of the angle,

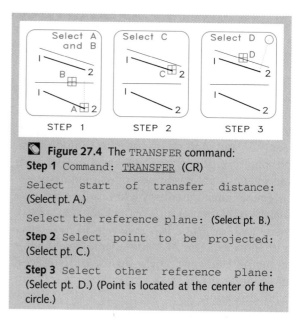

Figure 27.4 The TRANSFER command:
Step 1 Command: <u>TRANSFER</u> (CR)

Select start of transfer distance: (Select pt. A.)

Select the reference plane: (Select pt. B.)

Step 2 Select point to be projected: (Select pt. C.)

Step 3 Select other reference plane: (Select pt. D.) (Point is located at the center of the circle.)

the first line, and the second line. The second line must be picked counterclockwise from the first line. Locate the endpoint when prompted, and the bisector is drawn.

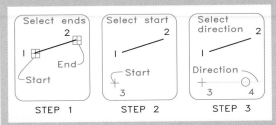

Figure 27.5 The COPYDIST command:
Step 1 Command: COPYDIST (CR)

Select start point of line distance to be copied: (Select end 1.)

End point?: (Select end 2.)

Step 2 Start point of new distance location: (Select pt. 3.)

Step 3 Which direction?: (Select with cursor, and endpoint 4 is located at the center of the circle.)

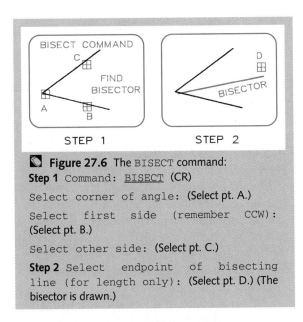

Figure 27.6 The BISECT command:
Step 1 Command: BISECT (CR)

Select corner of angle: (Select pt. A.)

Select first side (remember CCW): (Select pt. B.)

Select other side: (Select pt. C.)

Step 2 Select endpoint of bisecting line (for length only): (Select pt. D.) (The bisector is drawn.)

27.3 True-Length Lines

Primary Auxiliary View

Figure 27.7 shows the top and front views of line 1–2 pictorially and orthographically. Line 1–2 is not a princiʻal line, so it is not true length in a

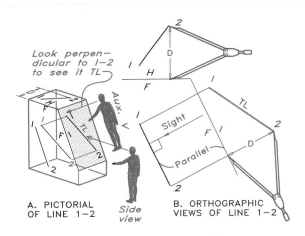

Figure 27.7

A A pictorial of line 1–2 is shown inside a projection box where an auxiliary plane is parallel to the line and perpendicular to the frontal plane.

B The auxiliary view is projected from the front orthographic view to find 1–2 true length.

principal view. Therefore a primary auxiliary view is required to find its true-length view.

In **Fig. 27.7A**, the line of sight is perpendicular to the front view of the line and reference line F1 is parallel to the line's frontal view. The auxiliary plane is parallel to the line and perpendicular to the frontal plane, accounting for its label, F1, where F and 1 are abbreviations for frontal and primary planes, respectively.

Projecting parallel to the line of sight and perpendicular to the F1 reference line yields the auxiliary view. Transferring distance D with dividers to the auxiliary view locates point 2 because the frontal plane appears as an edge in both the top and auxiliary views. Point 1 is located in the same manner, and the points are connected to find the true-length view of the line.

Figure 27.8 summarizes the steps of finding the true-length view of an oblique line. Letter all reference planes using the notation suggested in Chapter 26 and as shown in the examples

TRUE LENGTH OF A LINE

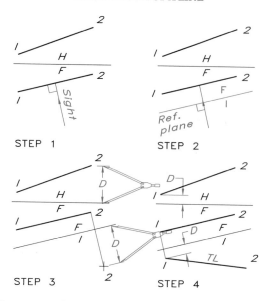

STEP 1

STEP 2

STEP 3

STEP 4

Figure 27.8 The true length of a line:

Step 1 To find the true length of line 1–2 the line of sight must be perpendicular to one of its views, the front view here.

Step 2 Draw the F1 reference line parallel to the line and perpendicular to the line of sight.

Step 3 Project point 2 perpendicularly from the front view. Transfer distance D from the top view to locate point 2 in the auxiliary view.

Step 4 Locate point 1 in the same manner to find line 1–2 true length in the auxiliary view.

throughout this chapter, with the exception of noted dimensions such as D. Use your dividers to transfer dimensions.

A primary auxiliary view projected parallel from a true-length view of a line in a principal view shows the point view of the line (**Fig. 27.9**). The auxiliary view projected from the line's front view gives a foreshortened view, not a point view, because the line is not true length in the front view.

⬧ **Computer Method** **Figure 27.10** shows how to use the PARALLEL and TRANSFER commands

POINT VIEW OF A LINE

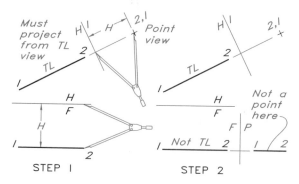

STEP 1

STEP 2

Figure 27.9 The point view of a line:

Step 1 The point view of a line is found in a primary auxiliary view projected from the true-length view of the line.

Step 2 An auxiliary view projected from a foreshortened view of a line gives a foreshortened view of the line, not its point view.

to find a line's true length by an auxiliary view. Draw a line parallel to the top view of line 1–2, and TRANSFER the endpoints of the line. Use the CENTER option of the OSNAP command to draw line 1–2 from the centers of the circles. Erase the circles afterward.

True Length by Analytical Geometry

Figure 27.11 shows how to find the true length of a frontal line (line 3–4) mathematically. The Pythagorean theorem states that the hypotenuse of a right triangle is equal to the square root of the sum of the squares of the other two sides. Because the line is true length in the front view, measuring that length provides a check on the mathematical solution.

The true length of a line shown pictorially in **Fig. 27.12A** (line 1–2) is determined by analytical geometry from its length in the front view where the X and Y distances form a right triangle. **Figure 27.12B** shows a second right triangle, 1–0–2, whose hypotenuse is the true length of line 1–2. Thus the true length of an oblique line is the square root of the sum of the squares of the X, Y, and Z distances that correspond to the width,

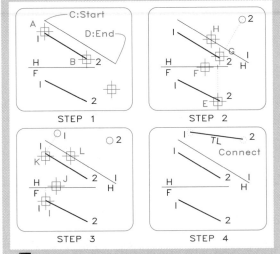

Figure 27.10 True length by computer:

Step 1 Command: <u>PARALLEL</u> (CR) (Follow the prompts in the caption of Fig. 27.3 to produce the reference line parallel to line 1–2 with PARALLEL.)

Step 2 Command: <u>TRANSFER</u> (CR) (Follow the prompts in the caption of Fig. 27.4 to locate point 2 in the auxiliary view with TRANSFER.)

Step 3 Locate point 1 in the auxiliary view by using TRANSFER.

Step 4 Draw a line from the centers of the circles obtained in the preceding steps by using the CENTER option of the OSNAP command. ERASE the circles after this step.

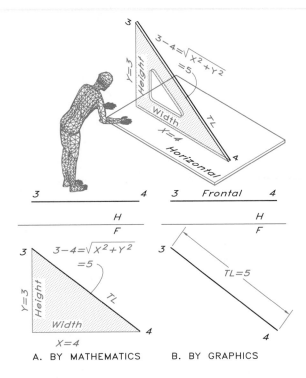

Figure 27.11 Apply the Pythagorean theorem to calculate the length of a line that appears true length in a view, the front view here. Because line 3–4 is true length in the front view, it can be measured to find its length graphically.

height, and depth of the triangles shown orthographically in **Fig. 27.12C**.

True-Length Diagram

A true-length diagram utilizes two perpendicular lines to find a line of true length (**Fig. 27.13**). This method does not give the line's direction, only its true length.

The two measurements laid out on the true-length diagram may be transferred from any two adjacent orthographic views. One measurement is the distance between the endpoints in one of the views. The other measurement, from the adjacent

view, is the distance between the endpoints perpendicular to the reference line between the two views. Here, these dimensions are vertical, V, and horizontal, H, between points 1 and 2.

27.4 Angles between Lines and Principal Planes

To measure the angle between a line and a plane, the line must appear true length and the plane as an edge in the same view (Fig. 27.14). A principal plane appears as an edge in a primary auxiliary view projected from it, so the angle a line makes with this principal plane can be measured if the line is true length in this auxiliary view.

TRUE-LENGTH LINES BY THE PYTHAGOREAN THEOREM

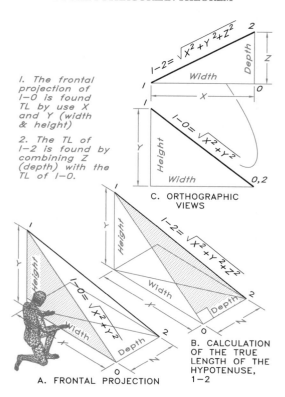

I. The frontal projection of 1–O is found TL by use X and Y (width & height)

2. The TL of 1–2 is found by combining Z (depth) with the TL of 1–O.

C. ORTHOGRAPHIC VIEWS

B. CALCULATION OF THE TRUE LENGTH OF THE HYPOTENUSE, 1–2

A. FRONTAL PROJECTION

Figure 27.12 To calculate the true length of a 3D line that is not true length in principal view, find (A) the frontal projection, line 1–O, by using the X and Y distances, and (B) the hypotenuse of the right triangle 1–O–2 by using the length of line 1–O and the Z distance. Then apply the Pythagorean theorem to find its length of 5. Orthographic views are shown in (C).

TRUE-LENGTH DIAGRAM

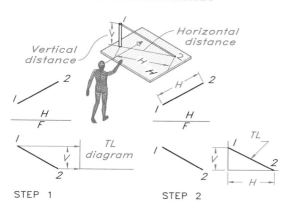

STEP 1   STEP 2

Figure 27.13 Using a true-length diagram:
Step 1 Transfer the vertical distance between the ends of line 1–2 to the vertical leg of the TL diagram.

Step 2 Transfer the horizontal length of the line in the top view to the horizontal leg of the TL diagram. The diagonal is the true length of line 1–2.

ANGLES WITH PRINCIPAL PLANES

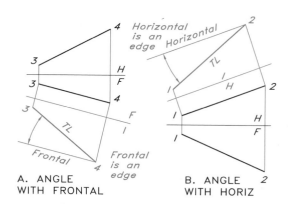

A. ANGLE WITH FRONTAL   B. ANGLE WITH HORIZ

Figure 27.14 Angles between lines and principal planes:
A An auxiliary view projected from the front view that shows the line true length will show the frontal plane as an edge, where its angle with the frontal plane can be measured.

B An auxiliary view projected from the top that shows the line true length will show the horizontal plane as an edge, where its angle with the horizontal plane can be measured.

27.5 Sloping Lines

Slope is the angle that a line makes with the horizontal plane when the line is true length and the plane is an edge. Figure 27.15 shows the three methods for specifying slope: slope angle, percent grade, and slope ratio.

The **slope angle** of line AB in **Fig. 27.16** is 31°. It can be measured in the front view where the line is true length.

MEASURING SLOPES OF LINES

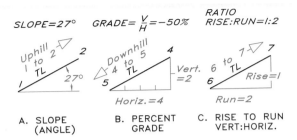

A. SLOPE (ANGLE)   B. PERCENT GRADE   C. RISE TO RUN VERT:HORIZ.

Figure 27.15 The inclination of a line with the horizontal may be measured and expressed as: (A) slope angle, (B) percent grade, and (C) slope ratio.

PERCENT GRADE OF A LINE

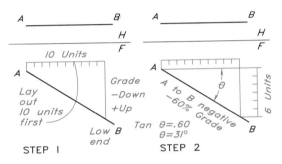

STEP I   STEP 2

Figure 27.16 The percent grade of a line:
Step 1 The percent grade of a line may be measured in the view where the horizontal appears as an edge and the line is true length (here, the front view). Lay off ten units parallel to the horizontal from the end of the line.

Step 2 A vertical distance from the end of the 10 units to the line measures 6 units. The percent grade is 6 divided by 10, or 60%. This is a negative grade from A to B because the line slopes downward from A. The tangent of this slope angle is 6/10, or 0.60, which can be used to find the slope of 31° from trigonometric function tables.

The **percent grade** is the ratio of the vertical (rise) divided by the horizontal (run) between the ends of a line, expressed as a percentage. The percent grade of line AB is determined in the front view of Fig. 27.16 where the line is true

SLOPE RATIO OF A LINE

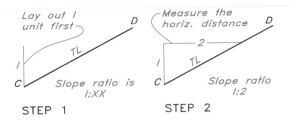

STEP 1   STEP 2

Figure 27.17 Slope ratio:
Step 1 Slope ratio always begins with 1, so lay out a vertical distance of 1 from end C.

Step 2 Lay off a horizontal distance from the end of the vertical line and measure it. It is 2, so the slope ratio (always expressed as 1:XX) of this line is 1:2.

length and the horizontal plane is an edge. Line AB has a -60% grade from A to B because the line slopes downward; it would be positive (upward) from B to A. Trigonometric tables reveal a slope angle of 31° for the angle's tangent of 0.60 (6/10).

The **slope ratio** is the ratio of a rise of 1 to the run. The rise is always written as 1, followed by a colon and the run (for example, 1:10, 1:200). **Figure 27.17** illustrates the graphical method of finding the slope ratio. The rise of 1 unit is laid off on the true-length view of CD. The corresponding run measures 2 units, for a slope ratio of 1:2.

The slope of oblique lines are found true length in an auxiliary view projected from the top view so that the horizontal reference plane will appear as an edge (**Fig. 27.18A**). The slope is expressed as an angle, or 26°.

To find the percent grade of an oblique line (**Fig. 27.18B**) lay off 10 units horizontally, parallel to the H1 reference line. The corresponding vertical distance measures 4.5 units, for a -45% grade from point 3 to point 4.

The principles of true-line length and angles between lines and planes are useful in applications such as the design of aggregate conveyors (**Fig. 27.19**) where slope is crucial to optimal operation of the equipment.

SLOPE OF OBLIQUE LINE

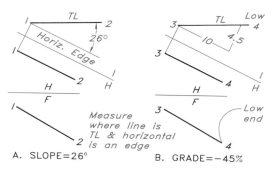

A. SLOPE=26°   B. GRADE=-45%

Figure 27.18

A Find the slope angle of an oblique line: (26° in this case) in a view where the horizontal appears as an edge and the line is true length;

B Find the percent grade in an auxiliary view projected from the top view where line 3–4 is true length (-45% from 3 to 4, the low end, in this case).

Figure 27.19 The design of these aggregate conveyors required the application of sloping-line principles in order to obtain their optimal slopes. (Courtesy of Smith Engineering Works.)

27.6 Bearings and Azimuths of Lines

Two types of bearings of a line's direction are compass bearings and azimuths. **Compass bearings** are angular measurements from north or

COMPASS DIRECTIONS

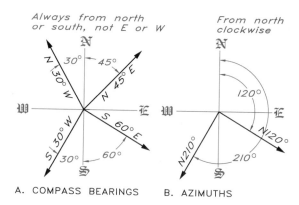

A. COMPASS BEARINGS   B. AZIMUTHS

Figure 27.20

A Compass bearings are measured with respect to north and south.

B Azimuths are measured clockwise from north up to 360°.

south. The line in **Fig. 27.20A** that makes a 30° angle with north has a bearing of N 30° W. The line making a 60° angle with south toward the east has a bearing of south 60° east, or S 60° E. Because a compass can be read only when held level, bearings of a line must be found in the top, or horizontal, view.

Azimuths are measured clockwise from north through 360° **(Fig. 27.20B)**. Azimuth bearings are written N 120°, N 210°, and so on, indicating that they are measured from north.

The bearing of a line is toward the low end of the line unless otherwise specified. For example, line 2–3 in **Fig. 27.21** has a bearing of N 45° E because the line's low end is point 3 in the front view.

Figure 27.22 shows how to find the bearing and slope of a line. This information may be used verbally to describe the line as having a bearing of S 60° E and a slope of 26° from point 5 to point 6. This information and the location of one point in the top and front views is sufficient to complete a 3D drawing of a line **(Fig. 27.23)**.

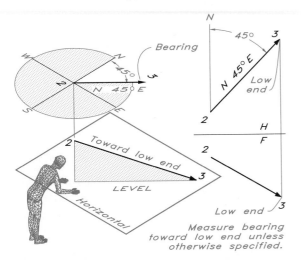

Figure 27.21 Measure the compass bearing of a line in the top view toward its low end (unless otherwise specified). Line 2–3 has a bearing of N 45° E toward the low end at 3.

SLOPE AND BEARING OF A LINE

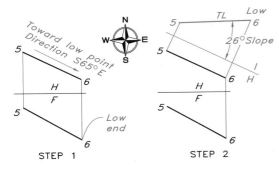

Figure 27.22 The slope and bearing of a line:
Step 1 Measure the bearing in the top view toward its low end, or S 65° E in this case.

Step 2 Measure the slope angle of 26° from the H1 reference line in an auxiliary view projected from the top view where the line is true length.

27.7 Application: Plot Plans

Figure 27.24 shows a typical plot plan for a tract of land. The boundary lines, their bearings, and the interior angles legally define the property.

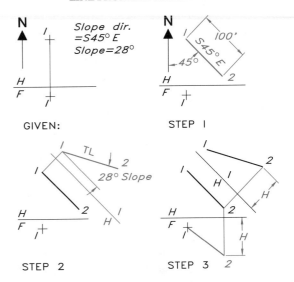

Figure 27.23 A line from slope specifications:
Required Draw a line through point 1 that bears S 45° E for 100 ft horizontally and slopes 28°.

Step 1 Draw the bearing and the horizontal distance in the top view.

Step 2 Project an auxiliary view from the top view and draw the line at a slope of 28°.

Step 3 Find the front view of line 1–2 by locating point 2 in the front view.

◈ **Computer Method** AutoCAD provides a `Surveyor's Units` option under the `UNITS` command for producing a plot plan from surveyor's notes and aligning it with the north arrow. This command prompts you in the following manner.

System of angle measure:	(Example)
1. Decimal degrees	45.0000
2. Degree/minutes/seconds	45d0'0"
3. Grads	50.0000g
4. Radians	0.7854r
5. Surveyor's units	N45d0'0"E

PLOT PLANS

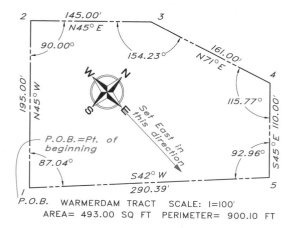

P.O.B. WARMERDAM TRACT   SCALE: 1=100'
AREA= 493.00 SQ FT   PERIMETER= 900.10 FT

Figure 27.24 This typical plot plan shows the lengths and bearings of each side of a tract of land, the interior angles, and the north arrow (produced with AutoCAD's Surveyor's Units option).

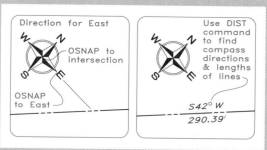

Figure 27.25 Compass direction by computer: **Step 1** Insert the compass arrow on the plot plan. Then use the UNITS command and select 5, Surveyor's units. It will prompt you to give the direction for east. Press F1, and with the cursor, select two points on the east line (west to east) to calibrate the compass direction.

Step 2 Using the DIST command, select the endpoints of each side clockwise about the plot from the point of beginning (P.O.B.). Label the sides with direction inside and lengths outside the plot.

```
Enter choice, 1 to 5 <default>: 5
```
By entering 5, you obtain surveyor's units that give directional angles with respect to north or south in east or west directions, such as N 30d45'10", where d = degrees, ' = minutes, and " = seconds. Interior angles measured with the DIM command give the angles in the same form, but without reference to compass directions, such as 152d34'17".

Next, the UNITS command gives a prompt for setting the direction of east to establish the relationship of the drawing to the directional north arrow:

```
    Direction for angle 0:

East     3 o'clock    =    0

North    12 o'clock   =    90

West     9 o'clock    =    180

South    6 o'clock    =    270

Enter direction for angle E <current>:
F1 (and align.)
```
Press F1 and the screen returns to the drawing you were working on. With the cursor, locate

west-to-east points on the eastward line of the compass arrow **(Fig. 27.25)**. By picking these two points, you establish the direction of north. The last prompt reads:

```
Do you want angles measured clockwise?
<N>: N (CR)
```
By selecting NO (N) you obtain angles measured in the standard direction (Fig. 27.24). Use the DIST command to find the lengths and directions of lines by selecting endpoints of the lines clockwise about the plot from the point of beginning (P.O.B.).

27.8 Contour Maps and Profiles

A **contour map** depicts variations in elevation of the earth in two dimensions **(Fig. 27.26)**. Three-dimensional representations involve (1) conventional orthographic views of the contour map combined with profiles and (2) contoured surface views (often in the form of models).

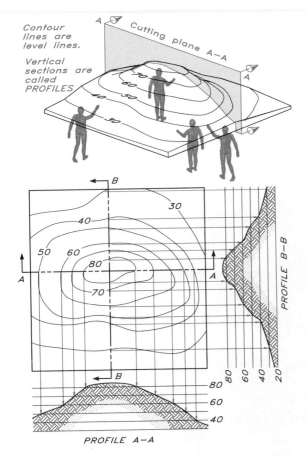

Contour lines are level lines.

Vertical sections are called PROFILES

Figure 27.26 A contour map shows variations in elevation on a surface. A profile is a vertical section through the contour map. To construct a profile, draw elevation lines parallel to the cutting planes, spacing them equally to show the difference in elevations of the contours (10 ft in this case). Then project crossing points of contours and the cutting plane to their respective elevations in the profile and connect them.

Contour lines are horizontal (level) lines that represent constant elevations from a horizontal datum such as sea level. The vertical interval of spacing between the contours shown in **Fig. 27.26** is 10 feet. Contour lines may be thought of as the intersection of horizontal planes with the surface of the earth as shown in the model in **Fig. 27.27**.

Figure 27.27 This model illustrates the use of contour lines in 3D space. (Courtesy of Burlington Industries.)

Contour maps contain contour lines that connect points of equal elevation on the earth's surface and therefore are continuous **(Fig. 27.26)**. The closer the contour lines are to each other, the steeper the terrain is.

Profiles are vertical sections through a contour map that show the earth's surface at any desired location **(Fig. 27.26)**. Contour lines represent edge views of equally spaced horizontal planes in profiles. True representation of a profile involves use of a vertical scale equal to the scale of the contour map; however, **the vertical scale usually is drawn larger to emphasize changes in elevation that often are slight compared to horizontal dimensions.**

Contoured surfaces also are depicted in drawings with contour lines **(Fig. 27.26)** or on models. When applied to objects other than the earth's surface—such as airfoils, automobile bodies, ship hulls, and household appliances—this technique of showing contours is called **lofting**.

Station numbers identify distances on a contour map. Surveyors use a chain (metal tape) 100 feet long, so primary stations are located 100 feet apart **(Fig. 27.28)**. For example, station 7 is 700

STATION NUMBERS

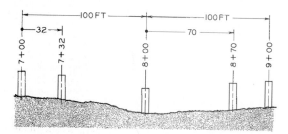

Figure 27.28 Primary station points are located 100 ft apart; for example, station 7 is 700 ft from station 0 (not shown). A point 32 ft beyond station 7 is labeled station 7 + 32. A point 870 ft from the origin is labeled station 8 + 70.

feet from the beginning point, station 0; a point 32 feet beyond station 7 is station 7 + 32.

A **plan–profile** combines a section of a contour map (a plan view) and a vertical section (a profile view). Engineers use plan–profile drawings extensively for construction projects such as pipelines, roadways, and waterways.

Application: Vertical Sections

In **Fig. 27.29**, a vertical section passed through the top view of an underground pipe gives a profile view. The pipe is known to have elevations of 90 feet at point 1 and 60 feet at point 2. Project an auxiliary view perpendicularly from the top view, locate contour lines, and draw the top of the earth over the pipe in profile. To measure the true lengths and angles of slope in the profile, use the same scale for both the contour map and the profile.

Pipeline installation (**Fig. 27.30**) requires major outlays for engineering design and construction. The use of profiles, found graphically, is the best way to make cost estimations for constructing ditches for laying underground pipe.

27.9 Plan–Profiles

A plan–profile drawing shows an underground drainage system from manhole 1 to manhole 3 in

VERTICAL SECTIONS

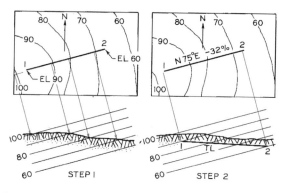

Figure 27.29 Drawing profiles (vertical sections):
Step 1 An underground pipe has elevations of 90 ft and 60 ft at its ends. Project an auxiliary view perpendicularly from the top view and draw contours at 10-ft intervals corresponding to their elevations in the plan view. Locate the ground surface by projecting from the contour lines in the plan view.

Step 2 Locate points 1 and 2 at elevations of 90 ft and 60 ft in the profile. Line 1–2 is TL in the section, so measure its slope (percent grade) here and label its bearing and slope in the top view.

Figure 27.30 Pipeline construction applies on the principles of descriptive geometry, true-length lines, and slopes of 3D lines. (Courtesy of Consumers Power Company.)

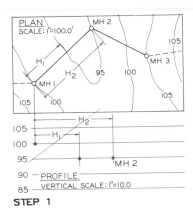

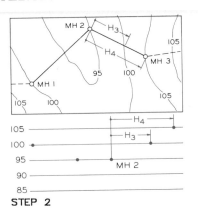

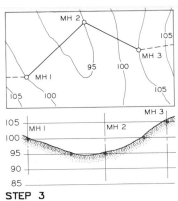

STEP 1   STEP 2   STEP 3

Figure 27.31 Plan–profile:

Required Find the profile of the earth's surface over the pipeline.

Step 1 Transfer distances H_1 and H_2 from manhole 1 in the plan view to their respective elevations in the profile view. This is not a true orthographic projection from the top view.

Step 2 Measure distances H_3 and H_4 from manhole 2 in plan and transfer them to their respective elevations in profile. These points represent elevations of points on the earth above the pipe.

Step 3 Connect these points with a freehand line and crosshatch the drawing to represent the earth's surface. Draw centerlines to show the locations of the three manholes.

Figs. **27.31** and **27.32**. The profile has a larger vertical scale to emphasize variations in the earth's surface and the grade of the pipe, although the vertical scale may be drawn at the same scale as the plan if desired.

The location of manhole 1 is projected to the profile orthographically, **but the remaining points are not (Fig. 27.31)**. Instead, transfer the distances where the contour lines cross the top view of the pipe to their respective elevations in the profile with your dividers to show the surface of the ground over the pipe.

Figure. 27.32 shows the manholes, their elevations, and the bottom line of the pipe. Find the drop from manhole 1 to manhole 2 (5.20 ft) by multiplying the horizontal distance of 260.00 ft by a -2.00% grade. The pipes intersect at manhole 2 at an angle, so the flow of the drainage is disrupted at the turn. A drop of 0.20 ft across the bottom of the

manhole compensates for the loss of pressure (head) through the manhole.

You cannot measure the true lengths of the pipes in the profile view when the vertical scale is different from the horizontal scale; you have to use trigonometry to calculate them.

27.10 Edge Views of Planes

Edge View

The edge view of a plane appears in a view where any line on the plane appears as a point. Recall that you can find a line as a point by projecting from its true-length view **(Fig. 27.33)**. You may obtain a true-length line on any plane by drawing a line parallel to one of the principal planes and projecting it to the adjacent view **(Fig. 27.34)**. You then get the edge view of the plane in an auxiliary view by finding the point view of line 3–4 on its surface.

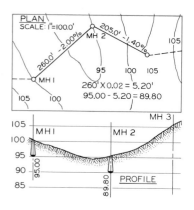

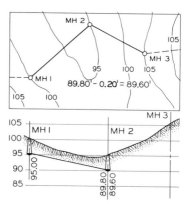

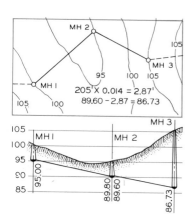

Figure 27.32 Determining manhole location in plan–profile:

Step 1 Multiply the horizontal distance from manhole 1(MH1) to manhole 2 (MH2) by the percent grade. Calculate the elevation of the bottom of MH2 by subtracting the amount of fall from the elevation of MH1.

Step 2 The lower side of MH2 is 0.20 ft lower than the inlet side to compensate for loss of head (pressure) because of the turn in the pipeline. Find the elevation on the lower side (89.60') and label it.

Step 3 Calculate the elevation of MH3 (86.73 ft) from the 1.40% grade from MH2 to MH3. Draw the flow line of the pipeline from manhole to manhole and label the elevations.

Dihedral Angle

The angle between two planes, a dihedral angle, is found in a view where the line of intersection between two planes appears as a point. In this view, both planes appear as edges and the angle between them is true size. The line of intersection, line 1–2, between the two planes shown in **Fig. 27.35** is true length in the top view. Project an auxiliary view from the top view to find the point view of line 1–2 and the edge views of both planes.

27.11 Planes and Lines

Piercing Points

By Projection **Figure 27.36** shows how to use projection to find the piercing point where line 1–2 passes through the plane. Pass cutting planes

POINT VIEW OF A LINE

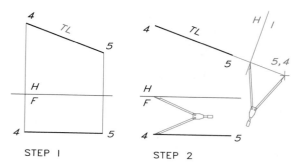

Figure 27.33 The point view of a line:

Step 1 Line 4–5 is horizontal in the front view and therefore is true length in the top view.

Step 2 Find the point view of line 4–5 by projecting parallel to its true length to the auxiliary view.

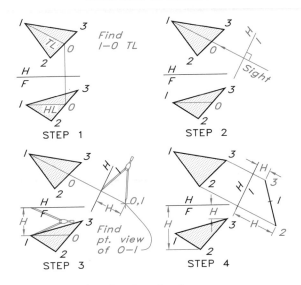

STEP 1

STEP 2

STEP 3

STEP 4

Figure 27.34 The edge view of a plane:

Step 1 To find the edge view of plane 1–2–3, draw horizontal line 1–O on its front view of the plane and project it to the top view, where it is true length.

Step 2 Draw a line of sight parallel to the true-length line 1–O. Draw reference line H1 perpendicular to the line of sight.

Step 3 Find the point view of 1–O in the auxiliary view by transferring height (H) from the front view.

Step 4 Locate points 2 and 3 in the same manner to find the edge view of the plane.

through the line and plane in the top view. Then project the trace of this cutting plane, line DE, to the front view to find piercing point P. Locate the top view of P and determine the visibility of the line.

By Auxiliary View You may also find the piercing point of a line and a plane by auxiliary view in which the plane is an edge (**Fig. 27.37**). The location of piercing point P in step 2 is where line AB crosses the edge view of the plane. Project point P to AB in the top view from the auxiliary view, and then to the front view. To verify the location of

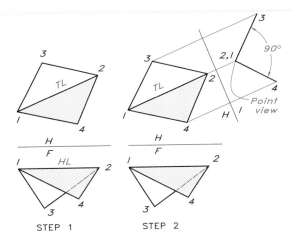

STEP 1

STEP 2

Figure 27.35 A dihedral angle:

Step 1 The line of intersection between the planes, line 1–2, is true length in the top view.

Step 2 The angle between the planes (the dihedral angle) is found in the auxiliary view where the line of intersection appears as a point and both planes are edges.

point P in the front view, transfer dimension H from the auxiliary view with dividers.

You can easily determine visibility for the top view because you see in the auxiliary view that line AP is higher than the plane and therefore is visible in the top view. Similarly, the top view shows that endpoint A is the forward-most point and line AP therefore is visible in the front view.

Perpendicular to a Plane

A perpendicular line appears true length and perpendicular to a plane where the plane appears as an edge. In **Fig. 27.38**, a line is to be drawn from point O perpendicular to the plane. Obtain an edge view of the plane and draw the true-length perpendicular to locate piercing point P. Locate point P in the top view by drawing line OP parallel to the H1 reference line. (This principle is reviewed in **Fig. 27.39**.) Line OP also is perpendicular to a true-length line in the top view of the

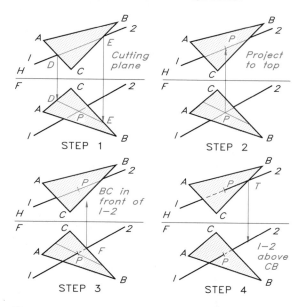

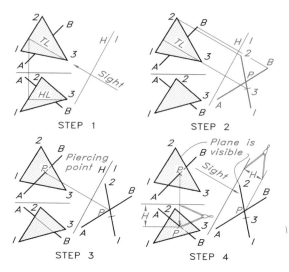

Figure 27.36 A piercing point by projection:

Step 1 Pass a vertical cutting plane through the top view of line 1–2, which cuts the plane along line DE. Project line DE to the front view to locate piercing point P.

Step 2 Project point P to the top view of line 1–2.

Step 3 Determine visibility in the front view by projecting the crossing point of lines CB and 1–2 to the top view. Because CB is encountered first, it is visible in front of line 1–2, making segment PF hidden in the front view.

Step 4 Determine visibility in the top view by projecting the crossing point of lines CB and 1–2 to the front view. Because line 1–2 is encountered first, it is above line CB, making TP visible in the top view.

Figure 27.37 A piercing point by auxiliary view:

Step 1 Draw a horizontal line on the plane in the front view and project it to the top view where it is true length on the plane.

Step 2 Find the edge view of the plane in an auxiliary view and project line AB to this view. Point P is the piercing point.

Step 3 Project point P to line AB in the top view. Line AP is nearest the H1 reference line, so it is the highest end of the line and is visible in the top view.

Step 4 Project point P to line AB in the front view. Line AP is visible in the front view because line AP is in front of 1–2.

plane. Obtain the front view of point P, along with its visibility, by projection.

Intersection

Auxiliary View Method To find the intersection between planes find the edge view of one of the planes **(Fig. 27.40)**. Then project piercing points L and M from the auxiliary view to their respective lines, lines 5–6 and 4–6, in the top view. Plane 4–5–L–M is visible in the top view because sight line 1 has an unobstructed view of the 4–5–L–M portion of the plane in the auxiliary view. Plane 4–5–L–M is visible in the front view because sight line 2 has an unobstructed view of the top view of this portion of the plane.

27.12 Sloping Planes

Slope and Direction of Slope
The slope of a plane is described as follows:

Angle of Slope: the angle that the plane's edge view makes with the edge of the horizontal plane.

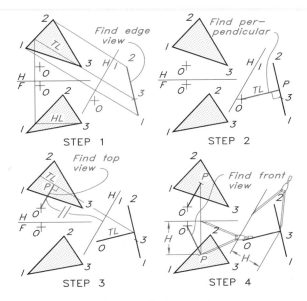

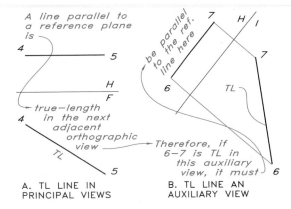

Figure 27.39 In the top view of Fig. 27.38, line OP is parallel to the H1 reference line because it is true length in the auxiliary view. Here, lines 4–5 and 6–7 are examples of this principle: Both are true length in one view and parallel to the reference line in the preceding view.

Figure 27.38 A line perpendicular to a plane:
Step 1 Find the edge view of the plane by finding the point view of a line on it in an auxiliary view. Project point O to this view, also.

Step 2 Draw line OP perpendicular to the edge view of the plane, which is true length in this view.

Step 3 Because line OP is true length in the auxiliary view, it must be parallel to the H1 reference line in the preceding view. Line OP is visible in the top view because it appears above the plane in the auxiliary view.

Step 4 Project point P to the front view and locate it by transferring height H from the auxiliary view with dividers.

Direction of Slope: the compass bearing of a line perpendicular to a true-length line in the top view of a plane toward its low side (the direction in which a ball would roll on the plane).

As **Fig. 27.41** shows, a ball would roll perpendicular to all horizontal lines on the roof toward the low side. **Figure 27.42** shows how to find a plane's direction of slope and slope angle.

You can establish a plane in three dimensions by working from slope and direction specifica-

tions **(Fig. 27.43)**. Draw the direction of slope in the top view to locate a perpendicular true-length line on the plane. Find the edge view of the plane by locating point 1 and constructing a slope of 30° through it in an auxiliary view. Transfer points 3 and 2 to the front view from the auxiliary view.

Application: Cut and Fill

A level roadway through irregular terrain and the embankment for an earthen dam **(Fig. 27.44)** involve the principles of cut and fill. **Cut and fill** is the process of cutting away high ground and filling low areas, generally of equal volumes.

In **Fig. 27.45**, a level roadway at an elevation of 60 feet is to be constructed along a specified centerline with specified angles of cut and fill. First, draw the roadway in the top view. Use contour intervals in the profile view of 10 feet to match those in the top view.

Next, measure and draw the cut angles on both sides of the roadway. Project the points where the cut angles cross each elevation line to the respective contour lines in the plan view to find the limits of cut.

INTERSECTION BETWEEN PLANES

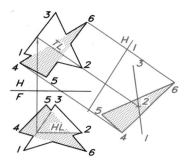
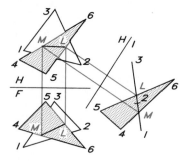
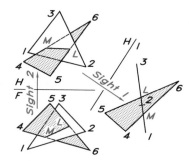

Figure 27.40 The intersection of planes by auxiliary view:

Step 1 Locate the edge view of one of the planes in an auxiliary view and project the other plane to this view.

Step 2 Piercing points L and M appear on the edge view of the plane. Project line LM to the top and front views.

Step 3 The line of sight from the top view strikes line L–5 first in the auxiliary view, indicating that line L–5 is visible in the top view. Line 4–5 is farthest forward in the top view and is visible in the front view.

SLOPE OF PLANE: DEFINITION

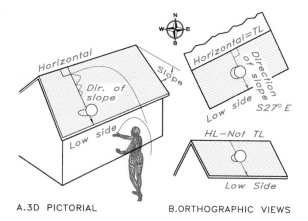

A.3D PICTORIAL   B.ORTHOGRAPHIC VIEWS

Figure 27.41 Slope definition:

A The direction of slope of a plane is the compass bearing of the direction in which a ball on the plane will roll.

B Slope direction is measured in the top view toward the low side of the plane and perpendicular to a horizontal line on the plane.

SLOPE AND BEARING OF A PLANE

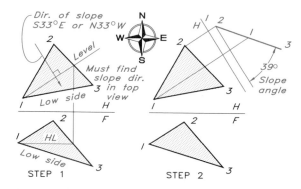

STEP 1   STEP 2

Figure 27.42 The slope and bearing of a plane:

Step 1 Slope direction is perpendicular to a true-length, level line in the top view toward the low side of the plane, or S 33° E in this case.

Step 2 Find the slope in an auxiliary view where the horizontal is an edge and the plane is an edge, or 39° in this case.

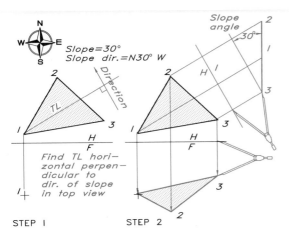

Figure 27.43 A plane from slope specifications:
Step 1 If the top view of a plane, the front view of point 1, and slope specifications are given, you can complete the front view. Draw the direction of slope in the top view and a true-length horizontal line on the plane perpendicular to the slope direction.

Step 2 Find a point view of the TL line in the auxiliary view to locate point 1. Draw the edge view of the plane through point 1 at a slope of 30°, according to the specifications. Find the front view by transferring height dimensions from the auxiliary view to the front view.

Then, measure and draw the fill angles in the profile view. Project the points where the fill angles cross each elevation line to the respective contour lines in the plan view to find the limits of fill. Finally, draw new contour lines inside the areas of cut parallel the centerline.

Application: Design of a Dam

Some of the terms used in designing a dam are (1) *crest,* or the top of the dam; (2) *water level;* and (3) *freeboard,* or the height of the crest above the water level (**Fig. 27.46**). The earthen dam shown in **Fig. 27.47** is an arc in plan view, with its center at C. The finished plan–profile drawing depicts both the dam and the level of the water held by the dam.

Figure 27.44 This dam was built by applying the principles of cut and fill. (Courtesy of the Bureau of Reclamation, U.S. Department of the Interior.)

Figure 27.48 is a photo of the 726-foot-high Hoover Dam, built in the 1930s. Because it was constructed of reinforced concrete instead of earth, the dam was designed in the shape of an arch bowed toward the water, taking advantage of the compressive strength of concrete.

Strike and Dip

Strike and dip are terms used in geology and mining engineering to describe the location of strata of ore under the surface of the earth:

> **Strike:** the compass bearings (two are possible) of a level line in the top view of a plane.

> **Dip:** the angle that the edge view of a plane makes with the horizontal and its general compass direction, such as NW or SW.

The dip angle lies in the primary auxiliary view projected from the top view. The dip direction is perpendicular to the strike and toward its low side.

Figure 27.49 demonstrates how to find the strike and dip of a plane. Here, the true-length line in the top view of the plane has a strike of

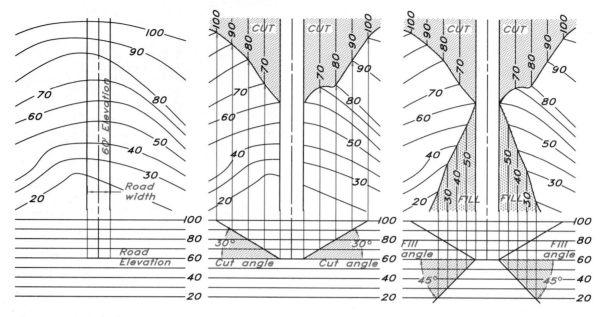

Figure 27.45 Cut and fill for a level roadway:

Step 1 Draw and label a series of elevation planes in the front view at the same scale as the contour map. Draw the width and elevation (60' in this case) of the roadway in the top and front views.

Step 2 Draw the cut angles on the higher sides of the road in the front view. Project the points of intersection between the cut angles and the contour planes in the front view to their respective contour lines in the top view to determine the limits of cut.

Step 3 Draw the fill angles on the lower sides of the road in the front view. Project the points in the front view where the fill angles cross the contour lines to their respective contour lines in the top view to give the limits of fill. Draw contour lines parallel to the centerline in the cut-and-fill areas to indicate the new contours.

TERMINOLOGY OF DAMS

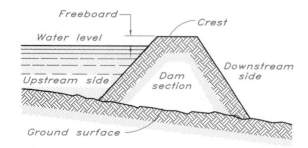

Figure 27.46 These terms and symbols are used in the design of a dam.

N 66° W or S 66° E. The dip angle appears in an auxiliary view projected from the top view that shows the horizontal (H1) and the plane as edges.

You may construct a plane from strike and dip specifications (**Fig. 27.50**). First, draw the strike as a true-length horizontal line on the plane and the dip direction perpendicular to the strike. Then find the edge view of the plane in the auxiliary view through point 1 at a dip of 30°. Locate points 2 and 3 in the front view by transferring them from the auxiliary view.

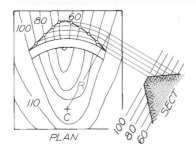

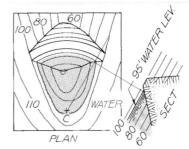

Figure 27.47 An arc-shaped dam:
Step 1 An arc-shaped dam with its center at C has a crest elevation of 100'. Draw radius R from C, project perpendicularly from this line, and draw a section (profile) through the dam from the specifications. Project the downstream side of the dam to radial line R. Use the radii from C to locate points on their respective contours.

Step 2 Project the elevations of the dam on the upstream side of the section to the radial line R. Use center C and your compass to locate points on their respective contour lines in the plan view as you project them from the profile.

Step 3 Draw the water level (95') in the profile. Project the points where the water intersects the dam to the radial line in the plan view as an arc from center C. Draw the upper limit of the water level between the 90' and 100' contour lines in the top view.

Figure 27.48 The Hoover Dam and Lake Mead, built from 1931 to 1935, is a classic dam design and a major civil engineering achievement. (Courtesy of the Bureau of Reclamation, U.S. Department of the Interior.)

27.13 Ore-Vein Applications

The principles of descriptive geometry can be applied to find the distance from a point to a plane. Techniques of finding such distances often are used to solve mining and geological problems. For example, test wells are drilled into coal seams to learn more about them (**Fig. 27.51**).

Application: Underground Ore Veins Geologists and mining engineers usually assume that strata of ore veins have upper and lower planes that are parallel. In **Fig. 27.52**, point O is on the upper surface of the earth and plane 1–2–3 is an underground ore vein. Point 4 is on the lower plane of the vein.

Find the edge view of plane 1-2-3 by projecting from the top view and then draw the lower plane through point 4 parallel to the upper plane. Draw the horizontal distance from point O to the plane parallel to the H1 reference line and the vertical distance perpendicular to line H1. The shortest distance is perpendicular to the ore vein. These

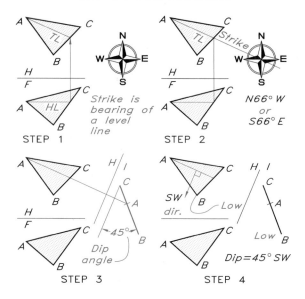

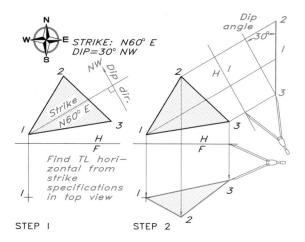

Figure 27.49 The strike and dip of a plane:

Step 1 Draw a horizontal line on the plane in the front view and project it to the top view, where it is true length.

Step 2 Strike is the compass direction of a level line on the plane in the top view—here, either N 66° W or S 66° E.

Step 3 Find the edge view of the plane in the auxiliary view. The dip angle of 45° is the angle between the H1 reference line and the edge view of the plane.

Step 4 The general compass direction of dip is toward the low side, and perpendicular to a strike in the top view or SW in this case. Dip direction in this example is 45°SW.

Figure 27.50 Strike and dip specifications:

Step 1 Draw the strike in the top view of the plane as a true-length horizontal line. Draw the direction of dip perpendicular to the strike toward the NW as specified.

Step 2 Find the point view of strike in the auxiliary view to locate point 1 where the edge view of the plane passes through it at a 30° dip, as specified. Complete the front view by transferring height (H) dimensions from the auxiliary to the front view.

three lines from point O are true length in the auxiliary view where the ore vein appears as an edge.

Application: Ore Vein Outcrop The same assumption regarding parallel planes is made in analyzing the orientation of underground ore veins that are inclined to the earth's surface and outcrop on its surface. When ore veins outcrop, open-pit mining can be used to reduce costs.

Figure 27.53 shows how to find the outcrop of an ore vein from the locations of sample drillings given on a contour map. Points A, B, and C are located on the upper plane of the ore vein, and

point D is located on the lower plane of the vein. Draw them in the front view at their determined elevations.

Find the edge view of the ore vein in an auxiliary view projected from the top view. Then project points on the upper surface where the vein crosses elevation lines back to their respective contour lines in the top view. Also project points on the lower surface of the vein (through point D) to the top. If the ore vein extends uniformly at its angle of inclination to the earth's surface, the area between these two lines will be the outcrop of the vein.

27.14 Intersections between Planes

Cutting Plane Method In **Fig. 27.54**, top and front views of two planes are given and the line of

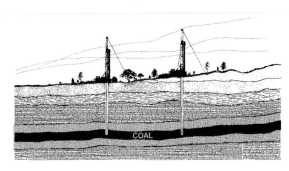

Figure 27.51 Test wells are drilled into coal zones to determine the elevations of coal seams that may contribute to the exploration for gas. (Courtesy of *Texas Eastern News*.)

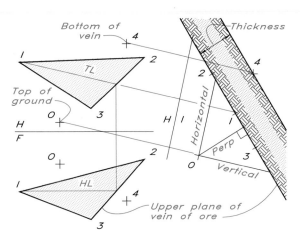

Figure 27.52 To find the vertical, horizontal, and perpendicular distances from a point to an ore vein, project an auxiliary view from the top view, where the vein appears as an edge. The thickness of an ore vein is perpendicular to the upper and lower planes of the vein.

ORE VEIN OUTCROP

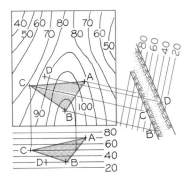

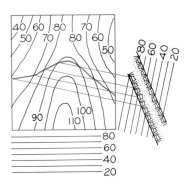

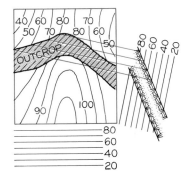

Figure 27.53 Locating an ore vein outcrop:

Step 1 Use points A, B, and C on the upper surface of the ore vein to find its edge view by projecting an auxiliary from the top view. Draw the lower surface of the vein parallel to the upper plane through point D.

Step 2 Project points of intersection between the upper plane of the vein and the contour lines in the auxiliary view to their respective contours in the top view to find a line of the outcrop.

Step 3 Project points from the lower plane in the auxiliary view to their respective contours in the top view to find the second line of outcrop. Crosshatch the area between the lines to depict the outcrop of the vein.

INTERSECTION BETWEEN PLANES

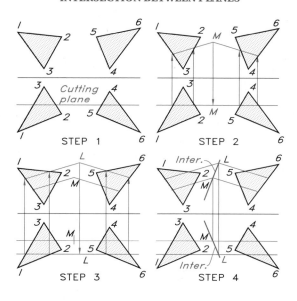

STEP 1

STEP 2

STEP 3

STEP 4

Figure 27.54 The intersection of planes by the cutting plane method:

Step 1 Draw a cutting plane that passes through both planes in the front view in any convenient direction.

Step 2 Project the intersections of the cutting plane to the top views of the planes. Find intersection point M in the top view and project it to the front view.

Step 3 Draw a second cutting plane through the front view of the planes and project it to the top view. Find intersection point L in the top view and project it to the front view.

Step 4 Connect point L and M in the top and front views to represent the 3D line of intersection of the extended planes.

INTERSECTION BETWEEN PLANES: STRIKE AND DIP METHOD

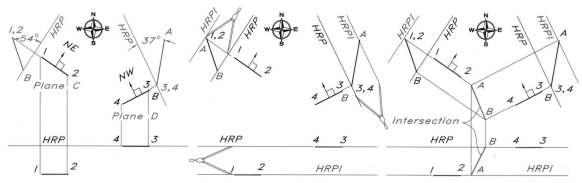

Figure 27.55 The intersection of planes by the strike and dip method:

Step 1 Lines 1–2 and 3–4 are strike lines and are true length in the top view. Use a common reference plane, HRP, to find the point view of each strike line by auxiliary views. Find the edge views by drawing the dip angles with the HRP line through the point views. The low side is the side of the dip arrow.

Step 2 Draw a supplementary horizontal plane, HRP1, at a convenient location in the front view. This plane, shown in both auxiliary views, is located H distance from HRP. The HRP1 cuts through each edge in both auxiliary views, locating A and B on each plane.

Step 3 Project points A from both auxiliary views of HRP1 in the top view to their intersection at A. Project points B and HRP1 to their intersection in the top view. Project points A and B to their respective planes in the front view. Line AB is the line of intersection between the two planes.

intersection between them, if they were extended, is to be determined. Draw cutting planes through both planes in either view at any angle and project them to the top view. Find points L and M in the top view to establish the line of intersection. Find the front view of line LM, the line of intersection, by projecting its endpoints from the top view to their respective planes in the front view.

Strike and Dip Method **Figure 27.55** shows how to locate the intersection of two planes located with strike and dip specifications. The given strike lines are true-length level lines in the top view, so the edge view of the planes appear in the auxiliary views where the strikes appear as points. Draw the edge views using the given dip angles and directions.

Use the additional horizontal datum plane HRP1 to find lines on each plane at equal elevations that intersect when projected to the top view from their auxiliary views. Connect points A and B as the line of intersection between the two planes in the top view and project it to the front view to establish line AB in three dimensions.

Problems

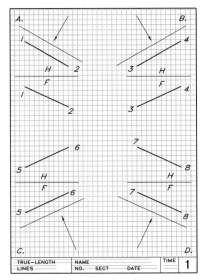

Figure 27.56 Problem 1 (A–D).

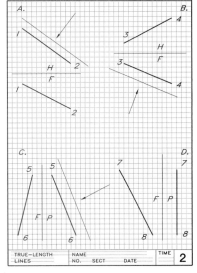

Figure 27.57 Problem 2 (A–D).

Use size A sheets for your solutions to the following problems and lay out the solutions using instruments. Each square on the grid is equal to 0.20 inch (5 mm). You may use either grid or plain paper. Label all reference planes and points with $1/8$-in. (3 mm) letters or numbers, using guidelines.

1. (Fig. 27.56)

(A–D) Find the true-length views of the lines by auxil-

iary view as indicated by the given lines of sight. Alternative method: Find the true-length of the lines by the Pythagorean theorem.

2. (Fig. 27.57)

(A–D) Find the true-length views of the lines by auxiliary view as indicated by the lines of sight. Alternative method: Find the true length of the lines by the Pythagorean theorem.

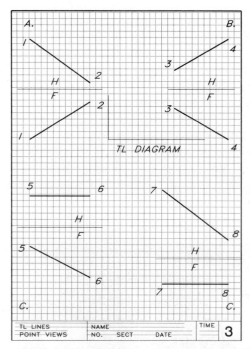

Figure 27.58 Problem 3 (A–D).

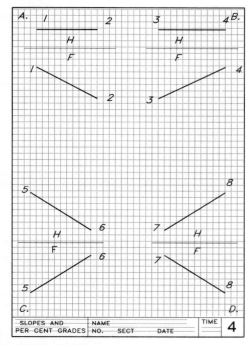

Figure 27.59 Problem 4 (A–D).

3. (Fig. 27.58)

(A–B) Find the true length of the lines with a true-length diagram.

(C–D) Find the point views of the lines.

4. (Fig. 27.59)

(A–D) Find the slope angle, tangent of the slope angle, and the percent grade of the four lines.

5. (Fig. 27.60)

(A–B) Find the edge views of the planes.

6. (Fig. 27.61)

(A) Find the angle between the planes.

(B) By projection, find the projection point between the line and plane and show visibility.

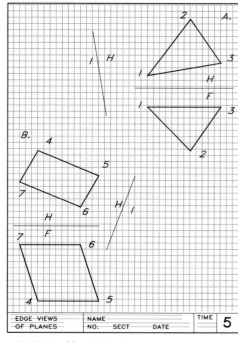

Figure 27.60 Problem 5 (A–B).

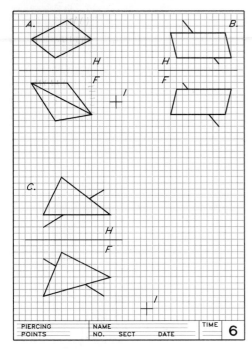

Figure 27.61 Problem 6 (A–C).

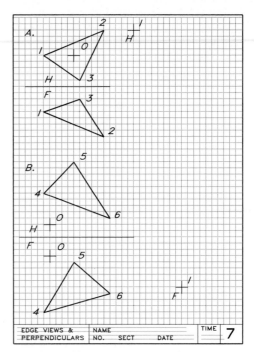

Figure 27.62 Problem 7 (A–B).

(C) By the auxiliary view method, find the point of intersection between the line and plane and show visibility.

7. **(Fig. 27.62)**

(A) Construct a 1 in. line perpendicular from point O on the plane and show it in all views.

(B) Construct a line perpendicular to the plane from point O, find the piercing point, and show visibility.

8. **(Fig. 27.63)**

(A) Find the line of intersection between the intersecting planes by using an auxiliary view.

(B) Find the angle between the planes by using an auxiliary view.

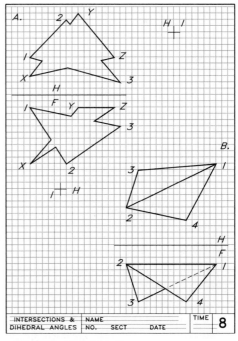

Figure 27.63 Problem 8 (A–B).

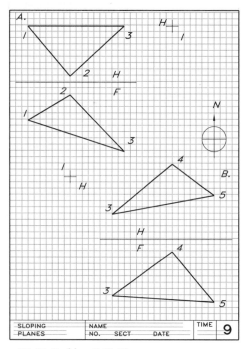

Figure 27.64 Problem 9 (A–B).

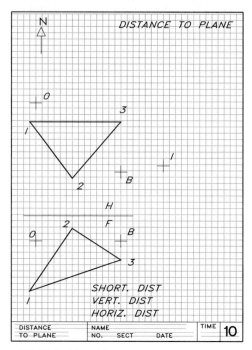

Figure 27.65 Problem 10.

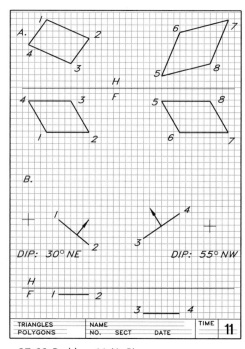

Figure 27.66 Problem 11 (A–B).

9. (Fig. 27.64)

(A–B) Find the direction of slope and the slope angle of the planes. Alternative method: Find the strike and dip of the planes.

10. (Fig. 27.65) Find the shortest distance, the horizontal distance, and the vertical distance from point O to the underground ore vein represented by plane 1–2–3. Point B is on the lower plane of the vein. Find the thickness of the vein.

11. (Fig. 27.66)

(A) Find the line of intersection between the two planes by the cutting plane method.

(B) Find the line of intersection between the two planes indicated by strike lines 1–2 and 3–4. The plane with strike line 1–2 has a dip of 30°, and the plane with strike line 3–4 has a dip of 55°.

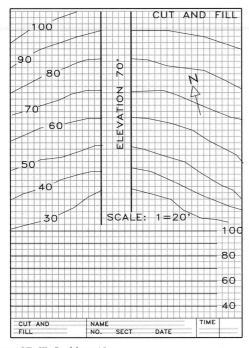

Figure 27.67 Problem 12.

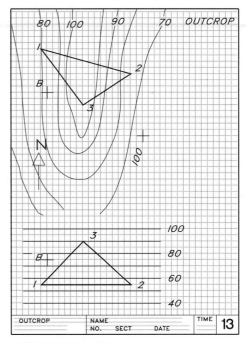

Figure 27.68 Problem 13.

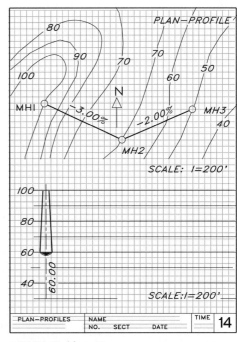

Figure 27.69 Problem 14.

12. (Fig. 27.67) Find the limits of cut and fill in the plan view of the roadway. Use a cut angle of 35° and fill angle of 40°.

13. (Fig. 27.68) Find the outcrop of the ore vein represented by plane 1–2–3 on its upper surface. Point B is on the lower surface.

14. (Fig. 27.69) Complete the plan–profile drawing of the drainage system from manhole 1 through manhole 2 to manhole 3, using the grades indicated. Allow a drop of 0.20 ft across each manhole to compensate for loss of pressure.

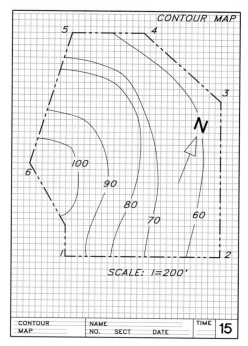

Figure 27.70 Problem 15.

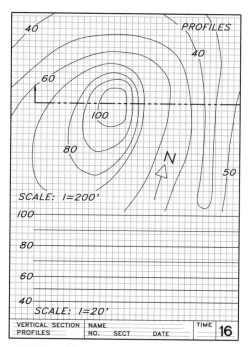

Figure 27.71 Problem 16.

15. (Fig. 27.70) Draw the contour map with its contour lines. Give the lengths of the sides, their compass directions, interior angles, scale, and north arrow.

16. (Fig. 27.71) Draw the contour map and construct the profile (vertical section) as indicated by the cutting plane line in the plan view. Note that the profile scale is different from the plan scale.

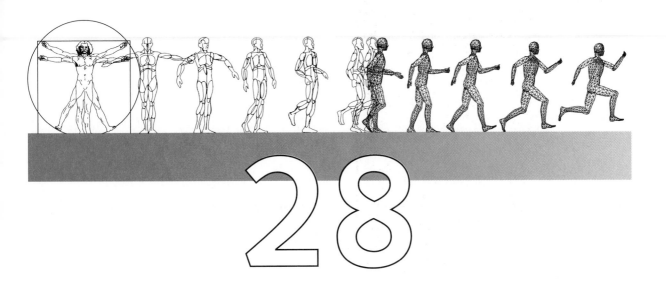

28

Successive Auxiliary Views

28.1 Introduction

A detailed drawing and specifications for a design cannot be completed without determining its geometry which usually requires the application of descriptive geometry. The structural supports for the shopping mall shown in **Fig. 28.1** are examples of complex descriptive geometry problems in which lengths must be determined, angles between lines and planes calculated, and 3D connectors designed.

The process of determining the 3D geometry of a design requires the use of secondary and successive auxiliary views. **Secondary auxiliary views are views projected from primary auxiliary views, and successive auxiliary views are views projected from secondary auxiliary views.**

28.2 Point View of a Line

Recall that, when a line appears true length, you can find its point view in a primary auxiliary view projected parallel from it. In **Fig. 28.2**, line 1–2 is true length in the top view because it is horizontal

Figure 28.1 This structural support of the roof system of a shopping mall was designed and fabricated through the application of descriptive geometry. (Courtesy of Lorna Stuckgold, Kaiser Engineers.)

POINT VIEW OF A LINE

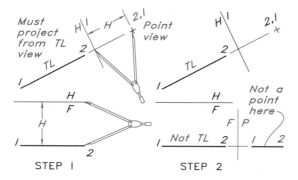

Figure 28.2 To find the point of view of a line, project an auxiliary view from the true-length view of the line.

POINT VIEW OF AN OBLIQUE LINE

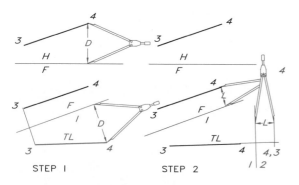

Figure 28.3 A point view of an oblique line:
Step 1 Draw a line of sight perpendicular to one of the views, the front view in this case. Line 3–4 is found true length in an auxiliary view projected perpendicularly from the front view.

Step 2 Draw a secondary reference line, 1–2, perpendicular to the true-length view of line 3–4. Find the point view by transferring dimension L from the front view to the secondary auxiliary view.

in the front view. To find its point view in the primary auxiliary view, first construct reference line H1 perpendicular to the true-length line. Transfer the height dimension, H, to the auxiliary view to locate the point view of 1–2.

DIHEDRAL ANGLE

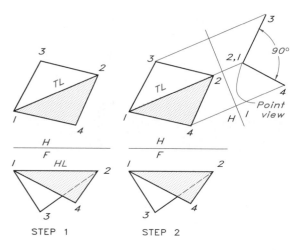

Figure 28.4 The angle between planes (the dihedral angle) appears in the view where their line of intersection projects as a point. The line of intersection, 1–2, is true length in the top view, so it can be found as a point in a view projected from the top view.

Line 3–4 in **Fig. 28.3** is not true length in either view. Finding the line's true length by a primary auxiliary view allows you to find its point view. To obtain a true-length view of line 3–4, project an auxiliary view from the front view (or from the top view). Projecting parallel from the true-length view to a secondary auxiliary view yields the point view of line 3–4. Label the line 4–3 because you see point 4 first in the secondary auxiliary view. Label the reference line between the primary and secondary planes 1–2 to represent the primary (1) and secondary (2) planes.

28.3 Dihedral Angles

Recall that the angle between two planes is called **a dihedral angle and can be found in a view where the line of intersection appears as a point.** The line of intersection lies on both planes, so both appear as edges when the intersection is a point view.

The planes shown in **Fig. 28.4** represent a special case because their line of intersection, line

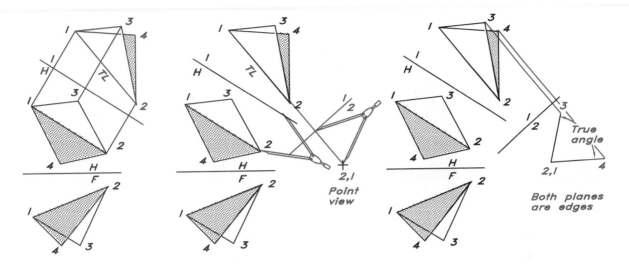

Figure 28.5 The angle between two planes:
Step 1 The angle between two planes is found in a view where the line of intersection (1–2) appears as a point. Find the true-length view of the intersection in a primary auxiliary view projected perpendicularly from it in the top view.

Step 2 Obtain the point view of the line of intersection in the secondary auxiliary view by projecting parallel to the true-length view of line 1–2.

Step 3 Complete the edge views of the planes in the secondary auxiliary view by locating points 3 and 4. Measure the angle between the planes (the dihedral angle) in this view.

1–2, is true length in the top view. This condition permits you to find the line's point view in a primary auxiliary view and measure the true angle between the planes.

Figure 28.5 presents a more typical case. Here, the line of intersection between the two planes is not true length in either view. The line of intersection, line 1–2, is true length in a primary auxiliary view, and the point view of the line appears in the secondary auxiliary view, where you measure the dihedral angle.

This principle was applied to determine the angles between wall panels of the control tower shown in **Fig. 28.6**. That allowed the corner braces to be designed and the structure to be assembled correctly.

28.4 True Size of a Plane

A plane can be found true size in a view projected perpendicularly from an edge view of a plane. The front view of plane 1–2–3 in **Fig. 28.7** appears as an edge in the front view as a special case. The plane's true size is in a primary auxiliary view projected perpendicularly from the edge view.

Figure 28.8 depicts a general case in which you can find the true-size view of plane 1–2–3 by finding the edge view of the plane and constructing a secondary auxiliary view projected perpendicularly from the edge view to find the plane true size.

This principle can be applied to find the angle between lines such as bends in an automobile

DESIGN APPLICATION

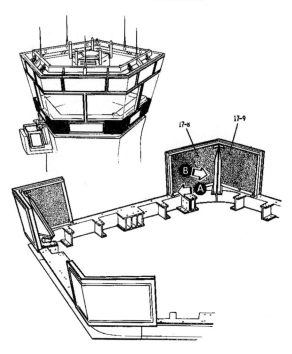

Figure 28.6 These control tower wall panels that intersect at compound angles illustrate the application of methods used to measure dihedral angles. The designer had to determine the angles between the wall panels in order to design corner connectors for securing the panels at their joints.

exhaust pipe **(Fig. 28.9)**. **Figure 28.10** shows how to solve this type of problem. The top and front views of intersecting centerlines are given, the angles of bend and the radii of curvature must be found. Angle 1–2–3 is an edge in the primary auxiliary view and true size in the secondary view, where it can be measured and the radius of curvature drawn. The structural base of the Unisphere® **(Fig. 28.11)** is an example of a design involving planes in many complex geometric combinations and was designed with applications of auxiliary views.

TRUE-SIZE PLANE: SPECIAL CASE

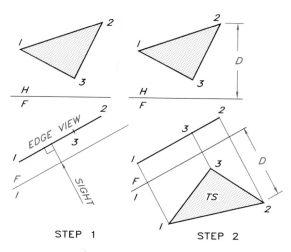

STEP 1   STEP 2

Figure 28.7 The true size of a plane (special case):
Step 1 Because plane 1–2–3 appears as an edge in the front view, it is a special case. Draw the line of sight perpendicular to its edge and the F1 reference line parallel to the edge.

Step 2 Find the true size of plane 1–2–3 in the primary auxiliary view by locating the vertex points with the depth D dimension.

TRUE-SIZE VIEW OF A PLANE

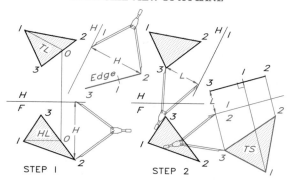

STEP 1   STEP 2

Figure 28.8 The true size of a plane (general case):
Step 1 Find the edge view of plane 1–2–3 by obtaining the point view of true-length line 1–O in a primary auxiliary view.

Step 2 Find a true-size view by projecting a secondary auxiliary view perpendicularly from the edge view of the plane.

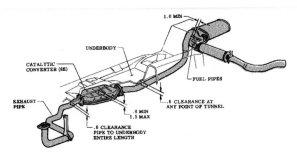

Figure 28.9 The designer determined the angles of bend in this automobile exhaust pipe by applying the principles of the angle between two lines. (Courtesy of General Motors Corporation.)

ANGLE BETWEEN LINES

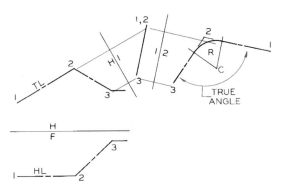

Figure 28.10 The angle between two lines is obtained by finding the plane of the two lines true size.

28.5 Shortest Distance from a Point to a Line: Line Method

The shortest distance from a point to a line can be measured in the view where the line appears as a point. The shortest distance from point 3 to line 1–2 that appears in a primary auxiliary view in **Fig. 28.12** (Step 1) is a special case. The distance from point 3 to the line is true length in the auxiliary view where the line is a point, so it is parallel to reference line F1 in the front view.

Figure 28.13 shows how to solve a general-case problem of this type, where neither line

Figure 28.11 This structural base for the Unisphere®, the trademark of New York's 1965 World's Fair, is an example of the application of methods to determine the size and angles of intersection of planes. (Courtesy of U.S. Steel Corporation.)

SHORTEST DISTANCE TO A LINE

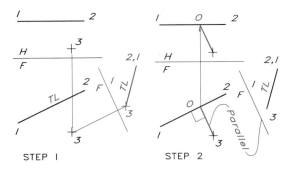

Figure 28.12 The shortest distance from a point to a line:
Step 1 The shortest distance from a point to a line is the true length where the line (1–2) appears as a point. The true-length view of the connecting line appears in the primary auxiliary view.

Step 2 When the connecting line is projected back to the front view, it must be parallel to the F1 reference line in the front view. Project line 3–O back to the top view.

appears true length in the given views. Line 1–2 is true length in the primary auxiliary projected perpendicularly from the front view. The point view of line 1–2 lies in the secondary auxiliary view,

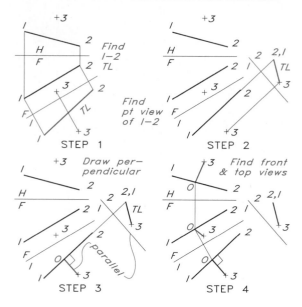

Tee

90° Elbow

Figure 28.14 The shortest distance between two lines or two planes is a line perpendicular to both. Thus the most economical and functional connection of pipes is by 90° tees and elbows, which are standard connectors.

Figure 28.13 The shortest distance from a point to a line:
Step 1 The shortest distance from a point to a line is found in the view where the line appears as a point. Find the true length of line 1–2 by projecting from the front view.

Step 2 Line 1–2 is a point in a secondary auxiliary view projected from the true-length view of line 1–2. The shortest distance to it is true length in this view.

Step 3 Since 3–O is true length in the secondary auxiliary view, it is parallel to the 1–2 reference line in the primary auxiliary view and perpendicular to the line.

Step 4 Find the front and top views of 3–O by projecting from the primary auxiliary view in sequence.

where the distance from point 3 is true length. Because line O–3 is true length in this view, it will be parallel to reference line 1–2 in the preceding view, the primary auxiliary view. It is also perpendicular to the true-length view of line 1–2 in the primary auxiliary view.

28.6 Shortest Distance between Skewed Lines: Line Method

Randomly positioned (nonparallel) lines are called **skewed lines**. **The shortest distance**

between two skewed lines is found in the view where one of the lines appears as a point.

The shortest distance between two lines is a line perpendicular to both lines. The location of the shortest distance between lines is both functional and economical. **Figure 28.14** shows standard 90° pipe connectors (tees and elbows) that are used to make the shortest connections between skewed pipes.

Figure 28.15 illustrates how to find the shortest distance between skewed lines with the line method. Find the true length of line 3–4 and then its point view in the secondary auxiliary view, where the shortest distance is perpendicular to line 1–2. Because the distance between the lines is true length in the secondary auxiliary view, it is parallel to reference line 1–2 in the primary auxiliary view. Find point O by projection and draw OP perpendicular to line 3–4. Project the line back to the given principal views.

28.7 Shortest Distance between Skewed Lines: Plane Method

You may also determine the shortest distance between skewed lines by the plane method, which requires construction of a plane through one of

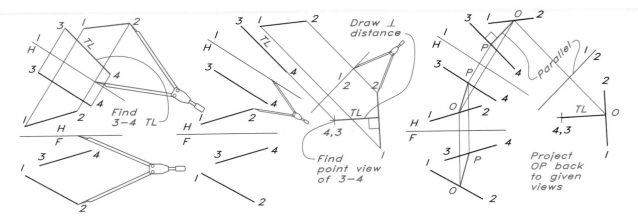

Figure 28.15 The shortest distance between skewed lines (line method):

Step 1 The shortest distance between two skewed lines appears in the view where one of the lines is a point. Find the true length of line 3–4, by projecting it from the top view along with line 1–2.

Step 2 Find the point view of line 3–4 in a secondary auxiliary view projected from the true-length view of line 3–4. The shortest distance between the lines is perpendicular to line 1–2.

Step 3 The shortest distance is true length in the secondary auxiliary view, so it must be parallel to the 1–2 reference line in the preceding view. Project line OP back to the given views.

the lines parallel to the other (**Fig. 28.16**). The top and front views of line O–2 are parallel to their respective views of line 3–4. Therefore plane 1–2–O is parallel to line 3–4. Both lines will appear parallel in an auxiliary view, where plane 1–2–0 appears as an edge.

Figure 28.17 demonstrates this principle. First construct plane 3–4–O. When its edge view is found in a primary auxiliary view, the lines appear parallel. To find the secondary auxiliary view where both lines are true length and cross, project a secondary auxiliary view perpendicularly from these parallel lines. The crossing point is the point view of the shortest distance between the lines. That distance is true length and perpendicular to both lines when projected to the primary auxiliary view as line LM. Project LM back to the given views to complete the solution.

The principle of the shortest distance between two skewed lines is used to determine the separation of power lines, where clearance is critical (**Fig. 28.18**).

PLANE PARALLEL TO A LINE

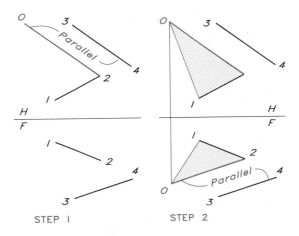

Figure 28.16 A plane through a line and parallel to another line:

Step 1 Draw line O–2 parallel to line 3–4 to a convenient length.

Step 2 Draw the front view of line O–2 parallel to the front view of line 3–4. Find the length of line O–2 in the front view by projecting from the top view of O. Plane 1–2–O is parallel to line 3–4.

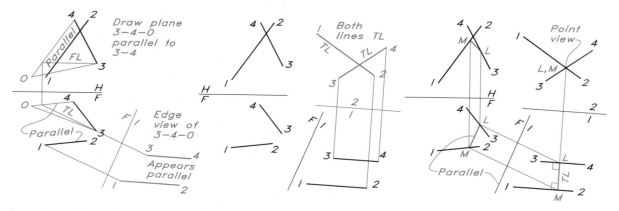

Figure 28.17 The shortest distance between skewed lines (plane method):

Step 1 Construct a plane through line 3–4 parallel to line 1–2. Find plane 3–4–O as an edge by projecting it from the front view, and the two lines will appear parallel.

Step 2 The shortest distance is true length in the primary auxiliary view and perpendicular to both lines. Project the secondary auxiliary view perpendicularly from the lines in the primary auxiliary to find both lines true length.

Step 3 The crossing point of the two lines is the point view of the perpendicular distance (LM) between them. Project LM to the primary auxiliary view, where it is true length, and back to the given views.

28.8 Shortest Level Distance between Skewed Lines

The shortest level (horizontal) distance between two skewed lines can be found by the plane method but not by the line method. In **Fig. 28.19**, plane 3–4–O is constructed parallel to line 1–2, and its edge view is found in the primary auxiliary view. Lines 1–2 and 3–4 appear parallel in this view, and the horizontal reference plane H1 appears as an edge.

A line of sight parallel to H1 is used, and the secondary reference line, 1–2, is drawn perpendicular to H1. The crossing point of the lines in the secondary auxiliary view locates the point view of the shortest horizontal distance between the lines. This line, LM, is true length in the primary auxiliary view and parallel to the H1 plane. Line LM is projected back to the given views. As a check on construction, LM must be parallel to the HF line in the front view, verifying that it is a level or horizontal line.

Figure 28.18 The determination of clearance between power lines is an electrical engineering problem that involves the application of methods of finding the distance between skewed lines. (Courtesy of the Tennessee Valley Authority.)

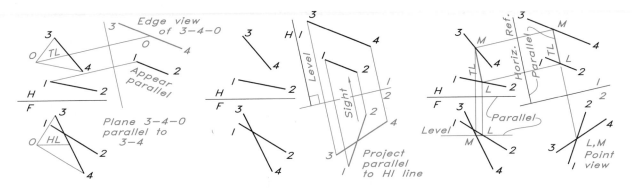

Figure 28.19 The shortest level distance between skewed lines (plane method):

Step 1 Construct plane O–3–4 parallel to line 1–2 by drawing line O–4 parallel to line 1–2. Find the edge view of plane O–3–4 by projecting off the top view, where the lines appear parallel. Project the auxiliary view from the top view to find the horizontal plane as an edge.

Step 2 Infinitely many horizontal (level) lines may be drawn parallel to reference line H1 between the lines in the auxiliary view, but the shortest one appears true length. Construct the secondary auxiliary view by projecting parallel to H1 to find the point view of the shortest level line.

Step 3 The crossing point of the two lines in the secondary auxiliary view is the point view of the level connector, LM. Project line LM back to the given views. Line LM is parallel to the horizontal reference plane in the front view, verifying that it is a level line.

Figure 28.20 The design of these conveyors involved the principles of skewed lines on a grade. (Courtesy of the American Aggregate Corporation.)

28.9 Shortest Grade Distance between Skewed Lines

Features of many applications (such as highways, power lines, or conveyors) are connected to other features at specified grades other than horizontal or perpendicular. For example, the design of conveyors, such as the one shown in **Fig. 28.20,** for transporting aggregate or grain involves the specification of slopes for optimum operation.

If you need to find a 40% grade connector between two lines **(Fig. 28.21)**, use the plane method. To obtain an edge view of the horizontal plane from which the 40% grade is constructed, you must project the primary auxiliary view from the top view. Construct a view in which the lines appear parallel and draw a 40% grade line from the edge view of the horizontal (H1) by laying off rise and run units of 4 and 10, respectively. The grade line may be constructed in two directions from the H1 reference line, but the shortest distance is the direction most nearly perpendicular to both lines.

Project the secondary auxiliary view parallel to this 40% grade line to find the crossing point of the lines to locate the shortest connector, LM. Project line LM back to all views; it is true length

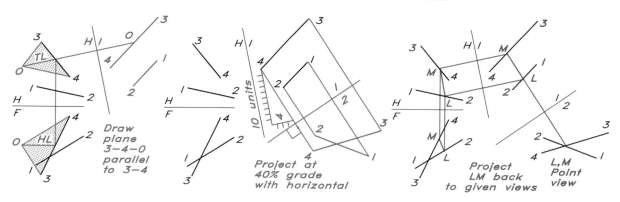

Figure 28.21 The grade distance between skewed lines:

Step 1 To find a level line or a line on a grade between two skewed lines, the primary auxiliary must be projected from the top view. Construct plane 3–4–O parallel to line 1–2. Find the edge view of the plane, and the lines appear parallel.

Step 2 Construct a 40% grade line from the edge view of the H1 reference line in the primary auxiliary view that is most nearly perpendicular to the lines. Project the secondary auxiliary view parallel to the grade line. The shortest grade distance appears true length in the primary auxiliary.

Step 3 The point of crossing of the two lines in the secondary auxiliary view establishes the point view of the 40% grade line, LM. Project LM back to the previous views in sequence.

in the primary auxiliary view, where the given lines appear parallel.

The shortest distances between skewed lines—perpendicular, horizontal, and perpendicular—are true length in the view where the lines appear parallel. The plane method is the general-case method that can be used to find any shortest connector between two lines.

28.10 Angular Distance to a Line

Standard connectors used to connect pipes and structural members are available in two standard angles: 90° (see Fig. 28.14) and 45°. Specifying these standard connectors in a design is far more economical than calling for fabrication of specially made connectors.

In **Fig. 28.22**, a line from point O that makes an angle of 45° with line 1–2 is to be found. Connect point O with the line's endpoints, 1 and

2, to find plane 1–2–O in the top and front views, and find its edge view in a primary auxiliary view. Find the true-size view of plane 1–2–O by projecting perpendicularly from its edge view. Measure the angle of the line from point O in this view, **where the plane of the line and point is true size**.

Draw the 45° connector from point O toward point 2 (the low point) if it slopes downward or toward point 1 if it slopes upward. Determine the upper and lower ends of line 1–2 by referring to the front view, where the height is easily seen. Project the 45° line, OP, back to the given views.

28.11 Angle between a Line and a Plane: Plane Method

The angle between a line and a plane can be measured in the view **where the plane appears as an edge and the line appears true length**. In **Fig. 28.23**, the edge view of plane 1–2–3 lies in a

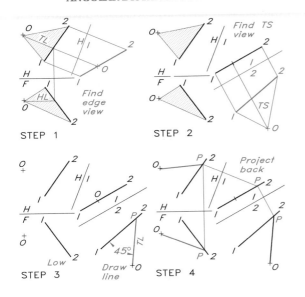

STEP 1 STEP 2

STEP 3 STEP 4

Figure 28.22 A line through a point with a given angle to a line:

Step 1 Connect O to each end of the line to form plane 1–2–O in both views. Draw a horizontal line in the front view of the plane and project it to the top view, where it is true length. Find the point view of AO and the edge view of the plane.

Step 2 Find the true size of plane 1–2–O in the primary auxiliary view by projecting perpendicularly from its edge view. Omit the outline of the plane in this view and only show line 1–2 and point O.

Step 3 Construct Line OP at an angle of 45° with line 1–2. If you draw the angle toward point 2 (the low end), the line slopes downward; if toward point 1, it slopes upward.

Step 4 Project line OP back to the other views in sequence.

ANGLE BETWEEN A LINE AND A PLANE

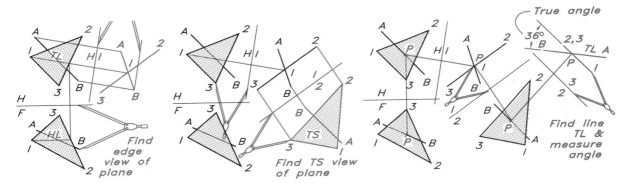

Figure 28.23 The angle between a line and plane (plane method):

Step 1 The angle between a line and a plane in the view where the plane is an edge and the line is true length. Find the plane as an edge by projecting it from the top view along with the line AB.

Step 2 To find the plane's true size, project perpendicularly from the edge view of the plane. A view projected in any direction from a true-size plane will show the plane as an edge.

Step 3 Project a third successive auxiliary view perpendicularly from line AB. The line appears true length and the plane as an edge in this view, where the angle is true size. Project line AB back in sequence to the given views and determine the piercing points and visibility in each view.

primary auxiliary view projected from the top view and is true size (step 2) where the line appears foreshortened. Line AB is true length in a third successive auxiliary view projected perpen-

dicularly from the secondary auxiliary view of line AB. The line appears true length and the plane appears as an edge in the third successive auxiliary view.

28.12 Angle between a Line and a Plane: Line Method

An alternative method (not illustrated here) of finding the angle between a line and a plane is the line method. In that method, the line, rather than the plane, is the primary geometric element that is projected. The line is found as true length in a primary auxiliary view; it is found as a point in the secondary auxiliary view; and the plane is found as an edge in the third successive auxiliary view. Because this last view is projected from the true-length view of the line, the line appears true length where the plane is an edge. When you project the piercing point back to the views in sequence, you can determine visibility for each view.

Problems

Use size A sheets for the following problems and lay out your solutions with instruments on grid or plain paper. Each square on the grid is equal to 0.20 in. (5 mm). Label all reference planes and points in each problem with $^1/_8$-in. letters or numbers, using guidelines.

Use the crosses marked "1" and "2" for positioning the primary and secondary reference lines. Primary reference lines should pass through "1" and secondary reference lines through "2."

1. **(Fig. 28.24)**

(A–B) Find the point views of the lines.

(C–D) Find the angles between the planes.

2. **(Fig. 28.25)**

(A–B) Find the true-size views of the planes.

3. **(Fig. 28.26)**

(A–B) Find the angles between the lines.

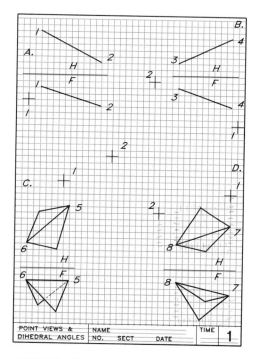

Figure 28.24 Problem 1 (A–D).

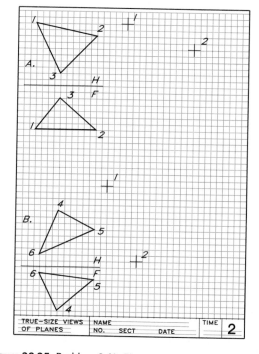

Figure 28.25 Problem 2 (A–B).

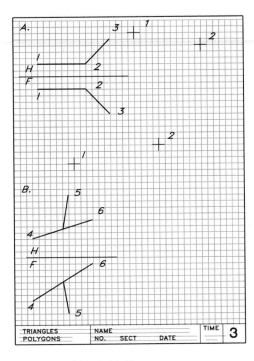

Figure 28.26 Problem 3 (A–B).

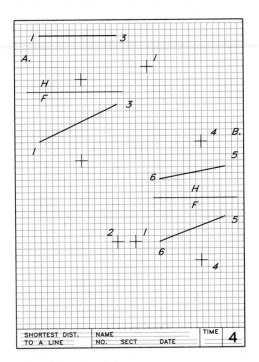

Figure 28.27 Problem 4 (A–B).

4. (Fig. 28.27)

(A–B) Find the shortest distances from the points to the lines. Show this distance in all views.

5. (Fig. 28.28)

(A–B) Find the shortest distances between the lines by the line method. Show this distance in all views.

6. (Fig. 28.29) Find the shortest distance between the lines by the plane method. Show the line in all views. Alternative problem: Find the shortest horizontal distance between the two lines and show the distance in all views.

7. (Fig. 28.30) Find the shortest 20% grade distance between the two lines. Show this distance in all views.

8. (Fig. 28.31) Find the connector from point O that intersects intersect line 1–2 at 60°. Show this line in all views. Project from the top view. Scale: full size.

9. (Fig. 29.32) Find the angle between the line and the plane by the plane method. Show visibility in all views.

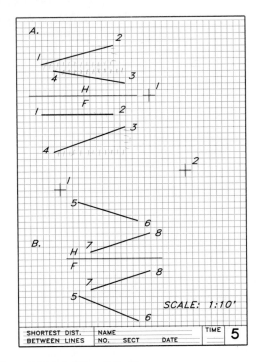

Figure 28.28 Problem 5 (A–B).

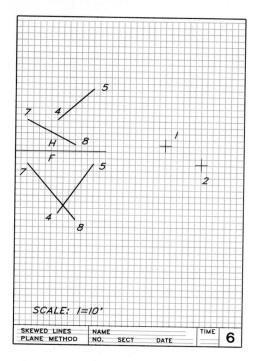

Figure 28.29 Problem 6.

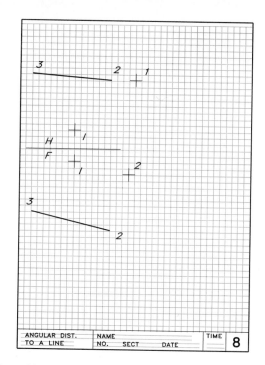

Figure 28.31 Problem 8.

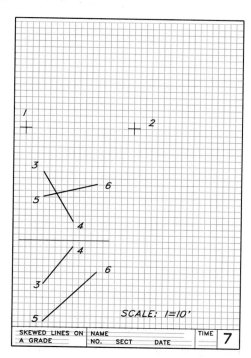

Figure 28.30 Problem 7.

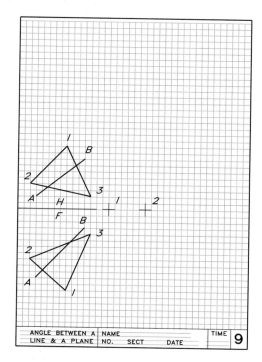

Figure 28.32 Problem 9.

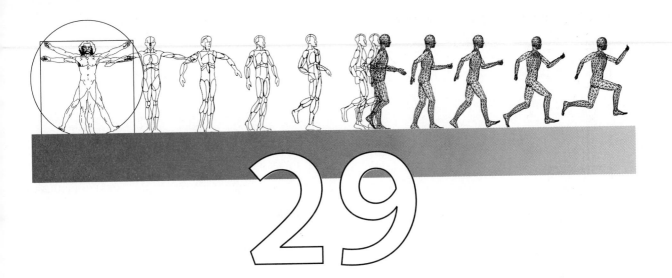

Revolution

29.1 Introduction

Figure 29.1 shows an automobile's front suspension, which was designed to revolve about several axes at each wheel. This design is just one of many based on the principles of revolution. **Revolution** is a technique of revolving an orthographic view into a new position to yield a true-size view of a surface or a line. Revolution was used to solve descriptive geometry problems before the introduction of the auxiliary-view method.

29.2 True-Length Lines

In the Front View

Figure 29.2 compares the auxiliary view and revolution methods of obtaining the true size of inclined surfaces. In the auxiliary view method, the observer changes position to an auxiliary vantage point and looks perpendicularly at the object's inclined surface. In the revolution method, the top view of the object is revolved

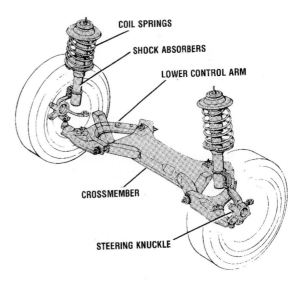

COIL SPRINGS

SHOCK ABSORBERS

LOWER CONTROL ARM

CROSSMEMBER

STEERING KNUCKLE

Figure 29.1 The front suspension of an automobile is an example of the application of the principles of revolution. From steering wheel to tires, revolution is a major part of the design. (Courtesy of Chrysler Corporation.)

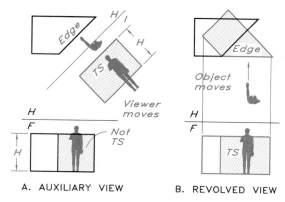

A. AUXILIARY VIEW   B. REVOLVED VIEW

Figure 29.2 Auxiliary views vs. revolved views:
A The surface is found true size in an auxiliary view.

B The surface is revolved to be seen true size in the front view.

TRUE LENGTH IN FRONT VIEW

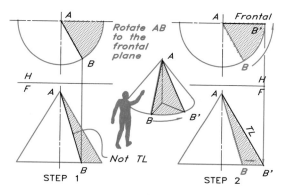

Figure 29.3 Determining true length in the front view:
Step 1 Use the top view of line AB as a radius to draw the base of a cone with point A as the apex. Draw the front view of the cone with a horizontal base through point B.

Step 2 Revolve the top view of line AB to be parallel to the frontal plane. When projected to the front view, frontal line AB' is the outside element of the cone and is true length.

about the axis until the edge view of the inclined plane is parallel to the frontal plane and perpendicular to the standard line of sight from the front view. In other words, the observer's line of sight

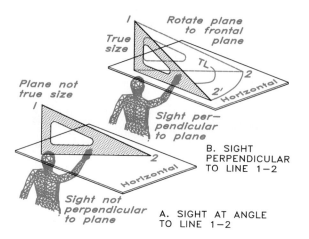

B. SIGHT PERPENDICULAR TO LINE 1–2

A. SIGHT AT ANGLE TO LINE 1–2

Figure 29.4
A Line 1–2 of the triangle does not appear true length in the front view because the observer's line of sight is not perpendicular to it.

B When the triangle is revolved into the frontal plane, his line of sight is perpendicular to it and line 1–2' is seen true length.

does not change, but the object is revolved until the plane appears true size in the observer's normal line of sight.

To find a true-length line in the front view by revolution (Fig. 29.3), revolve line AB into the frontal plane. The top view represents the circular base of a right cone, and the front view is the triangular view of a cone. Line AB' is the outside element of the cone's frontal line and is true length in the front view.

Figure 29.4 illustrates the technique of finding line 1–2 true length in the front view. The observer's line of sight is not perpendicular to the triangle containing line 1–2 in its first position, and line 1–2 is not seen true length. When the triangle is revolved into the frontal plane, the observer's line of sight is perpendicular to the plane, and line 1–2 is seen true length.

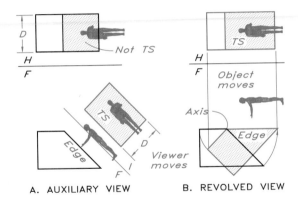

A. AUXILIARY VIEW   B. REVOLVED VIEW

Figure 29.5 Auxiliary views vs. revolved views:
A The inclined plane is true size in an auxiliary view with a line of sight perpendicular to its edge in the front view.

B The surface is true size in the top view after the front view is revolved parallel to the horizontal plane.

In the Top View

A surface that appears as an edge in the front view may be found true size in the top view by a primary auxiliary view or by a single revolution **(Fig. 29.5)**. The axis of revolution is a point in the front view and true length in the top view. Revolving the edge view of the plane into the horizontal in the front view and projecting it to the top view yields the surface's true size. As in the auxiliary view method, the depth dimension, D, does not change.

In **Fig. 29.6, revolving line CD into the horizontal gives its true length in the top view.** The arc of revolution in the front view represents the base of the cone of revolution. Line CD' is true length in the top view because it is an outside element of the cone. Note that the depth in the top view does not change.

In the Profile View

In **Fig. 29.7, revolving the front view of line EF into the profile plane gives its true length.** Projecting the circular view of the cone to the side view yields a triangular view of the cone. Because

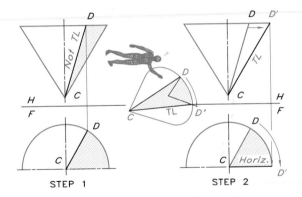

STEP 1   STEP 2

Figure 29.6 Obtaining true length of a line in the top view:
Step 1 Use the front view of line CD as a radius to draw the base of a cone with point C as the apex. Draw the top view of the cone with the base shown as a frontal plane.

Step 2 Revolve the front view of line CD into a horizontal position, CD'. When projected to the top view, CD' is the outside element of the cone and is true length.

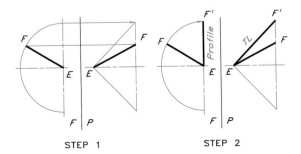

STEP 1   STEP 2

Figure 29.7 Finding true length of a line in the side view:
Step 1 Use the front view of line EF as a radius to draw the circular view of the base of a cone. Draw the side view of the cone with its base through point F, which gives a frontal edge.

Step 2 Revolve line EF in the frontal view to position EF' where it is a profile line. Line EF' in the profile view is true length because it is a profile line and the outside element of the cone.

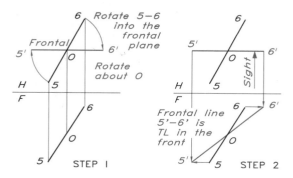

Figure 29.8 True-length views of lines may be obtained by revolving them about any point on the lines, not just their endpoints (here, the midpoint). Line 5–6 is revolved about its midpoint into the frontal plane in the top view and appears true length in the frontal view.

line EF' is a profile line, it is true length in the side view, where it is the outside element of the cone.

Alternative Points of Revolution

In the preceding examples, each line is revolved about one of its ends. However, a line may be revolved about any point on its length. **Figure 29.8** shows how to find line 5–6 true length by revolving it about point O.

29.3 Angles with a Line and Principal Planes

The angle between a line and plane appears true size in the view where the plane is an edge and the line is true length. In orthographic projection, two principal planes appear as edges in any principal view. Therefore, when a line appears true length in a principal view, the angle between the line and the two principal planes can be measured. The angle between the horizontal and the profile planes can be measured in **Fig. 29.9A** in the front view. The angle between horizontal and profile planes can be measured in the top view in **Fig. 29.9B**.

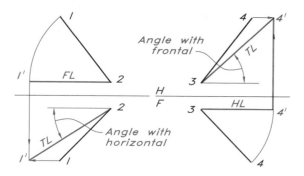

Figure 29.9 Angles with principal planes:
A The angle with the horizontal plane may be measured in the front view if the line appears true length.

B The angle with the frontal plane may be measured in the top view if the line appears true length.

TRUE SIZE OF A PLANE

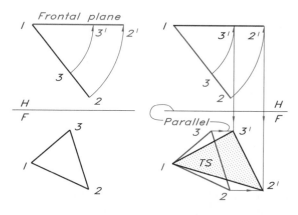

Figure 29.10 Determining true size of a plane:
Step 1 Revolve the edge view of the plane until it is parallel to the frontal plane.

Step 2 Project points 2' and 3' to the horizontal projectors from points 2 and 3 in the front view.

29.4 True Size of a Plane

When a plane appears as an edge in a principal view (the top view in **Fig. 29.10**), it can be

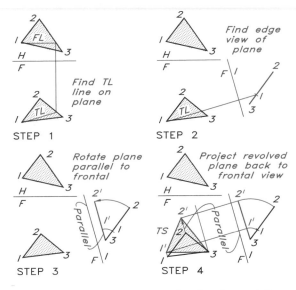

STEP 1   STEP 2

STEP 3   STEP 4

Figure 29.11 Obtaining true size of a plane by revolution:
Step 1 To find the edge view of the plane by revolution, draw a frontal line on the plane that is true length in the front view.

Step 2 Find the edge view of the plane by finding the point view of the frontal line.

Step 3 Revolve the edge view of the plane until it is parallel to the F1 reference line.

Step 4 Project the revolved points 1' and 2' to the front view to the projectors from points 1 and 2 that are parallel to the F1 reference line.

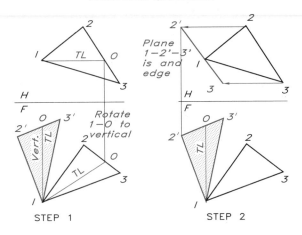

STEP 1   STEP 2

Figure 29.12 Finding the edge view of a plane:
Step 1 Draw a true-length frontal line on the front view of the plane. Revolve the front view until the true-length line is vertical.

Step 2 The true-length line, 1–A' is vertical, so it appears as a point in the top view, and the plane appears as edge 1–2'-3'.

revolved parallel to the frontal reference plane. The new front view is true size when projected horizontally across from its original front view.

The combination of an auxiliary view and a single revolution finds the plane in **Fig. 29.11** true size. After finding the plane as an edge by finding the point of view of a true-length line in the plane, revolve the edge view to be parallel to the F1 reference line. To find the true size of the plane, project the original points (1, 2, and 3) in the front view parallel to the F1 line to intersect the projectors from 1' and 2'. The true size of the plane

also may be found by projecting from the top view to find the edge view.

By Double Revolution

You may find the edge view of a plane by revolution without using auxiliary views **(Fig. 29.12)**. Draw a frontal line on plane 1–2–3, and project it to the front view where it is true length. Revolve the plane until the true-length line is vertical in the front view. The true-length line projects as a point in the top view; therefore the plane appears as an edge in this view. Projectors from points 2 and 3 from the top view are parallel to the HF reference line.

A second revolution, called a **double revolution**, positions this edge view of the plane parallel to the frontal plane, as shown in step 1 of **Fig. 29.13**. Projecting the top views of points 1" and 2" to the front view gives a true-size plane 1"–2"–3"

TRUE-SIZE VIEW OF A PLANE

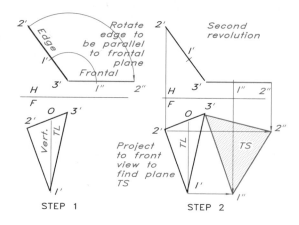

STEP 1

STEP 2

Figure 29.13 Finding true size of a plane:

Step 1 When a plane appears as an edge in the principal view (as in Fig. 29.12), you may revolve it to a position parallel to a reference line, or the frontal line in this case (3′–1″–2″).

Step 2 Project points 1″ and 2″ to the front view to intersect with the horizontal projectors from the original points 1′ and 2′. The plane 1″–2″–3′ is true size in this view.

DOUBLE REVOLUTION

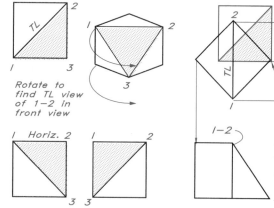

Figure 29.14 Obtaining true size of a plane by double revolution:

Given Three views of a block with an oblique plane across one corner.

Required Find the true size of the plane by revolution.

Step 1 Line 1–2 is horizontal in the frontal view and true length in the top view. Revolve the top view so that line 1–2 appears as a point in the front view.

Step 2 Plane 1–2–3 is an edge in step 1, so you can revolve this plane into a vertical position in the front view to obtain its true size in the side view. The depth dimension does not change.

(step 2). We could have shown this second revolution in Fig. 29.12, but it would have resulted in overlapping views, making observation of the separate steps difficult.

Figure 29.14 shows how to use double revolution to find the true size of the oblique plane

(1–2–3) of the object. Revolve the true-length line 1–2 on the plane in the top view until it is perpendicular to the frontal plane. Line 1–2 appears as a point in the front view, and the plane appears as an edge. This revolution changes the width and depth but not the height. Then revolve the edge

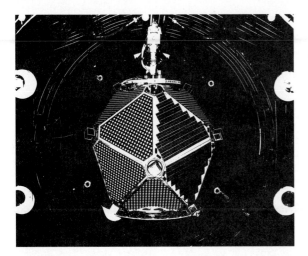

Figure 29.15 The use of revolution principles allows determination of the angles between the planes of this nuclear detection satellite. (Courtesy of TRW Space Technology Laboratories.)

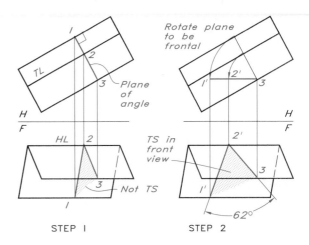

Figure 29.16 Finding the angle between planes:
Step 1 Draw a plane of the angle perpendicular to the true-length line of intersection between the planes in the top view and project it to the front view. The angle is not true size in the front view.

Step 2 Revolve the edge view of the plane of the angle to position angle 1'–2'–3 in the top view parallel to the frontal plane. Project this angle to the frontal view, where it is true size.

view of the plane into a vertical position parallel to the profile plane. To find the plane in true size, project to the profile view, where the depth remains unchanged but the height is greater.

29.5 Angle between Planes

The angle between the planes of the nuclear detection satellite shown in **Fig. 29.15** may be found by revolution instead of by auxiliary views. This application of geometry was essential to the design of the satellite.

In **Fig. 29.16**, finding the **dihedral** angle involves drawing its edge view perpendicular to the line of intersection and projecting the plane of the angle to the front view. Revolving the edge view of the angle until it is a frontal plane in the top view and projecting it to the front view yields its true-size view.

Figure 29.17 shows how to solve a similar problem. Here, the line of intersection does not appear true length in the given views; therefore an auxiliary view is needed to find its true length.

Draw the plane of the dihedral angle as an edge perpendicular to the true-length line of intersection. Project the foreshortened view of plane 1–2–3 to the top view. Then revolve the edge view of plane 1–2–3 in the primary auxiliary view until it is parallel to the H1 reference line. Project the revolved edge view of the angle back to the top view, where it is true size.

29.6 Determining Direction

To solve more advanced problems of revolution, you must be able to locate the basic directions of up, down, forward, and backward in any given view. In **Fig. 29.18A**, directional arrows in the top and front views identify the directions of backward and up. Pointing backward in the top view, line 4–5 appears as a point in the front view. Projecting arrow 4–5 to the auxiliary view as you would any

ANGLE BETWEEN OBLIQUE PLANES

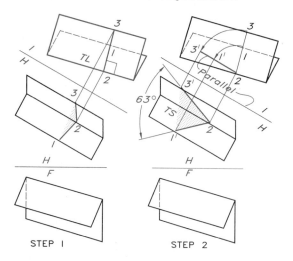

STEP 1   STEP 2

Figure 29.17 Determining the angle between oblique planes:

Step 1 To find a true-length view of the line of intersection, project it to an auxiliary view from the top view. Draw the plane of the angle perpendicular to the true length of the line of intersection and project it to the top view.

Step 2 Revolve the edge view of the plane of the angle until it is parallel to the H1 reference line so that the plane appears true size in the top view. The angle between the planes is 2–1'–3'.

DIRECTIONS OF FORWARD, BACK, UP, AND DOWN

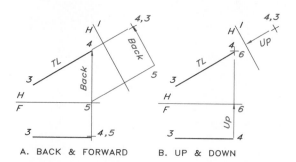

A. BACK & FORWARD   B. UP & DOWN

Figure 29.18 The directions of backward, forward, up, and down may be identified in the given views with arrows pointing in these directions. These directional arrows may be projected to successive auxiliary views. This drawing shows the directions of (A) backward and (B) up.

DIRECTIONS IN A SECONDARY AUXILIARY VIEW

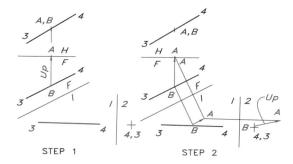

STEP 1   STEP 2

Figure 29.19 Finding direction in a secondary auxiliary view:

Step 1 To find the direction of up in the secondary auxiliary view, draw arrow AB pointing up in the front view. It appears as a point in the top view.

Step 2 Project arrow AB to the primary and secondary auxiliary views like any other line. Locate the direction of up in the secondary auxiliary view.

other line determines the direction of backward. By drawing the arrow on the other end of the line, you would find the direction of forward.

Locate the direction of up in **Fig. 29.18B** by drawing line 4–6 in the direction of up in the front view and as a point in the top view. Then find the arrow in the primary auxiliary by the usual projection method. The direction of down is in the opposite direction.

You find the location of directions in secondary auxiliary views in the same way. To determine the direction of up in **Fig. 29.19**, begin with an arrow that points up in the front view and appears as a point in the top view. Project the

arrow AB from the front view to the primary auxiliary view and then to a secondary auxiliary view to show the direction of up. Identify the other directions in the same way by beginning with the two principal views of a known direction.

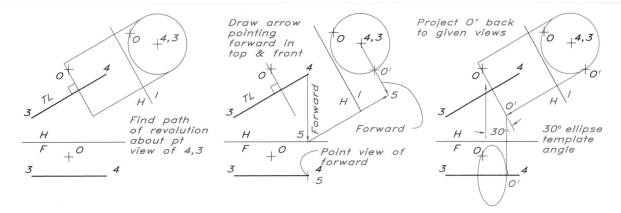

Figure 29.20 Revolving a point about an axis:

Step 1 To rotate point O about axis 3–4 to its most forward position, find the point view of the axis in a primary auxiliary view. The circular path of revolution appears as a circle in the auxiliary views and as an edge perpendicular to the axis in the top view.

Step 2 Locate the most forward position of point O by drawing an arrow pointing forward in the top view. It appears as a point in the front view. Find the arrow, 4–5, in the auxiliary view to locate point O'.

Step 3 Project point O' back to the given views. The path of revolution appears as an ellipse in the front view because the axis is not true length in this view. Draw a 30° ellipse because this is the angle of your line of sight with the circular path in the front view.

29.7 Revolution About an Axis

A Point

In **Fig. 29.20**, point O is to be revolved about axis 3–4 to its most forward position. **Draw the circular path of revolution in the primary auxiliary view, where the axis is a point.** Draw the direction of forward and find the new location of point O at O'. Project back through the successive views to find point O' in each view. Note that point O' lies on the line in the front view, verifying that point O' is in its most forward position.

In **Fig. 29.21** an additional auxiliary view is needed in order to rotate a point about an axis because axis 3–4 is not true length in the given views. You must find the true length of the axis before you can find it as a point in the secondary auxiliary view, where the path of revolution appears as a circle. Revolve point O into its highest position, O', and locate the up arrow, 3–5, in the secondary auxiliary view.

Project back to the given views to locate O in each view. Its position in the top view is over the axis, which verifies that the point is at its highest position.

The paths of revolution appear as edges when their axes are true length and as ellipses when their axes are not true length. The angle of the ellipse template for drawing the ellipse in the front view is the angle the projectors from the front view make with the edge view of the revolution in the primary auxiliary view. To find the ellipse in the top view, project an auxiliary view from the top view to obtain the path of revolution as an edge perpendicular to the true-length axis.

The handcrank of a casement window (**Fig. 29.22**) is an example of the application of revolution techniques. The designer must determine the clearances between the sill and the window frame when designing the crank.

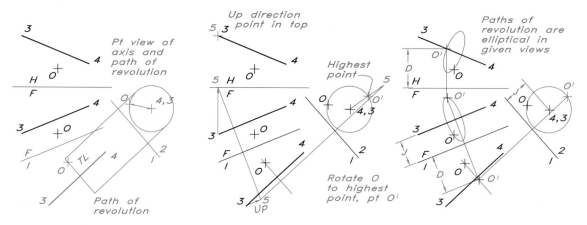

Figure 29.21 Revolving a point about an oblique axis:

Step 1 To rotate O about axis 3–4 to its highest position, find the axis as a point in a secondary auxiliary view and draw the circular path. The path of revolution appears as an edge in the primary auxiliary perpendicular to the true-length view of the axis.

Step 2 To locate the highest position on the path of revolution, draw arrow 3–5 pointing up in the top view and project it to the secondary auxiliary view to find O'.

Step 3 Project point O' back to the given views by transferring the dimensions J and D with your dividers. The highest point lies over the line in the top view, verifying its position. The path of revolution is elliptical wherever the axis is not true length.

HANDCRANK REVOLUTION

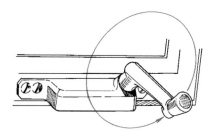

Figure 29.22 The handcrank on a casement window is an example of a problem solved by applying revolution principles. The handle must be properly positioned so as not to interfere with the windowsill or wall.

A Right Prism

The coal chute shown in **Fig. 29.23** conveys coal continuously between two buildings. The sides of the enclosed chute must be vertical and the

bottom of the chute's right section must be horizontal. Design of the chute required application of the technique of revolving a prism about its axis.

In **Fig. 29.24**, the right section is to be positioned about centerline AB so that two of its sides will be vertical. To do so, find the point view of the axis and project the direction of up to this view. Draw the right section about the axis so that two of its sides are parallel to the up arrow. Find the right section in the other views. Then construct the sides of the chute parallel to the axis. The bottom of the chute's right section will be horizontal and properly positioned for conveying coal.

29.8 A Line at a Specified Angle with Two Principal Planes

In **Fig. 29.25**, a line is to be drawn through point O that makes angles of 35° with the frontal plane

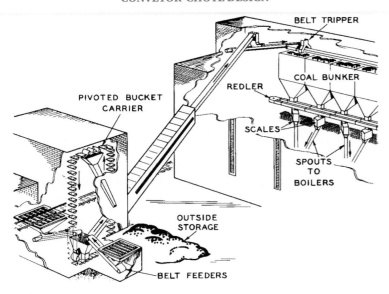

Figure 29.23 A conveyor chute must be installed so that two edges of its right section are vertical for the conveyors to function properly. (Courtesy of Stephens-Adamson Manufacturing Company.)

REVOLVING A PRISM ABOUT ITS AXIS

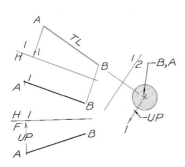

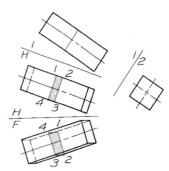

Figure 29.24 Revolving a prism about its axis:

Step 1 To draw a chute with two of its square sides vertical, locate the point view of centerline AB in the secondary auxiliary view and draw a circle about the axis with a diameter equal to the square section. Draw a vertical arrow in the front and top views and project it to the secondary auxiliary view to indicate the direction of vertical.

Step 2 Draw the right section, 1–2–3–4, in the secondary auxiliary view with two sides parallel to the vertical directional arrow. Project this section back to the successive views by transferring measurements with dividers. The edge view of the section may be anywhere along centerline AB in the primary auxiliary view.

Step 3 Draw the lateral edges of the prism through the corners of the right section parallel to the centerline in all views. Terminate the ends of the prism in the primary auxiliary view where they appear as edges perpendicular to the centerline. Project the corner points of the ends to the top and front views to establish the ends in these views.

29.8 A LINE AT A SPECIFIED ANGLE WITH TWO PRINCIPAL PLANES • 549

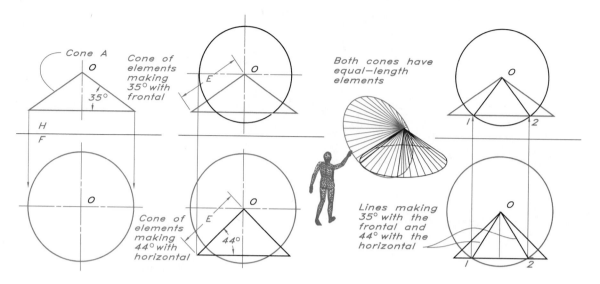

Figure 29.25 Constructing a line at specified angles:

Step 1 To draw lines making angles of 35° and 44° with the frontal and horizontal planes, respectively, begin by drawing a cone in the top view with outside elements making a 35° angle with the frontal plane. Construct the circular view of the cone in the front view with point O as the apex. All elements of this cone make an angle of 35° with the frontal plane.

Step 2 Draw a second cone in the front view with outside elements that make an angle of 44° with the horizontal plane. Draw the elements of this cone equal in length to element E of cone A. All elements of cone B make an angle of 44° with the horizontal plane.

Step 3 Because elements A and B are equal in length, two elements lie on the surface of each cone: lines O–1 and O–2. Locate points 1 and 2 at the point where the bases of the cones intersect in both views. These lines slope forward and down from this point at the specified angles.

and 44° with the horizontal plane and slopes forward and down.

First, draw the cone containing elements making 35° with the frontal plane and then the cone with elements making 44° with the horizontal

plane. The length of the elements of both cones must be equal so that the cones will intersect with equal elements. Finally, find lines O–1 and O–2, which are elements that lie on each cone and make the specified angles with the principal planes.

Problems

Use size A sheets for the following problems and lay out your solutions using instruments on grid or plain paper. Each square on the grid is equal to 0.20 in. (5mm). Label all reference planes and points in each problem with $^{1}/_{8}$-in. letters or numbers, using guidelines.

Use the crosses marked "1" and "2" for positioning primary and secondary reference lines. The primary reference line should pass through "1" and the secondary reference line through "2."

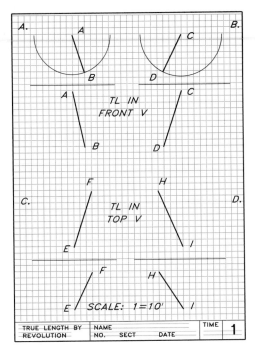

Figure 29.26 Problem 1 (A–D).

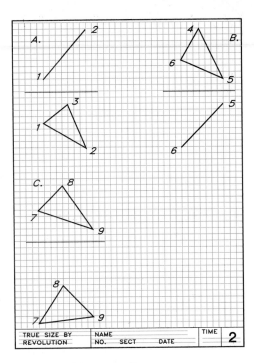

Figure 29.27 Problem 2 (A–C).

1. (Fig. 29.26)

(A–B) Find the true-length views of the lines in their front views by revolution.

(C–D) Find the true-length views of the lines in their top views by revolution.

2. (Fig. 29.27)

(A–B) By revolution, find the true-size views of plane 1–2–3 in the front view and plane 4–5–6 in the top view.

(C) By using an auxiliary view projected from the top view and one revolution, find the true-size view of plane 7–8–9.

3. (Fig. 29.28)

(A–B) Find the angles between the planes by revolution. Show construction.

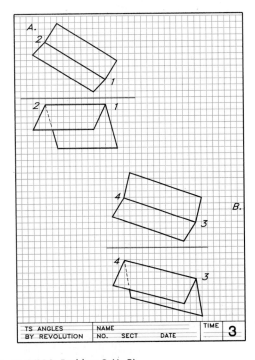

Figure 29.28 Problem 3 (A–B).

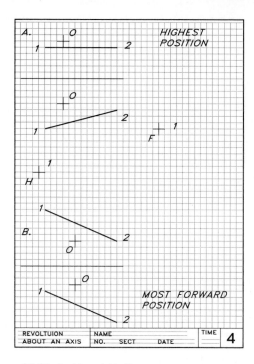

Figure 29.29 Problem 4 (A–B).

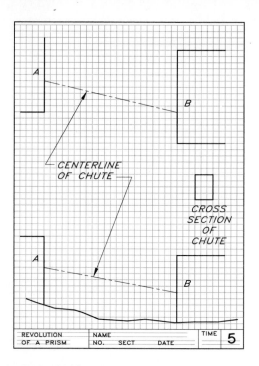

Figure 29.30 Problem 5.

4. (Fig. 29.29)

    (A) Perform the construction necessary to show point O revolved into its highest position.

    (B) Repeat (A) but show point O in its most forward position.

5. (Fig. 29.30) Construct a chute from A to B that has the cross section shown. The longer sides are to be vertical sides.

6. (Fig. 29.31) Draw the views of the line that is 3.2 in. long that makes a 30° angle with the frontal plane and a 52° angle with the horizontal plane.

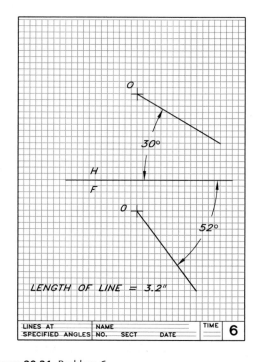

Figure 29.31 Problem 6.

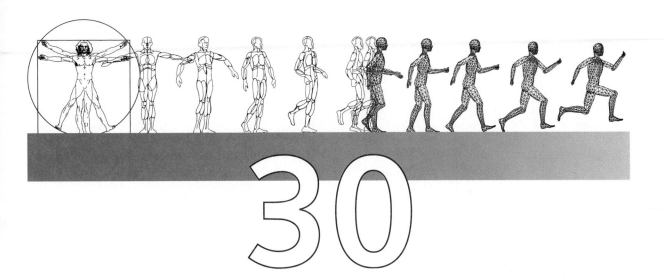

30

Vector Graphics

30.1 Introduction

Design of a structural system requires analysis of each member to determine the loads they must support and whether those loads are in tension or compression. Forces may be represented graphically by **vectors** and their magnitudes and directions determined in 3D space. Graphical methods are useful in the solution of vector problems as alternatives to conventional trigonometric and algebraic methods. Quantities such as distance, velocity, and electrical properties also may be represented as vectors for graphical solution.

30.2 Definitions

To help you understand more easily the discussion of vectors in this chapter, we define the following terms.

    Force: is a push or pull tending to produce motion. All forces have (1) magnitude, (2) direction, and (3) a point of application. The

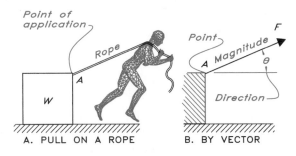

Figure 30.1 A force applied to an object (A) may be represented by vectors depicting the magnitude and direction of the force (B).

person shown pulling the rope in **Fig 30.1A** is applying a force to the weight W.

    Vector: a graphical representation of a force drawn to scale and depicting magnitude, direction, and point of application. The vector

RESULTANT: PARALLELOGRAM METHOD

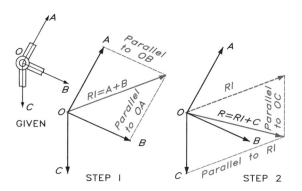

Figure 30.2 Determining the resultant by the parallelogram method:

Step 1 Draw a parallelogram with its sides parallel to vectors A and B. The diagonal R1 is the resultant of forces A and B.

Step 2 Draw a parallelogram using vectors R1 and C to find diagonal R, or the overall resultant that can replace forces A, B, and C.

in **Fig 30.1B** represents the force applied through the rope to pull the weight W.

Magnitude: the amount of push or pull represented by the length of the vector line, usually measured in pounds or kilograms.

Direction: the inclination of a force (with respect to a reference coordinate system) indicated by a line with an arrow at one end.

Point of application: the point through which the force is applied on the object or member (point A in **Fig 30.1A**).

Compression: the state created in a member by pushing forces that tend to shorten it. Compression is represented by the letter C or a plus sign (+).

Tension: the state created in a member by pulling forces that tend to stretch it. Tension is represented by the letter T or a minus sign (–).

System of forces: the combination of all forces acting on an object as shown (forces A, B, and C in **Fig 30.2**).

Resultant: a single force that can replace all the forces of a force system and have the same effect as the combined forces (force R1 in **Fig 30.2**).

Equilibrant: the opposite of a resultant; the single force that can be used to counterbalance all forces of a force system.

Components: separate forces that, if combined, would result in a single force; forces A and B are components of resultant R1 in **Fig 30.2**.

Space diagram: a diagram depicting the physical relationship between structural members, as given in **Fig 30.2**.

Vector diagram: a diagram of vectors representing the forces in a system and used to solve for unknown vectors in the system.

Metric units: standard units of weights and measures; the kilogram (kg) is the unit of mass (load), and one kilogram is approximately 2.2 pounds.

30.3 Coplanar, Concurrent Force Systems

When several forces, represented by vectors, act through a common point of application, the system is **concurrent**. In **Fig 30.2** vectors A, B, and C act through a single point; therefore this system is concurrent. When all vectors lie in the same plane, the system is **coplanar** and only one view is necessary to show them true length.

The **resultant** is the single vector that can replace all forces acting on the point of application. Resultants may be found graphically by (1) the parallelogram method and (2) the polygon method.

An **equilibrant** has the same magnitude, orientation, and point of application as the resultant in a system of forces, but in the opposite direction. The resultant of the system of forces shown

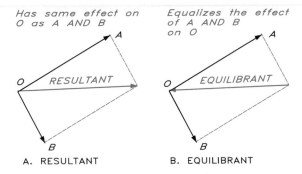

Figure 30.3 The (A) resultant and (B) equilibrant are equal in all respects except in direction (shown by arrowhead).

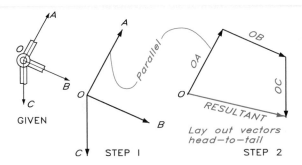

Figure 30.4 The resultant of a coplanar, concurrent system may be determined by the polygon method, in which the vectors are drawn head-to-tail. The vector that closes the polygon is the resultant.

in **Fig 30.3** is balanced by the equilibrant applied at point O, thereby causing the system to be in equilibrium.

Resultant: Parallelogram Method

In **Fig 30.2**, the vectors lie in the same plane, act through a common point, and are scaled to their known magnitudes. Use of the parallelogram method to determine resultants requires that the vectors be drawn to scale. Vectors A and B form two sides of a parallelogram. Constructing parallels to these vectors completes the parallelogram. Its diagonal, R1, is the resultant of forces A and B; that is, resultant R1 is the *vector sum* of vectors A and B.

Replaced by R1, vectors A and B now may be disregarded. Resultant R1 and vector C are two sides of a second parallelogram. Its diagonal, R, is the vector sum of R1 and C and the resultant of the entire system. Resultant R may be thought of as the only force acting on the point, thereby simplifying further analysis.

Resultant: Polygon Method

Figure 30.4 shows the same system of forces, but here the resultant is determined by the **polygon method**. Again, the vectors are drawn to scale but in this case head-to-tail, in their true directions to

form the polygon. The vectors are laid out in a clockwise sequence beginning with vector A. The polygon does not close, so the system is not in equilibrium but tends to be in motion. The resultant R (from the tail of vector A to the head of vector C) closes the polygon.

30.4 Noncoplanar, Concurrent Force Systems

When vectors lie in more than one plane of projection, they are **noncoplanar**, requiring 3D views for analysis of their spatial relationships. The resultant of a system of noncoplanar forces may be obtained by the parallelogram method if their projections are given in two adjacent orthographic views. Otherwise, it must be determined by the polygon method.

Resultant: Parallelogram Method

In **Fig 30.5** vectors 1 and 2 were used to construct the top and front views of a parallelogram and its diagonal R1 in both views. The front view of R1 must be an orthographic projection of its top view.

Then resultant R1 and vector 3 are resolved to form the overall resultant in both views. The top and front views of the resultant must project

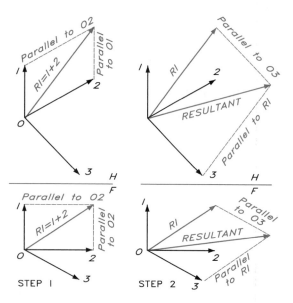

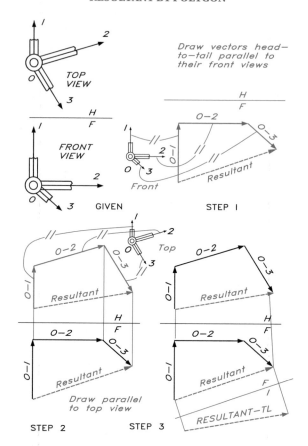

Figure 30.5 Finding the resultant by the parallelogram method:

Step 1 Use vectors 1 and 2 to construct a parallelogram in the top and front views. Diagonal R1 is the resultant of vectors 1 and 2.

Step 2 Use vectors 3 and R1 to construct a second parallelogram to find the overall resultant, R.

Figure 30.6 Obtaining the resultant by the polygon method:

Required Find the resultant of this system of forces by the polygon method.

Step 1 Lay off each vector head-to-tail in the front view parallel to the given view and find the front view of the resultant.

Step 2 Draw the same vectors head-to-tail in the top view, a 3D polygon projected above the front view.

Step 3 The resultant is found true length in an auxiliary view projected from the front view.

orthographically. The overall resultant replaces vectors 1, 2, and 3. However, it is an oblique line, so an auxiliary view (**Fig 30.6**) or revolution must be used to obtain its true length.

Resultant: Polygon Method

Figure 30.6 shows the solution of the same system of forces for the resultant by the polygon method. **Each vector is laid head-to-tail clockwise, beginning with vector 1 in the front view.**

Then the vectors are projected orthographically from the front view to the top view of the vector polygon. The vector polygon does not close, so the system is not in equilibrium. In both views the resultant (from the tail of vector 1 to the head of

vector 3) closes the polygon. However, the resultant is an oblique line, requiring an auxiliary view to obtain its true length.

Figure 30.7 The giant cranes on this offshore drilling rig are examples of coplanar, concurrent force systems. (Courtesy of Tennessee Gas Transmission Company.)

30.5 Forces in Equilibrium

The giant cranes on the offshore drilling platform shown in **Fig. 30.7** are examples of coplanar, concurrent structures in equilibrium. **A structure in equilibrium is one that is static with no motion taking place; the members balance each other.**

The coplanar, concurrent structure depicted in **Fig. 30.8** is designed to support a load of W = 165 kg. The maximum loading of each structural member determines the material and size of the members to be used in the design.

A single view of a vector polygon in equilibrium allows you to find only two unknown values. (Later, we show how to solve for three unknowns by using descriptive geometry.) Lay off the only known force, W = 165 kg, parallel to its given direction (here pointing vertically downward). Then draw the unknown forces A and B parallel to the supports to form the force polygon and scale (or calculate) the magnitude of these forces.

Analyze vectors A and B to determine whether they are in tension or compression and thus find their direction. Vector B points upward to the left, which is toward point O when transferred to the

structural diagram shown in the small drawing. Vectors that act toward their point of application are in compression. Vector A points away from point O when transferred to the structural diagram and therefore is in tension.

Figure 30.9 is a similar example involving determination of the loads in the structural members caused by the weight of 110 pounds acting through a pulley. The only difference between this solution and the previous one is the construction of two equal vectors at the outset to represent the cable loads on both sides of the pulley.

30.6 Coplanar Truss Analysis

Designers use vector polygons to determine the loads in each member of a truss by two graphical methods: (1) joint-by-joint analysis and (2) Maxwell diagrams.

Joint-by-Joint Analysis

In the Fink truss shown in **Fig. 30.10**, 3000-lb loads are applied at its joints. This method of designating forces is called **Bow's notation**. The exterior forces on the truss are labeled with letters placed between them, and numerals are placed between the interior members. Each vector is referred to by the number on each of its sides clockwise about its joint. For example, the vertical load at the left is denoted AB, with A at the tail and B at the head of the vector.

First analyze the joint at the left end with a reaction of 4500 lb. When you read clockwise about the joint, the force is EA, where E is the tail and A is the head of the vector. Continuing clockwise, the next forces are A–1 and 1–E, which close the polygon at E, the beginning letter. Place arrowheads in a head-to-tail sequence beginning with the known vector EA.

Determine tension and compression by relating the directions of each vector to the original joint. For example, A–1 points toward the joint and is in compression, whereas 1–E points away and is in tension. The truss is symmetrical and

COPLANAR FORCES IN EQUILIBRIUM

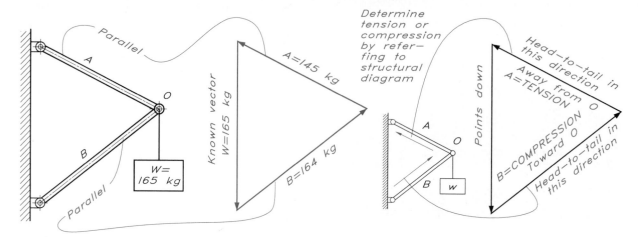

Figure 30.8 Determining coplanar forces in equilibrium:

Required Find the forces in the structural members supporting the 165 kg load.

Step 1 Draw the load of 165 kg as a vector. Draw vectors A and B parallel to their directions from the ends of the load and head-to-tail.

Step 2 Vector A points away from point O when transferred to the structural diagram and thus is in tension. Vector B points toward point O and thus is in compression.

FORCES IN EQUILIBRIUM WITH PULLEY

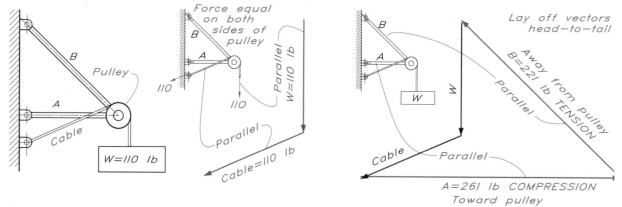

Figure 30.9 Obtaining forces in equilibrium (pulley application):

Required Find the forces in the members supporting the load of 110 lb on the pulley.

Step 1 The force in the cable is equal to 110 lb on both sides of the pulley. Draw these two forces as vectors head-to-tail and parallel to their directions in the space diagram.

Step 2 Draw A and B head-to-tail to close the polygon. Vector A points toward the point of application and thus is in compression. Vector B points away from the point and is in tension.

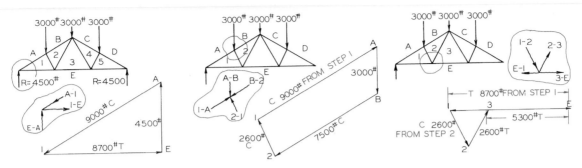

Figure 30.10 Analyzing a truss (joint-by-joint method):

Step 1 Use Bow's notation to label the truss, with letters between the exterior loads and numbers between interior members. Because it has only two unknowns, A–1 and 1–E, you may analyze the left joint. Find vectors A–1 and 1–E, by drawing them parallel to their directions from both ends of EA in a head-to-tail sequence.

Step 2 Using vector 1–A from step 1 and load AB, find the two unknowns, B–2 and 2–1. Draw the known vectors beginning with vector 1–A and then draw vectors B–2 and 2–1 to close the polygon, moving clockwise about the joint. A vector pointing toward the point of application is in compression. A vector pointing away from the point of application is in tension.

Step 3 Analyze the third joint by laying out vectors E–1 and 1–2 from the preceding steps. Vectors 2–3 and 3–E close the polygon and are parallel to their directions in the space diagram. Vectors 2–3 and 3–E point away from the point of application and thus are in tension.

equally loaded, so the loads in the members on the right will be equal to those on the left.

Analyze the other joints in the same way. The directions of the vectors are opposite at each end. For example, vector A–1 is toward the left in step 1 and toward the right in step 2.

Maxwell Diagrams

The Maxwell diagram is virtually the same as the joint-by-joint analysis, with the exception that the polygons overlap, with some vectors common to more than one polygon. In **Fig. 30.11** (step 1) the exterior loads are laid out head-to-tail in clockwise sequence—AB, BC, CD, DE, and EA—with a letter placed at each end of each vector. The forces are parallel so this force diagram is a vertical line.

Vector analysis begins at the left end where the force EA of 4500 lb is known. A free-body diagram

is sketched to isolate this joint. The two unknowns, A–1 and 1–E, are drawn parallel to their directions in the truss, with A–1 beginning at point A, 1–E beginning at point E, and both extended to point 1.

Because resultant EA points upward, A–1 must have its tail at A and its direction toward point 1. The free-body diagram shows that the direction is toward the point of application, which means that A–1 is in compression. Vector 1–E points away from the joint, which means that it is in tension. The vectors are coplanar and may be scaled to determine their magnitudes.

In step 2, where vectors 1–A and AB are known, the unknown vectors, B–2 and 2–1 may be determined. Vector B–2 is drawn parallel to its structural member through point B in the Maxwell diagram, and the line of vector 2–1 is extended from point 1 to intersect with B–2 at

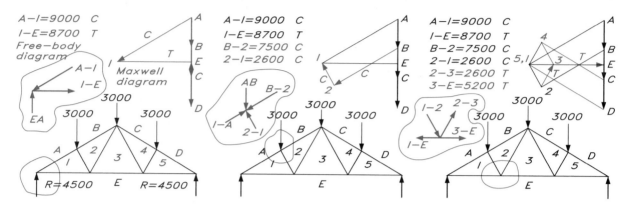

Figure 30.11 Analyzing a truss (Maxwell diagram method):

Step 1 Label the spaces between the outer loads on the truss with letters and the internal spaces with numbers using Bow's notation. Add the vertical loads end-to-end in a Maxwell diagram, and sketch a free-body diagram of the first joint. Use vectors EA, A–1, and 1–E (head-to-tail) to draw a vector diagram to find their magnitudes. Vector A–1 is in compression because it points toward the joint, and 1-E is in tension because it points away from the joint.

Step 2 Sketch the next joint to be analyzed. Because AB and A–1 are known, only 2–1 and B–2 are unknown. Draw them parallel to their direction (head-to-tail) in the Maxwell diagram using the previously found vector. Vectors B–2 and 2–1 are in compression, as each points toward the joint. Vector A–1 becomes vector 1–A when read in a clockwise direction.

Step 3 Sketch a free-body diagram of the next joint to be analyzed, where the unknowns are 2–3 and 3–E. Draw their vectors in the Maxwell diagram parallel to their given members to find point 3. Vectors 2–3 and 3–E are in tension because they point away from the joint. Repeat this process to find the vectors on the opposite side.

point 2. The arrows of each vector are drawn head-to-tail. Vectors B–2 and 2–1 point toward the joint in the free-body diagram; and therefore are in compression.

In step 3, the next joint is analyzed to find the forces in 2–3 and 3–E. The truss and its loading are symmetrical, so the Maxwell diagram will be symmetrical when completed.

If the last force polygon in the series does not close perfectly, an error in construction has occurred. A slight error may be disregarded, as a rounding error may be disregarded in mathematics. Arrowheads are unnecessary and usually are omitted on Maxwell diagrams because each vector will have the opposite direction when applied to a different joint.

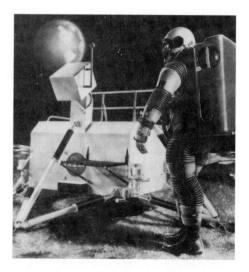

Figure 30.12 The structural members of this tripod support for a moon vehicle may be analyzed graphically to determine design load requirement. (Courtesy of NASA.)

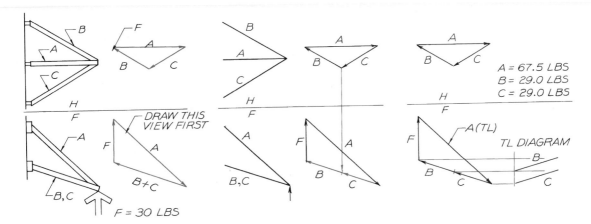

Figure 30.13 Noncoplanar structural analysis (special case):
Step 1 Forces B and C coincide in the front view, where there are only two unknowns. Draw vector F (30 lb) and the two unknown forces parallel to their front view in the front view of the vector polygon. To find the top view of A, project from the front and draw vectors B and C parallel to their top views.

Step 2 Project the point of intersection of vectors B and C in the top view to the front view to separate the head-to-tail vectors. Vectors B and C are in tension because they act away from the point in the space diagram, whereas vector A is in compression.

Step 3 The lengths of vectors B and C are not true length in their top and front views. Construct a true-length diagram and scale the lengths of these lines to obtain their magnitudes.

30.7 Noncoplanar Vector Analysis

Special Case

The solution of 3D vector systems requires the use of descriptive geometry because the system must be analyzed in 3D space. An example is the manned flying system (MFS) shown in **Fig 30.12**, which was analyzed to determine the loads on its support members. Weight on the moon is 0.165 of earth weight. Thus a tripod that must support 182 lb on earth needs to support only 30 lb on the moon.

In general, only two unknown vectors can be determined in a single view of a vector polygon that is in equilibrium. However, the system shown in **Fig 30.13 is a special case because members B and C lie in the same edge view of the plane in the front view**. Therefore solving for three unknowns is possible in this case.

Construct a vector polygon in the front view by drawing force F as a vector and using the other vectors as the sides of the polygon. Draw the top view using vectors B and C to form the polygon that closes at each end of vector A. Then find the front view of vectors B and C.

A true-length diagram gives the lengths of the vectors; measure them to determine their magnitudes. Vector A is in compression because it points toward the point of application. Vectors B and C are in tension because they point away from the point.

General Case

The structural frame shown in **Fig 30.14** is attached to a vertical wall to support a load of W = 1200 lb. There are three unknowns in each of the views, so begin by projecting an auxiliary view

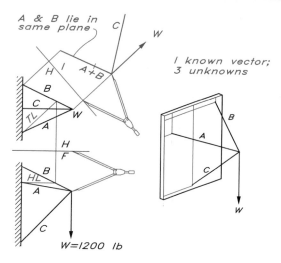

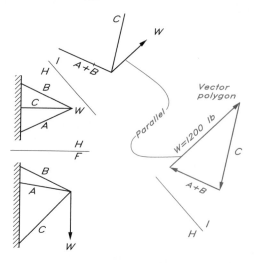

Figure 30.14 Noncoplanar structural analysis (general case):

Given The top and front views of a structural frame that supports a load of 1200 lb.

Required Find the loads in the structural members.

Step 1 To limit the unknowns to two, draw an auxiliary view where vectors A and B lie in the edge view of a plane. Draw a vector polygon parallel to the members in the auxiliary view in which W = 1200 lb is the only known vector.

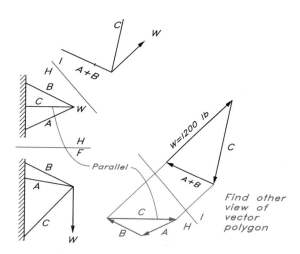

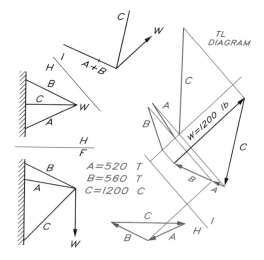

Step 2 Construct an orthographic projection of the vector polygon from step 1, with its vectors parallel to the members in the top view. The reference plane between the two views is parallel to the H1 plane. (This portion of the solution is closely related to that in Fig 30.13.)

Step 3 Project the intersection of A and B in the top view of the vector polygon to the auxiliary view polygon. Find the magnitudes of vectors A, B, and C in a true-length diagram and analyze them for tension or compression by referring to the top and auxiliary views.

Figure 30.15 Tractor sidebooms represent noncoplanar, concurrent systems of forces that can be solved graphically. (Courtesy of Trunkline Gas Company.)

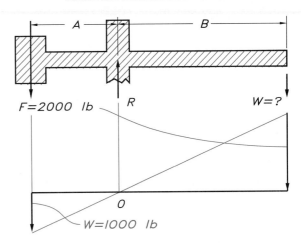

Figure 30.16 The load W required to balance the beam can be found by this diagram for a system of parallel, nonconcurrent forces.

from the top view to obtain the edge view of a plane containing vectors A and B, thereby reducing the number of unknowns to two. You no longer need refer to the front view.

Draw a vector polygon with vectors parallel to their members in the auxiliary view. Then draw an adjacent orthographic view of the vector polygon with vectors parallel to their members in the top view. Use a true-length diagram to find the true length of the vectors and measure their magnitudes.

An application of a 3D vector system is the side-boom tractors used to lower pipe into ditches during pipeline construction **(Fig 30.15)**.

30.8 Resultant of Parallel, Nonconcurrent Forces

The beam depicted in **Fig 30.16** is part of a rotational crane used to move building materials. The counterbalance weight is 2000 lb and the magnitude of the weight W is to be found, assuming that the support cables have been omitted.

To obtain the graphical solution, construct a line to represent the distance between forces F

and W. Project point O, the pivot point of balance, from the space diagram. Draw vectors F and W by transposing them to the opposite ends of the beam. Draw a line from the end of vector F through point O and extend it to intersect the extension of vector W to locate the end of vector W. When you scale it, you find that its magnitude is 1000 lb.

The beam depicted in **Fig 30.17** must carry the three loads shown. The requirement is to determine the magnitude of supports R_1 and R_2, the resultant of the loads, and the resultant's location. Begin by labeling the spaces between all vectors clockwise with Bow's notation and draw a vector diagram.

Extend the lines of force in the space diagram and draw the strings from the vector diagram in their respective spaces, parallel to their original directions. For example, string oa is parallel to string oA in space A between forces EA and AB, and string ob is in space B, beginning at the intersection of oa with vector AB. The last string, oe, closes the diagram, called a **funicular diagram**.

Transfer the direction of string oe to the force diagram, and lay it off through point O to intersect

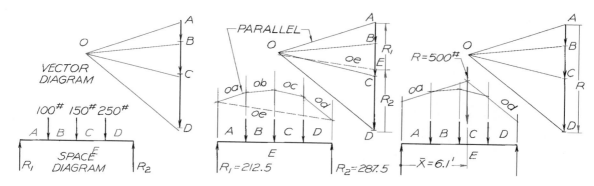

Figure 30.17 Analyzing parallel, nonconcurrent loads:

Step 1 Letter the spaces between the loads using Bow's notation. Find the sum of the vertical loads by drawing them head-to-tail in a vector diagram. Locate pole point O at a convenient location and draw strings from O to each end of the vectors.

Step 2 Extend the lines of vertical loads and draw a funicular diagram with string oa in the A space, ob in the B space, oc in the C space, and so on. The last string drawn, oe, closes the diagram. Transfer oe to the vector polygon to locate point E, thus establishing R₁ and R₂, which are EA and DE, respectively.

Step 3 The resultant of the three downward forces equals their graphical summation, line AD. Locate the resultant by extending strings oa and od in the funicular diagram to a point of intersection. The resultant, R = 500 lb, acts through this point in a downward direction at distance X̄ from the left end.

the load line at E. Vector DE represents R_2 (refer to Bow's notation as it was applied in step 1), and vector EA represents R_1.

The magnitude of the resultant of the loads is the sum of the downward forces, or the distance from A to D. To find the location of the resultant, extend the outside strings of the funicular diagram, oa and od, to intersect. The resultant has a magnitude of 500 lb, a vertical downward direction, and a point of application at X̄.

Problems

Draw your solutions to these problems with instruments on size A grid or plain sheets. Each grid square represents 0.20 in. (5mm). Make all notes, sketches, drawings, and graphical work neatly and use good design practices. Letter written matter legibly, using ¹/₈-in. guidelines.

1. (Fig. 30.18)

(A–B) Find the resultants of the force systems by the parallelogram and polygon methods. Scale: 1" = 100 lb.

2. (Fig. 30.19)

(A–B) Find the resultants of the force systems by the parallelogram and polygon methods. Scale: 1" = 100 lb.

3. (Fig. 30.20)

(A–B) Find the forces in the coplanar force systems. Label the members, assign the forces in each of them, and indicate whether the forces are compression or tension.

4. (Fig. 30.21) Find the forces in each member of the truss by using a Maxwell diagram. Make a table of forces and indicate compression and tension.

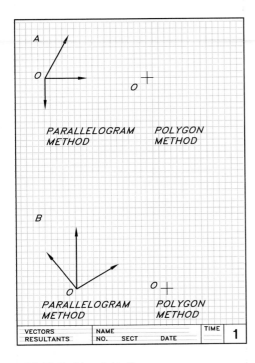

Figure 30.18 Problem 1 (A–B).

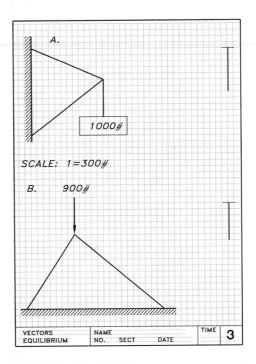

Figure 30.20 Problem 3 (A–B).

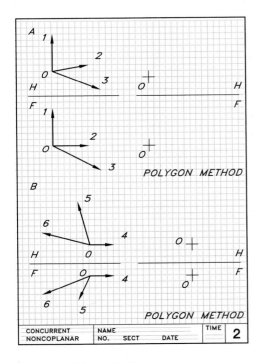

Figure 30.19 Problem 2 (A–B).

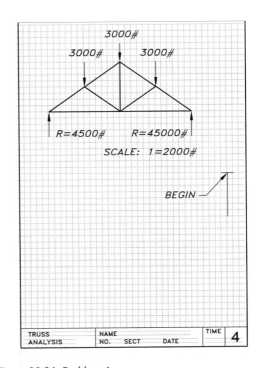

Figure 30.21 Problem 4.

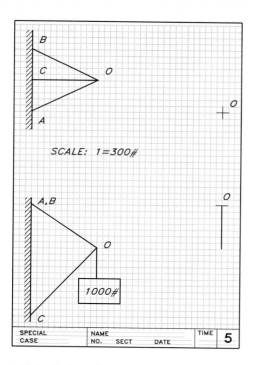

Figure 30.22 Problem 5.

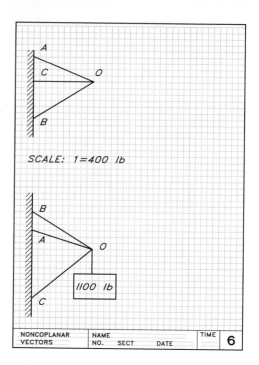

Figure 30.23 Problem 6.

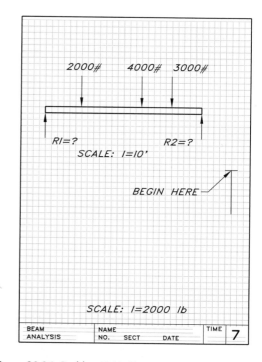

Figure 30.24 Problem 7 (A–B).

5. (Fig. 30.22) Find the forces in the members of the concurrent noncoplanar force system. Make a table of forces and indicate compression and tension.

6. (Fig. 30.23) Find the forces in the members of the concurrent noncoplanar force system. Make a table of forces and indicate compression and tension.

7. (Fig. 30.24)

  (A) Find the forces in reactions R1 and R2 necessary to equalize the loads applied to the beam.

  (B) Find the value and location of the single support that could replace both R1 and R2.

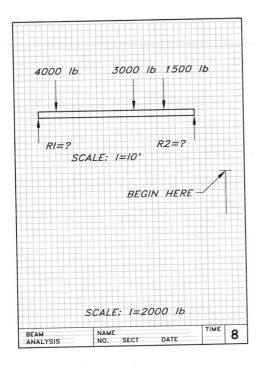

Figure 30.25 Problem 8 (A–B).

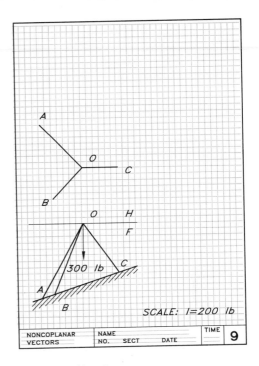

Figure 30.26 Problem 9.

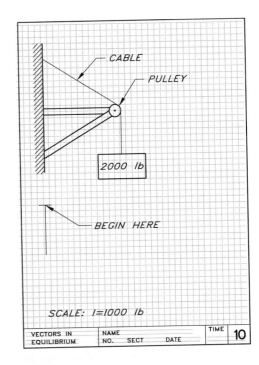

8. (Fig. 30.25) Repeat Problem 7 for this configuration.

9. (Fig. 30.26) Find the forces in the support members. Make a table of forces and indicate compression and tension.

10. (Fig. 30.27) Find the forces in the coplanar force system. Label the members, make a table of forces, and indicate compression and tension.

11. Repeat Problem 4, but use the joint-by-joint analysis instead of the Maxwell diagram method.

Figure 30.27 Problem 10.

31

Intersections and Developments

31.1 Introduction

Several methods may be used to find lines of intersection between parts that join. Usually such parts are made of sheet metal, or of plywood if used as forms for concrete. After **intersections** are found, **developments**, or flat patterns, can be laid out on sheet metal and cut to the desired shape. The refinery shown in **Fig. 31.1** illustrates many examples of intersections and developments.

31.2 Intersections of Lines and Planes

Figure 31.2 illustrates how to find the intersection between a line and a plane. This example is a special case in which the point of intersection clearly shows in the view where the plane appears as an edge. Projecting the piercing point, P, to the front view completes the visibility of the line.

This same principle is applied to finding the line of intersection between two planes (**Fig. 31.3**). By locating the piercing points of lines AB and DC and connecting these points, the line of intersection is found.

The angular intersection of two planes at a corner gives a line of intersection that bends around the corner (**Fig. 31.4**). First, find piercing points 2' and 1'. Then project corner point 3 from the side view where the vertical corner pierces the plane to the front view of the corner. Point 2' is hidden in the front view because it is on the back side.

Figure 31.5 shows how to find the intersection between a plane and prism where the plane appears as an edge. Obtain the piercing points for each corner line and connect them to form the line of intersection. Show visibility to complete the intersection.

Figure 31.6 depicts a more general case of an intersection between a plane and prism. Passing vertical cutting planes through the planes of the prism in the top view yields traces (cut lines) on the front view of the oblique plane on which the piercing points of the vertical corner lines lie. Connect the points and determine visibility to complete the solution.

In **Fig. 31.7**, finding the intersection between a foreshortened plane and an oblique prism

Figure 31.1 The design of this refinery in Borger, Texas, involved the application of many principles of intersections and developments. (Courtesy of Phillips Petroleum Company.)

LINE AND A PLANE

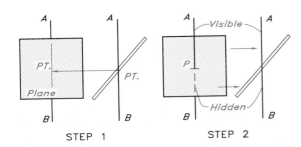

STEP 1 STEP 2

Figure 31.2 Intersection of a line and a plane:
Step 1 Find the point of intersection in the view where the plane appears as an edge, the side view in this case, and project it to the front view.

Step 2 Determine visibility in the front view by looking from the front view to the right-side view.

BETWEEN PLANES

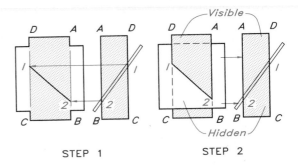

STEP 1 STEP 2

Figure 31.3 Intersection of planes:
Step 1 Find the piercing points of lines AB and DC with the plane where the plane appears as an edge and project them to the front view.

Step 2 Line 1–2 is the line of intersection. Determine visibility by looking from the front view to the right-side view.

PLANE AT A CORNER

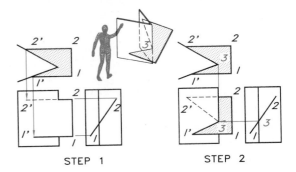

STEP 1 STEP 2

Figure 31.4 The intersection of a plane at a corner:
Step 1 The intersecting plane appears as an edge in the side view. Project intersection points 1' and 2' from the top and side views to the front view.

Step 2 The line of intersection from 1' to 2' must bend around the vertical corner at 3' in the top and side views. Project point 3' to the front view to locate line 1'–3'–2'.

involves finding an auxiliary view to obtain the edge view of the plane and simplify the problem. The piercing points of the corner lines of the prism lie in the auxiliary view and project back to the given views. Points 1, 2, and 3, projected from

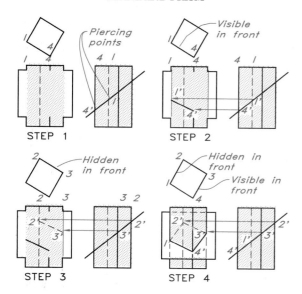

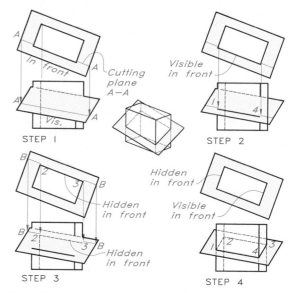

Figure 31.5 The intersection of a plane and a prism:
Step 1 Vertical corners 1 and 4 intersect the edge view of the plane in the side view at points 1′ and 4′.

Step 2 Project points 1′ and 4′ from the side view to lines 1 and 4 in the front view. Connect them to form a visible line of intersection.

Step 3 Vertical corners 2 and 3 intersect the edge view of the plane at points 2′ and 3′ in the side view. Project points 2′ and 3′ to the front view to form a hidden line of intersection.

Step 4 Connect points 1′, 2′, 3′, and 4′ and determine visibility by analyzing the top and side views.

Figure 31.6 The intersection of an oblique plane and a prism:
Step 1 Pass vertical cutting plane A–A through corners 1 and 4 in the top view and project endpoints to the front view.

Step 2 Locate piercing points 1′ and 4′ in the front view where line A–A crosses lines 1 and 4.

Step 3 Pass vertical cutting plane B–B through corners 2 and 3 in the top view and project them to the front view to locate piercing points 2′ and 3′.

Step 4 Connect the four piercing points and determine visibility by analysis of the top view.

the auxiliary view to the given views, are shown as examples. Analysis of crossing lines determines visibility to complete the line of intersection in the top and front views.

31.3 Intersections between Prisms

The techniques used to find the intersection between planes and lines also apply to finding the intersection between two prisms (**Fig. 31.8**). Project piercing points 1, 2, and 3 from the side and top views to the front view. Point X lies in the

side view where line of intersection 1–2 bends around the vertical corner of the vertical prism. Connect points 1, X, and 2 and determine visibility.

Figure 31.9 illustrates how to find the line of intersection between an inclined prism and a vertical prism. An auxiliary view reveals the end view of the inclined prism where its planes appear as edges. In the auxiliary view, plane 1–2 bends around corner AB at point P. Project points of intersection 1′ and 2′ from the top and auxiliary to their intersections in the front view. Then draw the line of intersection 1′–P–2′ for this portion of

OBLIQUE PLANE AND PRISM

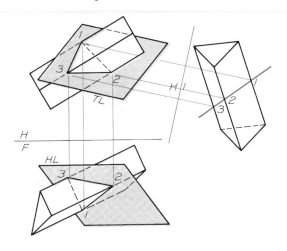

Figure 31.7 To find the intersection between a plane and a prism, construct a view in which the plane appears as an edge. Project piercing points 1, 2, and 3 back to the top and front views.

PERPENDICULAR PRISMS

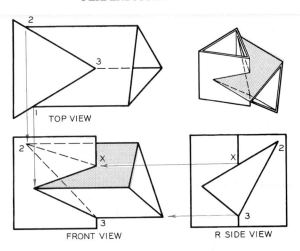

Figure 31.8 These are three views of intersecting prisms. The points of intersection are best found where intersecting planes appear as edges.

INTERSECTION BETWEEN PRISMS

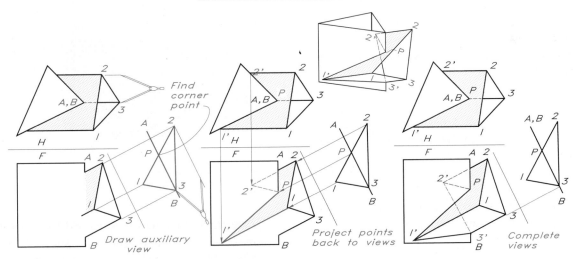

Figure 31.9 The intersection of prisms by auxiliary view:
Step 1 Construct the end view of the inclined prism by projecting an auxiliary view from the front view. Show line AB of the vertical prism in the auxiliary view.

Step 2 Locate piercing points 1' and 2' in the top and front views. Intersection line 1'–2' bends around corner AB at point P (projected from the auxiliary view).

Step 3 Intersection lines from 2' and 1' to 3' do not bend around the corner and are straight lines. Line 1'–3' is visible, and line 2'–3' is invisible.

INCLINED PRISM

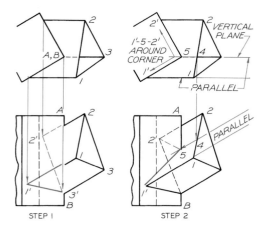

Figure 31.10 The intersection of prisms by projection:
Step 1 Project the piercing points of lines 1, 2, and 3 from the top view to the front view to locate piercing points 1', 2', and 3'.

Step 2 Pass a cutting plane through corner AB in the top view to locate point 5, where intersection line 1'–2' bends around the vertical prism. Find point 5 in the front view and draw line 1'–5–2'.

DESIGN APPLICATION: CONDUIT CONNECTOR

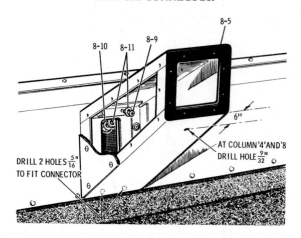

Figure 31.11 The design of this conduit connector involved the use of the principles of the intersection of a plane and prism. (Courtesy of the Federal Aviation Administration.)

the line of intersection. Connect the remaining lines, 1'–3' and 2'–3' to complete the solution.

Figure 31.10 shows an alternative method of solving this type of problem. Piercing points 1' and 2' appear in the front view as projections from the top view. Point 5 is the point where line 1'–5–2' bends around vertical corner AB. To find point 5 in the front view, pass a cutting plane through corner AB in the top view and project its trace to the front view. Draw the lines of intersection, 1'-5-2'.

The conduit connector shown in **Fig. 31.11** is an application of the principles of intersecting plane and prism.

31.4 Intersections between Planes and Cylinders

The sheet–metal flashing shown in **Fig. 31.12** is an example of a cylinder intersecting a plane.

DESIGN APPLICATION: FLASHING

Figure 31.12 This piece of sheet metal flashing illustrates the line of intersection between a plane and a cylinder.

Figure 31.13 shows how to find the intersection between a plane and a cylinder. Cutting planes passed vertically through the top view of the cylinder establish pairs of elements on the cylinder and their piercing points. Space the cutting planes conveniently apart by eye. Then project the piercing points to each view and draw the elliptical line of intersection.

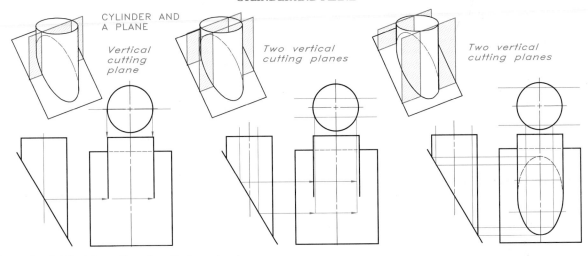

Figure 31.13 The intersection of a cylinder and a plane:
Step 1 Pass a vertical cutting plane through the cylinder parallel to its axis to find two points of intersection.

Step 2 Use two more cutting planes to find four additional points in the top and left side views. Project these points to the front view.

Step 3 Use additional cutting planes to find more points. Connect these points to give an elliptical line of intersection.

Figure 31.14 shows the solution of a more general problem. Here, the cylinder is vertical and the plane is oblique and does not appear as an edge. Passing vertical cutting planes through the cylinder and the plane in the top view gives elements on the cylinder and their piercing points on the plane. Projecting these points to the front view completes the elliptical line of intersection. The more cutting planes used, the more accurate the line of intersection will be.

Figure 31.15 demonstrates the general case of the intersection between a plane and cylinder, where both are oblique in the given views. An auxiliary view is used to show the edge view of the plane. Cutting planes passed through the cylinder parallel to its axis in the auxiliary view determine elements on the cylinder and their piercing points. An elliptical line of intersection is found

when the points are connected and projected back to the given views.

31.5 Intersections between Cylinders and Prisms

An inclined prism intersects a vertical cylinder in **Fig. 31.16**. A primary auxiliary view is drawn to show the end view of the inclined prism where its planes appear as edges. A series of vertical cutting planes in the top view establish lines lying on the surfaces of the cylinder and prism. The cutting planes, also shown in the auxiliary view, are the same distance apart as in the top view.

Projecting the line of intersection from 1 to 3 from the auxiliary view to the front view yields an elliptical line of intersection. The visibility of this line changes from visible to hidden at point X,

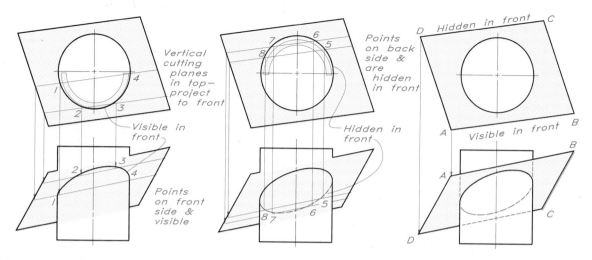

Figure 31.14 Intersection of a cylinder and an oblique plane:

Step 1 Pass vertical cutting planes through the cylinder in the top view to establish elements on its surface and lines on the oblique plane. Project piercing points 1, 2, 3, and 4 to the front view of their respective lines and connect them with a visible line.

Step 2 Use additional cutting planes to find other piercing points—5, 6, 7, and 8—and project them to the front view of their respective lines on the oblique plane. Connect these points with a hidden line.

Step 3 Determine visibility of the plane and cylinder in the front view. Line AB is visible by inspection of the top view, and line CD is hidden.

OBLIQUE CYLINDER AND OBLIQUE PLANE

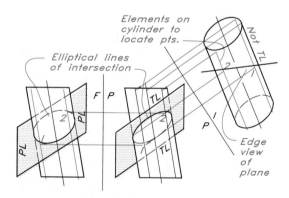

Figure 31.15 To find the intersection between an oblique cylinder and an oblique plane, construct a view that shows the plane as an edge. Cutting planes passed through the cylinder locate points on the line of intersection.

which appears in the auxiliary view and is projected to the front view. Continuing this process gives the lines of intersection of the other two planes of the prism.

31.6 Intersections between Two Cylinders

To find the line of intersection between two perpendicular cylinders, pass cutting planes through them parallel to their centerlines (**Fig. 31.17**). Each cutting plane locates a pair of elements on both cylinders that intersect at a piercing point. Connecting the points and determining visibility completes the solution. The design and fabrication of the vessels and pipes shown in **Fig. 31.18** required application of these principles.

Figure 31.19 illustrates how to find the intersection between nonperpendicular cylinders. This

CYLINDER AND A PRISM

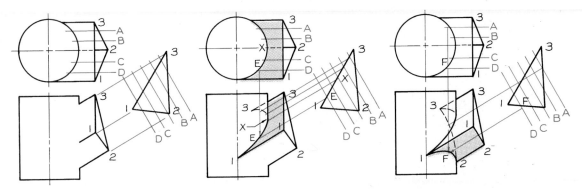

Figure 31.16 The intersection of a cylinder and a prism:

Step 1 Project an auxiliary view of the triangular prism from the front view to show the edge views of its planes. Draw frontal cutting planes through the top view of the cylinder and locate them in the auxiliary view with your dividers.

Step 2 Locate points along intersection line 1–3 in the top view and project them to the front view. For example, find point E on cutting plane D in the top and auxiliary views and project it to the front view where the projectors intersect. Visibility changes in the front view at point X.

Step 3 Determine the remaining points of intersection by using the other cutting planes. Project point F, shown in the top and auxiliary views, to the front view of line 1–2. Connect the points and determine visibility.

INTERSECTION BETWEEN CYLINDERS

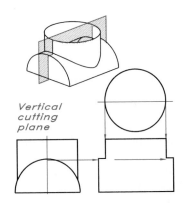

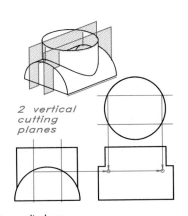

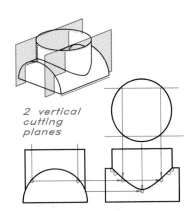

Figure 31.17 Obtaining the intersection of two cylinders:

Step 1 Pass a cutting plane through the cylinders parallel to their axes, locating two points of intersection.

Step 2 Use two more cutting planes to find four additional points of intersection.

Step 3 Use two more cutting planes to locate four more points. Connect the points to find the line of intersection.

Figure 31.18 These vessels and pipes are examples of intersections between cylinders. (Courtesy of Lone Star Gas Company.)

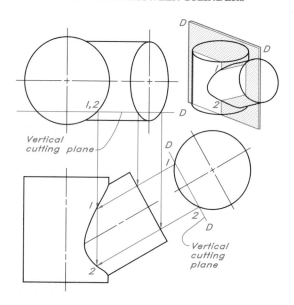

Figure 31.19 To find the intersection of these cylinders, locate the end view of the inclined cylinder in an auxiliary view. Then use vertical cutting planes to find the piercing points of the cylindrical elements and the line of intersection.

method involves passing a series of vertical cutting planes through the cylinders parallel to their centerlines. Points 1 and 2, labeled on cutting plane D, are typical of points on the line of intersection. Other points may be found in the same manner. Although the auxiliary view is not essential to the solution, it is an aid in visualizing the problem. Projecting points 1 and 2 on cutting plane D in the auxiliary view to the front view provides a check on the projections from the top view.

31.7 Intersections between Planes and Cones

To find points of intersection on a cone, use cutting planes that are **(1) perpendicular to the cone's axis** or **(2) parallel to the cone's axis**. The vertical planes in the top view of **Fig. 31.20A** cut radial lines on the cone and establish elements on its surface. The horizontal planes in **Fig. 31.20B** cut circular sections that appear true size in the top view of a right cone.

A series of **radial cutting planes** define elements on a cone (**Fig. 31.21**). These elements cross the edge view of the plane in the front view to locate piercing points of each element that, when projected to the top view of the same elements, lie on the line of intersection.

A series of **horizontal cutting planes** may be used to determine the line of intersection between a cone and an oblique plane (**Fig. 31.22**). The sections cut by these imaginary planes are circles in the top view. The cutting planes also locate lines on the oblique plane that intersect the circular sections cut by each respective cutting plane. The

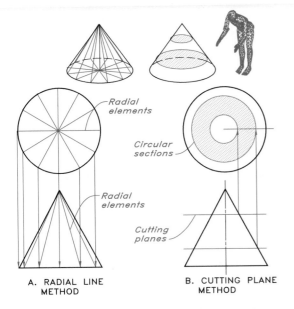

Figure 31.20 To find intersections on conical surfaces, you may use (A) radial cutting planes that pass through the cone's centerline and are perpendicular to its base, or (B) cutting planes that are parallel to the cone's base.

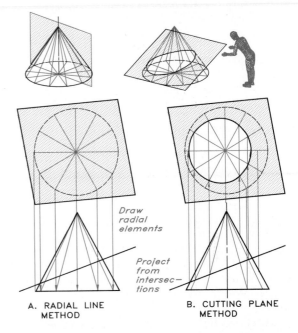

Figure 31.21 The intersection of a plane and a cone:
Step 1 Divide the base evenly in the top view and connect these points with the apex to establish elements on the cone. Project these elements to the front view.

Step 2 Project the piercing point of each element on the edge view of the plane to the top view of the same elements, and connect them to form the line of intersection.

points of intersection found in the top view project to the front view. We could have used the horizontal cutting-plane method in Fig. 31.21 to obtain the same results.

31.8 Intersections between Cones and Prisms

A primary auxiliary view gives the end view of the inclined prism that intersects the cone in **Fig. 31.23**. Cutting planes that radiate from the apex of the cone in the top view locate elements on the cone's surface that intersect the prism in the auxiliary view. These elements project to the front view.

Wherever the edge view of plane 1–3 intersects an element in the auxiliary view, the piercing points project to the same element in the front

and top views. Passing an extra cutting plane through point 3 in the auxiliary view locates an element that projects to the front and top views. Piercing point 3 projects to this element in sequence from the auxiliary view to the top view.

This same procedure yields the piercing points of the other two planes of the prism. All projections of points of intersection originate in the auxiliary view, where the planes of the prism appear as edges.

In **Fig. 31.24**, horizontal cutting planes passed through the front view of the cone and cylinder give a series of circular sections in the top view. Points 1 and 2, shown on cutting plane C in the

OBLIQUE PLANE AND A CONE

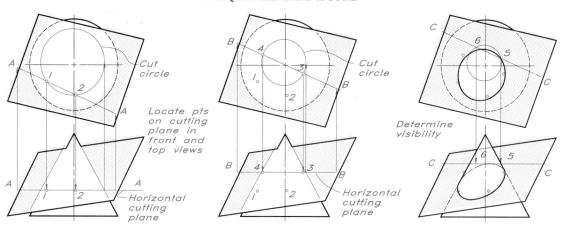

Figure 31.22 Determining the intersection of an oblique plane and a cone:

Step 1 Pass a horizontal cutting plane through the front view to establish a circular section on the cone and a line on the plane in the top view. The piercing points of this line are on the circular section. Project piercing points 1 and 2 to the front view.

Step 2 Pass horizontal cutting plane B–B through the front view in the same manner to locate piercing points 3 and 4 in the top view. Project these points to the horizontal plane in the front view from the top view.

Step 3 Use additional horizontal planes to find a sufficient number of points to complete the line of intersection and then determine visibility.

CONE AND PRISM

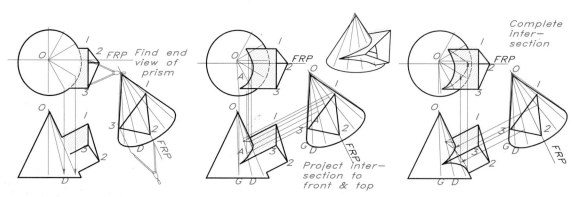

Figure 31.23 The intersection of a cone and a prism:

Step 1 Construct an auxiliary view to obtain the edge views of the lateral surfaces of the prism. In the auxiliary view, pass cutting planes through the cone that radiate from the apex to find elements on the cone. Project the elements to the front and auxiliary views.

Step 2 Locate the piercing points of the cone's elements with the edge view of plane 1–3 in the primary view and project them to the front and top views. For example, point A lies on element OD in the primary auxiliary view, so project it to the front and top views of OD.

Step 3 Locate the piercing points where the conical elements intersect the edge views of the planes of the prism in the auxiliary view. For example, find point B on O in the primary auxiliary view and project it to the front and top views of OE.

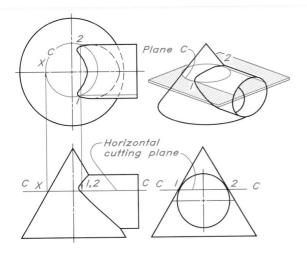

Figure 31.24 Horizontal cutting planes are used to find the intersection between the cone and the cylinder. The cutting planes cut circles in the top view. Only one cutting plane is shown here as an example.

Figure 31.25 This electrically operated distributor is an application of intersections between a cone and a series of cylinders. (Courtesy of GATX.)

top view, are typical and project to the front view. The same method produces other points.

This method is feasible only when the centerline of the cylinder is perpendicular to the axis of the cone, producing circular sections in the top view (rather than elliptical sections, which would be difficult to draw). The distributor housing in **Fig. 31.25** is an example of an intersection between cylinders and a cone.

31.9 Intersections between Prisms and Pyramids

Figure 31.26 shows how to find the intersection of an inclined prism with a pyramid. An auxiliary view shows the end view of the inclined prism and the pyramid. The radial lines OB and OA drawn through corners 1 and 3 in the auxiliary view project back to the front and top views. Projection locates intersecting points 1 and 3 on lines OB and OA in each view. Point P is the point where line 1–3 bends around corner OC. Finding lines of

intersection 1–4 and 4–3 and determining visibility completes the solution.

Figure 31.27 shows a horizontal prism that intersects a pyramid. An auxiliary view depicts the end view of the horizontal prism with its planes as edges. Passing a series of horizontal cutting planes through the corner points of the horizontal prism and the pyramid in the auxiliary view gives the lines of intersection, which form triangular sections in the top view.

The cutting plane through corner point P in the auxiliary view is an example of a typical cutting plane. At point P the line of intersection of this plane bends around the corner of the pyramid. Other cutting planes are passed through the corner lines of the prism in the auxiliary and front views. Each corner line extends in the top view to intersect the triangular section formed by the cutting plane, as shown at P.

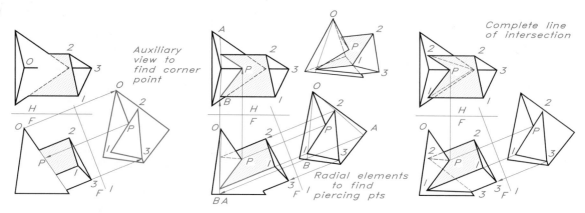

Figure 31.26 The intersection of a prism and a pyramid:

Step 1 Find the edge view of the surfaces of the prism by projecting an auxiliary view from the front view. Project the pyramid into this view also, showing only the visible surfaces.

Step 2 Pass planes A and B through apex O and points 1 and 3 in the auxiliary view. Project lines OA and OB to the front and top views and project points 1 and 3 to them. Point 2 lies on line OC. Connect points 1, 2, and 3 to give the intersection of the upper plane.

Step 3 Point 3 lies on line OC in the auxiliary view. Project this point to the principal views. Connect point 3 to points 1 and 2 to complete the intersections and show visibility. Assume that these geometric shapes are constructed of sheet metal.

31.10 Principles of Developments

The sheet metal mixing unit shown in **Fig. 31.28** was designed for fabrication from flat, sheet-metal stock. The part's flat pattern was developed and laid out in the plane of the stock and then folded and joined to form the 3D shape.

Figure 31.29 illustrates some of the standard edges and joints for sheet metal. The application determines the type of seam that is used.

Figure 31.30 shows the development of patterns for three typical shapes. The sides of a box are unfolded into a common plane. The cylinder is rolled out along a stretch-out line equal in length to its circumference. The pattern of a right cone is developed with the length of an element serving as a radius for drawing the base arc.

The construction of patterns for geometric shapes with parallel elements, such as the prisms and cylinders shown in **Fig. 31.31A** and **B**, begins with drawing **stretch-out lines parallel to the edge views of the shapes' right sections**. The distance around the right section becomes the length of the stretch-out line. The prism and cylinder in **Fig. 31.31C** and **D** are inclined, so their right sections are perpendicular to their sides, not parallel to their bases.

In development, **an inside pattern is preferable to an outside pattern** for two reasons: (1) most bending machines are designed to fold metal inward, and (2) markings and scribings will be hidden. The designer labels patterns with a series of lettered or numbered points on the layouts. **All lines on developments must be true length. Patterns should be laid out so that the seam line (a line where the pattern is joined) is the shortest line in order to reduce the expense of riveting or welding the seams**.

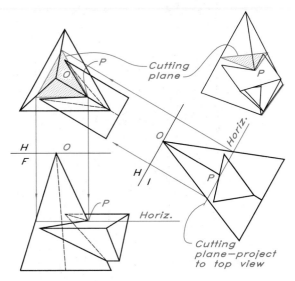

Figure 31.27 To locate the intersection of this pyramid and prism, obtain the end view of the prism in an auxiliary view. Pass horizontal cutting planes through the fold lines of the prism to find the piercing points and the line of intersection. A typical cutting plane is shown to find a corner point where a plane bends around a corner of the pyramid.

Figure 31.28 The design of this mixing unit required the application of the principles of intersections and developments. Its flat pattern was developed on flat sheet-metal stock that was bent into shape. These flat patterns are called developments.

EXAMPLES OF EDGES AND SEAMS (SHEET METAL)

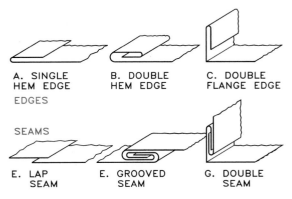

Figure 31.29 These are examples of several types of edges and seams used to join sheet-metal developments. Other seams are joined by riveting and welding.

31.11 Development of Rectangular Prisms

Figure 31.32 illustrates the development of a flat pattern for a rectangular prism. The edges of the prism are vertical and true length in the front view. The right section is perpendicular to these sides and the right section is true size in the top view. The stretch-out line begins with point 1 and is drawn parallel to the edge view of the right section.

If an inside pattern is to be laid out to the right, you must determine which point is to the right of the beginning point, point 1. Let's assume that you are standing inside the top view and are looking at point 1: You will see point 2 to the right of point 1.

To locate the fold lines of the pattern, transfer lines 2–3, 3–4, and 4–1 with your dividers from the right section in the top view to the stretch-

out line. The length of each fold line is its projected true length from the front view. Connect the ends of the fold lines to form the boundary of the developed surface. Draw the fold lines as thin

TYPES OF DEVELOPMENTS

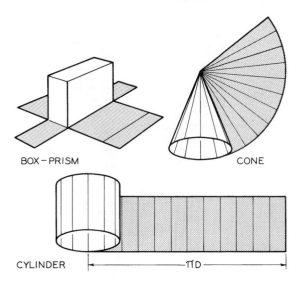

BOX – PRISM

CONE

CYLINDER |← ——— πD ——— →|

Figure 31.30 Three standard types of developments are the box, cylinder, and cone.

RECTANGULAR PRISM

2 to right of pt. 1

TS Right Section

Inside pattern

R Section

Stretch–out line

Thin fold lines

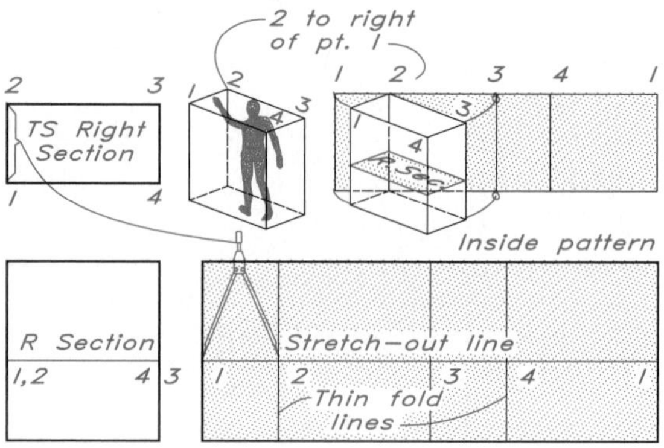

Figure 31.32 To develop a rectangular prism for an inside pattern, draw the stretch-out line parallel to the edge view of the right section. Transfer the distances between the fold lines from the true-size right section to the stretch-out line.

STRETCH-OUT LINES

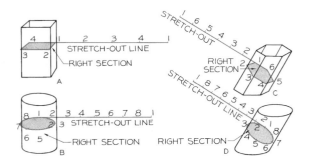

STRETCH-OUT LINE
RIGHT SECTION

A

STRETCH-OUT LINE
RIGHT SECTION

B

STRETCH-OUT
RIGHT SECTION

C

STRETCH-OUT LINE
RIGHT SECTION

D

Figure 31.31

A and **B** To obtain the developments of right prisms and right cylinders, roll out the right sections along a stretch-out line.

C and **D** Draw stretch-out lines parallel to the edge views of the right section of cylinders and prisms, or perpendicular to their true-length elements.

dark lines and the outside lines as thicker, visible object lines.

The can-crushing machine shown in **Fig. 31.33** comprises prisms and pyramids and was designed using the principles of development.

Development of the prism depicted in **Fig. 31.34** is similar to that shown in Fig. 31.32. Here, though, one of its ends is beveled (truncated) rather than square. The stretch-out line is parallel to the edge view of the right section in the front view. Lay off the true-length distances around the right section along the stretch-out line (beginning with the shortest one) and locate the fold lines. Find the lengths of the fold lines by projecting from the front view of these lines.

31.12 Development of Oblique Prisms

The prism shown in **Fig. 31.35** is inclined to the horizontal plane, but its fold lines are true length in the front view. The right section is an edge perpendicular to these fold lines, and the stretch-out line is parallel to the edge of the right section. A true-size view of the right section is found in the auxiliary view.

Transfer the distances between the fold lines from the true-size right section to the stretch-out

Figure 31.33 The principles of intersections and developments were applied to the design and construction of this can-crushing machine hopper.

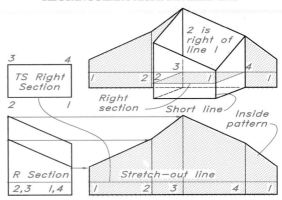

Figure 31.34 To develop an inside pattern of a rectangular prism with a beveled end, draw the stretch-out line parallel to the right section. Then find the fold lines by transferring distances between the fold lines from the true-size right section to the stretch-out line.

OBLIQUE PRISM

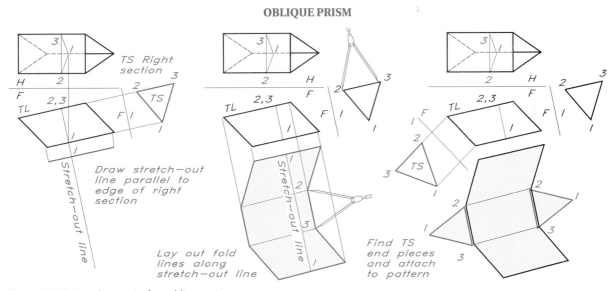

Figure 31.35 Development of an oblique prism:

Step 1 The edge view of the right section is perpendicular to the true-length axis of the prism in the front view. Find the true-size view of the right section by an auxiliary view. Draw the stretch-out line parallel to the edge view of the right section. The line through point 1 is the first line of the development.

Step 2 Because the pattern is to be laid out to the right from line 1, the next point is line 2 (from the auxiliary view). Transfer true-length lines 1–2, 2–3, and 3–1 from the right section to the stretch-out line to locate fold lines. Determine the lengths of bend lines by projection.

Step 3 Find true-size views of the end pieces by projecting auxiliary views from the front view. Connect these ends to the development to form the completed pattern. Draw fold lines as thin dark lines and outside lines as thicker, visible object lines.

OBLIQUE CHUTE

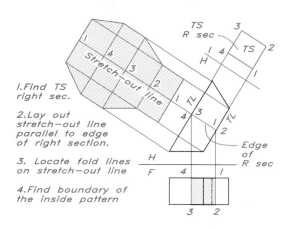

1.Find TS right sec.

2.Lay out stretch−out line parallel to edge of right section.

3. Locate fold lines on stretch−out line

4.Find boundary of the inside pattern

Figure 31.36 To develop this oblique chute, locate the right section true size in the auxiliary view. Draw the stretch-out line parallel to its right section. Find fold lines by transferring their spacing from the true-size right section to the stretch-out line.

GENERAL-CASE PRISM

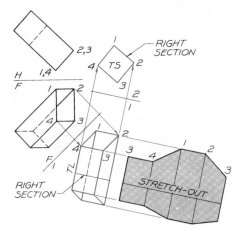

Figure 31.37 To develop an oblique prism, draw a primary auxiliary view in which the fold lines are true length and a secondary auxiliary view in which the right section appears true size. Use these views to develop the pattern the same way as in Fig. 31.36.

line. Find the lengths of the fold lines by projecting from the front view. Determine the ends of the prism and attach them to the pattern so that they can be folded into position.

In **Fig. 31.36**, the fold lines of the prism are true length in the top view, and the edge view of the right section is perpendicular to them. The stretch-out line is parallel to the edge view of the right section, and the true size of the right section appears in an auxiliary view projected from the top view. Transfer the distances about the right section to the stretch-out line to locate the fold lines, beginning with the shortest line. Find the lengths of the fold lines by projecting from the top view. Attach the end portions to the pattern to complete the construction.

A prism that does not project true length in either view may be developed as shown in **Fig. 31.37**. The fold lines are true length in an auxiliary view projected from the front view. The right section appears as an edge perpendicular to the fold lines in the auxiliary view and true size in a secondary auxiliary view.

Draw the stretch-out line parallel to the edge view on the right section. Locate the fold lines on the stretch-out line by measuring around the right section in the secondary auxiliary view, beginning with the shortest one. Then project the lengths of the fold lines to the development from the primary auxiliary view.

31.13 Development of Cylinders

Figure 31.38 illustrates how to develop a flat pattern of a right cylinder. The elements of the cylinder are true length in the front view, so the right section appears as an edge in this view and true size in the top view. **The stretch-out line is parallel to the edge view of the right section,** and point 1 is the beginning point because it lies on the shortest element.

Let's assume that you are standing inside the cylinder in the top view and are looking at point 1: You will see that point 2 is to the right of point 1. Therefore lay off point 2 to the right. of point 1 for developing an inside pattern.

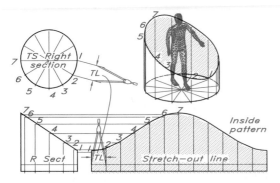

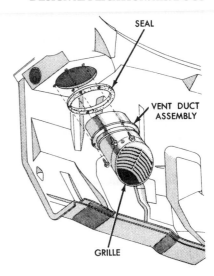

Figure 31.38 To develop an inside pattern for a truncated right cylinder, draw the stretch-out line parallel to the right section. Transfer points 1–7 from the top view to the stretch-out line that is parallel to the right section. Point 2 is to the right of point 1 for an inside pattern.

Figure 31.39 The design of this automobile ventilator air duct involved the use of development principles. (Courtesy of Ford Motor Company.)

By drawing radial lines at 15° or 30° intervals you can equally space the elements in the top view and conveniently lay them out along the stretch-out line as equal measurements. To complete the pattern, find the lengths of the elements by projecting from the front view. An application of a developed cylinder with a beveled end is the automobile air-conditioning duct shown in **Fig. 31.39**.

31.14 Development of Oblique Cylinders

The pattern for an oblique cylinder (**Fig. 31.40**) involves the same determinations as the preceding cases, but with the additional step of finding a true-size view of the right section in an auxiliary view. First, locate a series of equally spaced elements around the right section in the auxiliary view and project them back to the true-length view. Draw the stretch-out line parallel to the edge view of the right section in the front view.

Lay out the spacing between the elements along the stretch-out line, and draw the elements through these points perpendicular to the stretch-out line. Find the lengths of the elements

by projecting from the front view and complete the pattern.

A more general case is the oblique cylinder shown in **Fig. 31.41**, where the elements are not true length in the given views. A primary auxiliary view gives the elements true length, and a secondary auxiliary view yields a true-size view of the right section. Draw the stretch-out line parallel to the edge view of the right section in the primary auxiliary view. Transfer the elements to the stretch-out line from the true-size right section.

Draw the elements perpendicular to the stretch-out line and find their lengths by projecting from the primary auxiliary view. Connect the endpoints with a smooth curve to complete the pattern.

31.15 Development of Pyramids

All lines used to draw patterns must be true length, but pyramids have few lines that are true length in the given views. For this reason you

OBLIQUE CYLINDER

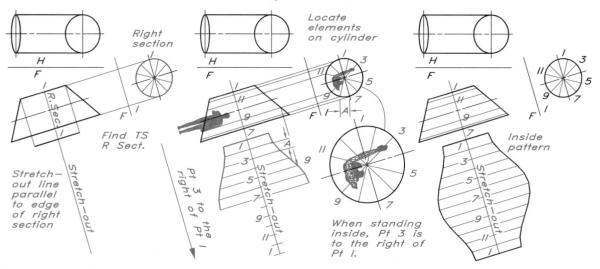

Figure 31.40 Development of an oblique cylinder:

Step 1 The right section is an edge perpendicular to the true-length axis in the front view. Draw an auxiliary view to find the right section true size. Divide the right section into equal chords. Draw a stretch-out line parallel to the edge view of the right section. Locate the shortest element at 1.

Step 2 Project elements from the right section to the front view. Transfer the chordal measurements in the auxiliary view to the stretch-out line to locate cylindrical elements and determine their lengths by projection.

Step 3 Locate the remaining elements to complete the inside pattern.

GENERAL-CASE CYLINDER

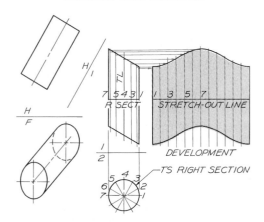

Figure 31.41 To develop an oblique cylinder, construct a primary auxiliary view in which the elements appear true length. The right section is found true size in a secondary auxiliary view. Complete the drawing as in Fig. 31.40.

must find the sloping corner lines true length before drawing a development.

Figure 31.42 shows the method of finding the corner lines of a pyramid true length by revolution. Revolve line O–5 into the frontal plane to line O–5' in the top view so that it will be true length in the front view. An application of the development of a pyramid is the sheet metal hopper shown in **Fig. 31.43**.

Figure 31.44 shows the development of a right pyramid. Line O–1 is revolved into the frontal plane in the top view to find its true length in the front view. Because it is a right pyramid, all corner lines are equal in length. Line O–1' is the radius for the base circle of the development. When you transfer distance 1–2 from the base in the top view to the development, it forms a chord on the base circle. Find lines 2–3, 3–4, and 4–1 in the same

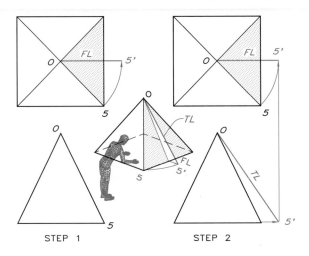

STEP 1   STEP 2

Figure 31.42 True length of a pyramid by revolution:
Step 1 To find the true length of corner line O–5 of a pyramid revolve it into the frontal plane in the top view, to O–5'.

Step 2 Project point 5' to the front view, where frontal line O–5' is true length.

DESIGN APPLICATION: HOPPER

Figure 31.43 This sheet-metal hopper is an application of a design involving development of a pyramid. (Courtesy of Gar–Bro.)

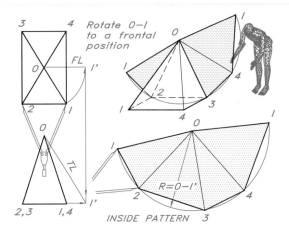

Figure 31.44 To develop this right pyramid, lay out an arc by using the true length of corner line O–1' as the radius. Transfer true-length distances around the base in the top view to the arc and darken the lines.

manner and in sequence. Draw the fold lines as thin lines from the base to the apex, point O.

A variation of this case is the truncated pyramid (**Fig. 31.45**). Development of the inside pattern proceeds as in the preceding case, but establishing the upper lines of the development requires an additional step. Revolution yields the true-length lines from the apex to points 1', 2', 3', and 4'. Lay off these distances along their respective lines on the pattern to find the upper boundary of the pattern.

31.16 Development of Cones

All elements of a right cone are equal in length (**Fig. 31.46**). Revolving element O–6 into its frontal position at O–6' gives its true length. When projected to the front view, line O–6' is true length and is the outside element of the cone. Projecting point 7 horizontally to element O–6' locates point 7'.

To develop the right cone depicted in **Fig. 31.47**, divide the base into equally spaced elements in the top view and project them to the

DEVELOPMENT OF A RIGHT PYRAMID

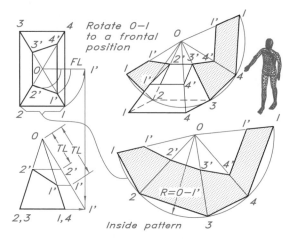

Figure 31.45 To develop an inside pattern of a truncated right pyramid, use the method shown in Fig. 31.42. Find the true lengths of elements O–1', O–2', O–3' and O–4' in the front view by revolution. Lay them off along their respective elements to find the upper boundary of the pattern.

front view, where they radiate to the apex at O. The outside elements in the front view, O–10 and O–4, are true length.

Using element O–10 as a radius, draw the base arc of the development. The spacing of the elements along the base circle is equal to the chordal distances between them on the base in the top view. Inspection of the top view from the inside, where point 2 is to the right of point 1, indicates that this is an inside pattern. The sheet-metal chambers in **Fig. 31.48** are applications of conical development.

Figure 31.49 shows the development of a truncated cone. To find its pattern, lay out the entire cone by using the true-length element O–1 as the radius, ignoring the portion removed from it. Locate the hyperbolic section formed by the inclined plane through the front view of the cone in the top view by projecting points on each element of the cone to the top view of these elements. For example, determine the true length of line O–3' by projecting point 3' horizontally to the

TRUE LENGTH BY REVOLUTION (CONE)

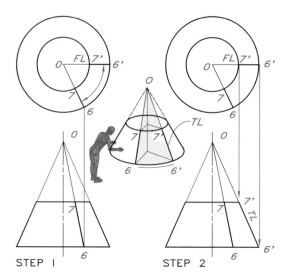

Figure 31.46 True length by revolution (cone):
Step 1 Revolve an element of a cone, O–6, into a frontal plane in the top view.

Step 2 Project point 6' to the front view, where it is a true-length outside element of the cone. Find the true length of line O–7' by projecting point 7' to the outside element in the front view.

DEVELOPMENT OF A RIGHT CONE

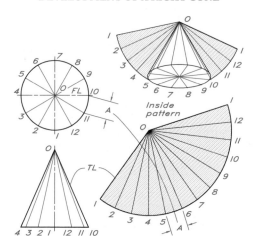

Figure 31.47 To develop an inside pattern of a right cone, use a true-length element (O–4 or O–10 in the front view) as the radius. Transfer chordal distances from the true-size base in the top view and mark them off along the arc.

Figure 31.48 This conical shape was formed from metal panels by applying principles of developments.

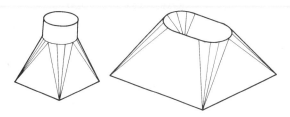

Figure 31.50 These are examples of transition pieces that connect parts having different cross sections.

Figure 31.51 Transition pieces were used to join a circular shape with a rectangular section in this application. (Courtesy of Western Precipitation Group, Joy Manufacturing Company.)

DEVELOPMENT OF A RIGHT CONE

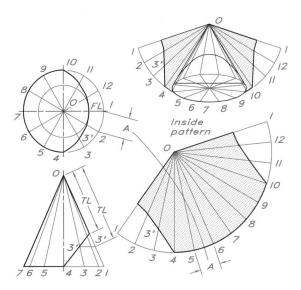

Figure 31.49 To develop a conical surface with a side opening, begin by laying it out as in Fig. 31.47. Find true-length elements by revolution in the front view and transfer them to their respective elements in the pattern.

true-length element O–1 in the front view. Lay off these distances, and others, along their respective elements to establish a smooth curve.

31.17 Development of Transition Pieces

A transition piece changes the shape of a section at one end to a different shape at the other end (Fig. 31.50). In industrial applications, transition pieces may be huge (**Fig. 31.51**) or relatively small (Fig. 31.48).

Figure 31.52 shows development of a transition piece. Radial elements extend from each corner to the equally spaced points on the circular end of the piece. Revolution gives the true length of each line. True-length lines 2–D, 3–D, and 2–3 yield the inside pattern of 2–3–D.

DEVELOPMENT OF A TRANSITION PIECE

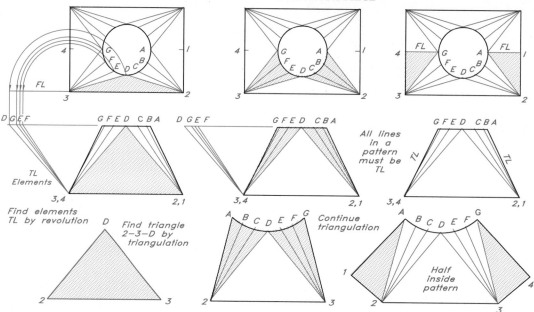

Figure 31.52 Development of a transition piece:

Step 1 Divide the circular end into equal parts in the top view and connect these points with lines to corner points 2 and 3. Find the true length of these lines by revolving and projecting them to the front view. Using true-length lines, draw triangle 2–3–D.

Step 2 Using other true-length lines and the chord distances on the circular end in the top view, draw a series of triangles joined at common sides. For example, draw arcs 2–C from point 2. To find point C, draw arc DC from point D. Chord DC is true length in the top view.

Step 3 Construct the remaining planes, A–1–2 and G–3–4, by triangulation to complete the inside half-pattern of the transition piece. Draw the fold lines, where the surface is to be bent, as thin lines. The seam line for the pattern is line A–1, the shortest line.

The true-length radial lines, used in combination with the true-length chordal distance in the top view, give a series of adjacent triangles to form the pattern beginning with element D2. Adding the triangles A–1–2 and G–3–4 at each end of the pattern completes the development of a half-pattern.

Problems

The scale of these problems allows you to fit two solutions on a size A sheet for a grid size of 0.20 in. (5 mm). For a grid size of 0.40 in. (10 mm), you can fit only one solution on a size A sheet.

Intersections

1–24. **(Fig. 31.53)** Lay out the problems and find the intersections that are necessary to complete the views.

Developments

25–48. **(Fig. 31.54)** Lay out the problems and draw the developments. Orient the long side of the sheet horizontally to allow space at the right of the given views for the development.

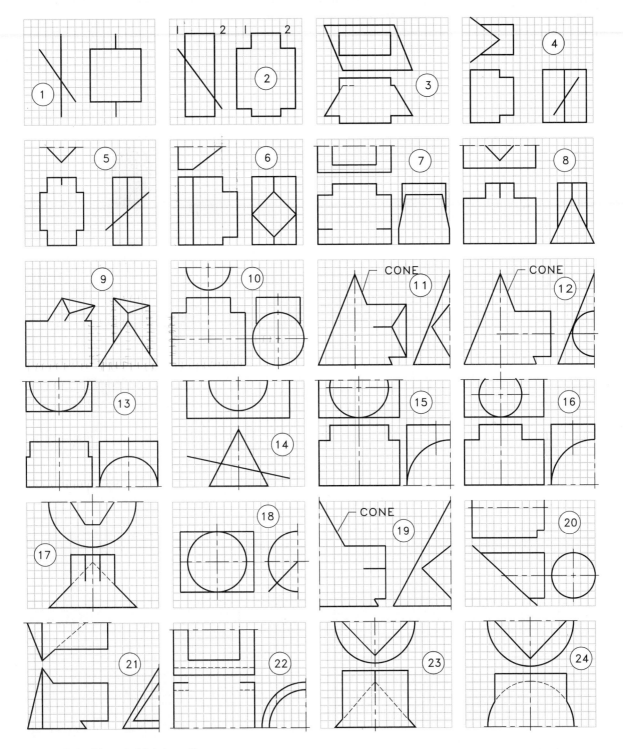

Figure 31.53 Problems 1–24. Intersections.

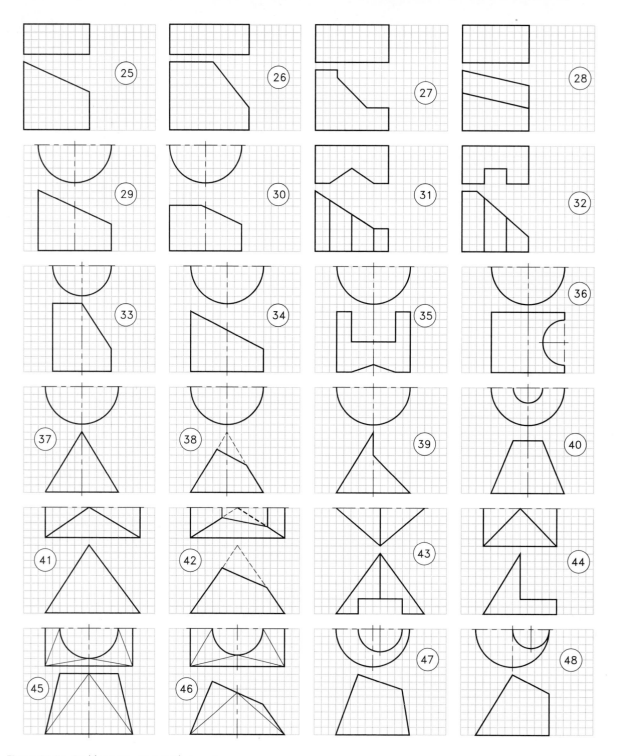

Figure 31.54 Problems 25–48. Developments.

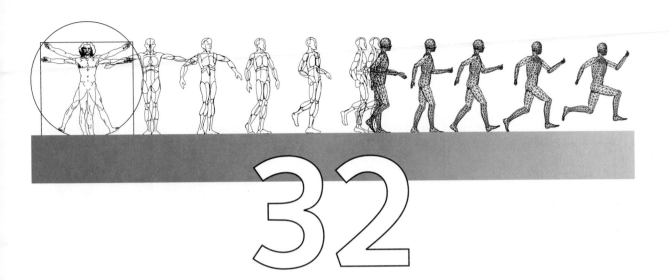

32

Graphs

32.1 Introduction

Data and information expressed as numbers and words are usually difficult to analyze or evaluate unless transcribed into graphical form, or as a **graph**. The term **chart** is an acceptable substitute for graph, but it is more appropriate when applied to maps, a specialized form of graphs.

Graphs are especially useful in presenting data at briefings where the data must be interpreted and communicated quickly to those in attendance **(Fig. 32.1)**. Graphs are a convenient way to condense and present data visually, allowing the data to be grasped much more easily than when presented as tables of numbers or verbally.

Several different types of graphs are widely used. Their application depends on the nature of the presentation required. The most common types of graphs are:

1. Pie graphs

2. Bar graphs

3. Linear coordinate graphs

Figure 32.1 Graphs are helpful in organizing and presenting technical data in briefings. (Courtesy of Compaq Computers.)

4. Logarithmic coordinate graphs

5. Semilogarithmic coordinate graphs

6. Polar graphs

7. Schematics and diagrams

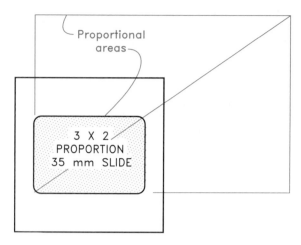

Figure 32.2 This diagonal-line method may be used to lay out drawings that are proportional to the area of a 35-mm slide.

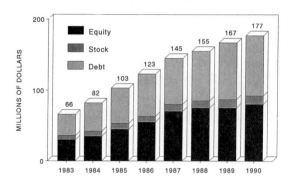

Figure 32.3 This computer-produced bar graph was generated by a program that converts numerical data into various types of graphs. (Courtesy of SlideWrite.)

Proportions

Graphs are used on large display boards, and in technical reports, as slides for a projector, or as transparencies for an overhead projector. Consequently, the proportion of the graph must be determined before it is constructed to match the page, slide, or transparency.

A graph that is to be photographed with a 35-mm camera must be drawn to the proportions of the film, or approximately 3 × 2 **(Fig. 32.2)**. This area may be enlarged or reduced proportionally by using the diagonal-line method.

The proportions of an overhead projector transparency are approximately 10 × 8. The image size should not exceed 9.5 inches × 7.5 inches to allow adequate margin for mounting the transparency on a frame (usually of cardboard).

Graphing by Computer

Many computer programs are available for converting numerical data into various types of graphs to improve comprehension and interpretation. These programs vary from data representation as bar graphs, coordinate graphs, and pie graphs to 3D mathematical models.

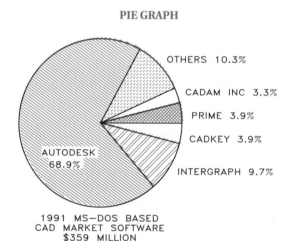

Figure 32.4 A pie graph shows the relationship of parts to a whole. It is most effective when there are only a few parts. (Courtesy of Autodesk.)

Computer-produced graphs are especially useful for preparing visual aids for projection on a screen for a presentation. They are equally valuable as graphs for technical reports. **Figure 32.3** is an example of a 3D bar graph printed by a laser printer from data that was input in tabular form. It

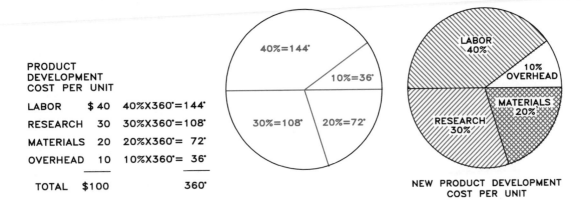

PRODUCT DEVELOPMENT COST PER UNIT		
LABOR	$ 40	40%X360°=144°
RESEARCH	30	30%X360°=108°
MATERIALS	20	20%X360°= 72°
OVERHEAD	10	10%X360°= 36°
TOTAL	$100	360°

NEW PRODUCT DEVELOPMENT COST PER UNIT

Figure 32.5 Pie graphs:
Step 1 Find the sum of the parts and the percentage that each is of the total. Multiply each percentage by 360° to obtain the angle of each sector of the graph.

Step 2 Draw the circle and construct each sector using the degrees of each from step 1. Place small sectors as nearly horizontal as possible.

Step 3 Label sectors with their proper names and percentages. Exact numbers also may be included in each sector.

could have also been plotted with a pen plotter or an impact printer.

32.2 Pie Graphs

Pie graphs compare the relationship of parts to a whole. For example, **Fig. 32.4** shows a pie graph that compares industry's use of various types of computer graphics software as percentages of the total, or their market share.

    Figure 32.5 illustrates the steps involved in drawing a pie graph. The data in this example, as simple as it is, are not as easily compared in numerical form as when drawn as a pie graph. Position thin sectors of a pie graph as nearly horizontal as possible to provide more space for labeling. When space is not available within the sectors, place labels outside the pie graph and, if necessary, use leaders (see Fig. 32.4). Showing the percentage represented by each sector is important and giving the actual numbers or values as part of the label is also desirable.

SINGLE BAR GRAPH

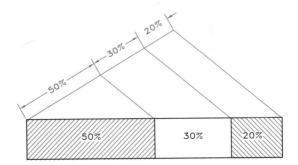

Figure 32.6 The diagonal-line method may be used to find the percentages of the parts to the whole, where the total bar represents 100%.

32.3 Bar Graphs

Bar graphs are widely used for comparing values because the general public understands them. A bar graph may be a single bar **(Fig. 32.6)** where the length of the bar representing 100% is divided

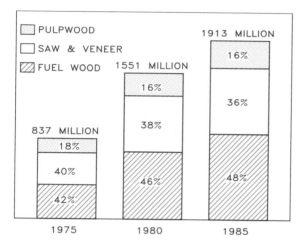

WORLDWIDE TIMBER PRODUCTION

Figure 32.7 In this bar graph, each bar represents 100% of the total amount and shows the percentages of the parts to the total.

into lengths proportional to the percentages of its three parts. In **Fig. 32.7** the bars not only show the overall production of timber (the total heights of the bars), but also the percentages of the total devoted to three uses of the timber.

Figure 32.8 shows how to convert data into a bar graph that can be used in a report or briefing. The axes of the graph carry labels, and its title appears inside the graph where space is available.

The bars of a bar graph should be sorted in ascending or descending order unless there is an overriding reason not to, such as a chronological sequence. An arbitrary arrangement of the bars, such as in alphabetical or numerical order, makes a graph difficult to evaluate (**Fig. 32.9A**). However, ranking the categories by bar length allows easier comparisons from smallest to largest (**Fig. 32.9B**). If the data are sequential and involve time, such as sales per month, a better arrangement of the bars is chronologically, to show the effect of time.

Bars in a bar graph may be horizontal (**Fig. 32.10**) or vertical. Data cannot be compared accu-

rately unless each bar is full length and originates at zero. Also, bars should not extend beyond the limits of the graph (giving the impression that the data were "too hot" to hold). Another form of bar graph shows plus and minus changes from a base value (**Fig. 32.11**).

32.4 Linear Coordinate Graphs

Figure 32.12 shows a typical **linear coordinate graph**, with notes explaining its important features. Divided into equal divisions, the axes are referred to as **linear scales**. Data points are plotted on the grid by using measurements, called **coordinates**, along each axis from zero. The plotted points are marked with symbols such as circles or squares that may be easily drawn with a template.

The horizontal scale of the graph is called the **abscissa or *x* axis**. The vertical scale is called the **ordinate or *y* axis**.

When the points have been plotted, a curve is drawn through them to represent the data. The line drawn to represent data points is called a **curve** regardless of whether it is a smooth straight line, curve, or broken line. The curve should not extend through the plotted points; rather, the points should be **left as open circles or other symbols**.

The curve is the most important part of the graph, so it should be drawn as the most prominent (thickest) line. If there are two curves in a graph, they should be drawn as different line types and labeled. The title of the graph is placed in a box inside the graph and units are given along the *x* and *y* axes with labels identifying the scales of the graph.

Broken-Line Graphs

Figure 32.13 shows the steps required to draw a linear coordinate graph. Because the data points represent sales, which have no predictable pattern, the data do not givg a smooth progression from point to point. Therefore the points are connected with a **broken-line curve** drawn as an angular line from point to point.

BAR GRAPH CONSTRUCTION

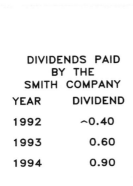

DIVIDENDS PAID
BY THE
SMITH COMPANY

YEAR	DIVIDEND
1992	⌐0.40
1993	0.60
1994	0.90

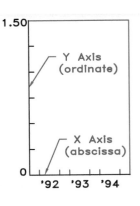

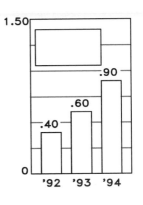

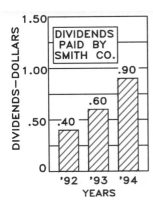

Figure 32.8 Constructing a bar graph:

Given These data are to be plotted as a bar graph.

Step 1 Scale the vertical and horizontal axes so that the data will fit on the grid. Bars should begin at zero on the ordinate.

Step 2 Draw and label the bars. The width of the bars should be greater than the space between them. Grid lines should not cross the bars.

Step 3 Strengthen lines, title the graph, label the axes, and cross-hatch the bars.

ARRANGING BARS BY LENGTH

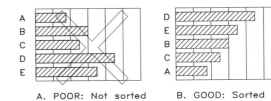

A. POOR: Not sorted   B. GOOD: Sorted

Figure 32.9

A When bars are arbitrarily arranged, such as alphabetically, the bar graph is difficult to interpret.

B When the bars are sorted by length, the graph is much easier to interpret.

Again, leave the symbols used to mark the data points open rather than extending grid lines or the data curve through them (**Fig. 32.14**). Each circle or symbol used to plot points should be about 1/8 in. (3 mm) in diameter. **Figure 32.15** shows typical data-point symbols and lines.

⬛ **Computer Method** Data points may be produced as CIRCLEs, DONUTs (open and closed), or

HORIZONTAL BAR GRAPH

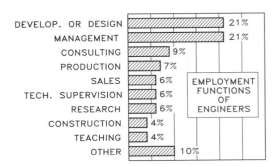

Figure 32.10 The horizontal bars of this graph (showing employment of engineers) are arranged in descending order. Note that the final category "Other" is made up of variations that represent less than 4% each of the total.

POLYGONs. The grid lines and curves that pass through the open symbols may be removed easily with the TRIM command (**Fig. 32.16**).

Titles The title of a graph may be located in any of the positions shown in **Fig. 32.17**. A graph's title

32.4 LINEAR COORDINATE GRAPHS • 597

BAR GRAPH: PLUS AND MINUS BARS

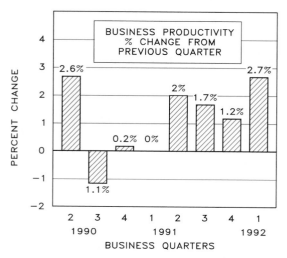

Figure 32.11 Bar graphs may be drawn with bars in both negative and positive directions.

MAJOR FEATURES OF A GRAPH

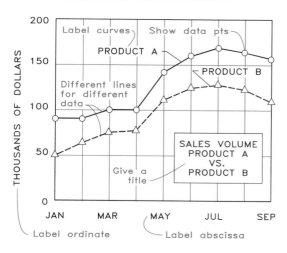

Figure 32.12 This basic linear coordinate graph illustrates the important features on a graph.

BROKEN-LINE GRAPH CONSTRUCTION

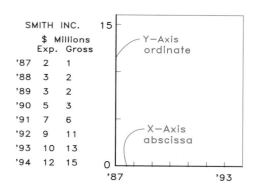
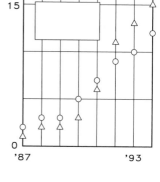
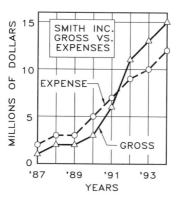

Figure 32.13 A broken-line graph:

Given A record of the Smith Company's gross income and expenses.

Step 1 Lay off the vertical (ordinate) and horizontal (abscissa) axes to provide space for the largest values.

Step 2 Draw division lines and plot data points for their respective years, using different symbols for each curve.

Step 3 Connect data points with straight lines, label the axes, title the graph, darken the lines, and label the curves.

should never be as meaningless as "graph" or "coordinate graph." Instead, it should identify concisely what the graph shows.

Scale and Labeling The calibration and labeling of the axes affects the appearance and readability

of a graph. **Figure 32.18A** shows a properly calibrated and labeled axis. **Figure 32.18B** and **32.18C** illustrates common mistakes: placing the grid lines too close together and labeling too many divisions along the axis. In Fig. 32.18C, the choice of the interval between the labeled values

DATA POINTS

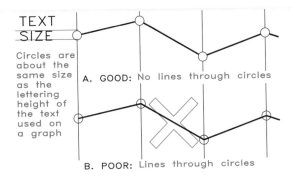

Figure 32.14 The curve of a graph drawn from point to point should not extend through the symbols used to represent data points.

SYMBOLS FOR DATA POINTS

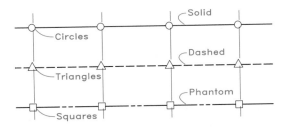

Figure 32.15 Symbols and lines such as these may be used to represent different curves on a graph. The data-point symbols should be drawn about the same size as the letter height being used, usually a little less than $1/8$ in. (3 mm) in diameter.

(9 units) makes interpolation between them difficult. For example, locating the value 22 by eye is more difficult on this scale than on the one shown in Fig. 32.18A.

Smooth-Line Graphs

The strength of concrete related to its curing time is plotted in **Fig. 32.19**. The strength of concrete changes gradually and continuously in relation to curing time. Therefore the data points are connected with a **smooth-line curve** rather than a broken-line curve. These relationships are repre-

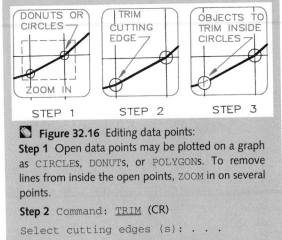

◆ **Figure 32.16** Editing data points:
Step 1 Open data points may be plotted on a graph as CIRCLEs, DONUTs, or POLYGONs. To remove lines from inside the open points, ZOOM in on several points.

Step 2 Command: <u>TRIM</u> (CR)

Select cutting edges (s): . . .

Select objects: (Select the circle.) (CR)

Step 3 Select object to trim: (Select the lines inside the circle, and they will be removed.) Continue this process for all points.

PLACEMENT OF TITLES ON GRAPHS

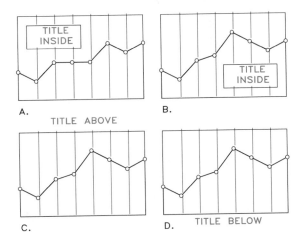

Figure 32.17
A and **B** The title of a graph may be placed inside a box within the graph. Box perimeter lines should not coincide with grid lines.

C Titles may be placed over the graph.

D Titles may be placed under the graph.

CALIBRATION OF GRAPH SCALES

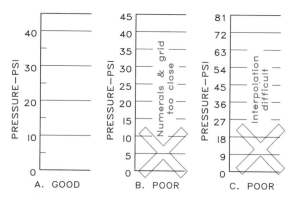

Figure 32.18

A The scale is properly labeled and calibrated. It has about the right number of grid lines and divisions, and the numbers are well spaced and easy to interpolate.

B The numbers are too close together, and there are too many grid lines.

C The increments selected make interpolation difficult.

SMOOTH-LINE GRAPH

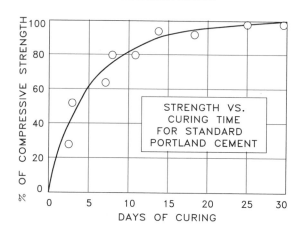

Figure 32.19 When the data being graphed involve gradual, continuous changes in relationships, the curve is drawn as a smooth line.

sented by the **best-fit curve**, a smooth curve that is an average representation of the points.

There is a smooth-line curve relationship between miles per gallon and the speed at which a

BEST-FIT CURVES APPROXIMATE DATA

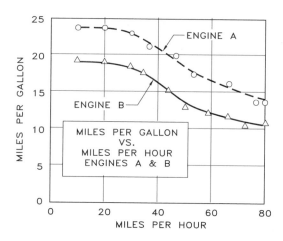

Figure 32.20 These data are represented by best-fit curves that approximate the data without necessarily passing through each point. The relationship of these data indicates that this curve should be a smooth-line curve rather than a broken-line curve.

car is driven. **Figure 32.20** compares the results for two engines.

A smooth-line curve on a graph implies that interpolations between data points can be made to estimate other values. Data points connected by a broken-line curve imply that interpolations between the plotted points cannot be made.

Straight-Line Graphs

Some graphs have neither broken-line curves nor smooth-line curves, but **straight-line curves (Fig. 32.21)**. On this graph, a third value can be determined from the two given values. For example, if you are driving 70 miles per hour and you take 5 seconds to react and apply your brakes, you will have traveled 550 feet in that time.

Two-Scale Coordinate Graphs

Graphs may contain different scales in combination, as shown in **Fig. 32.22**, where the vertical scale at the left is in units of pounds and the one

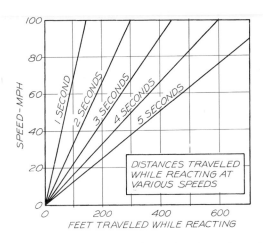

Figure 32.21 This graph may be used to determine a third value from the other two variables. For example, select a speed of 70 mph and a time of 5 seconds to find a distance traveled of 550 ft.

TWO-SCALE GRAPH

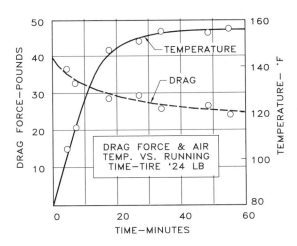

Figure 32.22 A two-scale graph has different scales along each y axis, and labels identify which scale applies to which curve.

at the right is in degrees of temperature. Both curves are drawn with respect to their *y* axes and each curve is labeled. Two-scale graphs of this type may be confusing unless they are clearly

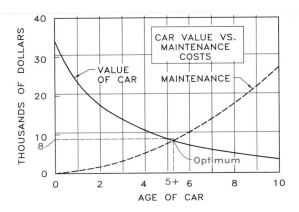

Figure 32.23 This graph shows the optimum time to sell a car based on the intersection of curves representing the depreciating value of the car and its increasing maintenance costs.

labeled. Two-scale graphs are effective for comparing related variables, as shown here.

Optimization Graphs

Figure 32.23 depicts the optimization of an automobile's depreciation in terms of maintenance cost increases. These two sets of data cross at an *x*-axis value of slightly more than five years, or the optimum point. At that time the cost of maintenance is equal to the value of the car, indicating that it might be a desirable time to buy a new car.

Figure 32.24 illustrates the steps involved in drawing an optimization graph. Here, the manufacturing cost per unit reduces as more units are made, causing warehousing costs to increase. Adding the two curves to get a third (total) curve indicates that the optimum number to manufacture at a time is about 8000 units (the low point on the total curve). When more or fewer units are manufactured, the total cost per unit is greater.

Composite Graphs

The graph shown in **Fig. 32.25** is a composite (or combination) of an area graph and a coordinate graph. The upper curve is a plot of the company's gross income. The lower curve is a plot of the

OPTIMIZATION GRAPH CONSTRUCTION

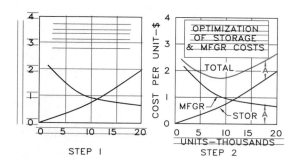

STEP 1

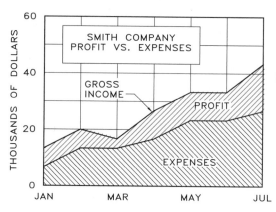

Figure 32.24 Constructing an optimization graph:
Step 1 Lay out the graph and plot the curves from the data given.

Step 2 Graphically add the two curves to find a third curve. For example, transfer distance A to locate a point on the third curve. The lowest point of the "total" curve is the optimum point, or 8000 units.

COMPOSITE GRAPH

Figure 32.25 This composite graph is a combination of a coordinate graph and an area graph. The upper area represents the difference between the two plotted curves.

BREAK-EVEN GRAPH CONSTRUCTION

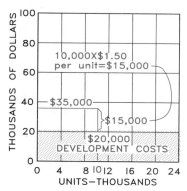

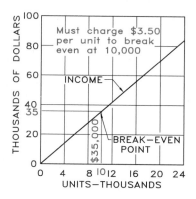

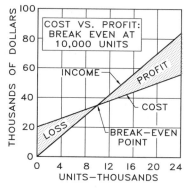

Figure 32.26 A break-even graph:
Step 1 Draw the graph and plot the development cost ($20,000 in this case). If each unit costs $1.50 to manufacture, the total investment would be $35,000 for 10,000 units, the break-even point.

Step 2 To break even at 10,000, the manufacturer must sell each unit for $3.50. Draw a line from zero through the break-even point of $35,000.

Step 3 The manufacturer incurs a loss of $20,000 at zero units, but it becomes progressively less until the break-even point is reached. Profit is the difference between cost and income to the right of the break-even point.

company's expenses. The difference between the two is the company's profit. Cross-hatching the areas emphasizes the relative sizes of expenses and profits.

Break-Even Graphs
Break-even graphs help in evaluating marketing and manufacturing costs to determine the selling price of a product. As **Fig. 32.26** shows, if the

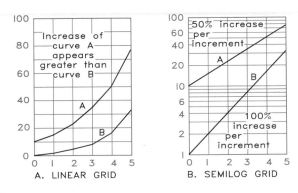

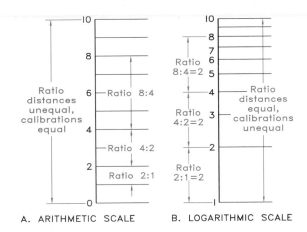

Figure 32.27 When plotted on a linear grid, curve A appears to be increasing at a greater rate than curve B. However, plotting the data on a semilogarithmic grid reveals the true rate of change.

Figure 32.28 The divisions on an arithmetic scale are equal and represent unequal ratios between points. The divisions on logarithmic scales are unequal and represent equal ratios.

desired break-even point for a product is 10,000 units, it must sell for $3.50 each to cover the costs of manufacturing and development.

32.5 Semilogarithmic Coordinate Graphs

Semilogarithmic graphs are called **ratio graphs** because they **graphically represent ratios**. One scale, usually the vertical scale, is logarithmic, and the other is linear (divided into equal divisions).

Figure 32.27 shows the same data plotted on a linear grid and on a semilogarithmic grid. The semilogarithmic graph reveals that the percentage change from 0 to 5 is greater for curve B than for curve A because here curve B is steeper. The plot on the linear grid appears to show the opposite result.

Figure 32.28 shows the relationship between the linear scale and the logarithmic scale. Equal divisions along the linear scale have unequal ratios, but equal divisions along the log scale have equal ratios.

Log scales may have one or many cycles. Each cycle increases by a factor of 10. For example, the scale shown in **Fig. 32.29A** is a three-cycle scale, and the one shown in **Fig. 32.29B** is a two-cycle scale. When scales must be drawn to a certain

length, commercially printed log scales may be used to transfer graphically the calibrations to the scale being used (**Fig. 32.29C**).

Figure 32.30 illustrates an application of a semilogarithmic graph for presenting industrial data. People who do not realize that semilog graphs are different from linear coordinate graphs may misunderstand them. Also, zero values cannot be shown on log scales.

Percentage Graphs

The percentage that one number is of another, or the percentage increase of one number to a greater number can be determined on a semilogarithmic graph (Fig. 32.31). Data plotted in **Fig. 32.31A** are used to find the percentage that 30 is of 60 (two points on the curve) by arithmetic. The vertical distance between them is the difference between their logarithms, so the percentage can be found graphically in **Fig 32.31B**. The distance from 30 to 60 is transferred to the log scale at the right of the graph and subtracted from the log of 100 to find the value of 50% as a direct reading of percentage.

LAYING OUT LOG SCALES

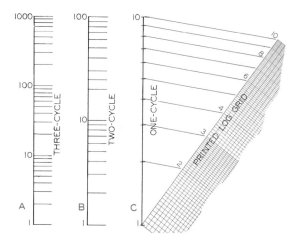

Figure 32.29 Logarithmic paper (either bought or drawn) may have several cycles: (A) three-cycle scales; (B) two-cycle scales, and (C) one-cycle scales. Calibrations may be projected to a scale of any length from a printed scale, as shown in (C).

SEMILOGARITHMIC GRAPH

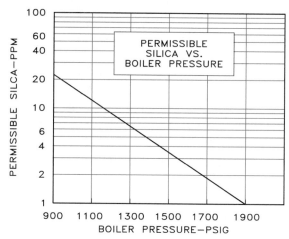

Figure 32.30 This semilogarithmic graph relates permissible silica (parts per million) to boiler pressure.

PERCENTAGE GRAPHS

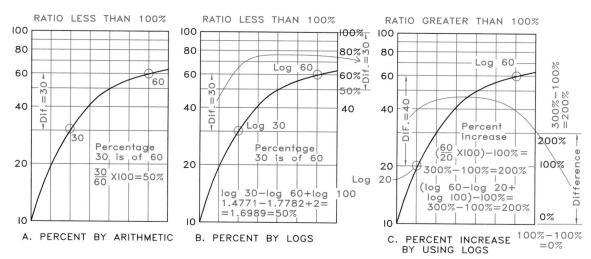

A. PERCENT BY ARITHMETIC   B. PERCENT BY LOGS   C. PERCENT INCREASE BY USING LOGS

Figure 32.31 Percentage graphs:

A To find the percentage that one data point is of another point (the percentage that 30 is of 60, for example) you may calculate it mathematically: (30/60)(100) = 50%.

B You may also find the percentage that 30 is of 60 mathematically by using the logarithms of the numbers. Or you may find it graphically by transferring the distance between 30 and 60 to the scale at the right, which shows that 30 is 50%.

C To find a percentage increase greater than 100%, divide the smaller number into the larger number. Find the difference between the logs of 60 and 20 with dividers and measure upward from 100% to find the percentage increase of 200%.

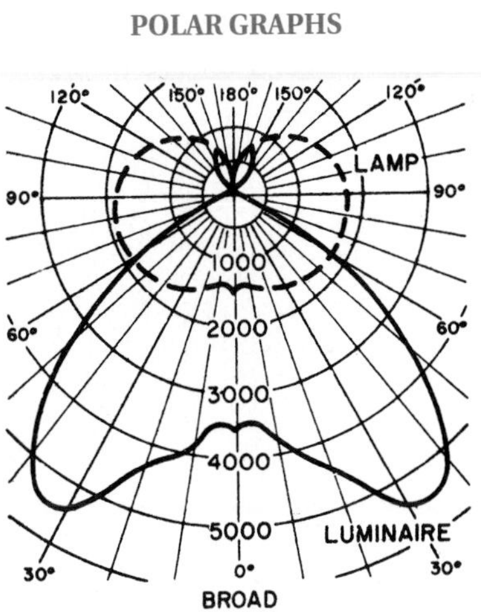

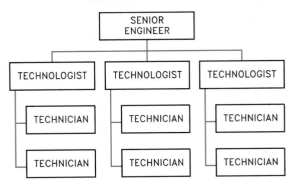

Figure 32.32 A polar graph is used to show the illumination characteristics of luminaires.

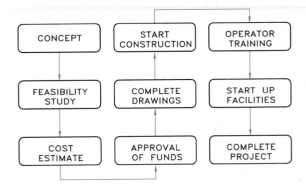

Figure 32.33 This block diagram shows the steps required to complete a project.

ORGANIZATIONAL CHART

Figure 32.34 This block diagram shows the organization of a design team.

In **Fig 32.31C**, the percentage increase between two points is transferred from the grid to the lower end of the log scale and measured upward because the increase is greater than zero. These methods may be used to find percentage increases or decreases for any set of points on the grid.

32.6 Polar Graphs

Polar graphs comprise a series of concentric circles with the origin at the center. Lines drawn from the center toward the perimeter of the graph allow data to be plotted through 360° by measuring values from the origin. For example, **Fig. 32.32** shows the maximum illumination of a lamp of 5500 lumens at 35° from the vertical and lesser amounts elsewhere. Printed graph paper may be purchased for use in drawing polar graphs.

32.7 Schematics

The block diagram shown in **Fig. 32.33** depicts the steps required to complete a construction project.

Each step is blocked in and connected with arrows to give the sequence.

The organization of a group of people can be represented in an organizational graph (usually called an organization chart), as shown in **Fig. 32.34**. The positions represented by the blocks in the lower part of the graph are responsible to those represented by the blocks above them. Lines of authority connecting the blocks give the routes for communication from one position to another, both upward and downward.

Geographical charts are used to combine maps and other relationships, such as weather (**Fig. 32.35**). Here, different hatching symbols represent the annual rainfall in various parts of the nation.

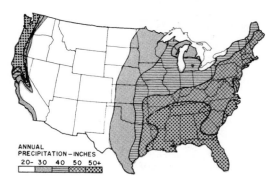

ANNUAL
PRECIPITATION—INCHES
20- 30 40 50 50+

Figure 32.35 This map chart shows annual precipitation for various parts of the United States. (Courtesy of the Structural Clay Products Institute.)

Problems

Draw your solutions to these problems on size A sheets in pencil or ink, as specified. Utilize the techniques presented and the examples given in this chapter.

Pie Graphs

1. Draw a pie graph that shows the comparative sources of retirees' income: investments, 34%; employment, 24%; social security, 21%; pensions, 19%; other, 2%.

2. Draw a pie graph that shows the number of members of the technological team: engineers, 985,000; technicians, 932,000; scientists, 410,000.

3. Construct a pie graph of the employment status of graduates of two-year technician programs a year after graduation: employed, 63%; continuing full-time study, 23%; considering job offers, 6%; military, 6%; other, 2%.

4. Construct a pie graph that shows the types of degrees held by engineers in aeronautical engineering: bachelor's, 65%; master's, 29%; Ph.D., 6%.

Bar Graphs

5. Draw a bar graph that shows expected job growth in Texas in 1993: Austin, 1.3%; San Antonio, 0.8%; Houston, 1.6%; Fort Worth, 0%; Dallas, 0.4%; all of Texas, 0.8%.

6. Draw a single-bar bar graph that represents 100% of a die casting alloy. The proportional parts of the alloy are: tin, 16%; lead, 24%; zinc, 38.8%; aluminum, 16.4%; copper, 4.8%.

7. Draw a bar graph that compares the number of skilled workers employed in various occupations. Use the following data and arrange the graph for ease of comparing occupations: carpenters, 82,000; all-round machinists, 310,000; plumbers, 350,000; bricklayers, 200,000; appliance servicers, 185,000; automotive mechanics, 760,000; electricians, 380,000; painters, 400,000.

8. Draw a bar graph that shows the characteristics of a typical U.S. family's spending: housing, 29.6%; food, 15.1%; transportation, 16.7%; clothing, 5.9%; retirement, 8.6%; entertainment, 4.9%; insurance, 5.2%; health care, 3.0%; charity, 3.1%; other, 7.9%.

9. Draw a bar graph that compares the corrosion resistance of the materials listed in the following table.

	Loss in Weight (%)	
	In Atmosphere	*In Sea Water*
Common Steel	100	100
10% nickel steel	70	80
25% nickel steel	20	55

Table 32.1

Angle with vertical	0°	10°	20°	30°	40°	50°	60°	70°	80°	90°
Candlepower (thous.) 2–400W	37	34	25	12	5.5	2.5	2	0.5	0.5	0.5
Candlepower (thous.) 1–1000W	22	21	19	16	12.3	7	3	2	0.5	0.5

10. Draw a bar graph using the data from Problem 1.

11. Draw a bar graph using the data from Problem 2.

12. Draw a bar graph using the data from Problem 3.

13. Construct a bar graph comparing sales and earnings of Apple Computer from 1980 through 1989. Data are by year for sales and earnings (profit) in billions of dollars: 1980, 0 and 0; 1981, 0.33 and 0.05; 1982, 0.60 and 0.07; 1983, 1.00 and 0.09; 1984, 1.51 and 0.08; 1985, 1.90 and 0.07; 1986, 1.85 and 0.12; 1987, 2.70 and 0.25; 1988, 4.15 and 0.40; 1989, 5.50 and 0.45.

Linear Coordinate Graphs

14. Draw a linear coordinate graph to show the estimated population growth in the U.S. from 1992 through 2050 in millions of people: 1992, 255; 2000, 270; 2010, 295; 2020, 325; 2030, 348; 2040, 360; 2050, 383.

15. Plot the data shown in Table 32.1 as a linear coordinate graph to assist in deciding which lamps should be selected to provide economical lighting for an industrial plant. The table gives the candlepower directly under the lamps (0°) and at various angles from the vertical for lamps mounted at a height of 25 ft.

16. Construct a linear coordinate graph that shows the relationship of energy costs (mills per kilowatt-hour) on the y axis to the percent capacity of a nuclear power plant and a gas-or oil-fired power plant on the x axis. Gas- or oil-fired plant data: 17 mills, 10%; 12 mills, 20%; 8 mills, 40%; 7 mills, 60%; 6 mills, 80%; 5.8 mills, 100%. Nuclear plant data: 24 mills, 10%; 14 mills, 20%; 7 mills, 40%; 5 mills, 60%; 4.2 mills, 80%; 3.7 mills, 100%.

17. Plot the data from Problem 13 as a linear coordinate graph.

18. Construct a linear coordinate graph to show the relationship between the transverse resilience in inch-pounds (ip) on the y axis and the single-blow impact in foot-pounds (fp) on the x axis of gray iron. Data: 21 fp, 375 ip; 22 fp, 350 ip; 23 fp, 380 ip; 30 fp, 400 ip; 32 fp, 420 ip; 33 fp, 410 ip; 38 fp, 510 ip; 45 fp, 615 ip; 50 fp, 585 ip; 60 fp, 785 ip; 70 fp, 900 ip; 75 fp, 920 ip.

19. Draw a linear coordinate graph to illustrate the trends in the export of U.S. services and products from 1980 through 1993: 1980, +8.5%; 1981, +2.5%; 1982, –7.5%; 1983, –4.5%; 1984, +7%; 1985, –2%; 1986, +7%; 1987, +12.5%; 1988, +17.5%; 1989, +10%; 1990, +6%; 1991, +3%; 1992, +5%; 1993, +7%.

20. Draw a linear coordinate graph that shows the voltage characteristics for a generator as given in the following table of values abscissa, armature current in amperes, (I_a); ordinate, terminal voltage in volts, (E_t):

I_a	E_t	I_a	E_t	I_a	E_t
0	288	31.1	181.8	41.5	68
5.4	275	35.4	156	40.5	42.5
11.8	257	39.7	108	39.5	26.5
15.6	247	40.5	97	37.8	16
22.2	224.5	40.7	90	13.0	0
26.2	217	41.4	77.5		

21. Draw a linear coordinate graph for the centrifugal pump test data in the following table. The units along the x axis are to be gallons per minute. Use four curves to represent the variables given.

Gallons per Minute	Discharge Pressure	Water HP	Electric HP	Efficiency (%)
0	19.0	0.00	1.36	0.0
75	17.5	0.72	2.25	32.0
115	15.0	1.00	2.54	39.4
154	10.0	1.00	2.74	36.5
185	5.0	0.74	2.80	26.5
200	3.0	0.63	2.83	22.2

Table 32.2

F°	Ultimate Strength	Elastic Limit
400	257,500	208,000
500	247,000	224,500
600	232,500	214,000
700	207,500	193,500
800	180,500	169,000
900	159,500	146,500
1000	142,500	128,500
1100	126,500	114,000
1200	114,500	96,500
1300	108,000	85,500

22. Draw a linear coordinate graph that compares two of the values shown in Table 32.2—ultimate strength and elastic limit—with degrees of temperature (x axis).

Break-Even Graphs

23. Construct a break-even graph that shows the earnings for a new product that has a development cost of $12,000. The break-even point is at 8000 units, and each costs $0.50 to manufacture. What would be the profit at volumes of 20,000 and 25,000?

24. Repeat Problem 23 except that the development costs are $80,000, the manufacturing cost of the first 10,000 units is $2.30 each, and the desired break-even point is 10,000 units. What would be the profit at volumes of 20,000 and 30,000?

Logarithmic Graphs

25. Use the data given in Table 32.3 to construct a logarithmic graph. Plot the vibration amplitude (A) as the ordinate and the vibration frequency (F) as the abscissa. The data for curve 1 represent the maximum limits of machinery in good condition with no danger from vibration. The data for curve 2 are the lower limits of machinery that is being vibrated excessively to the danger point. The vertical scale should be three cycles, and the horizontal scale should be two cycles.

26. Plot the following data on a two-cycle log graph to show the current in amperes (y axis) versus the voltage in volts (x axis) of precision temperature-sensing resistors. Data: 1 volt, 1.9 amps; 2 volts, 4 amps; 4 volts, 8 amps; 8 volts, 17 amps; 10 volts, 20 amps; 20 volts, 30 amps; 40 volts, 36 amps; 80 volts, 31 amps; 100 volts, 30 amps.

27. Plot the data from Problem 14 as a logarithmic graph.

28. Plot the data from Problem 20 as a logarithmic graph.

Semilogarithmic Graphs

29. Construct a semilogarithmic graph with the y axis a two-cycle log scale from 1 to 100 and the x axis a linear scale from 1 to 7. The object of the graph is to show the survivability of a shelter at varying distances from the atmospheric detonation of a one-megaton thermonuclear bomb. Plot overpressure in psi along the y axis, and distance from ground zero in miles along the x axis. The data points represent an 80% chance of survival of the shelter. Data: 1 mile, 55 psi; 2 miles, 11 psi; 3 miles, 4.5 psi; 4 miles, 2.5 psi; 5 miles, 2.0 psi; 6 miles, 1.3 psi.

30. The growth of two divisions of a company, Division A and Division B, is represented by the following data. Plot the data on a rectilinear graph and on a semilog graph with a one-cycle log scale on the y axis for sales in thousands of dollars and a linear scale on the x axis for years. Data: first year, A = $11,700 and B = $44,000; second year, A = $19,500 and B = $50,000; third year, A = $25,000 and B = $55,000; fourth year, A = $32,000 and B = $64,000; fifth year, A = $42,000 and B = $66,000; sixth year, A = $48,000 and B = $75,000. Which division has the better growth rate?

31. Draw a semilog chart showing probable engineering progress based on the following indices: 40,000 B.C., 21; 30,000 B.C., 21.5; 20,000 B.C., 22; 16,000 B.C., 23; 10,000 B.C., 27; 6000 B.C., 34; 4000 B.C., 39; 2000 B.C., 49; 500 B.C., = 60; A.D. 1900, 100. Use a horizontal scale of 1 in. = 10,000 years, a height of about 5 in., and two-cycle printed paper, if available.

32. Plot the data from Problem 20 as a semilogarithmic graph.

33. Plot the data from Problem 22 as a semilogarithmic graph.

Percentage Graphs

34. Plot the data from Problem 14 as a semilog graph to determine percentages and ratios for the data. What is the percentage increase in the demand for water from 1890 to 1920? What percentage of demand is the supply for 1900, 1930, and 1970?

Table 32.3

F	100	200	500	1000	2000	5000	10,000
$A(1)$	0.0028	0.002	0.0015	0.001	0.0006	0.0003	0.00013
$A(2)$	0.06	0.05	0.04	0.03	0.018	0.005	0.001

35. Using the graph plotted in Problem 30, determine the percentage of increase of Division A and Division B growth from year 1 to year 4. What percentage of sales of Division A are the sales of Division B at the end of year 2? At the end of year 6?

36. Plot the values for water horsepower and electric horsepower from Problem 22 on semilog paper against gallons per minute along the x axis. What percentage of electric horsepower is water horsepower when 75 gallons per minute are being pumped? What percentage increase in electric horsepower corre-

sponds to an increase in pumping volume from 0 to 185 gallons per minute?

Polar Graphs

37. Construct a polar graph from the data in Problem 15.

38. Construct a polar graph of the following illumination, in lumens at various angles, emitted from a luminaire. The 0° position is vertically under the overhead lamp. Data: 0°, 12,000; 10°, 15,000; 20°, 10,000; 30°, 8000; 40°, 4200; 50°, 2500; 60°, 1000; 70°, 0. The illumination is symmetrical about the vertical.

33

Nomography

33.1 Introduction

An additional aid in analyzing data is a graphical computer called a **nomogram**, **nomograph**, or "number chart." Basically, it is any graphical arrangement of calibrated scales and lines that may be used for calculations, usually those of a repetitive nature.

The term *nomogram* frequently denotes a specific type of scale arrangement called an **alignment chart** (which we call an *alignment graph*). **Figure 33.1** shows two examples of alignment graphs. Other types have curved scales or different scale arrangements for use with more complex problems.

An alignment graph usually is constructed to solve for one or more unknowns in a formula or empirical relationship between two or more quantities. For example, such a graph can be used to convert degrees Celsius to degrees Fahrenheit or to find the size of a structural member to sustain a certain load. To read an alignment graph, **place a straightedge or draw a line, called an isopleth, across the scales of the graph and read corre-

EXAMPLES OF NOMOGRAPHS

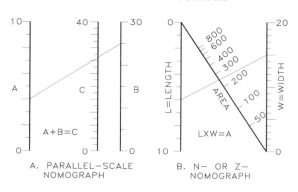

Figure 33.1 These two types of alignment graphs are typical nomographs.

sponding values from the scale on this line. **Figure 33.2** shows readings for the formula $W = U + V$.

33.2 Alignment Graph Scales

To construct any alignment graph, you must first determine the graduations of the scales that give

610

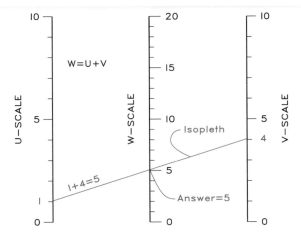

Figure 33.2 This graph illustrates the use of an isopleth to solve graphically for unknowns in an equation, $W = U + V$.

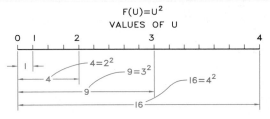

Figure 33.3 This functional scale contains units of measurement proportional to $F(U) = U2$.

the desired relationships. Alignment graph scales, called **functional scales**, are graduated according to values of some function of a variable. **Figure 33.3** illustrates a functional scale for $F(U) = U^2$. Substituting a value of U = 2 into the equation gives the position of U on the functional scale 4 units from zero, or $2^2 = 4$. Repeating this procedure for as many values of U as required yields a scale for finding all corresponding values of the function.

The Scale Modulus

Because the graduations on a functional scale are spaced in proportion to values of the function, a proportionality, or **scaling factor**, is needed. This constant of proportionality is called the **scale modulus**, m:

$$m = \frac{L}{F(U2) - F(U1)} \qquad (1)$$

where

m = scale modulus, in inches per functional unit,

L = desired length of the scale, in inches,

$F(U2)$ = functional value at the end of the scale, and

$F(U1)$ = functional value at the start of the scale.

For example, suppose that you are to construct a functional scale for $F(U) = \sin U$ with $0° \leq U \leq 45°$ and a scale 6 inches in length. Thus $L = 6$ in., $F(U2) = \sin 45° = 0.707$, $F(U1) = \sin 0° = 0$. Substituting these values into Eq. (1) gives

$$m = \frac{6}{0.707 - 0} = 8.49 \text{ in.}$$

per (sine) unit.

The Scale Equation

A **scale equation** makes possible graduation and calibration of functional scales. The general form of this equation is a variation of Eq. (1):

$$X = m[F(U) - F(U1)], \qquad (2)$$

where

X = distance from the measuring point of the scale to any graduation point,

m = scale modulus,

$F(U)$ = functional value at the graduation point,

$F(U1)$ = functional value at the measuring point of the scale.

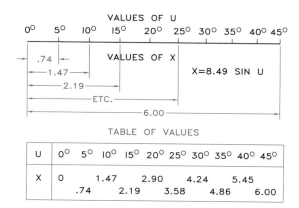

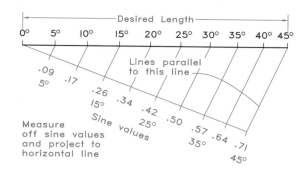

Figure 33.4 This drawing depicts a functional scale calibrated with values from the table, which were derived from the scale equation, $X = 8.49 \sin U$.

Figure 33.5 This functional scale shows the sine of the angles from 0° to 45° drawn graphically by the proportional line method. Draw the scale to a desired length and lay off the sine values of angles at 5° intervals along a construction line passing through the 0° end of the scale.

For example, let's construct a functional scale for the equation, $F(U) = \sin U$ $(0° \le U \le 45°)$. We have already determined that $m = 8.49$, $F(U) = \sin U$, and $F(U1) = \sin 0° = 0$. Thus, by substitution, Eq. (2) becomes

$$X = 8.49(\sin U - 0) = 8.49 \sin U.$$

Using this equation, we can substitute values of U and construct a table of positions. In this case, we calibrated the scale at 5° intervals, as shown in **Fig. 33.4**. The values of X give the positions, in inches, for the corresponding graduations measured from the start of the scale $(U = 0°)$. The initial measuring point does not have to be at one end of the scale, but an end is usually the most convenient point, especially if the functional value is zero at that point.

Figure 33.5 shows graphically how to locate functional values along a scale with the proportional line method. Measure the sine functions along a line at 5° intervals, with the end of the line passing through the 0° end of the scale. Transfer these functions from the inclined line with parallel lines back to the scale and label the functions.

33.3 Concurrent Scales

Concurrent scales aid in the rapid conversion of terms in one system of measurement into terms of a second system of measurement. Formulas of the type $F1 = F2$, which relate two variables, may be adapted to the concurrent scale format. A typical example is the Fahrenheit-Celsius temperature relation:

$$°F = \frac{9}{5} °C + 32.$$

Another is the area of a circle:

$$A = \pi r^2$$

Construction of a concurrent scale chart involves determining a functional scale for each side of the mathematical formula so that the position and lengths of each scale coincide. For example, to construct a conversion chart 5 inches long that gives the areas of circles whose radii range from 1 to 10, we first write $F1(A) = A$, $F2(r) = \pi r^2$, $r1 = 1$, and $r2 = 10$. The scale modulus for r is

$$m_r = \frac{L}{F2\,(r2) - F2\,(r1)}$$

$$= \frac{5}{\pi(10)^2 - \pi(1)^2} = 0.0161.$$

CALIBRATION OF RADIUS SCALE

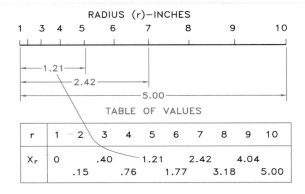

Figure 33.6 Calibrate one scale of a concurrent scale chart by using values from the table that have been calculated.

COMPLETED CONCURRENT SCALE GRAPH

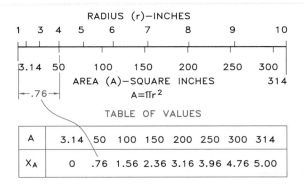

Figure 33.7 This completed concurrent scale chart is for the formula $A = \pi r^2$. Values for the A scale are from table.

Thus the scale equation for r becomes

$$X_r = m_r[F2(r) - F2(r1)]$$

$$= 0.0161[\pi r^2 - \pi(1)^2]$$

$$= 0.0161\pi(r^2 - 1)$$

$$= 0.0505(r^2 - 1).$$

Figure 33.6 shows a table of values for X_r and r. The r-scale values come from this table. From the original formula, $A = \pi r^2$, the limits of A are found to be $A1 = \pi = 3.14$ and $A2 = 100\pi = 314$. The scale modulus for concurrent scales is always the same for equal-length scales; therefore $m_A = m_r = 0.0161$, and the scale equation for A becomes

$$X_A = m_A[F1(A) - F1(A1)]$$

$$= 0.0161(A - 3.14).$$

We then compute the corresponding table of values for selected values of A, as shown in **Fig. 33.7**. We superimposed the A scale on the r scale and placed its calibrations on the other side of the line to facilitate reading.

To expand or contract one of the scales, simply use the alternative arrangement shown in **Fig. 33.8**. Draw the two scales parallel at any convenient distance and calibrate them in opposite directions. You must calculate a different scale

CONCURRENT GRAPH—UNEQUAL SCALES

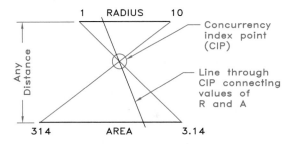

Figure 33.8 This construction is used to draw a concurrent graph with unequal scales.

modulus and corresponding scale equation for each scale if they are not the same length.

To construct concurrent scales graphically, use the proportional line method, as shown in **Fig. 33.9**. There are 101.6 mm in 4 in. Project millimeter units to the upper side of the inch scale with a series of parallel projectors.

33.4 Alignment Graphs with Three Variables

For a formula containing three functions (of one variable each), draw a nomograph by selecting the lengths and positions of two scales according to

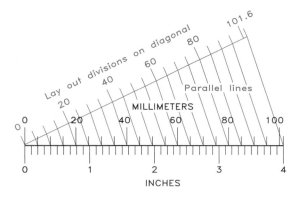

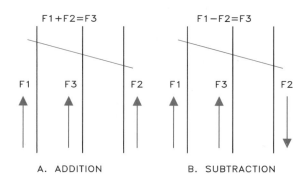

Figure 33.9 The proportional line method may be used to construct an alignment graph that converts inches to millimeters. The units at each end of the scales must be known. For example, there are 101.6 mm in 4 in.

Figure 33.10 These two common forms of parallel scale alignment nomographs show the directions in which the scales increase for addition and subtraction.

the range of variables and size of the chart desired. Calibrate these scales by using the scale equations presented in Section 33.3. Although mathematical relationships may be used to locate the third scale, graphical constructions are simpler and usually less subject to error. Examples of the various forms of these nomographs are presented in the following sections.

33.5 Parallel Scale Graphs

Linear Scales

Any formula of the type $F1 + F2 = F3$ may be represented as a parallel scale alignment graph **(Fig. 33.10)**. For addition, the three scales increase (functionally) in the same direction, and the function of the middle scale represents the sum of the other two. Reversing the direction of any scale changes the sign of its function in the formula, as for $F1 - F2 = F3$.

We can use the formula $Z = X + Y$ to illustrate this type of alignment graph **(Fig. 33.11)**. First, draw and calibrate the outer scales for X and Y. Then use two sets of data that yield a Z of 8 to locate the parallel Z scale. Divide the Z scale into

16 equal units. Add various values of X and Y to the finished nomograph to find their sums along the Z scale.

Figure 33.12 illustrated how the outer scales of a parallel-scale nomograph are calibrated for solving the equation $U + 2V = 3W$. The scales are placed any distance apart and are divided into linear divisions from 0 to 14 and 0 to 8 for U and V, respectively. These scales are used to complete the nomograph explained in **Fig. 33.13**.

Obtain the limits of calibration for the middle scale by connecting the endpoints of the outer scales and substituting these values into the formula. Here, W is 0 and 10 at the extreme ends. Select two pairs of corresponding values of U and V that give the same value of W. For example, $U = 0$ and $V = 7.5$ give $W = 5$. Also $W = 5$ when $U = 14$ and $V = 0.5$.

Because the W scale is linear ($3W$ is a linear function), you may subdivide it into uniform intervals by equal parts. For a nonlinear scale, determine the scale modulus (and the scale equation) by substituting length and its two end values into Eq. (1). You may use the scales to obtain infinitely many problem solutions.

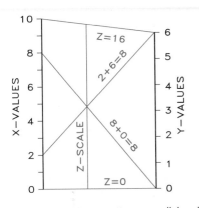 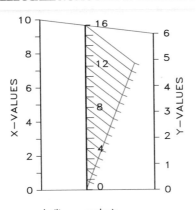 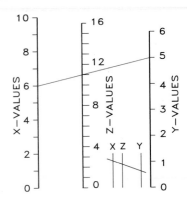

Figure 33.11 Constructing a parallel scale nomograph (linear scales):

Step 1 Draw two parallel scales at any length and calibrate them. Locate the parallel Z scale by selecting two sets of values that give the same value of Z (8 in this case). The ends of the Z scale have values of 0 and 16, or the sum of the end values of X and Y.

Step 2 Draw the Z scale through the point located in step 1 parallel to the other scales. Calibrate the scale from 0 to 16 by using the proportional line method.

Step 3 Calibrate and label the Z scale. Draw a key to show how to use the nomograph. If the Y scale were calibrated with 0 at the upper end instead of at the bottom, a different Z scale could be computed and the nomograph could be used for Z = X − Y.

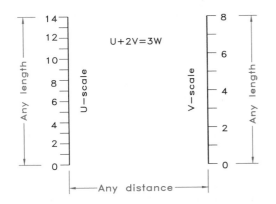

Figure 33.12 This calibration is of the outer scales for the formula $U + 2V = 3W$, where $0 \leq U \leq 14$ and $0 \leq V \leq 8$.

Logarithmic Scales

You may solve problems involving formulas of the type $(F1)(F2) = F3$ in a manner similar to that described in Fig. 33.11 by using logarithmic scales instead of linear scales (as shown in Fig. 33.15).

The first step in drawing a nomograph with logarithmic scales is to transfer the logarithmic functions to the scales. **Figure 33.14** shows the graphical method, which involves projecting units from a printed logarithmic scale to the nomographic scale.

Figure 33.15 illustrates the conversion of the formula $Z = XY$ into a nomograph. The desired limits for the X and Y scales are 1 and 10. Sets of values of X and Y that yield the same value of Z, 10 in this case, are used to locate the Z axis. The limits of the Z axis are 1 and 100.

The Z axis is drawn and calibrated as a two-cycle log scale. A key explains how to use an isopleth to add the logarithms of X and Y to give the log of Z. Recall that the addition of logarithms performs the operation of multiplication. Had the Y axis been calibrated in the opposite direction, with 1 at the upper end and 10 at the lower end, a new Z axis could have been calibrated. The nomograph then could be used for the formula $Z = Y/X$, involving the subtraction of logarithms.

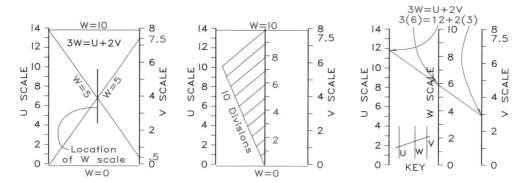

Figure 33.13 Constructing parallel scale nomograph (linear scales):

Step 1 Substitute the end values of the U and V scales into the formula to find the end values of the W scale: $W = 10$ and $W = 0$. Select two sets of U and V that will give the same value of W. For example, When $U = 0$ and $V = 7.5$, $W = 5$, and when $U = 14$ and $V = 0.5$, $W = 5$. Connect these sets of values to locate the W scale.

Step 2 Draw the W scale parallel to the outer scales to the limit lines of $W = 10$ and $W = 0$. This scale is 10 linear divisions long, so divide it graphically into 10 units. The W scale is a linear scale; construct it as shown in Fig. 33.11.

Step 3 Use the nomograph as illustrated by selecting any two known variables and connecting them with an isopleth to determine the third unknown. The key shows how to use the nomograph. The values of $U = 12$ and $V = 3$ verify the graph's accuracy.

CALIBRATION OF A LOGARITHMIC SCALE

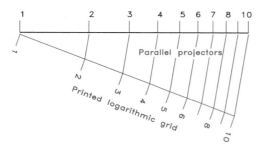

Figure 33.14 Here, a scale is calibrated graphically by projection from a printed logarithmic scale.

33.6 N or Z Nomographs

Whenever $F2$ and $F3$ are linear functions, we can avoid using logarithmic scales for formulas of the type

$$F1 = \frac{F2}{F3}$$

To do so, we use an **N graph (Fig. 33.16)**. The outer scales, or "legs," of the N are functional scales and therefore are linear if $F2$ and $F3$ are linear. If a parallel-scale graph were used for the same formula, all its scales would be logarithmic.

Some main features of the N graph are:

1. The outer scales are parallel functional scales of $F2$ and $F3$.

2. The outer scales increase (functionally) in opposite directions.

3. The diagonal scale connects the (functional) zeros of the outer scales.

4. In general, the diagonal scale is not a functional scale for the function $F1$ and is nonlinear.

Construction of an N nomograph is simplified because locating the middle (diagonal) scale is usually less of a problem than it is for a parallel scale graph. Calibration of the diagonal scale is most easily accomplished by graphical methods.

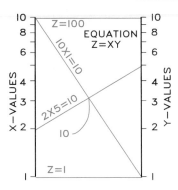
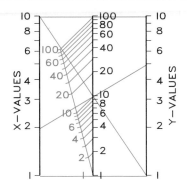
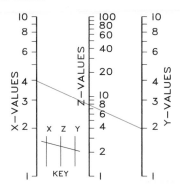

Figure 33.15 Parallel scale nomograph (logarithmic scales):

Step 1 For the equation $Z = XY$, draw parallel log scales. Construct sets of X and Y points that yield the same value of Z (10 in this case). Their intersection locates the Z scale with end values of 1 and 100.

Step 2 Graphically calibrate the Z axis as a two-cycle logarithmic scale from 1 to 100. This scale is parallel to the X and Y scales.

Step 3 Draw a key. An isopleth confirms that $4 \times 2 = 8$. By reversing the Y value scale from 1 downward to 10 and computing a different Z scale, the nomograph could be redrawn for the equation $Z = Y/X$.

N NOMOGRAPH FOR DIVISION

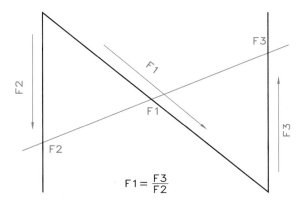

Figure 33.16 This N graph solves an equation of the form $F1 = F2/F3$.

Figure 33.17 shows how to construct a basic N graph of the equation $Z = Y/X$. Draw the diagonal to connect the zero ends of the scales. Then locate whole values along the diagonal by using combinations of X and Y values. Be sure to locate whole-value units along the diagonal that are easy to interpolate between. Label the diagonal and give a key explaining how to use the nomograph. Note that a sample **isopleth** verifies the correctness of the graphical relationship between the scales.

We can construct a more advanced N graph for the equation

$$A = \frac{B + 2}{C + 5},$$

where $0 \le B \le 8$ and $0 \le C \le 15$. This equation takes the form:

$$F1 = \frac{F2}{F3},$$

where

$$F1(A) = A,$$
$$F2(B) = B + 2, \text{ and}$$
$$F3(C) = C + 5.$$

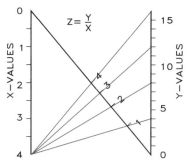

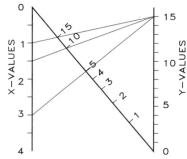

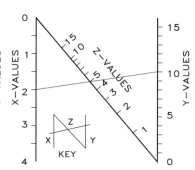

Figure 33.17 Constructing an N graph:
Step 1 Draw an N graph for the equation of $Z = Y/X$ with two parallel scales. Connect the zero ends of each scale with a diagonal scale. Use isopleths to locate units along the diagonal.

Step 2 Draw additional isopleths to locate other units along the diagonal. The units on the diagonal should be whole units to make interpolation easy.

Step 3 Label the diagonal scale and draw a key. An isopleth confirms that $10/2 = 5$. The accuracy of the N graph is greatest at the 0 end of the diagonal. The diagonal approaches infinity at the other end.

Thus the outer scales will be for $B + 2$ and $C + 5$, and the diagonal scale will be for A.

Begin the construction in the same manner as for a parallel scale graph by selecting the layout of the outer scales (**Fig. 33.18**). Determine the limits of the diagonal scale by connecting the endpoints on the outer scales, giving $A = 0.1$ for $B = 0$, $C = 15$ and $A = 2.0$ for $B = 8$, and $C = 0$. **Figure 33.19** shows these relationships and also gives the remainder of the construction.

Locate the diagonal scale by finding the **function zeros** of the outer scales (that is the points where $B + 2 = 0$ or $B = -2$ and $C + 5 = 0$ or $C = -5$). Then draw the diagonal scale by connecting these points.

Calibrating the diagonal scale is most easily accomplished by substituting into the formula. Selecting the upper limit of an outer scale, say, $B = 8$, gives the formula:

$$A = \frac{10}{C + 5}.$$

CALIBRATION OF N NOMOGRAPH

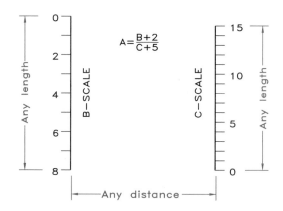

Figure 33.18 This calibration of the outer scales of an N graph is for the equation $A = (B + 2)/(C + 5)$.

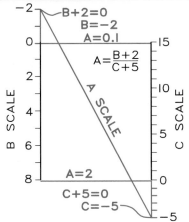

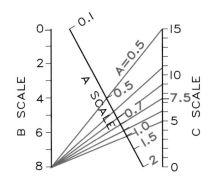

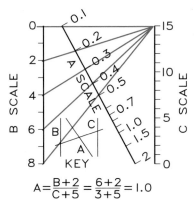

$$A = \frac{B+2}{C+5} = \frac{6+2}{3+5} = 1.0$$

Figure 33.19 Constructing an N graph:

Step 1 Locate the diagonal scale by finding the functional zeros of the outer scales. Do so by setting $B + 2 = 0$ and $C + 5 = 0$, which gives a zero value for A.

Step 2 Select the upper limit of one of the outer scales ($B = 8$ in this case) and substitute it into the equation to obtain a series of values of C for the desired values of A (see Table 33.1). Draw isopleths from $B = 8$ to the values of C to calibrate the A-scale.

Step 3 Calibrate the remainder of the A scale in the same manner by substituting $C = 15$ into the equation to determine a series of values on the B scale for desired values on the A scale (see Table 33.2). Draw isopleths from $C = 15$ to calibrate the A scale, as shown.

Then solve this equation for the other outer scale variable:

$$C = \frac{10}{A - 5}.$$

Using it as a "scale equation," make a table of values for the desired values of A and corresponding values of C (up to the limit of C in the graph), as shown in Table 33.1. Connect isopleths from $B = 8$ to the tabulated values of C. Their intersections with the diagonal scale give the required calibrations for approximately half the diagonal scale.

Calibrate the remainder of the diagonal scale by substituting the end value of the other outer scale ($C = 15$) into the formula, giving

$$A = \frac{B + 2}{20}.$$

Solving for B yields

$$B = 20A - 2.$$

Construct a table for the desired values of A as shown in Table 33.2. Isopleths connecting $C = 15$ with the tabulated values of B locate the remaining calibrations on the A scale.

Table 33.1

A	2.0	1.5	1.0	0.9	0.8	0.7	0.6	0.5
C	0	1.67	5.0	6.11	7.50	9.28	11.7	15.0

Table 33.2

A	0.5	0.4	0.3	0.2	0.1
B	8.0	6.0	4.0	2.0	0

Use the principles covered in this chapter to solve the following problems; use size A sheets. For problems involving geometric construction and mathematical calculations, show both the construction and calculations in the solutions. If the calculations are extensive, use a separate sheet for them.

Concurrent Scales

Construct concurrent scales for converting one type of unit to the other. The range of units for the scales is given for each relationship.

1. Kilometers and miles:

    1.609 km = 1 mile; from 10 to 100 miles.

2. Liters and U.S. gallons:

    1 L = 0.2692 U.S. gal; from 1 to 10 L.

3. Knots and miles per hour:

    1 knot = 1.15 mph; from 0 to 45 knots.

4. Horsepower and British thermal units:

    1 hp = 42.4 Btu; from 0 to 1200 hp.

5. Centigrade and Fahrenheit:

    °F = $\frac{9}{5}$°C + 32; from 32°F to 212°F.

6. Radius and area of a circle:

    $A = \pi r^2$; from r = 0 to 10.

7. Inches and millimeters:

    1 in. = 25.4 mm; from 0 to 5 in.

8. Numbers and their logarithms:

    Use logarithm tables; numbers from 1 to 10.

Addition and Subtraction Nomographs

Construct parallel scale nomographs to solve the following addition and subtraction problems.

9. $A = B + C$, where B = 0 to 10 and C = 0 to 5.

10. $Z = X + Y$, where X = 0 to 8 and Y = 0 to 12.

11. $Z = Y - X$, where X = 0 to 6 and Y = 0 to 24.

12. $A = C - B$, where C = 0 to 30 and B = 0 to 6.

13. $W = 2V + U$, where U = 0 to 12 and V = 0 to 9.

14. $W = 3U + V$, where U = 0 to 10, and V = 0 to 10.

15. Electrical current at a circuit junction:

$$I = I1 + I2,$$

where

I = current entering the junction in amperes (amps),

$I1$ = current leaving the junction, varying from 2 to 15 amps, and

$I2$ = current leaving junction, varying from 7 to 36 amps.

16. Pressure change in fluid flowing in a pipe:

$$\Delta P = P2 - P1,$$

where

ΔP = pressure change between two points in pounds per square inch (psi),

$P1$ = pressure upstream, varying from 3 psi to 12 psi, and

$P2$ = pressure downstream, varying from 10 psi to 15 psi.

Multiplication and Division: Parallel Scales

Construct parallel scale nomographs having logarithmic scales for performing the following multiplication and division operations.

17. Area of a rectangle: $A = H \times W$, where H = 1 to 10 and W = 1 to 12.

18. Area of a triangle: $A = \frac{1}{2}B \times H$, where B = 1 to 10, and H = 1 to 5.

19. Electrical potential between terminals of a conductor:

$$E = IR,$$

where

E = electrical potential in volts,

I = current, varying from 1 to 10 amps, and

R = resistance, varying from 5 to 30 ohms.

20. Pythagorean theorem:

$$C^2 = A^2 + B^2,$$

where

C = hypotenuse of a right triangle in centimeters (cm),

A = one leg of the right triangle, varying from 5 to 50 cm, and

B = second leg of the right triangle, varying from 20 to 80 cm.

21. Allowable pressure on a shaft bearing:

$$P = \frac{ZN}{100},$$

where

P = pressure in psi,

Z = viscosity of lubricant, varying from 15 to 50 (cp), and centipoise

N = angular velocity of shaft, varying from 10 to 1000 rpm.

22. Miles per gallon (mpg) an automobile gets: Miles vary from 1 to 500; gallons vary from 1 to 24.

23. Cost per mile (cpm) of an automobile: Miles vary from 1 to 500; cost varies from $1 to $28.

24. Angular velocity of a rotating body:

$$W = \frac{V}{R},$$

where

W = angular velocity, in radians per second,

V = peripheral velocity, varying from 1 to 100 m.$^{-1}$, and

R = radius, varying from 0.1 to 1 m.

N Nomographs

Construct N graphs that will solve the following equations.

25. Stress = P/A, where P varies from 0 to 1000 psi and A varies from 0 to 15 in^2.

26. Volume of a cylinder:

$$V = \pi r^2 h,$$

where

V = volume in in^3,

r = radius, varying from 5 to 10 ft, and

h = height, varying from 2 to 20 in.

27. Repeat Problem 17.

28. Repeat Problem 18.

29. Repeat Problem 19.

30. Repeat Problem 20.

31. Repeat Problem 21.

32. Repeat Problem 22.

33. Repeat Problem 23.

34. Repeat Problem 24.

Empirical Equations and Calculus

34.1 Introduction

Graphical methods are useful supplements to mathematical techniques of solving problems dealing with experimental data, especially when the data do not fit a mathematical equation. Data from laboratory experiments or from field tests are called **empirical data**. Empirical data often are expressed as one of three types of equations: linear, power, or exponential.

Analysis of empirical data begins with plots of the data on linear, logarithmic, or semilogarithmic grids. Fitting curves to the data determines which of the grids gives a straight-line relationship **(Fig. 34.1).** Data that yield a straight-line plot on one of these grids may be written as a mathematical equation.

34.2 Curve Equations

Two methods of finding the equation of a curve are (1) the selected-points method and (2) the slope-intercept method **(Fig. 34.2).**

LINEAR, LOGARITHMIC, AND SEMILOG GRIDS

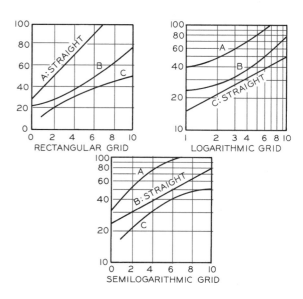

Figure 34.1 Plotting empirical data on each type of grid determines which yields a straight-line plot. A straight-line plot on one of these grids means that a mathematical equation can be written to describe the data.

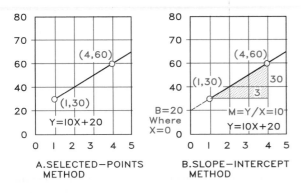

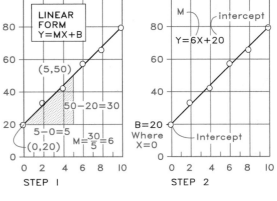

Figure 34.2
A One way to determine the equation of a straight line on a grid is by selecting any two points on the line.

B A second way is to use the slope-intercept equation. It requires finding the Y intercept (where $X = 0$), involving the extension of the curve to the Y axis.

Figure 34.3 The linear equation, $Y = MX + B$:
Step 1 A straight line on an arithmetic grid has an equation that takes the form $Y = MX + B$. Here, the slope, M, is 6.
Step 2 The intercept, $B = 20$, is substituted in the equation to obtain $Y = 6X + 20$.

Selected-Points Method

Two widely separated points on the line in **Fig. 34.2A**, (1, 30) and (4, 60), are selected and substituted into the equation:

$$\frac{Y - 30}{X - 1} = \frac{60 - 30}{4 - 1}$$

The resulting equation is

$$Y = 10X + 20.$$

Slope-Intercept Method

The slope-intercept method may be used when the Y intercept of the curve (where $X = 0$) is known. If the X axis is logarithmic, the log of $X = 1$ is 0, and the intercept is at $X = 1$.

In **Fig. 34.2B**, the curve does not intercept the Y axis, so it must be extended to find the intercept at $X = 0$, or $B = 20$ in this case. The slope of the curve, $M = 10$, is determined ($\Delta Y/\Delta X$ or 30/3), and it is substituted into the linear equation form to obtain $Y = 10X + 20$.

34.3　The Linear Equation: $Y = MX + B$

The curve shown in **Fig. 34.3** representing experimental data is a straight line on a linear graph, which identifies the data as **linear**, meaning that each measurement along the Y axis is directly proportional to the measurement along the X axis. We may use either the slope-intercept method or the selected-points method to find the equation for the data.

By using the slope-intercept method first, we select two known points along the curve. The vertical and horizontal differences between the coordinates of each point establish the adjacent sides of a right triangle. In the slope-intercept equation, $Y = MX + B$, the slope, M, is the tangent of the angle between the curve and the horizontal, B is the Y intercept of the curve (where $X = 0$), and X and Y are variables. Here, $M = 30/5 = 6$, and the intercept is at 20.

Substituting these values into the slope-intercept equation, we obtain $Y = 6X + 20$, from

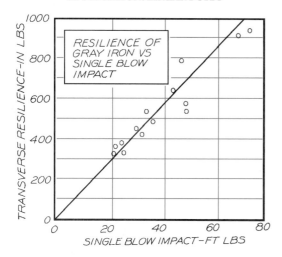

Figure 34.4 The relationship between the transverse strength of gray iron and impact resistance plots as a straight line yielding an equation of the form $Y = MX + B$ or $Y = 13.3X$.

which we may determine values of Y by substituting any value of X into the equation. If the curve had sloped downward to the right, the slope would have been negative.

Let's now use the selected-points method, assuming that we do not know the Y intercept. Select two widely separated points, say (2, 32) and (10, 80), and write the equation in the form:

$$\frac{Y - 32}{X - 2} = \frac{80 - 32}{10 - 2}$$

or

$$Y = 6X + 20,$$

Thus both methods give the same equation for the curve, or $Y = MX + B$.

Figure 34.4 shows the relationship between the transverse strength and impact resistance of gray iron obtained from empirical data. Plotted on a linear grid, the data are best represented by a straight line, so its equation takes the linear form.

34.4 The Power Equation: $Y = BX^M$

Data plotted on a logarithmic grid that yields a straight line (**Fig. 34.5**) may be expressed in the power equation form in which Y is a function of X raised to a power, or $Y = BX^M$. We obtain the equation of the data by using the Y intercept as B and the slope of the curve as M.

Select two points on the curve to form the slope triangle and use an engineer's scale to measure its slope. If you draw the horizontal side of the right triangle as 1 or a multiple of 10, you may read the vertical distance directly. Here, the slope M (tangent of the angle) is 0.54 and the Y intercept, B, is 7; thus the equation is $Y = 7X^{0.54}$, which in logarithmic form is

$$\log Y = \log B + M \log X,$$

$$= \log 7 + 0.54 \log X.$$

We used base-10 logarithms in these examples, but natural logs may be used with e (2.718) as the base.

In **Fig. 34.6** the intercept, B, lies on the Y axis where $X = 1$ because the curve is plotted on a logarithmic grid. The Y intercept ($B = 80$) is found where $X = 1$. Recall that the intercept at this point is analogous to the linear form of the equation because the log of 1 is 0. The curve slopes downward to the right making the slope, M, negative.

Figure 34.7 shows plots of empirical data that relate the specific weight (pounds per horsepower) of generators and hydraulic pumps to horsepower. The plots of the data are represented by straight lines on a logarithmic grid, which means that their equations take the power form.

34.5 The Exponential Equation: $Y = BM^X$

When data plotted on a semilogarithmic grid (**Fig. 34.8**) yield a straight line, the equation takes the form $Y = BM^X$, where B is the Y intercept, and M is the slope of the curve. Select two points along the curve, draw a right triangle, and find its slope, M. The slope of the curve is

$$\log M = \frac{\log 40 - \log 6}{8 - 3} = 0.1648,$$

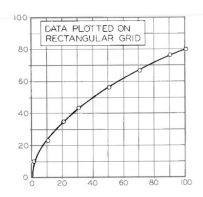

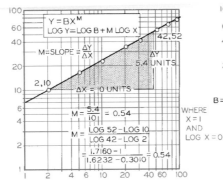

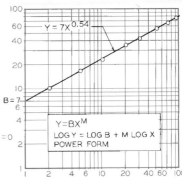

Figure 34.5 The power equation, $Y = BX^M$:
Given Plotting these data on a rectangular grid gives a parabolic curve. Because the curve is not a straight line on the rectangular grid, its equation is not linear.

Step 1 The data plots as a straight line on a logarithmic grid. Find the slope, M, graphically with an engineer's scale by setting ΔX at 10 units and measuring ΔY with the same scale to get 5.4 units. Thus $M = 5.4/10$, or 0.54.

Step 2 The intercept $B = 7$ lies at $X = 1$. Substitute the slope and intercept into the equation, which becomes $Y = 7X^{0.54}$.

APPLICATION: POWER EQUATION

INTERCEPT AT X = 1

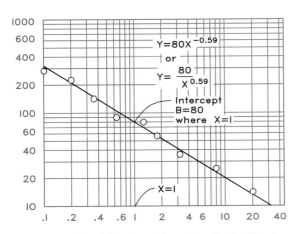

Figure 34.6 Use of the slope-intercept method with a logarithmic grid means that the intercept lies at $X = 1$. Here, the intercept lies at 80 (near the middle of the graph).

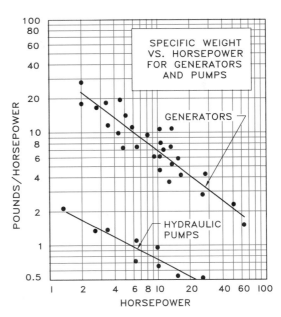

Figure 34.7 Empirical data plotted on a logarithmic grid showing the specific weight versus horsepower of electric generators and hydraulic pumps are straight lines. Their equations take the power form, $Y = BX^M$.

THE EXPONENTIAL EQUATION: $Y = BM^X$

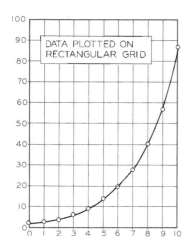

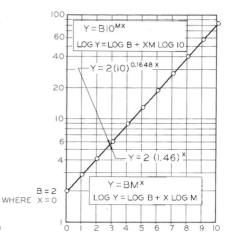

Figure 34.8 The exponential equation, $Y = BM^X$:

Given These data give a straight line on a semilogarithmic grid. Therefore the data take the equation form $Y = BM^X$.

Step 1 You must find the slope mathematically, not graphically, because the X and Y scales are unequal. Write the slope equation in either of the forms shown.

Step 2 Find the intercept, $B = 2$, where $X = 0$. Substitute the values for M and B into the equation to get $Y = 2(10)^{0.1648X}$ or $Y = 2(1.46)^X$.

or

$$M = (10)^{0.1648} = 1.46.$$

Substitute this value of M into the exponential equation:

$$Y = BM^X \quad \text{or} \quad Y = 2(1.46)^X,$$
$$Y = B(10)^{MX} \quad \text{or} \quad Y = 2(10)^{0.1648X},$$

where X is a variable that can be substituted into the equation to give infinitely many values for Y. In logarithmic form, the equation is

$$\log Y = \log B + X \log M,$$

or

$$\log Y = \log 2 + X \log 1.46.$$

The same methods give the negative slope of a curve. The curve shown in **Fig. 34.9** slopes downward to the right and therefore has a negative

NEGATIVE SLOPE (M)

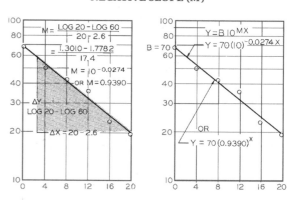

Figure 34.9

A When a curve slopes downward to the right, its slope is negative.

B Substitution gives these two forms of the equation for the curve.

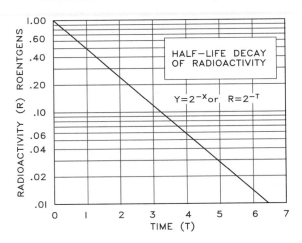

Figure 34.10 The decay of radioactivity is represented by a straight line on a semilog grid, indicating that its equation takes the exponential form, $Y = BM^X$ or $R = 2^{-T}$.

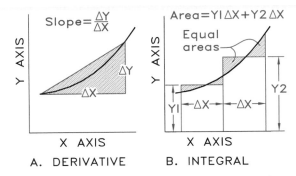

Figure 34.11
A The derivative of a curve is the rate of change at any point on the curve, or its slope, $\Delta Y/\Delta X$.

B The integral of a curve is the cumulative area enclosed by the curve, or the summation of the incremental areas comprising the whole.

slope. The slope, M, is the antilog of –0.0274. The intercept of 70 and the slope yield the equation for the curve.

The half-life decay of radioactivity plotted in **Fig. 34.10** compares decay to time. The half-life of different isotopes varies, so different values would be assigned along the X axis for them. However, the curves for all isotopes would be straight lines and take the exponential equation form.

34.6 Graphical Calculus

If the equation of a curve is known, calculus may be used to perform certain types of calculations. However, experimental data often do not fit standard mathematical equations, making impossible the application of calculus mathematically. In these cases, graphical calculus must be used. The two basic forms of calculus are (1) differential calculus and (2) integral calculus.

Differential calculus is used to determine the rate of change of one variable with respect to another (**Fig. 34.11A**). The rate of change at any instant along the curve is the slope of a line tangent to the curve at that point. Constructing a chord at any interval allows approximation of this slope. The tangent, $\Delta Y/\Delta X$, may represent miles per hour, weight versus length, or various other rates of change important in the analysis of data.

Integral calculus is the reverse of differential calculus. Integration is used to find the area under a curve (the product of the variables plotted on the X and Y axes). The area under a curve is approximated by dividing one of the variables into a number of very small rectangular bars under the curve (**Fig. 34.11B**). Each bar is drawn so that equal areas lie above and below the curve and the average height of the bar is near its midpoint.

SCALES FOR DIFFERENTIATION

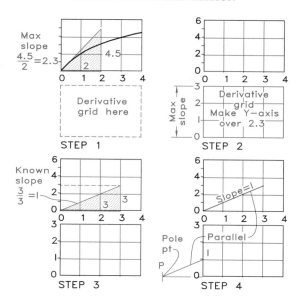

Figure 34.12 Scales for graphical differentiation:
Step 1 Estimate the maximum slope of the curve (here, 2.3) by drawing a line tangent to the curve where it is steepest.

Step 2 Draw the derivative grid with a maximum ordinate of 3.0 to accommodate the slope of 2.3.

Step 3 Find a known slope of 1 on the given grid. The slope has no relationship to the data curve.

Step 4 Draw a line from 1 on the Y axis of the derivative grid parallel to the slope of the triangle drawn in the given grid to locate the pole point on the extension of the X axis.

34.7 Graphical Differentiation

Graphical differentiation is used to determine the rate of change of two variables with respect to each other at any given point. **Figure 34.12** illustrates the preliminary construction of a derivative scale and pole point to be used to plot a derivative curve.

Figure 34.13 shows the graphical differentiation process. The maximum slope of the data

curve is estimated to be slightly less than 12, so an ordinate scale long enough to accommodate the maximum slope is selected. To locate the pole point, a line is drawn from point 12 on the ordinate axis of the derivative grid parallel to the known slope on the given curve grid and the line is extended to the X axis.

Next, a series of chords is constructed on the given curve. The interval between 0 and 1, where the curve is steepest, is divided in half to obtain a more accurate plot. After other bars are found, a smooth curve is drawn through the top of them so that the area above and under the top of each bar is equal. The rate of change, $\Delta Y/\Delta X$, can be found for any value of X in the derivative graph.

Applications

The mechanical handling shuttle shown in **Fig. 34.14** converts rotational motion into linear motion. The drawing of the linkage shows the end positions of point P, which is the zero point for plotting travel versus degrees of revolution. Rotation is constant at one revolution every three seconds, so the degrees of revolution may be converted to time **(Fig. 34.15)**. The drive crank, R1, is revolved at 30° intervals, and the distance that point P travels from its end position is plotted on the graph to give the distance-time relationship.

The ordinate scale of the derivative grid is scaled with an end value of 100 in./sec, or slightly larger than the estimated maximum slope of the curve. A slope of 40 is drawn on the given data grid. Pole point P is found by drawing a line from 40 on the derivative ordinate scale parallel to the slope of 40 in the given data graph to the extension of the X axis of the derivative graph.

Chords drawn on the curve approximate the slope at various points. Lines drawn from point P of the derivative scale are parallel to the chordal lines to the ordinate scale. These intersections on the Y axis are projected horizontally to their respective intervals to form vertical bars. A smooth curve is drawn through the tops of the bars to give an average of the bars. This curve is

GRAPHICAL DIFFERENTIATION

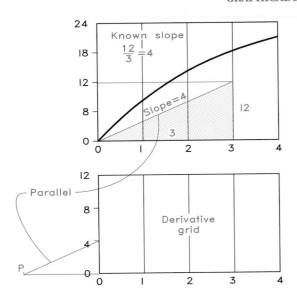

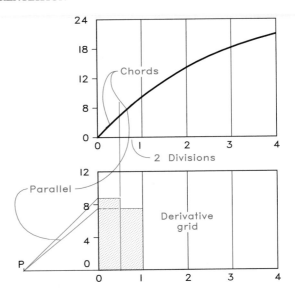

Figure 34.13 Graphical differentiation:

Required Find the derivative curve of the given data.

Step 1 Find the derivative grid and the pole point by using the steps presented in Fig. 34.12.

Step 2 Construct chords between intervals on the given curve and draw lines parallel to them through point P on the derivative grid. These lines locate the heights of bars in their respective intervals. Divide the first interval in half where the curve is sharpest.

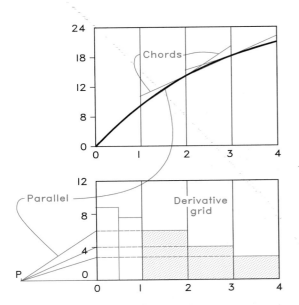

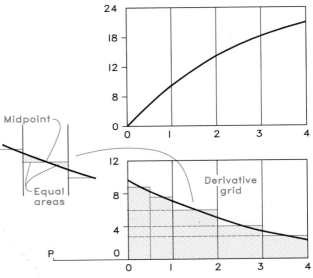

Step 3 Draw additional chords in the last three intervals. Draw lines parallel to these chords from the pole point to the Y axis to find additional bars in their respective intervals.

Step 4 The vertical bars represent the slopes of the curve at different intervals. Draw the derivative curve through the midpoints of the bars so that the areas below and above the bars are equal.

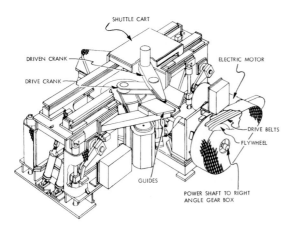

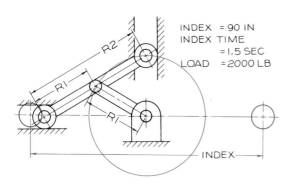

Figure 34.14 These pictorial and scale drawings are of an electrically powered mechanical handling shuttle used to move automobile parts on an assembly line. (Courtesy of General Motors Corporation.)

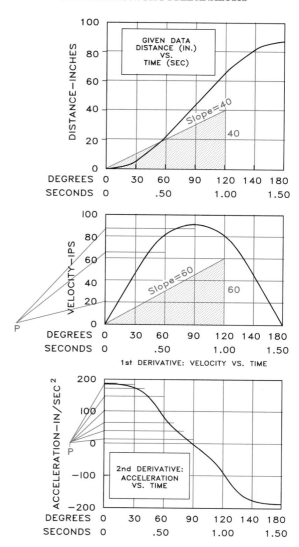

Figure 34.15 Velocity and acceleration of the mechanical handling shuttle may be obtained with graphical differential calculus.

used to find the velocity of the shuttle in inches per second at any time interval.

The second derivative curve, acceleration versus time, is constructed in the same manner as the first derivative curve. Inspection of the first derivative curve reveals that the maximum slope is an estimated 200 in./sec/sec. An easily measured scale for the ordinate scale is chosen. A new pole point P is found in the same manner as for the previous derivative.

Chords are drawn at intervals on the first derivative curve. Lines drawn from point P paral-

lel to these chords intersect the Y axis of the second derivative graph, where they are projected horizontally to their respective intervals, establishing the heights of the bars. A smooth curve

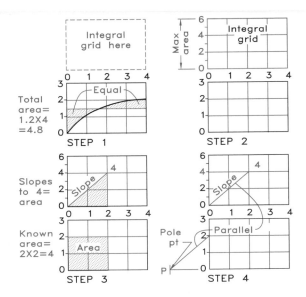

Figure 34.16 Scales for graphical integration:
Step 1 To determine the maximum value on the Y axis, draw a line to approximate the area under the given curve (4.8 square units here).

Step 2 Draw the integral with a Y axis of 6 to accommodate the maximum area.

Step 3 Find a known area of 4 on the given grid. Draw a slope from 0 to 4 on the integral grid directly above the known area to establish the integral.

Step 4 Draw a line from 2 on the Y axis of the given grid parallel to the slope line in the integral grid. Locate the pole point on the extension of the X axis.

drawn through the tops of the bars gives a close approximation of the average areas of the bars. The minus scale indicates deceleration.

These velocity and acceleration graphs show that parts being handled by the shuttle accelerate at a rapid rate until the maximum velocity is attained at 90°, at which time deceleration begins and continues until the parts come to rest.

34.8 Graphical Integration

Graphical integration is used to determine the area (product of two variables) under a curve. For example, if the Y axis represented pounds and the X axis represented feet, the integral curve would give the product of the variables, foot-pounds, at any interval of feet along the X axis. **Figure 34.16** illustrates how the scales and pole point for graphical integration are determined.

In **Fig. 34.17**, the total area under the curve is estimated to be less than 80 units. Therefore the maximum height of the Y axis on the integral curve is 80, drawn at a convenient scale. Pole point P is found by the steps shown in Fig. 34.16.

A series of vertical bars is constructed to approximate the areas under the curve. The narrower the bars, the more accurate will be the resulting plotted curve. The interval between 2 and 3 is divided into two bars to obtain more accuracy where the curve is sharpest. The tops of the bars are extended horizontally to the Y axis and these intersections are connected to point P.

Lines are drawn parallel to AP, BP, CP, DP, and EP in the integral grid to correspond to the respective intervals in the given grid. The intersection points of the chords are connected by a smooth curve—the integral curve—to give the cumulative product of the X and Y variables along the X axis.

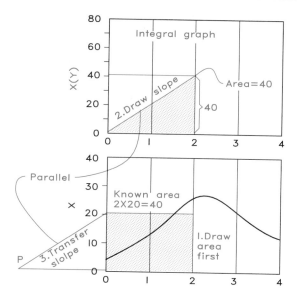

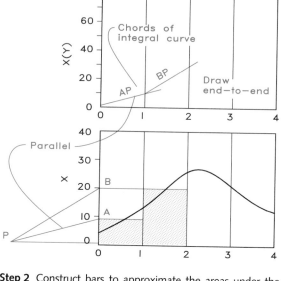

Figure 34.17 Performing graphical integration:
Required Plot the integral curve of the given data.

Step 1 Find pole point P by using the technique described in Fig. 34.16.

Step 2 Construct bars to approximate the areas under the curve. Project the heights of the bars to the Y axis and draw lines to the pole point. Draw sloping lines AP and BP at their respective intervals parallel to the lines drawn to P.

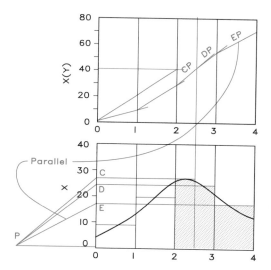

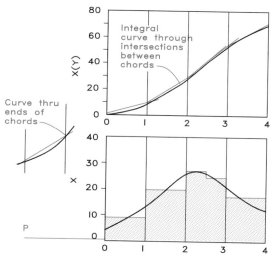

Step 3 Draw additional bars from 2 to 4 on the X axis. Project the heights of the bars to the Y axis and draw rays to P. Draw lines CP, DP, and EP at their respective intervals and parallel to their rays in the integral grid.

Step 4 The straight lines connected in the integral grid represent chords of the integral curve. Construct the integral curve to pass through the points where the chords intersect. An ordinate value on the integral curve represents the cumulative area under the given curve from zero to that point on the X axis.

Problems

Use size A sheets for your solutions to the following problems. For solutions involving mathematical calculations, show them on separate sheets if space is not available on the same sheet with the graphical solutions.

Empirical Equations—Logarithmic

1. Find the equation of the data shown in the following table. The empirical data compare input voltage, V, with input current in amperes, I, to a heat pump.

Y axis	V	0.8	1.3	1.75	1.85
X axis	I	20	30	40	45

2. Find the equation of the data presented in the following table. The empirical data give the relationship between peak allowable current in amperes, I, with the overload operating time in cycles at 60 cycles per second, C.

Y axis	I	2000	1840	1640	1480	1300	1200	1000
X axis	C	1	2	5	10	20	50	100

3. Find the equation of the data contained in the following table. The empirical data for a low-voltage circuit breaker used on a welding machine give the maximum loading during welding in amperes, rms, for the percentage of duty, pdc.

Y axis	rms	7500	5200	4400	3400	2300	1700
X axis	pdc	3	6	9	15	30	60

4. Construct a three-cycle by three-cycle logarithmic graph to find the equation of a machine's vibration displacement in mills (Y axis) and vibration frequency in cycles per minute, cpm (X axis). Data: 100 cpm, 0.80 mills; 400 cpm, 0.22 mills; 1000 cpm, 0.09 mills; 10,000 cpm, 0.009 mills; 50,000 cpm, 0.0017 mills.

5. Find the equation of the data shown in the following table that compares the velocities of air moving over a plane surface in feet per second, v at different heights in inches, y above the surface. Plot y values on the Y axis.

Y	0.1	0.2	0.3	0.4	0.6	0.8	1.2	1.6	2.4	3.2
V	18.8	21.0	22.6	24.1	26.0	27.3	29.2	30.6	32.4	33.7

6. Find the equation of the data presented in the following table that shows the distance traveled in feet, s, at various times in seconds, t, of a test vehicle. Plot s on the Y axis and t on the X axis.

t	1	2	3	4	5	6
s	15.8	63.3	146.0	264.0	420.0	580.0

Empirical Equations—Linear

7. Construct a linear graph to determine the equation for the annual cost of a compressor in relation to the compressor's size in horsepower. Plot cost on the Y axis and compressor size on the X axis. Data: 0 hp, $0; 50 hp, $2100; 100 hp, $4500; 150 hp, $6700; 200 hp, $9000; 250 hp, $11,400. Write the equation for these data.

8. Construct a linear graph on which the X axis is the mat depth from 0 to 4 in. and the Y axis is tons per hour per foot of width from 0 to 70 for a conveyor traveling at a rate of 50 feet per minute. The conveyor actually is a moving shaker that screens particles of coal by size. $X = 0$, $Y = 0$; $X = 1$, $Y = 13$; $X = 2$, $Y = 25$; $X = 3$, $Y = 38$; and $X = 4$, $Y = 50$.

9. Write the equation for the empirical data plotted in Fig. 34.8.

10. Plot the data contained in the following table on a linear graph and determine its equation. The empirical data show the deflection in centimeters of a spring, d, when it is loaded with different weights in kilograms, w. Plot w along the X axis and d along the Y axis.

w	0	1	2	3	4	5
d	0.45	1.10	1.45	2.03	2.38	3.09

11. Plot the data shown in the following table on a linear graph and determine the graph's equation. The empirical data show temperatures read from a Fahrenheit scale thermometer, °F, and a Celsius scale thermometer, °C. Plot the °C values along the X axis and the °F values along the Y axis.

°C	−6.8	6.0	16.0	32.2	52.0	76.0
°F	20.0	43.0	60.8	90.0	125.8	169.0

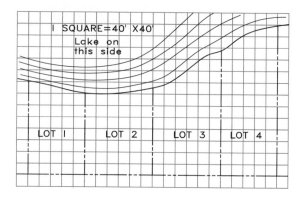

Figure 34.18 Problem 30. The tract of land shown here has been subdivided into four lots.

Empirical Equations—Semilogarithmic

12. Construct a semilog graph of the following data to determine their equation. Make the Y axis a two-cycle log scale and the X axis a 10-unit linear scale. Plot the voltage, V, along the Y axis and time, t, in sixteenths of a second along the X axis to represent resistor voltage during capacitor charging. Data: 0 sec, 10 V; 2 sec, 6 V; 4 sec, 3.6 V; 6 sec, 2.2 V; 8 sec, 1.4 V; 10 sec, 0.8 V.

13. Write the equation for the data plotted in Fig. 34.10.

14. Construct a semilog graph of the following data to determine their equation. Make the Y axis a three-cycle log scale and the X axis a linear scale with values from 0 to 250. These data relate the reduction factor, R (Y axis), to the mass thickness per square foot MT (X axis), of a nuclear protection barrier. Data: $0MT$, $1.0R$; $100MT$, $0.9R$; $150MT$, $0.028R$; $200MT$, $0.009R$; $300MT$, $0.0011R$.

15. An engineering firm considering expansion is reviewing its past income as shown in the following table. Their years of operation are represented by x, and their annual income (in tens of thousands of dollars) by N. Plot x along the X axis and N along the Y axis and determine the equation for the data.

x	1	2	3	4	5	6	7	8	9	10
N	0.05	0.08	0.12	0.20	0.32	0.51	0.80	1.30	2.05	3.25

Empirical Equations—General Types

16–21. Plot the experimental data shown in Table 34.1 on a grid that allows the data to appear as straight-line curves. Determine the equations of the data.

Calculus—Graphical Differentiation

22. Plot the equation $Y = X^3/6$ as a rectangular graph. Graphically differentiate the curve to determine the first and second derivatives.

23. Plot the following equation on a graph and find the derivative curve of the data on a second graph placed below the first: $Y = 2X^2$.

24. Plot the following equation on a graph and find the derivative curve of the data on a second graph placed below the first: $4Y = 8 - X^2$.

25. Plot the following equation on a graph and find the derivative curve of the data on a second graph placed below the first: $3Y = X^2 + 16$.

26. Plot the following equation on a graph and find the derivative curve of the data on a second graph placed below the first: $X = 3Y^2 - 5$.

Calculus—Graphical Integration

27. Plot the following equation on a graph and find the integral curve of the data on a second graph placed above the first: $Y = X^2$.

28. Plot the following equation on a graph and find the integral curve of the data on a second graph placed above the first: $Y = 9 - X^2$.

29. Plot the following equation on a graph and find the integral curve of the data on a second graph placed above the first: $Y = X$.

30. A plot plan shows that a tract of land is bounded by a lake front **(Fig. 34.18)**. By graphical integration, represent the cumulative area of the land (from left to right). What is the total area? What is the area of each lot?

Table 34.1

16	X	0	40	80	120	160	200	240	280			
	Y	4.0	7.0	9.8	12.5	15.3	17.2	21.0	24.0			
17	X	1	2	5	10	20	50	100	200	500	1000	
	Y	1.5	2.4	3.3	6.0	9.0	15.0	23.0	24.0	60.0	85.0	
18	X	1	5	10	50	100	500	1000				
	Y	3	10	19	70	110	400	700				
19	X	2	4	6	8	10	12	14				
	Y	6.5	14.0	32.0	75.0	115.0	320.0	710.0				
20	X	0	2	4	6	8	10	12	14			
	Y	20	34	53	96	115	270	430	730			
21	X	0	1	2	3	4	5	6	7	8	9	10
	Y	1.8	2.1	2.2	2.5	2.7	3.0	3.4	3.7	4.1	4.5	5.0

The Computer in Design and Graphics

35.1 Introduction

Computers are widely used in engineering and related fields and their use is expected to grow even more rapidly than in the past **(Fig. 35.1)**. Engineering and technology students must become computer literate, to understand the applications and implications of computers and their advantages. Not to do so will place students at a serious disadvantage in pursuing their careers.

35.2 Computer-Aided Design

Computer-aided design (CAD) involves solving design problems with the help of computers: to make graphic images on a video display, print out these images on paper with a plotter or printer, analyze design data, and store design information for easy retrieval. Many CAD systems perform these functions in an integrated manner, greatly increasing the designer's productivity.

Computer-aided design drafting (CADD), an offshoot of CAD, is the process of generating engi-

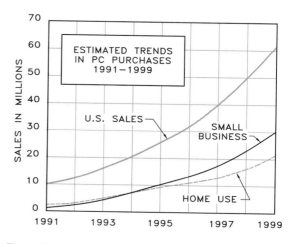

Figure 35.1 Annual sales of personal computers are expected to increase to more than 60 million by 1999. (Source: *The Dallas Morning News.*)

neering drawings and other technical documents by computer and is more directly related to drafting than is CAD. The CADD user inputs data by keyboard and/or mouse to produce illustrations on the monitor screen that can be reproduced as paper copies with a plotter or printer.

Engineers generally agree that the computer does not change the nature of the design process but is a significant tool that improves efficiency and productivity. The designer and the CAD system may be described as a design team: The designer provides knowledge, creativity, and control; the computer generates accurate, easily modifiable graphics, performs complex design analysis at great speed, and stores and recalls design information. Occasionally, the computer may augment or replace many of the engineer's other tools, but it cannot replace the design process, which is controlled by the designer.

Advantages of CAD and CADD

Depending on the nature of the problem and the sophistication of the computer system, computers offer the designer or drafter some or all of the following advantages.

1. **Easier creation and correction of drawings**. Working drawings may be created more quickly than by hand and making changes and modifications is more efficient than correcting drawings made by hand.

2. **Better visualization of drawings**. Many systems allow different views of the same object to be displayed and 3D pictorials to be rotated on the CRT screen.

3. **Database of drawing aids**. Creation and maintenance of design databases (libraries of designs) permits storing designs and symbols for easy recall and application to the solution of new problems.

4. **Quick and convenient design analysis**. Because the computer offers ease of analysis, the designer can evaluate alternative designs, thereby considering more possibilities while speeding up the process at the same time.

5. **Simulation and testing of designs**. Some computer systems make possible the simulation of a product's operation, testing the design under

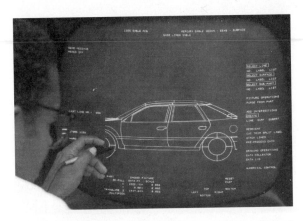

Figure 35.2 The Ford Taurus and Mercury Sable received more computer-aided design and engineering than any other car in the company's history. (Courtesy of Ford Motor Company.)

a variety of conditions and stresses. Computer testing may improve on or replace construction of models and prototypes.

6. **Increased accuracy**. The computer is capable of producing drawings with more accuracy than is possible by hand. Many CAD systems are even capable of detecting errors and informing the user of them.

7. **Improved filing**. Drawings can be more conveniently filed, retrieved, and transmitted on disks and tapes.

Computer Graphics

Computer graphics has an almost limitless number of applications in engineering and other technical fields. Most graphical solutions that are possible with a pencil can be done on a computer—and usually more productively. Applications vary from 3D modeling and finite element analysis to 2D drawings and mathematical calculations.

Automobile bodies are designed at the computer, as shown in **Fig. 35.2**. A scanning device records measurements of the life-size tape drawing of a new vehicle (**Fig. 35.3**). Once recorded and stored in the computer, the data can be manipulated to form the 3D shape of the body

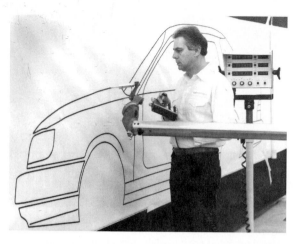

Figure 35.3 A scanning device records measurements from a life-size tape drawing of a new vehicle for storage in a computer and use for other design functions. (Courtesy of Ford Motor Company.)

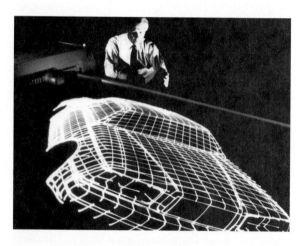

Figure 35.4 Design data stored in the computer's memory may be manipulated to obtain variations in body design of an automobile. (Courtesy of Ford Motor Company.)

style on the computer monitor from which 3D clay models are milled **(Fig. 35.4)**.

Once the domain of large computer systems, advanced applications can now be done on microcomputers. **Figure 35.5** depicts the modeling and 3D illustration of a site's topography by

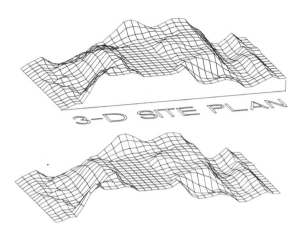

Figure 35.5 In this 3D representation of a site plan, the lower version is slightly different from the upper version. Hidden lines on the surface have been hidden by the computer. (Courtesy of LANDCADD, Incorporated.)

microcomputer. With simple keyboard commands, the user can view this site plan from infinitely many positions to obtain the best vantage points.

In **Fig. 35.6**, a perspective drawing of an urban area was plotted using microcomputer software. This program lets the viewer be positioned at any height and location in the site, and even "walk through" the area by successively changing positions.

35.3 Computer-Aided Design/ Computer-Aided Manufacturing

An important extension of CAD is its application to manufacturing. **Computer-aided design/computer-aided manufacturing** (CAD/CAM) systems may be used to design a part or product, devise the essential production steps, and electronically communicate this information to and control the operation of manufacturing equipment, including robots **(Fig. 35.7)**. These systems offer many advantages over traditional design and manufacturing systems, including less design effort, more efficient material use, reduced lead time, greater accuracy, and improved inventory maintenance.

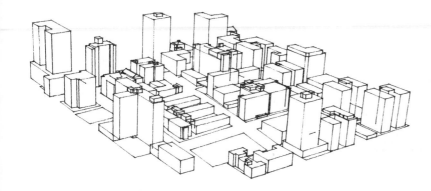

Figure 35.7 This automatic welding system for Chrysler LeBaron GTS and Dodge Lancer car bodies uses computer-controlled robots for consistent welds of all components in the unitized body structure. This assembly plant also features energy-efficient electric robot welders that require low maintenance and provide a high degree of accuracy. (Courtesy of Chrysler Corporation.)

An extension of CAD/CAM is **computer-integrated manufacturing** (CIM), a computer or system of computers that coordinates all stages of manufacturing. It enables manufacturers to custom design and produce products efficiently and economically.

35.4 Hardware Systems

Computer-aided design systems have three major components: the **designer, hardware,** and **software**. Hardware comprises the physical components of a computer system, and software is the programmer's instructions to the computer. The hardware of a computer graphics system includes the **computer, monitor, terminal, input devices** (keyboard, digitizers, and light pens), and **output devices** (plotters and printers).

Computer

The computer receives input from the user, executes instructions, and produces output. A sequence of instructions called a **program** controls the computer's activities. The part of the computer that follows the program's instructions is the **central processing unit** (CPU).

   Mainframe computers are large, fast, powerful, and expensive. Smaller and less costly than mainframes, **minicomputers** are used by many businesses.

   The smallest computers, **microcomputers**, are widely used for both personal and business applications **(Fig. 35.8)**. Advances in hardware technology and software capability, along with continually falling prices, have brought microcom-

Figure 35.8 The Hewlett-Packard Vectra 486U computer, keyboard, and monitor provide unprecedented graphics and system performance for Microsoft Windows and the CAD environment. (Courtesy of Hewlett-Packard Company.)

Figure 35.9 The user can take Compaq's Contoura laptop computer to the field for on-site work. (Courtesy of Compaq Computer Corporation.)

puters into wide use for engineering graphics applications in industry, government, and education. Moreover, the microcomputer is no longer bound to the office; the user can take it into the field for on-site work (**Fig. 35.9**).

Terminal

The terminal allows the user to communicate with the computer. It typically consists of a keyboard, a cathode-ray tube (monitor), and the interconnections between them and the computer (**Fig. 35.10**).

The **keyboard** allows the user to communicate with the computer through a set of alphanumeric and function keys. Keyboards resemble typewriters but include many other function keys, some of which may be user-defined.

The **cathode-ray tube** (CRT) is a video display tube and screen that is phosphor-coated. An electron gun emits a beam that sweeps out rows of raster lines on the screen. Each raster line consists of dots called **pixels**. Turning pixels on and off generates images on the screen. Raster-scanned CRTs refresh the picture display many times per second. A measure of the quality of a monitor is **resolution**, or the number of pixels per inch that can be produced on the screen. As the number of

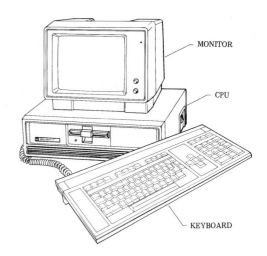

Figure 35.10 The basic components of a desktop computer system are the CPU, keyboard, and monitor (CRT).

pixels per square inch on the screen increases, the quality of the graphics obtained improves.

Raster-scan technology has largely replaced the vector-refreshed tube used in early video displays, in which each line in the picture is continuously redrawn by the computer. Because the vector-refreshed images are collections of lines instead of a set of individual pixels, raster-scanned displays are more realistic.

Figure 35.11 The digitizer tablet's set of coordinates correspond to points on the CRT screen. The user selects them by puck or stylus. (Courtesy of Compaq Computer Corporation.)

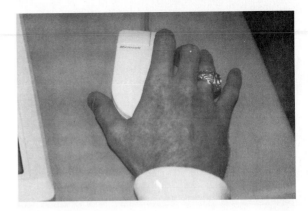

Figure 35.13 A mouse moves the cursor around a CRT screen, and the user executes commands or makes changes by pressing its buttons.

Figure 35.12 A light pen allows the user to modify drawings by touching the CRT screen with it. (Courtesy of Ford Motor Company.)

Input Devices

The **digitizer** is a graphics input device that can communicate information from an image into the computer in digital form for display, storage, or modification. The user attaches a drawing to a digitizer tablet and "traces" it with a stylus to convert the picture to a digital format based on the x and y coordinates of individual points **(Fig. 35.11)**.

The user may create or modify digitized pictures on a screen point-by-point with a **light pen**. The user touches the CRT screen with the light pen, telling the computer the position of the pen on the screen **(Fig. 35.12)**.

A **mouse** is a tabletop device that the user can move to position the cursor on the screen. The user may execute commands or change information on the screen with the mouse **(Fig. 35.13)**.

The **joy stick** allows the user to "steer" a cursor around the screen by tilting a lever in appropriate directions.

The **Spaceball®** enables the user to manipulate 3D objects on the screen **(Fig. 35.14)**. By pushing, pulling, and twisting the ball, the user may move 3D images on the screen.

Output Devices

The **plotter** is an output device directed by the computer to plot a drawing on paper or film with a pen in much the same manner as a drawing is made by hand. Three types of plotters are the **flatbed plotter**, **drum plotter**, and **sheet-fed plotter**. The paper of the flatbed plotter is attached to the bed and remains stationary while the pen moves about the paper in raised or lowered positions to produce the drawing. In the drum plotter,

Figure 35.14 The user manipulates the Spaceball® digitizer to move an object in 3D space on the computer screen by pushing, pulling, and twisting it. (Courtesy of CalComp Corporation.)

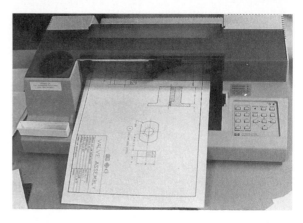

Figure 35.16 The Hewlett-Packard 7475 plotter produces pen-made drawings on size A and size B sheets.

Figure 35.15 The DesignMate® plotter is a pen plotter that can produce multicolored drawings that are accurate within 0.01 inch. (Courtesy of CalComp Corporation.)

the paper is held between grit wheels and rolled back and forth over a rotating drum as the pen moves left or right along the drum (**Fig. 35.15**). The sheet-fed plotter holds the paper by grit

wheels in a relatively flat position. The paper moves forward and backward and the pen moves right and left to plot the drawing (**Fig. 35.16**).

The **printer** transfers images to paper or film from the computer, but not with a pen. **An impact printer works like a typewriter, forming characters by striking typefaces against an inked ribbon and paper. Dot-matrix printers** are impact printers that have printheads composed of a rectangle (matrix) of pins that are raised or lowered to form characters. These pin patterns strike against a ribbon to make dotted characters on the paper. Dot-matrix printers are better than impact printers for printing graphics because their dot patterns correspond to lighted pixels on a CRT screen.

Nonimpact printers form characters by using ink sprays, laser beams, photography, or heat. One example, the thermal transfer plotter, uses electrical fields to direct jets of ink to appropriate spots on the paper (**Fig. 35.17**). Like the dot-matrix printer, the ink-jet printer forms a pattern of dots, thus limiting its accuracy and resolution. Multiple jet nozzles allow generation of multicolor graphics (**Fig. 35.18**).

Laser printers give an excellent resolution of dense, accurately drawn lines in text and draw-

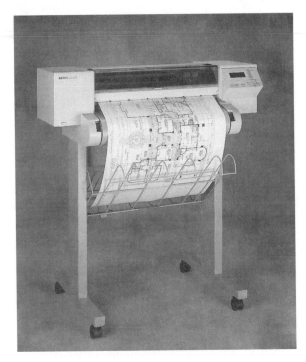

Figure 35.17 The PlotMaster® thermal transfer plotter/ printer produces fast, high-resolution color hard copy for computer-aided drawings. (Courtesy of CalComp Corporation.)

Figure 35.18 The Hewlett-Packard DesignJet 600® uses inkjet technology to make drawings with inkjet instead of pens. (Courtesy of Hewlett-Packard Company.)

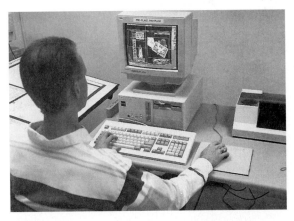

Figure 35.19 This typical computer-graphics workstation comprises a 486 IBM clone with a color monitor, mouse, and AutoCAD software.

ings, in color as well as in black and white. Although more costly than other types of printers, laser printers are being used in much greater numbers because of the quality of their output.

35.5 CAD Software for the Microcomputer

The availability of CAD software for the microcomputer continues to increase and each advance is more powerful than the previous version. Most of these packages require a computer with at least a 20 MB hard disk and 4 MB of RAM. Most computer graphics probably will be done on microcomputers in the future (**Fig. 35.19**).

The major advantages of CAD microcomputer software packages are their low cost and ease of upgrading. Among the more popular computer graphics software packages for the microcomputer are AutoCAD, VersaCAD, Professional Designer, MegaCADD, DynaPerspective, Silver Screen, and Cadkey.

35.6 Disk Operating Systems

A **disk operating system**, or DOS (pronounced "doss"), operates a computer by storing, retrieving, and managing data files. In this section we cover the fundamentals of DOS for IBM-compatible computers.

Random access memory (RAM) is the computer's memory in which information inserted through the keyboard is stored temporarily before it is saved on a disk. When the computer is turned off, RAM does not retain anything. The capacity of RAM for a DOS system varies from 640 KB to 16 MB and more.

Read-only memory (ROM) is information that is built into your computer at the factory and is never erased or changed. It contains the information necessary for your computer to begin operating.

Either hard disks or floppy disks may be used for **disk storage**. Hard disks are located within the computer and have capacities ranging from about 10 MB to several gigabytes. Hard disks store both programs and data. Floppy disks (5.25 inches square) have capacities of 360 KB and 1.2 MB. These disks store data and may be removed from the computer. Small, plastic-encased floppy disks (3.5 inches square) have storage capacities of 1.4 and 1.6 MB.

A **file** cannot exceed the capacity of the disk on which it is stored. It may be several million bytes in size when saved on a hard disk but far smaller when saved on a floppy disk.

Some commonly used **file extensions** are:

.BAK	a backup of another file
.BAT	a DOS batch file
.COM	a DOS command file
.DBF	a dBASE database file
.DOC	a document file
.EXE	a DOS file that can be executed
.TXT	a text file

Typical **AutoCAD files**, identified by their extensions, are:

.DWG	a drawing file
.BAK	a backup file
.SLD	a slide file

A **directory** is like a file drawer in which folders or files are stored. A directory may contain many files or only one.

American Standard Code for Information Interchange (ASCII) provides a means for converting files to a standard form so that they can be transferred to a different application. An ASCII file is a text file that can be listed or printed on the screen with the TYPE command.

Using DOS

The ">" symbol is the **DOS prompt** that asks for a command from the keyboard and appears as C>. The letter C represents the default drive, usually the hard disk where DOS executes the command.

The active drive may be changed to the A drive by typing A: (CR). The designation (CR) represents the "Carriage Return" key or the "Enter" key. The active drive may be changed to the B drive by typing B: (CR). The drive designation can be added as part of a DOS command such as DIR A: (CR) to obtain the directory of files on the A drive. Both upper and lowercase letters or a combination of the two may be used.

File names may contain one to eight characters followed by a period and an optional extension of one to three characters. Characters that can be used in file names are: any letter of the alphabet, any number, and characters & @ # $ % ( ) ' _. Characters that cannot be used are: * ? ", :.

Do not use the following file names: CON, AUX, COM1, COM2, LPT1, LPT2, LPT3, PRN, and NUL, which are names reserved for other DOS applications. Examples of acceptable file names are: FILENAME.EXT, CAR.DWG, 1234.#60, and ABCDEF.DOC.

Both DOS and AutoCAD use keys in combination (multiple-key commands) such as the Ctrl and Alt keys. The first key is held down while the second key is pressed. To stop a command that is in progress and exit to the DOS prompt, press Ctrl and C at the same time. Press Ctrl and S to stop scrolling information on the screen, and press Ctrl and S in combination to resume scrolling. When a computer locks up, restart it by typing `Ctrl`, `Alt`, and `Del` simultaneously.

Directories

The hard disk of a computer is divided into a series of partitions called **directories,** which may be divided into subdirectories, to hold files. A disk must be formatted before files can be stored on it. Formatting initializes the disk, organizes the tracks and sectors, and removes any existing files. The steps of formatting a disk are:

1. Insert the disk in drive A or B.

2. Type `C:` and (CR) to access the `FORMAT` command on the hard disk.

3. Type `FORMAT A:` and (CR) and formatting of the A drive begins. When completed, a prompt will ask whether another disk is to be formatted. Type `Y` or `N` and (CR) to make your selection.

Making a Directory Type `C:` (CR) to access the hard disk and obtain the `C>` prompt on the screen. Type `CD\` and (CR) to ensure that you are in the main directory. Type `MD\` ("Make Directory") followed by the name of the new directory (eight characters or less). For example, type `MD\XYZ` and (CR) to create a new directory named `XYZ` on the C drive.

Accessing a Directory If a floppy disk is in drive A, type `A:` (CR) to change the screen prompt `C>` to `A>`. Then type `DIR` (CR) to obtain the directory of files on the disk. To obtain a directory of the disk in drive A from a `C>` prompt, type `DIR A:` (CR).

For hard-disk directories, if the `C>` prompt appears on the screen, type `DIR` (CR) to obtain a listing of files on the C drive. If the `A>` prompt appears on the screen, type `DIR C:` (CR) to obtain a directory of files from the C drive.

Access a directory such as `CD\ABC` and (CR) and type `DIR` and (CR) to obtain a list of the files in directory `ABC` on the screen. You could type `DIR/P`, which would cause the scrolling to pause when the screen is full. By pressing (CR) the screen continues scrolling and pausing until all the files have been displayed. If you type `DIR/W` and (CR), the directory will display file names in a wide, multiple-column format across the screen, rather than the single-column listing obtained otherwise.

To gain access to the new directory created in the preceding example, `XYZ`, type `CD\XYZ` and (CR) and `C:\XYZ` will appear on the screen. The entry CD is an abbreviation for "Change Directory."

Removing a Directory To remove a directory, `ERASE` the files in it and type `RD JUNK` and (CR). The term `RD` is short for "Remove Directory."

Files

The fundamental rule of working with a computer is **Backup! Backup!** Backing up a file means copying it onto the same disk and onto different disks for protection in case a file goes bad or is lost. Disks can be protected from overwriting by covering the notch at the edge of the disk with adhesive tape or by sliding the gate on 3.5-inch disks.

COPY Command If the `A>` prompt is on the screen and a file from drive A is to be copied to the hard disk, drive C, type `COPY FILE1.XYZ C:` and (CR). If the `C>` prompt is on the screen, a file on drive A can be copied to drive C by typing `COPY A:FILE2.ABC C:` and (CR). In this case, the file will have the same name because no new name was assigned. By typing `COPY A:FILE2.ABC C:NEWFILE.XYZ` and (CR), you rename the file and copy it to the C drive.

Copying Wild Cards If `C:` appears on the screen, by typing `COPY *.DWG A:` (CR) you may copy all drawings to the floppy disk in drive A having the extension `.DWG`. The * designates the **wild card**, which means any name with the extension `.DWG`.

Similarly, by typing `COPY FILE.* A:` and (CR) you may copy all files with the name `FILE` but with different extensions. To copy all files from a floppy disk to a hard drive, type `COPY A:*.* C:`

and (CR). This command will copy files by any name with any extension to the C drive.

To copy a file onto the same disk, the new file must have a different name. Type COPY A:DRAW1.DWG A:DRAW1A.DWG and (CR) to make a new file called A:DRAW1A.DWG. To copy from one floppy disk to another when you have only one floppy disk drive, type COPY A:DRAW1.DWG B: and (CR). The program will prompt you to change the original and destination disks as the copying process proceeds. To copy an entire floppy onto a hard disk and into the directory of AutoCAD, type COPY A:*.* C:\ACAD and (CR). The files will be listed on the screen as they are copied.

Renaming a File To change the name of a file type REN DRAW1.DWG DRAW2.DWG and (CR) to change the file name DRAW1.DWG to a file named DRAW2.DWG. To check this operation type DIR and (CR) to get a list of the files.

Erasing a File Either the DEL or ERASE commands will remove a file, but be careful not to remove files accidentally. Type ERASE C:DRAW1.DWG and (CR) to erase file DRAW1.DWG from drive C. By typing DEL A:*.* and (CR), you can erase all files from drive A.

Printing What the Screen Displays

To print what is being scrolled on the screen, turn on your printer, press Ctrl and P at the same time, and type DIR A: and (CR). To deactivate the printing feature, press Ctrl-P again. To print the screen's display, turn on your printer and press Shift and PrtSc. The data are printed and the print-screen feature automatically shuts off.

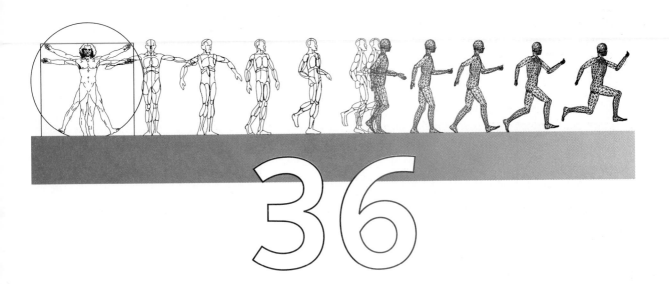

36

AutoCAD Computer Graphics

36.1 Introduction

In these last two chapters, we introduce computer graphics AutoCAD, Release 12, which runs on a 386 or faster computer, with at least a 40 Mb hard disk, a mouse (or tablet), and an A-B plotter (**Fig. 36.1**). We selected AutoCAD as the software for demonstrating computer graphics because it is the most widely used.

Our coverage of AutoCAD is necessarily brief, and we omitted some operations entirely because of space limitations. AutoCAD's concisely written reference manual is 685 pages long on 8.5 × 11-inch pages. However, the coverage here is sufficient to guide you through its applications for beginning engineering design and graphics.

Release 12 may be used in the same 2D drawing and 3D modeling manner as Release 11, but it contains 178 changes. Chapter 37 presents a portion of the Advanced Modeling Extension Solid Modeling (AME) software package as an addition to Release 12 (386 version or higher) for solid modeling.

Figure 36.1 This view is of one of the engineering design graphics computer laboratories at Texas A&M University.

36.2 Getting Started

If this is your first session with a computer, you are anxious to turn the computer on, make a drawing on the screen, and plot it without having

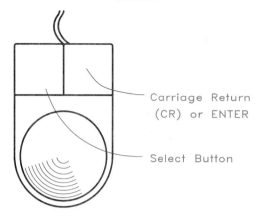

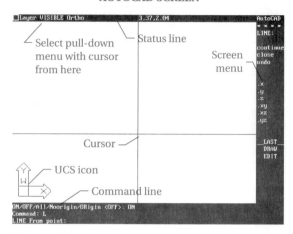

 Figure 36.2 The left-hand button on the mouse is the select button, and the right-hand button is the carriage return (CR) button.

Figure 36.3 After you boot up the system, the screen will appear ready for drawing.

to read instructions. This section helps you do just that.

Much of your interaction with the computer will be by means of a **mouse (Fig. 36.2)**, but you may enter many commands more quickly at the keyboard after you learn the commands. Press the left button to "click on" to select or pick. In some cases a "double click" is needed.

Begin by turning the computer on and **booting up** the system by typing ACAD12 (or the command used by your system) to ready the screen for drawing **(Fig. 36.3)**. Move the **cursor** (the crosshairs controlled by the mouse) about the screen with your mouse, select items on the side menu, and try the pull-down menu.

To create a new **file** on a disk, place your formatted disk in its slot, pick FILES from the pull-down menu and select NEW from the dialogue box **(Fig. 36.4)**. Then type the name of your new drawing, B:NEW-DWG, in the Edit box of the Create New Drawing dialogue box **(Fig. 36.5)**. The light at the B:drive where your disk is inserted will light up while a new file, B:NEW-DWG, is being saved.

Draw some lines on the screen just for fun **(Fig. 36.6)**. Move the cursor to the top bar to activate

CREATING A NEW FILE

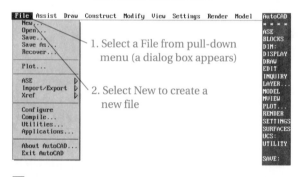

Figure 36.4 To begin the procedure for creating a new file, select File and New.

the pull-down menu and select DRAW, LINES, and SEGMENTS in succession to obtain the prompt From Point: in the COMMAND line at the bottom of the screen. Select a point with the mouse by clicking the left button, move the cursor to a second point, and select other points. To disconnect the rubber band that stretches from the last endpoint, press the right button on your mouse, which is the same as the Carriage Return (CR) or Enter key. We use (CR) as the abbreviation for this key. Try drawing a circle and other shapes on the screen.

CREATE NEW DRAWING BOX

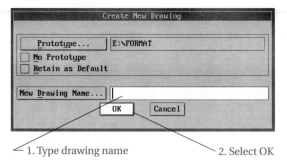

1. Type drawing name   2. Select OK

◈ **Figure 36.5** The `Create New Drawing` dialogue box lets you assign a prototype file (`E:\Format`) to a new file (`B:NEW-DWG`); press OK to proceed.

DRAWING WITH LINE

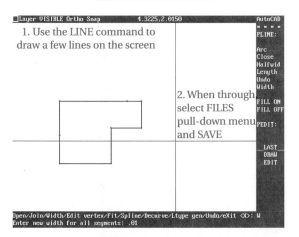

◈ **Figure 36.6** By selecting `LINE` from the side menu, you may produce lines on the screen by selecting points with the mouse.

You may **repeat** a command by pressing (CR) (the right button) twice at the end of the previous command. For example, `LINE` appears in the command line at the bottom of the screen after you press (CR) twice, if that was the previous command used.

To **update** a drawing file, `B:NEW-DWG`, click on `FILES` from the pull-down menu and `SAVE` from its dialogue box (**Fig. 36.7**). The command `_QSAVE`

SAVING A FILE

Select SAVE from pull-down menu

◈ **Figure 36.7** Select `SAVE` under `FILE` on the pull-down menu to save your drawing on a disk.

PLOTTING A FILE

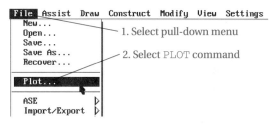

1. Select pull-down menu

2. Select PLOT command

◈ **Figure 36.8** To prepare to plot a drawing, select `PLOT` from the `FILE` option of the pull-down menu.

appears in the command line at the bottom of the screen, the light over drive `B:` blinks briefly, and the changes are added to `B:NEW-DWG` to update the file.

To **plot** a drawing, use the pull-down menu, select `FILES` and `PLOT` from the dialogue boxes (**Fig. 36.8**) which fills the screen with the `Plot Configuration` dialogue box (**Fig. 36.9**). Select only one setting, `Scaled to Fit`, to make the plotted drawing fill the sheet.

Load a size A paper sheet in the plotter. If the plotter is a Hewlett-Packard 7475, it will plot both size A and size B sheets with ink pens. From the dialogue box, select the OK button to send the drawing data to the plotter.

To end your first session with the computer, use the pull-down menu to select `FILES` and `EXIT AUTOCAD` from their dialogue boxes (**Fig. 36.10**) to return to the operating system.

Warning! When working from a floppy disk, do not remove the disk from its drive until you

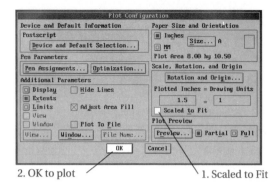

2. OK to plot   1. Scaled to Fit

Figure 36.9 The `Plot Configuration` dialogue box will appear on the screen for plotting instructions. Select `Scaled to Fit` and `OK` to plot.

EXITING AUTOCAD

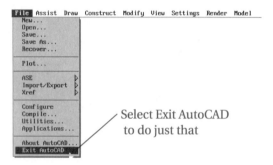

Select Exit AutoCAD to do just that

Figure 36.10 Select `Exit AutoCAD` from the `File` pull-down menu to exit AutoCAD.

have SAVEd the file and have performed either `Quit` or `End`.

That's how it works. Now, let's get into the details.

36.3 Basic Operations and Commands

Using Dialogue Boxes

AutoCAD 12 has many **dialogue boxes** with names beginning with DD (DDLMODES, for example) that interact with you. You may use the FILEDIA command to turn off (zero = off and 1 =

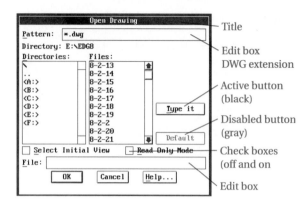

— Title
— Edit box
  DWG extension
— Active button
  (black)
— Disabled button
  (gray)
— Check boxes
  (off and on
— Edit box

Figure 36.11 To open a saved drawing, select `Open` from the `File` pull-down menu, and the `Open Drawing` dialogue box will appear on the screen. You may select the file to be opened from the list or type its name into the `Edit Box`.

on) the dialogue boxes if you prefer to not use them. When a command on a menu is followed by three dots (. . .) it will respond with a dialogue box when selected; some dialogue boxes have subdialogue boxes.

Double clicking (quickly pressing the `Select` button twice) with the mouse selects a file from a file list. You may also use double clicking in other applications as an alternative to making the selection and following it with `OK`. By experimentation you will soon learn when to apply double clicking most efficiently.

The `Open Drawing` dialogue box shown in **Fig. 36.11** illustrates the features of a typical dialogue box. The title of a dialogue box appears at the top. The box contains a series of **buttons** from which you may select by using the cursor. The button with the darkest outline is the `Default` button, such as the `OK` button, and disabled buttons are shown in gray.

Check boxes are for toggling the options `On` (an X is shown) or `Off` (the box is empty). **Edit boxes** are long boxes into which you type a

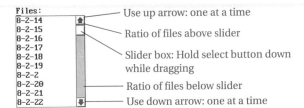

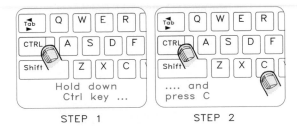

Figure 36.12 You may use Scroll Bars to scroll through a list of files by selecting the slider box while pressing the select button of the mouse. Directional arrows at each end move up or down the list one at a time.

Figure 36.14 To abort the current operation, hold down the Ctrl button and press the C key.

SUBDIALOGUE BOXES

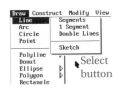

Figure 36.13 Dialogue boxes followed by triangles will give action boxes for further selections.

response, such as the name of file to be saved (**Fig. 36.11**). **Radio buttons** are square boxes with squares inside them that you may turn on or off with the cursor.

Scroll bars allow you to move through the lists in some dialogue boxes (**Fig. 36.12**). You may get subdialogue boxes with a command followed by a triangle (**Fig. 36.13**). After selecting LINE, you must define its type in a second box. Use the OK button to accept and activate the selections made in the dialogue box. We cover other features of dialogue boxes later in this chapter.

Control C (press Ctrl and C at the same time) exits the current command and gives the Command: prompt at the bottom of the screen, indicating that the program is ready for a new command (**Fig. 36.14**). Control C aborts a plot operation but does not take effect until the data have had time to exit the buffer.

Drawing Aids

You must become familiar with the drawing aids that are available. Use of the LIMITS command establishes the size of the drawing area to be covered with dots when GRID is on. Use of the GRID command sets the spacing of the dots within the LIMITS. A drawing that fills a size A sheet (11 × 8.5 inches) has a plotting area of about 10.1 × 7.8 inches (257 × 198 mm). The LIMITS command appears under SETTINGS of the pull-down and side menus:

Command: LIMITS (CR)

ON/OFF/<Lower left corner> <0.00,0.00>: (CR) (Accept default value.)

Upper right corner <12.00,9.00>: 11,8.5 (CR)

You may reset LIMITS at any time while producing a drawing with these steps.

Use the UNITS command to set the format of numerals and their fractional parts. Selecting UNITS under the pull-down heading, SETTINGS, gives the Units Control dialogue box (DDUNITS command) (**Fig. 36.15**). Set the UNITS format by selecting one of the following options for, say, 15 1/2 inches:

1. Scientific	1.55E+01
2. Decimal	15.50
3. Engineering	1'-3.50"
4. Architectural	1'-3 1/2"
5. Fractional	15 1/2

UNITS CONTROL BOX

Select form of units

Select number of decimal places for units and angles

Select form for angles

Figure 36.15 The `Units Control` dialogue box, under `SETTINGS` in the pull-down menu, lets you set the number of decimal places and the form of numbers and angles to the formats shown in Fig. 36.16.

DRAWING AIDS BOX

Figure 36.16 The `Drawing Aids` dialogue box is used to set `Modes`, `Snap`, and `Grid`.

To obtain the number of decimal places for fractions, pick the `Precision` edit box for a list of options. Select angular `UNITS` the same way. We presented applications of Surveyor units in Chapter 27.

The `Drawing Aids` dialogue box (`DDRMODES`) appears under the `SETTINGS` heading of the pull-down menu, from which you may select the following aids (**Fig. 36.16**).

Ortho The `Ortho` mode forces all lines to be either horizontal or vertical (not angular). To set `ORTHO`, check its box (Fig. 36.16) by pressing function key `F6`, or by typing `ORTHO` and typing `ON` or `OFF`. When set to `ON`, `ORTHO` appears in the status line at the top of the screen.

Solid Fill You may turn `Solid Fill On` or `Off` to show areas produced with the `SOLID` command as filled or outlined.

Quick Text The `Quick Text` mode saves regeneration time by showing text as boxes rather than as lines of text; restore text by turning `QTEXT Off` and typing `REGEN`.

Blips You may select the `Blips` mode from the dialogue box or use it by typing `BLIPMODE` and `On`

or `Off`. Blips are temporary markers that appear on the screen during drawing that you may remove by refreshing the screen (pressing `F7`).

Snap The `SNAP` mode forces the cursor to stop at points on an imaginary grid that you may set to any spacing. You may type the `X` and `Y` spacings of `SNAP` in the edit boxes of the pull-down dialogue box. Alternatively, you may begin from the side menu or the command line and type:

```
Command: SNAP (CR)

Snap spacing or ON/OFF/Aspect/ Rotate/
Style<0.25> .20
```

This response sets the `SNAP` grid to 0.20 units. When on, `SNAP` appears in the status line at the top of the screen. You may toggle `SNAP` on and off by pressing `F9`.

Select the `ASPECT` option by typing `A` to assign X and Y values if they are different. The `ROTATE` option prompts for the angle of rotation of the `SNAP` grid and its origin point, obtained by typing X and Y values in the edit boxes.

You may set `Snap Style` to `I` (isometric) or `S` (the standard rectangular pattern). The isometric option is for isometric drawings with axes 120° apart.

Grid Use `GRID` to fill the `LIMITS` area with the dot spacing assigned by typing values in the X and Y

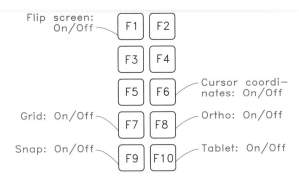

Flip screen: On/Off

Grid: On/Off

Snap: On/Off

Cursor coordinates: On/Off

Ortho: On/Off

Tablet: On/Off

Figure 36.17 These function keys may be used to turn settings (Fig. 36.16) Off or On.

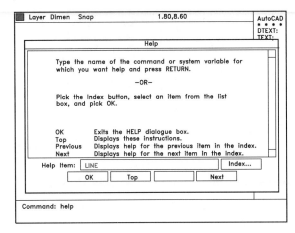

Figure 36.18 When you select HELP under the ASSIST heading, this dialogue box will appear, with a slider box for scrolling through the available HELP menus.

spacing boxes. Typing GRID and selecting the Snap option sets the grid equal to the SNAP spacing.

Drag You may use DRAGMODE to show the "dragging" movement of an entity as you move it. You may set it to On, Off, or Auto, which drags objects automatically without prompting. By pressing F6, you may obtain **running coordinates** at the top of the screen in the status line to show the cursor's position.

Function Keys Figure 36.17 shows the function keys used to toggle ON and OFF several of the drawing aids described.

Utility Commands

Utility commands control the operation of AutoCAD and make changes in files.

Help The HELP (?) dialogue box lists commands for which explanations are available. Press Ctrl C to abort the listing process and press F1 to return to the graphics mode. To obtain information on a command (LINE, for example), respond to the prompts as follows:

```
Command: HELP (or ?) (CR)

Command name (Return for list): LINE (CR)
```

To obtain the dialogue box shown in **Fig. 36.18**, select the HELP option under the ASSIST/Help! heading of the pull-down menu.

Shell The SHELL, or SH, command gives you access to the DOS operating system while remaining in the Drawing Editor as follows:

```
Command: SHELL (CR)

DOS command: DIR A: (Or similar command.)
(CR)

Command: (Reappears on screen.)
```

Purge You may use the PURGE command only if it is the first command utilized after you have loaded a file or immediately after it has been SAVEd.

```
Command: PURGE (CR)

Purge unused Blocks/Layers/

LTypes/SHapes/STyles? All: ALL (CR)
```

The ALL response lets you eliminate any unused entities, styles, layers, or features in the drawing file as the program prompts you for each, one at a

ALIASES (ABBREVIATIONS)

A	ARC	M	MOVE	
C	CIRCLE	MS	MSPACE	
CP	COPY	P	PAN	
DV	DVIEW	PS	PSPACE	
E	ERASE	PL	PLINE	
L	LINE	R	REDRAW	
LA	LAYER	Z	ZOOM	

◼ Figure 36.19 The ALIASES (abbreviations for commands) may be typed at the keyboard in their shortened form to speed up the entry process.

time. The other options purge specific categories of features of a drawing.

Aliases The use of ALIASES (abbreviations of command names) increases typing speed. Note how much faster typing R instead of REDRAW is **(Fig. 36.19)**.

Files The FILES dialogue box under the UTILITY heading of the pull-down or the side menu lists the following options:

Option 1: List drawing files gives the files on a specified drive. Enter B: and you will get a list of all files on the disk in the B drive with .DWG extensions.

Option 2: List user specified files displays the files that you specify. For example, by responding to ENTER FILE SEARCH SPECIFICATIONS: with B:*.BAK, you will obtain a list of all files with .BAK extensions on the disk in drive B.

Option 3: Delete files eliminates files. Specify a file such as B:DRAW1.DWG or use wild cards such as B:*.DWG to delete all files with a .DWG extension. When using wild cards, you may get a list of files meeting your specifications, one at a time, for deletion by entering Y or N.

Option 4: Rename files changes the name of a file by responding to the prompts as follows:

```
Enter current file name: B:DRAW1.DWG (CR)

Enter new file name: B:DRAW2.DWG (CR)
```

LAYER CONTROL BOX

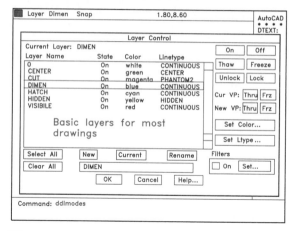

◼ Figure 36.20 Set Layers by using the Layer Control dialogue box from SETTINGS in the pull-down menu. The layers shown here are sufficient for making most working drawings.

Option 5: Copy file duplicates a file as follows:

```
Enter name of Source file name:
B:DRAW1.DWG (CR)

Enter name of Destination file name:
C:DRAW1A.DWG (CR)
```

Option 6: Unlock file unlocks a file as follows:

```
Enter locked file(s) specifications:
\DIR1\DRAW1.DWG

The file: \DIR1\DRAW1.DWG was locked
by J. Doe at

18:81 on 10/26/1994

Do you still wish to unlock it? <Y> Y

Lock was successfully removed.

1 file unlocked. Press RETURN to con-
tinue: (CR)
```

Layers

You may name an almost infinite number of layers on which to draw and assign each a name, color, and line type. One layer, for example, may be a layer named HIDDEN for drawing dashed lines in yellow.

Architects use separate copies of the same floor plan for different applications: dimensions, furniture arrangement, floor finishes, electrical details, and so on. Turning on the needed layers and turning off others allows use of the same basic plan for these applications.

Setting Layers The layers shown in the Layer Control dialogue box (DDLMODES) in **Fig. 36.20** are sufficient for most working drawings. Assigning different line types and colors to different layers distinguishes them from each other. The 0 (zero) layer is the default layer, which can be turned off but not deleted. Select LAYERs from the side menu and LAYER from the submenu and develop layers in the following manner.

New To create new layers type NEW:

```
Command: LAYER (CR)

?/Set/New/On/Off/Color/Ltype/

Freeze/Thaw: New or N (CR)

Layer name(s): VISIBLE,

HIDDEN,CENTER,HATCH,DIMEN,

CUT,WINDOW (CR)
```

Figure 36.20 shows the assignment of seven new layers by name, each having defaults of a white color and a continuous line type. You may assign new layers by using the dialogue box, typing a layer name in the edit box, or selecting NEW and then OK.

Color Set the COLOR of each layer as follows:

```
Command: LAYER

?/Set/NEW/ON/OFF/COLOR/Ltype/

Freeze/Thaw: COLOR or C (CR)

Color: RED (or 1) 1 (CR)

Layer name(s) for color 1

(red) <VISIBLE>: VISIBLE (CR)
```

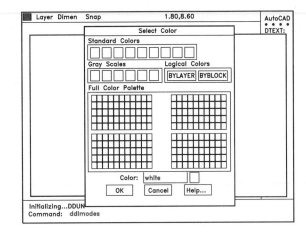

Figure 36.21 Use the Select Color dialogue box to assign colors to drawing layers by mouse or by typing in the Edit Box.

You may also pick colors from the Select Color dialogue box **(Fig. 36.21)**, obtained by selecting a layer from the Layer Control dialogue box and using the Set Color button (Fig. 36.20). The standard, most often used colors (1–9) appear at the top of the box. Assign a color to each layer and use the OK button to save the settings. Note that the color red was assigned to the VISIBLE layer in Fig. 36.20, which could have been typed as 1 or RED.

Line Types Assign line types to layers as follows:

```
Command: LAYER (CR)

?/Set/New/On/Off/Color/Ltype/

Freeze/Thaw: Ltype (or L) (CR)

Linetype (or ?) <CONTINUOUS>: HIDDEN (CR)

Layer name(s) for linetype

HIDDEN <0>: HIDDEN (CR)
```

Lines on the HIDDEN layer are generated as dashed (hidden) lines. By typing LINETYPE and pressing (CR) twice yields a display of the available line types **(Fig. 36.22)**. If you type LINETYPE and select LOAD, the list of these lines will appear in the subdialogue box Select Linetype **(Fig. 36.23)** of the

LINETYPES

CONTINUOUS	_____
BORDER	__ __ __ __ . __ __ __ .
BORDER2	_ _ _ . _ _ _ . _ _ .
BORDERX2	____ ____ __ . ____ ____ __.
CENTER	____ _ ____ _ ____ _ ____
CENTER2	__ _ __ _ __ _ __ _ __ _ __
CENTERX2	_____ __ _____ __ _____
DASHDOT	__ . __ . __ . __ . __ . __
DASHDOT2	_._._._._._._._._._.
DASHDOTX2	____ . ____ . ____ . ____
DASHED	__ __ __ __ __ __ __ __
DASHED2	_ _ _ _ _ _ _ _ _ _ _ _ _ _
DASHEDX2	____ ____ ____ ____ ____
DIVIDE	____ _ . ____ _ . ____ ..
DIVIDE2	_ . _ . _ . _ . _ . _ . _ ..
DIVIDEX2	_____ . . _____ . . _____
DOT	. . . .
DOT2	....
DOTX2	. . . .
HIDDEN	__ __ __ __ __ __ __ __
HIDDEN2	_ _ _ _ _ _ _ _ _ _ _ _ _ _
HIDDENX2	____ ____ ____ ____ ____
PHANTOM	_____ _ _ _____ _ _ _____
PHANTOM2	____ _ _ ____ _ _ ____ .. _
PHANTOMX2	_____ __ __ _____

Figure 36.22 To obtain a list of line types on the screen, type LINETYPE and press (CR) twice.

`Layer Control` dialogue box (Fig. 36.20). To select the type of line desired pick the corresponding layer line type and then use the `Set Ltype` button.

Set A layer must be `SET` for it to be the current layer and for you to be able to use it:

    Command: LAYER (CR)

    ?/Set/New/On/Off/Ltype/Freeze/

    Thaw: SET (or S) (CR)

    New current layer <0>:VISIBLE (CR) (CR)

You may also set the current layer by selecting it from the dialogue box (Fig. 36.20) and then using the `Current` button.

Rename You may rename a layer by selecting it from the dialogue box, editing its name in the edit box, and then picking the `RENAME` button.

SELECT LINETYPES BOX

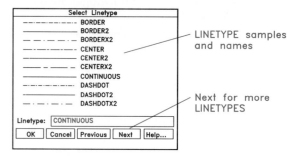

Figure 36.23 Assign line types to a layer by using the `Set Ltype...` dialogue box from the `Layer Control` box.

On/Off Turn defined layers on or off with the `ON/OFF` command in the following manner:

    Command: LAYER (CR)

    ?/Set/New/On/Off/Ltype/Freeze/

    Thaw: ON (or OFF)

    Layer name(s) to turn on: VISIBLE

(or * for all layers)

(Or `VISIBLE, HIDDEN, CENTER,` to turn

these layers ON or OFF.) (CR)

Using Layers All layers may be on, but only the current layer can be drawn on. Selecting the question-mark option (?) of LAYER, displays a list of the layers, their line types, colors, and on/off status. Press key F1 to change the screen back to the graphics editor. Obtain the `Layer Control` dialogue box to show the layers (Fig. 36.20) by selecting `Layer Control` under `Settings` of the pull-down menu.

Freeze and Thaw The `Freeze` and `Thaw` options under the LAYER command are similar to ON and OFF. When you FREEZE a layer, the program will ignore it until you THAW it. Therefore regeneration is much faster than using OFF. The current layer cannot be frozen.

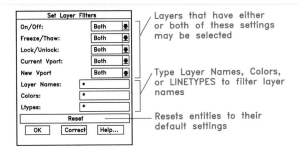

Figure 36.24 Set layer filters by using the Set `Layer Filter` box from the `Layer Control` box.

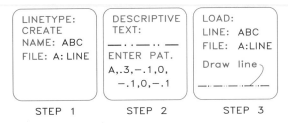

Figure 36.25 Producing custom-designed `Linetypes`:

Step 1 Command: LINETYPE (CR)

?/Create/Load/Set: <u>C</u> (CR)

Name of line type to create: <u>ABC</u> (CR)

File for storage of line type <default>:
<u>A:LINE</u> (CR)

Step 2 Descriptive text: (Type pattern.) (CR)

Enter pattern (on next line):

<u>A, .3,-.1,0,-.1,0,-.1</u> (CR)

Step 3 To load the new line type:

Command: <u>LOAD</u> (CR)

Name of line type to load: <u>ABC</u> (CR)

File to search <default>: <u>A:LINE</u> (CR)

(The new line is assigned to a layer and used.)

Filters You may use FILTERS to list layers—based on their specified properties—in the `Layer Name` list (Fig. 36.20). When you select FILTERS, the `Set Layer Filters` subdialogue box appears **(Fig. 36.24)**, from which you may specify the properties for filtering (or sorting) the layers. Pop-up lists show the options for `On/Off`, `Thaw/Freeze`, `Lock/Unlock` or `Both`. **Layer** Names, Colors and Ltypes can be typed in the edit boxes of a certain color, name, or linetype to specify the categories to be filtered and listed in the `Layer Control` dialogue box.

Select OK after setting filter options and the list of layers that have been sorted appears in the `Layer Name` dialogue box (Fig. 36.20). To turn filters off, select the On box and toggle it off.

LTSCALE Use LTSCALE to modify the length and spacing of noncontinuous lines, such as hidden lines. A large LTSCALE factor makes dashes and spaces longer and smaller factors do the opposite.

Save Layers The process of setting drawing aids takes time and effort, so you should save them by using the SAVE AS command under the FILE heading. You may save UNITS, GRID, SNAP, ORTHO, and other drawing aids as part of B:FORMAT for use as a PROTOTYPE having these settings. When you open B:Format, it will have no drawings on it, but it will

only contain the layers and settings previously made and ready for making drawings.

Custom-Designed Lines

By typing LINETYPE, (CR), ?, (CR) you may display the types and names of the lines available. You may assign these types of lines to layers by selecting the LTYPE option of the LAYER command and typing the preferred line type, such as hidden, center, or dashed.

If you want to custom-design types of lines, use the LINETYPE command and its CREATE option **(Fig. 36.25)**. The program will prompt you for the line's name and the file in which it is to be stored. Type in a description as dashes, spaces, and dots to represent the line. When prompted to enter the line pattern, begin with a dash (a positive length) or a dot, represent a space by minus

OPENING A FILE

Select OPEN from pull-down menu

🖰 **Figure 36.26** To open a file, select Open under File from the pull-down menu.

CREATING A NEW DRAWING

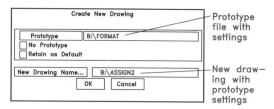

🖰 **Figure 36.27** To create a new file, select Create under File from the pull-down menu, enter the proto-type file, and type the file name (B:\ASSIGN2).

values, and separate spaces, dots, and dashes by commas. You need to show only one typical pattern (for example, .4, −.2, .2, −.2).

Use the LINETYPE command and its LOAD option to gain access to the newly designed line style. Then assign it to a LAYER by using the LTYPE option of the Layer Control dialogue box.

36.4 Beginning a New Drawing

Loading Files

There are two ways of loading B:FORMAT, which contains the layers and settings assigned to it, for a new drawing. One is with the OPEN option and the other is the NEW option under FILE of the pull-down menu (**Fig. 36.26**).

Open Option We can OPEN a file as an existing drawing, draw on it, and save it with SAVE AS to a different name (B:DWG1, for example), creating a new file with the settings of B:FORMAT. SAVE AS makes B:DWG1 the current drawing on the screen and B:Format remains in the background on the disk. You should periodically back up B:DWG1 with the SAVE command from the pull-down menu, the side menu, or from the keyboard.

New Option Select NEW, assign B:FORMAT as the PROTOTYPE file, type the new drawing's name (B:DWG1) in the Edit box, and select OK (**Fig. 36.27**). This procedure assigns the settings of B:FORMAT as the PROTOTYPE to a drawing named

SAVING A DRAWING

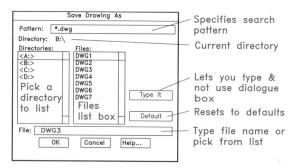

🖰 **Figure 36.28** To SAVE a file, select SAVE under. File from the pull-down menu and select or enter the name of the file to be saved.

B:DWG1, making the LAYERS and Drawing Aids available to it without having to set them.

Saving and Exiting

The Files portion of the pull-down menu gives the following commands for saving a drawing: SAVE, SAVE AS, and Exit AutoCAD. The side menu has these same commands and the command END. You may use either menu or type the commands at the keyboard.

Commands from the Pull-Down Menu The SAVE command displays a Save Drawing As dialogue box and requests a file name if you have not previously saved the drawing on the screen (**Fig. 36.28**). By selecting the directory and naming the

Select DISCARD CHANGES
to exit without saving a file

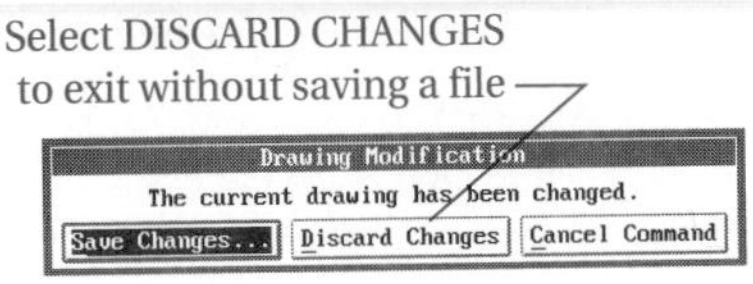

Figure 36.29 When EXITing or QUITting AutoCAD, you will be prompted to Save or Discard Changes, if you have not already saved them.

file B:NEW-DWG you save it on the floppy disk in drive B.

The QSAVE command does not request a file name and gives a "quick save" if the file has been previously saved during the session. When you type SAVE, the program prompts you for a file name or verification of the current file name.

The SAVEAS command displays a Save Drawing As dialogue box that requests a name for the drawing on the screen. The named drawing becomes the current file.

The Exit AutoCAD command gives the dialogue box shown in **Fig. 36.29**, which prompts you for the file name unless you have just saved it. To exit without saving your current drawing, select Discard Changes and you will exit without saving the drawing.

Commands from the Side Menu and by Typing The commands for saving, SAVE AS, QSAVE, and END, are listed under the UTILITY heading of the side menu for selection by use of the cursor. The same dialogue boxes given by the pull-down menu appear on the screen when you type a command (QSAVE, for example).

Exiting Options By selecting Exit AutoCAD from the pull-down menu or END or QUIT from the side menu, you will cause the screen to exit AutoCAD if you have saved your work.

END saves the current drawing and exits AutoCAD whether or not you have made changes. If you have not previously named the drawing, the

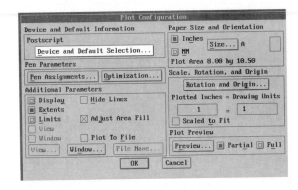

Figure 36.30 To plot, select PLOT under File from the pull-down menu to obtain the Plot Configuration dialogue box.

Create Drawing File dialogue box will prompt you for a file name. When the file name is typed, the previous drawing is automatically saved as a backup file with a .BAK extension and the current drawing is saved as a file with a .DWG extension.

QUIT (from the side menu) exits AutoCAD if there are no unsaved changes, or it gives a Drawing Modification dialogue box (Fig. 36.29) if there are unsaved changes. Select Cancel the command, Discard the changes, or Save Changes to obtain a dialogue box and name the file to be saved. When you name the file, the program will exit AutoCAD. The Exit AutoCAD command under FILE of the pull-down menu has the same function as the QUIT command.

36.5 Plotting a Drawing

Plotting Parameters

You can plot a drawing by selecting PLOT under the FILE heading of the pull-down menu to obtain the Plot Configuration dialogue box **(Fig. 36.30)**. Use the Device and Default Selection button to obtain the subdialogue box **(Fig. 36.31)** for selecting PLOTTER to make a pen drawing.

Press the Save Defaults to File button to display the Obtain from File subdialogue box

DEVICE AND DEFAULT SELECTION BOX

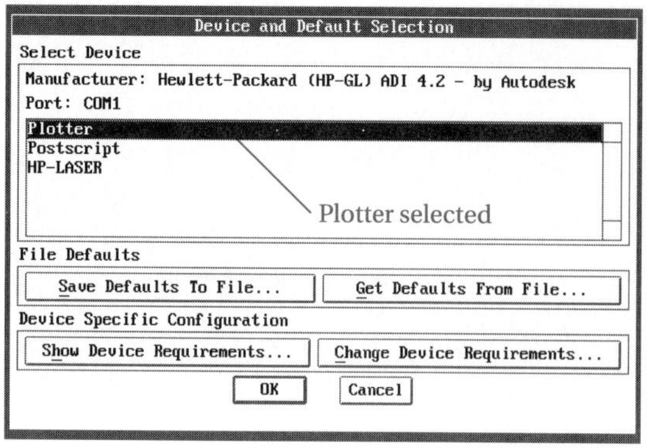

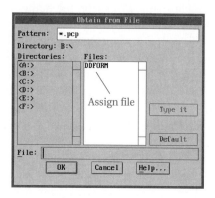

Figure 36.31 Use the Device and Default Selection box to select the plotter or printer to be used during
plotting.

PCP FILE (PLOT CONFIGURATION)

Figure 36.32 You may save the settings for a plot as a
PCP file by using the Save Defaults to File option
of the Device and Default Selection box.

to create a file with a .PCP extension called a Plot
Configuration Parameters file **(Fig. 36.32)**.
Naming this file B:FORMAT is convenient (with
an automatic .PCP extension) because you will
use it in conjunction with the prototype file,
B:FORMAT. DWG. Before plotting, add more parameters to B:FORMAT.PCP and save them.

Show Device Requirements (Fig. 36.31):
gives the dialogue box shown in **Fig. 36.33**,

DEVICE REQUIREMENTS

Figure 36.33 The Show Device Requirements
box gives Plotter port time-out in this case.

PEN ASSIGNMENTS DIALOGUE BOX

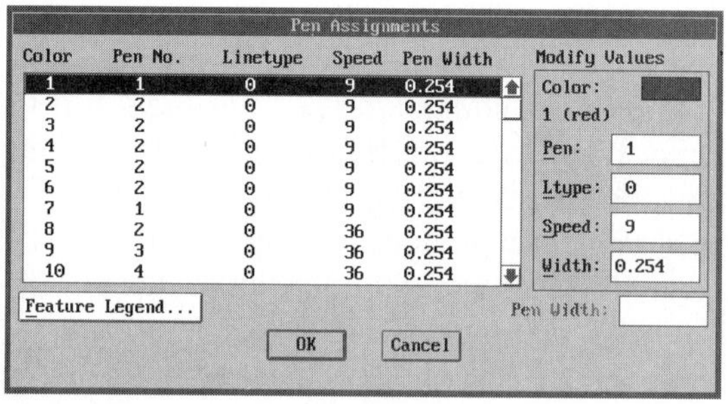

Figure 36.34 Use the Pen Assignments dialogue
box to assign colors to pen numbers, which represent the
slots of the pen holder on the plotter.

which indicates the length of time the plotter
will wait for plotter port time-out (or other
requirements). Pick OK to leave this box.

Change Device Requirements: allows selection of the plotting device. Pick OK.

Pen Assignments: show the assignment of
colors to pen numbers (slots in pen holder on
the plotter) **(Fig. 36.34)**. For example, the pen
in slot 1, a thick pen (P.7), will plot a red line
on the screen (VISIBLE layer).

Ltype: set to 0 because you have created layers with assigned linetypes.

Pen: use to change the pen for plotting a color
on the screen.

Speed: assigns the rate at which the plotter
pen moves while plotting. A speed of 9 is average and 36 is maximum.

FEATURE LEGEND BOX

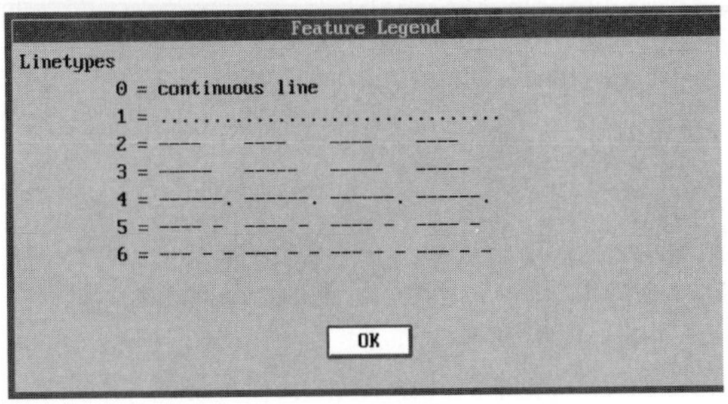

Figure 36.35 The Feature Legend box sets Linetypes at plot time. Setting lines at Continuous in order to use the Linetypes assigned by layer is the best procedure.

OPTIMIZING PEN MOTION

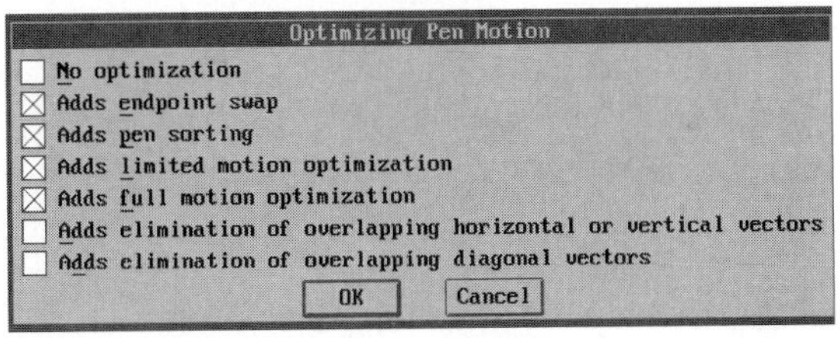

Figure 36.36 Use this dialogue box to optimize Pen Motion. Each selection activates all the features listed above it.

Pen Width: gives the line thicknesses based on pen tip widths for drawing polylines and traces (use Default).

WIDTH: option is grayed out if your plotter does not support this option.

Feature Legend: gives the dialogue box showing linetypes that you may choose **(Fig. 36.35)**. Do not use this option when you have assigned linetypes to layers.

Pen Optimization: shows the dialogue box that you may use to reduce wasted pen motion **(Fig. 36.36)**. When you select a check box, those above it (except the No optimization

ADDITIONAL PARAMETERS

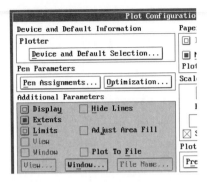

Figure 36.37 Additional Parameters may be selected to set the relationship between the screen and the plotted drawing.

box) become active. The last two boxes pertain to 3D figures and eliminate multiple strokes where lines coincide.

Additional Parameters of the Plot Configuration dialogue box **(Fig. 36.37)** offers the following settings.

Display: plots the portion of the drawing shown on the screen.

Extents: plots to the extent of a drawing, provided that the scale selected fits the plot. Good practice requires initializing the plot with Zoom/Extents prior to plotting.

Limits: plots the portion of the drawing defined by its LIMITS (grid pattern).

View: lists the saved Views that you can select from and plot **(Fig. 36.38)**. This box is grayed out if you do not save any views.

Window...: specifies the portion of your drawing to be plotted when you choose Pick by defining the window with the cursor or coordinates from the keyboard **(Fig. 36.39)**.

Hide Lines: removes hidden lines from 3D drawings.

SELECT VIEW TO PLOT

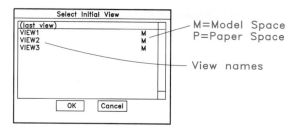

M=Model Space
P=Paper Space

View names

Figure 36.38 Views that have been saved may be recalled by selecting the `View` option. This box will be gray if no views have been saved.

WINDOW SELECTION

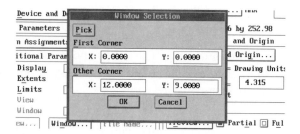

Figure 36.39 Use `Window Selection` to assign the portion of the screen to be plotted.

PAPER SIZE AND ORIENTATION

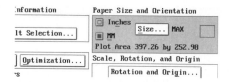

Figure 36.40 Use `Paper Size and Orientation` to size a plot, either in inches or millimeters. The size is listed as `Plot Area`.

PAPER SIZE AND ORIENTATION

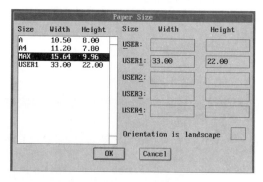

Figure 36.41 Standard and user-defined plotting sizes may be selected from the `Paper Size` box.

`Adjust Area Fill`: pulls in the boundaries of the filled area by one-half pen width for a more precisely drawn fill area.

`Plot to File`: sends the plot to a file with a .PLT extension for plotting with a utility program instead of to the plotter.

`File Name...`: gives a dialogue box from which you select a plot file.

`Paper Size and Orientation`: allows you to specify inches or millimeters as units for a plot **(Fig. 36.40)**. A rectangle in this area denotes whether the page has a portrait (vertical) or landscape (horizontal) orientation.

`Size...`: lists the standard sizes and your specified plot sizes **(Fig. 36.41)**. MAX is the

largest size that a plotter can handle. When you select a size and pick OK, the dialogue box leaves, and the dimensions appear at `Plot Area` (Fig. 36.40).

The `Plot Rotation and Origin` dialogue box offers the following options of plot orientation **(Fig. 36.42)**.

`Plot Rotation`: gives radio buttons for rotating plots 0°, 90°, 180°, or 270°.

`Plot Origin`: usually (0, 0), the lower-left corner of the sheet but may be offset by giving X and Y values in the `Edit` boxes.

`Scale`: is typed in the edit boxes, `Plotted Inches = Drawing Units:` **(Fig. 36.43)**. If you have selected metric units, the label would

PLOT ROTATION AND ORIGIN DIALOGUE BOX

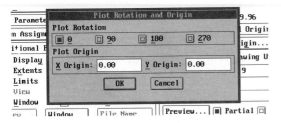

Figure 36.42 Set `Plot Rotation and Origin` with this box to position a drawing on the plot.

PLOT CONFIGURATION DIALOGUE BOX

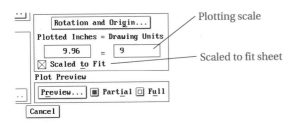

Figure 36.43 The `Scale to Fit` option lets you select a scale that will make the plot fill the available space.

read `Plotted MM`. Here, 1 = 1 is full size, 1 = 2 is half size, and 2 = 1 is double size.

`Scaled to Fit`: makes the drawing fill the plotting area and as large as possible.

`Plot Preview`: gives partial and full previews.

`Partial Preview`: shows two rectangles representing paper size and the drawing area **(Fig. 36.44)**. You cannot plot in the part of the drawing area that exceeds the paper size unless you adjust the scale, origin, or both. The triangular rotation icon shows in the lower left corner for 0° rotation, upper left for 90° rotation, upper right for 180° rotation, and lower right for 270° rotation.

`Full Preview`: shows the entire drawing on the screen and its relationship to the paper limits when plotted **(Fig. 36.45)**.

PREVIEW EFFECTIVE PLOTTING AREA

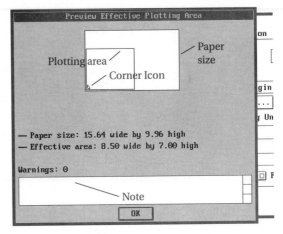

Figure 36.44 A `Partial Plot Preview` shows the boundaries of the drawing, plotting area, and the origin triangle in one of the corners.

FULL PREVIEW DISPLAY

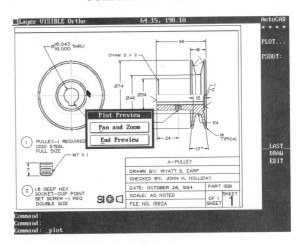

Figure 36.45 A `Full Plot Preview` shows the entire drawing as it will appear when plotted.

`Pan and Zoom`: allow you to check a drawing for details during preview.

`End Preview`: returns to the `Plot Configuration` dialogue box and the drawing when you pick OK.

1. Load paper, lower lever
2. Pens: P.7 (slot 1); P.3 (slot 2)
3. Set for size A
4. Press P1 and P2

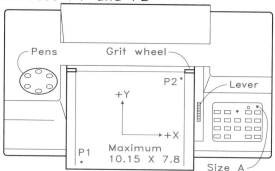

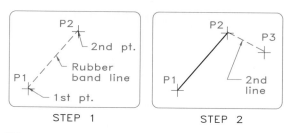

STEP 1   STEP 2

Figure 36.47 Drawing a LINE:

Step 1 Command: LINE or L

From point: P1

To point: P2 (The line is drawn.)

Step 2 To point: P3 (The line is drawn.)

((CR) to disengage the rubber band.)

Figure 36.46 This Hewlett-Packard 7475A plotter is typical of those used to plot A and B sheet sizes.

Update the .PCP file by picking Save Defaults to File... as shown in Fig. 36.31 to save these settings and make them available from the Get Defaults from File option when you make a plot with the same settings.

Readying the Plotter

After you set the parameters and approve the preview, pick OK. The command line then gives the following messages at the bottom of the screen:

    Effective plotting area: XX wide by
    YY high (CR)

    Position paper in plotter. (CR)

    Press RETURN to continue of S to Stop
    for hardware setup

Load your paper into the plotter as shown in **Fig. 36.46**, with the thick pen (P.7) in slot 1 and the thin pen (P.3) in slot 2 as previously specified by the Pen Assignments in Fig. 36.34. Press (CR) and the plot will begin. You may cancel plotting by pressing Ctrl C in combination which may require almost a minute to take effect. When plotting has been completed, select Exit AutoCAD

under File on the pull-down menu. You may use laser printing as well as pen plotters to make plots.

Now that you know how to set drawing aids, save files, and plot, you are ready to learn how to make drawings.

36.6 Drawing Lines, Points, and Figures

To begin, open your B:FORMAT file and make a new drawing file by using SAVE AS. Name the drawing (say, B:NO1), which becomes the current drawing and contains the settings and parameters of B:FORMAT.

Lines

You may draw a line by using the side menu, the keyboard, or the pull-down menu. If you use the side menu, select DRAW and LINE and respond to the command line, as shown in **Fig. 36.47**, to draw the lines by picking endpoints with the cursor and pressing the select button of the mouse (the left button). The current line will rubber band (stretch) from the last point and lines will be drawn in succession until you press (CR) or the right button of the mouse. If you use the pull-

LINE COMMAND FROM THE PULL-DOWN MENU

Figure 36.48 The LINE command appears under Draw on the pull-down menu. An action box offers four types of lines for use.

LINES: 2D SPACE

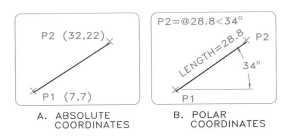

A. ABSOLUTE COORDINATES

B. POLAR COORDINATES

Figure 36.49

A Absolute coordinates may be typed to establish ends of a line.

B Polar coordinates are relative to the current point and are specified by a length and the angle, with the horizontal measured clockwise.

down menu, select DRAW, LINE, and SEGMENTS to begin drawing lines by using a dialogue box (**Fig. 36.48**).

Figure 36.49 compares absolute and polar coordinates. Use keyboard input to draw the types of lines shown in **Fig. 36.50** in absolute, polar, spherical, or cylindrical coordinate formats.

Absolute Cartesian Coordinates Type 3D coordinates in the form 2,3,1.5, to give the coordinates of a point from the origin of 0,0,0.

Polar Coordinates Type 2D coordinates as 3.6<56 to draw a line from 0,0,0 at an angle of 56° with the X axis in the X–Y plane.

LINES: 3D SPACE

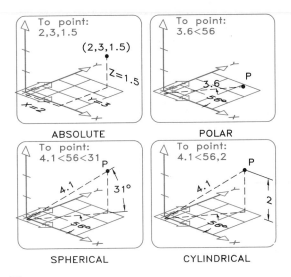

ABSOLUTE

POLAR

SPHERICAL

CYLINDRICAL

Figure 36.50 To locate a point in a drawing, use any of the formats shown here.

LINES: 3D SPACE

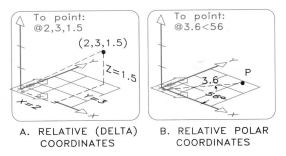

A. RELATIVE (DELTA) COORDINATES

B. RELATIVE POLAR COORDINATES

Figure 36.51 A point in a drawing may be located relative to the last point on the screen by either of these methods.

Spherical Coordinates Type 3D coordinates as 4.1<56<31 to locate a point 4.1 units from 0,0,0, at an angle of 56° with the X axis in the X–Y plane, and at an angle of 31° with the X-Y plane (Fig. 36.50).

Cylindrical Coordinates Type 3D coordinates as 4.1<56,2 to locate a point 4.1 units from 0,0,0 at

CLOSING A POLYGON

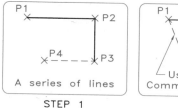 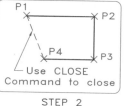

| STEP 1 | STEP 2 |

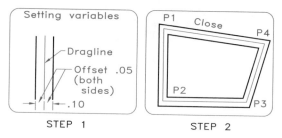

STEP 1 STEP 2

◆ **Figure 36.52** Closing a polygon:

Step 1 Draw a series of lines with the LINE command from P1 through P3. (The last one is a rubber-band line until the next point is selected.)

Step 2 After choosing P4, enter CLOSE, (CR) and the program will produce a closing line to the first point of the series.

an angle of 56° with the X axis in the X–Y plane and 2 units higher in the Z direction.

Other Types of Coordinates Relative (delta) **coordinates** locate points with respect to the current location of the cursor **(Fig. 36.51)**. Relative coordinates are X, Y, and Z values of a 3D point from the current point on the screen. Precede each coordinate with the @ symbol (@2,3,1.5, for example). Type **relative polar coordinates** as @3.6<56 to generate a line 3.6 units long, and making a 56° angle with the X axis in the X–Y plane.

To find the **last coordinates** used, type @ while in the line command to move the cursor back to the last point. **World coordinates** locate points in the **world coordinate system** regardless of the **user coordinate system** that you are using if you precede the coordinates with an asterisk (*). Examples are *4,3,5; *90<44,2; and @*1,3,4.

Closure The status line at the top of the screen shows the length of the line and its angle from the last point as it is being rubber banded from point to point. When you are drawing a continuous series of lines, use the CLOSE command to draw the last line to the beginning point **(Fig. 36.52)**.

DOUBLE LINES

◆ **Figure 36.53** Drawing double lines:

Step 1 Command: DLINE (CR)

Break/Caps/Dragline/Offset/Snap/Undo/
Width/<start point>: D (CR)

Set dragline position to Left/Center/Right/
<Offset from center=0.00>: .20 (CR)

Step 2 Arc/Break/ . . . /Width/<next point>:
P1 (CR)

Arc/Break/ . . . /Width/<next point>: P2 (CR)

Arc/Break/ . . . /Width/<next point>: P3 (CR)

Arc/Break/ . . . /Width/<next point>: P4 (CR)

Arc/Break/ . . . /Width/<next point>: CLOSE
(CR)

Double Lines Use the Line command Double-Line (DLINE) option to draw double lines **(Fig. 36.53)**. Options include Caps for connecting the lines at the ends; Dragline for specifying the line about which the Offset is made; Width for assigning the distance between the lines; Snap for snapping to objects; and Undo to reverse the last command.

Error Correction You may correct errors by typing one of the following commands.

CTRL X (Deletes the line.)

CTRL C (Cancels the current command and returns the Command: prompt.)

BACKSPACE (Deletes one character at a time.)

POINT STYLE

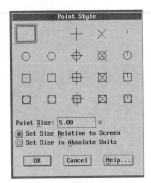

Figure 36.54 You may set the symbol used to draw a point (`DDPTYPE` dialogue box) and specify its size from this dialogue box.

DRAWING A CIRCLE

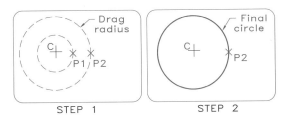

Figure 36.55 Using the `CIRCLE` command:
Step 1 `Command:` `CIRCLE` or `C` (CR)
`3P/2P/<Center point>:` `C` (with cursor)
`Diameter/<Radius>:` (Drag radius to P1, and then to P2 to dynamically change it, if `DRAGMODE` is ON.)
Step 2 Select the final radius to draw the circle.

Points

The `POINT` command places the symbols shown in **Fig. 36.54** on a drawing for plotting. Select a symbol by using one of two options that sets the point size `Relative` to the screen or in `Absolute` `units`. Use the point size box to assign the size of the marker on the drawing.

Circles

Use the `CIRCLE` command from `DRAW` of the pull-down menu to draw circles with a center and

CIRCLE: TANGENT TO CIRCLE AND LINE

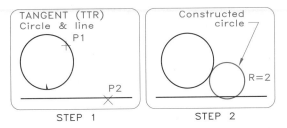

Figure 36.56 Producing a tangent to a circle and line:
Step 1 `Command:` `CIRCLE` or `C` (CR)
`3P/TTR/<Center Pt>:` `TTR`
`Enter Tangent spec:` `P1`
`Enter second Tangent spec:` `P2`
Step 2 `Radius:` `2` (CR)
(The circle with radius = 2 is drawn tangent to the line and circle.)

radius, a center and diameter, or three points. You may drag a circle to its final size with the mouse when `DRAGMODE` is ON (**Fig. 36.55**).

The `TTR` (tangent, tangent, radius) option of the `CIRCLE` command draws a circle tangent to a circle and a line, three lines, three circles, or two lines and a circle. To draw a circle tangent to a line and a circle, select the circle and line and give the radius (**Fig. 36.56**).

The `3P` option calculates the radius length and draws a circle tangent to three lines (**Fig. 36.57**). It also draws a circle tangent to two lines and a circle (**Fig. 36.58**) or to three circles.

Arcs

The `ARC` command from the `DRAW` menu has nine combinations of the variables starting point, center, angle, ending point, length of chord, and radius. The S, C, E version requires that you locate the starting point, S, the center, C and the ending point, E (**Fig. 36.59**). The arc begins at point S, but point E need not lie on the arc. Arcs are drawn counterclockwise by `Default`.

You may continue a line with an arc drawn from the last point of the line and tangent to it, as

ARC: TANGENT TO 3 LINES

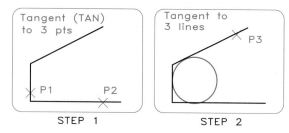

STEP 1   STEP 2

◆ **Figure 36.57** Constructing a tangent to three lines:

Step 1 Command: `CIRCLE` or `C` (CR)

`3P/2/TTR/ <Center point>: `3P`

`First point: `TAN` (CR) to `P1

`Second point: `TAN` (CR) to `P2

Step 2 `Third point: `TAN` (CR) to `P3

(The circle is drawn tangent to three lines.)

ARC: TANGENT TO CIRCLE AND LINES

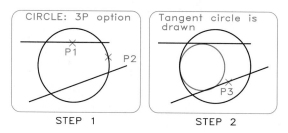

STEP 1   STEP 2

◆ **Figure 36.58** Constructing a tangent to two lines and a circle:

Step 1 Command: `CIRCLE` or `C` (CR)

`3P/2P/TTR/<Center point>: `3P` (CR)

`First point: `TAN` (CR) to `P1

`Second point: `TAN` (CR) to `P2

Step 2 `Third point: `TAN` `P3

(A circle is drawn tangent to the lines and circle.)

shown in **Fig. 36.60**. You will find this technique useful for drawing runouts when using fillets and rounds. You may also use it for drawing a line from an arc and tangent to it with the opposite sequence of the commands. With `DRAGMODE` on, you may observe the dragging of the arc or line to its final size.

ARC: SCE OPTION

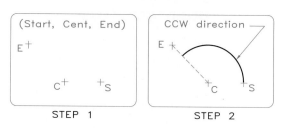

STEP 1   STEP 2

◆ **Figure 36.59** Using the `ARC` command:

Step 1 Command: `ARC` or `A` (CR)

`Arc Center/<Start point>: S`

`Center/End/<Second point>: `CENTER` or `C` (CR)

Step 2 `Angle/Length of chord/<Endpoint>: `DRAG to E`

(The arc is drawn counterclockwise (ccw) to an imaginary line from S to E.)

ARC: END OF LINE

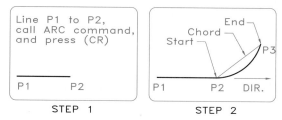

STEP 1   STEP 2

◆ **Figure 36.60** Drawing an arc tangent to the end of a line:

Step 1 Command: `LINE` or `L` (CR)

`From point: `P1

`To point: `P2` (CR)

Step 2 Command: `ARC` (CR)

`Center/<Start point>: (CR)`

`End point: `P3

Fillets

You may draw fillets between two nonparallel lines by using the `FILLET` command. After generating the fillet, trim the lines, as shown in **Fig. 36.61**.

Once you assign the radius it remains in memory for drawing additional fillets. By setting the

FILLET COMMAND

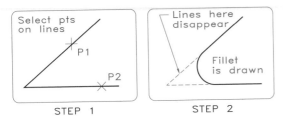

Figure 36.61 Using the FILLET command.

Step 1 Command: <u>FILLET</u> (CR)

Polyline Radius/<Select two objects>: <u>R</u> (CR)

Enter fillet radius <0.0000>: <u>1.5</u> (CR)

Command: (CR)

Step 2 FILLET Polyline Radius/<Select two objects>: <u>P1</u> and <u>P2</u> (The fillet is drawn, and the lines trimmed.)

radius to 0, you can extend nonparallel lines to a perfect intersection. To draw fillets tangent to circles or arcs, specify the radius and select points on each circle (**Fig. 36.62**).

Figure 36.63 shows examples of fillets that connect lines and arcs. The points selected on the two entities (see Section 36.8) determine the position of the fillet.

Chamfers

The CHAMFER command draws angular bevels at the intersections of lines or polylines. After assigning chamfer distances, select two lines and trim or extend the lines, and draw the CHAMFER. Press (CR) to repeat this command using the previous settings to chamfer other corners (**Fig. 36.64**).

Polygons

To draw a multisided figure composed of equal sides, use the POLYGON command. Locate the center, give the number of sides, and choose either the inscribe (I) or circumscribing (C) options (**Fig. 36.65**), as follows:

Command: <u>POLYGON</u> (CR)

Number of sides: <u>5</u> (CR)

FILLET: CIRCLES

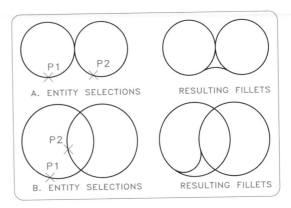

Figure 36.62 Obtaining tangent arcs:

Step 1 Command: <u>FILLET</u> (CR)

Polyline/Radius/<Select two objects>: <u>R</u> (CR)

Enter fillet radius <0.0000>: <u>1.2</u> (Example.) (CR)

Step 2 Command: (CR)

FILLET Polyline/Radius/<Select two objects>: (Select a point on each circle, and the fillet is drawn.)

FILLET: LINES

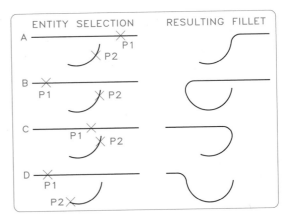

Figure 36.63 After choosing a fillet radius, select points on each line or arc for filleting. Each fillet is determined by the location of the points selected.

Edge/<Center of polygon>: (Locate center.)

Inscribed in circle\Circumscribed about circle (I/C): <u>I</u> (CR)

CHAMFER COMMAND

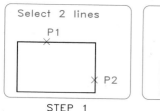

STEP 1   STEP 2

Figure 36.64 Using the CHAMFER command:

Step 1 Command: <u>CHAMFER</u> (CR)

Polyline/Distance/<Select first line>: D
(Used for nonpolylines.)

Enter first chamfer distance <0.4>: <u>1.40</u>
(CR)

Enter second chamfer distance <0.4>:
<u>1.00</u> (CR)

Step 2 Command: (CR)

CHAMFER Polyline/Distance/<Select first
line>: <u>P1</u>

Select second line: <u>P2</u>

ELLIPSES

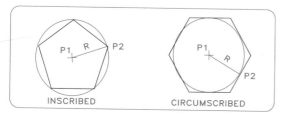

INSCRIBED   CIRCUMSCRIBED

Figure 36.65 To produce polygons inscribed in or circumscribed about a circle, use the POLYGON command.
Select Inscribed or Circumscribed, specify the
number of sides, and pick points P1 and P2 to define the
radius.

Radius of circle: (Type length and (CR) or
select length with pointer.)

When you use the EDGE option the next prompt
asks:

First endpoint of edge: (Select point.)

Second endpoint of edge: (Select point.)

ELLIPSE: AXIS ENDPOINTS

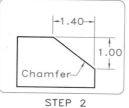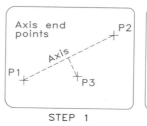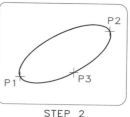

STEP 1   STEP 2

Figure 36.66 Constructing an ellipse—endpoints
method:

Step 1 Command: <u>ELLIPSE</u> (CR)

<Axis endpoint 1>/Center: <u>P1</u>

Axis endpoint 2: <u>P2</u>

<Other axis distance>/Rotation: <u>P3</u>

Step 2 The ellipse is drawn through P1 and P2 and a distance measured perpendicularly from P1–P2 by the location of P3.

The program draws the polygon counterclockwise
about the center point. This command can produce a maximum of 1024 sides.

Ellipses

You may use the ELLIPSE command to draw
ellipses by several methods. **Figure 36.66** shows
an ellipse drawn by selection of the endpoints of
the major axis and a third point, P3, to give the
length of the minor radius.

Figure 36.67 shows the center of the ellipse at
P1, selection of one axis endpoint at P2, and the
second axis length to P3 given. The program
draws the ellipse through point P2 and the endpoint of the minor diameter specified by P3.

Figure 36.68 shows the specified endpoints of
the axis and rotation angle. A rotation angle of 0°
gives a full circle, and a rotation of 90° gives an
edge. Construction of ellipses in isometric drawings is covered in Section 36.27.

TRACE

You may draw wide, multiple-stroke lines with the
TRACE command as follows:

ELLIPSE: CENTER AND AXES

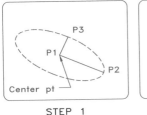
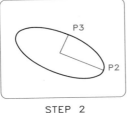

STEP 1   STEP 2

Figure 36.67 Constructing an ellipse—center and axes method:

Step 1 Command: <u>ELLIPSE</u> (CR)
(Axis endpoint 1)/Center: <u>C</u> (CR)
Center of ellipse: <u>P1</u>
Axis endpoint: <u>P2</u>
Step 2 <Other axis distance>/Rotation: <u>P3</u>
(The ellipse is drawn.)

ELLIPSE: CENTER AND ANGLE

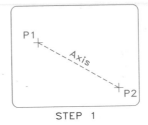
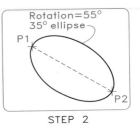

STEP 1   STEP 2

Figure 36.68 Constructing   ellipse—rotation   angle method:

Step 1 Command: <u>ELLIPSE</u> (CR)
<Axis endpoint>/Center: <u>P1</u>
Axis endpoint 2: <u>P2</u>
Step 2 <Other axis distance>/
Rotation: <u>R</u> (CR)
Rotation around major axis: <u>55</u> (CR)
(The ellipse is drawn.)

Command: <u>TRACE</u> (CR)

Width: <u>0.4</u> (CR)

From point: <u>2, 3</u> (CR)

To point: <u>4, 6</u> (CR)

To point: <u>6, 2</u> (CR)

With FILL on, the line will be drawn as shown in **Fig. 36.69**. When FILL is off, the lines will be drawn as parallel lines with miter angles.

36.7 Zooming and Panning

To enlarge or reduce parts of a drawing use the ZOOM command in the submenu under the DIS-PLAY menu. **Figure 36.70** shows the use of a ZOOM/WINDOW to select the portion to enlarge. From the side menu, instead of the WINDOW option, you could choose ALL, CENTER, DYNAMIC, EXTENTS, LEFT, PREVIOUS, VMAX, SCALE(X/XP), or type in a magnification factor as a number.

   All: expands the drawing's LIMITS to fill the screen.

TRACE COMMAND

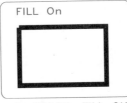

A. TRACE: FILL ON   B. TRACE: FILL OFF

Figure 36.69 Using the TRACE command:
A When you use the TRACE command with FILL ON, lines are drawn solid to the width specified.
B When FILL is OFF, parallel lines are drawn.

   Center: lets you pick a point in the drawing to be positioned at the center of the screen and specify its degree of magnification or reduction.

   Dynamic: lets you ZOOM and PAN by selecting points with the cursor.

   Extents: enlarges the drawing to bump against the edge of the screen.

ZOOM COMMAND

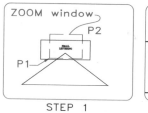
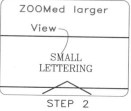

PAN COMMAND

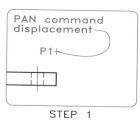
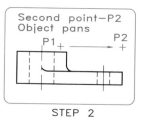

STEP 1	STEP 2

Figure 36.70 Using the `ZOOM` command:

Step 1 `Command: ZOOM or Z (CR)`

`All/Center/Dynamic/Extents/Left/Previous /Vmax/`

`Window/<Scale (X/XP)>:` `W (CR)` (Window is default.)

`First corner: P1`

Step 2 `Other corner: P2`

(The window is enlarged to fill the screen.)

Figure 36.71 Using the PAN command:

Step 1 `Command: PAN (CR)`

`Displacement: P1`

Step 2 `Second point: P2`

(The view point is moved to new position.)

`Left:` establishes the lower left corner and the height of the drawing to be `ZOOM`ed.

`Previous:` displays the last `ZOOM`ed view. You may `ZOOM` views almost infinitely many times.

`Vmax:` makes the current screen images as large as possible without forcing a complete regeneration.

`Scale X/XP:` magnifies a drawing relative to paper space <Scale `X/XP`> where `X` represents a fraction, say 1/4 or 0.25. By typing `1/4XP` or `.25XP` you scale the drawing to 1/4 size.

The `PAN` command **(Fig. 36.71)** lets you pan the view across the screen. The first point you select is a handle for dragging the view of the drawing to its new view. The drawing is not being relocated; your view of it is being changed.

36.8 Selecting Entities

A reoccurring prompt, `Select objects:`, asks you to select an entity, or entities, that you may want to `ERASE`, `CHANGE`, or otherwise modify.

Selection options include: `point, multiple, win- dow, window polygon, crossing window, crossing polygon, fence, box, all, last,` or `previous`. **Figure 36.72** illustrates how to select a single entity with the cursor and how to make multiple selections by typing `M` (multiple) when prompted to do so.

The `Window` (`W`) option is for selecting the diagonals of a window. Only entities totally within the window will be selected (Fig. 36.72). The `Crossing` window (`C`) shown in Fig. 36.72 selects the entities within the window and those crossed by it. When prompted to select objects, you can type `BOX` to obtain a window or a crossing window by selecting the second diagonal point to the right or left, respectively (Fig. 36.72).

When you use `LAST` (`L`), it picks the most recently drawn object. The `PREVIOUS` (`P`) option recalls the previously selected set of objects for editing. Enter `MOVE` and type `P` to recall and move the last group of selected entities.

When selecting a set of objects, you may remove them in reverse order one at a time with `Undo` by typing `U` repetitively. When selecting objects, you may remove a selected entity with `Remove` by typing `R` and picking those to be removed. To add entities to the set, use `Add` by

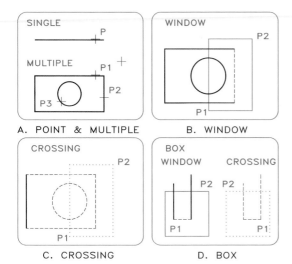

Figure 36.72 Object selection options:

A When prompted by the command `Select objects:`, select one or more entities individually.

B Use a `window` (W) to select objects lying completely within the window area.

C Use a `crossing window` (C) to select objects within or crossed by the window.

D You may use the `BOX` option as either a window or crossing option, determined by the sequence in which the diagonal corners are selected.

typing `A` and select those to be added. When finished, press (CR) to end the `Select/remove` prompt.

    When you use `SIngle` (SI), the program acts on the object or sets of objects without pausing for a response.

    `WPolygon` (WP) forms a solid-line polygon that has the same effect as a window **(Fig. 36.73)**. `CPolygon` (CP) forms a dotted-line polygon that has the same effect as a crossing window.

    `Fence` selects corner points of a polyline that will erase any object it crosses the same as a crossing window. `All` selects everything on the screen.

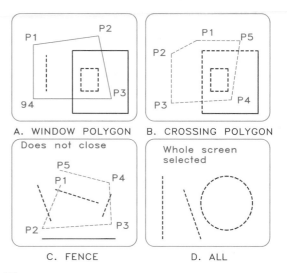

Figure 36.73

A Use a `window polygon` (WP) to select objects lying within it.

B Use a `crossing polygon` (CP) to select objects crossed by it.

C Use a `fence` (F), a nonclosing polyline, to select entities that it crosses.

D Type `ALL` to select everything on the screen.

36.9 Editing Commands

ERASE and BREAK

The `ERASE` command, a subcommand under `EDIT` (or `MODIFY`), deletes parts of a drawing. You may use the selection techniques described in the preceding section to select the entities to be erased **(Fig. 36.74)**. However, only entities completely within the window may be selected. Use a `CROSSING` (C) window **(Fig. 36.75)** to erase entities crossed by or lying within it.

    The `Default` of the `ERASE` command, `Select Objects`, allows you to pick entities and delete them one at a time by pressing (CR) **(Fig. 36.76)**. With the pull-down menu, you erase entities without confirmation as you select them. You may use `Oops` to restore the last erasure, but only the last one.

ERASING: WINDOW

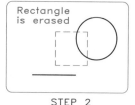

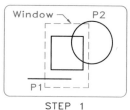

Figure 36.74 ERASE—the window option:
Step 1 Command: ERASE or E (CR)

Select objects: WINDOW or W (CR)

First corner: P1

Other corner: P2

Step 2 Select objects: (CR)

(Entities entirely within window are erased.)

ERASING: CROSSING WINDOW

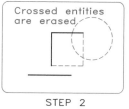

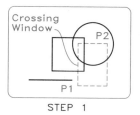

Figure 36.75 ERASE—the crossing option:
Step 1 Command: ERASE or E (CR)

Select objects: C (Crossing) (CR)

First corner: P1

Other corner: P2

Step 2 Select objects: (CR)

(Entities crossed by or in window are erased.)

The BREAK command removes part of a line, pline, arc, or circle **(Fig. 36.77)**. By using the F-option, ensure that you have chosen the desired line for breaking. If the breaks do not begin and end at intersections of the lines, omit the F response. To delete a portion of a line, use the BREAK command **(Fig. 36.78)** and window the line by selecting two points at the ends of the break.

ERASING: ENTITIES

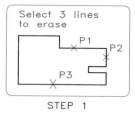

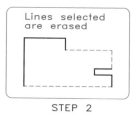

Figure 36.76 Erasing lines:
Step 1 Command: ERASE

Select Objects P1, P2, and P3: (Select lines.) (CR)
Step 2 Press (CR) again, and the lines are erased.

Use OOPS (CR) to recall erased entities.

BREAK COMMAND: F OPTION

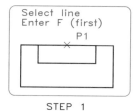

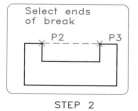

Figure 36.77 Using the BREAK command:
Step 1 Command: BREAK (CR)

Select object: P1 (On line to be broken.)

Enter second point or F: F (CR)

Step 2 Enter first point: P2

Enter second point: P3 (The line is broken from P2 to P3.)

MOVE and COPY

The MOVE command is used to reposition a drawing **(Fig. 36.79)** and the COPY command to duplicate it. When copied, the drawing remains in its original position and the duplicate is located where specified. Applied identically, the COPY and MOVE commands let you drag drawings into position (with DRAGMODE=AUTO) **(Fig. 36.80)**. The COPY command has a MULTIPLE option for positioning multiple copies of drawings in different places.

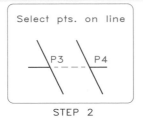

STEP 1   STEP 2

Figure 36.78 BREAK command—window option:
Step 1 Command: <u>BREAK</u> (CR)

Select object: <u>WINDOW</u> or <u>W</u> (CR)

(Window the line P1, P2.)

Step 2 Enter first point: <u>P3</u>

Enter second point: <u>P4</u>

(Line P4–P5 is removed. Use OSNAP to select intersection points.)

MOVE COMMAND

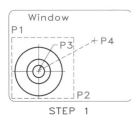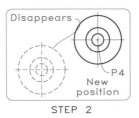

STEP 1   STEP 2

Figure 36.79 Using the MOVE command:
Step 1 Command: <u>MOVE</u> or <u>M</u> (CR)

Select objects <u>W</u>: (CR)

First corner: <u>P1</u>

Other corner: <u>P2</u> 5 found.

Select objects: (CR)

Base point or displacement: <u>P3</u>

Step 2 Second point of displacement: <u>P4</u>
(The object is moved to P4.)

TRIM

The TRIM command selects cutting edges for trimming selected lines, arcs, or circles that cross them **(Fig. 36.81)**. The cutting edges trim the entities to perfect intersections at their crossing points. The CROSSING option of TRIM selects four cutting edges and trims the lines **(Fig. 36.82)**. For

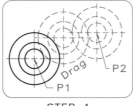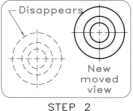

STEP 1   STEP 2

Figure 36.80 Using DRAGmode:
Step 1 Command: <u>MOVE</u> (CR)

Select objects: <u>W</u> (CR) (Window object.)

Base point or displacement: <u>DRAG</u> (CR)

Base point or displacement: <u>P1</u> (Or X, Y distance.)

Second point of displacement: <u>P2</u> (The drawing is DRAGged as the cursor is moved.)

Step 2 When moved to P2, press the Select button to draw the circle. If DRAGMODE is AUTO or On, dragging is automatic.

TRIM COMMAND

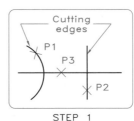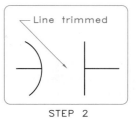

STEP 1   STEP 2

Figure 36.81 TRIMming edges:
Step 1 Command: <u>TRIM</u> (CR)

Select cutting edge(s) . . .

Select objects: <u>P1</u>

Select objects: <u>P2</u> (CR)

Step 2 Select object to trim: <u>P3</u>

(Line between cutting edges is removed.)

TRIM to work with a 3D drawing, the entities selected must be parallel to the current user coordinate system (UCS).

EXTEND

To lengthen lines and arcs to intersect an entity, use the EXTEND command **(Fig. 36.83)**. The

TRIMMING: CROSSING OPTION

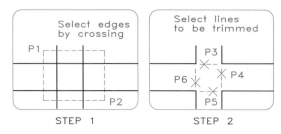

Figure 36.82 TRIMming by crossing:
Step 1 Command: TRIM (CR)

Select cutting edge(s) . . .

Select objects: <u>Crossing</u> Window (CR)

First corner: <u>P1</u>

Other corner: <u>P2</u> (CR)

Step 2 Select object to trim: <u>P3, P4, P5, P6</u>

(The lines are trimmed one at a time.)

prompt asks you to select the boundary entity and then the line or arc to be extended. You may extend more than one entity to the boundary entity. EXTEND will not work on closed PLINES, such as a polygon. For EXTEND to work with a 3D drawing, the entities selected must be parallel to the current UCS.

UNDO

The UNDO (U) command can undo the previous commands one at a time—all the way back to the beginning point of the current session. The REDO command reverses the last UNDO command if you use it immediately after applying that UNDO; OOPS will not work. The UNDO command works as follows:

Command: <u>UNDO</u> (CR)

Auto/Back/Control/End/Group/

Mark/<Number>: <u>4</u> (CR)

Entering 4 has the same effect as using the U command four separate times.

MARK identifies a point in the drawing process to which you may undo subsequent additions

EXTEND COMMAND

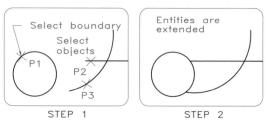

Figure 36.83 Using the EXTEND command:
Step 1 Command: <u>EXTEND</u> (CR)

Select boundary edge(s) . . .

Select objects: <u>P1</u> (CR)

Step 2 Select object to extend: <u>P2</u> and <u>P3</u>

(The line and arc are extended to the boundary.)

with the BACK option. Only that part of the drawing added after you place the MARK will be undone at the prompt:

This will undo everything.

OK? <Y>.

By responding Y, you remove the mark, allowing the next UNDO to proceed backward past the mark.

The UNDO GROUP, UNDO END, and AUTO subcommands remove groups of entities at one time. These options are meant to be used with menus.

The CONTROL subcommand has three options: ALL, NONE, and ONE. ALL turns on the full features of the UNDO command, and NONE turns them off. ONE uses UNDO commands for single operations and requires the least disk space.

CHANGE

The CHANGE command permits modification of single entities, including LINES, CIRCLES, TEXT, ATTRIBUTE DEFINITIONS, and BLOCKS. It also allows COLOR, LAYER, LINETYPE, and THICKNESS property changes.

To change the length of a line and the position of one of its ends, choose one end of the line and locate its new endpoint (**Fig. 36.84**). To change

CHANGING A LINE

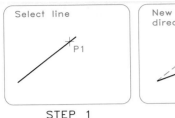
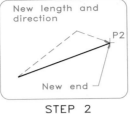

STEP 1   STEP 2

⬧ **Figure 36.84** Changing a line:
Step 1 Command: <u>CHANGE</u> (CR)

Select objects: <u>P1</u>

Select objects: (CR)

Step 2 Properties/<Change point>: <u>P2</u> (CR)

(End of line is moved.)

CHANGING CIRCLE SIZE

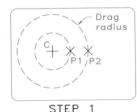
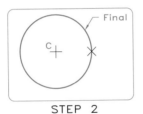

STEP 1   STEP 2

⬧ **Figure 36.85** Changing a circle:
Step 1 Command: <u>CHANGE</u> (CR)

Select objects: <u>P1</u> (CR)

Step 2 Properties/<Change point>: <u>P2</u> (To
change radius.)

the size of a circle, pick a point on its circumference and drag it to a new size (**Fig. 36.85**).

To change TEXT, press (CR) until the prompts STYLE, HEIGHT, ROTATION ANGLE, and TEXT STRING appear in sequence (**Fig. 36.86**). You may use the CHANGE command to revise ATTRIBUTE DEFINITIONS, including TAG, PROMPT STRING, and DEFAULT VALUE. Move or rotate BLOCKS with the CHANGE command, as shown in **Fig. 36.87**.

To change LAYER, COLOR, LTYPE, and THICK-NESS select entities when prompted and then type

CHANGING TEXT

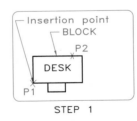

STEP 1   STEP 2

⬧ **Figure 36.86** Changing text:
Step 1 Command: <u>CHANGE</u>

Select objects: (Select text.)

Select objects: (CR)

Properties/<Change point> (CR)

Enter text insertion point: (Select point.)

Step 2 Text Style: <u>Standard</u>

New style or RETURN for no change: <u>RT</u>

New height <.50>: <u>0.20</u>

New rotation angle <0>: (CR)

New text <WORD>: <u>words</u> (CR)

CHANGING BLOCKS

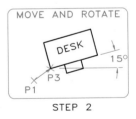

STEP 1   STEP 2

⬧ **Figure 36.87** Changing blocks:
Step 1 Command: <u>CHANGE</u> (CR)

Select objects: <u>P1</u>

Select objects: (CR)

Properties/<Change point>: (CR)

Enter block insertion point: <u>P3</u>
(Select new position.)

Step 2 New rotation angle <0>: <u>15</u> (CR)

(The block is moved and rotated 15°.)

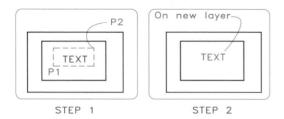

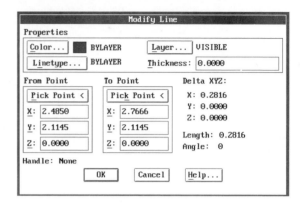

Figure 36.88 Changing layers:
Step 1 Command: CHANGE (CR)

Select objects: W (CR)

First corner: P1 (CR)

Other corner: P2 (CR)

Step 2 Properties < Change point >: P

Change what property (Color/LAyer/Thickness)? LA (CR)

New layer: VISIBLE (An existing layer.) (CR)

Figure 36.89 The Entity dialogue box under MODIFY on the pull-down menu is selected. It lets you CHANGE Color, Layer, Linetype, Thickness, and Point locations on entities.

P (Properties), as shown in **Fig. 36.88**. Type LA (layer) and the name of the layer that the text is to be moved to.

Although you may assign multiple COLORS and LTYPES to entities on the same layer with the CHANGE command, each layer should have only one layer and line type.

The THICKNESS property is the height of the extrusion in the Z direction of a 3D object drawn with the ELEV and THICKNESS commands. Changing the THICKNESS in a 2D drawing from 0 to a nonzero value converts it to an extruded 3D drawing. The THICKNESS option extrudes objects parallel to the Z axis of the current UCS.

You may also use the Modify Line dialogue box (DDMODIFY) under Modify/Entity of the pull-down menu to change COLOR, LINETYPE, LAYER, and THICKNESS properties and ends of lines (**Fig. 36.89**).

CHPROP

The CHPROP command also allows you to change color, linetype, layer, and thickness regardless of their extrusion direction, which is useful in modifying 3D drawings. Use CHPROP as follows:

Command: CHPROP

Select objects: (Select entities.)

Change what property (Color/LAyer/LType/Thickness)?

Make the same choice as when you are using the CHANGE command.

36.10 Working with Polylines

POLYLINE (PLINE)

The PLINE command draws a series of lines and arcs of varying widths to form a POLYLINE, which is a continuous line instead of a series of segmented lines, as follows:

Command: PLINE: (CR)

From point: (Select starting point.)

Current line-width is 0.0000

Arc/Close/Halfwidth/Length/ Undo/Width/<Endpoint of line>: W (For width.) (CR)

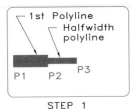 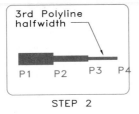

STEP 1   STEP 2

Figure 36.90 Using the PLINE command:
Step 1 Command: <u>PLINE</u> (CR)

From point: <u>P1</u>

Current line width is 0.0000

Arc/Close/Halfwidth/Length/Undo/
Width/<Endpoint of line>: <u>WIDTH</u> (CR)

Starting width <0.0000>: <u>.12</u> (CR)

Ending width <.12>: (CR)

Arc/Close/ . . . /<Endpoint of line>: <u>P2</u>

Arc/Close/ . . . /<Endpoint of line>:
<u>Halfwidth</u> (CR)

Starting half-width <0.3000>: <u>.3</u> (CR)

Ending half-width <0.3000>: (CR)

Arc/Close/ . . . /<Endpoint of line>: <u>P3</u>

Step 2 Arc/Close/ . . . /<Endpoint of line>:
<u>Halfwidth</u>

Starting half-width <0.2>: <u>.1</u> (CR)

Ending half-width <0.1000>: (CR)

Arc/Close/ . . . ?<Endpoint of line>: <u>P4</u>

Starting width <0.000>: <u>0.4</u> (CR)

Ending width <1.000>: <u>0.4</u> (CR)

(Select endpoint.)

The Default **response**, End point of line
(shown in brackets), asks for the first point of the
polyline. By typing W (width) you call the prompt
for the beginning width of the line (0.4) and its
ending width (0.4). Enter a width of 0 to obtain
single-stroke lines, the thinnest lines possible.
CLOSE automatically connects the last end of the
polyline with its beginning point and terminates
the command.

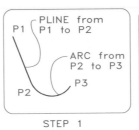 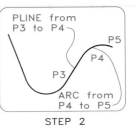

STEP 1   STEP 2

Figure 36.91 Producing polylines and arcs:
Step 1 Command: <u>PLINE</u>

From point: <u>P1</u> (CR)

Arc/Close/ . . . /Width/<Endpoint of
line>: <u>P2</u> (CR)(CR)

Command: <u>Arc</u> (CR)

Center/<Start point>: <u>P3</u> (CR)(CR)

Step 2 Command: <u>Line</u> (CR)

LINE From point: (CR)

Length of line: <u>P4</u> (CR)(CR)

Command: <u>ARC</u> (CR)

Center/<Start point>: <u>P5</u>

LENGTH continues a PLINE in the same direction
if you type the length of the segment. If the first line
was an arc, a line is drawn tangent to the arc. UNDO
erases the last segment of the polyline, and you
may repeat it to continue erasing segments.
HALFWIDTH specifies the width of the line measured
on both sides of its center line **(Fig. 36.90)**.

Arcs Selecting the Arc option of PLINE results in
the prompts listed in **Fig. 36.91**. When selected,
an arc will be drawn tangent to and from the last
line and pass through the next point selected.

ANGLE gives the prompt, Included angle:, to
which you give a positive or negative value. The
next prompt asks for Center/Radius/<End
point>: to draw an arc tangent to the previous
line segment. Select Center and you will be
prompted for the center of the next arc segment.
The next prompt is Angle/Length/<End

PEDIT: FIT CURVE

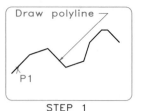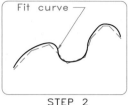

| STEP 1 | STEP 2 |

Figure 36.92 Using the PEDIT command:
Step 1 Command: <u>PEDIT</u> (CR)

Select Polyline: <u>P1</u>

Step 2 Close/Join/Width/Edit vertex/Fit/
Spline/Decurve/Ltype gen/Undo/eXit <X>:
<u>Fit</u> (CR)

(The curve is smoothed.)

PEDIT: FITS AND SPLINES

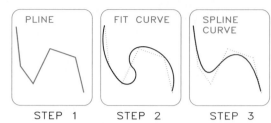

| STEP 1 | STEP 2 | STEP 3 |

Figure 36.93 The SPLINE curve:
Step 1 Draw a polyline PLINE.

Step 2 Command: <u>PEDIT</u>

Select polyline: (Select line.)

Close/Join/Width/Edit vertex/Fit/
Spline/Decurve/Ltype gen/Undo/eXit <X>: <u>Fit</u>

Step 3 (Use PEDIT with the SPLINE option for mathematical curve.)

point>:, where Angle is the included angle, and Length is the chordal length of the arc. CLOSE causes PLINE's arc segment to close to its beginning point.

DIRECTION lets you override the default, which draws the next arc tangent to the last PLINE segment. When you see the prompt, Direction from starting point:, pick the beginning point and respond to the next prompt, Endpoint, by picking a second point to give the direction of the arc.

LINE switches the PLINE command back to the straight-line mode. RADIUS gives the prompt, Radius:, for specifying the size of the next arc. The next prompt, Angle/Length/<End point>:, lets you specify the included angle or the arc's chordal length. SECOND PT gives two prompts, Second point: and Endpoint:, for selecting points on a three-point arc.

PEDIT

To modify polylines, use the PEDIT command as follows:

Select polyline: (Select line with cursor.)

Entity selected is not a polyline

Do you want it turned into one? <u>Y</u>

Close/Join/Width/Edit vertex/Fit/
Spline/Decurve/Ltype gen/Undo
/exit<X>: <u>CLOSE</u> (Used to close PLINE.)

If the PLINE is already closed, the OPEN option replaces the CLOSE command.

Join The JOIN option gives the prompt, Select objects Window or Last:, for selecting segments to join as a polyline. Segments must have exact meeting points for joining to occur.

Width gives the prompt, Enter new width for all segments:, for assigning a new width to a PLINE. FIT (F) converts the polyline into a line composed of circular arcs passing through each vertex **(Fig. 36.92)**. You may use the EDIT VERTEX command discussed shortly to change the resulting curve.

SPLINE (S) modifies the polyline the same as FIT CURVE did. However, it draws connecting cubic curves passing through the first and last points, but not necessarily through the other points **(Fig. 36.93)**. SPLINE is good for drawing data curves on graphs. DECURVE (D) converts FIT or

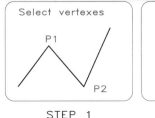

◾ Figure 36.94 PEDIT—break:
Step 1 Command: PEDIT (CR)

Select polyline: (Select line.)

Close/Join/Width/Edit vertex/Fit/

Spline/Decurve/Ltype gen/Undo/eXit <X>: E

Next/Previous/Break/Insert/Move/Regen/
Straighten/

Tangent/Width/eXit<N>: Next to P1 (CR)

Next/Previous/. . ./Width/eXit<N>: Break (CR)

Next/Previous/Go/eXit <N>: Next to P2

Step 2 Next/Previous/Go/eXit <N>: Go (CR)
(The polyline between P1 and P2 is erased.)

SPLINE curves to their original straight-line forms.

LTYPE GEN (L) applies broken lines (such as hidden lines) in a continuous pattern around vertices. Without application of this option, the broken lines you produce may ignore gaps in splined and fit curves. By setting the system variable PLINEGEN to On, you generate LINETYPE as PLINES are drawn.

UNDO (U) reverses the most recent PEDIT editing step. EDIT VERTEX (E) selects vertexes of the PLINE for editing by marking the first vertex with an X when you pick the polyline. An arrow and the following prompt will appear if you specified a tangent direction for the vertex:

Next/Previous/Break/Insert/
Move/Regen/Straighten/Tangent/
Width/eXit/<N>: Next (CR)

You move the X marker to the next or previous vertexes with the NEXT (N) and PREVIOUS (P)

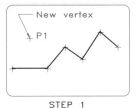

◾ Figure 36.95 PEDIT—add vertex:
Step 1 Use EDIT VERTEX option and place X on vertex before the new vertex.

Close/Join/Width/Edit vertex/Fit/Spline/

Decurve/Ltype gen/Undo/eXit <X>: Insert (CR)

Enter location of new vertex: P1

Step 2 Press (CR), the new vertex is inserted, and the polyline passes through it.

options by pressing (CR). BREAK (B) gives the prompts in **Fig. 36.94**. Move the X marker to a selected vertex, use NEXT or PREVIOUS to move to a second point and pick GO to erase and the line between the vertexes. Select EXIT to leave the BREAK command and return to EDIT VERTEX. INSERT adds a new vertex to the polyline between a selected vertex and the next vertex **(Fig. 36.95)**. MOVE (M) relocates a selected vertex **(Fig. 36.96)**.

STRAIGHTEN (S) converts the polyline between two selected points into a straight line from the current X-marker position on the line when you select S. The prompt, Next/Previous/ Go/eXit/<N>, is obtained. To straighten the line between the vertices move the X marker to a new vertex and select GO **(Fig. 36.97)**. Enter X to eXit and return to the EDIT VERTEX prompt.

TANGENT (T) allows you to select a tangent direction at the vertex marked by the X for curve fitting by responding to the prompt Direction of tangent. Enter the angle from the keyboard, or with the cursor. WIDTH (W) lets you set the beginning and ending widths of an existing line segment from the X-marked vertex. Use the NEXT and PREVIOUS options to confirm which direction

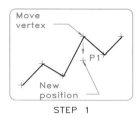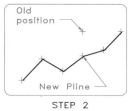

Figure 36.96 PEDIT—move vertex:
Step 1 Command: PEDIT (CR)

Select polyline: (Select line.)

Close/Join/Width/Edit vertex/Fit/

Spline/Decurve/Ltype gen/Undo/eXit <X>: E

((CR) X to vertex to be moved.)

Next/Previous/Break/Insert/Move/

Regen/Straighten/Tangent/Width/eXit <N>:
Move or M (CR)

Enter location of new vertex: P1

Step 2 Press (CR), and the polyline will be changed to pass through the moved vertex.

the line will be drawn from the X marker. The polyline will change to its new thickness when you regenerate the screen with REGEN (R). Use Exit to exit from the PEDIT command.

36.11 Grips

Grips are small squares that appear on entities at midpoints and ends of a line and at the insertion point of text. You may select grips to perform several modifications: STRETCH, MOVE, ROTATE, SCALE, and MIRROR. The GRIPS dialogue box (DDGRIPS) appears under SETTINGS of the pull-down menu **(Fig. 36.98)**.

Enable Grips is a check box that turns on grips for all objects. Enable Grips Within Blocks turns on grips for entities within a BLOCK; when turned off, it gives a single grip at the insertion point.

Grip Colors assigns colors to selected and unselected grips. Unselected grips are not filled

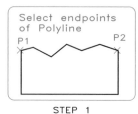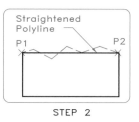

Figure 36.97 PEDIT—straighten line:
Step 1 Use EDIT VERTEX option of PEDIT, and place the X at the vertex where straightening is to begin, P1.

Next/Previous/Break/Insert/Move/Regen/
Straighten/

Tangent/Width/eXit<N>: Straighten (CR)

Step 2 Next/Previous/Go/eXit <N>: Next (CR)
to P2

Next/Previous/Go/eXit <N>: Go (CR) (Line P1–P2 is straightened.)

GRIPS DIALOGUE BOX (DDGRIPS)

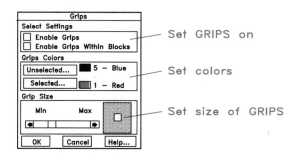

Figure 36.98 The Grips dialogue box is under Settings of the pull-down menu. Use it to enable or disable grips, assign grip colors, and set grip sizes.

in. When you select this option, the Color Dialogue box is displayed for your use in selecting colors. To vary the size of grip boxes, use Grip Size to move the Slider box.

Using Grips When you select entities with the cursor, grips will appear on them as open squares. When you pick a grip and make it a "hot" point, it

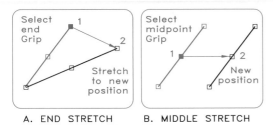

A. END STRETCH   B. MIDDLE STRETCH

Figure 36.99 Stretching with `Grips`:
A Select the line and grips appear; select the endpoint to be stretched (P1) and move to the new position (P2).

B Select the midpoint grip (P1) pick a second point (P2); the line is moved to it.

GRIPS: MOVE OPTION

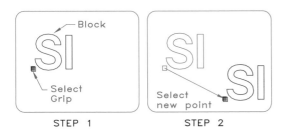

STEP 1   STEP 2

Figure 36.100 Moving text with `Grips`:
Step 1 Select the text and a grip appears at the insertion point. Pick the grip.

Step 2 Move the cursor to a new position, which moves the text.

will be filled with color. By holding down `SHIFT`, you may pick more than one grip as "hot" points, but you must select the last grip of a series without pressing `SHIFT`. To remove the grips, press `Ctrl C` twice.

By turning the grips on and by successively pressing (CR), you sequentially activate the options `STRETCH`, `MOVE`, `ROTATE`, `SCALE`, and `MOVE`, each with its own subcommands.

`STRETCH` lets you select the endpoint grip of a line as a "hot" point, and a second point can be picked as the new end of the line (**Fig. 36.99**). To

GRIPS: ROTATE OPTION

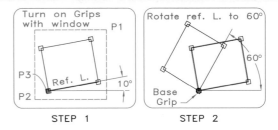

STEP 1   STEP 2

Figure 36.101 Rotating with `Grips`:
Step 1 (Turn on grips by windowing; select P3 as the base; press (CR) until `**ROTATE**` appears.)

`<Rotation angle>/Base point/Copy/Undo/ Reference/eXit:` R (CR)

`Reference angle <0>:` 10 (CR)

Step 2 `**ROTATE**`

`<New angle>/Base point/Copy/Undo/Reference/ eXit:` 60 (CR)

(Object is rotated 50° from first position.)

move the line to a new position, select the midpoint grip.

`MOVE` relocates objects without changing their orientation or size. To move the `BLOCK` shown in **Fig. 36.100**, relocate its grip (its insertion point) icon. Use the options `Base point`, `Copy`, `Undo`, and `eXit` for their obvious purposes.

`ROTATE` revolves an object about a selected grip. The `Reference` option rotates an object about a selected grip to its rotation angle, by dragging or when you insert a number. To rotate the object shown in **Fig. 36.101** 60° from the reference line simply requires typing 60.

`SCALE` sizes objects from the grip selected as a base point (**Fig. 36.102**). You may assign the scale factor by typing, dragging, or selecting a reference dimension and giving it a new dimension.

`MIRROR` makes a mirror image of an object when you select two grips as the mirror line (**Fig. 36.103**). By holding down `SHIFT` while selecting the second grip on the mirror line, you will not remove the initial drawing.

GRIPS: SCALE OPTION

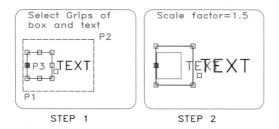

Figure 36.102 Scaling with Grips:
Step 1 (Turn on Grips by windowing; select P3 as the base; press (CR) until **SCALE** appears.)
Step 2 <Scale factor>/Base point/Copy/Undo/
Reference/eXit: 1.5 (CR)
(Objects are scaled.)

GRIPS: MIRROR OPTION

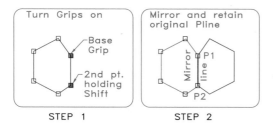

Figure 36.103 Mirroring with Grips:
Step 1 (Turn on Grips by windowing; select P1 as the base; press (CR) until **MIRROR** appears.)
Step 2 **MIRROR**
<Second point>/Base point/Copy/Undo/eXit:
P2 (While holding Shift) (CR) (Object is mirrored and original is retained.)

36.12 Hatching

Recall that hatching is a pattern of lines drawn to fill sectioned areas, to accent areas in graphs, and for similar purposes. The Boundary Hatch (BHATCH) dialogue box appears under the DRAW heading of the pull-down menu **(Fig. 36.104)**.

HATCH OPTIONS gives a subdialogue box **(Fig. 36.105)**. Use Pattern to obtain a display of pat-

BOUNDARY HATCH (BHATCH)

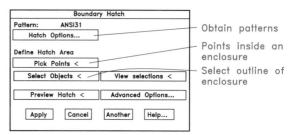

Figure 36.104 Use the Boundary Hatch (BHATCH) dialogue box under the Draw heading of the pull-down menu to make settings for hatching.

HATCH OPTIONS BOX

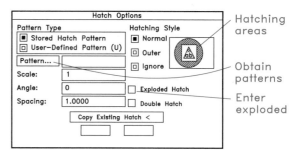

Figure 36.105 The Hatch Options dialogue box gives access to hatch patterns (Fig. 36.104) or sets user-defined patterns. Styles of hatching are Normal, Outer, or Ignore, as shown in the circular icon.

terns for selection **(Fig. 36.106)**. Then use the radio box, Stored Hatch Pattern (Fig. 36.105), and select Pattern... to obtain the various patterns to select from. After you have selected a pattern and pressed OK, the Hatch Options dialogue box returns to the screen (Fig. 36.105). Use Scale to set the spacing between the lines of a pattern and Angle to assign the angle of the hatch pattern. When the User-Defined Pattern radio box is selected, you are prompted for Angle and Spacing for equally spaced hatch lines.

You may choose Hatching Style as Normal, Outer, or Ignore from the dialogue box (Fig. 36.105), which shows a circle with a square inside it to illustrate the effect of each choice. Normal

CHOOSE HATCH PATTERN BOX

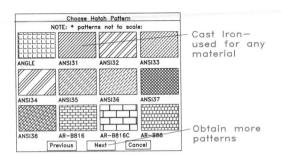

Figure 36.106 The first screen of hatch patterns from which you may make selections is shown here.

HATCHING AN AREA

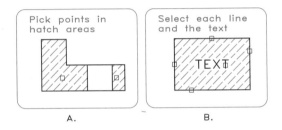

Figure 36.107 Hatching an area:

A The `Pick Points` option of hatching prompts for points inside of boundaries for hatching.

B The `Select Object` option of hatching requires that you select the boundary lines of an enclosure.

hatches every other nested area beginning with the outside. `Outer` hatches the outside area. `Ignore` hatches the entire area from the outer boundary. Text (if selected) will not be hatched, but a window within the hatching will appear around it.

`Exploded Hatch` inserts a hatch pattern as a group of separate entities as if it were inserted as *NAME (*ANSI31, for example). `Copy Existing Hatch` allows you to select a hatch pattern from an existing pattern drawn on the screen and apply it to the next area to be hatched. Use `Pick Points` (Fig. 36.104) to select areas inside boundaries to be hatched; you may pick multiple points. `Select`

BOUNDARY DEFINITION ERRORS

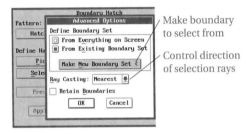

A. OUTSIDE   **B. BOUNDARY OPEN**

Figure 36.108 Boundary definition errors:

A If you select a point outside an enclosed area, this message will appear.

B If the boundary you select is not closed perfectly, this message will appear.

ADVANCED OPTIONS

Figure 36.109 The `Advanced Option` dialogue box lets you assign boundary areas and ray castings for hatching, as shown in Fig. 36.110.

`Objects` requires that you choose the boundaries of the area to be hatched.

Hatching by picking points and selecting objects is illustrated in **Fig. 36.107.** When you select points outside the boundary or if the boundary is not closed, error messages (**Fig. 36.108**) will indicate the cause of the error.

`View Selections` shows all the boundaries and selected entities. `Preview Hatch` shows the result of your hatch selections before you apply them. You may make changes in hatch and preview them until you are satisfied. `Apply` generates hatching as a part of the drawing. `Another` allows multiple hatching operations to be performed.

`ADVANCED OPTIONS` gives the dialogue box shown in **Fig. 36.109** with options for making hatching more efficient. `From everything on`

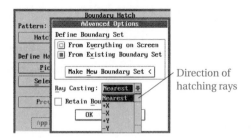

Direction of hatching rays

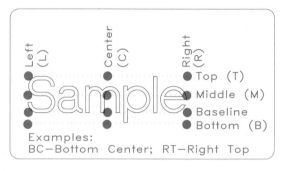

Figure 36.110 To set the direction of ray castings (the direction that hatching rays will be cast), use this action box.

Figure 36.111 You may add text to a drawing by using any of the insertion points shown. For example, BC means the bottom center of a word or sentence that will be located at the cursor point.

the `Screen` requires AutoCAD to look at all entities on the screen during hatching. Specifying a boundary increases speed.

`Make New Boundary Set` returns the drawing to the screen so that you may select entities on the drawing as the boundary set. Hatching will be performed only in this area of the drawing. Select `From Existing Boundary Set` after making a new boundary set, which becomes the default until `BHATCH` is completed. `Retain Boundaries` keeps the defined boundary as polyline in your drawing.

`Ray Casting` is the direction of rays cast to seek a hatching boundary when you use `Pick Points` to select a point within boundaries (**Fig. 36.110**). `Nearest`, the default, casts a ray to the nearest hatchable entity and turns left to trace the boundary. When you want to be sure that you have picked the correct boundary, use the `+X`, `-X`, `+Y`, or `-Y` options to define the ray's direction as it searches for the entity to be hatched.

We covered `HATCH`, the command used before `BHATCH` was available, in Chapter 16. You may use the `BPOLY` command to form a polyline by selecting loosely connected boundary entities, making this procedure less rigorous to apply than the `HATCH` command.

36.13 Text and Numerals

Use the `DTEXT` (dynamic text) and `TEXT` commands to insert text in a drawing. `DTEXT` displays

the text on the screen as you type it, and the `TEXT` command shows the text after you have finished typing a line; therefore `DTEXT` is used almost exclusively. `DTEXT` begins by asking for the left starting point (the default) for the text line:

```
Command: DTEXT
Justify/Style/<Start point>:
```
(Select point with cursor.)
```
Height <.18>: .125
Rotation angle <0>: (CR)
Text: TYPE WORDS
```

`JUSTIFY` prompts you for the insertion point for a string of text (**Fig. 36.111**). The abbreviation `BC` means bottom center, `RT` means right top, and so on (**Fig. 36.112**).

Figure 36.113 shows how to type multiple lines of text with `DTEXT`. It automatically spaces the lines when you press (CR). You may insert the special characters shown in **Fig. 36.114** by typing a double percent sign, `%%`, in front of them.

To reduce screen regeneration time, draw the text as boxes by turning `ON` the `QTEXT` command (**Fig. 36.115**). When you turn `OFF` `QTEXT`, the full text will be redrawn after regeneration (`REGEN`).

DTEXT: JUSTIFY OPTIONS

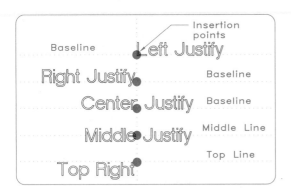

Figure 36.112 The JUSTIFY option allows you to insert text aligned with the insertion points, as shown.

DTEXT: MULTIPLE LINES

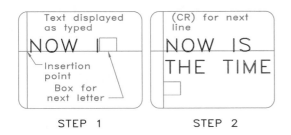

STEP 1   STEP 2

Figure 36.113 Inserting DTEXT:

Step 1 Command: DTEXT

Justify/Style/<Start point>: (Select point.)

Height <.18>: .125

Rotation angle <0>: (CR)

Text: NOW IS

Step 2 Press (CR) and box advances to next line of Text, THE TIME (CR) (Box spaces down with each (CR).)

The STYLE Command

Figure 36.116 shows many of AutoCAD's text fonts and their names. The default style is STANDARD, which uses the TXT font. To obtain the fonts shown in Fig. 36.116, first get the STYLE command by selecting Draw from the pull-down menu; then choose Style and Set Style. After you have

DTEXT: SPECIAL CHARACTERS

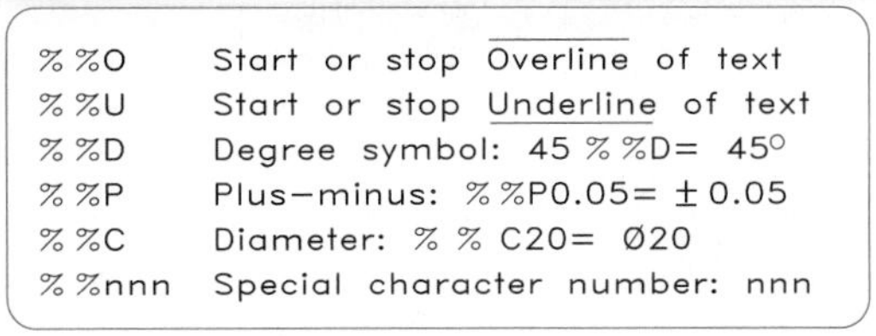

Figure 36.114 These special characters beginning with %% are used to obtain these symbols while typing DTEXT and TEXT.

QTEXT COMMAND

QTEXT OFF   QTEXT ON

⌀24 CBORE
16 DEEP

After REGEN

STEP 1   STEP 2

Figure 36.115 Using the QTEXT command:

Step 1 Command: QTEXT (CR) ON/OFF <CURRENT>: ON (CR)

Step 2 Command: REGEN (CR) (Text is shown as boxes.)

selected a font, the following prompts appear on the screen:

Command: STYLE (CR)

Text style name (or?)<current>: PRETTY (CR)

Font file <TXT>: COMPLEX (CR)

Height <0.20>: 0 (CR)

Width factor <default>: 1.25 (CR)

Obliquing angle <45>: 0 (CR)

Backwards? <Y/N>: N (CR)

Upside-down? <Y/N>: N (CR)

Vertical? <Y/N>: N (CR)

PRETTY is now the current text style.

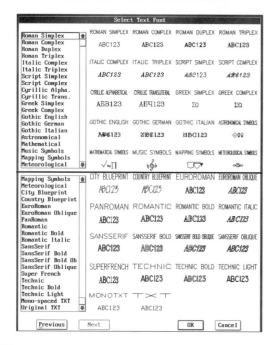

Figure 36.116 These fonts can be accessed under DRAW: TEXT: set style.

The style named PRETTY will retain these settings until you change them. If you respond to the first prompt with ?, you get a list of the defined text STYLES. To obtain examples of fonts that you may select by cursor for a new style, select OPTIONS from the pull-down menu, then the DTEXT option, and finally Text Font.

If you later create a STYLE called PRETTY and select a different font, such as ROMAN, the text previously produced with the STYLE called PRETTY (which used the COMPLEX font) will be redone with the ROMAN font. This technique changes the TXT and MONOTXT fonts to more attractive fonts at plot time. Meanwhile, you have saved time by using TXT and MONOTXT fonts because they regenerate quickly.

The VERTICAL font draws vertical lines of text with letters stacked one under the other. Use a rotation angle of 270° with this font. To select a line of text from the screen and display it in a dialogue box for editing, use the DDEDIT command

Figure 36.117 The Edit Text box (DDEDIT) appears under the heading of the EDIT side menu. By activating this command, you may select lines of text for modification.

MIRROR COMMAND

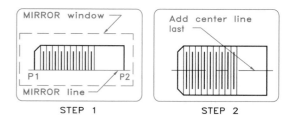

Figure 36.118 Using the MIRROR command:
Step 1 Command: MIRROR (CR)
Select objects: W (CR) (Window drawing to be mirrored.)
First point or mirror line: P1
Second point: P2
Step 2 Delete old objects? <N>: N (CR) (The object is mirrored about the mirror line.)

(Fig. 36.117). After making corrections, press the OK button to revise the text on the screen.

36.14 Mirroring

Objects

You may draw symmetrical objects as partial figures and then mirror them about an axis with the MIRROR command **(Fig. 36.118)**. If a line such as P1–P2 coincides with the MIRROR line, the line will be drawn twice when mirrored. Therefore you should draw parting lines of this type after the view has been mirrored.

MIRRTEXT SETVAR

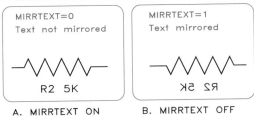

Figure 36.119 Obtaining mirrored text:

A When you give the variable MIRRTEXT of the SETVAR command a value of 0, the text will not be mirrored.

B When you give MIRRTEXT a value of 1, the text will be mirrored.

OSNAP FROM THE PULL-DOWN MENU

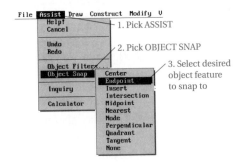

Figure 36.120 The ASSIST column under the pull-down menu gives you access to OSNAP options for picking entities on the screen.

Text

MIRRTEXT is a system variable under the SETVAR command for mirroring text. By setting MIRRTEXT to 0 (off), the text will not be mirrored; if set to 1 (on) the text, along with the drawing, will be mirrored **(Fig. 36.119)**.

36.15 Object Snap

By using Object Snap (OSNAP), you may snap to object features rather than to a grid. Activate OSNAP by selecting the ASSIST box of the pull-down menu to obtain the options in **Fig. 36.120**.

OSNAP: ENDPOINT OPTION

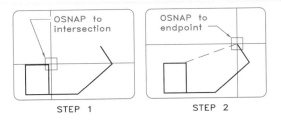

Figure 36.121 OSNAP option—intersection and endpoint:

Step 1 Command: LINE (CR)

From point: (Select the stars in the side menu; select INTERSEC.)

INTERSEC of: (Select intersection.)

Step 2 To point: (Select ENDPOINT option of OSNAP.)

To point: ENDPOINT of (Select endpoint of line and the line is drawn.)

OSNAP: TANGENT OPTION

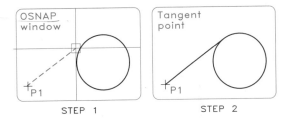

Figure 36.122 OSNAP tangent—point to a circle:

Step 1 Command: LINE (CR)

From point: P1

To point: (Select OSNAP from the menu, and select TANGENT.)

Step 2 To point: P2 (The line is drawn from P1 tangent to the circle near the selected point.)

Or you may use the line of stars, *****, in the side menu to activate OSNAP. **Figure 36.121** shows how to produce a line from an intersection to the endpoint of a line. OSNAP is not a primary command, but it is an accessory to other commands, such as LINE, MOVE, and BREAK.

Figure 36.122 shows a line drawn from P1 tangent to the circle by using TANGENT to select a

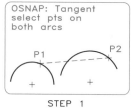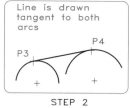

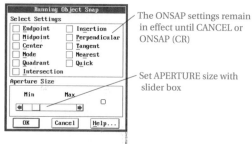

Figure 36.123 OSNAP—tangent to two arcs:
Step 1 Command: <u>LINE</u> (CR)

From point: (Select the stars in the side menu;
select the TAN option.)

TANGENT to: <u>P1</u>

Step 2 To point: (Select TAN option of OSNAP.)

TANGENT to: <u>P2</u> (The line is drawn.)

point on the circle near the tangent point. The program automatically locates the tangent point and produces the line. **Figure 36.123** shows how to use the TANGENT option of OSNAP to produce a line tangent to two arcs and trim the tangent line to the tangent points.

The options of OSNAP are NEAREST, ENDPOINT, MIDPOINT, CENTER, NODE, QUADRANT, INTERSECTION, INSERT, PERPENDICULAR, TANGENT, QUICK, and NONE. The NODE option snaps to a POINT, the QUADRANT option snaps to one of the four compass points on a circle, and the INSERT option snaps to the insertion point of a BLOCK. The QUICK option reduces searching time by selecting the first object encountered within the target area rather than searching for the one closest to the target's center. The NONE option turns off OSNAP for the next selection.

You may lock in OSNAP settings as running OSNAPS when you will be using them many times. For example, to snap repetitively to ENDPOINTS and circle CENTERS, do the following:

Command: <u>OSNAP</u> (CR)

Object snap modes: <u>ENDPOINT, CENTER</u> (CR)

Figure 36.124 You may set a Running OSNAP with this dialogue box from SETTINGS of the pull-down menu.

Now your cursor has an aperture target at its intersection for picking endpoints and centers of arcs and circles. Remove this OSNAP setting in the following manner:

Command: <u>OSNAP</u> (CR)

Object snap modes: (**Select** NONE or OFF, or press (CR)

To set running OSNAPS, you may also use the dialogue box under the Settings/Object Snap headings of the pull-down menu (**Fig. 36.124**).

The APERTURE command assigns a size to the target box that appears at the cursor for OSNAP applications. Its size can vary from 1 to 50 pixels square.

36.16 Miscellaneous Commands

ARRAY

You may generate circular or rectangular patterns (rows and columns) of selected objects with the ARRAY command. For example, **Fig. 36.125** shows how to locate a series of holes on a bolt circle by drawing the first hole and using ARRAY to transform it into a polar array.

To produce a rectangular ARRAY, begin by making the drawing in the lower left corner and following the steps shown in **Fig. 36.126**. You may terminate an array in progress with Ctrl C.

ARRAY: POLAR OPTION

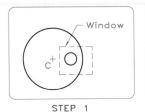 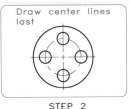

Figure 36.125 ARRAY—polar:
Step 1 Drawing the figure to array.

`Command:` `ARRAY` (CR)

`Select objects:` `W` (Window the hole.) (CR)

`Select objects:` (CR)

`Rectangular or Polar array (R/P):` `P` (CR)

`Center point of array:` `C` (CR)

Step 2 `Number of items:` `4`

`Angle to fill (+=ccw, -=cw)<360>:` `360` (CR)

`Rotate objects as they are copied? <Y>` (CR)

(The array is drawn.)

ARRAY: RECTANGULAR OPTION

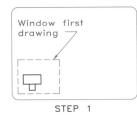 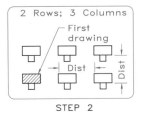

Figure 36.126 ARRAY—rectangular:
Step 1 Draw desk in lower left of ARRAY.

`Command:` `ARRAY` (CR)

`Select Objects:` `W` (Window desk.) (CR)

`Rectangular or Circular array (R/P):` `R`

Step 2 `Number of rows (---) <1>:` `2` (CR)

`Number of columns (<vb><vb><vb>) <1>:` `3` (CR)

`Unit cell distance between rows (---):` `4` (CR)

`Distance between columns (<vb><vb><vb>):` `3.5` (CR)

(CR) (The array is drawn.)

ARRAY: ANGULAR OPTION

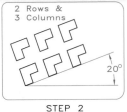

Figure 36.127 ARRAY—at an angle:
Step 1 Rotate the grid at the desired angle using the SNAP command. Draw the object in the lower left corner of the array.

`Command:` `ARRAY` (CR)

`Select objects.` (Window the object.)

`Rectangular or Polar array <R/P>:` `R` (CR)

Step 2 `Number of rows(---) <1>:` `2` (CR)

`Number of columns (<vb><vb><vb>) <1>:` `3` (CR)

`Unit cell or distance between rows (---):` `2` (CR)

`Distance between columns (<vb><vb><vb>):` `3` (CR)

You may produce rectangular arrays at angles by rotating the SNAP mode (**Fig. 36.127**). Draw the first object in the lower left corner of the array and specify the number of rows and columns and the cell distances when prompted by the ARRAY command.

DONUT

To draw a circle with an opening at its center, use the DONUT command by assigning outside and inside diameters (**Fig. 36.128**). However, if you set the inside diameter to 0, DONUT will give a solid circle.

SCALE

The SCALE command reduces or enlarges previously drawn objects. The desk shown in **Fig. 36.129** was enlarged by windowing it, selecting a base point, and typing a scale factor of `1.6`. Both the labels and the objects were enlarged in the X and Y directions.

DONUT COMMAND

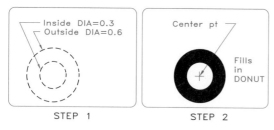

Figure 36.128 Using the DONUT command:
Step 1 Command: <u>DONUT</u> (CR)

Inside diameter <0.0000>: <u>.30</u> (CR)

Outside diameter <0.0000>: <u>.60</u> (CR)

Step 2 Center of donut: (Select point, and donut is drawn.)

SCALING: BY NUMBER

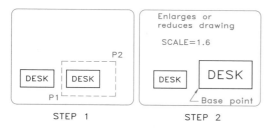

Figure 36.129 Using the SCALE command:
Step 1 Command: <u>SCALE</u> (CR)

Select objects: Window with <u>P1</u> and <u>P2</u>

Step 2 Base point: (Select base point.)

<Scale factor>/Reference: <u>1.6</u> (CR)

(The desk is drawn 60% larger.)

A second option of the SCALE command lets you select the length of an object and then specify its new length as a ratio of the first dimension (**Fig. 36.130**). You may enter the lengths numerically with the cursor or the keyboard.

STRETCH

The STRETCH command expands or contracts a portion of a drawing, leaving one of its ends sta-

SCALING: BY REFERENCE

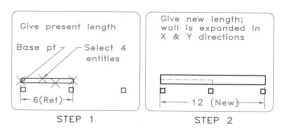

Figure 36.130 Scaling—reference dimensions:
Step 1 Command: <u>SCALE</u> (CR)

Select objects: (Select points on each line.)

Base point: (Select point.)

<Scale factor>/Reference: <u>R</u> (CR)

Reference length <1>: <u>6</u> (CR)

Step 2 New length: <u>12</u> (CR)

(The drawing is enlarged in all directions.)

STRETCH COMMAND

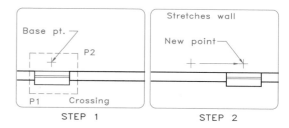

Figure 36.131 Using the STRETCH command:
Step 1 Command: <u>STRETCH</u> (CR)

Select objects: <u>C</u> (Crossing window) (CR)

First corner: <u>P1</u>

Other corner: <u>P2</u>

Select objects: (CR)

Base point: (Select base point.)

Step 2 New point: (Select new point.)

(The symbol is repositioned.)

tionary. The window symbol in the floor plan in **Fig. 36.131** was STRETCHed to a new position, leaving the lines of the wall unchanged. You must use a CROSSING window to select lines for stretching.

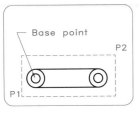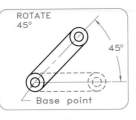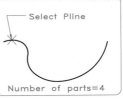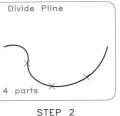

STEP 1   STEP 2   STEP 1   STEP 2

Figure 36.132 Using the ROTATE command:
Step 1 Command: <u>ROTATE</u> (CR)
Select objects: <u>W</u> (CR) <u>P1</u> and <u>P2</u>
Step 2 Base point: <u>P3</u>
<Rotation angle>/Reference: <u>45</u> (CR)
(Object is rotated 45° ccw.)

Figure 36.133 DIVIDE—a line:
Step 1 Command: <u>DIVIDE</u> (CR)
Select object to divide: (Select PLINE.)
Step 2 <Number of segments>/Block: <u>4</u> (CR)
(PDMODE symbols are placed along the line, dividing it.)

ROTATE

To rotate an object about a base point, use the ROTATE command (**Fig. 36.132**). Window the object, select a base point, and type the rotation angle at the keyboard or select it with the mouse. You may also rotate drawings made on multiple layers.

SETVAR

You may inspect many modes, sizes, limits, and variables in AutoCAD's memory under the SETVAR command by typing SETVAR and ?. The SETVAR command can change these variables if they are not read-only commands. To change one or more variables (TEXTSIZE, for example) respond as follows:

    Command: <u>SETVAR</u> (CR)

    Variable name or ?: <u>TEXTSIZE</u> (CR)

    New value for TEXTSIZE <0.18>: <u>0.125</u> (CR)

To change the default value given in brackets, simply type a new value, say 0.125. By entering the SETVAR command with an apostrophe in front of it('SETVAR), you may use it transparently while another command is in progress.

DIVIDE

The DIVIDE command places markers on an entity that divides it into a specified number of equal segments. **Figure 36.133** depicts an entity selected with the cursor to be divided into four segments. The markers are equally spaced along the line. They are POINTS the type and size currently set for the PDMODE and PDSIZE variables under the SETVAR command.

The BLOCK option of the DIVIDE command allows you to use previously saved blocks (rectangles in this example) as markers on the line (**Fig. 36.134**). The blocks may be either ALIGNED or NOT ALIGNED.

MEASURE

You may use the MEASURE command to place markers along an arc, circle, polyline, or line at specified distances apart (**Fig. 36.135**). To respond to the Select object to measure prompt, which prompts for the segment length, pick a point on the entity near its end where measuring is to begin. The program places POINT markers along the line (or entity) beginning with the selected end. The last segment is usually a shorter length.

DIVIDE: BLOCK MARKER

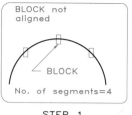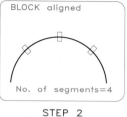

STEP 1   STEP 2

◈ **Figure 36.134** `DIVIDE`—an arc:
Step 1 `Command: ` <u>`DIVIDE`</u> (CR)

`Select object to divide:` (Select arc.)

`<Number of segments>/Block:` <u>`B`</u> (CR)

`Block name to insert:` <u>`RECT`</u> (CR)

`Align block with object? <Y>` <u>`N`</u> (CR)

`Number of segments:` <u>`4`</u> (CR)

Step 2 `Align block with object? <Y>` (CR)
(The blocks radiate from the arc's center.)

MEASURE COMMAND

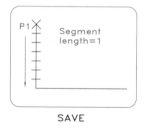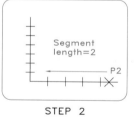

SAVE   STEP 2

◈ **Figure 36.135** Using the `MEASURE` command:
Step 1 `Command:` <u>`MEASURE`</u> (CR)

`Select object to measure:` <u>`P1`</u>

`<Segment length>/Block:` <u>`1`</u> (CR)

(Six divisions are measured along the line starting at the end nearest P1.)

Step 2 `Command:` <u>`MEASURE`</u> (CR)

`Select object to measure:` <u>`P1`</u>

`<Segment length>/Block:` <u>`2`</u> (CR)

(Four equal divisions are measured along the line.)

OFFSET COMMAND

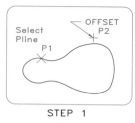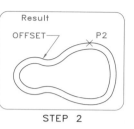

STEP 1   STEP 2

◈ **Figure 36.136** Using the `OFFSET` command:
Step 1 `Command:` <u>`OFFSET`</u> (CR)

`Offset distance or Through <Through>:` <u>`T`</u> (CR)

`Select object to offset:` <u>`P1`</u>

`Through point:` <u>`P2`</u>

Step 2 An enlarged `PLINE` is drawn that passes through P2.

OFFSET

You may produce an entity parallel to and offset from another entity, such as the polyline, with the `OFFSET` command **(Fig. 36.136)**. `OFFSET` prompts you for the offset distance or the point through which the offset line must pass and then prompts you to select the entity and the side of the offset.

The `OFFSET` command is helpful in drawing parallel lines to represent walls of a floor plan **(Fig. 36.137)**. You may set the offset distance precisely with the keyboard, repetitively select lines, and pick the offset side with the cursor. To trim the corners of the walls to perfect intersections, set a radius of 0 to the `FILLET` command. For `OFFSET` to work in a 3D drawing, the selected entities must be parallel to the current UCS.

36.17 BLOCKS

One of the more productive features of computer graphics is the capability of building a file of drawings called `BLOCKS` for repetitive use. Objects such as the SI symbol are drawn in the conventional manner, made into `BLOCKS`, and inserted into drawings as needed **(Fig. 36.138)**. Select the

OFFSET: PARALLEL LINES

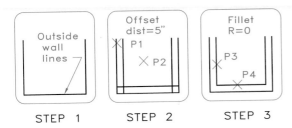

STEP 1   STEP 2   STEP 3

◈ **Figure 36.137** OFFSET—parallel lines:
Step 1 Command: <u>OFFSET</u> (CR)

Offset distance or Through <last>: <u>5</u>

Step 2 Select object to offset: <u>P1</u>

Side to offset: <u>P2</u>

Step 3 Use FILLET (R=0) to trim corners by selecting P3 and P4. The TRIM command also may be used.

BLOCK AND INSERT COMMANDS

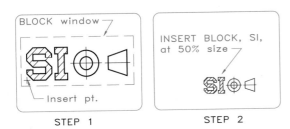

STEP 1   STEP 2

◈ **Figure 36.138** Using the BLOCK command:
Step 1 Make a drawing to be BLOCKed, and respond to the BLOCK command.

BLOCK name (or ?): <u>SI</u> (CR)

Insertion base point: (Select insertion point.)

Select objects: <u>WINDOW</u> or W (CR)

(Window the drawing, and it disappears into memory.)

Step 2 Command: <u>INSERT</u> (CR)

Block name (or ?): <u>SI</u> (CR)

Insertion point: (Select point with cursor.)

X-scale factor <1>/Corner/XYZ: <u>0.5</u> (CR)

Y-scale factor <default=X>: (CR)

Rotation angle <0>: (CR)

(Block SI is inserted at 50% size.)

INSERT DIALOGUE BOX (DDINSERT)

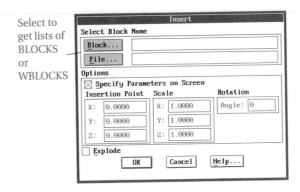

◈ **Figure 36.139** Use the Insert dialogue box (DDINSERT) to insert Blocks and Wblocks into a drawing exploded or unexploded. Selecting Block or File yields a list of the Block or Files.

SELECT A DRAWING FILE

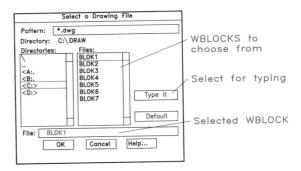

◈ **Figure 36.140** Use the list of WBLOCKS given to choose one for insertion.

INSERT command, which appears under the DRAW portion of the pull-down menu **(Fig. 36.139)**, to obtain a dialogue box for inserting both blocks and files (BLOCK and WBLOCK). When you select Block the Select a Drawing File dialogue box appears for selection of a BLOCK **(Fig. 36.140)**.

You cannot erase a portion of a BLOCK; you must erase the entire BLOCK. However, you may edit BLOCK if it is inserted with a star typed in front of its name—for example, *SI (CR). Moreover, the dialogue box (Fig. 36.139) contains a check box for

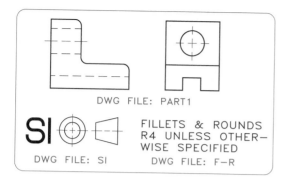

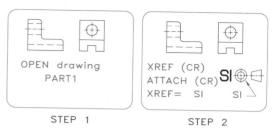

Figure 36.141 We utilize these three drawing files to illustrate how to use `External References` (XREFs) to produce a single drawing.

Figure 36.142 Producing a drawing with one XREF:
Step 1 OPEN the drawing, PART1.
Step 2 Command; <u>XREF</u> (CR)
?/Bind/Detach/Path/Reload/<Attach>: <u>A</u> (CR)
Xref to Attach <Default>: <u>SI</u> (CR)
Attach Xref SI: <u>SI</u> (SI is loaded.)
Insertion point: (Select.)
X scale factor <1>: (CR)
Y scale factor <1>: (CR)
Rotation angle <0>: (CR) (SI symbol is inserted.)

inserting an exploded `BLOCK` into a drawing. The `EXPLODE` command separates an inserted `BLOCK` into individual entities that you may erase one at a time by typing `EXPLODE` and then selecting the `BLOCK`.

You may use a `BLOCK` only on the current drawing file unless you convert it to a `WBLOCK` (`Write Block`), which becomes an independent file (not part of a file). Perform this conversion as follows:

Command: <u>WBLOCK</u> (CR)

File Name: <u>B:SI</u> (CR) (This assigns the name of the `WBLOCK` to drive B.)

Block Name: <u>SI</u> (CR) (This is the name of the `BLOCK` that is being changed to a `WBLOCK`.)

To redefine a `BLOCK` select a previously used `BLOCK` name to receive the prompt, `Redefine it? <N>:`. Type `Y` for "yes" and select the new drawing to be blocked. The redefined `BLOCK` is updated automatically by replacing it with the new `BLOCK`.

36.18 External References

`External References` are existing drawing files that are temporarily attached to a current file to form a combination drawing. When the drawing session ends, the XREF is discarded, leaving only the name of the XREF and its path as part of the drawing. If the XREF is updated, the latest version will be attached to the current drawing.

Figure 36.141 shows three drawings: a two-view drawing of a part, an SI symbol, and a fillet and round note. To combine the three drawings, load the PART1 file **(Fig. 36.142)**, and use the XREF command and the ATTACH option to place the SI file. Use the ATTACH option of the XREF command again to place the F-R file.

By using the ? option of the XREF command, you may obtain a list of the names and paths of the XREFs **(Fig. 36.143)**. Use the ? option of the BLOCK command to obtain a list of the XREFs that were attached to the current file **(Fig. 36.144)**.

When inserted in the current drawing, the XREFs will accept the settings of the current drawing if there is a conflict in assignment of colors to layers with the same name. **Figure 36.145** displays how AutoCAD managed a named object definition conflict by typing the command LAYER and ?.

ATTACHING TWO XREF

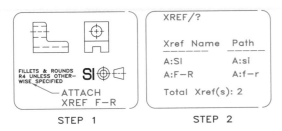

```
XREF/?

Xref Name    Path

A:SI         A:si
A:F-R        A:f-r

Total Xref(s): 2
```

STEP 1   STEP 2

Figure 36.143 Producing a drawing with two XREFs:
Step 1 Command: <u>XREF</u>

?/Bind/Detach/Path/Reload/<Attach>: <u>A</u>

Xref to attach: <u>F-R</u>

Attach Xref F-R: <u>F-R</u> (F–R is loaded.)

Insertion point: (Select.)

X scale factor <1>: (CR)

Y scale factor <1>: (CR)

Rotation angle <0>: (CR)

Step 2 Command: <u>XREF</u>

?/Bind/Detach/Path/Reload/<Attach>: <u>?</u>

Xref(s) to list <*>: (CR) (XREFs are listed.)

BLOCKS AS XREFS

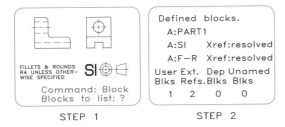

```
Defined blocks.
  A:PART1
  A:SI      Xref:resolved
  A:F-R     Xref:resolved
User Ext.  Dep Unamed
Blks Refs.Blks Blks
  1     2     0    0
```

STEP 1   STEP 2

Figure 36.144 Using BLOCKs as XREFs:
Step 1 Attached XREFs are listed under the BLOCK ?
command.

Step 2 BLOCKS (A:PART1, A:SI, and A:F-R) are
listed. There are two XREFs and one user BLOCK.

36.19 Important Commands

Transparent Commands

You may use **transparent commands** when another
command is in progress. Transparent commands are

LISTING XREFS

```
Layer name  State Color  Linetype
----------   --    --    ------
0            ON    7     CONTIN.
VISIBLE      ON    1     CONTIN.
HIDDEN       ON    2     CENTER
CENTER       ON    3     CONTIN.
HATCH        ON    4     CONTIN.
DIMEN        ON    5     PHANTOM
CUT |VISIBLE ON    6     CONTIN. Xdep: SI
WINDOW|VISIBLE ON  7     CONTIN. Xdep: F-R

                          SI and F-R
                          dependent
       Combination        upon CUT and
       layers             WINDOW layers
```

Figure 36.145 Colors of layers being changed as they
are used in XREFs will retain the color of the parent drawing into which they are placed. The ? option under the
LAYER command gives a table showing that XREFs, SI,
and F-R are dependent on the A:PART1 file.

typed with apostrophes in front of them. For example, if you are dimensioning a part and want to use
the HELP command, type ´HELP and (CR) to complete the dimensioning command. Available transparent commands are ´GRAPHSCR, ´HELP, ´PAN,
´REDRAW, ´RESUME, ´SETVAR, ´TEXTSCR, ´VIEW, and
´ZOOM. Commands selected from the pull-down
menu are transparent commands by default.

VIEW

You may save parts of a drawing as separate views
with the VIEW command. It is an effective way to
save separate "snapshots" of parts of a large drawing, such as a house, for viewing one at a time.

Figure 36.146 illustrates how to create VIEWs.
You may make the entire image on the screen into
a VIEW with the Save option and name it when
prompted. The Window option saves as a VIEW the
portion of a drawing that you select with a window. Type Restore and give the VIEW's name to
display it. Type ? to get a list of the saved VIEW.
The Delete subcommand removes a VIEW from
the list.

VIEW COMMAND

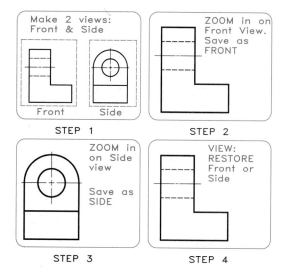

Figure 36.146 Using the VIEW command:
Step 1 The two-view drawing can be saved as separate VIEWs.

Step 2 ZOOM the front view to fill the screen.

Command: VIEW (CR)

?/Delete/Restore/Save/Window: S (CR)

View name to save: FRONT (CR)

Step 3 ZOOM the side view to fill the screen.

Command: VIEW (CR)

?/Delete/Restore/Save/Window: S (CR)

View name to save: SIDE (CR)

Step 4 To display a view: Command: VIEW (CR)

?/Delete/Restore/Save/Window: R (CR)

View name to save: FRONT (CR) (View is displayed on screen.)

Inquiry Commands

The **Inquiry commands** used to obtain information about entities and files are LIST, DBLIST, STATUS, TIME, HELP, DIST, ID, and AREA. Typing LIST and selecting a circle when prompted gives you information about the circle **(Fig. 36.147)**. You may obtain information about any entity in this manner. DBLIST lists all the entities of a draw-

LIST COMMAND

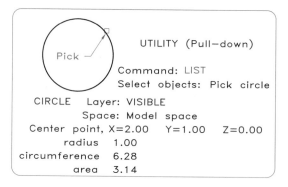

Figure 36.147 Use the LIST command from the UTILITY option of the pull-down menu to obtain information about entities on the screen, such as a circle.

TIME COMMAND

```
TIME
Current time:          27 Apr 1994 at 19:0949.270
Drawing created:        1 Apr 1994 at 20:11:37.950
Drawing last updated:  19 Apr 1994 at 19:24:42.360
Time in drawing editor: 0 days 00:06:18.660
Elapsed timer:          0 days 00:06:18.660
```

Figure 36.148 You may use the TIME command to inspect the time spent on a drawing and for setting the time for beginning an assignment.

ing. Ctrl S stops and starts scrolling the list, and Ctrl C aborts the command.

STATUS gives information about settings, layers, coordinates, and disk space. TIME displays information about the time spent on a drawing **(Fig. 36.148)**. RESET the timer and turn it ON to record the time of a drawing session, but you cannot erase the cumulative time without destroying the drawing file. The DISPLAY option displays the time opposite the heading, Elapsed Time:, showing the length of the current session after RESETting.

HELP gives a screen of options from which you can obtain an Index of topics that have help explanations. Or, you can type the name of the

AREA COMMAND

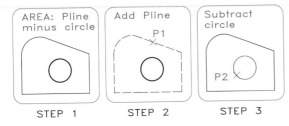

STEP 1   STEP 2   STEP 3

Figure 36.149 Calculating areas:

Step 1 You may determine the area of an object drawn as a PLINE having a circular hole with the AREA command.

Step 2 Command: AREA (CR)

<First point>/Entity/Add/Subtract: A (CR)

<First point>/Entity/Subtract: E (CR)

(ADD mode) Select circle or polyline: P1 (CR)

Step 3 (ADD mode) Select circle or polyline: (CR)

<First point>/Entity/Subtract: S (CR)

<First point>/Entity/Add: E (CR)

(SUBTRACT mode) Select circle or polyline: P2 (CR) (Circle's area and total area are given.)

command in Help Item: to obtain help information about it.

DIST measures the distance between selected points without drawing a line. When prompted, select the first and second points with the cursor or type coordinates at the keyboard to obtain the distance, its angle, and the ΔX and ΔY distances. ID gives the X, Y, and Z coordinates of any point that you pick on the screen.

AREA gives the perimeter and area of an enclosed area. Prompts request First point, Next point:, Next point:, and so on until you have picked all the points; then press (CR). To compute areas add and subtract areas, as shown in **Fig. 36.149**.

By typing FILES when in the Drawing Editor, you can display the File Utility Menu **(Fig. 36.150)**. Selecting 0 and the F1 key returns it to the Drawing Editor.

FILES COMMAND

```
0.  Exit File Utility Menu
1.  List Drawing files
2.  List user specified files
3.  Delete files
4.  Rename files
5.  Copy file
6.  Unlock file
Enter selection (0 to 6) <0>:
```

Figure 36.150 The FILES command lets you leave the current drawing to make changes in existing files without leaving AutoCad.

TYPES OF DIMENSIONS

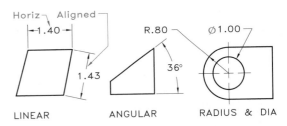

LINEAR   ANGULAR   RADIUS & DIA

Figure 36.151 These are common types of dimensions that appear on drawings.

36.20 Dimensioning

Figure 36.151 shows the types of dimensions that you may apply. Because measurements for dimensions are made from drawings on the screen, the drawings must be full size on the screen and scaled when plotted.

You may apply dimensions as **associative dimensions** or **exploded dimensions**. Insert associative dimensions as if the dimension and extension lines, text, and arrows were a BLOCK by setting the dimensioning variable DIMASO to ON. Apply exploded dimensions as individual entities, which you can modify independently (DIMASO-OFF).

Before you can use dimensioning, you must set many variables. These include sizing arrow-

DIM VARS	DEFAULT	DEFINITION
DIMALT	OFF	Alternate units selected
DIMALTD	2	Alternate units decimal pls.
DIMALTF	25.4	Alternate units scale factor
DIMAPOST	NONE	Alternate units text suffix
DIMASO	ON	Create associative dimens.
DIMASZ	.125	Arrow size
DIMBLK	NONE	Arrow block
DIMBLK1	NONE	Separate arrow block 1
DIMBLK2	NONE	Separate arrow block 2
DIMCEN	−.05	Center mark size
DIMCLRD	BYBLOCK	Dimension line color
DIMCLRE	BYBLOCK	Extension line color
DIMCLRT	BYBLOCK	Dimension text color
DIMDLE	0	Dimension line extension
DIMDLI	.38	Dimension line increment
DIMEXE	.12	Extension line extension
DIMEXO	.06	Extension line offset
DIMGAP	.05	Dimension line gap
DIMLFAC	1	Length factor
DIMLIM	OFF	Limits dimensioning
DIMPOST	NONE	Dimension text suffix
DIMRND	0	Rounding value
DIMSAH	OFF	Separate arrow blocks
DIMSCALE	1	Dimension scale factor
DIMSE1	OFF	Suppress extension line 1
DIMSE2	OFF	Suppress extension line 2
DIMSHO	OFF	Show dragged dimensions
DIMSOXD	OFF	Suppress outside dim. lines
DIMSTYLE	UNNAMED	Dimension style
DIMTAD	OFF	Text above dimension line
DIMTFAC	1	Tolerance text scale factor
DIMTIH	ON	Text inside horizontal
DIMTIX	OFF	Text inside extension lines
DIMTM	0	Minus tolerance value
DIMTP	0	Plus tolerance value
DIMTOFL	OFF	Tolerance dimensioning
DIMTOH	ON	Text outside horizontal
DIMTOL	OFF	Tolerance dimensioning
DIMTSZ	0	Tick size
DIMTVP	0	Text vertical position
DIMTXT	.125	Text size
DIMZIN	0	Zero suppression

Figure 36.152 These dimensioning variables are listed when you type DIM (CR) and Status.

heads and numerals, specifying extension line offsets, assigning text fonts, and adopting units, to name a few.

Basic Dimensioning Variables
Dimensioning variables (DIM VARS), their assigned values, and definitions appear in the STATUS option under the DIM: command of the side menu

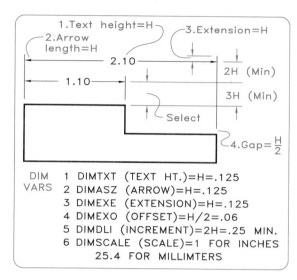

DIM VARS		
	1	DIMTXT (TEXT HT.)=H=.125
	2	DIMASZ (ARROW)=H=.125
	3	DIMEXE (EXTENSION)=H=.125
	4	DIMEXO (OFFSET)=H/2=.06
	5	DIMDLI (INCREMENT)=2H=.25 MIN.
	6	DIMSCALE (SCALE)=1 FOR INCHES
		25.4 FOR MILLIMTERS

Figure 36.153 Dimensioning variables are based on the height of the lettering (text), usually about 1/8 inch.

(Fig. 36.152). Sizes of dimensioning variables are based on the letter height, 0.125 inch, for example **(Fig. 36.153).**

To set and save a few variables needed for basic applications, OPEN the file B:FORMAT so that you can specify and save variables to it. To set each variable, obtain DIM: in the side menu, select the DIM VARS command to get a list of the variables, and assign values to them. The basic DIM VARS and their values are DIMTXT, DIMASZ, DIMEXE, DIMEXO, DIMTAD, DIMDLI, DIMASO (set to Off), and DIMSCALE, as shown in Fig. 36.153; they apply to most drawings. The UNITS command sets decimal fractions to two decimal places for inches.

SAVE these settings to B:FORMAT with no drawings on it. Now, use the SAVEAS command to name a new file (B:BASIC, for example), which becomes the current file with the same settings as B:FORMAT to preserve the previously made settings.

We discuss dimensioning variables throughout this section, but for now use the B:BASIC file to explore the fundamentals of dimensioning.

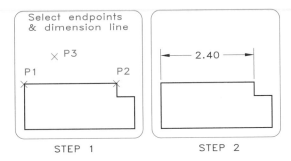

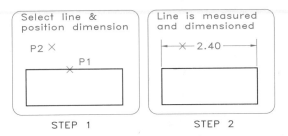

Figure 36.154 Dimensioning a line:
Step 1 Command: <u>DIM</u> (CR)

Dim: <u>HOR</u> (CR)

First extension line origin or (CR) to select: <u>P1</u>

Second extension line origin: <u>P2</u>

Dimension line location: <u>P3</u>

Step 2 Press (CR) and 2.40 appears at the command line. Press (CR) to accept 2.40, or type a different value and press (CR).

Figure 36.155 Semiautomatic dimensioning of a line:
Step 1 Command: <u>DIM</u> (CR)

Dim: <u>HOR</u> (CR)

First extension line origin or (CR) to select: (CR)

Select entity: <u>P1</u>

Dimension line location: <u>P2</u>

Step 2 Press (CR) and 2.40 appears at the command line. Press (CR) to accept this value (or give a different value) and the dimension is drawn.

Linear Dimensions

Begin dimensioning by selecting DRAW from the pull-down menu, picking DIMENSIONS, and specifying the type of dimension line, such as horizontal **(Fig. 36.154)**. Prompts at the command line ask for the endpoints of the line. When you respond to the prompt, the program locates the dimension line and automatically produces the dimension.

You may also apply a dimension semiautomatically **(Fig. 36.155)** by pressing (CR) when prompted for endpoints, selecting the line to be dimensioned, and locating its dimension line. The program measures the line and produces the dimension.

To continue a chain of dimensions, select (or type) the CONTINUE option to obtain the next endpoint **(Fig. 36.156)**. Similarly, you may apply BASELINE dimensions from the first endpoint requested and offset each dimension incremen-

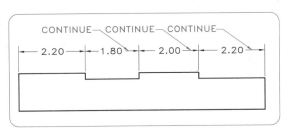

Figure 36.156 When dimensions are to be placed end to end, use the CONTINUE option to specify the origin of the second extension line after the first dimension line has been drawn.

tally with the dimension line increment variable (DIMDLI) **(Fig. 36.157)**.

In these examples we placed the dimension text in a gap (DIMGAP) at the center of the dimension line because DIMTAD (text above dimension lines) was Off **(Fig. 36.158)**. When DIMTAD is On, the text is placed above the dimension line.

BASELINE DIMENSIONS

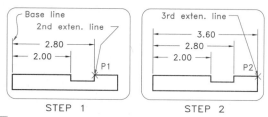

STEP 1   STEP 2

Figure 36.157 Using the BASELINE option:
Step 1 Command: DIM (CR)

Dimension the first line (2.00) as shown in Fig. 36.154. When prompted for the next dimension, respond with BASELINE (CR) and you will be prompted for the second extension line origin: P1

Step 2 The second dimension line will be drawn. Respond to Dim: with BASELINE (CR) again, and select P2 when asked for second extension line origin. Continue in this manner for other dimensions.

TEXT ABOVE DIMENSION LINE: DIMTAD

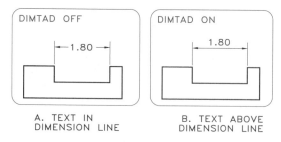

A. TEXT IN   B. TEXT ABOVE
DIMENSION LINE   DIMENSION LINE

Figure 36.158 The DIMTAD variable:
A When DIMTAD is OFF, text is inserted within the dimension line.

B When DIMTAD is ON, text is placed above the dimension line.

Ordinate Dimensions

Ordinate dimensions are used to dimension surfaces from an origin point where X=0 and Y=0. The UCSICON must be set to OR (for ORIGIN) and moved to the 0,0 point on the drawing, as shown in **Fig. 36.159**. The X-ordinate dimension of 2.30

ORDINATE DIMENSIONS: XDATUM

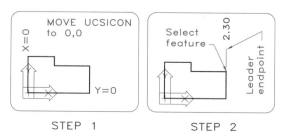

STEP 1   STEP 2

Figure 36.159 ORDINATE dimensions—Xdatum:
Step 1 Move the UCSICON to the origin where X=0 and Y=0.

Step 2 Command: ORDINATE

Select Feature: (Select line.)

Leader endpoint (Xdatum/Ydatum): X (CR)

Leader endpoint: (Select.) (CR) (The dimension is drawn.)

ORDINATE DIMENSIONS: YDATUM

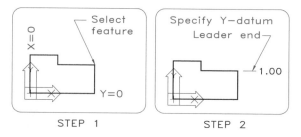

STEP 1   STEP 2

Figure 36.160 ORDINATE dimensions—Ydatum:
Step 1 Command: DIM

Dim: ORDINATE

Select Feature: (Select line.)

Step 2 Leader endpoint (Xdatum/Ydatum): Y (CR)

Leader endpoint: (Select.) (CR) (The dimension is drawn.)

means that the vertical plane is 2.30 units along the X axis from the origin. The Y-ordinate dimension of 1.00 **(Fig. 36.160)** means that the horizontal plane is 1.00 unit in the Y direction from Y = 0.

DIMENSIONING ANGLES

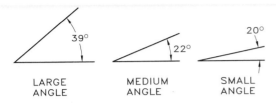

Figure 36.161 Angles may be dimensioned in any of these three formats using AutoCAD.

ANGULAR DIMENSION

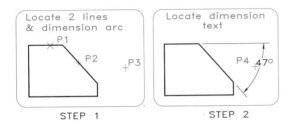

Figure 36.162 Using the ANGULAR command:.
Step 1 Command: <u>DIM</u> (CR)

Dim: <u>ANGULAR</u> (CR)

Select first line: <u>P1</u>

Second line: <u>P2</u>

Enter dimension line arc location: <u>P3</u>

Dimension text <47>: (CR)

Step 2 Enter text location: <u>P4</u> or (CR)

Dimensioning Angles

Figure 36.161 shows variations for dimensioning angles based on the space available. Begin by selecting two lines of the angle and locate the dimension line arc (**Fig. 36.162**). Where space permits, center the dimension value in the arc between the arrows.

You may dimension an angle by beginning with its vertex and locating the endpoints of each line (**Fig. 36.163**). For angles of more than 90°, select a point on the sides of the angle and move counterclockwise to the second point (**Fig. 36.164**). Locate the dimension arc and position the text.

ANGLE: ENDPOINT OPTION

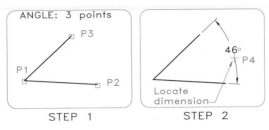

Figure 36.163 ANGULAR command—endpoint option:
Step 1 DIM: ANGULAR

Select arc, circle, line, or RETURN: (CR)

Angle Vertex: <u>P1</u>

First angle endpoint: <u>P2</u>

Second angle endpoint: <u>P3</u>

Step 2 Enter dimension line arc location: <u>P4</u>

Dimension text <46>: (CR)

Enter text location: (CR) (Accepts 46.)

ANGLE: OBTUSE ANGLE

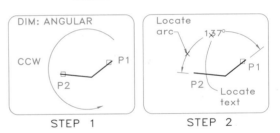

Figure 36.164 ANGULAR command—large angles:
Step 1 Dim: <u>ANGULAR</u>

Select arc, circle, line, or RETURN: <u>P1</u>

Second line: <u>P2</u>

Step 2 Enter dimension line arc location:
(Select point.)

Dimension text <137>: (CR)

Enter text location: (CR)

Dimensioning Circles and Arcs

Dimensions of circles may be positioned as shown in **Fig. 36.165**. Changing the variables DIMTIX and DIMTOFL lets you dimension circles as shown in **Fig. 36.166** and **Fig. 36.167**.

DIMENSIONING CIRCLES

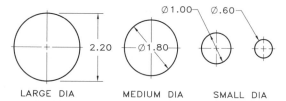

Figure 36.165 These types of dimensions are available for dimensioning circles with AutoCAD.

DIAMETER FORMATS

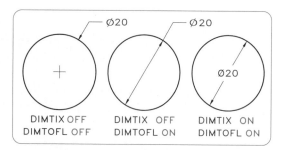

Figure 36.166 These are examples of dimensions with various DIM VARS settings for circles.

The DIMTIX (On) variable forces the text inside the extension lines regardless of the available space. The DIMTOFL (On) variable produces a dimension line between extension lines when the text is located outside.

When you dimension arcs, place an R in front of the text (R1.00, for example), as shown in **Fig. 36.168. Figure 36.169** shows how to dimension an arc with a radius and leader. Use the RADIUS command to add a dimension or a note to an arc. This command measures the arc and inserts this measurement. When the LEADER command is used, you must know the circle's diameter and type it in to override that measurement **(Fig. 36.170)**. LEADER does not measure radii or diameters.

DIMENSIONING DIAMETERS

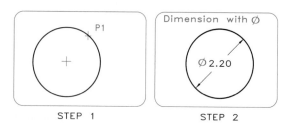

Figure 36.167 Dimensioning a diameter:
Step 1 Command: Dim (CR)

Dim: DIAMETER (CR)

Select arc or circle: P1

Step 2 Dimension text <2.20>: (CR) (Accept 2.20 and the dimension is drawn from P1 through the center.)

DIMENSIONING ARCS

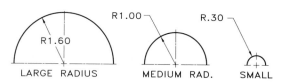

Figure 36.168 Dimension arcs by using one of the formats shown, depending on the size of the radius.

DIMENSIONING RADII

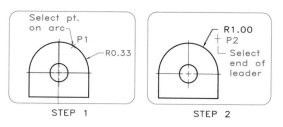

Figure 36.169 Using the LEADER option:
Step 1 Command: DIM (CR)

Select arc or circle: P1

Dimension text <0.33>: (CR)

Step 2 Enter leader length for text: P2
(The leader and R1.00 is drawn.)

DIAMETER BY LEADER

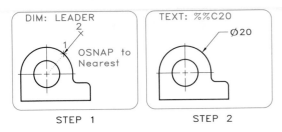

STEP 1   STEP 2

🔲 **Figure 36.170** Using the LEADER option with OSNAP:
Step 1 Dim: <u>LEADER</u>

Leader start point: (OSNAP to NEAREST)

To point: <u>P2</u>

To point: (CR)

Step 2 Dimension text <1>: <u>%%C20</u> (CR)

DIMENSIONING STYLES AND VARIABLES: DDIM

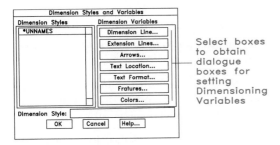

🔲 **Figure 36.171** Use the Dimension Styles dialogue box from Settings on the pull-down menu to set dimensioning variables and assign them to styles.

Dimension Variables

You may activate nearly all commands from the pull-down menu, side menu, or keyboard for associative and nonassociative dimensions. You may assign variables from the Dimension Styles and Variables (DDIM) box that appears under Settings of the pull-down menu to SAVE them as a Dimension Style file by typing its name, FULL-ENG, in the edit box **(Fig. 36.171)**. RESTORE Dimension Styles by selecting them from the Dimension Styles list. (We discuss Dimension Styles later in this section.)

DIMENSION LINE

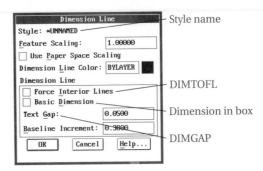

🔲 **Figure 36.172** Use the Dimension Line box to set scale and color, force interior lines, specify a basic dimension, size the text gap, and specify the baseline increment.

Use the subdialogue box Dimension Line to assign several dimensioning variables **(Fig. 36.172)**.

Feature Scaling (DIMSCALE): scales all variables assigned to dimensions.

Paper Space Scaling: computes the scaling ratio between the current model space viewport and paper space.

Dimension Line Color (DIMCLD): sets the color of the dimension line. BYLAYER is preferred.

Force Interior Lines (DIMTOFL): forces a dimension line between extension lines, even when text is placed outside.

Basic Dimension: places dimensioning text in a rectangular box.

Text Gap (DIMGAP): specifies the space around a numeral placed in a dimension line.

Baseline Increment (DIMDLI): sets the spacing between dimensioning lines when dimensioning is done from a common plane.

The Extension Lines subdialogue box allows the following settings **(Fig. 36.173)**.

Extension Above Line (DIMEXE): specifies the length of the extension lines beyond the dimension line.

EXTENSION LINES

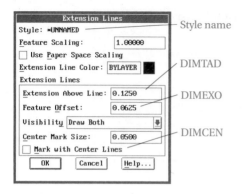

Figure 36.173 Use this box to assign settings to the extension lines.

ARROWS

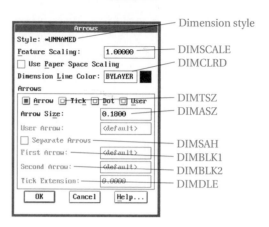

Figure 36.174 The Arrow box lets you assign settings to arrows on dimension lines.

Feature Offset (DIMEXO): assigns the gap size between the extension line and the object.

Visibility (DIMSE1 and DIMSE2): gives a pop-up list for suppressing the first, second, both, or neither extension line.

Center Mark Size (DIMCENTER): gives the dimensions of arcs and circles and places a plus sign at their centers when the CENTER, RADIUS, and DIAMETER commands are used. Enter zero for no center mark.

Mark with Center Lines: draws the center mark with center lines when the dimension line is placed outside the circle.

The Arrows subdialogue box gives the setting for arrows, as shown in **Fig. 36.174**:

Arrow, Dot, or User: specifies the type of arrowhead at the ends of dimension lines and sets the Tick value to zero.

Arrow Size (DIMASZ): sets the length of arrows or dots.

Tick (DIMTSZ): draws ticks (slash marks) instead of arrows on dimension lines at the size specified by Arrow Size.

User: allows application of user-defined arrowheads (DIMBLK) to dimensions and sepa-

rate arrowheads (DIMSAH) defined as DIMBLK1 and DIMBLK2. (We present more details later in this section.)

Tick Extension: available with Ticks and used to assign the length of the dimension line overhang beyond the extension line.

Use the Text Location subdialogue box to modify text location setting (**Fig. 36.175**).

Dimension Text Color (DIMCLT): assigns the color of dimensioning numerals; BYLAYER is preferred.

Text Height (DIMTXT): gives the height of the dimensioning numerals.

Tolerance Height (DIMTFAC): assigns heights of text used for tolerance dimensions; use the DIMTFAC variable to store the ratio of tolerance height divided by text height as a factor.

Horizontal Text Placement: sets text dimensions by Default, which places dimension text inside extension lines if space permits, or outside if it does not; text is placed outside for radius and diameter dimensions (DIMTIX Off and DIMSOXD Off).

TEXT LOCATION	TEXT FORMAT

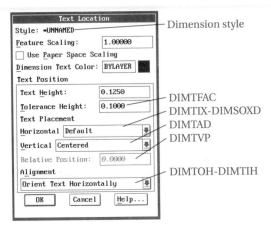

Figure 36.175 Use the Text Location box to make location and size assignments to text within dimension lines.

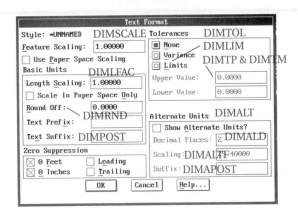

Figure 36.176 The Text Format box lets you make settings that pertain to the format of dimensioning and tolerancing text.

Force Text Inside option: turns DIMTIX On and DIMSOXD On to force text between extension lines and suppresses outside dimension lines. Text, Arrows Inside forces text inside extension lines and suppresses arrows and dimensions that would be placed outside extension lines (DIMTIX On and DIMSOXD Off).

Vertical: has three options for setting text in vertical dimension lines. Center places text inside the dimension line (DIMTAD Off); Above places it above the dimension line (DIMTAD On); and Relative places it with the variable (DIMTVP).

Relative Position of text: controlled by the value entered in the edit box, which is divided by the text height and stored as DIMTVP.

Alignment: offers four settings for the orientation of dimensioning text. Orient Text Horizontally inserts text horizontally regardless of dimension-line direction with a DIMTIH On and DIMTOH On setting; Align with Dimension Line aligns the text with the dimension line with a DIMTIH Off and DIMTOH

On setting; Aligned When Inside Only aligns text placed inside extension lines with the dimension line with a DIMTIH On and DIMTOH Off setting; and Aligned When Outside Only aligns text placed outside extension lines with the dimension line with a DIMTIH Off and DIMTOH On setting.

The Text Format subdialogue box gives the text settings shown in **Fig. 36.176**.

Length Scaling (DIMLFAC): applies a scale factor to all measurements.

Scale in Paper Space Only: applies the length scaling factor to dimensions created in paper space and stores it as negative value in the DIMLFAC variable.

Round Off (DIMRND): sets the round-off value of all except angular dimensions.

Text Prefix (DIMPOST): places a string of text in front of dimensioning numerals (APPROX. 36, for example).

Text Suffix (DIMPOST): places text following dimensioning numerals (24 mm, for example).

DIMENSIONING FEATURE BOX

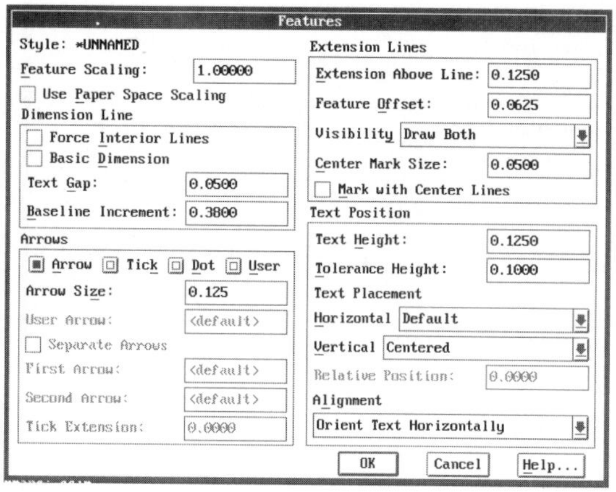

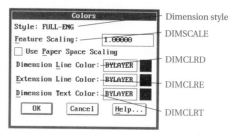

Figure 36.177 The Feature box summarizes the settings made.

Zero Suppression (DIMZIN): gives four check boxes for eliminating zeros. For example, check the 0 Feet box to change 0'-9" to 9"; check the 0 Inch box to change 5'-0 1/2" to 5'-1/2"; check Leading to change 0.22 to .22; and check Trailing to change 10.500 to 10.5. Tolerances: specified as Upper (DIMTP) and Lower Values (DIMTM); selecting Variance (DIMLIM Off), gives tolerance in plus–minus form after the basic dimension; check Limits (DIMLIM On) to give the tolerances as maximum and minimum values.

Alternate Units: modifies the second dimensioning values used in dual dimensioning (10" [254.00 mm], for example).

Decimal Places (DIMALTD): specifies the decimal places for alternate dimensions; Scaling (DIMALTF) specifies the scaling factor for alternate dimension in dual dimensioning; Suffix (DIMAPOST) gives text that follows the alternate dimension numerals (say 20 mm).

Use the Features subdialogue box to examine and modify the variables inserted by means of the previously described subdialogue boxes (**Fig. 36.177**).

COLORS DIALOGUE BOX

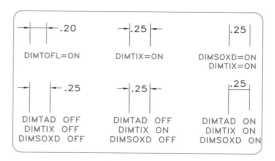

Figure 36.178 The Colors dialogue box assigns colors to dimension lines, extension lines, and text. The Bylayer setting is preferred rather than mixing colors on the same layer.

EFFECTS OF DIMSOXD

Figure 36.179 These are examples of applying the dimensioning variables DIMTOFL, DIMTIX, and DIMSOXD used in conjunction with DIMTAD.

The Color subdialogue box allows the assignment of colors to dimension lines, extension lines, and dimension text (**Fig. 36.178**). The BYLAYER setting is recommended, allowing you to use the colors assigned to each layer.

Text Placement
The effect of the variable settings selected in Fig. 36.175 on text and dimension placement is shown in **Fig. 36.179**. Alternate dimensions (dual dimensioning) assigned in Fig. 36.175 give the results shown in **Fig. 36.180**. Experiment with other variables to learn their effects.

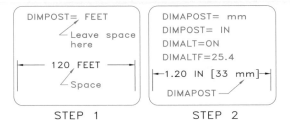

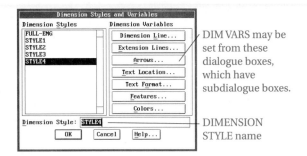

DIM VARS may be set from these dialogue boxes, which have subdialogue boxes.

DIMENSION STYLE name

Figure 36.180 The DIMPOST dimensioning variable adds text after dimension numerals. DIMAPOST adds text after alternate dimension numerals in dual dimensioning.

Figure 36.181 The Dimension Styles and Variables box lists the Dimension Styles after you have defined them.

36.21 DIMENSION STYLE

Appearing under SETTINGS, the Dimension Styles dialogue boxes permit you to set and save variables as Styles, which are listed as part of the drawing file **(Fig. 36.181)** and which you obtain with LIST. You may set dimension styles for both associative and nonassociative dimensions. By typing the Dimension Style name in the edit box and selecting OK, you save the current settings to this named file. For example, you may need a Dimension Style to locate diametric text outside small circles and another Dimension Style to place it inside larger circles. You may select these styles with a cursor to change from one set of variables to another.

When in Dim: you may save a Dimension Style by typing Save and the name of the style when prompted. Changing a single variable requires a new Dimension Style. If you change a variable while still in the current file, the new settings will be applied to all dimensions previously placed with Dimension Style.

For example, if you changed DIMTAD (text above dimension line) from Off to On, the text would be moved above the dimension lines. You cannot save the dimension variables DIMASO and DIMSHO as part of a Dimension Style setting. You must set them from the side menu or type them at the command line.

OVERRRIDE assigns different variable settings on a one-time basis without affecting the current variable settings when associative dimensions are being used. For example, OVERRIDE changes Textstyle to ROMAN and DIMTSZ from arrows to ticks **(Fig. 36.182)**. From the Dim: mode, type RESTORE, select the dimensions to be modified with the cursor, and type OVERRIDE at the command line. When prompted, type the names of the variables to be overridden and, when prompted again, press (CR) to change the Tick and Dimscale variables on the drawing.

When prompted with Modify dimension style "Style1"? <N>, press (CR) or enter No to leave Dimension Style unchanged. A Yes response updates Dimension Style. Previously drawn dimensions will be updated with this new variable. (We discuss other editing commands for associative dimensions shortly.)

Format File Settings

You may SAVE Dimensioning Variables to your prototype file, B:FORMAT, to preserve layers, text, and other settings as discussed in Section 36.16. In addition, you may save Dimension Styles to make them available for a new drawing, named by the SAVEAS command, or as a prototype file when CREATING a file.

OVERRIDE OPTION	STRETCHING DIMENSIONS

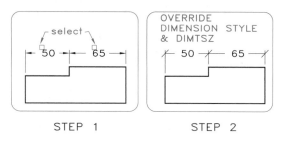

STEP 1   STEP 2

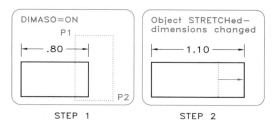

STEP 1   STEP 2

Figure 36.182 Associative dimensions—OVERRIDE option:

Step 1 Command: DIM

Dim: OVERRIDE

Dimension variable to override: STYLE (CR)

Current value <Standard> New value: ROMAN (CR)

Dimension variable to override: DIMTSZ (CR)

Current value (0.00) New value: .10 (CR)

Dimension variable to override: (CR)

Select objects: (Select both dimensions.)

Step 2 Press (CR) and the dimensioning variables are redrawn on the screen conforming to the changes. If any of the selected dimensions contain references to a dimension style, AutoCAD prompts: Modify dimension style "style name"? <N>: Enter Y or N.

36.22 Associative Dimensioning

The entities of associative dimensions (arrows, extension lines, dimension lines, and text) are a unit, which allows you to change dimensions as you STRETCH them. **Figure 36.183** shows a 0.80-wide box STRETCHed to a width of 1.00 and its corresponding dimension change.

By picking any part of the dimension, you will be selecting a BLOCK. You must then use the EXPLODE command to separate associative dimensions into individual entities for editing.

Set DIMASO to 1 (ON) to apply associative dimensions. Set DIMSHO to 1 (ON) to show changes in dimensioning values dynamically as they are stretched. You cannot store the DIMSHO value in a Dimension Style.

Figure 36.183 Associative dimensions—STRETCH option: **Step 1** Set DIMASO to ON and apply dimensions to the part. Use the STRETCH command and CROSSING WINDOW at the end of the part.

Step 2 Select a new endpoint for the part and it will be lengthened and new dimensions will be calculated.

Use the following DIM: command options to modify associative dimensions when DIMASO is On.

HOMETEXT: repositions text to its standard position at the center of the dimension line after being changed by the STRETCH command or the TEDIT subcommand (**Fig. 36.184**).

NEWTEXT: changes the dimensioning text within a dimension line (**Fig. 36.185**); a null response [pressing (CR)] to a prompt for new text, substitutes the new text for the measured distance.

OBLIQUE: creates linear dimensions with oblique extension lines (**Fig. 36.186**).

TEDIT: changes the position and orientation of dimensioning text.

Dim: TEDIT

Select dimension: (Select dimension.)

Enter text location (Left/Right/Home/Angle): L

(Drag to new location.)

Home: moves the text back to its original position.

Left and Right: moves the text to near the left and right ends of the dimension line.

HOMETEXT OPTION

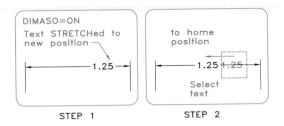

STEP 1   STEP 2

⬙ Figure 36.184 Associative dimensions—HOMETEXT command:

Step 1 Turn DIMASO ON and dimension the object. If the object and dimensions are STRETCHed, the dimension numerals will not be centered.

Step 2 Dim: <u>HOMETEXT</u> (CR)

Select objects: (Select text.) (The text will automatically center itself in the dimension line.)

NEWTEXT OPTION

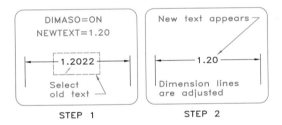

STEP 1   STEP 2

⬙ Figure 36.185 Associative dimensions—NEWTEXT command:

Step 1 Turn DIMASO ON and dimension the object.

Dim: <u>NEWTEXT</u> (CR)

Enter new dimension text: <u>1.20</u> (CR)

Select objects: (Select text to be changed.)

Step 2 The old text will be replaced with new text.

Angle: rotates the text to a specified angle.

TROTATE: sets the angle of rotation for dimensioning text as shown in **Fig. 36.187**; TROTATE works like the Angle option of the TEDIT command, except that it allows selection of more than one dimension at a time.

OBLIQUE OPTION

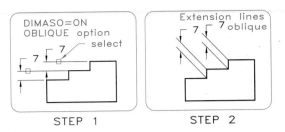

STEP 1   STEP 2

⬙ Figure 36.186 Associative dimensions—OBLIQUE option:

Step 1 Command: <u>DIM</u>

Dim: <u>OBLIQUE</u>

Select objects: (Select an associative dimension.)
2 selected, 2 found

Select objects: (CR)

Enter obliquing angle (RETURN for none): <u>135</u>

Step 2 Press (CR) and the extension lines are drawn at an angle and the dimension numerals are repositioned.

TROTATE OPTION

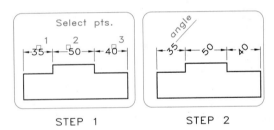

STEP 1   STEP 2

⬙ Figure 36.187 Associative dimensions—TROTATE option:

Step 1 Command: <u>DIM</u>

Dim: <u>TROTATE</u>

Enter new text angle: <u>45</u>

Select objects: (Select each associative dimension.)

Step 2 Press (CR) and the text is redrawn at an angle within the dimension lines.

UPDATE: reassigns dimension variables, dimension styles, text style, and units settings to the current drawing; for example, **Fig. 36.188** illustrates a change in the DIMSCALE

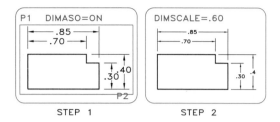

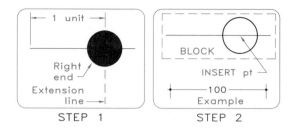

Figure 36.188 Associative dimensions—UPDATE command:
Step 1 Turn DIMASO ON and dimension the object. Change any dimensioning variable you wish.

Dim: UPDATE (CR)

Select objects: (Window drawing.)

Step 2 Changing the DIMSCALE variable to 0.60 changes all dimensioning factors to their new values.

Figure 36.189 Producing custom-designed arrowheads:
Step 1 An arrowhead, a DONUT in this case, is 1 unit long.

Step 2 It is BLOCKed with the INSERTion point at the extension line. Use the DIMBLK option under DIM VARS to assign the BLOCK by its name, which becomes the new arrowhead.

after application of the associative dimensions and updating of the drawing; a window may be used to select the dimensions of a drawing for updating.

VARIABLES: lists the variable settings of Dimension Style that is identical to the STATUS command list for your inspection.

Dim: VAR

Current dimension style: (Style name.)

?/Enter dimension style name or
RETURN to select dimension: (CR)

Select Dimension: (Select.)

You may choose a dimension style by typing Dimstyle at the command line. Or you may select a previously inserted dimension of Dimension Style that you want to use. The name of the style will appear at the command line, which means that this style has been set for the next application of dimensions.

36.23 Special Arrowheads

Instead of standard arrowheads, you may use custom arrowheads and other symbols at the ends of dimension lines. If you select the Dim: command DIMBLK, and type DOT, dimension lines will end with dots.

To create a symbol (a dot, in this case), draw a right-end dot as shown in **Fig. 36.189**. Its length (including the dimension line segment to where the extension line will cross) must be one drawing unit long. Make a BLOCK of the dot and dimension line, with the insertion point at the center of the dot where it will join an extension line. Enter the dimensioning command Dim: and type DIMBLK. When prompted, type the name of the block to be used at the ends of the dimension lines instead of arrowheads. To disable the special dot, set the DIMBLK variable to a period (.).

DIMSAH gives separate arrowheads at each end of the dimension line (**Fig. 36.190**). DIMBLK1 and DIMBLK2 define two right-end arrowheads. When you insert a dimension, place DIMBLK1 at the first extension line and DIMBLK2 at the second extension line. For this option to work, DIMTSZ must be set to zero, or ticks will be drawn instead.

SEPARATE ARROWHEADS

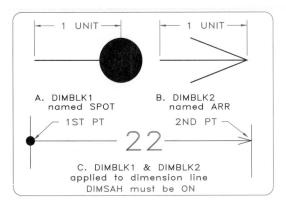

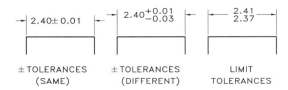

Figure 36.190 You may create special arrowhead BLOCKS (here named SPOT and ARR). When DIMSAH (separate arrowheads) is set to ON, DIMBLK1 is set to SPOT and DIMBLK2 is set to ARR. The first end of the dimension line will have SPOT as the arrowhead and the second end will have ARR as the arrowhead.

TOLERANCE FORMATS

Figure 36.191 You may tolerance dimensions in any of these three formats.

36.24 Toleranced Dimensions

To tolerance dimensions automatically, use the formats shown in **Fig. 36.191** by setting Dim Vars: DIMTOL to On and assigning a plus tolerance to DIMTP and a minus tolerance to DIMTM. When DIMLIM is On, the upper and lower limits of the tolerances will be given as shown in **Fig. 36.192**. DIMTFAC is a scale factor that controls the text height of the tolerance values, which is about 80% of the basic dimension.

TOLERANCE VARIABLES

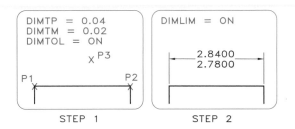

Figure 36.192 Producing toleranced dimensions:
Step 1 To tolerance a dimension, set Dim Vars DIMTOL to ON. Set DIMTP (plus tolerance) and DIMTM (minus tolerance) to the desired values. Set DIMLIM to ON to convert the tolerances into the limit form.

Step 2 Pick P1, P2, and P3 to specify the first extension line, second extension line, and the dimension location. The toleranced dimension is drawn.

OBLIQUE PICTORIAL

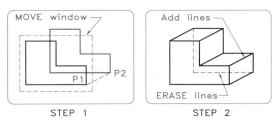

Figure 36.193 Producing an oblique pictorial:
Step 1 Draw the frontal surface of the oblique and COPY this view from P1 to P2.

Step 2 Connect the corner points and ERASE the invisible lines to complete the oblique.

36.25 Pictorials

Oblique

Figure 36.193 shows how to construct an oblique pictorial. Construct the front orthographic view and then COPY it behind the first view at the angle you have selected for the receding axis. Use OSNAP to connect the visible endpoints and erase invisi-

ORTHOGRAPHIC AND ISOMETRIC GRIDS

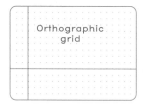
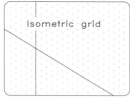

A. ORTHOGRAPHIC GRID   B. ISOMETRIC GRID

Figure 36.194
A The orthographic grid is called the STANDARD style of the SNAP mode.

B The ISOMETRIC style is that of the SNAP mode.

ISOMETRIC PICTORIAL

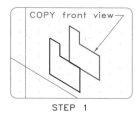
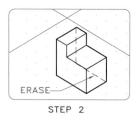

STEP 1   STEP 2

Figure 36.195 Drawing an isometric pictorial:
Step 1 Draw the front view of the isometric and COPY at its proper depth.

Step 2 Connect the corner points and ERASE the invisible lines. You may move the cursor lines into three positions by using Ctrl E or ISOPLANE.

ble lines. Circles appear as true circles on the true-size front surface, but you should avoid circular features on the receding planes because constructing them is complex.

Isometric

Use the STYLE subcommand of the Snap command to change the rectangular GRID (Standard or S) to ISOMETRIC (I), with the dots plotted in an isometric pattern, vertically and at 30° with the horizontal (**Fig. 36.194**). The lines of the cursor will align with two of the isometric axes, and you can

ISOMETRIC ELLIPSES

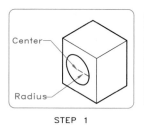
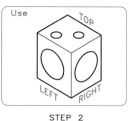

STEP 1   STEP 2

Figure 36.196 ELLIPSE command—isometric mode:
Step 1 When in the isometric SNAP mode, proceed as follows:

Command: ELLIPSE (CR)

<Axis endpoint 1>/Center/Isocircle: I (CR)

Center of circle: (Select center.)

<Circle radius>/Diameter: (Drag radius.)

Step 2 The isometric ellipse is drawn on the current ISOPLANE.

SNAP them to the grid. Use Ctrl E or the ISO-PLANE command to rotate the axes 120°. When ORTHO is ON, lines are forced parallel to the isometric axes. **Figure 36.195** shows how to construct an isometric drawing.

When you set SNAP to the isometric mode, the ELLIPSE command gives the following options:

<Axis endpoint 1>/Center/Isocircle: I (CR)

Selection of the ISOCIRCLE option and Ctrl E positions ellipses in one of the isometric orientations shown in **Fig. 36.196**. **Figure 36.197** shows an isometric with partial ellipses.

The oblique and isometric drawings covered here are 2D views that appear to be 3D views, but they cannot be rotated on the screen to obtain different viewpoints of them. We cover methods of true 3D computer-aided drawing in Chapter 37.

36.26 Grid Rotation

To draw lines parallel to a given line, rotate and align the grid on the screen with the existing lines.

STEP 1   STEP 2

Figure 36.197 Partial ellipses:
Step 1 DRAW isometric ELLIPSES to show the semicircular feature. DRAW the tangent lines.

Step 2 TRIM the lines and supply the lines needed to complete the isometric.

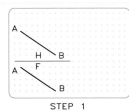
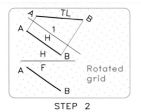

STEP 1   STEP 2

Figure 36.198 Solving a descriptive geometry problem:
Step 1 Command: SNAP (CR)

Snap spacing or ON/OFF/Aspect/Rotate/
Style <0.1250>: R (CR)

Base point <0.0>: A (In top view.)

Rotation angle <0>: B (In top view.)

Step 2 Draw H1 reference line. OSNAP from points A and B perpendicular to H1. Extend the projectors to locate the true-length view.

Use the ROTATE command, an option under the SNAP command, as follows:

```
Command: SNAP (CR)

ON/OFF/Value/Aspect/Rotate/
Style: R (Rotate) (CR)

Base point <0, 0>: (Select a point.)

Rotation angle <0>:
```

(Type the angle or specify the angle on the screen.)

The grid is aligned with the selected points. **Figure 36.198** shows a descriptive geometry problem in which an auxiliary view gives a true-length view of line AB. Return the grid to its original position by selecting the SNAP command and ROTATE as follows:

```
Base point <2,3>: (CR)

Rotation angle <37>: 0 (CR) (Realigns grid
```
to its original position.)

36.27 Digitizing with a Tablet

You may digitize drawings on paper point-by-point by using a digitizing tablet. To calibrate a drawing for digitizing, tape it to the tablet and proceed as follows:

```
Command: TABLET

Option (ON/OFF/CAL/CFG): CAL (CR)

Calibrate tablet for use

Digitize first known point: (Select
point.)

Enter coordinates for first point:
1,1 (CR)

Digitize second known point: (Select
point.)

Enter coordinates for second point:
10,1
```

Digitize points from left to right or from the bottom to the top of the drawing. Use ON or OFF to turn the tablet mode on or off. Use function key F10 to turn the tablet off so the pointer can access the side menu at the right of the tablet. For example, select the LINE command, pick a beginning point on the tablet with your pointer, and select other points in sequence.

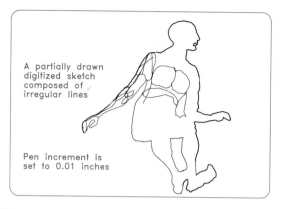

Figure 36.199 We made this drawing by using the SKETCH command with a tablet instead of a mouse. The drawing was taped to the tablet and traced with the cursor (using increments of 0.01″).

SKETCH Command

Use the SKETCH command with a tablet to trace drawings composed of irregular lines (**Fig. 36.199**). This type of digitizing requires a tablet; a mouse alone is unsatisfactory. Attach the drawing to the tablet, and calibrate it, and proceed as follows:

```
Command: SKETCH (CR)

Record increment <0.1>: 0.01 (CR)

Sketch. Pen eXit Quit Record Erase
Connect.
```

The Record Increment option specifies the distances between the endpoints of the connecting lines that are sketched. Other command options are the following.

Pen: raises or lowers pen.

eXit: records lines and exits.

Quit: discards temporary lines and exits.

Record: records temporary lines.

Erase: deletes selected lines.

Connect: joins current line to last endpoint.

. (period): line is drawn from the current location of cursor to the last endpoint.

Begin sketching by moving your pointer to the first point with the pen up, lower the pen (P), and trace the line with the pointer, which draws the line on the screen. Erase by raising the pen (P), enter ERASE (E), move the pointer backward from the current point, and select the point where you want the erasure to stop.

All lines are temporary until you select Record (R) or eXit (X) to record them. Begin new lines by repeating these steps. You may set the SKPOLY variable by typing SETVAR, entering SKPOLY, and typing 1 (on). When you use the SKETCH command, lines will be drawn as continuous polylines in contrast to SKETCHed lines, which are connected, separate segments.

36.28 Preparing a Slide Show

Commands for creating programming slides on the screen are SCRIPT, MSLIDE, RSCRIPT, DELAY, RESUME, and VSLIDE. To convert a drawing on the screen to a slide, ZOOM in on the drawing so that the image displayed represents how the slide will appear (**Fig. 36.200**). Type MSLIDE and name the file A:FRONTSL when prompted. Then ZOOM in on the side view, enter MSLIDE, and name it B:SIDESL. Make as many slides as desired and erase the last drawing when finished.

Enter the command VSLIDE and type B:FRONTSL to obtain this slide on the screen. Type REDRAW to remove it from the screen. Recall B:SIDESL with VSLIDE in the same manner. Now that you have checked your slides, QUIT and EXIT (or use the SHELL command to gain access to DOS).

To create a SCRIPT file, type SH to obtain the DOS prompt, C>, and use the EDLIN command to begin a script called A:DISPLAY.SCR as follows:

```
C> EDLIN A:DISPLAY.SCR

* I
```

MAKING SLIDES

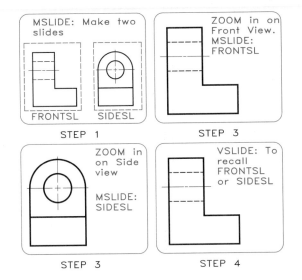

Figure 36.200 Preparing slide shows:
Step 1 Make a drawing that is to be saved as two separate slides.

Step 2 ZOOM in on the front view.

Command: MSLIDE (CR) Slide file <current>: FRONTSL (CR)

Step 3 ZOOM in on the side view.

Command: MSLIDE (CR) Slide file <current>: SIDESL (CR)

Step 4 Command: VSLIDE (CR) Slide file FRONTSL (CR)

(The slide is displayed.)

Type the following script file, placing the disk drive, B:, in front of the slide names:

```
1: VSLIDE B:FRONTSL (CR)

2: DELAY 2000 (CR)

3: VSLIDE B:SIDESL (CR)

4: DELAY 1000 (CR)

5: RSCRIPT (CR)

6: RESUME (CR)
```

7: (Ctrl Break)

*E (Saves file and ends session.)

The DELAY command specifies pauses in milliseconds, which is an approximation because of variations among computers.

To run the SCRIPT and show the slides, load AutoCAD's Drawing Editor by calling up any new drawing. Type SCRIPT and B:DISPLAY as the Filename when prompted. The slides will be shown in sequence according to the script and will recycle. Stop the show by pressing Ctrl C or Backspace.

Omitting the RSCRIPT command from the SCRIPT file, would have stopped the sequence after two slides. To repeat the show, type RESUME. When through, type QUIT to return to the Main Menu.

Many more advanced slides may be developed. For example, the SCRIPT file can include such variables as LIMITS, SNAP, GRID, UNITS, and TEXT, which are activated by the SCRIPT. By using your imagination, you can produce a series of slides to give animated action.

The EDLIN command is an MS-DOS line editor that can be used to write the SCRIPT file. Some of the commands used with EDLIN are:

Type the line number and (CR); goes to that line for editing

D: deletes lines; for example, 2D (CR) to delete line 2.

E: ends session and writes file to disk; for example: Ctrl Break and E.

I: inserts line into file; for example, 4I to insert above line 4.

L: lists lines; for example, L (CR).

Q: quits editing without saving file; for example, Ctrl Break and Q.

S: searches for a specified string; for example, ?S string to locate.

Ctrl Break ends insertion mode.

Ctrl Z ends insertion mode.

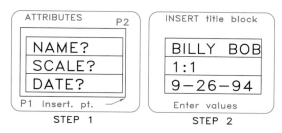

ATTRIBUTES   P2   INSERT title block

NAME?   BILLY BOB
SCALE?   1:1
DATE?   9-26-94

P1 Insert. pt.   Enter values
         STEP 1   STEP 2

◆ **Figure 36.201** Defining attributes:

Step 1 Command: <u>ATTDEF</u> (CR)

```
Attribute modes—Invisible:N constant:N
Verify:N Preset:N
```

```
Enter (ICVP) to change, RETURN when
done: V (CR)
```

```
Attribute modes—Invisible:N Constant:N
Verify: Y Preset: N
```

```
Enter (ICVP) to change, RETURN when
done: (CR)
```

```
Attribute tag: NAME (CR)
```

```
Attribute prompt: NAME? (CR)
```

```
Default attribute value: (Blank)
```

```
Start point or Align/Center/Fit/Middle/
Right/Style:
```

(Locate tag on title block.)

```
Height <0.2000>: .125 (CR)
```

```
Rotation angle <0>: (CR)
```

(Repeat for Scale and Date)

Step 2 BLOCK the title block with a P1 and P2 window.

```
Command: INSERT
```

```
Block name <or?>: TITLE
```

```
Insertion point: (Select.)
```

```
X scale factor <1>/Corner/XYZ: (CR)
```

```
Y scale factor <default=X>: (CR)
```

```
Rotation angle <0>: (CR) (Respond to the prompts
```
to complete the title block.)

36.29 Attributes

Definition

Attributes are text labels saved in combination with drawings as special types of BLOCKs. You may produce a drawing with attribute text defined by

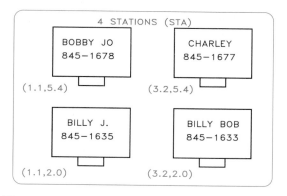

4 STATIONS (STA)

BOBBY JO   CHARLEY
845-1678   845-1677

(1.1,5.4)   (3.2,5.4)

BILLY J.   BILLY BOB
845-1635   845-1633

(1.1,2.0)   (3.2,2.0)

◆ **Figure 36.202** In this example, we inserted four blocks with attributes.

the ATTDEF command, save it as a BLOCK, and then insert it repetitively. With each insertion, prompts will ask for attribute values, which you insert on the drawing or in the file as invisible values for tabulation in an attribute report.

Figure 36.201 shows the method of defining attributes within a title block drawing. By entering I, C, or V, one at a time, you change the settings of INVISIBLE, CONSTANT, and VERIFY from On and Off and vice versa. If the INVISIBLE mode is on, attribute values will not appear on the drawing. If the CONSTANT mode is on, a fixed attribute value will be assigned to each BLOCK insertion. If the VERIFY mode is on, you will be able to verify the correctness of inserted attributes.

You may add multiple attributes to each drawing. After adding all the attributes needed, transform the drawing into a BLOCK. When you insert the BLOCK, you are prompted to enter attribute values through the keyboard:

```
Enter attribute values
```

```
Name?: <NONE>: BILLY BOB (CR)
```

```
Scale? <full size>: (CR) (Accepts full size
```
default.)

```
Date: <NONE>: 9-26-94 (CR)
```

ATTRIBUTE DEFINITION: DDATTDEF

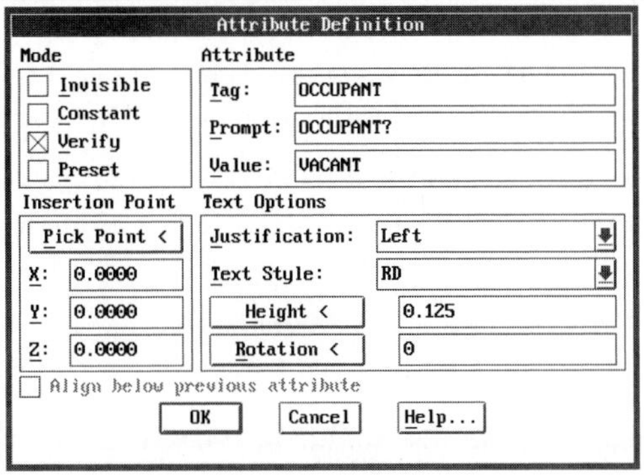

Figure 36.203 This `Attribute Definition` box (`DDATTDEF`) may be used to define attributes.

ENTER ATTRIBUTES WHEN ATTDIA=1

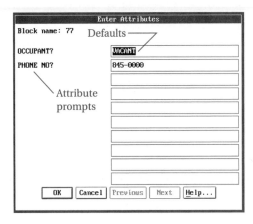

Figure 36.204 When `ATTDIA` is turned `ON` (set to `1`), you may enter attributes on the screen.

When finished, press (CR) to obtain a list of the attribute values for verification. When the end of the list is reached, the program will plot the values on the drawing as assigned. **Figure 36.202** illustrates multiple insertions of `BLOCKS` having different attribute values.

You may define attributes by using the `Attribution Definition` dialogue box (`DDATTDEF`) shown in **Fig. 36.203**. Enter attribute definitions in the response boxes and set other variables with their respective buttons. You may use the `Enter Attributes` dialogue box (`ATTDIA`) shown in **Fig. 36.204** to insert `BLOCKS` if you set `ATTDIA` to `1`. The values listed are the default values, which you may change or accept.

Visibility

To change Attribute visibility, use the `ATTDISP` command as follows:

`Command:` <u>`ATTDISP`</u> (CR)

`Normal/On/Off <current value>:` <u>`N`</u> (CR)

When you enter `N`, the attributes specified as hidden or visible will appear that way on the drawing. The `ON` option will display all values on the drawing, whether hidden or visible. The `OFF` option does not show any of the values on the drawing. To redisplay the values, you must REGENerate the drawing.

Editing

The `ATTEDIT` command edits attributes globally or one at a time. Use the command as follows:

`Command:` <u>`ATTEDIT`</u> (CR)

`Edit attributes one by one? <Y>` <u>`Y`</u> (CR)

By responding `Yes` (`Y`), you may edit the attributes currently visible on the screen. By responding `No` (`N`), you may edit all attributes globally. Select the attributes to be edited by responding as follows:

`Block name specification <*>:` (CR)

`Attribute tag specification <*>:` (CR)

`Attribute value specification <*>:` (CR)

If you press (CR) after the `BLOCK` name specification, attributes of `BLOCK` of all names will be selected for editing. If you specified the `BLOCK` name but entered (CR) after the tag specification, attributes for all tags will be selected. If you responded with `NAME` for the tag specification, only attributes with the tag `NAME` will be selected.

EDIT ATTRIBUTES (DDATTE)

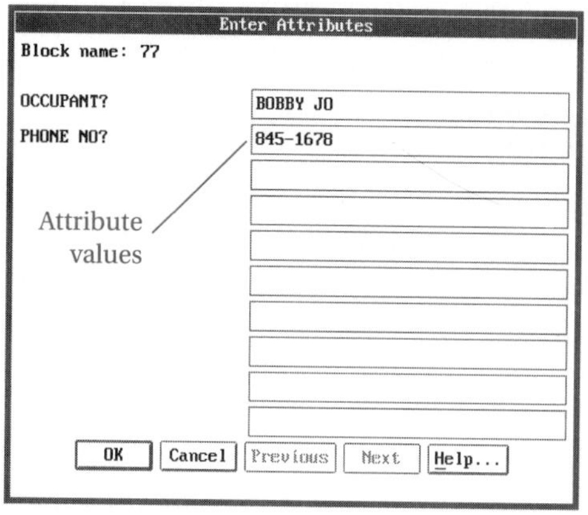

Figure 36.205 Use the `Edit Attribute` box to modify attributes.

EXTRACT: TEMPLATE FILE

```
Blk name   BL:STA   C008000 8 characters
X-coord.   BL:X     N006001 6 numbers
Y-coord.   BL:Y     N006001 6 numbers
Name       NAME     C011000 11 characters
Phone      PHONE    C008000 8 characters

N=numbers              N006001 means
006=field width        000000  6 spaces
001=1 decimal pl.      0000.0
```

Figure 36.206 We typed this template file with `EDLIN` so that attributes could be extracted from a file similar to the one shown in Fig. 36.202.

By entering a `\` you will be able to edit null-value attributes. You will now be prompted to

```
Select ATTRIBUTES:
```

Pick the attributes on the screen and each one eligible for editing will be marked with an `X`. The next prompt is

```
Value/Position/Height/Angle/
Style/Layer/Color/Next <N>:
```

By selecting the first letter of these commands, you may modify the attributes in accordance with these options. To do global editing, use the following responses:

```
Command: ATTEDIT
Edit attributes one by one? <Y> N
(Global option.)
Global edit of Attribute values.
Edit only Attributes visible on screen?
<Y> N
```

(Screen goes to Text mode if `N`.)

(Drawing must be regenerated afterward.)

These steps involve the selection of attributes in the same manner as when you do them one at a time, except that changes are applied uniformly to all selected blocks and attributes.

The `Edit Attributes` dialogue box (`DDATTE`) illustrated in **Fig. 36.205** gives a listing of the attributes of any block you select when prompted. You may change the attributes but cannot modify them with respect to `position`, `height`, or `style`.

Extract

Use the `ATTEXT` command to extract attributes from a drawing for printing in tabular form. Applications include bills of materials, parts lists, and inventories. Let's look at an elementary application.

The `Desk BLOCK` with attributes is inserted four times and the drawing is saved as `B:OFFICE` (Fig. 36.202). Next, you have to use a word processor to write a template file called `B:DESKS.TXT`. Use the `DOS` editor `EDLIN` by typing `SH` to return to `DOS`. Type `EDLIN B:DESKS.TXT` to begin a new file and type `I` to insert the lines of the template as shown in **Fig. 36.206**. End by typing `Ctrl Break` and `E` for end. Return to drawing file

CDF AND SDF TEMPLATE FORMATS

```
COMMA DELIMITED FORMAT (CDF)
'STA', 1.1, 5.4, 'BOBBY JO','845—1678'
'STA', 3.2, 5.4, 'CHARLEY','845—1677'
'STA', 1.1, 2.0, 'BILLY J.','845—1635'
'STA', 3.2, 2.0, 'BILLY BOB','845—1633'

SPACE DELIMITED FORMAT (SDF)
STA    1.1   5.4   BOBBY JO    845—1678
STA    3.2   5.4   CHARLEY     845—1677
STA    1.1   2.0   BILLY J.    845—1635
STA    3.2   2.0   BILLY BOB   845—1633
```

Figure 36.207 We produced these examples of CDF and SDF attribute extractions by using the template file shown in Fig. 36.208.

ATTRIBUTE EXTRACTION (DDATTEXT)

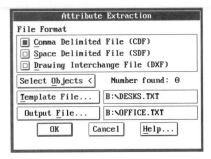

Figure 36.208 You may use this Attribute Extraction dialogue box (DDATTEXT) as an alternative method of extracting attributes.

B:OFFICE and type ATTEXT:

Command: ATTEXT (CR)

CDF, SDF, or DXF Attribute extract (or Entitles)? <C>: C (CR)

Template file <default>: B:DESKS (Omit .TXT.)

Extract file name <A:OFFICE>: B:OFFICE

8 entities in extract file:

Type SH to return to DOS; type EDLIN B:OFFICE (CR) and LIST, and the extract file will be displayed as a Comma Delimited Format (CDF) table **(Fig. 36.207)**.

An alternative method is to use the Space Delimited Format (SDF), which utilizes the same template file, B:DESKS.TXT, and gives the table of values shown in Fig. 36.207. To exchange data points with other software, use the Drawing Interchange File (DXF).

You may use the Attribute Extraction dialogue box (DDATTEXT) shown in **Fig. 36.208** to make the same selections by choosing buttons and typing responses in the edit boxes.

Problems

The problems at the ends of the preceding chapters may be drawn and plotted using CAD techniques as covered in this chapter and previous chapters. Additional problems are given in **Figs. 36.209–36.222** to develop computer graphics skills. Produce each for plotting on a size A sheet (11 in. × 8.5 in.).

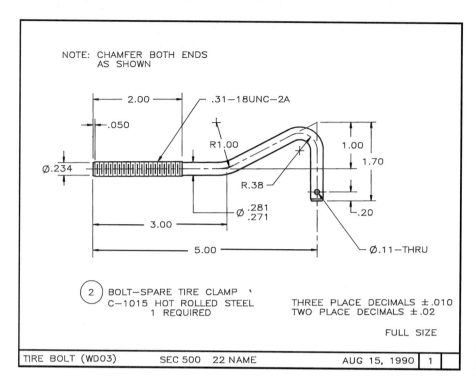

NOTE: CHAMFER BOTH ENDS
AS SHOWN

2.00

.050

.31−18UNC−2A

R1.00

Ø.234

1.00

1.70

R.38

Ø .281
.271

.20

3.00

5.00

Ø.11−THRU

(2) BOLT−SPARE TIRE CLAMP
C−1015 HOT ROLLED STEEL
1 REQUIRED

THREE PLACE DECIMALS ±.010
TWO PLACE DECIMALS ±.02

FULL SIZE

| TIRE BOLT (WD03) | SEC 500   22 NAME | AUG 15, 1990 | 1 | |

Figure 36.209 Problem 1. Bolt for spare tire clamp.

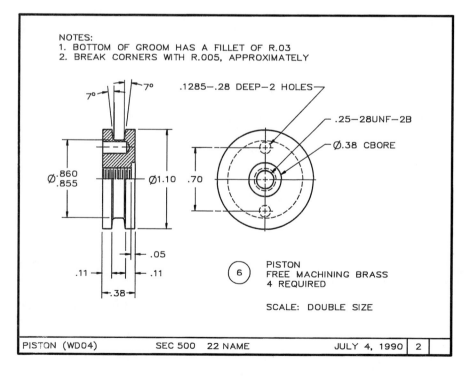

NOTES:
1. BOTTOM OF GROOM HAS A FILLET OF R.03
2. BREAK CORNERS WITH R.005, APPROXIMATELY

7°   7°

.1285−.28 DEEP−2 HOLES

.25−28UNF−2B

Ø.38 CBORE

Ø .860
.855

Ø1.10

.70

.05

.11   .11

.38

(6) PISTON
FREE MACHINING BRASS
4 REQUIRED

SCALE: DOUBLE SIZE

| PISTON (WD04) | SEC 500   22 NAME | JULY 4, 1990 | 2 | |

Figure 36.210 Problem 2. Piston.

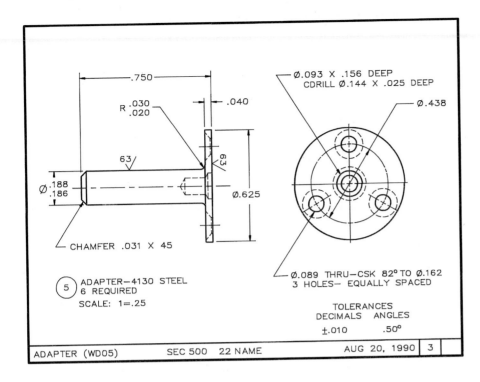

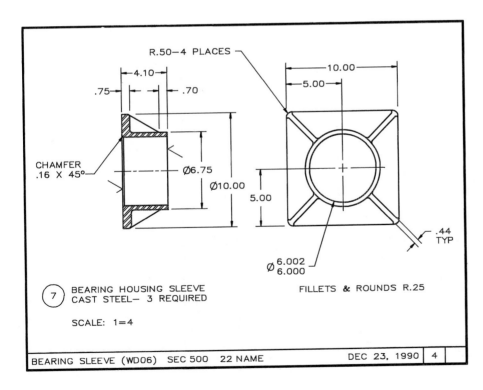

Figure 36.211 Problem 3. Adapter.

Figure 36.212 Problem 4. Bearing housing sleeve.

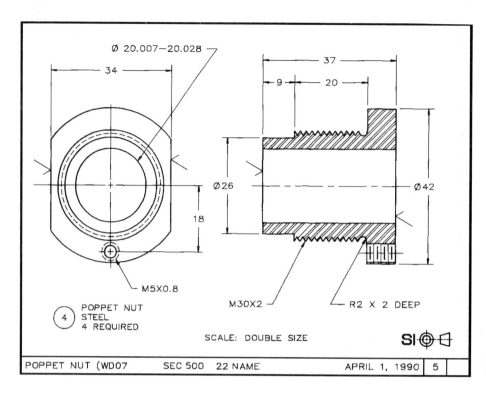

Figure 36.213 Problem 5.
Poppet nut.

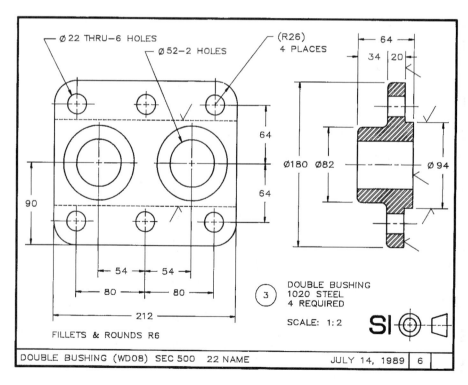

Figure 36.214 Problem 6.
Double bushing.

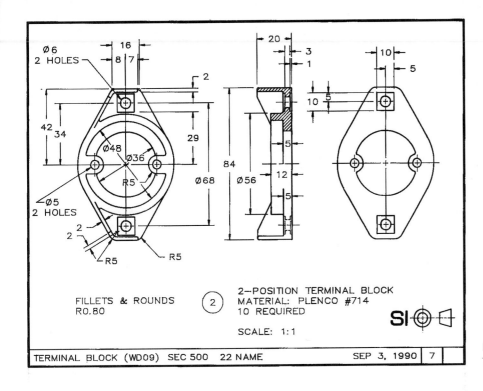

Ø6
2 HOLES

16
8 7

2

20
3
1

10
5

Ø48
Ø36

42
34

29

10
5

84

5

Ø68

R5

Ø56

12

Ø5
2 HOLES

5

2
2

R5

R5

FILLETS & ROUNDS
R0.80

(2)

2-POSITION TERMINAL BLOCK
MATERIAL: PLENCO #714
10 REQUIRED

SI⊕◁

SCALE: 1:1

TERMINAL BLOCK (WD09) SEC 500   22 NAME

SEP 3, 1990

7

Figure 36.215 Problem 7.
Terminal block.

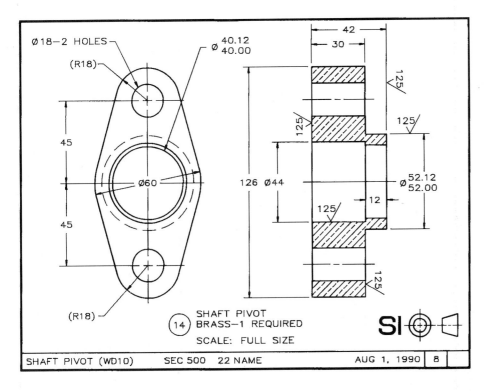

Ø18-2 HOLES

(R18)

Ø 40.12
40.00

42
30

45

125

125
125

Ø60

126 Ø44

Ø 52.12
52.00

45

125

12

(R18)

(14)

SHAFT PIVOT
BRASS-1 REQUIRED

SI⊕◁

125

SCALE: FULL SIZE

SHAFT PIVOT (WD10)   SEC 500   22 NAME

AUG 1, 1990

8

Figure 36.216 Problem 8.
Shaft pivot.

PROBLEMS • 725

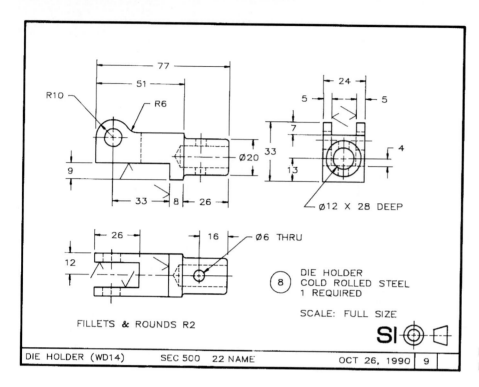

Figure 36.217 Problem 9. Die holder.

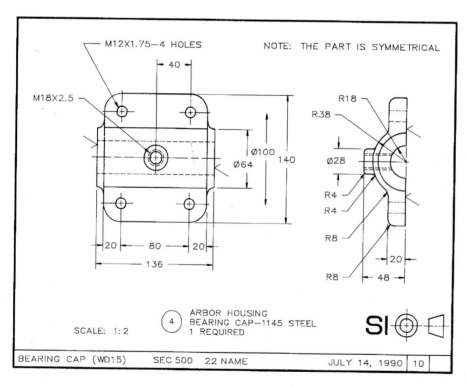

Figure 36.218 Problem 10. Bearing cap.

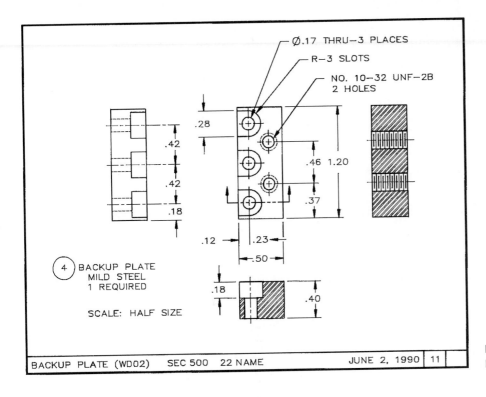

Ø.17 THRU–3 PLACES

R–3 SLOTS

NO. 10–32 UNF–2B
2 HOLES

.28

.42

.42

.18

.46 1.20

.37

.12 .23

.50

.18

.40

④ BACKUP PLATE
MILD STEEL
1 REQUIRED

SCALE: HALF SIZE

BACKUP PLATE (WD02)   SEC 500   22 NAME   JUNE 2, 1990   11

Figure 36.219 Problem 11.
Backup plate.

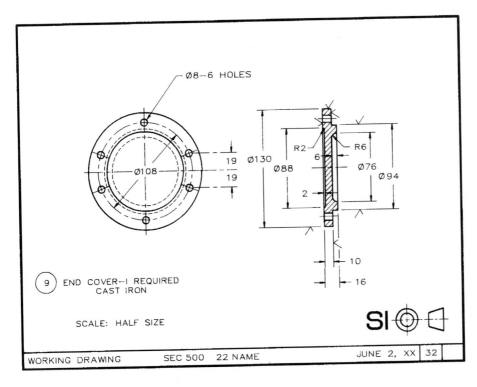

Ø8–6 HOLES

19

19

Ø130

Ø108

Ø88

R2

6

R6

Ø76

Ø94

2

10

16

⑨ END COVER–1 REQUIRED
CAST IRON

SCALE: HALF SIZE

SI ⊕ ◁

WORKING DRAWING   SEC 500   22 NAME   JUNE 2, XX   32

Figure 36.220 Problem 12.
End cover.

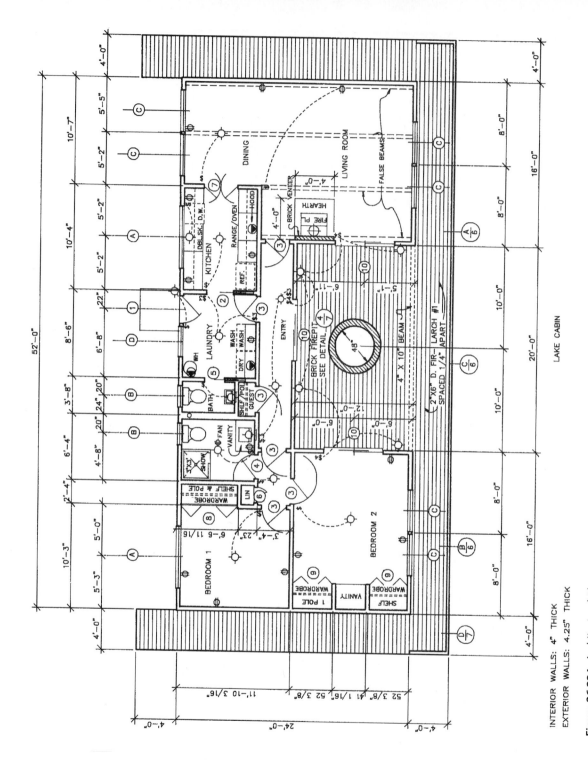

INTERIOR WALLS: 4" THICK
EXTERIOR WALLS: 4.25" THICK

Figure 36.221 Architectural Plan. Draw the house plan on a size D sheet at a scale of 1/4"=1'-0". Include a door schedule, window schedule, and legend on the same sheet. On a second sheet draw elevations of the house, which will require a degree of design on your part.

LEGEND

Symbol	Description	Height
⏀	110V DUPLEX	12" ABOVE FLR.
$	SWITCH	48" ABOVE FLR.
$3	3–WAY SWITCH	48" ABOVE FLR.
$4	4–WAY SWITCH	48" ABOVE FLR.
⊖	CEILING, WALL AND RECESSED LIGHTS	
▲	220V RANGE, DRYER, WATER HEATER	10" ABOVE COUNTER

WINDOW SCHEDULE

MARK	NO.	STOCK	TYPE	MFGR.	ROUGH OPENING
A	2	6030	HORIZONTAL SLIDING	ALENCO	72 3/4" X 36 1/2"
B	2	2030	HORIZONTAL SLIDING	ALENCO	24 3/4" X 36 1/2"
C	6	4040	HORIZONTAL SLIDING	ALENCO	48 3/4" X 48 1/2"
D	1	3030	HORIZONTAL SLIDING	ALENCO	36 3/4" X 36 1/2"

DOOR SCHEDULE

MARK	NO.	TYPE	MATERIAL	SIZE
1	1	1 LT. PANEL SASH DOOR	BIRCH	2'–8" X 6'–8" X 1 3/4"
2	1	FLUSH HOLLOW CORE — INTERIOR	BIRCH	2'–8" X 6'–8" X 1 3/8"
3	6	FLUSH HOLLOW CORE — INTERIOR	BIRCH	2'–6" X 6'–8" X 1 3/8"
4	1	FLUSH HOLLOW CORE — INTERIOR	BIRCH	2'–4" X 6'–8" X 1 3/8"
5	1	FLUSH HOLLOW CORE — INTERIOR	BIRCH	2'–0" X 6'–8" X 1 3/8"
6	1	FLUSH HOLLOW CORE — INTERIOR	BIRCH	1'–6" X 6'–8" X 1 3/8"
7	1 PR.	SWING. CAFE DOORS	BIRCH	1'–8" X 4'–0" X 1 1/8"
8	1	FLUSH HOLLOW CORE — FOLDING	BIRCH	6'–0" X 6'–8" X 1 3/8"
9	2	FLUSH HOLLOW CORE — INTERIOR	BIRCH	4'–0" X 6'–8" X 1 3/8"
10	3	SLIDING GLASS (6069 – 2V)	GLASS	6'–0" X 6'–8 1/2"

Figure 36.222 Schedules and legends for the architectural plan (sheet 1, Fig. 36.221).

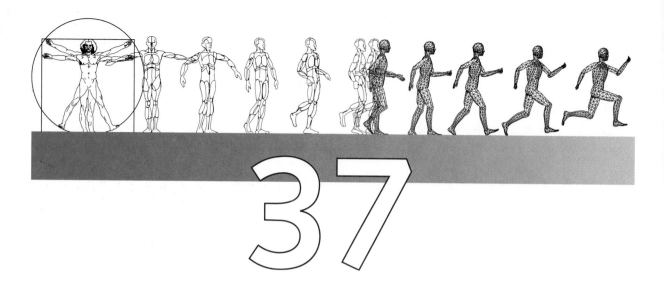

3D Drawing, Solid Modeling, and Rendering

37.1 Introduction

You may use AutoCAD to make three-dimensional (3D) pictorial drawings that can be rotated on the screen and viewed from any angle. Although the coverage of this topic here is necessarily brief, we do present the basic principles of 3D drawing. In addition, we discuss solid modeling and rendering.

37.2 Using Paper Space and Model Space

AutoCAD offers two distinctly different ways of obtaining views for drawing objects: with TILE-MODE=1 (ON) and with TILEMODE=0 (Off). When TILEMODE is on, you may divide the screen into viewports (VPORTS) that abut each other like tiles in standard arrangements (**Fig. 37.1A**). When TILEMODE is off, viewports (MVIEW) can be made in standard and nonstandard positions, as shown in **Fig. 37.1B**.

TILEMODE ON

When you set TILEMODE On, the screen enters model space by default, allowing you to produce

TILEMODE ON AND OFF

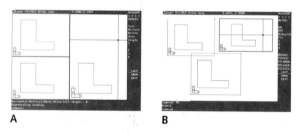

A   B

🔷 **Figure 37.1**

A When TILEMODE=1 (ON), the viewports are arranged to abut each other like tiles.

B When TILEMODE=0 (OFF), 3D viewports are not tiled and are made with MVIEW (make view), which can overlap.

drawings in both two dimensions (2D) and three dimensions (3D). The VPORTS command permits you to show four tiled viewports at a time, up to a total of sixteen on the screen. Any drawing made before application of the VPORTS command will be duplicated in each viewport, as though multi-

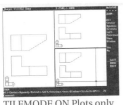

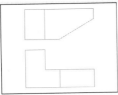

TILEMODE ON Plots only
the active viewport.

Plot of the active
viewport.

Figure 37.2 When TILEMODE=1 (ON), only the active viewport can be plotted to paper.

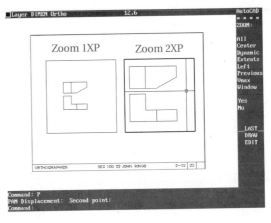

Figure 37.4 Type MS to enter model space and type ZOOM and 1XP to show the drawing full size in that viewport. Select the other viewport, type ZOOM and 2XP to obtain a double-size view in that viewport.

PAPER SPACE AND MODEL SPACE IN COMBINATION

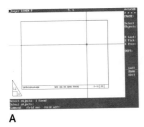

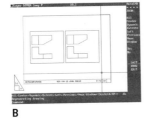

A

B

Figure 37.3

A When TILEMODE=0 (OFF) and you type PS, the screen is in paper space and a 2D border can be drawn.

B Use the MVIEW command to define the diagonal corners of 3D viewports.

ple monitors were wired to your computer (**Fig. 37.2**). Only one viewport at a time can be active—the one in which the cursor appears—and only the active viewport can be plotted.

TILEMODE OFF

When you set TILEMODE Off, type PS to enter paper space (2D space). The letter P appears in the pull-down menu bar at the top of the screen, and the triangle icon appears at the lower left of the screen. Drawings produced in this mode will be 2D drawings, so now would be a good time to insert a 2D border (**Fig. 37.3A**).

Type MVIEW to create two model-space windows by selecting their diagonals (**Fig. 37.3B**). Drawings made when TILEMODE was On will appear in the model-space viewports. Type MS to enter model space; the cursor appears in the active viewport. To move the cursor from one viewport to the next, select the desired viewport with the cursor. You may scale the contents of each model-space port by typing ZOOM and 1XP for a full-size drawing, 2XP for a double-size drawing, and 0.5XP for a half-size drawing (**Fig. 37.4**).

Type PS to return to paper space. The triangle icon reappears in the left corner, and P appears at the top of the screen. Plotting from paper space enables you to combine the 2D and 3D drawings into a single drawing (**Fig. 37.5**).

Paper Space versus Model Space
The following points summarize what you can do with PAPER SPACE (PS):

1. Use MVIEW to establish 3D viewports.

2. STRETCH, MOVE, and SCALE 3D viewports.

3. Erase 3D viewports.

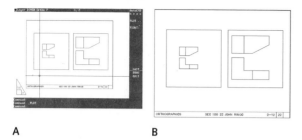

A   **B**

📐 **Figure 37.5**

A Type PS to enter paper space for plotting a drawing.

B Both paper space and model space are plotted.

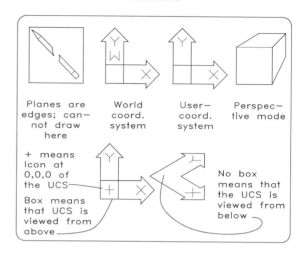

📐 **Figure 37.6** These are the various VPORTS icons that appear on the screen to show the X and Y axes.

4. Freeze model-space outlines.

5. Insert 2D drawings.

6. Add 2D text across 3D viewports.

The following points summarize what you can do with MODEL SPACE (MS):

1. Modify a 3D drawing.

2. Rotate the User Coordinate System (UCS).

3. PAN, ZOOM, SCALE, etc., 3D drawings.

4. Attach dimensions to the 3D drawing.

5. Use HIDEPLOT to remove lines from selected viewports.

6. Erase the contents of a 3D viewport.

37.3 3D Drawing: TILEMODE=1

Fundamentals

To experiment with 3D drawing, set TILEMODE On (=1), type UCSICON, and set to ON; and the XY icon appears **(Fig. 37.6)**. The broken-pencil icon warns that the projection plane appears as an edge, making drawing impractical in that viewpoint. When the UCSICON is at the origin, a plus sign appears in its corner box. When a W appears on the icon, it represents the World Coordinate

DETERMINATION OF THE Z AXIS

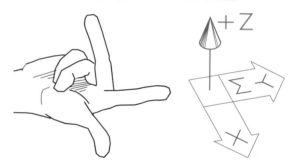

📐 **Figure 37.7** By pointing the thumb of your right hand in the positive X direction and your index finger in the positive Y direction, your middle finger points in the positive Z direction.

System (WCS); an icon without the W represents a User Coordinate System (UCS). The icon appearing as an oblique view of a box indicates that the current drawing is a perspective. The UCSICON does not show the Z axis; to find it, use the right-hand rule shown in **Fig. 37.7**.

Recall that you may obtain a maximum of 4 VPORTS at a time on the screen with the VPORTS command **(Fig. 37.8)**. You may divide these

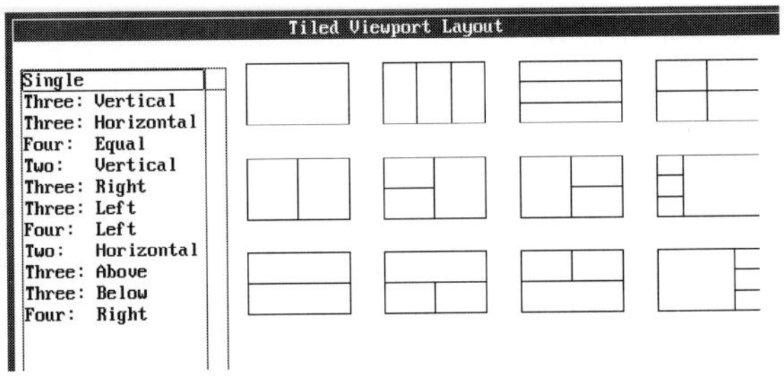

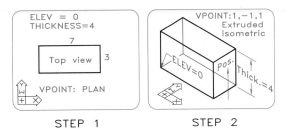

Figure 37.8 The `Tiled Viewports Layout` dialogue box may be used to set viewports on the screen when `TILEMODE=1`.

Figure 37.9 Extrusion of a box:
Step 1 Command: <u>ELEV</u>
`New current elevation <0>:` <u>0</u>
`New current thickness <0>:` <u>4</u>
`Command:` <u>LINE</u>
`From point:` (Draw 7 × 3 rectangle as a top view.)
Step 2 Command: <u>VPOINT</u>
`Rotate/<view point><current>:` <u>1,-1,1</u>
(An isometric view of the extruded box appears.)

`VPORTS` as well, up to a total of 16 on the screen. Use the `SAVE` option of `VPORTS` to save viewport arrangements.

Extrusion

The extrusion technique is an elementary method of drawing 3D objects with `ELEV` (elevation), `THICKNESS`, `PLAN`, and `HIDE`.

`ELEV`: sets the height of the plane of the `XY` icon, the plane of the drawing.

`THICKNESS`: the distance of the extrusion parallel to the Z axis.

`PLAN`: changes the `UCS` to give a true-size view of the `XY` icon and a true-size view of any surface parallel to it.

`HIDE`: removes the hidden lines of the extruded surfaces.

Turn the `UCSICON` On to obtain the `XY` icon, type `ELEV` and set it to 0, type `THICKNESS` and set it to 4, and generate the plan view of the object with the `LINE` command (**Fig. 37.9**). The X and Y axes are true size in the `PLAN` view (the top view), and the Z axis is perpendicular to them and points toward you.

Type `VPOINT` and specify a line of sight with settings of 1,-1,1 to obtain an isometric view of the extruded block (Fig. 37.9). When you type `PLAN`, the `UCSICON` is true size and the portion of the object parallel to it appears true size. You may use all the regular `DRAW` commands (such as `LINE`, `CIRCLE`, `TEXT`, and `ARC`) to draw features, all of which will be extruded 4 units in the Z direction in 2D or 3D views. The extrusion of 4 units will remain in effect until you reset it to another value with the `ELEV` command.

`HIDE` removes hidden lines from an extruded object, resulting in an empty-box look (**Fig. 37.10**). Type `3DFACE` and `OSNAP` to corner points on its upper surface, to make the top of the box opaque when `HIDE` command is applied (**Fig. 37.11**).

You may use the `SOLID` command also to opaque the upper surface by assigning it an `ELEV` equal to the extruded `THICKNESS` (4), setting its `THICKNESS` to 0, and applying a solid area to the part's top. Type `HIDE` and the surface appears opaque by this alternative method.

HIDING AN EXTRUDED BOX

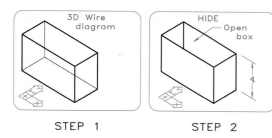

Figure 37.10 Extrusion of a box (HIDE option):
Step 1 (After the box is drawn, the box appears as a wire diagram.)

Step 2 Command: <u>HIDE</u> (CR)

(Vertical surfaces are opaque (solid) and the top appears open.)

3DFACE COMMAND

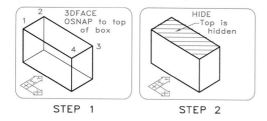

Figure 37.11 Extrusion of a box (3DFACE):
Step 1 Command: <u>OSNAP</u>

Object snap modes: <u>END</u>

Command: <u>3DFACE</u>

First point: <u>1</u>

Second point: <u>2</u>

Third point: <u>3</u>

Fourth point: <u>4</u>

Third point: (CR)

Step 2 Command: <u>HIDE</u> (CR)

(The top surface appears as an opaque surface.)

Coordinate Systems
Drawings in 3D are made in the plane of the coordinate system indicated by an xY icon. You must become familiar with the two coordinate systems.

UCS: 3POINT OPTION

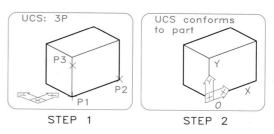

Figure 37.12 Using the 3Point option:
Step 1 Command: <u>UCS</u> (Select 3point option.)

Origin point <0, 0, 0>: <u>P1</u> (Origin.)

Point on positive portion of the X axis: <u>P2</u>

Point on positive Y portion of UCS XY plane: <u>P3</u>

Step 2 Transfer the UCS icon to the origin. The plus sign at its corner box indicates that it is at the origin.

The **World Coordinate System** (WCS) has its origin where X, Y, and Z all are 0. Usually, the X and Y axes are true length in the top, or plan, view. The UCSICON has a W on it when placed at the WCS origin.

The **User Coordinate System** (UCS) may be located within the WCS with its X and Y axes positioned in any direction and its origin at any selected point. Type UCSICON and select On to turn the icon on, select Origin to move it to the origin, and establish a UCS in the following manner:

Command: <u>UCS</u> (CR)

Origin/ZAxis/3point/Entity/
View/X/Y/Z/Prev/Restore/Save/
Del/?/<World>: <u>O</u> (CR)

Origin point <0,0,0>: (Select desired origin.) (CR)

To use the 3point (3P) option to set the UCS to the plane of an object, pick its origin, the X axis, and the Y axis, as shown in **Fig. 37.12**. The UCS command also has the following options.

ZAxis: Select a new origin and pick a point on the positive portion of the new Z axis to locate a new UCS.

Entity: Pick an entity (other than 3D `Polyline` or a polygon mesh) and the UCS will have the same positive Z axis as the entity selected.

View: Select `VIEW` to establish a UCS in which the X-Y plane is parallel to the screen, which allows the application of 2D text to a 3D drawing.

X/Y/Z: Specify X, Y, or Z as the axis about which to rotate the UCS and type the angle of rotation. Example: Type X (CR) and then 90 (CR) to rotate the icon 90° about the X axis.

Previous: Recalls the previous UCS.

Restore: Select R and type the name of the saved UCS, which then becomes the current UCS.

Save: Select S and name the current UCS to save it.

Delete: Select D and give the name of the UCS to delete it.

?: gives a listing of the current and saved coordinate systems; if unnamed, the current UCS is listed as *WORLD* or *NO NAME*.

The UCS Control dialogue box (DDUCS) lists the coordinate systems (**Fig. 37.13**). The *WORLD* coordinate system is listed first and is followed by other saved systems. The current UCS is indicated by CUR, but you may select a new one with the cursor (SETUP2, for example) or by picking the Current box. You may delete a UCS by picking the Delete box or rename it by picking the Rename To box and typing its new name. OK confirms any action made and Cancel closes the dialogue box. When you pick List the UCS Origin Point and Axis Vectors dialogue box appears and shows the coordinates of the origin and the endpoints of the three axes. Select OK to return to the DDUCS.

Activate the UCSICON in the following manner:

Command: UCSICON (CR)

ON/OFF/ALL/Noorigin/ORigin<On>: OR (CR)

These options perform the following functions:

ON/OFF: turns the icon on and off.

UCS CONTROL DIALOGUE BOX

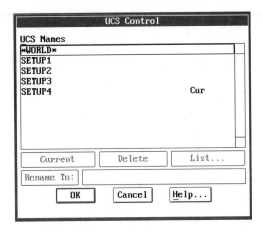

Figure 37.13 The UCS Control dialogue box located under SETTINGS in the pull-down menu lists the UCS names that you have saved and indicates the current one with Cur.

ALL: displays the icon in all viewports.

Noorigin: displays the icon at the lower left corner regardless of the location of the UCS origin.

ORigin: places the icon at the origin of the current coordinate system if space permits, or at the lower left if space is unavailable.

Setting VPOINTS

The VPOINT command controls the viewpoint of a UCS from the UCS Orientation dialogue box, the axes, or from typed coordinates. Use the UCS Orientation dialogue box from Settings of the pull-down menu, to select UCS orientations for principal orthographic views or select the UCSICON symbol for nonstandard views (**Fig. 37.14**). For example, select the top view icon, specify the origin when prompted, and type PLAN to obtain a true-size top view of the UCSICON, where the X and Y axes are true size.

When you select the UCSICON symbol, a set of X, Y, and Z axes appear on the screen (**Fig. 37.15**). Selecting a point on the *globe* (**Fig. 37.16**) gives a

UCS ORIENTATION DIALOGUE BOX

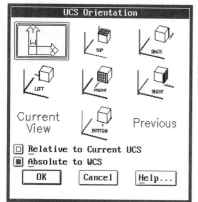

Standard views can be selected from this dialogue box

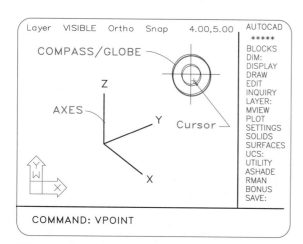

Figure 37.14 Use the UCS Orientation dialogue box to select standard views of objects.

THE COMPASS GLOBE

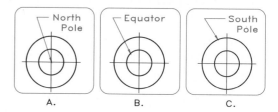

Figure 37.16

A The north pole locates the viewpoint at the crossing of the cross hairs.

B The small circle locates the viewpoint at the equator.

C The large circle locates the viewpoint at the south pole.

VPOINT AND AXES

Figure 37.15 When you type the VPOINT command and press (CR) twice, a globe and a set of axes will appear on the screen for your use in selecting a viewpoint.

THE COMPASS GLOBE

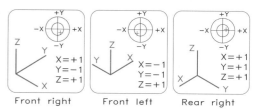

Figure 37.17 These views illustrate the relationships between points on the VPOINT globe and the VPOINT values typed at the keyboard.

gin (0,0,0) is seen from a point that is 1 unit in the X direction, 1 unit in the negative Y direction, and 1 unit in the positive Z direction from the origin (**Fig. 37.18**).

The VPOINT command's ROTATE option (**Fig. 37.19**) lets you specify the angle with the X axis in the X-Y plane and the angle with the X-Y plane. To select viewpoints including the Axes, Presets, and Set Vpoint options, use a pull-down menu under the heading VIEW/Set View/View Point. The Viewpoint Presets dialogue box in **Fig. 37.20** allows you to set the viewpoint angle with the X axis in the X-Y plane and the angle with the X-Y plane by using the cursor or by typing.

view of the object from that position. To recycle this command, press (CR) twice to select other views.

Figure 37.17 compares the viewpoints found on the globe with those entered as numbers at the keyboard. A VPOINT of 1,-1,1 means that the ori-

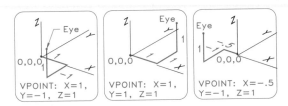

Figure 37.18 These graphical examples illustrate the selection of VPOINTs (eye) from the keyboard.

VPOINT COMMAND

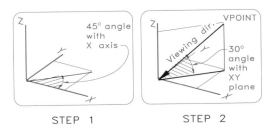

STEP 1   STEP 2

Figure 37.19 Using the VPOINT command:
Step 1 Command: <u>VPOINT</u>

Rotate/<View Point> <current>: <u>R</u> (CR)

Enter angle in XY plane from X axis <current>: <u>45</u> (CR)

Step 2 Enter angle from XY plane <current>: <u>30</u> (CR)

To find the six principal orthographic views, type the following values when using the VPOINT command:

Top view	0,0,1
Front view	0,-1,0
Right side view	1,0,0
Left side view	-1,0,0
Rear view	0,1,0
Bottom view	0,0,-1

VIEWPOINT PRESETS DIALOGUE BOX

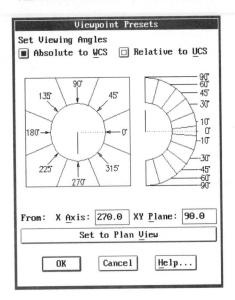

Figure 37.20 Use the Viewpoint Presets dialogue box to select the point of view in the XY plane and with the XY plane by selecting intervals with the cursor or by typing at the keyboard.

Applying Extrusions

To obtain an extruded box similar to the one shown in Fig. 37.9 in isometric **(Fig. 37.21)** type VPOINT (CR) and the coordinates 1,-1,1. Move the UCSICON to the object's lower left corner with the UCS command and rotate it 90° about the X axis to lie in the frontal plane of the box. Type PLAN to make the UCSICON true size. The front view of the box appears true size because it is parallel to the icon.

As **Fig. 37.22** shows, you may obtain a true-size drawing in the plane of the UCS (the frontal plane of the box in this case), in which a circle appears. To draw the circle as an extrusion, set ELEV to 0 and THICKNESS to -3 (the depth of the box). Apply 3DFACE to the upper and lower planes of the box and use HIDE to show visibility.

Basic 3D Forms

You may use the 3D Surfaces/3D Objects commands under Draw in the pull-down menu to

SETTING THE UCS IN THE FRONT VIEW

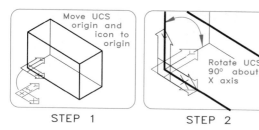

STEP 1   STEP 2

Figure 37.21 Extrusion of a box (setting UCS):
Step 1 `Command: UCSICON ON/OFF/ALL/Noorigin/` `ORigin <current ON/OFF state>:` OR `(CR)`

`Command:` UCS `(CR)`

`Origin/ZAxis/3point/Entity/View/X/Y/Z/` `Prev/Restore/Save/Del/?/`

`<World>:` O `(CR)`

`Origin point:` (OSNAP to END at corner of box.)

Step 2 `Command:` UCS `(CR)`

`Origin/ZAxis/3point/Entity/View/X/Y/Z/` `... <World>:` X `(CR)`

`Rotation angle about X axis <0>:` 90 `(CR)`

(The icon is placed in the frontal plane of the box.)

EXTRUDING A HOLE

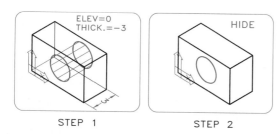

STEP 1   STEP 2

Figure 37.22 Extruding a hole:
Step 1 `Command:` ELEV `(CR)`

`New current elevation <0>:` 0 `(CR)`

`New current thickness <4>:` -3 `(CR)`

`Command:` CIRCLE `3P/2P/TTR/<Center point>:` `(CR)` (Select center with cursor.)

`Diameter/<Radius>:` `(CR)` .5 (A 1″ diameter cylinder is extruded 3″ deep into the box.)

Step 2 `Command:` HIDE (The outline of the hole is shown, but it cannot be seen through.)

THE BOX COMMAND

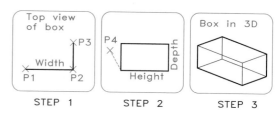

STEP 1   STEP 2   STEP 3

Figure 37.23 Applying the BOX command:
Step 1 `Command:` BOX `(CR)`

`Corner of box:` P1

`Length:` P2

Step 2 `Height:` P3

Step 3 `Rotation angle about Z axis:` 0 `(CR)` (Select a VPOINT of 1,-1,1 for an isometric of the 3D box.)

THE WEDGE COMMAND

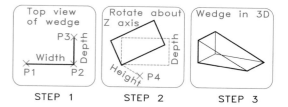

STEP 1   STEP 2   STEP 3

Figure 37.24 Utilizing the WEDGE command:
Step 1 `Command:` WEDGE `(CR)`

`Corner of wedge:` P1

`Length:` P2

`Width:` P3

Step 2 `Height:` P4

Step 3 `Rotation angle about Z axis:` -15 `(CR)` (Type VPOINT 1,-1,1 to get the wedge in 3D.)

select primitive 3D forms: Box, Pyramid, Dome, Sphere, Dish, Wedge, Cone, and Torus (toroid).

Box: produces a cube or a box when you select a corner, specify length, width, and height, and select the angle of rotation about the Z axis **(Fig. 37.23)**.

Wedge: produces a wedge when you follow the same steps used to draw the box **(Fig. 37.24)**.

THE PYRAMID COMMAND

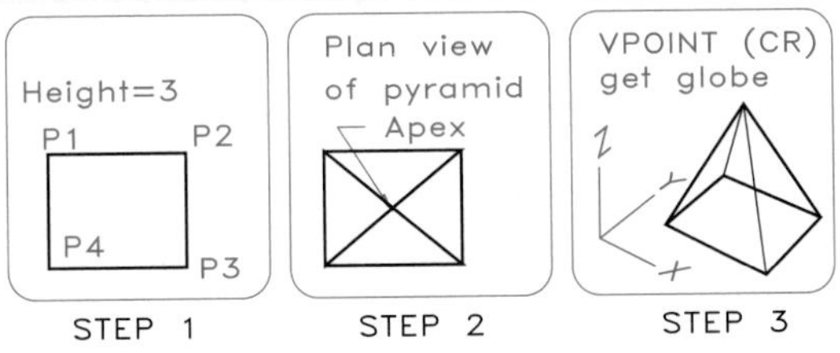

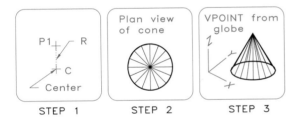

Figure 37.25 Using the PYRAMID command:
Step 1 Command: <u>PYRAMID</u> (CR)

First base point: <u>P1</u>

Second base point: <u>P2</u>

Third base point: <u>P3</u>

Tetrahedron/<Fourth base point>: <u>P4</u>

Step 2 Ridge/Top. <Apex point>: <u>.XY</u> of (need Z): <u>2</u> (CR)

Step 3 Set the VPOINT to 1,-1,1 to get a 3D view of the pyramid.

THE CONE COMMAND

Figure 37.26 Applying the CONE command:
Step 1 Command: <u>CONE</u> (CR)

Base center point: <u>C</u>

Diameter/<radius> of base: <u>P1</u>

Diameter/<radius> of top <0>: (CR)

Step 2 Height: <u>3</u> (CR)

Number of segments <16>: (CR)

(The plan view of the cone is generated.)

Step 3 Set a VPOINT of 1,-1,1 to obtain an isometric of the cone.

Pyramid: produces a pyramid extending to its apex **(Fig. 37.25)** or as a truncated pyramid with its apex removed.

THE SPHERE COMMAND

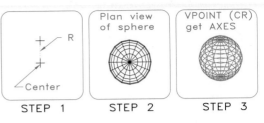

Figure 37.27 Utilizing the SPHERE command:
Step 1 Command: <u>SPHERE</u> (CR)

Center of sphere: <u>C</u>

Diameter/<radius>: (Select with pointer.)

Number of longitudinal segments<16>: (CR)

Number of latitudinal segments <16>: (CR)

Step 2 The plan view of the sphere is generated.

Step 3 Set VPOINT to 1,-1,1 to obtain an isometric of the sphere.

Cone: produces a cone to its apex **(Fig. 37.26)** or as a truncated cone with its apex removed.

Sphere: produces a ball when you select its center and radius **(Fig. 37.27)**.

Dish: produces the lower hemisphere of a sphere when you select its center and radius **(Fig. 37.28)**.

Dome: produces the upper hemisphere of a sphere when you select its center and radius **(Fig. 37.29)**.

Torus: produces a donut shape (a torus or toroid), as shown in **Fig. 37.30**.

Other selections under 3D Surfaces are Surface Revolution, Ruled Surface, Edge Defined Patch, Tabulated Surface, and 3D Face. You may set values for Surface Tabulation 1 (SURFTAB1) and Surface Tabulation 2 (SURFTAB2) by selecting from these boxes. After producing the drawing, use HIDE to remove hidden lines to make the 3D forms appear solid.

THE DISH COMMAND

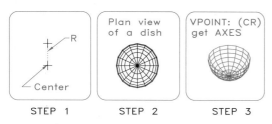

| STEP 1 | STEP 2 | STEP 3 |

Figure 37.28 Applying the DISH command:
Step 1 Command: <u>DISH</u> (CR)

Center of dish: (Pick center.)

Diameter/<radius>: (Select with cursor.)

Number of longitudinal segments <16>: (CR)

Number of latitudinal segments <8>: (CR)

Step 2 The plan view of the dish is generated.

Step 3 Select a VPOINT of 1,-1,1 for a 3D view of the dish.

THE DOME COMMAND

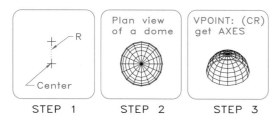

| STEP 1 | STEP 2 | STEP 3 |

Figure 37.29 Using the DOME command:
Step 1 Command: <u>DOME</u> (CR)

Center of dome: (Select with cursor.)

Diameter/<radius>: (Select with cursor.)

Number of longitudinal segments <16>: (CR)

Number of latitudinal segments <8>: (CR)

Step 2 The plan view of the dome is generated.

Step 3 Select a VPOINT of 1,-1,1 for a 3D view of the dome.

DVIEW Command

The DVIEW command is similar to the VPOINT command, but with it you may change viewpoints dynamically, showing new views of the object as

THE TORUS COMMAND

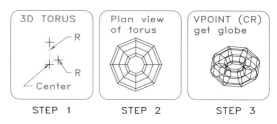

| STEP 1 | STEP 2 | STEP 3 |

Figure 37.30 Utilizing the TORUS command:
Step 1 Command: <u>TORUS</u> (CR)

Center of torus: (Select with pointer.)

Diameter/<radius> of torus: (Select with pointer.)

Diameter/<radius> of tube: (Select with pointer.)

Segments around tube circumference <16>: <u>8</u> (CR)

Segments around torus circumference <16>: <u>8</u> (CR)

Step 2 The plan view of the torus is generated.

Step 3 Set a VPOINT of 1,-1,1 for a 3D view of the torus.

VPOINT: DISTANCE OPTION

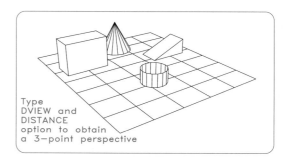

Type DVIEW and DISTANCE option to obtain a 3—point perspective

Figure 37.31 When you select the DISTANCE option of the VPOINT command, you get a perspective view.

you move the cursor. In addition to obtaining axonometric views (parallel projections), you may generate three-point perspectives in which parallel lines converge toward vanishing points for realistic pictorials (**Fig. 37.31**). The DVIEW command has the following options:

THE DVIEWBLOCK HOUSE

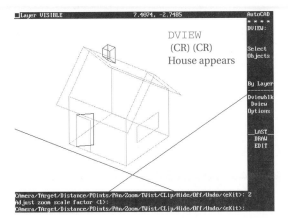

Figure 37.32 To obtain the DVIEWBLOCK house, type VPOINT and press (CR) twice.

DVIEW: TARGET OPTION

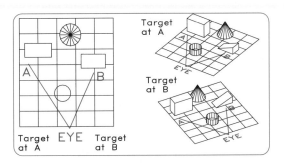

◇ **Figure 37.34** Use the TArget option of the DVIEW command to obtain different views of a scene by moving the target to different locations about a stationary camera.

DVIEW : CAMERA OPTION

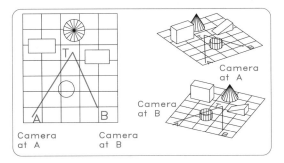

◇ **Figure 37.33** Use the CAmera option of the DVIEW command to obtain different views of a stationary target point by moving the position of the camera.

```
Command: DVIEW (CR)
CAmera/TArget?Distance/POints/PAn/Zoom/
TWist/CLip/Hide/Off/Undo/<eXit>:
```

When you type DVIEW and press (CR) twice, the top view of the DVIEWBLOCK house appears. You may experiment with the following DVIEW options **(Fig. 37.32)**.

CAmera rotates your viewpoint as if the camera were moving about the target **(Fig. 37.33)**. As you move the cursor, the viewpoint of the object changes dynamically until you select a view.

TArget rotates the specified target about the camera **(Fig. 37.34)**. Prompts are identical to those of the CAmera option, but the camera remains stationary as the target rotates about it.

DISTANCE positions the camera at its current distance from the object and turns it into a perspective. The perspective box icon replaces the XY icon. When prompted, give a distance to the target by typing the value or by using the slider bar with a range from 0X to 16X; 1X is the current distance to the target.

POints specifies the target point and the camera position for viewing a drawing **(Fig. 37.35)**. PAn moves the viewpoint of a drawing without changing its magnification or true position. With the Perspective mode, you must use the pointing device; otherwise, you may type the coordinates.

Zoom changes the magnification of a drawing in the same manner as the Zoom/Center command when Perspective is off. If Perspective is on, Zoom dynamically varies the magnification of the drawing as if the camera lens were changing from a wide-angle to a telescopic lens.

TWist rotates the drawing about an axis that is perpendicular to the screen. For example, you may twist a drawing from a vertical format to a horizontal format.

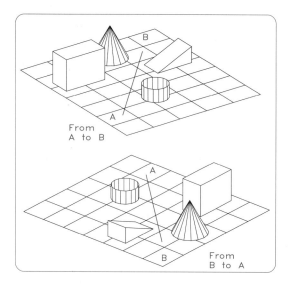

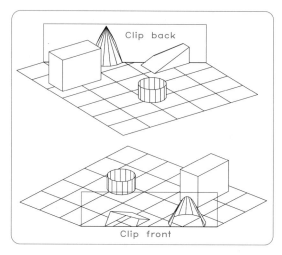

Figure 37.35 To locate the target and camera points (with X, Y, and Z coordinates) and obtain the desired view, use the POint option of the DVIEW command. You may find both perspective and axonometric views with this option.

Figure 37.36 The CLip option of the DVIEW command removes Back or Front portions of a drawing. The clipping plane is parallel to the drawing screen.

CLip places cutting planes perpendicular to the line of sight to remove portions of a drawing either in front or in back of the plane when you select one of the Back/Front/<Off> options **(Fig. 37.36)**. Off turns CLip off. When Perspective is turned on, the frontal clipping plane remains on at the camera position.

Hide suppresses the invisible lines of a drawing. OFF turns off the Perspective mode enabled by the Distance option. Undo reverses the previous DVIEW operations one at a time. eXit ends the DVIEW command and displays the resulting drawing.

The ZOOM, PAN, SKETCH, 'ZOOM, and 'PAN commands do not work in the perspective mode. The program gives you a message to this effect and cancels the command. You may use ZOOM and PAN as options under the DVIEW command.

3D Polygon Meshes

You may utilize **meshes** to define surfaces that you can later modify with the PEDIT command.

Commands for generating meshes are RULESURF, TABSURF, REVSURF, EDGESURF, and 3DMESH. We illustrate the use of the first four of these commands—first in general terms and then individually—to construct a drawing of a connector.

Load AutoCAD's file MESHES, turn off layers POINTS, SMOOTH-1, and SMOOTH-2, to get a 3D framework of the connector. Then specify a VIEWPOINT of 1,-1,1 to obtain an isometric view of the frame (Fig. 37.42) and apply a 3D skin to it with 3DMESH. After applying each mesh, move them 0,5,0 to form a hollow shell adjacent to the wire-frame diagram to complete the connector.

RULESURF Command Begin with the RULESURF command. It produces lines between two entities, which can be curves, arcs, polylines, lines, or points, even if one boundary is a point **(Fig. 37.37)**. The system variable, SURFTAB1, assigns vertices along each entity when you type SETVAR (CR) and give the number of divisions. For example, when you set SURFTAB1 to 6, it divides the entity into six divisions of mesh.

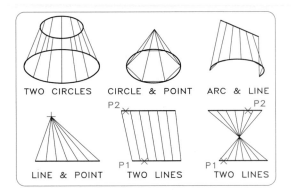

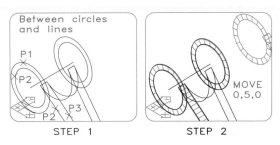

Figure 37.37 The RULESURF command connects entities with a series of 3DFACES. The positions of the selected points determine the direction of the ruled lines for connecting two lines.

Apply RULESURF to add faces between the circular ends, the parallel edges, and the arcs (**Fig. 37.38**). Set SURFTAB1 to 20 to give this number of faces for each RULESURF, and move each set of faces to 0,5,0 to form a hollow shell.

TABSURF Command Apply TABSURF to tabulate a surface from a curve parallel and equal in length to a directional vector (**Fig. 37.39**). Select the circle first and the vector next to tabulate a cylinder with elements equal in length to the vector. The SURFTAB1 system variable (20, in this case) controls the density of the tabulated surface. Move the tabulated surfaces to 0,5,0 to add them to the hollow shell.

REVSURF Command Then use REVSURF to revolve a surface about an axis (**Fig. 37.40**), move it to 0,5,0, and add it to the hollow shell. Lines, polylines, arcs, or circles may be revolved to form a surface of revolution, and SURFTAB1 controls its density. If a circle or a closed polyline is to be revolved about an axis (a torus, for example), system variable SURFTAB2 controls the density of the Line, Circle, Arc, or Pline that is being

Figure 37.38 Applying the RULESURF command:
Step 1 Command: RULESURF (CR)
Select entity??: P1
Select entity??: P2 (Circles are faced.)
Select entity??: P3
Select entity??: P4 (Ends are faced.) (CR)
Step 2 Command: MOVE (CR)
Select objects? (Select faces) (CR)
Base point or displacement: 0,5,0 (CR)
(Faces are moved.)

MESHES: TABSURF

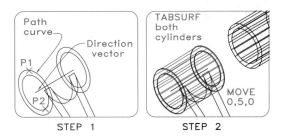

Figure 37.39 Using the TABSURF command:
Step 1 Command: TABSURF
Select path curve: P1
Select direction vector: P2
(The spacing between the tabulated vectors is determined by the SURFTAB1 variable.)
Step 2 Command: MOVE (CR)
Select objects? (Select faces) (CR)
Base point or displacement: 0,5,0 (CR)
(Faces are moved.)

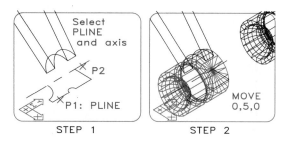

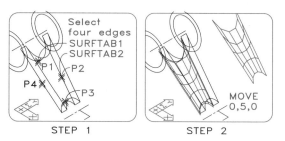

Figure 37.40 Applying the REVSURF command:
Step 1 Command: <u>REVSURF</u> (CR)

Select path of curve: <u>P1</u>

Select axis of revolution: <u>P2</u>

Start angle <0>: (CR)

Included angle (+=ccw,-=cw)<Full circle>:
<u>360</u> (CR)

Step 2 Command: <u>MOVE</u> (CR)

Select objects? (Select faces) (CR)

Base point or displacement: <u>0,5,0</u> (CR)
(Faces are moved.)

Figure 37.41 Utilizing the EDGESURF command:
Step 1 Command: <u>EDGESURF</u> (CR)

Select edge 1: <u>P1</u>

Select edge 2: <u>P2</u>

Select edge 3: <u>P3</u>

Select edge 4: <u>P4</u> (Mesh is drawn in the boundary. System variables SURFTAB1 and SURFTAB2 determine the density of the mesh.)

Step 2 Command: <u>MOVE</u> (CR)

Select objects? (Select faces) (CR)

Base point or displacement: <u>0,5,0</u> (CR)
(Faces are moved.)

revolved, and SURFTAB1 controls the density of the path of revolution.

EDGESURF Command   Select four edges (boundary lines) that intersect to be spanned with a mesh by using EDGESURF **(Fig. 37.41)**. System variables SURFTAB1 and SURFTAB2 control the density of the first and second entities selected, respectively, and you may select the edges in any order. Move the edge surfaces to 0,5,0 to complete the hollow shell of the connector.

Completion of the Connector   Figure 37.42 shows the beginning frame and two final views of the meshed connector, after we have applied the HIDE command. You may now obtain infinitely many views of the 3D connector drawing with the VIEW-POINT and DVIEW commands.

3DMESH Command   When you specify vertices in the M and N directions (the maximum number

COMPLETED CONNECTOR 3D DRAWING

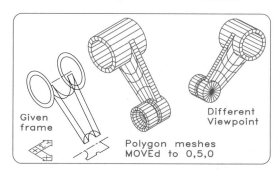

Figure 37.42 The given frame of the connector and the resulting 3D hollow shells from two viewpoints are shown here after use of the HIDE command.

of each direction is 256), the 3DMESH command defines a 3D mesh **(Fig. 37.43)**. Prompts will ask for the X, Y, and Z values of each vertex, starting at the origin. You may specify vertices in arrangements other than rectangular formats. When you

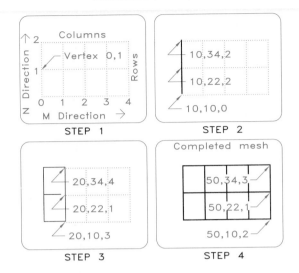

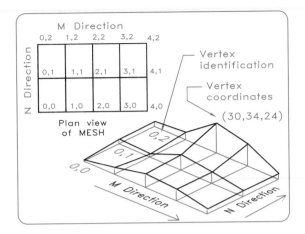

Figure 37.44 Type VPOINT and select a viewpoint for a 3D view of the 3DMESH.

Figure 37.43 Applying the 3DMESH command:
Step 1 Command: <u>3DMESH</u> (CR)

Mesh M size: <u>5</u> (CR)

Mesh N size: <u>3</u> (CR)

Step 2 Vertex (0,0): <u>10,10,0</u>

Vertex (0,1): <u>10,22,2</u> (CR)

Vertex (0,2): <u>10,34,2</u> (CR)

Step 3 Vertex (1,0): <u>20,10,3</u> (CR)

Vertex (1,1): <u>20,22,1</u> (CR)

Vertex (1,2): <u>20,34,4</u> (CR)

Step 4 (Moving to the last set.)

Vertex (4,0): <u>50,10,2</u> (CR)

Vertex (4,1): <u>50,22,1</u> (CR)

Vertex (4,2): <u>50,34,3</u> (CR)

(The mesh is drawn.)

have entered all vertices, the mesh appears as shown in **Fig. 37.44**. Select a VPOINT to give a 3D view of it.

Use PEDIT to change a mesh:

Command: <u>PEDIT</u> (CR)

Edit vertex/Smooth surface/
Desmooth/Mclose/Nclose/Undo/eXit <X>:

Smooth converts a mesh into a smooth 3D surface. Desmooth converts a smoothed surface to its straight-line form. The Undo command reverses the previous PEDIT option. If a mesh is closed, Mopen and Nopen replace the commands Mclose and Nclose as options, and eXit ends the PEDIT command.

Use Edit vertex to change individual vertices of the mesh:

Vertex (M,N): <u>M</u>

Next/Previous/Left/Right/Up/
Down/Move/REgen/eXit <N>

An X (pointer) appears at the origin for selecting a vertex to be changed. The Next and Previous options move the X forward and backward from its current position. The Left and Right options move the pointer in the N directions, and Up and Down options move it in the M directions.

Place the X on a vertex and select Move to receive the prompt Enter new location. Type the X, Y, and Z coordinates of the vertex's new position or select it with the cursor and type REgen to display the revised mesh.

3DPOLY: ABSOLUTE COORDINATES

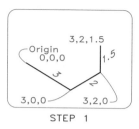
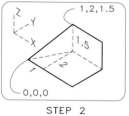

STEP 1   STEP 2

Figure 37.45 Applying 3DPOLY (absolute coordinates):
Step 1 Command: <u>3DPOLY</u> (or LINE) (CR)

From point: <u>0,0,0</u> (CR)

Close/Undo/<Endpoint of line>: <u>3,0,0</u> (CR)

Close/Undo/<Endpoint of line>: <u>3,2,0</u> (CR)

Close/Undo/<Endpoint of line>: <u>3,2,1.5</u> (CR)

Step 2 Close/Undo/<Endpoint of line>: <u>1,2,1.5</u> (CR)

Close/Undo/<Endpoint of line>: <u>C</u> (Back to origin.) (CR)

3DPOLY: RELATIVE COORDINATES

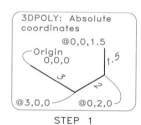
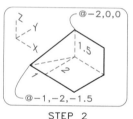

STEP 1   STEP 2

Figure 37.46 Using 3DPOLY (relative coordinates):
Step 1 Command: <u>3DPOLY</u> (or LINE) (CR)

From point: (Select.)

Close/Undo/<End of line>: <u>@3,0,0</u> (CR)

Close/Undo/<End of line>: <u>@0,2,0</u> (CR)

Close/Undo/<End of line>: <u>@0,0,1.5</u> (CR)

Step 2 Close/Undo/<End of line>: <u>@-2,0,0</u> (CR)

Close/Undo/<End of line>: <u>@-1,-2,-1.5</u> (CR)

LINE, PLINE, and 3DPOLY Commands

The LINE command draws 2D lines when you type X and Y values and 3D lines when you type X,

SPLFRAME VARIABLE

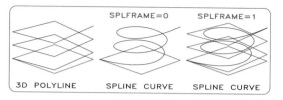

Figure 37.47 The PEDIT and the SPLFRAME system variable (set to 0) are used to spline a 3D spiral. Set SPLFRAME=1 to display both the spline and the defining polyline.

Y, and Z coordinates. Use PLINE to draw a 2D polyline and 3DPOLY to draw 3D polylines with X, Y, and Z coordinates.

To draw lines with absolute coordinates from 0,0,0 in 3D space use the 3DPOLY or LINE command (**Fig. 37.45**). **Figure 37.46** depicts relative coordinates that were typed in the form of @X,Y,Z or @3,0,0 to locate their endpoints with respect to the last point.

All OSNAP modes apply to LINEs, PLINEs, and the visible edges of polyface meshes. You may STRETCH 3D objects in their X and Y directions, but not in their Z direction. Change height dimensions with the ELEV command.

You may change 3DPOLY lines with the PEDIT command to form a spline curve (a helix in this case), as shown in **Fig. 37.47**. System variable SPLINEGS (SPLINE-Generated Segments) control the resolution of the spline curve, and system variable SPLFRAME controls the display of the polygon enclosing the spline curve. When you set SPLFRAME to 0 (off), only the spline curve is shown; when you set it to 1 (on), the curve and its defining polyline are displayed.

3DFACE Command

Use 3DFACE to obtain faces that you may make opaque with the HIDE command (**Fig. 37.48**). Apply 3DFACE to wire-diagram frames by snapping to their endpoints with OSNAPs END option.

3DFACE COMMAND

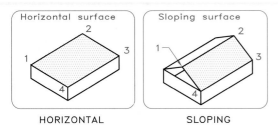

Figure 37.48 3DFACE opaques planes of wire diagrams by snapping to endpoints and produces opaque faces without a wire diagram.

3DFACE: MULTIPLE PLANES

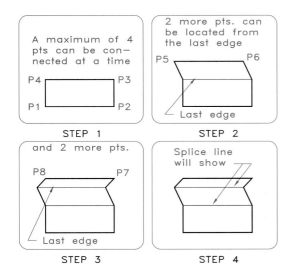

Figure 37.49 Utilizing the 3DFACE command:
Step 1 Command: <u>3DFACE</u> (CR)

First point: <u>P1</u>

Second point: <u>P2</u>

Third point: <u>P3</u>

Fourth point: <u>P4</u>

Step 2 Third point: <u>P5</u>

Fourth point: <u>P6</u>

Step 3 Third point: <u>P7</u>

Fourth point: <u>P8</u>

Step 4 Third point: (CR) (Splice lines will show.)

3DFACE: I OPTION

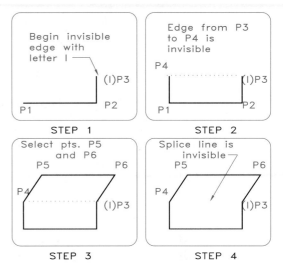

Figure 37.50 Applying the 3DFACE (I option):
Step 1 Command: <u>3DFACE</u> (CR)

First point: <u>P1</u>

Second point: <u>P2</u>

Third point: **I** (CR) <u>P3</u>

Step 2 Fourth point: <u>P4</u>

Step 3 Third point <u>P5</u>:

Fourth point: <u>P6</u>

Step 4 Third point: (CR) (Splice line P3–P4 is invisible.)

Figure 37.49 shows how to select corners of a 3DFACE to form an opaque plane. The maximum number of points that you may select to define a 3DFACE is four, after which prompts ask for points 3 and 4 (using the previous two points as points 1 and 2). Successively added areas are connected with splice lines giving a patchwork appearance. By typing I (invisible) and (CR) prior to selecting the beginning point of a splice line, you can eliminate the *patchwork* look by making that line invisible **(Fig. 37.50)**.

XYZ Filters

Use filters to pick points that adopt the coordinates of 3D points **(Fig. 37.51)**. When a command

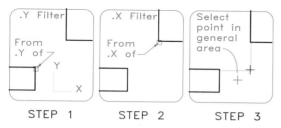

Figure 37.51 Using X and Y filters:
Step 1 Command: LINE (CR)
From point: .Y of (Select corner.) (need XZ)

Step 2 Close/Undo/<Endpoint of line>: .X of (Select corner.) (need YZ)

Step 3 Select a point in the general area of the desired position and the point appears at the intersection of the Y coordinate from the side view and the X coordinate from the top view.

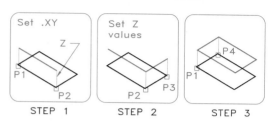

Figure 37.52 Utilizing 3D filters:
Step 1 Command: 3DPOLY (or LINE) (CR)

From point: .XY of P1

of (need Z) 2 (CR)

Close/Undo/<Endpoint of line>: .XY of P2
(need Z) 2 (CR)

Step 2 Close/Undo/<Endpoint of line>: .XY of P3 (need Z) 2 (CR)

Step 3 Close/Undo/<Endpoint of line>: .XY of P4 (need Z) 2 (CR)

Close/Undo/<Endpoint of line>: .XY of P1
(need Z) 2 (CR)

prompts you to select a point (as in the LINE command), type a period (.) followed by one or two coordinates (for example, .X or .XY) and select the point with the cursor. The program will prompt you for the missing coordinate or coordinates.

To locate the front view of a point projected from the left side and top views, select the .Y of the left point and the .X of the top point and the front view of the point SNAPs to the point sharing these coordinates.

Figure 37.52 shows a 3D drawing made with filters by selecting XY coordinates of a given point and specifying Z as 2 when prompted. Here, set OSNAP to END to select the endpoints. When you filter points in the XY plane, a prompt will ask for a Z coordinate to locate the point in 3D space.

37.4 A Wire Diagram Drawing in 3D

Figure 37.53 shows how to construct a 3D wire diagram of an object. First, turn UCSICON to On and draw the plan view of the object and set the UCS origin to 0,0,0 at the midpoint of a line. Specify a viewpoint (1,-1,1,) for an isometric

view of the surface. Copy the first drawing to 0,0,-4 to establish the lower plane of the object with a height of 4. Connect the corners of the planes with vertical lines that SNAP to them, forming a 3D wire diagram.

Figure 37.54 illustrates an arrangement chosen from VPORTS for displaying the wire diagram in four ports. Use the SAVE option of UCS to save the upper right viewport as ISO. Then, select the upper left viewport (step 2) and type PLAN to obtain a true-size top view. Type UCS and SAVE to save this UCS as TOP.

Next, move to the lower left viewport (step 3), rotate the UCS 90° about the X axis, and type PLAN to obtain the front view. SAVE this UCS as FRONT. Finally, rotate the UCS 90° about the Y axis and type PLAN to get the side view. SAVE it as SIDE.

This process gives top, front, and right side views and isometric views of the wire-frame drawing. Any lines added in any of the four viewports would be shown in the others simultaneously.

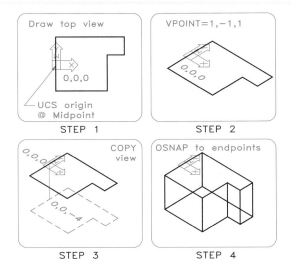

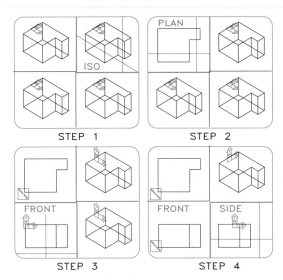

Figure 37.53 Constructing a wireframe drawing in 3D:
Step 1 Draw the plan view of the part. Use UCS to set the origin at the midpoint of the line. Type UCSICON and Origin to locate the XY icon at the UCS origin.

Step 2 Command: <u>VPOINT</u> (CR)

Rotate/<View point> <0,0,0>: <u>1,-1,1</u> (CR)
(A 3D view is displayed.)

Step 3 Command: <u>COPY</u> (CR)

Select objects: <u>W</u> (Window 3D view.)

<Base point or displacement>/Multiple: <u>0,0,-4</u> (CR)

Second point of displacement: (CR)

Step 4 Set OSNAP to END and connect the corner points with the LINE command.

To set the system variable UCSFOLLOW to ON (1), use the SETVAR command. When on, the UCSFOLLOW variable augments the PLAN option when you RESTORE a UCS, automatically showing the X and Y axes true size in the restored view. If you set UCSFOLLOW before selecting multiple VPORTS, it will apply to all VIEWPORTS and must be turned on or off one at a time.

Note: While working in VPORTS, you may find that the CHANGE command does not always prop-

Figure 37.54 Obtaining multiple views of a 3D object:
Step 1 Command: <u>VPORTS</u> (CR)

Save/Restore/Delete/Join/SIngle/?/2/<3>/ 4: <u>4</u> (CR)

(The 3D drawing appears in four ports.)

Step 2 (Select the upper left port.)

Command: <u>PLAN</u> (CR)

<Current UCS/UCS/World: (CR)

Command: <u>ZOOM</u>

All/Center/....<Scale(X)>: <u>.6X</u> (CR)

Command: <u>UCS</u> (CR)

Origin/Xaxis/...<World>: <u>SAVE</u> (CR)

?/Name of UCS: <u>TOP</u> (CR)

Step 3 (Select lower left port.)

Command: <u>UCS</u> (CR)

Origin/Xaxis/...<World>: <u>X</u> (CR)

Rotation angle about X axis: <u>90</u> (CR)

Command: <u>PLAN</u> (CR)

<Current UCS>/UCS/World: (CR) (Get front view.)

UCS (Command repeated.)

Origin/Xaxis...<World>: <u>SAVE</u> (CR)

?/Name of UCS: <u>FRONT</u> (CR)

Step 4 (Select lower right port.)

Use UCS and Y to rotate the UCS 90° about the Y axis; type PLAN to obtain the right side view and save as SIDE.

ANGLE BRACKET PROBLEM

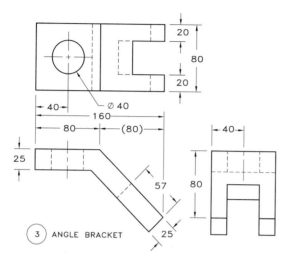

Figure 37.55 We use this angle bracket with an inclined surface to demonstrate how to produce a 3D view of the part in Figs. 37.56–37.60.

erly select entities in 3D. Therefore, you should use the CHPROP command as an alternative method of selecting and changing layers and 3D entities.

37.5 Object with an Inclined Surface

Figure 37.55 shows three orthographic views of a dimensioned angle bracket with an inclined plane. Your task is to produce a 3D view of the bracket with TILEMODE=1 (on).

Part 1: Fig. 37.56 Draw the top view of the bracket and set the UCS origin at the midpoint of one of its lines. Move the UCSICON to the UCS origin to obtain an isometric view (VPOINT=1,-1,1) and copy the surface to 0.25 inch below the upper surface. Connect the corner points by OSNAPping to them with the END option.

Part 2: Fig. 37.57 Select four VPORTS for displaying the object. Rotate the UCS of the lower left port 90° about the X axis, making the UCSICON parallel to the object's front view. Type PLAN to obtain the

ANGLE BRACKET: PART 1

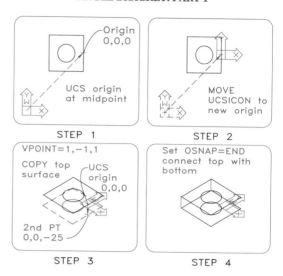

Figure 37.56 Connecting top and bottom surfaces:
Step 1 Draw the top view of the part and set the UCS origin at the midpoint of the line.

Step 2 Use UCSICON and Origin to place the icon at the UCS origin.

Step 3 Set a VPOINT to 1,-1,1 to obtain an isometric view of the plane. Copy the surface to a position of 0,0,-25 from its origin.

Step 4 Set the OSNAP to END and connect the corners of the two planes.

front view and type UCS and SAVE to preserve this UCS as FRONT.

Part 3: Fig. 37.58 RESTORE FRONT and rotate the UCS 90° about the Y axis to place the XY icon parallel to the side plane of the object. Activate the lower right panel, SAVE the UCS, and type PLAN to obtain the right side view. Obtain the top view in the upper right panel of this figure in the same manner as you did for the two previous orthographic views.

Part 4: Fig. 37.59 Activate the right side port to RESTORE the UCS named SIDE and show it in isometric by setting VPOINT to (1,-1,1). Establish

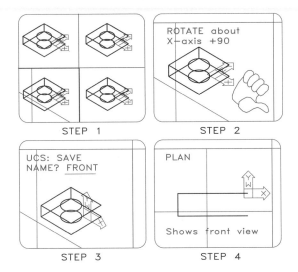

Figure 37.57 Using four-port construction:
Step 1 Command: <u>VPORTS</u> (Select four ports; and the 3D drawing is shown in each.)

Step 2 (Select lower left port.) Command: <u>UCS</u> (CR) (Select X option and rotate UCS 90° about the X axis for a front view.)

Step 3 Command: <u>UCS</u> (CR) (Enter <u>SAVE</u>.) ?/Name of UCS: <u>FRONT</u> (CR)

Step 4 Command: <u>PLAN</u> (CR) (Get front view and ZOOM to the desired size.)

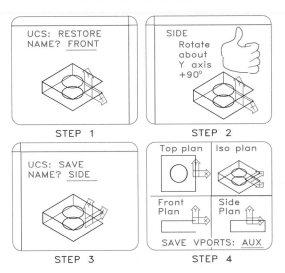

Figure 37.58 Producing top and side views:
Step 1 You may RESTORE the FRONT view with the UCS command, but it is unnecessary here.

Step 2 (Select the lower right port.) Command: <u>UCS</u> (CR) (Select Y and rotate the UCS 90°.)

Step 3 Command: <u>UCS</u> (CR)

SAVE ?/Name of UCS: <u>SIDE</u> (CR)

Step 4 Command: <u>PLAN</u> (CR) (Obtain the side view.) Use the same steps to obtain and save a top view in the upper left port. Leave the 3D view in the upper right corner.

the UCS origin with coordinates of 0,-80, 80 at the end of the sloping surface; the XY icon moves to it. Rotate the XY icon –45° about the X axis to make it lie in the plane of the inclined surface. SAVE this UCS as INCLINED. Produce the notch by using the coordinates taken from the given view of the object (Fig. 37.55). You may draw these points with the cursor while observing the polar coordinates at the top of the screen because the plane of the inclined surface lies in the UCS.

Part 5: Fig. 37.60 Use the OFFSET command with an offset of 0.25 inch to select lines in the isometric view of the inclined surface one at a time. Use the lower left port to indicate the side of the

offset (P). The offset lines appear in all ports. Apply FILLET (R=0) to the lines in the upper right port to make them intersect precisely. Use OSNAP and the END option to draw lines connecting the corner points of the wire diagram.

Use 3DFACE with OSNAP set to END to apply faces to the planes of the wire diagram, which will require the selection of several vpoints. When done, use VPOINT and HIDE, to obtain a 3D view of the bracket with hidden lines suppressed.

37.6 Drawing with TILEMODE=0

We produced all the previous drawings in 3D space with TILEMODE=1 (on). In the remainder of this sec-

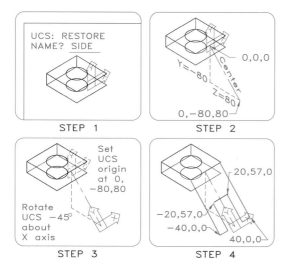

ANGLE BRACKET: PART 4

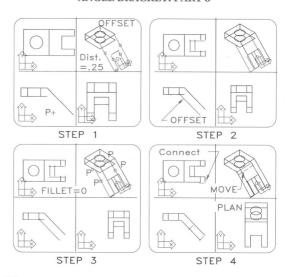

ANGLE BRACKET: PART 5

Figure 37.59 Generating the inclined surface:
Step 1 (Select the side view port.) Command: UCS (CR) (Select RESTORE.)

?/Name of UCS: SIDE (CR) Command: VIEW-PORTS (CR) (SAVE this viewport configuration as AUX1.)

Command: VPORTS (CR) (Select SIngle to obtain a full-screen image.)

Step 2 Command: UCS Origin/Zaxis/...<World>: OR (Select origin.)

Origin point: 0,-80,80 (CR)

Step 3 Command: UCSICON (CR) (Select ORigin to place icon at a new origin.)

Command: UCS (CR) (Select X and rotate UCS 45° about the X axis.) Use UCS command to SAVE as INCLINED.

Step 4 Use the LINE command and dimensions from the given view of the part to specify the corners of the notch.

Figure 37.60 Completion of the drawing:
Step 1 Command: VPORTS (RESTORE the four-port configuration AUX1.) Use OFFSET and select a point on the notch in the 3D view with an offset distance of 25, and locate the side of the offset in the front viewport.

Step 2 The lower surface of the inclined plane is displayed.

Step 3 Use FILLET to trim the ends of the intersecting lines. Pick the lines in the 3D view.

Step 4 With OSNAP set to END, connect the endpoints of lines. MOVE the horizontal intersection line to its new position. Any of the views may be used for developing the drawing.

tion, we produce drawings with TILEMODE=0 (off) in order to move between paper space and model space. (Recall the overview of paper space and model space in Section 37.2.)

Moving between Paper Space and Model Space

Type UCSICON and On so that the icon will appear to indicate which mode the screen is in. A triangle

appears when it is in paper space, and the XY icon appears when it is in model space. Type TILEMODE and set it to 0 (off) which turns the screen into paper space (blank screen) with a PSPACE icon in the lower left corner **(Fig. 37.61)**. Then set the paper-space LIMITS large enough to contain the border, about 12 × 9 inches, and insert a size A border.

Type MVIEW, select diagonal corners of a model space viewport with the cursor, COPY it, and type MSPACE (or MS) to enter model space and make one of the model spaces active **(Fig. 37.62)**. Set the LIMITS in this 3D viewport to 24,20, or large enough to contain the object to be drawn.

TILEMODE=0: PAPER SPACE

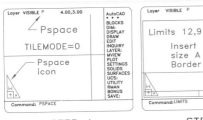

STEP 1   STEP 2

Figure 37.61 Using nontiled viewports (`PSPACE`):
Step 1 Command: `TILEMODE` (CR)

New value for `TILEMODE` <1>: `0` (CR) (Enters
`PSPACE`; 2D icon and P in status line appear.)

Step 2 Command: `INSERT` (CR)

Block name (or?) : `BORDER-A` (CR)

Insertion point: `0,0` (CR)

X scale factor <1>/Corner/XYZ: `1` (CR)

Y scale factor (default = X): `2` (CR)

Rotation angle <0>: `0` (CR) (Border is inserted in
`Pspace`. Set limits to `12,9` and `ZOOM/ALL`.)

MVIEW TO MAKE MS WINDOWS

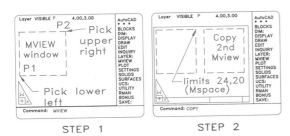

STEP 1   STEP 2

Figure 37.62 Using nontiled viewports (`MVIEW` option):
Step 1 Command: `MVIEW` (Switching to paper space.)

`ON/OFF/Hideplot/Fit/2/3/4/Restore/<First
Point>`: `P1`

Other corner: `P2` (CR) (A 3D model-space port is
drawn.)

Step 2 While still in paper space, make a copy of the 3D
model-space port, `Return` to MS, and set `VPORT` limits to
`24,20`.

Other options under `MVIEW`, which operates
only from paper space, are `ON`, `OFF`, `Hideplot`,
`Fit`, 2, 3, 4, and `Restore`.

DRAWING IN MODEL SPACE

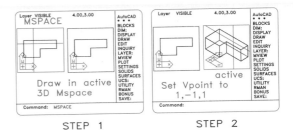

STEP 1   STEP 2

Figure 37.63 Using nontiled viewports (drawing in
model space):
Step 1 Command: `MSPACE` (3D icons appear in both
ports; the cursor is in the active port.)

(Use `ELEV` and `THICKNESS` to draw an extrusion of a
`BLOCK`; it will show in both views.)

Step 2 Pick the right viewport to activate it.
Command: `VPOINT` (CR)
`Rotate/<View point> <current>`: `1, -1, 1` (CR)
(An isometric view is obtained.)

`ON/OFF`: select and turn off viewports to save
regeneration time but leave at least one on.

`Hideplot`: when turned on, removes the hidden lines from the selected viewports.

`Fit`: makes a viewport that is the size of the
current screen.

`2/3/4`: lets you create the number of viewports needed and the areas to be filled.

`Restore`: recalls a viewport configuration
saved by `VPORTS`.

Figure 37.63 shows a 3D object produced by
the extrusion method in one of the 3D viewports
and shown simultaneously in a second port.
Select a `VPOINT` for an isometric view of the
object. Type `PS` to enter paper space **(Fig. 37.64)**.
The cursor spans the screen in paper space. Now
the drawing must be scaled.

The paper-space `LIMITS` are `12,9`, inside of
which will be two model-space viewports with
`LIMITS` of `24,20` each. These model-space ports
must be sized to fit, which requires calculations.

ZOOMING TO SCALE

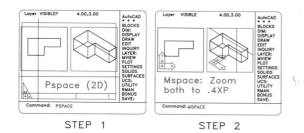

STEP 1   STEP 2

Figure 37.64 Using nontiled viewports (ZOOMing to scale):

Step 1 Command: <u>PSPACE</u> (CR) (The 2D border reappears and the cursor spans the screen. Because PS had limits of 12,9 and MS had limits of 24,20, you must ZOOM the MS ports to about 0.20 size for both to fit in border.)

Step 2 Command: <u>MSPACE</u> (CR) (One of the 3D ports becomes active.)

Command: <u>ZOOM</u> (CR)

All/Center/ ... /<Scale (X/XP)>: <u>.2XP</u> (Active port is scaled to 0.2 size to fit within 2D border. Select and ZOOM other view to 0.2XP.)

The combined width of the two 24-inch wide model spaces is 48 inches. When scaled to half size, their width is 24 inches, or too wide to fit. But if you use a 0.20 (two-tenths) scale, their widths are 9.6 inches, or small enough to fit inside the size A border (Fig. 37.64).

While in model space, type ZOOM and use the X/XP (2/10XP or 0.2XP) option to size the contents of each 3D port. This factor changes the width limit of each viewport from 24 to 4.8 inches, with both drawings having the same scale on the screen. In other words, you may scale viewports with width limits of 24 inches to a full-size width of 4.8 inches in paper space.

Type PS to enter paper space and use STRETCH to reduce the size of the MSPACE outlines on the screen (**Fig. 37.65**). Reposition the model-space views for plotting with MOVE.

You have to plot from PSPACE. Both the 2D and 3D drawings are plotted at the same time,

MOVING MS VIEWPORTS

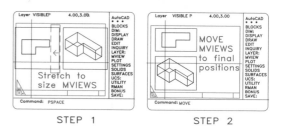

STEP 1   STEP 2

Figure 37.65 Using nontiled viewports (moving ports):

Step 1 Command: <u>PSPACE</u> (CR) (Switch to 2D paper space. Use the STRETCH command to reduce the 3D ports.)

Step 2 MOVE MVIEW ports to their final positions within the 2D border in paper space.

PLOTTING FROM PAPER SPACE

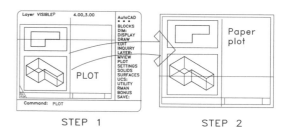

STEP 1   STEP 2

Figure 37.66 Using nontiled viewports (plotting):

Step 1 Command: <u>PLOT</u> (CR) (Give specification for plotting from paper space.)

Step 2 Plot the size A layout to show both 2D (paper space) and 3D (model space) in the same plot.

including the window outlines around the model-space ports (**Fig. 37.66**). If you do not want the window outlines, CHANGE them to a separate LAYER (such as WINDOW), FREEZE it before plotting, and PLOT in the usual manner.

To remove hidden lines from a plot, apply a mesh to the object. Use the HIDEPLOT option of MVIEW and the Remove hidden lines button from the PLOT dialogue box. From paper space, type MVIEW, select HIDEPLOT, and choose the outlines of the model-space viewports to be hidden when

DIMENSIONING IN MODEL SPACE

STEP 1   STEP 2

Figure 37.67 Dimensioning in model space:
Step 1 When you set the Dim Vars DIMSCALE to 0, the dimension variable in model space will match those specified for paper space. Move the UCSICON to the plane of the dimension. Use associative dimensions and SNAP to the endpoints of the object.

Step 2 When you apply a dimension in one viewport, it is shown in all 3D viewports and the size of the dimensioning variables are the same in MSPACE as in PSPACE.

plotted. Using VPLAYER, select 3D viewports from paper space in which layers may be turned off or frozen while remaining on or thawed in other viewports. (Figure 37.69 illustrates an application of this command.)

37.7 Dimensioning in 3D

You may add dimensions to objects in model space to match the dimensioning variables set for paper space. When DIMSCALE=0, the program sizes dimensioning variables in model space to match those set in paper space (**Fig. 37.67**). Use associative dimensions (DIMASO=On) and SNAP to the endpoints of the object to attach a horizontal dimension in the top view, which appears in both ports.

Select the isometric viewport and rotate the XY icon parallel to the frontal plane (UCS/X/90°), as demonstrated in **Fig. 37.68**. To attach a vertical dimension to the object, SNAP to the endpoints of the isometric view, and it appears as an edge in the top view.

You cannot see the vertical dimension in the top view, so select the top viewport and type

DIMENSIONING A 3D VIEW

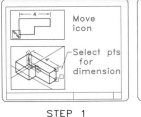

STEP 1   STEP 2

Figure 37.68 Dimensioning a 3D view:
Step 1 Move the UCSICON to the plane of the vertical dimension in the 3D view. Select the three points of the dimension.

Step 2 The dimension appears in the 3D view and in the top view. It is an edge and is not readable in the top view.

USING THE VPLAYER COMMAND

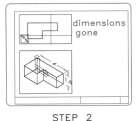

STEP 1   STEP 2

Figure 37.69 Dimensioning with the VPLAYER command:
Step 1 Because the vertical dimension is not readable in the top view, set VPLAYER to On, select the border of the 3D viewport (while in PS) and turn Off the Dimen Layer in that view.

Step 2 The dimensions are removed in the top view but remain in the 3D view.

VPLAYER and its FREEZE option to turn off the dimensioning layer (**Fig. 37.69**). That action freezes both the vertical and horizontal dimensions in the top view but retains them in the isometric view where they are readable. You may add other dimensions in the same manner.

Options under VPLAYER that operate from paper space are Freeze, Thaw, Reset, Newfrz, and Vpvisdflt.

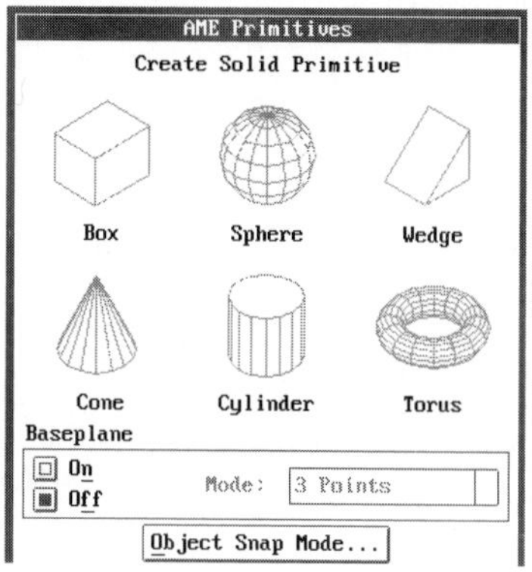

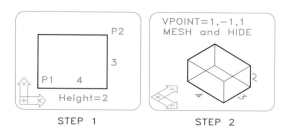

Basic solids that can be chosen for 3D modeling

Figure 37.70 These six primitives in the AME Primitives dialogue box may be selected and produced as solids.

Figure 37.71 Producing a box:
Step 1 Command: <u>SOLBOX</u> (CR)
Corner of box: <u>P1</u> (CR)
Cube/Length/<Other corner>: <u>P2</u> (CR)
Height <1>: <u>2</u> (CR)
Step 2 Command: <u>VPOINT</u> (CR)
Enter vpoint <0,0,0>: <u>1,-1,1</u> (CR)
(3D view of a solid box is obtained.)

Freeze/Thaw: allows layers in selected viewports to be frozen or thawed; Thaw does not work on globally frozen layers.

Reset: changes visibility of one or more layers in selected viewports to their current default setting.

Newfrz: creates a new layer that is visible in the current viewport and is frozen in all other viewports.

Vpvisdflt: used to set default visibility by viewport for any layer; the default setting freezes or thaws layers in new viewports.

37.8 Introduction to Solid Modeling

An auxiliary software package, **Advanced Modeling Extension (AME)**, is available from Autodesk and a portion of it comes with Release 12 to provide solid modeling capabilities. The following introduction to solid modeling is a brief overview of Autodesk's 340-page manual.

Solid modeling is a technique of creating 3D solids from solid primitives—boxes, cylinders, spheres, and others—that can be added or sub-

tracted from each other to form an object. In solid modeling, objects cannot be created on the screen that cannot actually be made.

The AME system allows you to extrude 2D planar drawings into 3D solids, to transform wire diagrams into solids, and to modify solids by adding fillets or chamfers and other features. Meshes and shading to render the solids add realism.

Solid Primitives

Solid primitives—Box, Sphere, Wedge, Cone, Cylinder, and Torus—appear in the AME Primitive dialogue box under Model of the pull-down menu and under Model of the side menu. You may also select a solid primitive by typing DDSOLPRM (**Fig. 37.70**). Solid primitives are inserted as wire diagrams that cannot be hidden by HIDE until they have been meshed. Type SOLMESH, select the wire diagrams that turn the primitives into solids, and type HIDE to obtain them as 3D models with hidden lines suppressed. You must follow these steps to create each primitive.

Box Use SOLBOX to create the most basic of all solids, the box (**Fig. 37.71**); its base is drawn in the

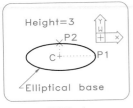
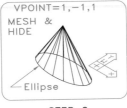

STEP 1   STEP 2

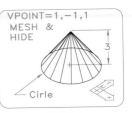 Figure 37.72 Producing a cone with an elliptical base:
Step 1 Command: <u>SOLCONE</u> (CR)

Elliptical/<Center point>: <u>E</u> (CR)

<Axis endpoint 1>/Center of base: <u>C</u> (CR)

Center of ellipse: <u>C</u>

Axis endpoint: <u>P1</u>

Other axis distance: <u>P2</u>

Height of cone: <u>3</u> (CR)

Step 2 Command: <u>VPOINT</u> (CR)

Enter vpoint <0,0,0>: <u>1,-1,1</u> (CR)

(3D view of the cone is drawn.)

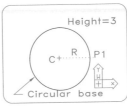
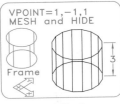

STEP 1   STEP 2

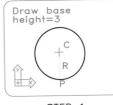 Figure 37.73 Producing a cone with a circular base:
Step 1 Command: <u>SOLCONE</u> (CR)

Elliptical/<Center point>: <u>C</u> (CR)

Diameter/<Radius>: <u>P1</u> (CR)

Height of cone: <u>3</u> (CR)

Step 2 Command: <u>VPOINT</u> (CR)

Enter vpoint <0,0,0>: <u>1,-1,1</u> (CR)

(3D view of the cone is drawn.)

THE SOLCYL COMMAND

STEP 1   STEP 2

Figure 37.74 Producing a cylinder:
Step 1 Command: <u>SOLCYL</u> (CR)

Elliptical/<Center point>: <u>P</u> (CR)

Height of cylinder: <u>3</u> (CR)

Step 2 Command: <u>VPOINT</u> (CR)

Enter Vpoint <0,0,0> <u>1,-1,1</u> (CR)

(3D view of a cylinder is obtained.)

---

plane of the current UCS. To generate the dimensions of the box with separate widths and depths, diagonal corners of the base, or as a cube, type values at the keyboard or use the cursor to select points on the screen.

Cone Use SOLCONE to construct cones with circular or elliptical bases. **Figure 37.72** illustrates how to produce a cone with an elliptical base by locating its center, axis endpoints, and height. **Figure 37.73** shows how to produce a cone with a circular base by specifying its center, radius (or center and diameter), and height. Type values at the keyboard or specify them with the cursor.

Cylinder Use SOLCYL to produce a solid cylinder in a manner similar to that used to obtain a cone. **Figure 37.74** illustrates the procedure for a cylinder with a circular base; cylinders with elliptical bases also may be drawn.

Sphere You may create a sphere with SOLSPHERE by responding to the prompts with the center and radius (or center and diameter) of a ball (**Fig. 37.75**). Position the sphere so that the axis connecting its north and south poles is parallel to the Z axis of the current UCS and its center is on the plane of the UCS.

THE SOLSPHERE COMMAND

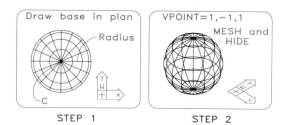

STEP 1 STEP 2

Figure 37.75 Producing a sphere:
Step 1 Command: <u>SOLSPHERE</u> (CR)
Center of sphere: <u>C</u> (CR)
Diameter/<Radius> of sphere: <u>4</u> (CR)
Step 2 Command: <u>VPOINT</u> (CR)
Enter vpoint <0,0,0>: <u>1,-1,1</u> (CR)
(3D view of a sphere is obtained.)

THE SOLTORUS COMMAND

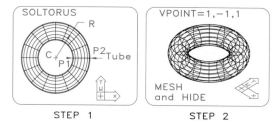

STEP 1 STEP 2

Figure 37.76 Producing a torus:
Step 1 Command: <u>SOLTORUS</u> (CR)
Center of torus: <u>C</u> (CR)
Diameter/<Radius> of torus: (Select.) (CR)
Diameter/<Radius> of tube: (Select.) (CR)
Step 2 Command: <u>VPOINT</u> (CR)
Enter vpoint <0,0,0>: <u>1,-1,1</u> (CR)
(3D view of a solid torus is obtained.)

Torus To create a torus (or toroid, a doughnut shape) respond to the prompts with its center, diameter or radius of the tube, and diameter or radius of revolution (**Fig. 37.76**). Position the diameter of the torus to lie in the plane of the current UCS.

THE SOLWEDGE COMMAND

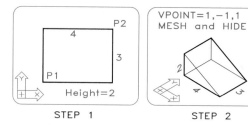

STEP 1 STEP 2

Figure 37.77 Producing a wedge:
Step 1 Command: <u>SOLWEDGE</u> (CR)
Corner of wedge: <u>P1</u> (CR)
Length/<Other corner>: <u>P2</u> (CR)
Height: <u>2</u> (CR)
Step 2 Command: <u>VPOINT</u> (CR)
Enter vpoint <0,0,0>: <u>1,-1,1</u> (CR)
(3D view of a solid wedge is obtained.)

SOLEXT COMMAND

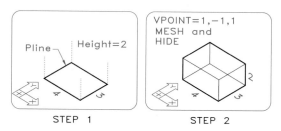

STEP 1 STEP 2

Figure 37.78 Generating extrusions:
Step 1 Command: <u>SOLEXT</u> (CR)
Select polylines and circles for extrusion.
Select objects: (Select rectangle.)
Select objects: (CR)
Height of extrusion: <u>2</u> (CR)
Extrusion taper angle from Z <0>: (CR)
Step 2 Command: <u>VPOINT</u> (CR)
Enter vpoint <0,0,0>: <u>1,-1,1</u> (CR)
(3D view of the extrusion is obtained.)

Wedge Use SOLWEDGE to generate a rectangular base of a wedge that lies in the plane of the current UCS and whose upper plane slopes toward a second point selected (**Fig. 37.77**). Prompts ask

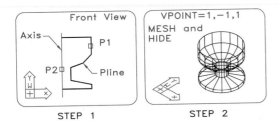

■ **Figure 37.79** Revolution of a solid:
Step 1 Command: <u>SOLREV</u> (CR)

(Select polyline or circle for revolution.)

Select objects: <u>P1</u>

Select axis: <u>P2</u> (CR)

Axis of revolution-Entity/X/Y/<Start point of axis>: <u>E</u> (CR)

Included angle <full circle>: (CR)

Step 2 Command: <u>VPOINT</u> (CR)

Enter vpoint <0,0,0>: <u>1,-1,1</u> (CR)

(3D view of the revolved solid is obtained.)

■ **Figure 37.80** Solidifying an extruded wire frame:
Step 1 Command: <u>SOLIDIFY</u> (CR)

Select objects: (Select base.)

Select objects: (CR)

Step 2 Command: <u>VPOINT</u> (CR)

Enter vpoint <0,0,0>: <u>1,-1,1</u> (CR)

(3D view of the solidified surface is obtained.)

for the length, width, and height of the wedge, each of which you may type at the keyboard or specify with the cursor.

Extrusions

Use SOLEXT to extrude polylines, polygons, circles, ellipses, and 3D entities drawn as 2D figures to a specified height and taper their sides inward from the base (if desired). **Figure 37.78** shows a rectangle drawn as a polyline extruded upward to a selected height. Polylines with crossing or intersecting segments cannot be extruded.

Solid Revolution

You may revolve (sweep) polylines, polygons, circles, ellipses, and 3D poly objects about an axis, if they have at least 3 and less than 300 vertices, with the SOLREV command. The path of revolution can start and end at any point between 0° and 360°. Do not revolve polylines that have been Fit or

Splined because of the extensive calculations required. **Figure 37.79** shows a polyline revolved a full 360° about an axis.

Solidify

You may convert objects extruded using the THICKNESS command with nonzero thicknesses (not as solids) into solids with the SOLIDIFY command **(Fig. 37.80)**. Polylines, polygons, circles, ellipses, traces, donuts, and 2D polygons are such shapes. Their corners must be perfect joints and segments cannot intersect for SOLIDIFY to work.

Subtracting Solids

To subtract one intersecting solid from another, use SOLSUB **(Fig. 37.81)**. This approach is especially helpful in subtracting a cylinder from a solid in order to represent a hole.

Adding Solids

You may join overlapping solids with SOLUNION to form a single composite solid model. **Figure 37.82** shows how to unify a box and cylinder into a single solid by selecting each in succession.

THE SOLSUB COMMAND

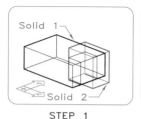 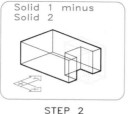

STEP 1   STEP 2

 **Figure 37.81** Subtracting solids:
Step 1 Command: <u>SOLSUB</u> (CR)

Source objects ...

Select objects: (Select solid 1.)

Select objects: (CR)

Objects to subtract from them ...

Select objects: (Select solid 2.)

Select objects: (CR)

Step 2 Command: <u>VPOINT</u> (CR)

Enter vpoint <0,0,0>: <u>1,-1,1</u> (CR)

(3D view of the resulting solid is obtained.)

THE SOLUNION COMMAND

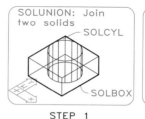

STEP 1   STEP 2

Figure 37.82 Joining of solids:
Step 1 Command: <u>SOLUNION</u> (CR)

Select objects: (Select box.) (CR)

Select objects: (Select cylinder.) (CR)

Select objects: (CR)

Step 2 Command: <u>VPOINT</u> (CR)

Enter vpoint <0,0,0>: <u>1,-1,1</u> (CR)

(3D view of the composite solid is obtained.)

Separating Solids

Use the SOLSEP command to separate solids previously combined with the SOLSUB and SOLUNION commands. You may undo the commands to edit or correct and then reenter them.

Chamfers

To chamfer solid objects, use the SOLCHAM command. Select the base surface, the adjoining surface, and the edges to be chamfered (**Fig. 37.83**). Respond to the prompts to enter the chamfer distances along the first and second surfaces. When you satisfy these prompts, the program automatically generates the chamfers.

Fillets

To use SOLFILL, select the intersecting edges of solids to be filleted or rounded and specify the diameter or radius of the fillet (**Fig. 37.84**). You may also use the SOLFILL command to round the ends of cylindrical objects.

THE SOLCHAM COMMAND

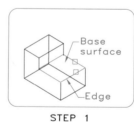 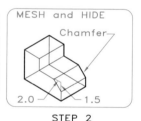

STEP 1   STEP 2

Figure 37.83 Creating a solid chamfer:
Step 1 Command: <u>SOLCHAM</u> (CR)

Select base surface: (Select.)

<OK>/ Next: (CR)

Select edges to be chamfered (Press ENTER when done): (Select.) (CR)

1 edge selected

Enter distance along base surface <default>: <u>2</u>

Enter distance along adjacent surface <default>: <u>1.5</u>

Step 2 Command: <u>VPOINT</u> (CR)

Enter vpoint <0,0,0>: <u>1,-1,1</u> (CR)

(3D view of chamfered object is obtained.)

THE SOLFILL COMMAND

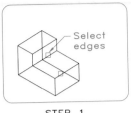
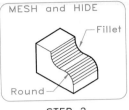

STEP 1   STEP 2

Figure 37.84 Producing a solid fillet:
Step 1 Command: <u>SOLFILL</u> (CR)

Select edges to be filleted (Press ENTER when done): (Select 2 edges.) (CR)

2 edges selected

Diameter/<Radius> of fillet <default>: <u>3</u> (CR)

Step 2 Command: <u>VPOINT</u> (CR)

Enter vpoint <0,0,0>: <u>1,-1,1</u> (CR)

(3D view of filleted object is obtained.)

THE SOLMOVE COMMAND

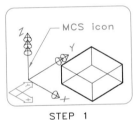
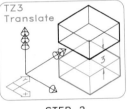

STEP 1   STEP 2

Figure 37.85 Moving a solid:
Step 1 Command: <u>SOLMOVE</u> (CR)

Select objects: (Select box.) (CR)

Select objects: (CR)

Redefining block SOLAXES

<Motion description>/?: <u>TZ3</u> (CR)

<Motion description>/?: (CR)

Step 2 (The solid box is translated 3 units in the Z direction.)

Solid Move

Use SOLMOVE to rotate and move solids. After you select the solids, an icon appears to indicate a temporary motion coordinate system (MCS) with

THE SOLHPAT COMMAND

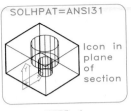
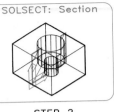

STEP 1   STEP 2

Figure 37.86 Obtaining a section view:
Step 1 Command: <u>SOLHPAT</u> (CR)

Hatch pattern <default>: <u>ANSI31</u> (CR)

(Position UCS icon in plane of section.)

Step 2 Command: <u>SOLSECT</u> (CR)

Select objects: (Select box.)

Select objects: (CR)

(Section is drawn through the object in plane of icon.)

one, two, and three arrowheads at each end **(Fig. 37.85)**. The single arrow defines the X axis, two arrows define the Y axis, and three arrows define the Z axis.

When prompted with <Motion description>/?:, type ? to obtain a list of 15 options for moving or rotating the selected solid. The meaning of some of these options is self-explanatory, but others require reference to HELP screens. Figure 37.85 shows a solid box translated (moved) along the Z axis a distance of 3 units. Here, the SOLMOVE option typed TZ3 means that the translation is parallel to the Z axis for a distance of 3 units.

Sections

Recall that a sectioning plane passed through a 3D solid enables you to visualize better its internal features. The use of SOLHPAT (solid hatch pattern) assigns a hatching pattern. Type SOLVAR (solid variables) to enter SOLHANGLE and SOLHSIZE, which control the angle and spacing, respectively, of the hatching. You may also enter variables at the command line. **Figure 37.86** shows SOLHPAT set to ANSI31 for a cast iron symbol.

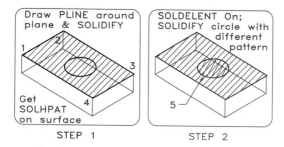

Figure 37.87 Region modeling (hatching):
Step 1 A PLINE is drawn to SNAP to the lines defining the region and it is hatched.

Command: SOLIDIFY (CR)

Select objects: (Pick Pline.)

(Change hatch pattern with SOLHPAT.)

Step 2 Command: SOLELENT (CR)

Delete the entity after extrusion, revolution or solidification? (1=never, 2=ask, 3=always)<3>: 2 (CR)

Command: SOLIDIFY (CR)

Delete the entities that are solidified? <N> Y (CR)

Select objects: Select circle (CR) (It is hatched with a different pattern.)

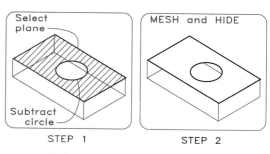

Figure 37.88 Region modeling (opaque):
Step 1 Command: SOLSUB (CR)

Source objects...

Select objects: (Select plane) (CR)

Objects to subtract from them: (Select circle.) (CR)

(Circle is subtracted.)

Step 2 Command: SOLMESH (CR)

Select objects: (Select plane) (CR)

Command: HIDE (CR)

Select objects: (Select plane) (CR)

(Plane appears opaque with hole removed.)

Place the UCS icon on the object to establish the plane of the section and type SOLSECT. The section with cast-iron symbols passing through the object appears on the screen. To find other sections through the object in the same manner, simply move the icon.

37.9 Region Modeling

Region modeling is a 2D version of solid modeling by which you may convert a surface into a solid plane (**Fig. 37.87**). Enclose the upper plane of a wire diagram with a PLINE by SNAPping to its corner points, hatch the surface, and make it solid with the SOLIDIFY command. Select the hatch symbol with the SOLHPAT command just as you did with the standard Hatch command. To delete the border of the area being solidified, turn

on SOLDELENT before applying the SOLIDIFY command.

Use SOLSUB to select the plane and then the circle to be subtracted from the plane (**Fig. 37.88**). Mesh the plane with SOLMESH and use HIDE to suppress the hidden lines of the plane.

SOLIDIFY cannot be applied to 3D wire diagrams or to solids that have been exploded.

37.10 Summary of Solid Commands

You may use the following commands in region and solid modeling, but not with both types in combination.

SOLAREA (solid area): computes the area of the region or solid.

SOLCHP (solid change property): used to change properties, move, copy, replace, delete, or resize solid primitives.

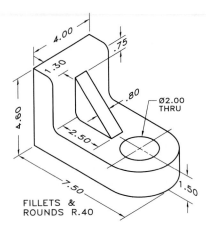

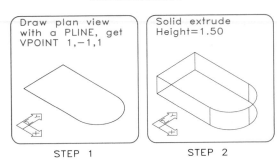

Figure 37.89 We use this bracket to demonstrate solid modeling in Figs. 37.90–37.96.

Figure 37.90 Bracket (SOLEXT command):
Step 1 Draw plan view of base of bracket with PLINE.

Command: VPOINT (CR)

Rotate/<Viewpoint>: 1,-1,1 (CR)

(Get isometric view of base.)

Step 2 Command: SOLEXT (CR)

(Select regions polylines and circles for extrusion.)

Select objects: P1

1 found

Select objects: (CR)

Height of extrusion: 1.50 (CR)

Extrusion taper angle <0>: (CR)

(Base is extruded to a 1.50 unit height.)

SOLFEAT (solid feature): creates standard entities from a region or solid model.

SOLINT (solid intersection): creates a region that is common to two intersecting regions.

SOLLIST (solid list): gives defining information about a region.

SOLMASSP (solid mass properties): displays the mass properties, moments of inertia, product of inertia, radii of gyration, principal moments, and principal direction.

SOLMAT (solid material): assigns material types to a solid that are used with SOLMASSP.

SOLMESH (solid mesh): applies a mesh to solid; necessary for HIDE and SHADE to work.

SOLMOVE (solid move): moves and rotates regions.

SOLPURGE (solid purge): erases unused entities of region to lighten the file.

SOLSEP (solid separate): separates solids and/or regions that were merged by SOLUNION, SOLINT, and SOLSUB.

SOLSUB (solid subtract): subtracts one region from another.

SOLUCS (solid user coordinate system): aligns the current UCS with an edge or face of a region.

SOLUNION (solid union): combines regions or solids into a single composite part.

SOLWIRE (solid wire): converts a solid into a wire diagram.

37.11 An Example of Solid Modeling

Let's use the bracket shown in **Fig. 37.89** to produce a solid model. As shown in **Fig. 37.90**, first create the plan view of the base in the plane of the XY icon as a PLINE and obtain an isometric view

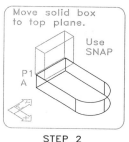

STEP 1 STEP 2

Figure 37.91 Bracket (SOLBOX command):
Step 1 Command: <u>SOLBOX</u> (CR)

Baseplane/Center/<Corner of box><0,0,0>: (CR)

Cube/Length/<Other corner>: <u>@1.30,4</u> (CR)

Height: <u>3.1</u> (CR) (Box is drawn.)

Step 2 (Set SNAP to END.)

Command: <u>MOVE</u> (CR)

Select objects: (Select box.)

Base point or displacement: <u>P1</u> (CR)

Second point of displacement: <u>A</u> (CR)

(Box is moved to base extrusion.)

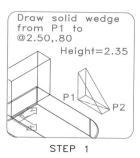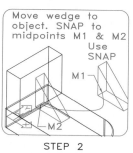

STEP 1 STEP 2

Figure 37.92 Bracket (SOLWEDGE command):
Step 1 Command: <u>SOLWEDGE</u> (CR)

Baseplane/<Corner of wedge>: <u>P1</u> (CR)

Length/<Other corner>: <u>@2.5,.8</u> (CR)

Height: <u>2.35</u> (CR) (Wedge is drawn.)

Step 2 (Set SNAP to MID.)

Command: <u>MOVE</u> (CR)

Select objects: (Select wedge.)

Base point or displacement: <u>M1</u> (CR)

Second point of displacement: <u>M2</u> (CR)

(Wedge is moved to midpoint of line.)

of it by VPOINT 1,-1,1. Then extrude the base to a height of 1.50 units.

Next use SOLBOX to generate a box 1.30 × 4.00 × 3.10 units at a convenient location (**Fig. 37.91**) and then move P1 of the box to point A of the base. Use SOLWEDGE to draw a wedge conforming to the measurements of bracket's rib (**Fig. 37.92**). Select and move the midpoint of a line on the wedge to the midpoint of the line on the upright box.

Next, create a cylinder with SOLCYL at a convenient position (**Fig. 37.93**) to represent the hole in the bracket. Then, move it to the center of the semicircular end of the base by using the Center option of OSNAP. **Figure 37.94** demonstrates how to use SOLSUB to subtract the hole from the base. Next, use SOLUNION to join the base, upright box, and wedge, making it a composite solid. Use SOLFILL (**Fig. 37.95**) to select the edges to be filleted and rounded, specify a radius of 0.40 unit,

and show fillets as wire diagrams. Finally, use SOLMESH to apply a mesh to the bracket and HIDE to suppress its hidden lines (**Fig. 37.96**). You may view the bracket with VPOINT or DVIEW from infinitely many viewpoints.

37.12 Solid Inquiry Commands

Inquiry commands give information in text form about selected solids. SOLLIST gives mathematical data defining edges, faces, solids, and trees, which provide sufficient definitions of solids.

SOLMASSP calculates the mass properties of selected solids: mass (weight), volume, centroid, moments of inertia, products of inertia, radii of gyration, principal moments (in X, Y, and Z directions), and the dimensions of the bounding box of the solid selected. SOLAREA computes the area of a surface or solid.

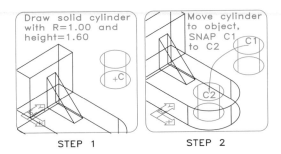

STEP 1   STEP 2

Figure 37.93 Bracket (SOCYL command):

Step 1 Command: <u>SOLCYL</u> or <u>CYL</u> (CR)

Elliptical/<Center point>: <u>C</u> (CR)

Diameter/<Radius>: <u>1.00</u> (CR)

Center of other end/<Height>: <u>1.60</u> (CR)

(Cylinder is drawn.)

Step 2 (Set SNAP to CENT.)

Command: <u>MOVE</u> (CR)

Select objects: (Select cylinder.)

Base point or displacement: <u>P1</u> (CR)

Second point of displacement: <u>P2</u> (CR)

(Cylinder is moved to center of arc.)

37.13 Solid Representations

SOLMESH and SOLWIRE place a skin on a wire diagram or convert it back to a wire diagram, respectively. SOLWDENS (solid wire density from 1 to 12), controls the density of the mesh applied to the frame of solid primitive or to a wire diagram.

SOLMESH applies faces to the surface of a solid to permit the use of HIDE to suppress hidden lines and converts curved surfaces to straight edges. SOLWIRE displays the solid as a wire-frame drawing, the opposite of SOLMESH.

37.14 Rendering

Rendering is the process of adding color, lighting, and surface textures to objects with mesh surfaces. Computer-aided rendering gives 3D draw-

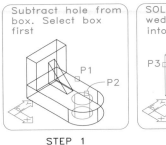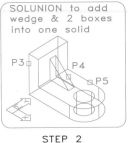

STEP 1   STEP 2

Figure 37.94 Bracket (SOLSUB and SOLUNION commands):

Step 1 Command: <u>SOLSUB</u>

Source objects...

Select objects: <u>P1</u>

Select objects: (CR)

1 solid selected.

Objects to subtract from them...

Select objects: <u>P2</u>

Select objects: (CR)

2 solids selected. (Hole is removed.)

Step 2 Command: <u>SOLUNION</u> (CR)

Select objects: <u>P3, P4, P5</u>

Select objects: (CR)

3 solids selected. (Bracket is unified.)

ings a realistic appearance by the application of shading—and much faster than an illustrator can do by hand.

Basics of Rendering

Let's use the bracket from Fig. 37.89 to demonstrate the rendering process. **Figure 37.97** displays the bracket from different viewpoints in two ports. From RENDER of the pull-down menu bar, select PREFERENCES to obtain the Rendering Preferences box **(Fig. 37.98)**. You may specify rendering as Full Render or Quick Render. Although Quick Render is slightly faster, Full Render gives a better final image. Select the Full Render button, select the OK box, pick RENDER, and the

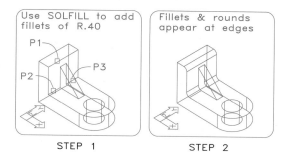

STEP 1   STEP 2

Figure 37.95 Bracket (SOLFILL command):
Step 1 Command: SOLFILL
`Pick edges of solids to be filled`
(Press ENTER when done): P1, P2, P3 (CR)
`3 edges selected.`
Step 2 `Diameter/<Radius> of fillet<0.00>:`
.40 (CR)
(Fillets and rounds are indicated on bracket.)

RENDER option from the pull-down menu to render the bracket **(Fig. 37.99).**

Let's keep several descriptive views of the bracket from different angles with the DVIEW and its CAMERA option and SAVE them as VIEWS. For example, pick the right viewport where the bracket is in perspective (Fig. 37.97); type VIEW, select the SAVE option, and name the part VIEW1 in response to the prompt. Select and name other VIEWs the same way. You may recall VIEWs with the VIEW command's RESTORE option.

Lights
Adding Lights Illumination from point lights, distant lights, and ambient lights that you control can enhance a rendering. **Point lights** emit rays in all directions from a point source. **Distant lights** emit parallel beams like those of sunlight. **Ambient light** comes from no particular source and provides constant illumination of all surfaces of an object.

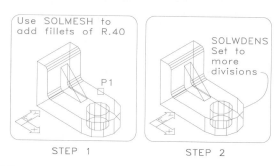

STEP 1   STEP 2

Figure 37.96 Bracket (SOLMESH and HIDE commands):
Step 1 Command: SOLMESH (CR)
`Select solids to be meshed...`
`Select objects:` P1 (CR)
`1 solid selected.`
`Surface meshing of current solid is completed.`
`Creating block for mesh representation...`
`Done.`
Step 2 Command: HIDE (CR)
`Regenerating drawing.`
`Hiding lines: done 100%` (Visibility is shown.)
(SOLWDENS can be set from 1 to 12 to adjust the density of the 3D faces on curving surfaces.)

BRACKET TO BE RENDERED

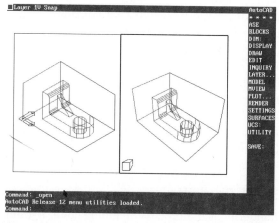

Figure 37.97 Use the bracket as the model to render.

RENDERING PREFERENCES DIALOGUE BOX

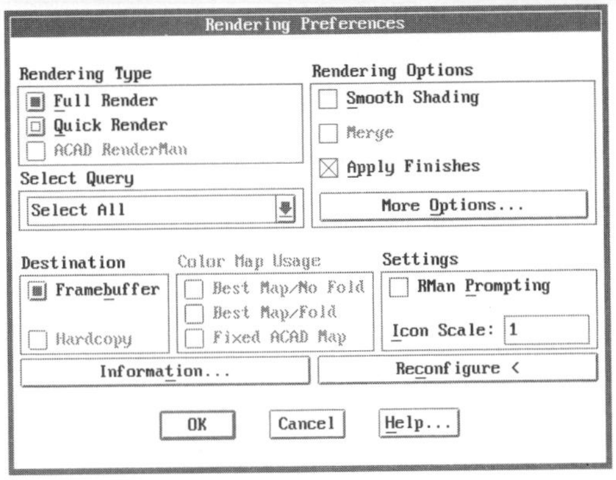

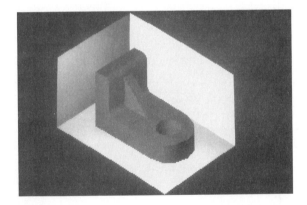

Figure 37.98 This dialogue box appears under `Render` in the pull-down menu. Select `Full Render` to make the best rendering of an object.

RENDERED BRACKET

Figure 37.99 This view shows a full rendering of the bracket.

You may add multiple point and distant light sources to a drawing and indicate that you have done so with symbols **(Fig. 37.100)**. From `Render` in the pull-down menu, select `Lights` to get its dialogue box **(Fig. 37.101)**, which lists three types of lights. Select `New` to get the `New Light Type` dialogue box, select the `Distant Light` button,

DISTANT LIGHTS AND POINT LIGHTS

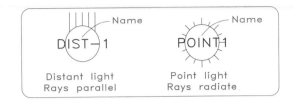

Figure 37.100 Place point light and distant light symbols on drawings to indicate the positions for lighting.

LIGHTS DIALOGUE BOX

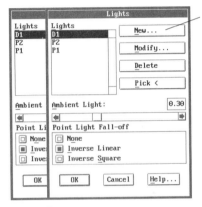

Select NEW to get dialogue box for picking light types

A   B

Figure 37.101

A The `Lights` dialogue box appears under `Render` of the pull-down menu.

B To obtain the `New Light Type` dialogue box, click on `New` of the `Lights` dialogue box.

and pick `OK`. When you place the light source at your viewpoint on the drawing, the `New Distant Light` dialogue box appears. Type `D2` in the `Light Name` box to name the light **(Fig. 37.102)**. To set `Light Intensity` type `0` for off or `1` for bright, or use the slider bar. Leave intensity set to `1`, select `OK` to return to `Lights` where `D2` is listed, and pick OK to exit. Select `Render` from the `Render` pull-down menu to obtain a new image of the bracket using distant light `D2`.

NEW DISTANT LIGHT DIALOGUE BOX

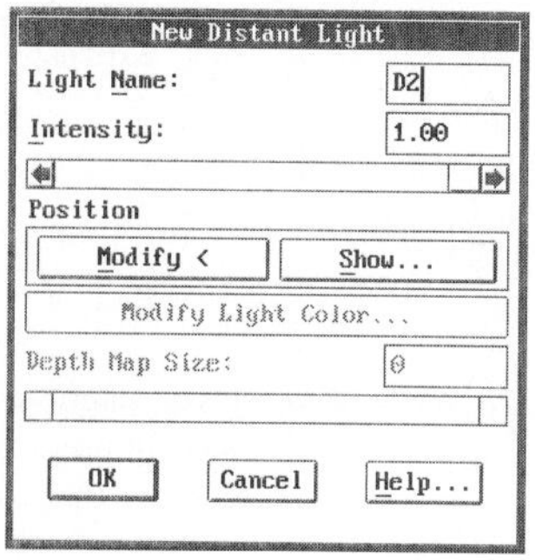

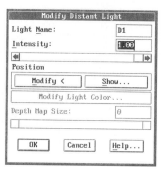

Box for setting New Distant Light (D2)

Figure 37.102 Type a new `Distant Light`, `D2`, in the `Light Name` box. Select `Modify` to obtain the `Modify Distant Light` dialogue box shown in Fig. 37.103.

Modifying Lights Select `Render` from the pull-down menu and `Lights` to obtain its dialogue box (Fig. 37.101). Then double click on `D1` (distant light) to obtain the `Modify Distant Light` dialogue box **(Fig. 37.103)**. Experiment with `Intensity` by changing to other settings (for example, `0.2` and `0.6`) and render the views to observe the brightness of the image.

Moving Lights To move distant light `D2`, type `PLAN` to get a top view of the bracket, select `Lights` from the `Render` pull-down menu, select `D2` from the `Lights` dialogue box, and pick `Modify` (Fig. 37.101). The `Modify Distant Light` dialogue box appears (Fig. 37.103). Select `Modify` under the heading `Position`, and the screen returns to the drawing where light `D2` and the target are shown **(Fig. 37.104)**.

The prompt `Enter light target <current>`: appears. Press (CR) to retain the current target and obtain a rubber-band line attached to it and the prompt, `enter light location <current>`. Type `.XY`, pick the light location with the cursor, and respond to the prompt `need Z` with `8` to locate the light in 3D space and establish the direction of the parallel light rays.

MODIFY DISTANT LIGHT DIALOGUE BOX

Box for setting light intensity and position for light D1

Figure 37.103 To change the intensity of `Distant Light D2`, type at the keyboard or use the slider bar.

TWO POSITIONS OF DISTANT LIGHT SOURCES

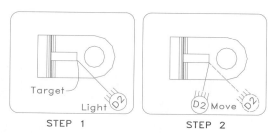

Figure 37.104 Moving a light:

Step 1 Select `Modify` from the `Modify Distant Light` dialogue box, which transfers you to the drawing. `Enter light target <current>`: (CR) (Accept current target.)

Step 2 `Enter light location <current>`: `.XY` (CR) `of` (need Z): 8 (CR) (The `Modify Distant Light` box reappears; click on OK to exit and `Render` the bracket.)

Type `VIEW`, select `Restore`, and enter `VIEW2` to return to the 3D view of the bracket. Select `Render` from the pull-down menu to compare the original lighting of the part with the new lighting **(Fig. 37.105)**.

Light Fall-Off The characteristic whereby a point light diminishes as it travels from its source is called **fall-off.** Only `point lights` have `fall-off`.

A.   B.

Figure 37.105 These rendered views of the bracket compare the effects of `Distant Light D2` in positions 1 and 2.

POINT LIGHT FALL-OFF

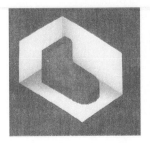 

A.   B.

Figure 37.107 This rendering compares `Fall-off` when (A) `Inverse Linear` lighting and (B) `Inverse Square` lighting are applied to the bracket.

Point Light Fall-off	
☐ None	Only Point Lights
■ Inverse Linear	have Fall-off;
☐ Inverse Square	Distant Lights do not

Figure 37.106 `Light Fall-off` settings in the `Light` dialogue box are `None`, `Inverse Linear`, and `Inverse Square`. With `Inverse Linear`, an object 4 units away from the light is 1/4 as bright as it is 1 unit away from the light. With `Inverse Square`, an object 4 units away from the light is 1/16th as bright as it is 1 unit away from the light.

`Distant light` and `ambient light` have uniform intensities at all distances. The `Lights` dialogue box **(Fig. 37.106)** shows the types of fall-off that may be specified.

None is a light with no fall-off, giving all objects the same brightness. **Inverse linear** makes an object 4 units from the light 1/4 as bright, and an object 8 units from the light 1/8th as bright. **Inverse square** makes an object 4 units from the light 1/16th as bright and an object 8 units from the light 1/64th as bright.

Type `VIEW`, select `Restore`, and obtain `VIEW 2`. Select `Lights` from `Render` in the pull-down

menu. Then pick `D2`, select `Modify` to get the `Modify Distant Light` dialogue box, and change the intensity to `0` to turn the light off. Pick `OK` to return to the Lights dialogue box. Turn off `P2` in the same manner, leaving only point light `Point1` on so that you can observe the effects of fall-off.

When you set `Fall-off` to `Inverse Linear`, the bracket appears as shown in **Fig. 37.107A**. When you set it to `Inverse Square` it appears as shown in **Fig. 37.107B**.

Ambient Light So far, ambient light has been set to `0.3`. Select `Lights` from the pull-down menu under `Render` to obtain the `Lights` dialogue box. Set ambient lighting to `0.85` by typing or using the slider bar and render the bracket to obtain the lighting shown in **Fig. 37.108**.

Scenes

You may save `VIEWS` and light settings in combination as `SCENES`. `SCENES` may have several lights (or no lights) but only one `VIEW`.

Open your file of the bracket used in the previous examples and select `SCENES` from the `Render` pull-down menu, to obtain the Scenes dialogue box **(Fig. 37.109)**. Select `New` to get the `New Scene`

A.   **B.**

■ **Figure 37.108**

A Ambient light set to `0.85` (bright).

B Ambient light set to `0.30` (dim).

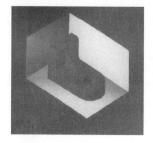 

A.   **B.**

■ **Figure 37.110**

A `SCENE1` is rendered as a combination of lights with the `VIEW` of the bracket.

B `SCENE2` is the same view with different lighting.

SCENES DIALOGUE BOXES

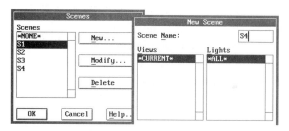

■ **Figure 37.109** The `Scenes` dialogue box appears under `Render` of the pull-down menu; select `New` to obtain the `New Scene` dialogue box.

dialogue box, which lists all model-space `Views` and `Lights`. Enter the name of the `SCENE` in the box (`S2`, for example). The program truncates the name to eight characters if it is longer than that. Select `VIEW2` as the `VIEW` in `Scene2` (`S2`), point light `P1`, distant light `D1`, and `OK`. The `Scenes` dialogue box reappears. You may establish other scenes by repeating these steps.

If you selected `*NONE*` as the scene to render, all lights would be on in the current view. If no lights existed an over-the-shoulder distant light would be given.

Now select `Scene2` (`S2`) from the dialogue box, `OK`, and `Render` to obtain a rendered `Scene2`. Compare `Scene1` (`S1`) **(Fig. 37.110A)** and `Scene2` **(Fig. 37.110B)** after they have been rendered.

To modify scenes, select a name from `Scenes` in the `Scenes` dialogue box (Fig. 37.109) and pick `Modify` to obtain the `Modify Scene` dialogue box, which is similar to the `New Scene` dialogue box. You may select different lights and delete previously selected lights. Select `OK` to keep your modifications, return to the `Scenes` dialogue box, select `OK` to exit, and render the modified `Scene`.

37.15 Finishes

The **finish** of an object is the reflective quality of its surfaces from dull to shiny. Let's use the file `PINS2` in AutoCAD's tutorial to illustrate the application of finishes **(Fig. 37.111)**. Three push pins are given: red, green, and blue, from left to right.

Select `FINISHES` from the pull-down menu under `Render` to get the `Finishes` dialogue box **(Fig. 37.112)**. Select `*GLOBAL*` to make settings that will apply to all three pins and to obtain the `Modify Finish` dialogue box **(Fig. 37.113)**. Default settings of four types of finish are given: `Ambient`, `Diffuse`, `Specular`, and `Roughness`.

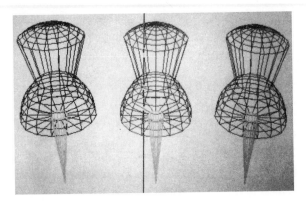

Figure 37.111 This illustration is a file named PINS2, which is available on AutoCAD's tutorial.

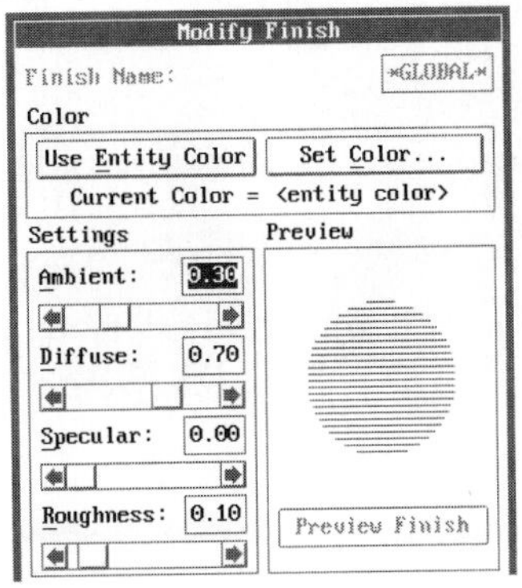

Finish types—Ambient, Diffuse, Specular, and Roughness—can be set from this box and a Preview obtained

Figure 37.113 Use the Modify Finish box to make finish settings. Select OK to accept the defaults and return to the Finish dialogue box.

FINISHES DIALOGUE BOX

Box for adding New finishes and Modifying existing finishes

Figure 37.112 The Finishes dialogue box appears under Render of the pull-down menu. Double click on *GLOBAL* to get the Modify Finish dialogue box shown in Fig. 37.113.

The settings of each type of finish must lie between the limits of 0 and 1.

Ambient: factors ranging from 0.7 to 1.0 give bright background lighting; no lighting at 0.

Diffuse: factors ranging from 0.7 to 1.0 yield dull finishes.

Specular: factors ranging from 0.7 to 1.0 give shiny finishes with highlights; finish is dull at 0, the default.

Roughness: factors increase the spread of specular highlights as their values increase.

Click on OK to exit the Modify Finish dialogue box, click on OK of the Finishes dialogue box to exit from it, and render the pins. Global finishes apply to all three pins, and each type is dull, with little highlighting (**Fig. 37.114**).

To give the blue pin on the right a shiny finish, select Finishes from the Render pull-down menu, click on New to obtain the New Finish dialogue box, which is similar to the Modify Finish dialogue box (Fig. 37.113). Enter RIGHT as the finish name and leave Current Color set to the entity color. (Finish colors may be assigned so that objects are rendered in colors different from their entity colors.) Use the following factors: Ambient=0.3, Diffuse=0.0, Specular=0.7, and Roughness=0.1.

When you select Preview Finish from the New Finish dialogue box, the program renders the sphere in the box (if your machine is configured to render to a viewport), similar to the one shown in Fig. 37.113. Click on OK to exit the New Finish dialogue box and return to the drawing. A cylindrical symbol appears at the UCS origin that

Figure 37.114 The pins have been rendered with `Global` finishes applied to each.

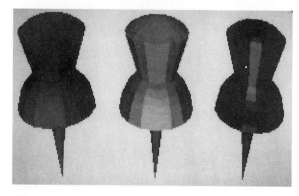

Figure 37.115 After selecting `Apply Finishes` from the `Rendering Preferences` dialogue box, turn `Diffuse Off` (=0) and set `Specular` to `0.7` to make the right pin shinier.

contains the finish information. A prompt asks you to `Enter New Finish location <current>,` which lets you locate the symbol on the drawing. After you select the position, the `Finishes` dialogue box returns.

To `Attach` the finishes to the right pin, select `RIGHT` from the list, and `Attach/Entities`. The program returns you to the drawing and gives the prompt `Select objects to attach "RIGHT" to:`. When you select the pin at the right, the `Finishes` dialogue box returns; click on `OK` to exit.

`Render` the object to get a shinier blue pin **(Fig. 37.115)**.

When the option `Apply Finishes` of the `Rendering Preferences` dialogue box (Fig. 37.97) is on, the specified finishes are applied to the pin that you selected. When `Apply Finishes` is off, settings are applied globally to all three pins with the `Global` finish settings. By setting `Smooth Shading` to `On` in the `Rendering Preferences` dialogue box, you will get smoother surfaces by removing the multifaceted look.

Problems

The problems at the ends of the previous chapters can be drawn and plotted using computer graphics techniques covered in this chapter and previous chapters.

Appendix Contents

APPENDIX 1 • Abbreviations (ANSI Z 32.13)

Word	Abbreviation	Word	Abbreviation	Word	Abbreviation
Allowance	ALLOW	Elevation	ELEV	Not to scale	NTS
Alloy	ALY	Equal	EQ	Number	NO.
Aluminum	AL	Estimate	EST	Octagon	OCT
Amount	AMT	Exterior	EXT	On center	OC
Anneal	ANL	Fahrenheit	F	Ounce	OZ
Approximate	APPROX	Feet	(') FT	Outside diameter	OD
Area	A	Feet per minute	FPM	Parallel	PAR.
Assembly	ASSY	Feet per second	FPS	Perpendicular	PERP
Auxiliary	AUX	Fillet	FIL	Piece	PC
Average	AVG	Fillister	FIL	Plastic	PLSTC
Babbitt	BAB	Finish	FIN.	Plate	PL
Between	BET.	Finish all over	FAO	Point	PT
Between centers	BC	Flat head	FH	Polish	POL
Bevel	BEV	Foot	(') FT	Pound	LB
Bill of material	B/M	Front	FR	Pounds per square inch	PSI
Both sides	BS	Gage	GA	Pressure	PRESS.
Bottom	BOT	Gallon	GAL	Production	PROD
Brass	BRS	Galvanize	GALV	Quarter	QTR
Brazing	BRZG	Galvanized iron	GI	Radius	R
Broach	BRO	General	GEN	Ream	RM
Bronze	BRZ	Gram	G	Rectangle	RECT
Cadmium plate	CD PL	Grind	GRD	Reference	REF
Cap screw	CAP SCR	Groove	GRV	Required	REQD
Case harden	CH	Hardware	HDW	Revise	REV
Cast iron	CI	Head	HD	Revolution	REV
Cast steel	CS	Heat treat	HT TR	Revolutions per minute	RPM
Casting	CSTG	Hexagon	HEX	Right	R
Center	CTR	Horizontal	HOR	Right hand	RH
Centerline	CL	Horsepower	HP	Rough	RGH
Center to center	C to C	Hot rolled steel	HRS	Screw	SCR
Centigrade	C	Hour	HR	Section	SECT
Centigram	CG	Hundredweight	CWT	Set screw	SS
Centimeter	cm	Inch	(") IN.	Shaft	SFT
Chamfer	CHAM	Inches per second	IPS	Slotted	SLOT.
Circle	CIR	Inside diameter	ID	Socket	SOC
Clockwise	CW	Interior	INT	Spherical	SPHER
Cold rolled steel	CRS	Iron	I	Spot faced	SF
Cotter	COT	Key	K	Spring	SPG
Counterclockwise	CCW	Kip (1000 lb)	K	Square	SQ
Counterbore	CBORE	Left	L	Station	STA
Counterdrill	CDRILL	Left hand	LH	Steel	STL
Counterpunch	CPUNCH	Length	LG	Symmetrical	SYM
Countersink	CSK	Light	LT	Taper	TPR
Cubic centimeter	cc	Machine	MACH	Temperature	TEMP
Cubic feet per minute	CFM	Malleable	MALL	Tension	TENS.
Cubic foot	CU FT	Manhole	MH	Thick	THK
Cubic inch	CU IN.	Manufacture	MFR	Thousand	M
Cylinder	CYL	Material	MATL	Thousand pound	KIP
Diagonal	DIAG	Maximum	MAX	Thread	THD
Diameter	DIA	Metal	MET.	Tolerance	TOL
Distance	DIST	Meter (Instrument or		Typical	TYP
Ditto	DO	measure of length)	M	Vertical	VERT
Down	DN	Miles	MI	Volume	VOL
Dozen	DOZ	Miles per gallon	MPG	Washer	WASH.
Drafting	DFTG	Miles per hour	MPH	Weight	WT
Drawing	DWG	Millimeter	MM	Width	W
Drill	DR	Minimum	MIN	Wrought iron	WI
Each	EA	Normal	NOR	Yard	YD

DECIMAL EQUIVALENTS—INCH-MILLIMETER CONVERSION TABLE

1/2	1/4	1/8	1/16	1/32	1/64	Decimals	Millimeters
					1	.015625	.396875
				1		.031250	.793750
					3	.046875	1.190625
			1			.062500	1.587500
					5	.078125	1.984375
				3		.093750	2.381250
					7	.109375	2.778125
		1				.125000	3.175000
					9	.140625	3.571875
				5		.156250	3.968750
					11	.171875	4.365625
			3			.187500	4.762500
					13	.203125	5.159375
				7		.218750	5.556250
					15	.234375	5.953125
	1					.250000	6.350000
					17	.265625	6.746875
				9		.281250	7.143750
					19	.296875	7.540625
			5			.312500	7.937500
					21	.328125	8.334375
				11		.343750	8.731250
					23	.359375	9.128125
		3				.375000	9.525000
					25	.390625	9.921875
				13		.406250	10.318750
					27	.421875	10.715625
			7			.437500	11.112500
					29	.453125	11.509375
				15		.468750	11.906250
					31	.484375	12.303125
1						.500000	12.700000

1/2	1/4	1/8	1/16	1/32	1/64	Decimals	Millimeters
					33	.515625	13.096875
				17		.531250	13.493750
					35	.546875	13.890625
			9			.562500	14.287500
					37	.578125	14.684375
				19		.593750	15.081250
					39	.609375	15.478125
		5				.625000	15.875000
					41	.640625	16.271875
				21		.656250	16.668750
					43	.671875	17.065625
			11			.687500	17.462500
					45	.703125	17.859375
				23		.718750	18.256250
					47	.734375	18.653125
	3					.750000	19.050000
					49	.765625	19.446875
				25		.781250	19.843750
					51	.796875	20.240625
			13			.812500	20.637500
					53	.828125	21.034375
				27		.843750	21.431250
					55	.859375	21.828125
		7				.875000	22.225000
					57	.890625	22.621875
				29		.906250	23.018750
					59	.921875	23.415625
			15			.937500	23.812500
					61	.953125	24.209375
				31		.968750	24.606250
					63	.984375	25.003125
	4	8	16	32	64	1.000000	25.400000

LENGTH

1 millimeter (mm) = 0.03937 inch
1 centimeter (cm) = 0.39370 inch
1 meter (m) = 39.37008 inches
1 meter = 3.2808 feet
1 meter = 1.0936 yards
1 kilometer (km) = 0.6214 miles
1 inch = 25.4 millimeters (mm)
1 inch = 2.54 centimeters
1 foot = 304.8 millimeters
1 foot = 0.3048 meters
1 yard = 0.9144 meters
1 mile = 1.609 kilometers

AREA

1 square millimeter = 0.00155 square inches
1 square centimeter = 0.155 square inches
1 square meter = 10.764 square feet
1 square meter = 1.196 square yards
1 square kilometer = 0.3861 square mile
1 square inch = 645.2 square millimeters
1 square inch = 6.452 square centimeters
1 square foot = 929 square centimeters
1 square foot = 0.0929 square meters
1 square yard = 0.836 square meters
1 square mile = 2.5899 square kilometers

DRY CAPACITY

1 cubic centimeter (cm3) = 0.061 cubic inches
1 liter = 0.0353 cubic foot
1 liter = 61.023 cubic inches
1 cubic meter (m3) = 35.315 cubic feet
1 cubic meter = 1.308 cubic yards
1 cubic inch = 16.38706 cubic centimeters
1 cubic foot = 0.02832 cubic meters
1 cubic foot = 28.317 liters
1 cubic yard = 0.7646 cubic meters

LIQUID CAPACITY

1 liter = 1.0567 U.S. quarts
1 liter = 0.2642 U.S. gallons
1 liter = 0.2200 Imperial gallons
1 cubic meter = 264.2 U.S. gallons
1 cubic meter = 219.969 Imperial gallons
1 U.S. quart = 0.946 liters
1 Imperial quart = 1.136 liters
1 U.S. gallon = 3.785 liters
1 Imperial gallon = 4.546 liters

WEIGHT

1 gram (g) = 15.432 grains
1 gram = 0.03215 ounce troy
1 gram = 0.03527 ounce avoirdupois
1 kilogram (kg) = 35.274 ounces avoirdupois
1 kilogram = 2.2046 pounds
1000 kilograms = 1 metric ton (t)
1000 kilograms = 1.1023 tons of 2000 pounds
1000 kilograms = 0.9842 tons of 2240 pounds
1 ounce avoirdupois = 28.35 grams
1 ounce troy = 31.103 grams
1 pound = 453.6 grams
1 pound = 0.4536 kilograms
1 ton of 2240 pounds = 1016 kilograms
1 ton of 2240 pounds = 1.016 metric tons
1 grain = 0.0648 grams
1 metric ton = 0.9842 tons of 2240 pounds
1 metric ton = 2204.6 pounds

N	0	1	2	3	4	5	6	7	8	9
1.0	.0000	.0043	.0086	.0128	.0170	.0212	.0253	.0294	.0334	.0374
1.1	.0414	.0453	.0492	.0531	.0569	.0607	.0645	.0682	.0719	.0755
1.2	.0792	.0828	.0864	.0899	.0934	.0969	.1004	.1038	.1072	.1106
1.3	.1139	.1173	.1206	.1239	.1271	.1303	.1335	.1367	.1399	.1430
1.4	.1461	.1492	.1523	.1553	.1584	.1614	.1644	.1673	.1703	.1732
1.5	.1761	.1790	.1818	.1847	.1875	.1903	.1931	.1959	.1987	.2014
1.6	.2041	.2068	.2095	.2122	.2148	.2175	.2201	.2227	.2253	.2279
1.7	.2304	.2330	.2355	.2380	.2405	.2430	.2455	.2480	.2504	.2529
1.8	.2553	.2577	.2601	.2625	.2648	.2672	.2695	.2718	.2742	.2765
1.9	.2788	.2810	.2833	.2856	.2878	.2900	.2923	.2945	.2967	.2989
2.0	.3010	.3032	.3054	.3075	.3096	.3118	.3139	.3160	.3181	.3201
2.1	.3222	.3243	.3263	.3284	.3304	.3324	.3345	.3365	.3385	.3404
2.2	.3424	.3444	.3464	.3483	.3502	.3522	.3541	.3560	.3579	.3598
2.3	.3617	.3636	.3655	.3674	.3692	.3711	.3729	.3747	.3766	.3784
2.4	.3802	.3820	.3838	.3856	.3874	.3892	.3909	.3927	.3945	.3962
2.5	.3979	.3997	.4014	.4031	.4048	.4065	.4082	.4099	.4116	.4133
2.6	.4150	.4166	.4183	.4200	.4216	.4232	.4249	.4265	.4281	.4298
2.7	.4314	.4330	.4346	.4362	.4378	.4393	.4409	.4425	.4440	.4456
2.8	.4472	.4487	.4502	.4518	.4533	.4548	.4564	.4579	.4594	.4609
2.9	.4624	.4639	.4654	.4669	.4683	.4698	.4713	.4728	.4742	.4757
3.0	.4771	.4786	.4800	.4814	.4829	.4843	.4857	.4871	.4886	.4900
3.1	.4914	.4928	.4942	.4955	.4969	.4983	.4997	.5011	.5024	.5038
3.2	.5051	.5065	.5079	.5092	.5105	.5119	.5132	.5145	.5159	.5172
3.3	.5185	.5198	.5211	.5224	.5237	.5250	.5263	.5276	.5289	.5302
3.4	.5315	.5328	.5340	.5353	.5366	.5378	.5391	.5403	.5416	.5428
3.5	.5441	.5453	.5465	.5478	.5490	.5502	.5514	.5527	.5539	.5551
3.6	.5563	.5575	.5587	.5599	.5611	.5623	.5635	.5647	.5658	.5670
3.7	.5682	.5694	.5705	.5717	.5729	.5740	.5752	.5763	.5775	.5786
3.8	.5798	.5809	.5821	.5832	.5843	.5855	.5866	.5877	.5888	.5899
3.9	.5911	.5922	.5933	.5944	.5955	.5966	.5977	.5988	.5999	.6010
4.0	.6021	.6031	.6042	.6053	.6064	.6075	.6085	.6096	.6107	.6117
4.1	.6128	.6138	.6149	.6160	.6170	.6180	.6191	.6201	.6212	.6222
4.2	.6232	.6243	.6253	.6263	.6274	.6284	.6294	.6304	.6314	.6325
4.3	.6335	.6345	.6355	.6365	.6375	.6385	.6395	.6405	.6415	.6425
4.4	.6435	.6444	.6454	.6464	.6474	.6484	.6493	.6503	.6513	.6522
4.5	.6532	.6542	.6551	.6561	.6571	.6580	.6590	.6599	.6609	.6618
4.6	.6628	.6637	.6646	.6656	.6665	.6675	.6684	.6693	.6702	.6712
4.7	.6721	.6730	.6739	.6749	.6758	.6767	.6776	.6785	.6794	.6803
4.8	.6812	.6821	.6830	.6839	.6848	.6857	.6866	.6875	.6884	.6893
4.9	.6902	.6911	.6920	.6928	.6937	.6946	.6955	.6964	.6972	.6981
5.0	.6990	.6998	.7007	.7016	.7024	.7033	.7042	.7050	.7059	.7067
5.1	.7076	.7084	.7093	.7101	.7110	.7118	.7126	.7135	.7143	.7152
5.2	.7160	.7168	.7177	.7185	.7193	.7202	.7210	.7218	.7226	.7235
5.3	.7243	.7251	.7259	.7267	.7275	.7284	.7292	.7300	.7308	7316
5.4	.7324	.7332	.7340	.7348	.7356	.7364	.7372	.7380	.7388	.7396
N	0	1	2	3	4	5	6	7	8	9

N	0	1	2	3	4	5	6	7	8	9
5.5	.7404	.7412	.7419	.7427	.7435	.7443	.7451	.7459	.7466	.7474
5.6	.7482	.7490	.7497	.7505	.7513	.7520	.7528	.7536	.7543	.7551
5.7	.7559	.7566	.7574	.7582	.7589	.7597	.7604	.7612	.7619	.7627
5.8	.7634	.7642	.7649	.7657	.7664	.7672	.7679	.7686	.7694	.7701
5.9	.7709	.7716	.7723	.7731	.7738	.7745	.7752	.7760	.7767	.7774
6.0	.7782	.7789	.7796	.7803	.7810	.7818	.7825	.7832	.7839	.7846
6.1	.7853	.7860	.7868	.7875	.7882	.7889	.7896	.7903	.7910	.7917
6.2	.7924	.7931	.7938	.7945	.7952	.7959	.7966	.7973	.7980	.7987
6.3	.7993	.8000	.8007	.8014	.8021	.8028	.8035	.8041	.8048	.8055
6.4	.8062	.8069	.8075	.8082	.8089	.8096	.8102	.8109	.8116	.8122
6.5	.8129	.8136	.8142	.8149	.8156	.8162	.8169	.8176	.8182	.8189
6.6	.8195	.8202	.8209	.8215	.8222	.8228	.8235	.8241	.8248	.8254
6.7	.8261	.8267	.8274	.8280	.8287	.8293	.8299	.8306	.8312	.8319
6.8	.8325	.8331	.8338	.8344	.8351	.8357	.8363	.8370	.8376	.8382
6.9	.8388	.8395	.8401	.8407	.8414	.8420	.8426	.8432	.8439	.8445
7.0	.8451	.8457	.8463	.8470	.8476	.8482	.8488	.8494	.8500	.8506
7.1	.8513	.8519	.8525	.8531	.8537	.8543	.8549	.8555	.8561	.8567
7.2	.8573	.8579	.8585	.8591	.8597	.8603	.8609	.8615	.8621	.8627
7.3	.8633	.8639	.8645	.8651	.8657	.8663	.8669	.8675	.8681	.8686
7.4	.8692	.8698	.8704	.8710	.8716	.8722	.8727	.8733	.8739	.8745
7.5	.8751	.8756	.8762	.8768	.8774	.8779	.8785	.8791	.8797	.8802
7.6	.8808	.8814	.8820	.8825	.8831	.8837	.8842	.8848	.8854	.8859
7.7	.8865	.8871	.8876	.8882	.8887	.8893	.8899	.8904	.8910	.8915
7.8	.8921	.8927	.8932	.8938	.8943	.8949	.8954	.8960	.8965	.8971
7.9	.8976	.8982	.8987	.8993	.8998	.9004	.9009	.9015	.9020	.9025
8.0	.9031	.9036	.9042	.9047	.9053	.9058	.9063	.9069	.9074	.9079
8.1	.9085	.9090	.9096	.9101	.9106	.9112	.9117	.9122	.9128	.9133
8.2	.9138	.9143	.9149	.9154	.9159	.9165	.9170	.9175	.9180	.9186
8.3	.9191	.9196	.9201	.9206	.9212	.9217	.9222	.9227	.9232	.9238
8.4	.9243	.9248	.9253	.9258	.9263	.9269	.9274	.9279	.9284	.9289
8.5	.9294	.9299	.9304	.9309	.9315	.9320	.9325	.9330	.9335	.9340
8.6	.9345	.9350	.9355	.9360	.9365	.9370	.9375	.9380	.9385	.9390
8.7	.9395	.9400	.9405	.9410	.9415	.9420	.9425	.9430	.9435	.9440
8.8	.9445	.9450	.9455	.9460	.9465	.9469	.9474	.9479	.9484	.9489
8.9	.9494	.9499	.9504	.9509	.9513	.9518	.9523	.9528	.9533	.9538
9.0	.9542	.9547	.9552	.9557	.9562	.9566	.9571	.9576	.9581	.9586
9.1	.9590	.9595	.9600	.9605	.9609	.9614	.9619	.9624	.9628	.9633
9.2	.9638	.9643	.9647	.9652	.9657	.9661	.9666	.9671	.9675	.9680
9.3	.9685	.9689	.9694	.9699	.9703	.9708	.9713	.9717	.9722	.9727
9.4	.9731	.9736	.9741	.9745	.9750	.9754	.9759	.9763	.9768	.9773
9.5	.9777	.9782	.9786	.9791	.9795	.9800	.9805	.9809	.9814	.9818
9.6	.9823	.9827	.9832	.9836	.9841	.9845	.9850	.9854	.9859	.9863
9.7	.9868	.9872	.9877	.9881	.9886	.9890	.9894	.9899	.9903	.9908
9.8	.9912	.9917	.9921	.9926	.9930	.9934	.9939	.9943	.9948	.9952
9.9	.9956	.9961	.9965	.9969	.9974	.9978	.9983	.9987	.9991	.9996
N	0	1	2	3	4	5	6	7	8	9

ANGLE in DEGREES	SINE	COSINE	TAN	COTAN	ANGLE in DEGREES
0	0.0000	1.0000	0.0000		90
1	.0175	.9998	.0175	57.290	89
2	.0349	.9994	.0349	28.636	88
3	.0523	.9986	.0524	19.081	87
4	.0698	.9976	.0699	14.301	86
5	.0872	.9962	.0875	11.430	85
6	.1045	.9945	.1051	9.5144	84
7	.1219	.9925	.1228	8.1443	83
8	.1392	.9903	.1405	7.1154	82
9	.1564	.9877	.1584	6.3138	81
10	.1736	.9848	.1763	5.6713	80
11	.1908	.9816	.1944	5.1446	79
12	.2079	.9781	.2126	4.7046	78
13	.2250	.9744	.2309	4.3315	77
14	.2419	.9703	.2493	4.0108	76
15	.2588	.9659	.2679	3.7321	75
16	.2756	.9613	.2867	3.4874	74
17	.2924	.9563	.3057	3.2709	73
18	.3090	.9511	.3249	3.0777	72
19	.3256	.9455	.3443	2.9042	71
20	.3420	.9397	.3640	2.7475	70
21	.3584	.9336	.3839	2.6051	69
22	.3746	.9272	.4040	2.4751	68
23	.3907	.9205	.4245	2.3559	67
24	.4067	.9135	.4452	2.2460	66
25	.4226	.9063	.4663	2.1445	65
26	.4384	.8988	.4877	2.0503	64
27	.4540	.8910	.5095	1.9626	63
28	.4695	.8829	.5317	1.8807	62
29	.4848	.8746	.5543	1.8040	61
30	.5000	.8660	.5774	1.7321	60
31	.5150	.8572	.6009	1.6643	59
32	.5299	.8480	.6249	1.6003	58
33	.5446	.8387	.6494	1.5399	57
34	.5592	.8290	.6745	1.4826	56
35	.5736	.8192	.7002	1.4281	55
36	.5878	.8090	.7265	1.3764	54
37	.6018	.7986	.7536	1.3270	53
38	.6157	.7880	.7813	1.2799	52
39	.6293	.7771	.8098	1.2349	51
40	.6428	.7660	.8391	1.1918	50
41	.6561	.7547	.8693	1.1504	49
42	.6691	.7431	.9004	1.1106	48
43	.6820	.7314	.9325	1.0724	47
44	.6947	.7193	.9657	1.0355	46
45	.7071	.7071	1.0000	1.0000	45

Substance	Weight Lb. per Cu. Ft.	Specific Gravity	Substance	Weight Lb. per Cu. Ft.	Specific Gravity
METALS, ALLOYS, ORES			**TIMBER, U. S. SEASONED**		
Aluminum, cast, hammered	165	2.55-2.75	**Moisture Content by Weight:**		
Brass, cast, rolled	534	8.4-8.7	Seasoned timber 15 to 20%		
Bronze, 7.9 to 14% Sn	509	7.4-8.9	Green timber up to 50%		
Bronze, aluminum	481	7.7	Ash, white, red	40	0.62-0.65
Copper, cast, rolled	556	8.8-9.0	Cedar, white, red	22	0.32-0.38
Copper ore, pyrites	262	4.1-4.3	Chestnut	41	0.66
Gold, cast, hammered	1205	19.25-19.3	Cypress	30	0.48
Iron, cast, pig	450	7.2	Fir, Douglas spruce	32	0.51
Iron, wrought	485	7.6-7.9	Fir, eastern	25	0.40
Iron, spiegel-eisen	468	7.5	Elm, white	45	0.72
Iron, ferro-silicon	437	6.7-7.3	Hemlock	29	0.42-0.52
Iron ore, hematite	325	5.2	Hickory	49	0.74-0.84
Iron ore, hematite in bank	160-180	....	Locust	46	0.73
Iron ore, hematite loose	130-160	....	Maple, hard	43	0.68
Iron ore, limonite	237	3.6-4.0	Maple, white	33	0.53
Iron ore, magnetite	315	4.9-5.2	Oak, chestnut	54	0.86
Iron slag	172	2.5-3.0	Oak, live	59	0.95
Lead	710	11.37	Oak, red, black	41	0.65
Lead ore, galena	465	7.3-7.6	Oak, white	46	0.74
Magnesium, alloys	112	1.74-1.83	Pine, Oregon	32	0.51
Manganese	475	7.2-8.0	Pine, red	30	0.48
Manganese ore, pyrolusite	259	3.7-4.6	Pine, white	26	0.41
Mercury	849	13.6	Pine, yellow, long-leaf	44	0.70
Monel Metal	556	8.8-9.0	Pine, yellow, short-leaf	38	0.61
Nickel	565	8.9-9.2	Poplar	30	0.48
Platinum, cast, hammered	1330	21.1-21.5	Redwood, California	26	0.42
Silver, cast, hammered	656	10.4-10.6	Spruce, white, black	27	0.40-0.46
Steel, rolled	490	7.85	Walnut, black	38	0.61
Tin, cast, hammered	459	7.2-7.5			
Tin ore, cassiterite	418	6.4-7.0			
Zinc, cast, rolled	440	6.9-7.2			
Zinc ore, blende	253	3.9-4.2	**VARIOUS LIQUIDS**		
			Alcohol, 100%	49	0.79
			Acids, muriatic 40%	75	1.20
			Acids, nitric 91%	94	1.50
			Acids, sulphuric 87%	112	1.80
VARIOUS SOLIDS			Lye, soda 66%	106	1.70
Cereals, oats ....bulk	32	....	Oils, vegetable	58	0.91-0.94
Cereals, barley ....bulk	39	....	Oils, mineral, lubricants	57	0.90-0.93
Cereals, corn, rye ....bulk	48	....	Water, 4°C. max. density	62.428	1.0
Cereals, wheat ....bulk	48	....	Water, 100°C.	59.830	0.9584
Hay and Straw ....bales	20	....	Water, ice	56	0.88-0.92
Cotton, Flax, Hemp	93	1.47-1.50	Water, snow, fresh fallen	8	.125
Fats	58	0.90-0.97	Water, sea water	64	1.02-1.03
Flour, loose	28	0.40-0.50			
Flour, pressed	47	0.70-0.80			
Glass, common	156	2.40-2.60			
Glass, plate or crown	161	2.45-2.72	**GASES**		
Glass, crystal	184	2.90-3.00			
Leather	59	0.86-1.02	Air, 0°C. 760 mm.	.08071	1.0
Paper	58	0.70-1.15	Ammonia	.0478	0.5920
Potatoes, piled	42	....	Carbon dioxide	.1234	1.5291
Rubber, caoutchouc	59	0.92-0.96	Carbon monoxide	.0781	0.9673
Rubber goods	94	1.0-2.0	Gas, illuminating	.028-.036	0.35-0.45
Salt, granulated, piled	48	....	Gas, natural	.038-.039	0.47-0.48
Saltpeter	67	....	Hydrogen	.00559	0.0693
Starch	96	1.53	Nitrogen	.0784	0.9714
Sulphur	125	1.93-2.07	Oxygen	.0892	1.1056
Wool.	82	1.32			

The specific gravities of solids and liquids refer to water at 4°C., those of gases to air at 0°C. and 760 mm. pressure. The weights per cubic foot are derived from average specific gravities, except where stated that weights are for bulk, heaped or loose material, etc.

(Courtesy of the American Institute of Steel Construction.)

APPENDIX 7 • Wire and Sheet Metal Gages (in decimals of an inch)

WIRE AND SHEET METAL GAGES
IN DECIMALS OF AN INCH

Name of Gage	United States Standard Gage*		The United States Steel Wire Gage	American or Brown & Sharpe Wire Gage	New Birmingham Standard Sheet & Hoop Gage	British Imperial or English Legal Standard Wire Gage	Birmingham or Stubs Iron Wire Gage	Name of Gage
Principal Use	Uncoated Steel Sheets and Light Plates		Steel Wire except Music Wire	Non-Ferrous Sheets and Wire	Iron and Steel Sheets and Hoops	Wire	Strips, Bands, Hoops and Wire	Principal Use
Gage No.	Weight Oz. per Sq. Ft.	Approx. Thickness Inches	Thickness, Inches					Gage No.
7/0's			.4900		.6666	.500		7/0's
6/0's			.4615	.5800	.625	.464		6/0's
5/0's			.4305	.5165	.5883	.432	.500	5/0's
4/0's			.3938	.4600	.5416	.400	.454	4/0's
3/0's			.3625	.4096	.500	.372	.425	3/0's
2/0's			.3310	.3648	.4452	.348	.380	2/0's
0			.3065	.3249	.3964	.324	.340	0
1			.2830	.2893	.3532	.300	.300	1
2			.2625	.2576	.3147	.276	.284	2
3	160	.2391	.2437	.2294	.2804	.252	.259	3
4	150	.2242	.2253	.2043	.250	.232	.238	4
5	140	.2092	.2070	.1819	.2225	.212	.220	5
6	130	.1943	.1920	.1620	.1981	.192	.203	6
7	120	.1793	.1770	.1443	.1764	.176	.180	7
8	110	.1644	.1620	.1285	.1570	.160	.165	8
9	100	.1495	.1483	.1144	.1398	.144	.148	9
10	90	.1345	.1350	.1019	.1250	.128	.134	10
11	80	.1196	.1205	.0907	.1113	.116	.120	11
12	70	.1046	.1055	.0808	.0991	.104	.109	12
13	60	.0897	.0915	.0720	.0882	.092	.095	13
14	50	0747	.0800	.0641	.0785	.080	.083	14
15	45	.0673	.0720	.0571	.0699	.072	.072	15
16	40	.0598	.0625	.0508	.0625	.064	.065	16
17	36	.0538	.0540	.0453	.0556	.056	.058	17
18	32	.0478	.0475	.0403	.0495	.048	.049	18
19	28	.0418	.0410	.0359	.0440	.040	.042	19
20	24	.0359	.0348	.0320	.0392	.036	.035	20
21	22	.0329	.0318	.0285	.0349	.032	.032	21
22	20	.0299	.0286	.0253	.0313	.028	.028	22
23	18	.0269	.0258	.0226	.0278	.024	.025	23
24	16	.0239	.0230	.0201	.0248	.022	.022	24
25	14	.0209	.0204	.0179	.0220	.020	.020	25
26	12	.0179	.0181	.0159	.0196	.018	.018	26
27	11	.0164	.0173	.0142	.0175	.0164	.016	27
28	10	.0149	.0162	.0126	.0156	.0148	.014	28
29	9	.0135	.0150	.0113	.0139	.0136	.013	29
30	8	.0120	.0140	.0100	.0123	.0124	.012	30
31	7	.0105	.0132	.0089	.0110	.0116	.010	31
32	6.5	.0097	.0128	.0080	.0098	.0108	.009	32
33	6	.0090	.0118	.0071	.0087	.0100	.008	33
34	5.5	.0082	.0104	.0063	.0077	.0092	.007	34
35	5	.0075	.0095	.0056	.0069	.0084	.005	35
36	4.5	.0067	.0090	.0050	.0061	.0076	.004	36
37	4.25	.0064	.0085	.0045	.0054	.0068		37
38	4	.0060	.0080	.0040	.0048	.0060		38
39			.0075	.0035	.0043	.0052		39
40			.0070	.0031	.0039	.0048		40

* U. S. Standard Gage is officially a weight gage, in oz. per sq. ft. as tabulated. The Approx. Thickness shown is the "Manufacturers' Standard" of the American Iron and Steel Institute, based on steel as weighing 501.81 lbs. per cu. ft. (489.6 true weight plus 2.5 percent for average over-run in area and thickness). The A.I.S.I. standard nomenclature for flat rolled carbon steel is as follows:

Widths, Inches	Thicknesses, Inch							
	0.2500 and thicker	0.2499 to 0.2031	0.2030 to 0.1875	0.1874 to 0.0568	0.0567 to 0.0344	0.0343 to 0.0255	0.0254 to 0.0142	0.0141 and thinner
To 3½ incl.	Bar	Bar	Strip	Strip	Strip	Strip	Sheet	Sheet
Over 3½ to 6 incl.	Bar	Bar	Strip	Strip	Strip	Sheet	Sheet	Sheet
" 6 to 12 "	Plate	Strip	Strip	Strip	Sheet	Sheet	Sheet	Sheet
" 12 to 32 "	Plate	Sheet	Sheet	Sheet	Sheet	Sheet	Sheet	Black Plate
" 32 to 48 "	Plate	Sheet	Sheet	Sheet	Sheet	Sheet	Sheet	Sheet
" 48	Plate	Plate	Plate	Sheet	Sheet	Sheet	Sheet	—

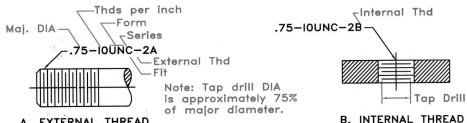

Note: Tap drill DIA is approximately 75% of major diameter.

A. EXTERNAL THREAD

B. INTERNAL THREAD

Nominal Diameter	Basic Diameter	Coarse NC & UNC		Fine NF & UNF		Extra Fine NEF/UNEF	
		Thds per In.	Tap Drill DIA	Thds per In.	Tap Drill DIA	Thds per In.	Tap Drill DIA
0	.060			80	.0469		
1	.073	64	No.53	72	No.53		
2	.086	56	No.50	64	No.50		
3	.099	48	No.47	56	No.45		
4	.112	40	No.43	48	No.42		
5	.125	40	No.38	44	No.37		
6	.138	32	No.36	40	No.33		
8	.164	32	No.29	36	No.29		
10	.190	24	No.25	32	No.21		
12	.216	24	No.16	28	No.14	32	No.13
1/4	.250	20	No.7	28	No.3	32	.2189
5/16	.3125	18	F	24	I	32	.2813
3/8	.375	16	.3125	24	Q	32	.3438
7/16	.4375	14	U	20	.3906	28	.4062
1/2	.500	13	.4219	20	.4531	28	.4688
9/16	.5625	12	.4844	18	.5156	24	.5156
5/8	.625	11	.5313	18	.5781	24	.5781
11/16	.6875	...	...	...	...	24	.6406
3/4	.750	10	.6563	16	.6875	20	.7031
13/16	.8125	...	...	...	...	20	.7656
7/8	.875	9	.7656	14	.8125	20	.8281
15/16	.9375	...	...	...	...	20	.8906

Nominal Diameter	Basic Diameter	Coarse NC & UNC		Fine NF & UNF		Extra Fine NEF/UNEF	
		Thds per In.	Tap Drill DIA	Thds per In.	Tap Drill DIA	Thds per In.	Tap Drill DIA
1	1.000	8	.875	12	.922	20	.953
1-1/16	1.063	...	...	...	...	18	1.000
1-1/8	1.125	7	.904	12	1.046	18	1.070
1-3/16	1.188	...	...	...	...	18	1.141
1-1/4	1.250	7	1.109	12	1,172	18	1.188
1-5/16	1.313	...	...	...	...	18	1.266
1-3/8	1.375	6	1.219	12	1.297	18	1.313
1-7/16	1.438	...	...	...	...	18	1.375
1-1/2	1.500	6	1.344	12	1.422	18	1.438
1-9/16	1.563	...	...	...	...	18	1.500
1-5/8	1.625	...	...	...	...	18	1.563
1-11/16	1.688	...	...	...	...	18	1.625
1-3/4	1.750	5	1.563	...	...	...	...
2	2.000	4.5	1.781	...	...	...	...
2-1/4	2.250	4.5	2.031	...	...	...	...
2-1/2	2.500	4	2.250	...	...	...	...
2-3/4	2.750	4	2.500	...	...	...	...
3	3.000	4	2.750	...	...	...	...
3-1/4	3.250	4	...	...	...	...	...
3-1/2	3.500	4	...	...	...	...	...
3-3/4	3.750	4	...	...	...	...	...
4	4.000	4	...	...	...	...	...

ANSI/ASME B1.1—1989

APPENDIX 9 • Screw Threads: American National and Unified (inches)
Constant-Pitch Threads

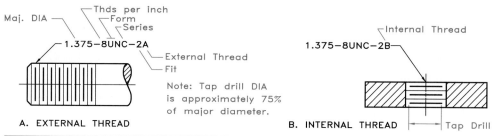

Maj. DIA ⌐Thds per inch
⌐Form
⌐Series

1.375−8UNC−2A

⌐External Thread
⌐Fit

Note: Tap drill DIA is approximately 75% of major diameter.

A. EXTERNAL THREAD

⌐Internal Thread

1.375−8UNC−2B

B. INTERNAL THREAD ⊢ Tap Drill

Nominal Diameter	8 Pitch 8N & 8UN		12 Pitch 12N & 12UN		16 Pitch 16N & 16UN		Nominal Diameter	8 Pitch 8N & 8UN		12 Pitch 12N & 12UN		16 Pitch 16N & 16UN	
	Thds per In.	Tap Drill DIA	Thds per In.	Tap Drill DIA	Thds per In.	Tap Drill DIA		Thds per In.	Tap Drill DIA	Thds per In.	Tap Drill DIA	Thds per In.	Tap Drill DIA
.500	...	...	12	.422	...	...	2.063	...	...	...	...	16	2.000
.563	...	...	12	.484	...	...	2.125	...	...	12	2.047	16	2.063
.625	...	...	12	.547	...	...	2.188	...	...	...	...	16	2.125
.688	...	...	12	.609	...	...	2.250	8	2.125	12	2.172	16	2.188
.750	...	...	12	.672	16	.688	2.313	...	...	...	...	16	2.250
.813	...	...	12	.734	16	.750	2.375	...	...	12	2.297	16	2.313
.875	...	...	12	.797	16	.813	2.438	...	...	...	...	16	2.375
.934	...	...	12	.859	16	.875	2.500	8	2.375	12	2.422	16	2.438
1.000	8	.875	12	.922	16	.938	2.625	...	...	12	2.547	16	2.563
1.063	...	...	12	.984	16	1.000	2.750	8	2.625	12	2.717	16	2.688
1.125	8	1.000	12	1.047	16	1.063	2.875	...	...	12	...	16	...
1.188	...	...	12	1.109	16	1.125	3.000	8	2.875	12	...	16	...
1.250	8	1.125	12	1.172	16	1.188	3.125	...	...	12	...	16	...
1.313	...	...	12	1.234	16	1.250	3.250	8	...	12	...	16	...
1.375	8	1.250	12	1.297	16	1.313	3.375	...	...	12	...	16	...
1.434	...	...	12	1.359	16	1.375	3.500	8	...	12	...	16	...
1.500	8	1.375	12	1.422	16	1.438	3.625	...	...	12	...	16	...
1.563	...	...	...	...	16	1.500	3.750	8	...	12	...	16	...
1.625	8	1.500	12	1.547	16	1.563	3.875	...	...	12	...	16	...
1.688	...	...	...	...	16	1.625	4.000	8	...	12	...	16	...
1.750	8	1.625	12	1.672	16	1.688	4.250	8	...	12	...	16	...
1.813	...	...	...	...	16	1.750	4.500	8	...	12	...	16	...
1.875	8	1.750	12	1.797	16	1.813	4.750	8	...	12	...	16	...
1.934	...	...	...	...	16	1.875	5.000	8	...	12	...	16	...
2.000	8	1.875	12	1.922	16	1.938	5.250	8	...	12	...	16	...

ANSI/ASME B1.1−1989

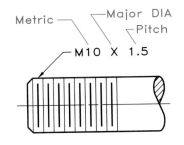

Metric
Major DIA
Pitch
M10 X 1.5

A. EXTERNAL THREAD

Note: Tap drill DIA
is approximately 75%
of major diameter.

M10 X 1.5

B. INTERNAL THREAD   Tap Drill

COARSE		FINE		COARSE		FINE	
MAJ. DIA & THD PITCH	TAP DRILL	MAJ. DIA & THD PITCH	TAP DRILL	MAJ. DIA & THD PITCH	TAP DRILL	MAJ. DIA & THD PITCH	TAP DRILL
M1.6 X 0.35	1.25			M20 X 2.5	17.5	M20 X 1.5	18.5
M1.8 X 0.35	1.45			M22 X 2.5	19.5	M22 X 1.5	20.5
M2 X 0.4	1.6			M24 X 3	21.0	M24 X 2	22.0
M2.2 X 0.45	1.75			M27 X 3	24.0	M27 X 2	25.0
M2.5 X 0.45	2.05			M30 X 3.5	26.5	M30 X 2	28.0
M3 X 0.5	2.5			M33 X 3.5	29.5	M33 X 2	31.0
M3.5 X 0.6	2.9			M36 X 4	32.0	M36 X 3	33.0
M4 X 0.7	3.3			M39 X 4	35.0	M39 X 3	36.0
M4.5 X 0.75	3.75			M42 X 4.5	37.5	M42 X 3	39.0
M5 X 0.8	4.2			M45 X 4.5	40.5	M45 X 3	42.0
M6 X 1	5.0			M48 X 5	43.0	M48 X 3	45.0
M7 X 1	6.0			M52 X 5	47.0	M52 X 3	49.0
M8 X 1.25	6.8	M8 X 1	7.0	M56 X 5.5	50.5	M56 X 4	52.0
M9 X 1.25	7.75			M60 X 5.5	54.5	M60 X 4	56.0
M10 X 1.5	8.5	M10 X 1.25	8.75	M64 X 6	58.0	M64 X 4	60.0
M11 X 1.5	9.5			M68 X 6	62.0	M68 X 4	64.0
M12 X 1.75	10.3	M12 X 1.25	10.5	M72 X 6	66.0	M72 X 4	68.0
M14 X 2	12.0	M14 X 1.5	12.5	M80 X 6	74.0	M80 X 4	76.0
M16 X 2	14.0	M16 X 1.5	14.5	M90 X 6	84.0	M90 X 4	86.0
M18 X 2.5	15.5	M18 X 1.5	16.5	M100 X 6	94.0	M100 X 4	96.0

ANSI/AME B1.1 (1989)

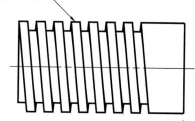

2.00−2.5 SQUARE

Typical thread note

Dimensions are in inches

Size	Size	Thds per inch
3/8	.375	12
7/16	.438	10
1/2	.500	10
9/16	.563	8
5/8	.625	8
3/4	.75	6
7/8	.875	5
1	1.000	5

Size	Size	Thds per inch
1-1/8	1.125	4
1-1/4	1.250	4
1-1/2	1.500	3
1-3/4	1.750	2-1/2
2	2.000	2-1/2
2-1/4	2.250	2
2-1/2	2.500	2
2-3/4	2.750	2

Size	Size	Thds per inch
3	3.000	1-1/2
3-1/4	3.125	1-1/2
3-1/2	3.500	1-1/3
3-3/4	3.750	1-1/3
4	4.000	1-1/3
4-1/4	4.250	1-1/3
4-1/2	4.500	1
Larger		1

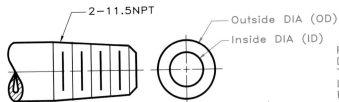

2−11.5NPT

Outside DIA (OD)

Inside DIA (ID)

PIPES THRU 12 INCHES IN DIA ARE SPECIFIED BY THEIR INSIDE DIAMETERS. LARGER PIPES ARE SPECIFIED BY THEIR OD,

$\frac{1}{16}$ DIA to $1\frac{1}{4}$ DIA   Dimensions in inches

Nominal ID	$\frac{1}{16}$	$\frac{1}{8}$	$\frac{1}{4}$	$\frac{3}{8}$	$\frac{1}{2}$	$\frac{3}{4}$	1	$1\frac{1}{4}$
Outside DIA	0.313	0.405	0.540	0.675	0.840	1.050	1.315	1.660
Thds/Inch	27	27	18	18	14	14	$11\frac{1}{2}$	$11\frac{1}{2}$

$1\frac{1}{2}$ DIA to 6 DIA

Nominal ID	$1\frac{1}{2}$	2	$2\frac{1}{2}$	3	$3\frac{1}{2}$	4	5	6
Outside DIA	1.900	2.375	2.875	3.500	4.000	4.500	5.563	6.625
Thds/Inch	$11\frac{1}{2}$	$11\frac{1}{2}$	8	8	8	8	8	8

8 DIA to 24 DIA

Nominal ID	8	10	12	14 OD	16 OD	18 OD	20 OD	24 OD
Outside DIA	8.625	10.750	12.750	14.000	16.000	18.000	20.000	24.000
Thds/Inch	8	8	8	8	8	8	8	8

ANSI B2.1

APPENDIX 13 • Square Bolts (inches)

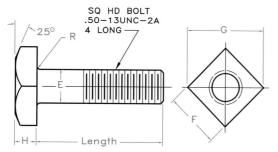

SQ HD BOLT
.50—13UNC—2A
4 LONG

STANDARD COMMERCIAL LENGTHS																		
	.50	.75	1.00	1.25	1.50	1.75	2.00	2.50	3.00	3.50	4.00	4.50	5.00	5.50	6.00	6.50	7.00	8.00 *PLUS
.25	●	●	●	●	●	●	●											
.31		●	●	●	●	●	●	●	●		●							
.375			●	●	●	●	●	●	●	●	●	●						
.500		●	●	●	●	●	●	●	●	●	●	●	●	●	●	●	●	*14
.625				●	●	●	●	●	●	●	●	●	●	●	●	●	●	*12
.750						●	●	●	●	●	●	●	●	●	●	●	●	*13
.875									●	●	●	●	●	●				
1.00															●	●	●	*12
1.25															●	●	●	

DIAMETER

*14 MEANS THAT LENGTHS ARE AVAILABLE
AT 1 INCH INCREMENTS UP 14 INCHES.

DIA	E Max.	F Max.	G Avg.	H Max.	R Max.
1/4	.250	.375	.530	.188	.031
5/16	.313	.500	.707	.220	.031
3/8	.375	.563	.795	.268	.031
7/16	.438	.625	.884	.316	.031
1/2	.500	.750	1.061	.348	.031
5/8	.625	.938	1.326	.444	.062

DIA	E Max.	F Max.	G Avg.	H Max.	R Max.
3/4	.750	1.125	1.591	.524	.062
7/8	.875	1.313	1.856	.620	.062
1	1.000	1.500	2.121	.684	.093
1—1/8	1.125	1.688	2.386	.780	.093
1—1/4	1.250	1.875	2.652	.876	.093
1—3/8	1.375	2.625	2.917	.940	.093
1—1/2	1.500	2.250	3.182	1.036	.093

APPENDIX 14 • Square Nuts

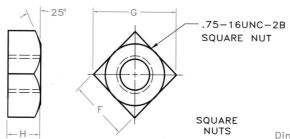

.75—16UNC—2B
SQUARE NUT

SQUARE
NUTS

Dimensions are in inches

DIA	DIA	F Max.	G Avg.	H Max.
1/4	.250	.438	.619	.235
5/16	.313	.563	.795	.283
3/8	.375	.625	.884	.346
7/16	.438	.750	1.061	.394
1/2	.500	.813	1.149	.458
5/8	.625	1.000	1.414	.569

DIA	DIA	F Max.	G Avg.	H Max.
3/4	.750	1.125	1.591	.680
7/8	.875	1.313	1.856	.792
1	1.000	1.500	2.121	.903
1—1/8	1.125	1.688	2.386	1.030
1—1/4	1.250	1.875	2.652	1.126
1—3/8	1.375	1.063	2.917	1.237
1—1/2	1.500	2.250	3.182	1.348

APPENDIX 15 • Hexagon Head Bolts

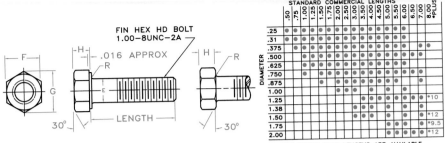

Dimensions are in inches

DIA	E Max.	F Max.	G Avg.	H Max.	R Max.
1/4	.250	.438	.505	.163	.025
5/16	.313	.500	.577	.211	.025
3/8	.375	.563	.650	.243	.025
7/16	.438	.625	.722	.291	.025
1/2	.500	.750	.866	.323	.025
9/16	.563	.812	.938	.371	.045
5/8	.625	.938	1.083	.403	.045
3/4	.750	1.125	1.299	.483	.045
7/8	.875	1.313	1.516	.563	.065
1	1.000	1.500	1.732	.627	.095

DIA	E Max.	F Max.	G Avg.	H Max.	R Max.
1-1/8	1.125	1.688	1.949	.718	.095
1-1/4	1.250	1.875	2.165	.813	.095
1-3/8	1.375	2.063	2.382	.878	.095
1-1/2	1.500	2.250	2.598	.974	.095
1-3/4	1.750	2.625	3.031	1.134	.095
2	2.000	3.000	3.464	1.263	.095
2-1/4	2.250	3.375	3.897	1.423	.095
2-1/2	2.500	3.750	4.330	1.583	.095
2-3/4	2.750	4.125	4.763	1.744	.095
3	3.000	4.500	5.196	1.935	.095

APPENDIX 16 • Hex Nuts and Hex Jam Nuts

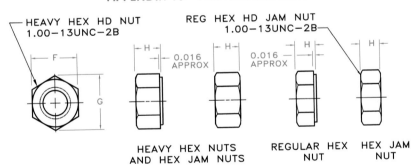

HEAVY HEX NUTS AND HEX JAM NUTS

REGULAR HEX NUT   HEX JAM NUT

MAJOR DIA		F Max.	G Avg.	H Max.	H Max.
1/4	.250	.438	.505	.226	.163
5/16	.313	.500	.577	.273	.195
3/8	.375	.563	.650	.337	.227
7/16	.438	.688	.794	.385	.260
1/2	.500	.750	.866	.448	.323
9/16	.563	.875	1.010	.496	.324
5/8	.625	.938	1.083	.559	.387

MAJOR DIA		F Max.	G Avg.	H Max.	H Max.
3/4	.750	1.125	1.299	.665	.446
7/8	.875	1.313	1.516	.776	.510
1	1.000	1.500	1.732	.887	.575
1-1/8	1.125	1.688	1.949	.899	.639
1-1/4	1.250	1.875	2.165	1.094	.751
1-3/8	1.375	2.063	2.382	1.206	.815
1-1/2	1.500	2.250	2.598	1.317	.880

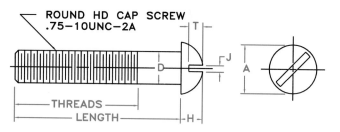

STANDARD COMMERCIAL LENGTHS											
DIAMETER	.50	.75	1.00	1.25	1.50	1.75	2.00	2.50	3.00	3.50	4.00
.25	●	●	●	●	●	●	●				
.31		●	●	●	●	●	●	●	●	●	●
.375			●	●	●	●	●	●	●	●	●
.500			●	●	●	●	●	●	●	●	●
.625				●	●	●	●	●	●	●	●
.750							●	●	●	●	●

OTHER LENGTHS AND DIAMETERS ARE
AVAILABLE, BUT THESE ARE THE
MORE STANDARD ONES.

Dimensions are in inches

DIA	D Max.	A Max.	H Avg.	J Max.	T Max.	DIA	D Max.	A Max.	H Avg.	J Max.	T Max.
1/4	.250	.437	.191	.075	.117	1/2	.500	.812	.354	.106	.218
5/16	.313	.562	.245	.084	.151	9/16	.563	.937	.409	.118	.252
3/8	.375	.625	.273	.094	.168	5/8	.625	1.000	.437	.133	.270
7/16	.438	.750	.328	.094	.202	3/4	.750	1.250	.546	.149	.338

FLAT HEAD CAP SCREWS

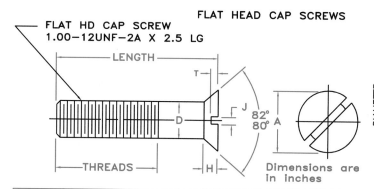

STANDARD COMMERCIAL LENGTHS											
DIAMETER	.50	.75	1.00	1.25	1.50	1.75	2.00	2.50	3.00	3.50	4.00
.25	●	●	●	●	●	●	●				
.31		●	●	●	●	●	●	●	●	●	●
.375			●	●	●	●	●	●	●	●	●
.500			●	●	●	●	●	●	●	●	●
.625				●	●	●	●	●	●	●	●
.750							●	●	●	●	●
.875								●	●	●	●
1.00									●	●	●
1.50									●	●	●

OTHER LENGTHS AND DIAMETERS ARE
AVAILABLE, BUT THESE ARE THE
MORE STANDARD ONES.

Dimensions are in inches

DIA	D Max.	A Max.	H Avg.	J Max.	T Max.	DIA	D Max.	A Max.	H Avg.	J Max.	T Max.
1/4	.250	.500	.140	.075	.068	3/4	.750	1.375	.352	.149	.171
5/16	.313	.625	.177	.084	.086	7/8	.875	1.625	.423	.167	.206
3/8	.375	.750	.210	.094	.103	1	1.000	1.875	.494	.188	.240
7/16	.438	.813	.210	.094	.103	1-1/8	1.125	2.062	.529	.196	.257
1/2	.500	.875	.210	.106	.103	1-1/4	1.250	2.312	.600	.211	.291
9/16	.563	1.000	.244	.118	.120	1-3/8	1.375	2.562	.665	.226	.326
5/8	.625	1.125	.281	.133	.137	1-1/2	1.500	2.812	.742	.258	.360

APPENDIX 19 • Fillister Head Cap Screws

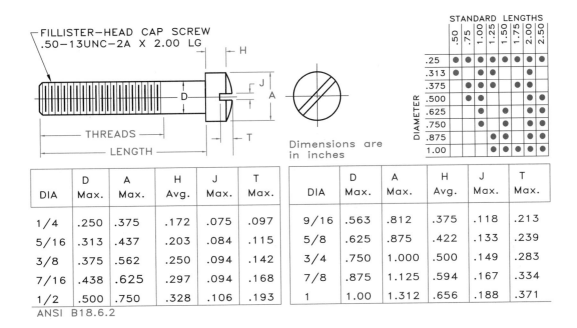

FILLISTER-HEAD CAP SCREW
.50-13UNC-2A X 2.00 LG

THREADS

LENGTH

Dimensions are in inches

DIA	D Max.	A Max.	H Avg.	J Max.	T Max.
1/4	.250	.375	.172	.075	.097
5/16	.313	.437	.203	.084	.115
3/8	.375	.562	.250	.094	.142
7/16	.438	.625	.297	.094	.168
1/2	.500	.750	.328	.106	.193

DIA	D Max.	A Max.	H Avg.	J Max.	T Max.
9/16	.563	.812	.375	.118	.213
5/8	.625	.875	.422	.133	.239
3/4	.750	1.000	.500	.149	.283
7/8	.875	1.125	.594	.167	.334
1	1.00	1.312	.656	.188	.371

ANSI B18.6.2

APPENDIX 20 • Flat Socket Head Cap Screws

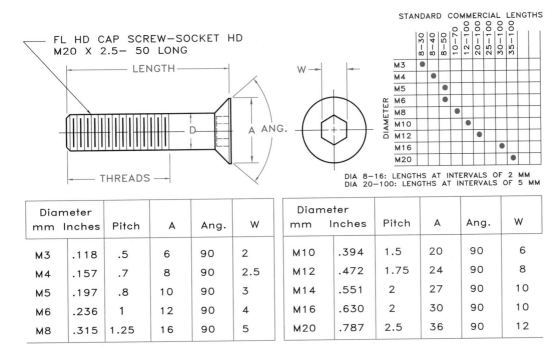

FL HD CAP SCREW-SOCKET HD
M20 X 2.5- 50 LONG

LENGTH

THREADS

DIA 8-16: LENGTHS AT INTERVALS OF 2 MM
DIA 20-100: LENGTHS AT INTERVALS OF 5 MM

Diameter mm	Inches	Pitch	A	Ang.	W
M3	.118	.5	6	90	2
M4	.157	.7	8	90	2.5
M5	.197	.8	10	90	3
M6	.236	1	12	90	4
M8	.315	1.25	16	90	5

Diameter mm	Inches	Pitch	A	Ang.	W
M10	.394	1.5	20	90	6
M12	.472	1.75	24	90	8
M14	.551	2	27	90	10
M16	.630	2	30	90	10
M20	.787	2.5	36	90	12

APPENDIX 21 • Socket Head Cap Screws

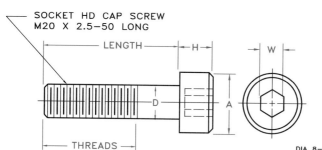

SOCKET HD CAP SCREW
M20 X 2.5–50 LONG

STANDARD COMMERCIAL LENGTHS

DIA 8–16: LENGTHS AT INTERVALS OF 2 MM
DIA 20–100: LENGTHS AT INTERVALS OF 5 MM

Diameter mm	Inches	Pitch	A	H	W
M3	.118	.5	6	3	2
M4	.157	.7	8	4	3
M5	.197	.8	10	5	4
M6	.236	1	12	6	6
M8	.315	1.25	16	8	6

Diameter mm	Inches	Pitch	A	H	W
M10	.394	1.5	20	10	8
M12	.472	1.75	24	12	10
M14	.551	2	27	14	12
M16	.630	2	30	16	14
M20	.787	2.5	36	20	17

APPENDIX 22 • Round Head Machine Screws

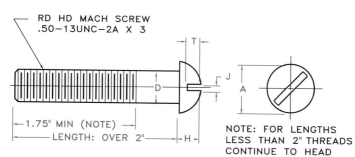

RD HD MACH SCREW
.50–13UNC–2A X 3

1.75" MIN (NOTE)
LENGTH: OVER 2"

NOTE: FOR LENGTHS LESS THAN 2" THREADS CONTINUE TO HEAD

STANDARD LENGTHS

OTHER LENGTHS AND DIAMETERS ARE AVAILABLE; THESE ARE THE MORE STANDARD ONES.

Dimensions are in inches

DIA	D Max.	A Max.	H Avg.	J Max.	T Max.
0	.060	.113	.053	.023	.039
1	.073	.138	.061	.026	.044
2	.086	.162	.069	.031	.048
3	.099	.187	.078	.035	.053
4	.112	.211	.086	.039	.058
5	.125	.236	.095	.043	.063
6	.138	.260	.103	.048	.068
8	.164	.309	.120	.054	.077
10	.190	.359	.137	.060	.087

DIA	D Max.	A Max.	H Avg.	J Max.	T Max.
12	.216	.408	.153	.067	.096
1/4	.250	.472	.175	.075	.109
5/16	.313	.590	.216	.084	.132
3/8	.375	.708	.256	.094	.155
7/16	.438	.750	.328	.094	.196
1/2	.500	.813	.355	.106	.211
9/16	.563	.938	.410	.118	.242
5/8	.625	1.000	.438	.133	.258
3/4	.750	1.250	.547	.149	.320

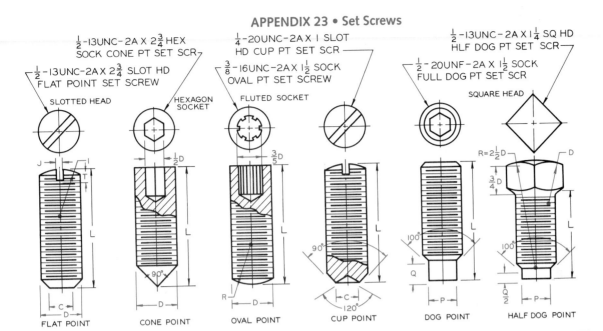

D	I	J	T	R	C		P		Q	q
Nominal Size	Radius of Headless Crown	Width of Slot	Depth of Slot	Oval Point Radius	Diameter of Cup and Flat Points		Diameter of Dog Point		Length of Dog Point	
					Max	Min	Max	Min	Full	Half
5 0.125	0.125	0.023	0.031	0.094	0.067	0.057	0.083	0.078	0.060	0.030
6 0.138	0.138	0.025	0.035	0.109	0.047	0.064	0.092	0.087	0.070	0.035
8 0.164	0.164	0.029	0.041	0.125	0.087	0.076	0.109	0.103	0.080	0.040
10 0.190	0.190	0.032	0.048	0.141	0.102	0.088	0.127	0.120	0.090	0.045
12 0.216	0.216	0.036	0.054	0.156	0.115	0.101	0.144	0.137	0.110	0.055
$\frac{1}{4}$ 0.250	0.250	0.045	0.063	0.188	0.132	0.118	0.156	0.149	0.125	0.063
$\frac{5}{16}$ 0.3125	0.313	0.051	0.076	0.234	0.172	0.156	0.203	0.195	0.156	0.078
$\frac{3}{8}$ 0.375	0.375	0.064	0.094	0.281	0.212	0.194	0.250	0.241	0.188	0.094
$\frac{7}{16}$ 0.4375	0.438	0.072	0.109	0.328	0.252	0.232	0.297	0.287	0.219	0.109
$\frac{1}{2}$ 0.500	0.500	0.081	0.125	0.375	0.291	0.270	0.344	0.344	0.250	0.125
$\frac{9}{16}$ 0.5625	0.563	0.091	0.141	0.422	0.332	0.309	0.391	0.379	0.281	0.140
$\frac{5}{8}$ 0.625	0.625	0.102	0.156	0.469	0.371	0.347	0.469	0.456	0.313	0.156
$\frac{3}{4}$ 0.750	0.750	0.129	0.188	0.563	0.450	0.425	0.563	0.549	0.375	0.188

Dimensions for the set screws shows in Fig. 18.44 (dimensions in inches)

Source: Courtesy of ANSI; B18.6.2.

APPENDIX 24 • Length of Thread Engagement Groups

Nominal Size Diam.		Pitch P	Length of Thread Engagement					Nominal Size Diam.		Pitch P	Length of Thread Engagement				
			Group S		Group N		Group L				Group S		Group N		Group L
Over	To and Incl		To and Incl	Over	To and Incl	Over		Over	To and Incl		To and Incl	Over	To and Incl	Over
1.5	2.8	0.2	0.5	0.5	1.5	1.5		22.4	45	1	4	4	12	12	
		0.25	0.6	0.6	1.9	1.9				1.5	6.3	6.3	19	19	
		0.35	0.8	0.8	2.6	2.6				2	8.5	8.5	25	25	
		0.4	1	1	3	3				3	12	12	36	36	
		0.45	1.3	1.3	3.8	3.8				3.5	15	15	45	45	
										4	18	18	53	53	
										4.5	21	21	63	63	
2.8	5.6	0.35	1	1	3	3		45	90	1.5	7.5	7.5	22	22	
		0.5	1.5	1.5	4.5	4.5				2	9.5	9.5	28	28	
		0.6	1.7	1.7	5	5				3	15	15	45	45	
		0.7	2	2	6	6				4	19	19	56	56	
		0.75	2.2	2.2	6.7	6.7				5	24	24	71	71	
		0.8	2.5	2.5	7.5	7.5				5.5	28	28	85	85	
										6	32	32	95	95	
5.6	11.2	0.75	2.4	2.4	7.1	7.1		90	180	2	12	12	36	36	
		1	3	3	9	9				3	18	18	53	53	
		1.25	4	4	12	12				4	24	24	71	71	
		1.5	5	5	15	15				6	36	36	106	106	
11.2	22.4	1	3.8	3.8	11	11		180	355	3	20	20	60	60	
		1.25	4.5	4.5	13	13				4	26	26	80	80	
		1.5	5.6	5.6	16	16				6	40	40	118	118	
		1.75	6	6	18	18									
		2	8	8	24	24									
		2.5	10	10	30	30									

All dimensions are given in millimeters. (Courtesy of ISO Standards.)

Number Size Drills

Size	Drill Diameter Inches	Drill Diameter mm	Size	Drill Diameter Inches	Drill Diameter mm	Size	Drill Diameter Inches	Drill Diameter mm	Size	Drill Diameter Inches	Drill Diameter mm
1	0.2280	5.7912	21	0.1590	4.0386	41	0.0960	2.4384	61	0.0390	0.9906
2	0.2210	5.6134	22	0.1570	3.9878	42	0.0935	2.3622	62	0.0380	0.9652
3	0.2130	5.4102	23	0.1540	3.9116	43	0.0890	2.2606	63	0.0370	0.9398
4	0.2090	5.3086	24	0.1520	3.8608	44	0.0860	2.1844	64	0.0360	0.9144
5	0.2055	5.2197	25	0.1495	3.7973	45	0.0820	2.0828	65	0.0350	0.8890
6	0.2040	5.1816	26	0.1470	3.7338	46	0.0810	2.0574	66	0.0330	0.8382
7	0.2010	5.1054	27	0.1440	3.6576	47	0.0785	1.9812	67	0.0320	0.8128
8	0.1990	5.0800	28	0.1405	3.5560	48	0.0760	1.9304	68	0.0310	0.7874
9	0.1960	4.9784	29	0.1360	3.4544	49	0.0730	1.8542	69	0.0292	0.7417
10	0.1935	4.9149	30	0.1285	3.2639	50	0.0700	1.7780	70	0.0280	0.7112
11	0.1910	4.8514	31	0.1200	3.0480	51	0.0670	1.7018	71	0.0260	0.6604
12	0.1890	4.8006	32	0.1160	2.9464	52	0.0635	1.6129	72	0.0250	0.6350
13	0.1850	4.6990	33	0.1130	2.8702	53	0.0595	1.5113	73	0.0240	0.6096
14	0.1820	4.6228	34	0.1110	2.8194	54	0.0550	1.3970	74	0.0225	0.5715
15	0.1800	4.5720	35	0.1100	2.7940	55	0.0520	1.3208	75	0.0210	0.5334
16	0.1770	4.4958	36	0.1065	2.7051	56	0.0465	1.1684	76	0.0200	0.5080
17	0.1730	4.3942	37	0.1040	2.6416	57	0.0430	1.0922	77	0.0180	0.4572
18	0.1695	4.3053	38	0.1015	2.5781	58	0.0420	1.0668	78	0.0160	0.4064
19	0.1660	4.2164	39	0.0995	2.5273	59	0.0410	1.0414	79	0.0145	0.3638
20	0.1610	4.0894	40	0.0980	2.4892	60	0.0400	1.0160	80	0.0135	0.3429

Metric Drill Sizes

Preferred sizes are in color type. Decimal-inch equivalents are for reference only.

mm	in.	mm	in.	mm	in.	mm	in.	mm	in.	mm	in.	mm	in.
.40	.0157	1.03	.0406	2.20	.0866	5.00	.1969	10.00	.3937	21.50	.8465	48.00	1.8898
.42	.0165	1.05	.0413	2.30	.0906	5.20	.2047	10.30	.4055	22.00	.8661	50.00	1.9685
.45	.0177	1.08	.0425	2.40	.0945	5.30	.2087	10.50	.4134	23.00	.9055	51.50	2.0276
.48	.0189	1.10	.0433	2.50	.0984	5.40	.2126	10.80	.4252	24.00	.9449	53.00	2.0866
.50	.0197	1.15	.0453	2.60	.1024	5.60	.2205	11.00	.4331	25.00	.9843	54.00	2.1260
.52	.0205	1.20	.0472	2.70	.1063	5.80	.2283	11.50	.4528	26.00	1.0236	56.00	2.2047
.55	.0217	1.25	.0492	2.80	.1102	6.00	.2362	12.00	.4724	27.00	1.0630	58.00	2.2835
.58	.0228	1.30	.0512	2.90	.1142	6.20	.2441	12.50	.4921	28.00	1.1024	60.00	2.3622
.60	.0236	1.35	.0531	3.00	.1181	6.30	.2480	13.00	.5118	29.00	1.1417		
.62	.0244	1.40	.0551	3.10	.1220	6.50	.2559	13.50	.5315	30.00	1.1811		
.65	.0256	1.45	.0571	3.20	.1260	6.70	.2638	14.00	.5512	31.00	1.2205		
.68	.0268	1.50	.0591	3.30	.1299	6.80	.2677	14.50	.5709	32.00	1.2598		
.70	.0276	1.55	.0610	3.40	.1339	6.90	.2717	15.00	.5906	33.00	1.2992		
.72	.0283	1.60	.0630	3.50	.1378	7.10	.2795	15.50	.6102	34.00	1.3386		
.75	.0295	1.65	.0650	3.60	.1417	7.30	.2874	16.00	.6299	35.00	1.3780		
.78	.0307	1.70	.0669	3.70	.1457	7.50	.2953	16.50	.6496	36.00	1.4173		
.80	.0315	1.75	.0689	3.80	.1496	7.80	.3071	17.00	.6693	37.00	1.4567		
.82	.0323	1.80	.0709	3.90	.1535	8.00	.3150	17.50	.6890	38.00	1.4961		
.85	.0335	1.85	.0728	4.00	.1575	8.20	.3228	18.00	.7087	39.00	1.5354		
.88	.0346	1.90	.0748	4.10	.1614	8.50	.3346	18.50	.7283	40.00	1.5748		
.90	.0354	1.95	.0768	4.20	.1654	8.80	.3465	19.00	.7480	41.00	1.6142		
.92	.0362	2.00	.0787	4.40	.1732	9.00	.3543	19.50	.7677	42.00	1.6535		
.95	.0374	2.05	.0807	4.50	.1772	9.20	.3622	20.00	.7874	43.50	1.7126		
.98	.0386	2.10	.0827	4.60	.1811	9.50	.3740	20.50	.8071	45.00	1.7717		
1.00	.0394	2.15	.0846	4.80	.1890	9.80	.3858	21.00	.8268	46.50	1.8307		

Cont.

Letter Size Drills

Size	Drill Diameter Inches	mm	Size	Drill Diameter Inches	mm	Size	Drill Diameter Inches	mm	Size	Drill Diameter Inches	mm
A	0.234	5.944	H	0.266	6.756	O	0.316	8.026	V	0.377	9.576
B	0.238	6.045	I	0.272	6.909	P	0.323	8.204	W	0.386	9.804
C	0.242	6.147	J	0.277	7.036	Q	0.332	8.433	X	0.397	10.084
D	0.246	6.248	K	0.281	7.137	R	0.339	8.611	Y	0.404	10.262
E	0.250	6.350	L	0.290	7.366	S	0.348	8.839	Z	0.413	10.490
F	0.257	6.528	M	0.295	7.493	T	0.358	9.093			
G	0.261	6.629	N	0.302	7.601	U	0.368	9.347			

(Courtesy of General Motors Corporation.)

APPENDIX 26 • Straight Pins

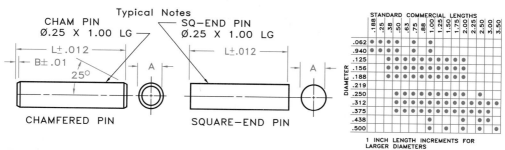

Nominal DIA	Diamter A Max	Min	Chamfer B
0.062	0.0625	0.0605	0.015
0.094	0.0937	0.0917	0.015
0.109	0.1094	0.1074	0.015
0.125	0.1250	0.1230	0.015
0.156	0.1562	0.1542	0.015
0.188	0.1875	0.1855	0.015

Nominal DIA	Diamter A Max	Min	Chamfer B
0.219	0.2187	0.2167	0.015
0.250	0.2500	0.2480	0.015
0.312	0.3125	0.3095	0.015
0.375	0.3750	0.3720	0.030
0.438	0.4345	0.4345	0.030
0.500	0.4970	0.4970	0.030

ANSI: B5.20

APPENDIX 27 • Taper Pins

TYPICAL NOTE:
NO. 2 TAPER PIN—1.500 LG

L=Length

Number	7/0	6/0	5/0	4/0	3/0	2/0	0	1	2	3	4	5	6	7	8	9	10
Size (large end)	0.0625	0.0780	0.0940	0.1090	0.1250	0.1410	0.1560	0.1720	0.1930	0.2190	0.2500	0.2890	0.3410	0.4090	0.4920	0.5910	0.7060
Length, L																	
0.375	X	X															
0.500	X	X	X														
0.625	X	X	X	X	X	X	X										
0.750		X	X	X	X	X	X	X									
0.875				X	X	X	X	X	X								
1.000			X	X	X	X	X	X	X	X							
1.250						X	X	X	X	X	X						
1.500						X	X	X	X	X	X	X					
1.750							X	X	X	X	X	X	X				
2.000								X	X	X	X	X	X	X	X		
2.250									X	X	X	X	X	X	X	X	
2.500										X	X	X	X	X	X	X	
2.750										X	X	X	X	X	X	X	X
3.000										X	X	X	X	X	X	X	X
3.250												X	X	X	X	X	X
3.500													X	X	X	X	X
3.750													X		X	X	X
4.000															X	X	X
4.250																X	X
4.500															X	X	X
4.750															X	X	X
5.000																X	X
5.250																X	X
5.500																X	X
5.750																X	X
6.000																	X

All dimensions are given in inches.

Standard reamers are available for pins given above the line.

Pins Nos. 11 (size 0.8600), 12 (size 1.032), 13 (size 1.241), and 14 (1.523) are special sizes—hence their lengths are special.

To find small diameter of pin, multiply the length by 0.02083 and subtract the result from the large diameter.

(Courtesy of ANSI; B5.20–1958.)

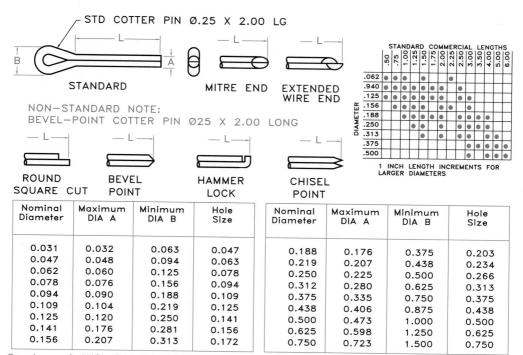

Courtesy of ANSI: B18.8.1–1983

Nominal Diameter	Maximum DIA A	Minimum DIA B	Hole Size
0.031	0.032	0.063	0.047
0.047	0.048	0.094	0.063
0.062	0.060	0.125	0.078
0.078	0.076	0.156	0.094
0.094	0.090	0.188	0.109
0.109	0.104	0.219	0.125
0.125	0.120	0.250	0.141
0.141	0.176	0.281	0.156
0.156	0.207	0.313	0.172

Nominal Diameter	Maximum DIA A	Minimum DIA B	Hole Size
0.188	0.176	0.375	0.203
0.219	0.207	0.438	0.234
0.250	0.225	0.500	0.266
0.312	0.280	0.625	0.313
0.375	0.335	0.750	0.375
0.438	0.406	0.875	0.438
0.500	0.473	1.000	0.500
0.625	0.598	1.250	0.625
0.750	0.723	1.500	0.750

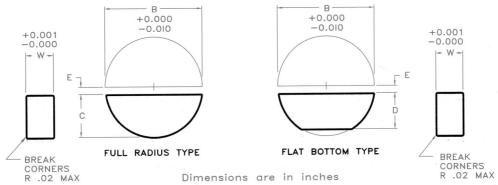

FULL RADIUS TYPE

FLAT BOTTOM TYPE

BREAK CORNERS R .02 MAX

Dimensions are in inches

Key No.	W X B	C Max.	D Max.	E		Key No.	W X B	C Max.	D Max.	E
204	1/16 X 1/2	.203	.194	.047		506	5/32 X 3/4	.313	.303	.063
304	3/32 X 1/2	.203	.194	.047		606	3/16 X 3/4	.313	.303	.063
404	1/8 X 1/2	.203	.194	.047		507	5/32 X 7/8	.375	.365	.063
305	3/32 X 5/8	.250	.240	.063		607	3/16 X 7/8	.375	.365	.063
405	1/8 X 5/8	.250	.240	.063		807	1/4 X 7/8	.375	.365	.063
505	5/32 X 5/8	.250	.240	.063		608	3/16 X 1	.438	.428	.063
406	1/8 X 3/4	.313	.303	.063		609	3/16 X 1-1/8	.484	.475	.078

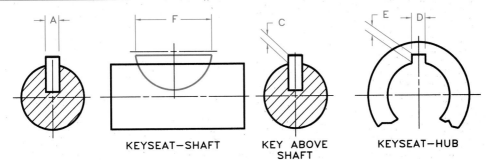

KEYSEAT—SHAFT

KEY ABOVE SHAFT

KEYSEAT—HUB

Key No.	A Min.	C +.005 −.000	F	D +.005 −.000	E +.005 −.000		Key No.	A Min.	C +.005 −.000	F	D +.005 −.000	E +.005 −.000
204	.0615	.0312	.500	.0635	.0372		506	.1553	.0781	.750	.1573	.0841
304	.0928	.0469	.500	.0948	.0529		606	.1863	.0937	.750	.1885	.0997
404	.1240	.0625	.500	.1260	.0685		507	.1553	.0781	.875	.1573	.0841
305	.0928	.0625	.625	.0948	.0529		607	.1863	.0937	.875	.1885	.0997
405	.1240	.0469	.625	.1260	.0685		807	.2487	.1250	.875	.2510	.1310
505	.1553	.0625	.625	.1573	.0841		608	.1863	.3393	1.000	.1885	.0997
406	.1240	.0781	.750	.1260	.0685		609	.1863	.3853	1.125	.1885	.0997

KEY SIZES VS. SHAFT SIZES

Shaft DIA	to .375	to .500	to .750	to 1.313	to 1.188	to 1.448	to 1.750	to 2.125	to 2.500
Key Nos.	204	304	404	505	606	807	810	1011	1211
		305	405	506	607	808	811	1012	1212
			406	507	608	809	812		
					609				

APPENDIX 30 • Standard Keys and Keyways

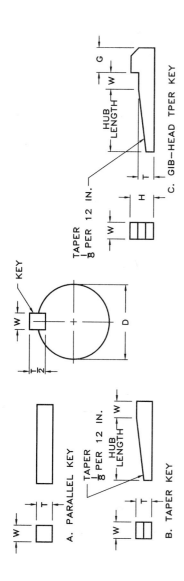

A. PARALLEL KEY

B. TAPER KEY

C. GIB-HEAD TPER KEY

Sprocket Bore (= Shaft Diam.) Inches D	Keyway Dimensions — Inches For Square Key Width W	Depth T/2	For Flat Key Width W	Depth T/2	Key Dimensions — Inches Square Width W	Height T	Flat Width W	Height T	Tolerance on W and T (−)	Gib Head Dimensions — Inches Square Key H	G	Flat Key H	G	Key Tolerances Taper and Gib Head W (−)	T (+)
1/2 — 9/16	1/8	1/16	1/8	3/64	1/8	1/8	1/8	3/32	0.002	1/4	7/32	3/16	1/8	0.002	0.002
5/8 — 7/8	3/16	3/32	3/16	1/16	3/16	3/16	3/16	1/8	0.002	5/16	9/32	1/4	3/16	0.002	0.002
13/16 — 1 1/4	1/4	1/8	1/4	3/32	1/4	1/4	1/4	3/16	0.002	7/16	11/32	5/16	1/4	0.002	0.002
1 3/16 — 1 3/8	5/16	5/32	5/16	1/8	5/16	5/16	5/16	1/4	0.002	9/16	13/32	3/8	5/16	0.002	0.002
1 7/16 — 1 3/4	3/8	3/16	3/8	1/8	3/8	3/8	3/8	1/4	0.002	11/16	15/32	7/16	3/8	0.002	0.002
1 13/16 — 2 1/4	1/2	1/4	1/2	3/16	1/2	1/2	1/2	3/8	0.0025	7/8	19/32	5/8	1/2	0.0025	0.0025
2 5/16 — 2 3/4	5/8	5/16	5/8	7/32	5/8	5/8	5/8	7/16	0.0025	1 1/16	23/32	3/4	5/8	0.0025	0.0025
2 7/8 — 3 1/4	3/4	3/8	3/4	1/4	3/4	3/4	3/4	1/2	0.0025	1 1/4	7/8	7/8	3/4	0.0025	0.0025
3 3/8 — 3 3/4	7/8	7/16	7/8	5/16	7/8	7/8	7/8	5/8	0.003	1 1/2	1	1 1/16	7/8	0.003	0.003
3 7/8 — 4 1/2	1	1/2	1	3/8	1	1	1	3/4	0.003	1 3/4	1 3/16	1 1/4	1	0.003	0.003
4 3/4 — 5 1/2	1 1/4	5/8	1 1/4	7/16	1 1/4	1 1/4	1 1/4	7/8	0.003	2	1 7/16	1 1/2	1 1/4	0.003	0.003
5 3/4 — 7 7/8	1 1/2	3/4	1 1/2	1/2	1 1/2	1 1/2	1 1/2	1	0.003	2 1/2	1 3/4	1 3/4	1 1/2	0.003	0.003
7 1/2 — 9 7/8	1 3/4	7/8	..	..	1 3/4	1 3/4	..	..	0.004	3	2	..	..	0.004	0.004
10 — 12 1/2	2	1	..	..	2	2	..	..	0.004	3 1/2	2 3/8	..	..	0.004	0.004

Standard Keyway Tolerances: Straight Keyway — Width (W) + .005 Depth (T/2) + .010
                                                                    − .000   − .000

Taper Keyway — Width (W) + .005 Depth (T/2) + .000
                                   − .000   − .010

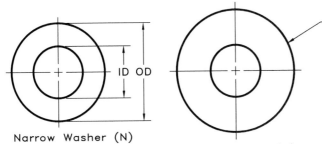

.938 X 2.25 X .165
TYPE A PLAIN WASHER

Dimensioned
Washer

In Screw Size Column
N= Narrow washer
W= Wide washer

Narrow Washer (N)
TYPE A PLAIN WASHERS

WIDE WASHER (W)

SCREW SIZE	ID SIZE	OD SIZE	THICK-NESS	SCREW SIZE	ID SIZE	OD SIZE	THICK-NESS
0.138	0.156	0.375	0.049	0.875 N	0.938	1.750	0.134
0.164	0.188	0.438	0.049	0.875 W	0.938	2.250	0.165
0.190	0.219	0.500	0.049	1.000 N	1.062	2.000	0.134
0.188	0.250	0.562	0.049	1.000 W	1.062	2.500	0.165
0.216	0.250	0.562	0.065	1.125 N	1.250	2.250	0.134
0.250 N	0.281	0.625	0.065	1.125 W	1.250	2.750	0.165
0.250 W	0.312	0.734	0.065	1.250 N	1.375	2.500	0.165
0.312 N	0.344	0.688	0.065	1.250 W	1.375	3.000	0.165
0.312 W	0.375	0.875	0.083	1.375 N	1.500	2.750	0.165
0.375 N	0.406	0.812	0.065	1.375 W	1.500	3.250	0.180
0.375 W	0.438	1.000	0.083	1.500 N	1.625	3.000	0.165
0.438 N	0.469	0.922	0.065	1.500 W	1.625	3.500	0.180
0.438 W	0.500	1.250	0.083	1.625	1.750	3.750	0.180
0.500 N	0.531	1.062	0.095	1.750	1.875	4.000	0.180
0.500 W	0.562	1.375	0.109	1.875	2.000	4.250	0.180
0.562 N	0.594	1.156	0.095	2.000	2.125	4.500	0.180
0.562 W	0.594	1.469	0.109	2.250	2.375	4.750	0.220
0.625 N	0.625	1.312	0.095	2.500	2.625	5.000	0.238
0.625 N	0.625	1.750	0.134	2.750	2.875	5.250	0.259
0.750 W	0.812	1.469	0.134	3.000	3.125	5.500	0.284
0.750 W	0.812	2.000	0.148				

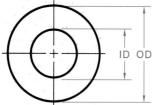

17 X 27 X 2
WROUGHT WASHER

Dimensioned
Washer

DIN= German Industrial
Standard (ISO)

FLAT WASHERS
DIN 9021

WROUGHT WASHERS
DIN 433

SCREW SIZE	ID SIZE	OD SIZE	THICK-NESS	SCREW SIZE	ID SIZE	OD SIZE	THICK-NESS
3	3.2	9	0.8	2.6	2.8	5.5	0.5
4	4.3	12	1	3	3.2	6	0.5
5	5.3	15	1.5	4	4.3	8	0.5
6	6.4	18	1.5	5	5.3	10	1.0
8	8.4	25	2	6	6.4	11	1.5
10	10.5	30	2.5	8	8.4	15	1.5
12	13	40	3	10	10.5	18	1.5
14	15	45	3	12	13	20	2.0
16	17	50	3	14	15	25	2.0
18	19	56	4	16	17	27	2.0
20	21	60	4	18	19	30	2.5
				20	21	33	2.5

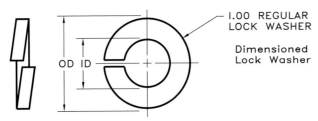

SCREW SIZE	ID SIZE	OD SIZE	THICK- NESS	SCREW SIZE	ID SIZE	OD SIZE	THICK- NESS
0.164	0.168	0.175	0.040	0.812	0.826	1.367	0.203
0.190	0.194	0.202	0.047	0.875	0.890	1.464	0.219
0.216	0.221	0.229	0.056	0.938	0.954	1.560	0.234
0.250	0.255	0.263	0.062	1.000	1.017	1.661	0.250
0.312	0.318	0.328	0.078	1.062	1.080	1.756	0.266
0.375	0.382	0.393	0.094	1.125	1.144	1.853	0.281
0.438	0.446	0.459	0.109	1.188	1.208	1.950	0.297
0.500	0.509	0.523	0.125	1.250	1.271	2.045	0.312
0.562	0.572	0.587	0.141	1.312	1.334	2.141	0.328
0.625	0.636	0.653	0.156	1.375	1.398	2.239	0.344
0.688	0.700	0.718	0.172	1.438	1.462	2.334	0.359
0.750	0.763	0.783	0.188	1.500	1.525	2.430	0.375

METRIC LOCK WASHERS—DIN 127 (Millimeters)

SCREW SIZE	ID SIZE	OD SIZE	THICK- NESS	SCREW SIZE	ID SIZE	OD SIZE	THICK- NESS
4	4.1	7.1	0.9	22	22.5	34.5	4
5	5.1	8.7	1.2	24	24.5	38.5	5
6	6.1	11.1	1.6	27	27.5	41.5	5
8	8.2	12.1	1.6	30	30.5	46.5	6
10	10.2	14.2	2	33	33.5	53.5	6
12	12.1	17.2	2.2	36	36.5	56.5	6
14	14.2	20.2	2.5	39	39.5	59.5	6
16	16.2	23.2	3	42	42.5	66.5	7
18	18.2	26.2	3.5	45	45.5	69.5	7
20	20.2	28.2	3.5	48	49	73	7

APPENDIX 34 • American Standard Running and Sliding Fits (hole basis)

Limits are in thousandths of an inch.
Limits for hole and shaft are applied algebraically to the basic size to obtain the limits of size for the parts.
Data in bold face are in accordance with ABC agreements.
Symbols H5, g5, etc., are Hole and Shaft designations used in ABC System.

Nominal Size Range Inches Over — To	Class RC 1 Limits of Clearance	Standard Limits Hole H5	Shaft g4	Class RC 2 Limits of Clearance	Standard Limits Hole H6	Shaft g5	Class RC 3 Limits of Clearance	Standard Limits Hole H7	Shaft f6	Class RC 4 Limits of Clearance	Standard Limits Hole H8	Shaft f7
0 — 0.12	0.1 / 0.45	+0.2 / 0	−0.1 / −0.25	0.1 / 0.55	+0.25 / 0	−0.1 / −0.3	0.3 / 0.95	+0.4 / 0	−0.3 / −0.55	0.3 / 1.3	+0.6 / 0	−0.3 / −0.7
0.12 — 0.24	0.15 / 0.5	+0.2 / 0	−0.15 / −0.3	0.15 / 0.65	+0.3 / 0	−0.15 / −0.35	0.4 / 1.12	+0.5 / 0	−0.4 / −0.7	0.4 / 1.6	+0.7 / 0	−0.4 / −0.9
0.24 — 0.40	0.2 / 0.6	0.25 / 0	−0.2 / −0.35	0.2 / 0.85	+0.4 / 0	−0.2 / −0.45	0.5 / 1.5	+0.6 / 0	−0.5 / −0.9	0.5 / 2.0	+0.9 / 0	−0.5 / −1.1
0.40 — 0.71	0.25 / 0.75	+0.3 / 0	−0.25 / −0.45	0.25 / 0.95	+0.4 / 0	−0.25 / −0.55	0.6 / 1.7	+0.7 / 0	−0.6 / −1.0	0.6 / 2.3	+1.0 / 0	−0.6 / −1.3
0.71 — 1.19	0.3 / 0.95	+0.4 / 0	−0.3 / −0.55	0.3 / 1.2	+0.5 / 0	−0.3 / −0.7	0.8 / 2.1	+0.8 / 0	−0.8 / −1.3	0.8 / 2.8	+1.2 / 0	−0.8 / −1.6
1.19 — 1.97	0.4 / 1.1	+0.4 / 0	−0.4 / −0.7	0.4 / 1.4	+0.6 / 0	−0.4 / −0.8	1.0 / 2.6	+1.0 / 0	−1.0 / −1.6	1.0 / 3.6	+1.6 / 0	−1.0 / −2.0
1.97 — 3.15	0.4 / 1.2	+0.5 / 0	−0.4 / −0.7	0.4 / 1.6	+0.7 / 0	−0.4 / −0.9	1.2 / 3.1	+1.2 / 0	−1.2 / −1.9	1.2 / 4.2	+1.8 / 0	−1.2 / −2.4
3.15 — 4.73	0.5 / 1.5	+0.6 / 0	−0.5 / −0.9	0.5 / 2.0	+0.9 / 0	−0.5 / −1.1	1.4 / 3.7	+1.4 / 0	−1.4 / −2.3	1.4 / 5.0	+2.2 / 0	−1.4 / −2.8
4.73 — 7.09	0.6 / 1.8	+0.7 / 0	−0.6 / −1.1	0.6 / 2.3	+1.0 / 0	−0.6 / −1.3	1.6 / 4.2	+1.6 / 0	−1.6 / −2.6	1.6 / 5.7	+2.5 / 0	−1.6 / −3.2
7.09 — 9.85	0.6 / 2.0	+0.8 / 0	−0.6 / −1.2	0.6 / 2.6	+1.2 / 0	−0.6 / −1.4	2.0 / 5.0	+1.8 / 0	−2.0 / −3.2	2.0 / 6.6	+2.8 / 0	−2.0 / −3.8
9.85 — 12.41	0.8 / 2.3	+0.9 / 0	−0.8 / −1.4	0.8 / 2.9	+1.2 / 0	−0.8 / −1.7	2.5 / 5.7	+2.0 / 0	−2.5 / −3.7	2.5 / 7.5	+3.0 / 0	−2.5 / −4.5
12.41 — 15.75	1.0 / 2.7	+1.0 / 0	−1.0 / −1.7	1.0 / 3.4	+1.4 / 0	−1.0 / −2.0	3.0 / 6.6	+ / 0	−3.0 / −4.4	3.0 / 8.7	+3.5 / 0	−3.0 / −5.2
15.75 — 19.69	1.2 / 3.0	+1.0 / 0	−1.2 / −2.0	1.2 / 3.8	+1.6 / 0	−1.2 / −2.2	4.0 / 8.1	+1.6 / 0	−4.0 / −5.6	4.0 / 10.5	+4.0 / 0	−4.0 / −6.5
19.69 — 30.09	1.6 / 3.7	+1.2 / 0	−1.6 / −2.5	1.6 / 4.8	+2.0 / 0	−1.6 / −2.8	5.0 / 10.0	+3.0 / 0	−5.0 / −7.0	5.0 / 13.0	+5.0 / 0	−5.0 / −8.0
30.09 — 41.49	2.0 / 4.6	+1.6 / 0	−2.0 / −3.0	2.0 / 6.1	+2.5 / 0	−2.0 / −3.6	6.0 / 12.5	+4.0 / 0	−6.0 / −8.5	6.0 / 16.0	+6.0 / 0	−6.0 / −10.0
41.49 — 56.19	2.5 / 5.7	+2.0 / 0	−2.5 / −3.7	2.5 / 7.5	+3.0 / 0	−2.5 / −4.5	8.0 / 16.0	+5.0 / 0	−8.0 / −11.0	8.0 / 21.0	+8.0 / 0	−8.0 / −13.0
56.19 — 76.39	3.0 / 7.1	+2.5 / 0	−3.0 / −4.6	3.0 / 9.5	+4.0 / 0	−3.0 / −5.5	10.0 / 20.0	+6.0 / 0	−10.0 / −14.0	10.0 / 26.0	+10.0 / 0	−10.0 / −16.0
76.39 — 100.9	4.0 / 9.0	+3.0 / 0	−4.0 / −6.0	4.0 / 12.0	+5.0 / 0	−4.0 / −7.0	12.0 / 25.0	+8.0 / 0	−12.0 / −17.0	12.0 / 32.0	+12.0 / 0	−12.0 / −20.0
100.9 — 131.9	5.0 / 11.5	+4.0 / 0	−5.0 / −7.5	5.0 / 15.0	+6.0 / 0	−5.0 / −9.0	16.0 / 32.0	+10.0 / 0	−16.0 / −22.0	16.0 / 36.0	+16.0 / 0	−16.0 / −26.0
131.9 — 171.9	6.0 / 14.0	+5.0 / 0	−6.0 / −9.0	6.0 / 19.0	+8.0 / 0	−6.0 / −11.0	18.0 / 38.0	+8.0 / 0	−18.0 / −26.0	18.0 / 50.0	+20.0 / 0	−18.0 / −30.0
171.9 — 200	8.0 / 18.0	+6.0 / 0	−8.0 / −12.0	8.0 / 22.0	+10.0 / 0	−8.0 / −12.0	22.0 / 48.0	+16.0 / 0	−22.0 / −32.0	22.0 / 63.0	+25.0 / 0	−22.0 / −38.0

(Courtesy of USASI; B4.1–1955.)

Cont.

Class RC 5			Class RC 6			Class RC 7			Class RC 8			Class RC 9			Nominal Size Range Inches	
Limits of Clearance	Standard Limits		Limits of Clearance	Standard Limits		Limits of Clearance	Standard Limits		Limits of Clearance	Standard Limits		Limits of Clearance	Standard Limits			
	Hole H8	Shaft e7		Hole H9	Shaft e8		Hole H9	Shaft d8		Hole H10	Shaft c9		Hole H11	Shaft	Over	To
0.6 / 1.6	+0.6 / −0	−0.6 / −1.0	0.6 / 2.2	+1.0 / −0	−0.6 / −1.2	1.0 / 2.6	+1.0 / 0	−1.0 / −1.6	2.5 / 5.1	+1.6 / 0	−2.5 / −3.5	4.0 / 8.1	+2.5 / 0	−4.0 / −5.6	0	0.12
0.8 / 2.0	+0.7 / −0	−0.8 / −1.3	0.8 / 2.7	+1.2 / −0	−0.8 / −1.5	1.2 / 3.1	+1.2 / 0	−1.2 / −1.9	2.8 / 5.8	+1.8 / 0	−2.8 / −4.0	4.5 / 9.0	+3.0 / 0	−4.5 / −6.0	0.12	0.24
1.0 / 2.5	+0.9 / −0	−1.0 / −1.6	1.0 / 3.3	+1.4 / −0	−1.0 / −1.9	1.6 / 3.9	+1.4 / 0	−1.6 / −2.5	3.0 / 6.6	+2.2 / 0	−3.0 / −4.4	5.0 / 10.7	+3.5 / 0	−5.0 / −7.2	0.24	0.40
1.2 / 2.9	+1.0 / −0	−1.2 / −1.9	1.2 / 3.8	+1.6 / −0	−1.2 / −2.2	2.0 / 4.6	+1.6 / 0	−2.0 / −3.0	3.5 / 7.9	+2.8 / 0	−3.5 / −5.1	6.0 / 12.8	+4.0 / −0	−6.0 / −8.8	0.40	0.71
1.6 / 3.6	+1.2 / −0	−1.6 / −2.4	1.6 / 4.8	+2.0 / −0	−1.6 / −2.8	2.5 / 5.7	+2.0 / 0	−2.5 / −3.7	4.5 / 10.0	+3.5 / 0	−4.5 / −6.5	7.0 / 15.5	+5.0 / 0	−7.0 / −10.5	0.71	1.19
2.0 / 4.6	+1.6 / −0	−2.0 / −3.0	2.0 / 6.1	+2.5 / −0	−2.0 / −3.6	3.0 / 7.1	+2.5 / 0	−3.0 / −4.6	5.0 / 11.5	+4.0 / 0	−5.0 / −7.5	8.0 / 18.0	+6.0 / 0	−8.0 / −12.0	1.19	1.97
2.5 / 5.5	+1.8 / −0	−2.5 / −3.7	2.5 / 7.3	+3.0 / −0	−2.5 / −4.3	4.0 / 8.8	+3.0 / 0	−4.0 / −5.8	6.0 / 13.5	+4.5 / 0	−6.0 / −9.0	9.0 / 20.5	+7.0 / 0	−9.0 / −13.5	1.97	3.15
3.0 / 6.6	+2.2 / −0	−3.0 / −4.4	3.0 / 8.7	+3.5 / −0	−3.0 / −5.2	5.0 / 10.7	+3.5 / 0	−5.0 / −7.2	7.0 / 15.5	+5.0 / 0	−7.0 / −10.5	10.0 / 24.0	+9.0 / 0	−10.0 / −15.0	3.15	4.73
3.5 / 7.6	+2.5 / −0	−3.5 / −5.1	3.5 / 10.0	+4.0 / −0	−3.5 / −6.0	6.0 / 12.5	+4.0 / 0	−6.0 / −8.5	8.0 / 18.0	+6.0 / 0	−8.0 / −12.0	12.0 / 28.0	+10.0 / 0	−12.0 / −18.0	4.73	7.09
4.0 / 8.6	+2.8 / −0	−4.0 / −5.8	4.0 / 11.3	+4.5 / −0	−4.0 / −6.8	7.0 / 14.3	+4.5 / 0	−7.0 / −9.8	10.0 / 21.5	+7.0 / 0	−10.0 / −14.5	15.0 / 34.0	+12.0 / 0	−15.0 / −22.0	7.09	9.85
5.0 / 10.0	+3.0 / 0	−5.0 / −7.0	5.0 / 13.0	+5.0 / 0	−5.0 / −8.0	8.0 / 16.0	+5.0 / 0	−8.0 / −11.0	12.0 / 25.0	+8.0 / 0	−12.0 / −17.0	18.0 / 38.0	+12.0 / 0	−18.0 / −26.0	9.85	12.41
6.0 / 11.7	+3.5 / 0	−6.0 / −8.2	6.0 / 15.5	+6.0 / 0	−6.0 / −9.5	10.0 / 19.5	+6.0 / 0	−10.0 / −13.5	14.0 / 29.0	+9.0 / 0	−14.0 / −20.0	22.0 / 45.0	+14.0 / 0	−22.0 / −31.0	12.41	15.75
8.0 / 14.5	+4.0 / 0	−8.0 / −10.5	8.0 / 18.0	+6.0 / 0	−8.0 / −12.0	12.0 / 22.0	+6.0 / 0	−12.0 / −16.0	16.0 / 32.0	+10.0 / 0	−16.0 / −22.0	25.0 / 51.0	+16.0 / 0	−25.0 / −35.0	15.75	19.69
10.0 / 18.0	+5.0 / 0	−10.0 / −13.0	10.0 / 23.0	+8.0 / 0	−10.0 / −15.0	16.0 / 29.0	+8.0 / 0	−16.0 / −21.0	20.0 / 40.0	+12.0 / 0	−20.0 / −28.0	30.0 / 62.0	+20.0 / 0	−30.0 / −42.0	19.69	30.09
12.0 / 22.0	+6.0 / 0	−12.0 / −16.0	12.0 / 28.0	+10.0 / 0	−12.0 / −18.0	20.0 / 36.0	+10.0 / 0	−20.0 / −26.0	25.0 / 51.0	+16.0 / 0	−25.0 / −35.0	40.0 / 81.0	+25.0 / 0	−40.0 / −56.0	30.09	41.49
16.0 / 29.0	+8.0 / 0	−16.0 / −21.0	16.0 / 36.0	+12.0 / 0	−16.0 / −24.0	25.0 / 45.0	+12.0 / 0	−25.0 / −33.0	30.0 / 62.0	+20.0 / 0	−30.0 / −42.0	50.0 / 100	+30.0 / 0	−50.0 / −70.0	41.49	56.19
20.0 / 36.0	+10.0 / 0	−20.0 / −26.0	20.0 / 46.0	+16.0 / 0	−20.0 / −30.0	30.0 / 56.0	+16.0 / 0	−30.0 / −40.0	40.0 / 81.0	+25.0 / 0	−40.0 / −56.0	60.0 / 125	+40.0 / 0	−60.0 / −85.0	56.19	76.39
25.0 / 45.0	+12.0 / 0	−25.0 / −33.0	25.0 / 57.0	+20.0 / 0	−25.0 / −37.0	40.0 / 72.0	+20.0 / 0	−40.0 / −52.0	50.0 / 100	+30.0 / 0	−50.0 / −70.0	80.0 / 160	+50.0 / 0	−80.0 / −110	76.39	100.9
30.0 / 56.0	+16.0 / 0	−30.0 / −40.0	30.0 / 71.0	+25.0 / 0	−30.0 / −46.0	50.0 / 91.0	+25.0 / 0	−50.0 / −66.0	60.0 / 125	+40.0 / 0	−60.0 / −85.0	100 / 200	+60.0 / 0	−100 / −140	100.9	131.9
35.0 / 67.0	+20.0 / 0	−35.0 / −47.0	35.0 / 85.0	+30.0 / 0	−35.0 / −55.0	60.0 / 110.0	+30.0 / 0	−60.0 / −80.0	80.0 / 160	+50.0 / 0	−80.0 / −110	130 / 260	+80.0 / 0	−130 / −180	131.9	171.9
45.0 / 86.0	+25.0 / 0	−45.0 / −61.0	45.0 / 110.0	+40.0 / 0	−45.0 / −70.0	80.0 / 145.0	+40.0 / 0	−80.0 / −105.0	100 / 200	+60.0 / 0	−100 / −140	150 / 310	+100 / 0	−150 / −210	171.9	200

(Courtesy of ANSI; B4.1–1955.)

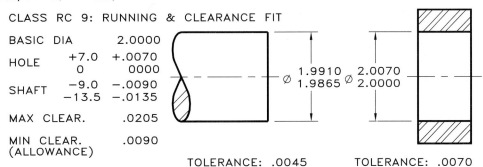

CLASS RC 9: RUNNING & CLEARANCE FIT

BASIC DIA	2.0000	
HOLE	+7.0 / 0	+.0070 / 0000
SHAFT	−9.0 / −13.5	−.0090 / −.0135
MAX CLEAR.	.0205	
MIN CLEAR. (ALLOWANCE)	.0090	

Ø 1.9910 / 1.9865   Ø 2.0070 / 2.0000

TOLERANCE: .0045   TOLERANCE: .0070

Limits are in thousandths of an inch.
Limits for hole and shaft are applied algebraically to the basic size to obtain the limits of size for the parts.
Data in bold face are in accordance with ABC agreements.
Symbols H9,f8, etc., are Hole and Shaft designations used in ABC System.

Nominal Size Range Inches Over	To	Class LC 1 Limits of Clearance	LC1 Hole H6	LC1 Shaft h5	Class LC 2 Limits of Clearance	LC2 Hole H7	LC2 Shaft h6	Class LC 3 Limits of Clearance	LC3 Hole H8	LC3 Shaft h7	Class LC 4 Limits of Clearance	LC4 Hole H10	LC4 Shaft h9	Class LC 5 Limits of Clearance	LC5 Hole H7	LC5 Shaft g6
0 —	0.12	0 / 0.45	+0.25 / -0	+0 / -0.2	0 / 0.65	+0.4 / -0	+0 / -0.25	0 / 1	+0.6 / -0	+0 / -0.4	0 / 2.6	+1.6 / -0	+0 / -1.0	0.1 / 0.75	+0.4 / -0	-0.1 / -0.35
0.12—	0.24	0 / 0.5	+0.3 / -0	+0 / -0.2	0 / 0.8	+0.5 / -0	+0 / -0.3	0 / 1.2	+0.7 / -0	+0 / -0.5	0 / 3.0	+1.8 / -0	+0 / -1.2	0.15 / 0.95	+0.5 / -0	-0.15 / -0.45
0.24—	0.40	0 / 0.65	+0.4 / -0	+0 / -0.25	0 / 1.0	+0.6 / -0	+0 / -0.4	0 / 1.5	+0.9 / -0	+0 / -0.6	0 / 3.6	+2.2 / -0	+0 / -1.4	0.2 / 1.2	+0.6 / -0	-0.2 / -0.6
0.40—	0.71	0 / 0.7	+0.4 / -0	+0 / -0.3	0 / 1.1	+0.7 / -0	+0 / -0.4	0 / 1.7	+1.0 / -0	+0 / -0.7	0 / 4.4	+2.8 / -0	+0 / -1.6	0.25 / 1.35	+0.7 / -0	-0.25 / -0.65
0.71—	1.19	0 / 0.9	+0.5 / -0	+0 / -0.4	0 / 1.3	+0.8 / -0	+0 / -0.5	0 / 2	+1.2 / -0	+0 / -0.8	0 / 5.5	+3.5 / -0	+0 / -2.0	0.3 / 1.6	+0.8 / -0	-0.3 / -0.8
1.19—	1.97	0 / 1.0	+0.6 / -0	+0 / -0.4	0 / 1.6	+1.0 / -0	+0 / -0.6	0 / 2.6	+1.6 / -0	+0 / -1	0 / 6.5	+4.0 / -0	+0 / -2.5	0.4 / 2.0	+1.0 / -0	-0.4 / -1.0
1.97—	3.15	0 / 1.2	+0.7 / -0	+0 / -0.5	0 / 1.9	+1.2 / -0	+0 / -0.7	0 / 3	+1.8 / -0	+0 / -1.2	0 / 7.5	+4.5 / -0	+0 / -3	0.4 / 2.3	+1.2 / -0	-0.4 / -1.1
3.15—	4.73	0 / 1.5	+0.9 / -0	+0 / -0.6	0 / 2.3	+1.4 / -0	+0 / -0.9	0 / 3.6	+2.2 / -0	+0 / -1.4	0 / 8.5	+5.0 / -0	+0 / -3.5	0.5 / 2.8	+1.4 / -0	-0.5 / -1.4
4.73—	7.09	0 / 1.7	+1.0 / -0	+0 / -0.7	0 / 2.6	+1.6 / -0	+0 / -1.0	0 / 4.1	+2.5 / -0	+0 / -1.6	0 / 10	+6.0 / -0	+0 / -4	0.6 / 3.2	+1.6 / -0	-0.6 / -1.6
7.09—	9.85	0 / 2.0	+1.2 / -0	+0 / -0.8	0 / 3.0	+1.8 / -0	+0 / -1.2	0 / 4.6	+2.8 / -0	+0 / -1.8	0 / 11.5	+7.0 / -0	+0 / -4.5	0.6 / 3.6	+1.8 / -0	-0.6 / -1.8
9.85—	12.41	0 / 2.1	+1.2 / -0	+0 / -0.9	0 / 3.2	+2.0 / -0	+0 / -1.2	0 / 5	+3.0 / -0	+0 / -2.0	0 / 13	+8.0 / -0	+0 / -5	0.7 / 3.9	+2.0 / -0	-0.7 / -1.9
12.41—	15.75	0 / 2.4	+1.4 / -0	+0 / -1.0	0 / 3.6	+2.2 / -0	+0 / -1.4	0 / 5.7	+3.5 / -0	+0 / -2.2	0 / 15	+9.0 / -0	+0 / -6	0.7 / 4.3	+2.2 / -0	-0.7 / -2.1
15.75—	19.69	0 / 2.6	+1.6 / -0	+0 / -1.0	0 / 4.1	+2.5 / -0	+0 / -1.6	0 / 6.5	+4 / -0	+0 / -2.5	0 / 16	+10.0 / -0	+0 / -6	0.8 / 4.9	+2.5 / -0	-0.8 / -2.4
19.69—	30.09	0 / 3.2	+2.0 / -0	+0 / -1.2	0 / 5.0	+3 / -0	+0 / -2	0 / 8	+5 / -0	+0 / -3	0 / 20	+12.0 / -0	+0 / -8	0.9 / 5.9	+3.0 / -0	-0.9 / -2.9
30.09—	41.49	0 / 4.1	+2.5 / -0	+0 / -1.6	0 / 6.5	+4 / -0	+0 / -2.5	0 / 10	+6 / -0	+0 / -4	0 / 26	+16.0 / -0	+0 / -10	1.0 / 7.5	+4.0 / -0	-1.0 / -3.5
41.49—	56.19	0 / 5.0	+3.0 / -0	+0 / -2.0	0 / 8.0	+5 / -0	+0 / -3	0 / 13	+8 / -0	+0 / -5	0 / 32	+20.0 / -0	+0 / -12	1.2 / 9.2	+5.0 / -0	-1.2 / -4.2
56.19—	76.39	0 / 6.5	+4.0 / -0	+0 / -2.5	0 / 10	+6 / -0	+0 / -4	0 / 16	+10 / -0	+0 / -6	0 / 41	+25.0 / -0	+0 / -16	1.2 / 11.2	+6.0 / -0	-1.2 / -5.2
76.39—	100.9	0 / 8.0	+5.0 / -0	+0 / -3.0	0 / 13	+8 / -0	+0 / -5	0 / 20	+12 / -0	+0 / -8	0 / 50	+30.0 / -0	+0 / -20	1.4 / 14.4	+8.0 / -0	-1.4 / -6.4
100.9 —	131.9	0 / 10.0	+6.0 / -0	+0 / -4.0	0 / 16	+10 / -0	+0 / -6	0 / 26	+16 / -0	+0 / -10	0 / 65	+40.0 / -0	+0 / -25	1.6 / 17.6	+10.0 / -0	-1.6 / -7.6
131.9 —	171.9	0 / 13.0	+8.0 / -0	+0 / -5.0	0 / 20	+12 / -0	+0 / -8	0 / 32	+20 / -0	+0 / -12	0 / 8	+50.0 / -0	+0 / -30	1.8 / 21.8	+12.0 / -0	-1.8 / -9.8
171.9 —	200	0 / 16.0	+10.0 / -0	+0 / -6.0	0 / 26	+16 / -0	+0 / -10	0 / 41	+25 / -0	+0 / -16	0 / 100	+60.0 / -0	+0 / -40	1.8 / 27.8	+16.0 / -0	-1.8 / -11.8

(Courtesy of USASI; B4.1-1955.)

Cont.

Class LC 6			Class LC 7			Class LC 8			Class LC 9			Class LC 10			Class LC 11			Nominal Size Range Inches	
Limits of Clearance	Standard Limits		Limits of Clearance	Standard Limits		Limits of Clearance	Standard Limits		Limits of Clearance	Standard Limits		Limits of Clearance	Standard Limits		Limits of Clearance	Standard Limits			
	Hole H9	Shaft f8		Hole H10	Shaft e9		Hole H10	Shaft d9		Hole H11	Shaft c10		Hole H12	Shaft		Hole H13	Shaft	Over	To
0.3 / 1.9	+1.0 / 0	−0.3 / −0.9	0.6 / 3.2	+1.6 / 0	−0.6 / −1.6	1.0 / 3.6	+1.6 / −0	−1.0 / −2.0	2.5 / 6.6	+2.5 / −0	−2.5 / −4.1	4 / 12	+4 / −0	−4 / −8	5 / 17	+6 / −0	−5 / −11	0	0.12
0.4 / 2.3	+1.2 / 0	−0.4 / −1.1	0.8 / 3.8	+1.8 / 0	−0.8 / −2.0	1.2 / 4.2	+1.8 / −0	−1.2 / −2.4	2.8 / 7.6	+3.0 / −0	−2.8 / −4.6	4.5 / 14.5	+5 / −0	−4.5 / −9.5	6 / 20	+7 / −0	−6 / −13	0.12	0.24
0.5 / 2.8	+1.4 / 0	−0.5 / −1.4	1.0 / 4.6	+2.2 / 0	−1.0 / −2.4	1.6 / 5.2	+2.2 / −0	−1.6 / −3.0	3.0 / 8.7	+3.5 / −0	−3.0 / −5.2	5 / 17	+6 / −0	−5 / −11	7 / 25	+9 / −0	−7 / −16	0.24	0.40
0.6 / 3.2	+1.6 / 0	−0.6 / −1.6	1.2 / 5.6	+2.8 / 0	−1.2 / −2.8	2.0 / 6.4	+2.8 / −0	−2.0 / −3.6	3.5 / 10.3	+4.0 / −0	−3.5 / −6.3	6 / 20	+7 / −0	−6 / −13	8 / 28	+10 / −0	−8 / −18	0.40	0.71
0.8 / 4.0	+2.0 / 0	−0.8 / −2.0	1.6 / 7.1	+3.5 / 0	−1.6 / −3.6	2.5 / 8.0	+3.5 / −0	−2.5 / −4.5	4.5 / 13.0	+5.0 / −0	−4.5 / −8.0	7 / 23	+8 / −0	−7 / −15	10 / 34	+12 / −0	−10 / −22	0.71	1.19
1.0 / 5.1	+2.5 / 0	−1.0 / −2.6	2.0 / 8.5	+4.0 / 0	−2.0 / −4.5	3.0 / 9.5	+4.0 / −0	−3.0 / −5.5	5 / 15	+6 / −0	−5 / −9	8 / 28	+10 / −0	−8 / −18	12 / 44	+16 / −0	−12 / −28	1.19	1.97
1.2 / 6.0	+3.0 / 0	−1.2 / −3.0	2.5 / 10.0	+4.5 / 0	−2.5 / −5.5	4.0 / 11.5	+4.5 / −0	−4.0 / −7.0	6 / 17.5	+7 / −0	−6 / −10.5	10 / 34	+12 / −0	−10 / −22	14 / 50	+18 / −0	−14 / −32	1.97	3.15
1.4 / 7.1	+3.5 / 0	−1.4 / −3.6	3.0 / 11.5	+5.0 / 0	−3.0 / −6.5	5.0 / 13.5	+5.0 / −0	−5.0 / −8.5	7 / 21	+9 / −0	−7 / −12	11 / 39	+14 / −0	−11 / −25	16 / 60	+22 / −0	−16 / −38	3.15	4.73
1.6 / 8.1	+4.0 / 0	−1.6 / −4.1	3.5 / 13.5	+6.0 / 0	−3.5 / −7.5	6 / 16	+6 / −0	−6 / −10	8 / 24	+10 / −0	−8 / −14	12 / 44	+16 / −0	−12 / −28	18 / 68	+25 / −0	−18 / −43	4.73	7.09
2.0 / 9.3	+4.5 / 0	−2.0 / −4.8	4.0 / 15.5	+7.0 / 0	−4.0 / −8.5	7 / 18.5	+7 / −0	−7 / −11.5	10 / 29	+12 / −0	−10 / −17	16 / 52	+18 / −0	−16 / −34	22 / 78	+28 / −0	−22 / −50	7.09	9.85
2.2 / 10.2	+5.0 / 0	−2.2 / −5.2	4.5 / 17.5	+8.0 / 0	−4.5 / −9.5	7 / 20	+8 / −0	−7 / −12	12 / 32	+12 / −0	−12 / −20	20 / 60	+20 / −0	−20 / −40	28 / 88	+30 / −0	−28 / −58	9.85	12.41
2.5 / 12.0	+6.0 / 0	−2.5 / −6.0	5.0 / 20.0	+9.0 / 0	−5 / −11	8 / 23	+9 / −0	−8 / −14	14 / 37	+14 / −0	−14 / −23	22 / 66	+22 / −0	−22 / −44	30 / 100	+35 / −0	−30 / −65	12.41	15.75
2.8 / 12.8	+6.0 / 0	−2.8 / −6.8	5.0 / 21.0	+10.0 / 0	−5 / −11	9 / 25	+10 / −0	−9 / −15	16 / 42	+16 / −0	−16 / −26	25 / 75	+25 / −0	−25 / −50	35 / 115	+40 / −0	−35 / −75	15.75	19.69
3.0 / 16.0	+8.0 / 0	−3.0 / −8.0	6.0 / 26.0	+12.0 / −0	−6 / −14	10 / 30	+12 / −0	−10 / −18	18 / 50	+20 / −0	−18 / −30	28 / 88	+30 / −0	−28 / −58	40 / 140	+50 / −0	−40 / −90	19.69	30.09
3.5 / 19.5	+10.0 / 0	−3.5 / −9.5	7.0 / 33.0	+16.0 / −0	−7 / −17	12 / 38	+16 / −0	−12 / −22	20 / 61	+25 / −0	−20 / −36	30 / 110	+40 / −0	−30 / −70	45 / 165	+60 / −0	−45 / −105	30.09	41.49
4.0 / 24.0	+12.0 / 0	−4.0 / −12.0	8.0 / 40.0	+20.0 / −0	−8 / −20	14 / 46	+20 / −0	−14 / −26	25 / 75	+30 / −0	−25 / −45	40 / 140	+50 / −0	−40 / −90	60 / 220	+80 / −0	−60 / −140	41.49	56.19
4.5 / 30.5	+16.0 / 0	−4.5 / −14.5	9.0 / 50.0	+25.0 / −0	−9 / −25	16 / 57	+25 / −0	−16 / −32	30 / 95	+40 / −0	−30 / −55	50 / 170	+60 / −0	−50 / 110	70 / 270	+100 / −0	−70 / −170	56.19	76.39
5.0 / 37.0	+20.0 / 0	−5 / −17	10.0 / 60.0	+30.0 / −0	−10 / −30	18 / 68	+30 / −0	−18 / −38	35 / 115	+50 / −0	−35 / −65	50 / 210	+80 / −0	−50 / −130	80 / 330	+125 / −0	−80 / −205	76.39	100.9
6.0 / 47.0	+25.0 / 0	−6 / −22	12.0 / 67.0	+40.0 / −0	−12 / −27	20 / 85	+40 / −0	−20 / −45	40 / 140	+60 / −0	−40 / −80	60 / 260	+100 / −0	−60 / −160	90 / 410	+160 / −0	−90 / −250	100.9	131.9
7.0 / 57.0	+30.0 / 0	−7 / −27	14.0 / 94.0	+50.0 / −0	−14 / −44	25 / 105	+50 / −0	−25 / −55	50 / 180	+80 / −0	−50 / −100	80 / 330	+125 / −0	−80 / −205	100 / 500	+200 / −0	−100 / −300	131.9	171.9
7.0 / 72.0	+40.0 / 0	−7 / −32	14.0 / 114.0	+60.0 / −0	−14 / −54	25 / 125	+60 / −0	−25 / −65	50 / 210	+100 / −0	−50 / −110	90 / 410	+160 / −0	−90 / −250	125 / 625	+250 / −0	−125 / −375	171.9	200

(Courtesy of ANSI; B4.1–1955.)

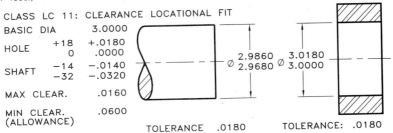

CLASS LC 11: CLEARANCE LOCATIONAL FIT

BASIC DIA		3.0000
HOLE	+18 / 0	+.0180 / .0000
SHAFT	−14 / −32	−.0140 / −.0320
MAX CLEAR.		.0160
MIN CLEAR. (ALLOWANCE)		.0600

Ø 2.9860 / 2.9680   Ø 3.0180 / 3.0000

TOLERANCE .0180   TOLERANCE: .0180

APPENDIX 36 • American Standard Transition Locational Fits (hole basis)

Limits are in thousandths of an inch.

Limits for hole and shaft are applied algebraically to the basic size to obtain the limits of size for the mating parts.

Data in bold face are in accordance with ABC agreements.

"Fit" represents the maximum interference (minus values) and the maximum clearance (plus values).

Symbols H7, js6, etc., are Hole and Shaft designations used in ABC System.

Nominal Size Range Inches Over	To	Class LT 1 Fit	LT 1 Hole H7	LT 1 Shaft js6	Class LT 2 Fit	LT 2 Hole H8	LT 2 Shaft js7	Class LT 3 Fit	LT 3 Hole H7	LT 3 Shaft k6	Class LT 4 Fit	LT 4 Hole H8	LT 4 Shaft k7	Class LT 5 Fit	LT 5 Hole H7	LT 5 Shaft n6	Class LT 6 Fit	LT 6 Hole H7	LT 6 Shaft n7
0 –	0.12	−0.10 / +0.50	+0.4 / −0	+0.10 / −0.10	−0.2 / +0.8	+0.6 / −0	+0.2 / −0.2							−0.5 / +0.15	+0.4 / −0	+0.5 / +0.25	−0.65 / +0.15	+0.4 / −0	+0.65 / +0.25
0.12 –	0.24	−0.15 / +0.65	+0.5 / −0	+0.15 / −0.15	−0.25 / +0.95	+0.7 / −0	+0.25 / −0.25							−0.6 / +0.2	+0.5 / −0	+0.6 / +0.3	−0.8 / +0.2	+0.5 / −0	+0.8 / +0.3
0.24 –	0.40	−0.2 / +0.8	+0.6 / −0	+0.2 / −0.2	−0.3 / +1.2	+0.9 / −0	+0.3 / −0.3	−0.5 / +0.5	+0.6 / −0	+0.5 / +0.1	−0.7 / +0.8	+0.9 / −0	+0.7 / +0.1	−0.8 / +0.2	+0.6 / −0	+0.8 / +0.4	−1.0 / +0.2	+0.6 / −0	+1.0 / +0.4
0.40 –	0.71	−0.2 / +0.9	+0.7 / −0	+0.2 / −0.2	−0.35 / +1.35	+1.0 / −0	+0.35 / −0.35	−0.5 / +0.6	+0.7 / −0	+0.5 / +0.1	−0.8 / +0.9	+1.0 / −0	+0.8 / +0.1	−0.9 / +0.2	+0.7 / −0	+0.9 / +0.5	−1.2 / +0.2	+0.7 / −0	+1.2 / +0.5
0.71 –	1.19	−0.25 / +1.05	+0.8 / −0	+0.25 / −0.25	−0.4 / +1.6	+1.2 / −0	+0.4 / −0.4	−0.6 / +0.7	+0.8 / −0	+0.6 / +0.1	−0.9 / +1.1	+1.2 / −0	+0.9 / +0.1	−1.1 / +0.2	+0.8 / −0	+1.1 / +0.6	−1.4 / +0.2	+0.8 / −0	+1.4 / +0.6
1.19 –	1.97	−0.3 / +1.3	+1.0 / −0	+0.3 / −0.3	−0.5 / +2.1	+1.6 / −0	+0.5 / −0.5	−0.7 / +0.9	+1.0 / −0	+0.7 / +0.1	−1.1 / +1.5	+1.6 / −0	+1.1 / +0.1	−1.3 / +0.3	+1.0 / −0	+1.3 / +0.7	−1.7 / +0.3	+1.0 / −0	+1.7 / +0.7
1.97 –	3.15	−0.3 / +1.5	+1.2 / −0	+0.3 / −0.3	−0.6 / +2.4	+1.8 / −0	+0.6 / −0.6	−0.8 / +1.1	+1.2 / −0	+0.8 / +0.1	−1.3 / +1.7	+1.8 / −0	+1.3 / +0.1	−1.5 / +0.4	+1.2 / −0	+1.5 / +0.8	−2.0 / +0.4	+1.2 / −0	+2.0 / +0.8
3.15 –	4.73	−0.4 / +1.8	+1.4 / −0	+0.4 / −0.4	−0.7 / +2.9	+2.2 / −0	+0.7 / −0.7	−1.0 / +1.3	+1.4 / −0	+1.0 / +0.1	−1.5 / +2.1	+2.2 / −0	+1.5 / +0.1	−1.9 / +0.4	+1.4 / −0	+1.9 / +1.0	−2.4 / +0.4	+1.4 / −0	+2.4 / +1.0
4.73 –	7.09	−0.5 / +2.1	+1.6 / −0	+0.5 / −0.5	−0.8 / +3.3	+2.5 / −0	+0.8 / −0.8	−1.1 / +1.5	+1.6 / −0	+1.1 / +0.1	−1.7 / +2.4	+2.5 / −0	+1.7 / +0.1	−2.2 / +0.4	+1.6 / −0	+2.2 / +1.2	−2.8 / +0.4	+1.6 / −0	+2.8 / +1.2
7.09 –	9.85	−0.6 / +2.4	+1.8 / −0	+0.6 / −0.6	−0.9 / +3.7	+2.8 / −0	+0.9 / −0.9	−1.4 / +1.6	+1.8 / −0	+1.4 / +0.2	−2.0 / +2.6	+2.8 / −0	+2.0 / +0.2	−2.6 / +0.4	+1.8 / −0	+2.6 / +1.4	−3.2 / +0.4	+1.8 / −0	+3.2 / +1.4
9.85 –	12.41	−0.6 / +2.6	+2.0 / −0	+0.6 / −0.6	−1.0 / +4.0	+3.0 / −0	+1.0 / −1.0	−1.4 / +1.8	+2.0 / −0	+1.4 / +0.2	−2.2 / +2.8	+3.0 / −0	+2.2 / +0.2	−2.6 / +0.6	+2.0 / −0	+2.6 / +1.4	−3.4 / +0.6	+2.0 / −0	+3.4 / +1.4
12.41 –	15.75	−0.7 / +2.9	+2.2 / −0	+0.7 / −0.7	−1.0 / +4.5	+3.5 / −0	+1.0 / −1.0	−1.6 / +2.0	+2.2 / −0	+1.6 / +0.2	−2.4 / +3.3	+3.5 / −0	+2.4 / +0.2	−3.0 / +0.6	+2.2 / −0	+3.0 / +1.6	−3.8 / +0.6	+2.2 / −0	+3.8 / +1.6
15.75 –	19.69	−0.8 / +3.3	+2.5 / −0	+0.8 / −0.8	−1.2 / +5.2	+4.0 / −0	+1.2 / −1.2	−1.8 / +2.3	+2.5 / −0	+1.8 / +0.2	−2.7 / +3.8	+4.0 / −0	+2.7 / +0.2	−3.4 / +0.7	+2.5 / −0	+3.4 / +1.8	−4.3 / +0.7	+2.5 / −0	+4.3 / +1.8

(Courtesy of ANSI; B4.1–1955.)

Limits are in thousandths of an inch.
Limits for hole and shaft are applied algebraically to the
basic size to obtain the limits of size for the parts.
Data in bold face are in accordance with ABC agreements,
Symbols H7, p6, etc., are Hole and Shaft designations
used in ABC System.

Nominal Size Range Inches Over	To	Class LN 1 Limits of Interference	Class LN 1 Standard Limits Hole H6	Class LN 1 Standard Limits Shaft n5	Class LN 2 Limits of Interference	Class LN 2 Standard Limits Hole H7	Class LN 2 Standard Limits Shaft p6	Class LN 3 Limits of Interference	Class LN 3 Standard Limits Hole H7	Class LN 3 Standard Limits Shaft r6
0	0.12	0 / 0.45	+ 0.25 / − 0	+0.45 / +0.25	0 / 0.65	+ 0.4 / − 0	+ 0.65 / + 0.4	0.1 / 0.75	+ 0.4 / − 0	+ 0.75 / + 0.5
0.12	0.24	0 / 0.5	+ 0.3 / − 0	+0.5 / +0.3	0 / 0.8	+ 0.5 / − 0	+ 0.8 / + 0.5	0.1 / 0.9	+ 0.5 / 0	+ 0.9 / + 0.6
0.24	0.40	0 / 0.65	+ 0.4 / − 0	+0.65 / +0.4	0 / 1.0	+ 0.6 / − 0	+ 1.0 / + 0.6	0.2 / 1.2	+ 0.6 / − 0	+ 1.2 / + 0.8
0.40	0.71	0 / 0.8	+ 0.4 / − 0	+0.8 / +0.4	0 / 1.1	+ 0.7 / − 0	+ 1.1 / + 0.7	0.3 / 1.4	+ 0.7 / − 0	+ 1.4 / + 1.0
0.71	1.19	0 / 1.0	+ 0.5 / − 0	+1.0 / +0.5	0 / 1.3	+ 0.8 / − 0	+ 1.3 / + 0.8	0.4 / 1.7	+ 0.8 / − 0	+ 1.7 / + 1.2
1.19	1.97	0 / 1.1	+ 0.6 / − 0	+1.1 / +0.6	0 / 1.6	+ 1.0 / − 0	+ 1.6 / + 1.0	0.4 / 2.0	+ 1.0 / − 0	+ 2.0 / + 1.4
1.97	3.15	0.1 / 1.3	+ 0.7 / − 0	+1.3 / +0.7	0.2 / 2.1	+ 1.2 / − 0	+ 2.1 / + 1.4	0.4 / 2.3	+ 1.2 / − 0	+ 2.3 / + 1.6
3.15	4.73	0.1 / 1.6	+ 0.9 / − 0	+1.6 / +1.0	0.2 / 2.5	+ 1.4 / − 0	+ 2.5 / + 1.6	0.6 / 2.9	+ 1.4 / − 0	+ 2.9 / + 2.0
4.73	7.09	0.2 / 1.9	+ 1.0 / − 0	+1.9 / +1.2	0.2 / 2.8	+ 1.6 / − 0	+ 2.8 / + 1.8	0.9 / 3.5	+ 1.6 / − 0	+ 3.5 / + 2.5
7.09	9.85	0.2 / 2.2	+ 1.2 / − 0	+2.2 / +1.4	0.2 / 3.2	+ 1.8 / − 0	+ 3.2 / + 2.0	1.2 / 4.2	+ 1.8 / − 0	+ 4.2 / + 3.0
9.85	12.41	0.2 / 2.3	+ 1.2 / − 0	+2.3 / +1.4	0.2 / 3.4	+ 2.0 / − 0	+ 3.4 / + 2.2	1.5 / 4.7	+ 2.0 / − 0	+ 4.7 / + 3.5
12.41	15.75	0.2 / 2.6	+ 1.4 / − 0	+2.6 / +1.6	0.3 / 3.9	+ 2.2 / − 0	+ 3.9 / + 2.5	2.3 / 5.9	+ 2.2 / − 0	+ 5.9 / + 4.5
15.75	19.69	0.2 / 2.8	+ 1.6 / − 0	+2.8 / +1.8	0.3 / 4.4	+ 2.5 / − 0	+ 4.4 / + 2.8	2.5 / 6.6	+ 2.5 / − 0	+ 6.6 / + 5.0
19.69	30.09		+ 2.0 / − 0		0.5 / 5.5	+ 3 / − 0	+ 5.5 / + 3.5	4 / 9	+ 3 / − 0	+ 9 / + 7
30.09	41.49		+ 2.5 / − 0		0.5 / 7.0	+ 4 / − 0	+ 7.0 / + 4.5	5 / 11.5	+ 4 / − 0	+11.5 / + 9
41.49	56.19		+ 3.0 / − 0		1 / 9	+ 5 / − 0	+ 9 / + 6	7 / 15	+ 5 / − 0	+15 / +12
56.19	76.39		+ 4.0 / − 0		1 / 11	+ 6 / − 0	+11 / + 7	10 / 20	+ 6 / − 0	+20 / +16
76.39	100.9		+ 5.0 / − 0		1 / 14	+ 8 / − 0	+14 / + 9	12 / 25	+ 8 / − 0	+25 / +20
100.9	131.9		+ 6.0 / − 0		2 / 18	+10 / − 0	+18 / +12	15 / 31	+10 / − 0	+31 / +25
131.9	171.9		+ 8.0 / − 0		4 / 24	+12 / − 0	+24 / +16	18 / 38	+12 / − 0	+38 / +30
171.9	200		+10.0 / − 0		4 / 30	+16 / − 0	+30 / +20	24 / 50	+16 / − 0	+50 / +40

(Courtesy of ANSI; B4.1–1955.)

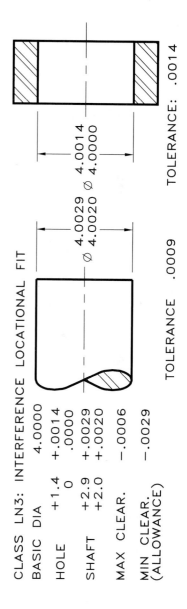

CLASS LN3: INTERFERENCE LOCATIONAL FIT

TOLERANCE: .0014

TOLERANCE .0009

BASIC DIA.		4.0000
HOLE	+1.4 / 0	+.0014 / .0000
SHAFT	+2.9 / +2.0	+.0029 / +.0020
MAX CLEAR.		−.0006
MIN CLEAR. (ALLOWANCE)		−.0029

Ø 4.0029 / Ø 4.0020

4.0014 / 4.0000

APPENDIX 38 • American Standard Force and Shrink Fits (hole basis)

Limits are in thousandths of an inch.
Limits for hole and shaft are applied algebraically to the basic size to obtain the limits of size for the parts.
Data in bold face are in accordance with ABC agreements.
Symbols H7, s6, etc., are Hole and Shaft designations used in ABC System.

Nominal Size Range Inches (Over — To)	FN1 Limits of Interference	FN1 Hole H6	FN1 Shaft	FN2 Limits of Interference	FN2 Hole H7	FN2 Shaft s6	FN3 Limits of Interference	FN3 Hole H7	FN3 Shaft t6	FN4 Limits of Interference	FN4 Hole H7	FN4 Shaft u6	FN5 Limits of Interference	FN5 Hole H8	FN5 Shaft x7
0 — 0.12	0.05 / 0.5	+0.25 / −0	+0.5 / +0.3	0.2 / 0.85	+0.4 / −0	+0.85 / +0.6				0.3 / 0.95	+0.4 / −0	+0.95 / +0.7	0.3 / 1.3	+0.6 / −0	+1.3 / +0.9
0.12 — 0.24	0.1 / 0.6	+0.3 / −0	+0.6 / +0.4	0.2 / 1.0	+0.5 / −0	+1.0 / +0.7				0.4 / 1.2	+0.5 / −0	+1.2 / +0.9	0.5 / 1.7	+0.7 / −0	+1.7 / +1.2
0.24 — 0.40	0.1 / 0.75	+0.4 / −0	+0.75 / +0.5	0.4 / 1.4	+0.6 / −0	+1.4 / +1.0				0.6 / 1.6	+0.6 / −0	+1.6 / +1.2	0.5 / 2.0	+0.9 / −0	+2.0 / +1.4
0.40 — 0.56	0.1 / 0.8	−0.4 / −0	+0.8 / +0.5	0.5 / 1.6	+0.7 / −0	+1.6 / +1.2				0.7 / 1.8	+0.7 / −0	+1.8 / +1.4	0.6 / 2.3	+1.0 / −0	+2.3 / +1.6
0.56 — 0.71	0.2 / 0.9	+0.4 / −0	+0.9 / +0.6	0.5 / 1.6	+0.7 / −0	+1.6 / +1.2				0.7 / 1.8	+0.7 / −0	+1.8 / +1.4	0.8 / 2.5	+1.0 / −0	+2.5 / +1.8
0.71 — 0.95	0.2 / 1.1	+0.5 / −0	+1.1 / +0.7	0.6 / 1.9	+0.8 / −0	+1.9 / +1.4				0.8 / 2.1	+0.8 / −0	+2.1 / +1.6	1.0 / 3.0	+1.2 / −0	+3.0 / +2.2
0.95 — 1.19	0.3 / 1.2	+0.5 / −0	+1.2 / +0.8	0.6 / 1.9	+0.8 / −0	+1.9 / +1.4	0.8 / 2.1	+0.8 / −0	+2.1 / +1.6	1.0 / 2.3	+0.8 / −0	+2.3 / +1.8	1.3 / 3.3	+1.2 / −0	+3.3 / +2.5
1.19 — 1.58	0.3 / 1.3	+0.6 / −0	+1.3 / +0.9	0.8 / 2.4	+1.0 / −0	+2.4 / +1.8	1.0 / 2.6	+1.0 / −0	+2.6 / +2.0	1.5 / 3.1	+1.0 / −0	+3.1 / +2.5	1.4 / 4.0	+1.6 / −0	+4.0 / +3.0
1.58 — 1.97	0.4 / 1.4	+0.6 / −0	+1.4 / +1.0	0.8 / 2.4	+1.0 / −0	+2.4 / +1.8	1.2 / 2.8	+1.0 / −0	+2.8 / +2.2	1.8 / 3.4	+1.0 / −0	+3.4 / +2.8	2.4 / 5.0	+1.6 / −0	+5.0 / +4.0
1.97 — 2.56	0.6 / 1.8	+0.7 / −0	+1.8 / +1.3	0.8 / 2.7	+1.2 / −0	+2.7 / +2.0	1.3 / 3.2	+1.2 / −0	+3.2 / +2.5	2.3 / 4.2	+1.2 / −0	+4.2 / +3.5	3.2 / 6.2	+1.8 / −0	+6.2 / +5.0
2.56 — 3.15	0.7 / 1.9	+0.7 / −0	+1.9 / +1.4	1.0 / 2.9	+1.2 / −0	+2.9 / +2.2	1.8 / 3.7	+1.2 / −0	+3.7 / +3.0	2.8 / 4.7	+1.2 / −0	+4.7 / +4.0	4.2 / 7.2	+1.8 / −0	+7.2 / +6.0
3.15 — 3.94	0.9 / 2.4	+0.9 / −0	+2.4 / +1.8	1.4 / 3.7	+1.4 / −0	+3.7 / +2.8	2.1 / 4.4	+1.4 / −0	+4.4 / +3.5	3.6 / 5.9	+1.4 / −0	+5.9 / +5.0	4.8 / 8.4	+2.2 / −0	+8.4 / +7.0
3.94 — 4.73	1.1 / 2.6	+0.9 / −0	+2.6 / +2.0	1.6 / 3.9	+1.4 / −0	+3.9 / +3.0	2.6 / 4.9	+1.4 / −0	+4.9 / +4.0	4.6 / 6.9	+1.4 / −0	+6.9 / +6.0	5.8 / 9.4	+2.2 / −0	+9.4 / +8.0
4.73 — 5.52	1.2 / 2.9	+1.0 / −0	+2.9 / +2.2	1.9 / 4.5	+1.6 / −0	+4.5 / +3.5	3.4 / 6.0	+1.6 / −0	+6.0 / +5.0	5.4 / 8.0	+1.6 / −0	+8.0 / +7.0	7.5 / 11.6	+2.5 / −0	+11.6 / +10.0
5.52 — 6.30	1.5 / 3.2	+1.0 / −0	+3.2 / +2.5	2.4 / 5.0	+1.6 / −0	+5.0 / +4.0	3.4 / 6.0	+1.6 / −0	+6.0 / +5.0	5.4 / 8.0	+1.6 / −0	+8.0 / +7.0	9.5 / 13.6	+2.5 / −0	+13.6 / +12.0
6.30 — 7.09	1.8 / 3.5	+1.0 / −0	+3.5 / +2.8	2.9 / 5.5	+1.6 / −0	+5.5 / +4.5	4.4 / 7.0	+1.6 / −0	+7.0 / +6.0	6.4 / 9.0	+1.6 / −0	+9.0 / +8.0	9.5 / 13.6	+2.5 / −0	+13.6 / +12.0
7.09 — 7.88	1.8 / 3.8	+1.2 / −0	+3.8 / +3.0	3.2 / 6.2	+1.8 / −0	+6.2 / +5.0	5.2 / 8.2	+1.8 / −0	+8.2 / +7.0	7.2 / 10.2	+1.8 / −0	+10.2 / +9.0	11.2 / 15.8	+2.8 / −0	+15.8 / +14.0
7.88 — 8.86	2.3 / 4.3	+1.2 / −0	+4.3 / +3.5	3.2 / 6.2	+1.8 / −0	+6.2 / +5.0	5.2 / 8.2	+1.8 / −0	+8.2 / +7.0	8.2 / 11.2	+1.8 / −0	+11.2 / +10.0	13.2 / 17.8	+2.8 / −0	+17.8 / +16.0
8.86 — 9.85	2.3 / 4.3	+1.2 / −0	+4.3 / +3.5	4.2 / 7.2	+1.8 / −0	+7.2 / +6.0	6.2 / 9.2	+1.8 / −0	+9.2 / +8.0	10.2 / 13.2	+1.8 / −0	+13.2 / +12.0	13.2 / 17.8	+2.8 / −0	+17.8 / +16.0
9.85 — 11.03	2.8 / 4.9	+1.2 / −0	+4.9 / +4.0	4.0 / 7.2	+2.0 / −0	+7.2 / +6.0	7.0 / 10.2	+2.0 / −0	+10.2 / +9.0	10.0 / 13.2	+2.0 / −0	+13.2 / +12.0	15.0 / 20.0	+3.0 / −0	+20.0 / +18.0
11.03 — 12.41	2.8 / 4.9	+1.2 / −0	+4.9 / +4.0	5.0 / 8.2	+2.0 / −0	+8.2 / +7.0	7.0 / 10.2	+2.0 / −0	+10.2 / +9.0	12.0 / 15.2	+2.0 / −0	+15.2 / +14.0	17.0 / 22.0	+3.0 / −0	+22.0 / +20.0
12.41 — 13.98	3.1 / 5.5	+1.4 / −0	+5.5 / +4.5	5.8 / 9.4	+2.2 / −0	+9.4 / +8.0	7.8 / 11.4	+2.2 / −0	+11.4 / +10.0	13.8 / 17.4	+2.2 / −0	+17.4 / +16.0	18.5 / 24.2	+3.5 / +0	+24.2 / +22.0
13.98 — 15.75	3.6 / 6.1	+1.4 / −0	+6.1 / +5.0	5.8 / 9.4	+2.2 / −0	+9.4 / +8.0	9.8 / 13.4	+2.2 / −0	+13.4 / +12.0	15.8 / 19.4	+2.2 / −0	+19.4 / +18.0	21.5 / 27.2	+3.5 / −0	+27.2 / +25.0
15.75 — 17.72	4.4 / 7.0	+1.6 / −0	+7.0 / +6.0	6.5 / 10.6	+2.5 / −0	+10.6 / +9.0	9.5 / 13.6	+2.5 / −0	+13.6 / +12.0	17.5 / 21.6	+2.5 / −0	+21.6 / +20.0	24.0 / 30.5	+4.0 / −0	+30.5 / +28.0
17.72 — 19.69	4.4 / 7.0	+1.6 / −0	+7.0 / +6.0	7.5 / 11.6	+2.5 / −0	+11.6 / +10.0	11.5 / 15.6	+2.5 / −0	+15.6 / +14.0	19.5 / 23.6	+2.5 / −0	+23.6 / +22.0	26.0 / 32.5	+4.0 / −0	+32.5 / +30.0

(Courtesy of ANSI; B4.1–1955.)

The international tolerance grades (ANSI B4.2)

Dimensions are in mm.

Basic sizes		Tolerance grades[3]																		
Over	Up to and including	IT01	IT0	IT1	IT2	IT3	IT4	IT5	IT6	IT7	IT8	IT9	IT10	IT11	IT12	IT13	IT14	IT15	IT16
0	3	0.0003	0.0005	0.0008	0.0012	0.002	0.003	0.004	0.006	0.010	0.014	0.025	0.040	0.060	0.100	0.140	0.250	0.400	0.600
3	6	0.0004	0.0006	0.001	0.0015	0.0025	0.004	0.005	0.008	0.012	0.018	0.030	0.048	0.075	0.120	0.180	0.300	0.480	0.750
6	10	0.0004	0.0006	0.001	0.0015	0.0025	0.004	0.006	0.009	0.015	0.022	0.036	0.058	0.090	0.150	0.220	0.360	0.580	0.900
10	18	0.0005	0.0008	0.0012	0.002	0.003	0.005	0.008	0.011	0.018	0.027	0.043	0.070	0.110	0.180	0.270	0.430	0.700	1.100
18	30	0.0006	0.001	0.0015	0.0025	0.004	0.006	0.009	0.013	0.021	0.033	0.052	0.084	0.130	0.210	0.330	0.520	0.840	1.300
30	50	0.0006	0.001	0.0015	0.0025	0.004	0.007	0.011	0.016	0.025	0.039	0.062	0.100	0.160	0.250	0.390	0.620	1.000	1.600
50	80	0.0008	0.0012	0.002	0.003	0.005	0.008	0.013	0.019	0.030	0.046	0.074	0.120	0.190	0.300	0.460	0.740	1.200	1.900
80	120	0.001	0.0015	0.0025	0.004	0.006	0.010	0.015	0.022	0.035	0.054	0.087	0.140	0.220	0.350	0.540	0.870	1.400	2.200
120	180	0.0012	0.002	0.0035	0.005	0.008	0.012	0.018	0.025	0.040	0.063	0.100	0.160	0.250	0.400	0.630	1.000	1.600	2.500
180	250	0.002	0.003	0.0045	0.007	0.010	0.014	0.020	0.029	0.046	0.072	0.115	0.185	0.290	0.460	0.720	1.150	1.850	2.900
250	315	0.0025	0.004	0.006	0.008	0.012	0.016	0.023	0.032	0.052	0.081	0.130	0.210	0.320	0.520	0.810	1.300	2.100	3.200
315	400	0.003	0.005	0.007	0.009	0.013	0.018	0.025	0.036	0.057	0.089	0.140	0.230	0.360	0.570	0.890	1.400	2.300	3.600
400	500	0.004	0.006	0.008	0.010	0.015	0.020	0.027	0.040	0.063	0.097	0.155	0.250	0.400	0.630	0.970	1.550	2.500	4.000
500	630	0.0045	0.006	0.009	0.011	0.016	0.022	0.030	0.044	0.070	0.110	0.175	0.280	0.440	0.700	1.100	1.750	2.800	4.400
630	800	0.005	0.007	0.010	0.013	0.018	0.025	0.035	0.050	0.080	0.125	0.200	0.320	0.500	0.800	1.250	2.000	3.200	5.000
800	1000	0.0055	0.008	0.011	0.015	0.021	0.029	0.040	0.056	0.090	0.140	0.230	0.360	0.560	0.900	1.400	2.300	3.600	5.600
1000	1250	0.0065	0.009	0.013	0.018	0.024	0.034	0.046	0.066	0.105	0.165	0.260	0.420	0.660	1.050	1.650	2.600	4.200	6.600
1250	1600	0.008	0.011	0.015	0.021	0.029	0.040	0.054	0.078	0.125	0.195	0.310	0.500	0.780	1.250	1.950	3.100	5.000	7.800
1600	2000	0.009	0.013	0.018	0.025	0.035	0.048	0.065	0.092	0.150	0.230	0.370	0.600	0.920	1.500	2.300	3.700	6.000	9.200
2000	2500	0.011	0.015	0.022	0.030	0.041	0.057	0.077	0.110	0.175	0.280	0.440	0.700	1.100	1.750	2.800	4.400	7.000	11.000
2500	3150	0.013	0.018	0.026	0.036	0.050	0.069	0.093	0.135	0.210	0.330	0.540	0.860	1.350	2.100	3.300	5.400	8.600	13.500

[3] IT Values for tolerance grades larger than IT16 can be calculated by using the following formulas:
IT17 = IT12 × 10; IT18 = IT13 × 10; etc.

APPENDIX 40 • Preferred Hole Basis Clearance Fits—Cylindrical Fits (ANSI B4.2)

AMERICAN NATIONAL STANDARD PREFERRED METRIC LIMITS AND FITS   ANSI B4.2 - 1978

Dimensions in mm.

BASIC SIZE		LOOSE RUNNING Hole H11	LOOSE RUNNING Shaft c11	LOOSE RUNNING Fit	FREE RUNNING Hole H9	FREE RUNNING Shaft d9	FREE RUNNING Fit	CLOSE RUNNING Hole H8	CLOSE RUNNING Shaft f7	CLOSE RUNNING Fit	SLIDING Hole H7	SLIDING Shaft g6	SLIDING Fit	LOCATIONAL CLEARANCE Hole H7	LOCATIONAL CLEARANCE Shaft h6	LOCATIONAL CLEARANCE Fit
1	MAX	1.060	0.940	0.180	1.025	0.980	0.070	1.014	0.994	0.030	1.010	0.998	0.018	1.010	1.000	0.016
	MIN	1.000	0.880	0.060	1.000	0.955	0.020	1.000	0.984	0.006	1.000	0.992	0.002	1.000	0.994	0.000
1.2	MAX	1.260	1.140	0.180	1.225	1.180	0.070	1.214	1.194	0.030	1.210	1.198	0.018	1.210	1.200	0.016
	MIN	1.200	1.080	0.060	1.200	1.155	0.020	1.200	1.184	0.006	1.200	1.192	0.002	1.200	1.194	0.000
1.6	MAX	1.660	1.540	0.180	1.625	1.580	0.070	1.614	1.594	0.030	1.610	1.598	0.018	1.610	1.600	0.016
	MIN	1.600	1.480	0.060	1.600	1.555	0.020	1.600	1.584	0.006	1.600	1.592	0.002	1.600	1.594	0.000
2	MAX	2.060	1.940	0.180	2.025	1.980	0.070	2.014	1.994	0.030	2.010	1.998	0.018	2.010	2.000	0.016
	MIN	2.000	1.880	0.060	2.000	1.955	0.020	2.000	1.984	0.006	2.000	1.992	0.002	2.000	1.994	0.000
2.5	MAX	2.560	2.440	0.180	2.525	2.480	0.070	2.514	2.494	0.030	2.510	2.498	0.018	2.510	2.500	0.016
	MIN	2.500	2.380	0.060	2.500	2.455	0.020	2.500	2.484	0.006	2.500	2.492	0.002	2.500	2.494	0.000
3	MAX	3.060	2.940	0.180	3.025	2.980	0.070	3.014	2.994	0.030	3.010	2.998	0.018	3.010	3.000	0.016
	MIN	3.000	2.880	0.060	3.000	2.955	0.020	3.000	2.984	0.006	3.000	2.992	0.002	3.000	2.994	0.000
4	MAX	4.075	3.930	0.220	4.030	3.970	0.090	4.018	3.990	0.040	4.012	3.996	0.024	4.012	4.000	0.020
	MIN	4.000	3.855	0.070	4.000	3.940	0.030	4.000	3.978	0.010	4.000	3.988	0.004	4.000	3.992	0.000
5	MAX	5.075	4.930	0.220	5.030	4.970	0.090	5.018	4.990	0.040	5.012	4.996	0.024	5.012	5.000	0.020
	MIN	5.000	4.855	0.070	5.000	4.940	0.030	5.000	4.978	0.010	5.000	4.988	0.004	5.000	4.992	0.000
6	MAX	6.075	5.930	0.220	6.030	5.970	0.090	6.018	5.990	0.040	6.012	5.996	0.024	6.012	6.000	0.020
	MIN	6.000	5.855	0.070	6.000	5.940	0.030	6.000	5.978	0.010	6.000	5.988	0.004	6.000	5.992	0.000
8	MAX	8.090	7.920	0.260	8.036	7.960	0.112	8.022	7.987	0.050	8.015	7.995	0.029	8.015	8.000	0.024
	MIN	8.000	7.830	0.080	8.000	7.924	0.040	8.000	7.972	0.013	8.000	7.986	0.005	8.000	7.991	0.000
10	MAX	10.090	9.920	0.260	10.036	9.960	0.112	10.022	9.987	0.050	10.015	9.995	0.029	10.015	10.000	0.024
	MIN	10.000	9.830	0.080	10.000	9.924	0.040	10.000	9.972	0.013	10.000	9.986	0.005	10.000	9.991	0.000
12	MAX	12.110	11.905	0.315	12.043	11.950	0.136	12.027	11.984	0.061	12.018	11.994	0.035	12.018	12.000	0.029
	MIN	12.000	11.795	0.095	12.000	11.907	0.050	12.000	11.966	0.016	12.000	11.983	0.006	12.000	11.989	0.000
16	MAX	16.110	15.905	0.315	16.043	15.950	0.136	16.027	15.984	0.061	16.018	15.994	0.035	16.018	16.000	0.029
	MIN	16.000	15.795	0.095	16.000	15.907	0.050	16.000	15.966	0.016	16.000	15.983	0.006	16.000	15.989	0.000
20	MAX	20.130	19.890	0.370	20.052	19.935	0.169	20.033	19.980	0.074	20.021	19.993	0.041	20.021	20.000	0.034
	MIN	20.000	19.760	0.110	20.000	19.883	0.065	20.000	19.959	0.020	20.000	19.980	0.007	20.000	19.987	0.000
25	MAX	25.130	24.890	0.370	25.052	24.935	0.169	25.033	24.980	0.074	25.021	24.993	0.041	25.021	25.000	0.034
	MIN	25.000	24.760	0.110	25.000	24.883	0.065	25.000	24.959	0.020	25.000	24.980	0.007	25.000	24.987	0.000
30	MAX	30.130	29.890	0.370	30.052	29.935	0.169	30.033	29.980	0.074	30.021	29.993	0.041	30.021	30.000	0.034
	MIN	30.000	29.760	0.110	30.000	29.883	0.065	30.000	29.959	0.020	30.000	29.980	0.007	30.000	29.987	0.000

APPENDIX 40 • Preferred Hole Basis Clearance Fits—Cylindrical Fits (cont.)

AMERICAN NATIONAL STANDARD PREFERRED METRIC LIMITS AND FITS ANSI B4.2 - 1978

Dimensions in mm.

BASIC SIZE		LOOSE RUNNING Hole H11	Shaft c11	Fit	FREE RUNNING Hole H9	Shaft d9	Fit	CLOSE RUNNING Hole H8	Shaft f7	Fit	SLIDING Hole H7	Shaft g6	Fit	LOCATIONAL CLEARANCE Hole H7	Shaft h6	Fit
40	MAX	40.160	39.880	0.440	40.062	39.920	0.204	40.039	39.975	0.089	40.025	39.991	0.050	40.025	40.000	0.041
	MIN	40.000	39.720	0.120	40.000	39.858	0.080	40.000	39.950	0.025	40.000	39.975	0.009	40.000	39.984	0.000
50	MAX	50.160	49.870	0.450	50.062	49.920	0.204	50.039	49.975	0.089	50.025	49.991	0.050	50.025	50.000	0.041
	MIN	50.000	49.710	0.130	50.000	49.858	0.080	50.000	49.950	0.025	50.000	49.975	0.009	50.000	49.984	0.000
60	MAX	60.190	59.860	0.520	60.074	59.900	0.248	60.046	59.970	0.106	60.030	59.990	0.059	60.030	60.000	0.049
	MIN	60.000	59.670	0.140	60.000	59.826	0.100	60.000	59.940	0.030	60.000	59.971	0.010	60.000	59.981	0.000
80	MAX	80.190	79.850	0.530	80.074	79.900	0.248	80.046	79.970	0.106	80.030	79.990	0.059	80.030	80.000	0.049
	MIN	80.000	79.660	0.150	80.000	79.826	0.100	80.000	79.940	0.030	80.000	79.971	0.010	80.000	79.981	0.000
100	MAX	100.220	99.830	0.610	100.087	99.880	0.294	100.054	99.964	0.125	100.035	99.988	0.069	100.035	100.000	0.057
	MIN	100.000	99.610	0.170	100.000	99.793	0.120	100.000	99.929	0.036	100.000	99.966	0.012	100.000	99.978	0.000
120	MAX	120.220	119.820	0.620	120.087	119.880	0.294	120.054	119.964	0.125	120.035	119.988	0.069	120.035	120.000	0.057
	MIN	120.000	119.600	0.180	120.000	119.793	0.120	120.000	119.929	0.036	120.000	119.966	0.012	120.000	119.978	0.000
160	MAX	160.250	159.790	0.710	160.100	159.855	0.345	160.063	159.957	0.146	160.040	159.986	0.079	160.040	160.000	0.065
	MIN	160.000	159.540	0.210	160.000	159.755	0.145	160.000	159.917	0.043	160.000	159.961	0.014	160.000	159.975	0.000
200	MAX	200.290	199.760	0.820	200.115	199.830	0.400	200.072	199.950	0.168	200.046	199.985	0.090	200.046	200.000	0.075
	MIN	200.000	199.470	0.240	200.000	199.715	0.170	200.000	199.904	0.050	200.000	199.956	0.015	200.000	199.971	0.000
250	MAX	250.290	249.720	0.860	250.115	249.830	0.400	250.072	249.950	0.168	250.046	249.985	0.090	250.046	250.000	0.075
	MIN	250.000	249.430	0.280	250.000	249.715	0.170	250.000	249.904	0.050	250.000	249.956	0.015	250.000	249.971	0.000
300	MAX	300.320	299.670	0.970	300.130	299.810	0.450	300.081	299.944	0.189	300.052	299.983	0.101	300.052	300.000	0.084
	MIN	300.000	299.350	0.330	300.000	299.680	0.190	300.000	299.892	0.056	300.000	299.951	0.017	300.000	299.968	0.000
400	MAX	400.360	399.600	1.120	400.140	399.790	0.490	400.089	399.938	0.208	400.057	399.982	0.111	400.057	400.000	0.093
	MIN	400.000	399.240	0.400	400.000	399.650	0.210	400.000	399.881	0.062	400.000	399.946	0.018	400.000	399.964	0.000
500	MAX	500.400	499.520	1.280	500.155	499.770	0.540	500.097	499.932	0.228	500.063	499.980	0.123	500.063	500.000	0.103
	MIN	500.000	499.120	0.480	500.000	499.615	0.230	500.000	499.869	0.068	500.000	499.940	0.020	500.000	499.960	0.000

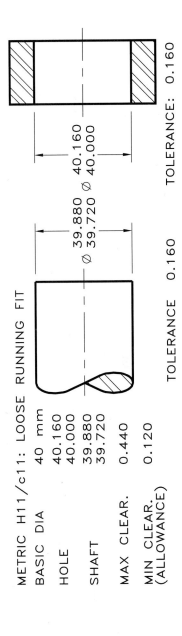

TOLERANCE: 0.160

Ø 40.160 / Ø 40.000

TOLERANCE 0.160

Ø 39.880 / Ø 39.720

METRIC H11/c11: LOOSE RUNNING FIT

BASIC DIA	40 mm
HOLE	40.160 / 40.000
SHAFT	39.880 / 39.720
MAX CLEAR.	0.440
MIN CLEAR. (ALLOWANCE)	0.120

APPENDIX 41 • Preferred Hole Basis Transition and Interference Fits—Cylindrical Fits (ANSI B4.2)

AMERICAN NATIONAL STANDARD PREFERRED METRIC LIMITS AND FITS   ANSI B4.2 - 1978

Dimensions in mm.

BASIC SIZE		LOCATIONAL TRANSN. Hole H7	Shaft k6	Fit	LOCATIONAL TRANSN. Hole H7	Shaft n6	Fit	LOCATIONAL INTERF. Hole H7	Shaft p6	Fit	MEDIUM DRIVE Hole H7	Shaft s6	Fit	FORCE Hole H7	Shaft u6	Fit
1	MAX	1.010	1.006	0.010	1.010	1.010	0.006	1.010	1.012	0.004	1.010	1.020	-0.004	1.010	1.024	-0.008
	MIN	1.000	1.000	-0.006	1.000	1.004	-0.010	1.000	1.006	-0.012	1.000	1.014	-0.020	1.000	1.018	-0.024
1.2	MAX	1.210	1.206	0.010	1.210	1.210	0.006	1.210	1.212	0.004	1.210	1.220	-0.004	1.210	1.224	-0.008
	MIN	1.200	1.200	-0.006	1.200	1.204	-0.010	1.200	1.206	-0.012	1.200	1.214	-0.020	1.200	1.218	-0.024
1.6	MAX	1.610	1.606	0.010	1.610	1.610	0.006	1.610	1.612	0.004	1.610	1.620	-0.004	1.610	1.624	-0.008
	MIN	1.600	1.600	-0.006	1.600	1.604	-0.010	1.600	1.606	-0.012	1.600	1.614	-0.020	1.600	1.618	-0.024
2	MAX	2.010	2.006	0.010	2.010	2.010	0.006	2.010	2.010	0.004	2.010	2.020	-0.004	2.010	2.024	-0.008
	MIN	2.000	2.000	-0.006	2.000	2.004	-0.010	2.000	2.006	-0.012	2.000	2.014	-0.020	2.000	2.018	-0.024
2.5	MAX	2.510	2.506	0.010	2.510	2.510	0.006	2.510	2.512	0.004	2.510	2.520	-0.004	2.510	2.524	-0.008
	MIN	2.500	2.500	-0.006	2.500	2.504	-0.010	2.500	2.506	-0.012	2.500	2.514	-0.020	2.500	2.518	-0.024
3	MAX	3.010	3.006	0.010	3.010	3.010	0.006	3.010	3.012	0.004	3.010	3.020	-0.004	3.010	3.024	-0.008
	MIN	3.000	3.000	-0.006	3.000	3.004	-0.010	3.000	3.006	-0.012	3.000	3.014	-0.020	3.000	3.018	-0.024
4	MAX	4.012	4.009	0.011	4.012	4.016	0.004	4.012	4.020	0.000	4.012	4.027	-0.007	4.012	4.031	-0.011
	MIN	4.000	4.001	-0.009	4.000	4.008	-0.016	4.000	4.012	-0.020	4.000	4.019	-0.027	4.000	4.023	-0.031
5	MAX	5.012	5.009	0.011	5.012	5.016	0.004	5.012	5.020	0.000	5.012	5.027	-0.007	5.012	5.031	-0.011
	MIN	5.000	5.001	-0.009	5.000	5.008	-0.016	5.000	5.012	-0.020	5.000	5.019	-0.027	5.000	5.023	-0.031
6	MAX	6.012	6.009	0.011	6.012	6.016	0.004	6.012	6.020	0.000	6.012	6.027	-0.007	6.012	6.031	-0.011
	MIN	6.000	6.001	-0.009	6.000	6.008	-0.016	6.000	6.012	-0.020	6.000	6.019	-0.027	6.000	6.023	-0.031
8	MAX	8.015	8.010	0.014	8.015	8.019	0.005	8.015	8.024	0.000	8.015	8.032	-0.008	8.015	8.037	-0.013
	MIN	8.000	8.001	-0.010	8.000	8.010	-0.019	8.000	8.015	-0.024	8.000	8.023	-0.032	8.000	8.028	-0.037
10	MAX	10.015	10.010	0.014	10.015	10.019	0.005	10.015	10.024	0.000	10.015	10.032	-0.008	10.015	10.037	-0.013
	MIN	10.000	10.001	-0.010	10.000	10.010	-0.019	10.000	10.015	-0.024	10.000	10.023	-0.032	10.000	10.028	-0.037
12	MAX	12.018	12.012	0.017	12.018	12.023	0.006	12.018	12.029	0.000	12.018	12.039	-0.010	12.018	12.044	-0.015
	MIN	12.000	12.001	-0.012	12.000	12.012	-0.023	12.000	12.018	-0.029	12.000	12.028	-0.039	12.000	12.033	-0.044
16	MAX	16.018	16.012	0.017	16.018	16.023	0.006	16.018	16.029	0.000	16.018	16.039	-0.010	16.018	16.044	-0.015
	MIN	16.000	16.001	-0.012	16.000	16.012	-0.023	16.000	16.018	-0.029	16.000	16.028	-0.039	16.000	16.033	-0.044
20	MAX	20.021	20.015	0.019	20.021	20.028	0.006	20.021	20.035	-0.001	20.021	20.048	-0.014	20.021	20.054	-0.020
	MIN	20.000	20.002	-0.015	20.000	20.015	-0.028	20.000	20.022	-0.035	20.000	20.035	-0.048	20.000	20.041	-0.054
25	MAX	25.021	25.015	0.019	25.021	25.028	0.006	25.021	25.035	-0.001	25.021	25.048	-0.014	25.021	25.061	-0.027
	MIN	25.000	25.002	-0.015	25.000	25.015	-0.028	25.000	25.022	-0.035	25.000	25.035	-0.048	25.000	25.048	-0.061
30	MAX	30.015	30.015	0.019	30.021	30.028	0.006	30.021	30.035	-0.001	30.021	30.048	-0.014	30.021	30.061	-0.027
	MIN	30.000	30.002	-0.015	30.000	30.015	-0.028	30.000	30.022	-0.035	30.000	30.035	-0.048	30.000	30.048	-0.061

APPENDIX 41 • Preferred Hole Basis Transition and Interference Fits—Cylindrical Fits (cont.)

AMERICAN NATIONAL STANDARD PREFERRED METRIC LIMITS AND FITS ANSI B4.2 - 1978

Dimensions in mm.

BASIC SIZE		LOCATIONAL TRANSN. Hole H7	Shaft k6	Fit	LOCATIONAL TRANSN. Hole H7	Shaft n6	Fit	LOCATIONAL INTERF. Hole H7	Shaft p6	Fit	MEDIUM DRIVE Hole H7	Shaft s6	Fit	FORCE Hole H7	Shaft u6	Fit
40	MAX	40.025	40.018	0.023	40.025	40.033	0.008	40.025	40.042	-0.001	40.025	40.059	-0.018	40.025	40.076	-0.035
	MIN	40.000	40.002	-0.018	40.000	40.017	-0.033	40.000	40.026	-0.042	40.000	40.043	-0.059	40.000	40.060	-0.076
50	MAX	50.025	50.018	0.023	50.025	50.033	0.008	50.025	50.042	-0.001	50.025	50.059	-0.018	50.025	50.086	-0.045
	MIN	50.000	50.002	-0.018	50.000	50.017	-0.033	50.000	50.026	-0.042	50.000	50.043	-0.059	50.000	50.070	-0.086
60	MAX	60.030	60.021	0.028	60.030	60.039	0.010	60.030	60.051	-0.002	60.030	60.072	-0.023	60.030	60.106	-0.057
	MIN	60.000	60.002	-0.021	60.000	60.020	-0.039	60.000	60.032	-0.051	60.000	60.053	-0.072	60.000	60.087	-0.106
80	MAX	80.030	80.021	0.028	80.030	80.039	0.010	80.030	80.051	-0.002	80.030	80.078	-0.029	80.030	80.121	-0.072
	MIN	80.000	80.002	-0.021	80.000	80.020	-0.039	80.000	80.032	-0.051	80.000	80.059	-0.078	80.000	80.102	-0.121
100	MAX	100.035	100.025	0.032	100.035	100.045	0.012	100.035	100.059	-0.002	100.035	100.093	-0.036	100.035	100.146	-0.089
	MIN	100.000	100.003	-0.025	100.000	100.023	-0.045	100.000	100.037	-0.059	100.000	100.071	-0.093	100.000	100.124	-0.146
120	MAX	120.035	120.025	0.032	120.035	120.045	0.012	120.035	120.059	-0.002	120.035	120.101	-0.044	120.035	120.166	-0.109
	MIN	120.000	120.003	-0.025	120.000	120.023	-0.045	120.000	120.037	-0.059	120.000	120.079	-0.101	120.000	120.144	-0.166
160	MAX	160.040	160.028	0.037	160.040	160.052	0.013	160.040	160.068	-0.003	160.040	160.125	-0.060	160.040	160.215	-0.150
	MIN	160.000	160.003	-0.028	160.000	160.027	-0.052	160.000	160.043	-0.068	160.000	160.100	-0.125	160.000	160.190	-0.215
200	MAX	200.046	200.033	0.042	200.046	200.060	0.015	200.046	200.079	-0.004	200.046	200.151	-0.076	200.046	200.265	-0.190
	MIN	200.000	200.004	-0.033	200.000	200.031	-0.060	200.000	200.050	-0.079	200.000	200.122	-0.151	200.000	200.236	-0.265
250	MAX	250.046	250.033	0.042	250.046	250.060	0.015	250.046	250.079	-0.004	250.046	250.169	-0.094	250.046	250.313	-0.238
	MIN	250.000	250.004	-0.033	250.000	250.031	-0.060	250.000	250.050	-0.079	250.000	250.140	-0.169	250.000	250.284	-0.313
300	MAX	300.052	300.036	0.048	300.052	300.066	0.018	300.052	300.088	-0.004	300.052	300.202	-0.118	300.052	300.382	-0.298
	MIN	300.000	300.004	-0.036	300.000	300.034	-0.066	300.000	300.056	-0.088	300.000	300.170	-0.202	300.000	300.350	-0.382
400	MAX	400.057	400.040	0.053	400.057	400.073	0.020	400.057	400.098	-0.005	400.057	400.244	-0.151	400.057	400.471	-0.378
	MIN	400.000	400.004	-0.040	400.000	400.037	-0.073	400.000	400.062	-0.098	400.000	400.208	-0.244	400.000	400.435	-0.471
500	MAX	500.063	500.045	0.058	500.063	500.080	0.023	500.063	500.108	-0.005	500.063	500.292	-0.189	500.063	500.580	-0.477
	MIN	500.000	500.005	-0.045	500.000	500.040	-0.080	500.000	500.068	-0.108	500.000	500.252	-0.292	500.000	500.540	-0.580

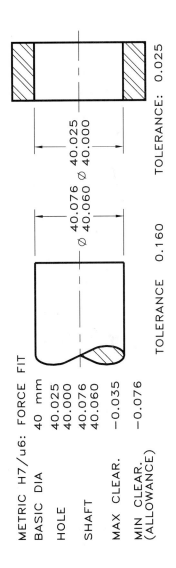

METRIC H7/u6: FORCE FIT

BASIC DIA	40 mm
HOLE	40.025 / 40.000
SHAFT	40.076 / 40.060
MAX CLEAR.	-0.035
MIN CLEAR. (ALLOWANCE)	-0.076

Ø 40.076 / Ø 40.060   Ø 40.025 / Ø 40.000

TOLERANCE 0.160   TOLERANCE 0.025   TOLERANCE: 0.025

APPENDIX 42 • Preferred Shaft Basis Clearance Fits—Cylindrical Fits (ANSI B4.2)

AMERICAN NATIONAL STANDARD PREFERRED METRIC LIMITS AND FITS   ANSI B4.2 - 1978

Dimensions in mm.

BASIC SIZE		LOOSE RUNNING			FREE RUNNING			CLOSE RUNNING			SLIDING			LOCATIONAL CLEARANCE		
		Hole C11	Shaft h11	Fit	Hole D9	Shaft h9	Fit	Hole F8	Shaft h7	Fit	Hole G7	Shaft h6	Fit	Hole H7	Shaft h6	Fit
1	MAX	1.120	1.000	0.180	1.045	1.000	0.070	1.020	1.000	0.030	1.012	1.000	0.018	1.010	1.000	0.016
	MIN	1.060	0.940	0.060	1.020	0.975	0.020	1.006	0.990	0.006	1.002	0.994	0.002	1.000	0.994	0.000
1.2	MAX	1.320	1.200	0.180	1.245	1.200	0.070	1.220	1.200	0.030	1.212	1.200	0.018	1.210	1.200	0.016
	MIN	1.260	1.140	0.060	1.220	1.175	0.020	1.206	1.190	0.006	1.202	1.194	0.002	1.200	1.194	0.000
1.6	MAX	1.720	1.600	0.180	1.645	1.600	0.070	1.620	1.600	0.030	1.612	1.600	0.018	1.610	1.600	0.016
	MIN	1.660	1.540	0.060	1.620	1.575	0.020	1.606	1.590	0.006	1.602	1.594	0.002	1.600	1.594	0.000
2	MAX	2.120	2.000	0.180	2.045	2.000	0.070	2.020	2.000	0.030	2.012	2.000	0.018	2.010	2.000	0.016
	MIN	2.060	1.940	0.060	2.020	1.975	0.020	2.006	1.990	0.006	2.002	1.994	0.002	2.000	1.994	0.000
2.5	MAX	2.620	2.500	0.180	2.545	2.500	0.070	2.520	2.500	0.030	2.512	2.500	0.018	2.510	2.500	0.016
	MIN	2.560	2.440	0.060	2.520	2.475	0.020	2.506	2.490	0.006	2.502	2.494	0.002	2.500	2.494	0.000
3	MAX	3.120	3.000	0.180	3.045	3.000	0.070	3.020	3.000	0.030	3.012	3.000	0.018	3.010	3.000	0.016
	MIN	3.060	2.940	0.060	3.020	2.975	0.020	3.006	2.990	0.006	3.002	2.994	0.002	3.000	2.994	0.000
4	MAX	4.145	4.000	0.220	4.060	4.000	0.090	4.028	4.000	0.040	4.016	4.000	0.024	4.012	4.000	0.020
	MIN	4.070	3.925	0.070	4.030	3.970	0.030	4.010	3.988	0.010	4.004	3.992	0.004	4.000	3.992	0.000
5	MAX	5.145	5.000	0.220	5.060	5.000	0.090	5.028	5.000	0.040	5.016	5.000	0.024	5.012	5.000	0.020
	MIN	5.070	4.925	0.070	5.030	4.970	0.030	5.010	4.988	0.010	5.004	4.992	0.004	5.000	4.992	0.000
6	MAX	6.145	6.000	0.220	6.060	6.000	0.090	6.028	6.000	0.040	6.016	6.000	0.024	6.012	6.000	0.020
	MIN	6.070	5.925	0.070	6.030	5.970	0.030	6.010	5.988	0.010	6.004	5.992	0.004	6.000	5.992	0.000
8	MAX	8.170	8.000	0.260	8.076	8.000	0.112	8.035	8.000	0.050	8.020	8.000	0.029	8.015	8.000	0.024
	MIN	8.080	7.910	0.080	8.040	7.964	0.040	8.013	7.985	0.013	8.005	7.991	0.005	8.000	7.991	0.000
10	MAX	10.170	10.000	0.260	10.076	10.000	0.112	10.035	10.000	0.050	10.020	10.000	0.029	10.015	10.000	0.024
	MIN	10.080	9.910	0.080	10.040	9.964	0.040	10.013	9.985	0.013	10.005	9.991	0.005	10.000	9.991	0.000
12	MAX	12.205	12.000	0.315	12.093	12.000	0.136	12.043	12.000	0.061	12.024	12.000	0.035	12.018	12.000	0.029
	MIN	12.095	11.890	0.095	12.050	11.957	0.050	12.016	11.982	0.016	12.006	11.989	0.006	12.000	11.989	0.000
16	MAX	16.205	16.000	0.315	16.093	16.000	0.136	16.043	16.000	0.061	16.024	16.000	0.035	16.018	16.000	0.029
	MIN	16.095	15.890	0.095	16.050	15.957	0.050	16.016	15.982	0.016	16.006	15.989	0.006	16.000	15.989	0.000
20	MAX	20.240	20.000	0.370	20.117	20.000	0.169	20.053	20.000	0.074	20.028	20.000	0.041	20.021	20.000	0.034
	MIN	20.110	19.870	0.110	20.065	19.948	0.065	20.020	19.979	0.020	20.007	19.987	0.007	20.000	19.987	0.000
25	MAX	25.240	25.000	0.370	25.117	25.000	0.169	25.053	25.000	0.074	25.028	25.000	0.041	25.021	25.000	0.034
	MIN	25.110	24.870	0.110	25.065	24.948	0.065	25.020	24.979	0.020	25.007	24.987	0.007	25.000	24.987	0.000
30	MAX	30.240	30.000	0.370	30.117	30.000	0.169	30.053	30.000	0.074	30.028	30.000	0.041	30.021	30.000	0.034
	MIN	30.110	29.870	0.110	30.065	29.948	0.065	30.020	29.979	0.020	30.007	29.987	0.007	30.000	29.987	0.000

md

Dimensions in mm.

BASIC SIZE		LOOSE RUNNING Hole C11	LOOSE RUNNING Shaft h11	LOOSE RUNNING Fit	FREE RUNNING Hole D9	FREE RUNNING Shaft h9	FREE RUNNING Fit	CLOSE RUNNING Hole F8	CLOSE RUNNING Shaft h7	CLOSE RUNNING Fit	SLIDING Hole G7	SLIDING Shaft h6	SLIDING Fit	LOCATIONAL CLEARANCE Hole H7	LOCATIONAL CLEARANCE Shaft h6	LOCATIONAL CLEARANCE Fit
40	MAX	40.280	40.000	0.440	40.142	40.000	0.204	40.064	40.000	0.089	40.034	40.000	0.050	40.025	40.000	0.041
	MIN	40.120	39.840	0.120	40.080	39.938	0.080	40.025	39.975	0.025	40.009	39.984	0.009	40.000	39.984	0.000
50	MAX	50.290	50.000	0.450	50.142	50.000	0.204	50.064	50.000	0.089	50.034	50.000	0.050	50.025	50.000	0.041
	MIN	50.130	49.840	0.130	50.080	49.938	0.080	50.025	49.975	0.025	50.009	49.984	0.009	50.000	49.984	0.000
60	MAX	60.330	60.000	0.520	60.174	60.000	0.248	60.076	60.000	0.106	60.040	60.000	0.059	60.030	60.000	0.049
	MIN	60.140	59.810	0.140	60.100	59.926	0.100	60.030	59.970	0.030	60.010	59.981	0.010	60.000	59.981	0.000
80	MAX	80.340	80.000	0.530	80.174	80.000	0.248	80.076	80.000	0.106	80.040	80.000	0.059	80.030	80.000	0.049
	MIN	80.150	79.810	0.150	80.100	79.926	0.100	80.030	79.970	0.030	80.010	79.981	0.010	80.000	79.981	0.000
100	MAX	100.390	100.000	0.610	100.207	100.000	0.294	100.090	100.000	0.125	100.047	100.000	0.069	100.035	100.000	0.057
	MIN	100.170	99.780	0.170	100.120	99.913	0.120	100.036	99.965	0.036	100.012	99.978	0.012	100.000	99.978	0.000
120	MAX	120.400	120.000	0.620	120.207	120.000	0.294	120.090	120.000	0.125	120.047	120.000	0.069	120.035	120.000	0.057
	MIN	120.180	119.780	0.180	120.120	119.913	0.120	120.036	119.965	0.036	120.012	119.978	0.012	120.000	119.978	0.000
160	MAX	160.460	160.000	0.710	160.245	160.000	0.345	160.106	160.000	0.146	160.054	160.000	0.079	160.040	160.000	0.065
	MIN	160.210	159.750	0.210	160.145	159.900	0.145	160.043	159.960	0.043	160.014	159.975	0.014	160.000	159.975	0.000
200	MAX	200.530	200.000	0.820	200.285	200.000	0.400	200.122	200.000	0.168	200.061	200.000	0.090	200.046	200.000	0.075
	MIN	200.240	199.710	0.240	200.170	199.885	0.170	200.050	199.954	0.050	200.015	199.971	0.015	200.000	199.971	0.000
250	MAX	250.570	250.000	0.860	250.285	250.000	0.400	250.122	250.000	0.168	250.061	250.000	0.090	250.046	250.000	0.075
	MIN	250.280	249.710	0.280	250.170	249.885	0.170	250.050	249.954	0.050	250.015	249.971	0.015	250.000	249.971	0.000
300	MAX	300.650	300.000	0.970	300.320	300.000	0.450	300.137	300.000	0.189	300.069	300.000	0.101	300.052	300.000	0.084
	MIN	300.330	299.680	0.330	300.190	299.870	0.190	300.056	299.948	0.056	300.017	299.968	0.017	300.000	299.968	0.000
400	MAX	400.760	400.000	1.120	400.350	400.000	0.490	400.151	400.000	0.208	400.075	400.000	0.111	400.057	400.000	0.093
	MIN	400.400	399.640	0.400	400.210	399.860	0.210	400.062	399.943	0.062	400.018	399.964	0.018	400.000	399.964	0.000
500	MAX	500.880	500.000	1.280	500.385	500.000	0.540	500.165	500.000	0.228	500.083	500.000	0.123	500.063	500.000	0.103
	MIN	500.480	499.600	0.480	500.230	499.845	0.230	500.068	499.937	0.068	500.020	499.960	0.020	500.000	499.960	0.000

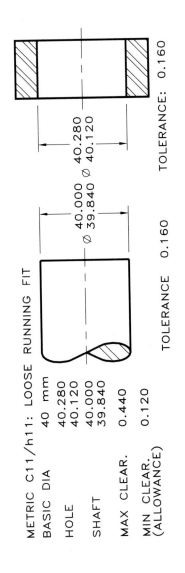

METRIC C11/h11: LOOSE RUNNING FIT

BASIC DIA	40 mm
HOLE	40.280 / 40.120
SHAFT	40.000 / 39.840
MAX CLEAR.	0.440
MIN CLEAR. (ALLOWANCE)	0.120

Ø 40.000   40.280 Ø 40.120   TOLERANCE: 0.160

Ø 40.000 / 39.840   TOLERANCE 0.160

APPENDIX 43 •Preferred Shaft Basis Transition and Interference Fits—Cylindrical Fits (ANSI B4.2)

AMERICAN NATIONAL STANDARD PREFERRED METRIC LIMITS AND FITS   ANSI B4.2 - 1978

Dimensions in mm.

BASIC SIZE		LOCATIONAL TRANSN. Hole K7	Shaft h6	Fit	LOCATIONAL TRANSN. Hole N7	Shaft h6	Fit	LOCATIONAL INTERF. Hole P7	Shaft h6	Fit	MEDIUM DRIVE Hole S7	Shaft h6	Fit	FORCE Hole U7	Shaft h6	Fit
1	MAX	1.000	1.000	0.006	0.996	1.000	0.002	0.994	1.000	0.000	0.986	1.000	-0.008	0.982	1.000	-0.012
	MIN	0.990	0.994	-0.010	0.986	0.994	-0.014	0.984	0.994	-0.016	0.976	0.994	-0.024	0.972	0.994	-0.028
1.2	MAX	1.200	1.200	0.006	1.196	1.200	0.002	1.194	1.200	0.000	1.186	1.200	-0.008	1.182	1.200	-0.012
	MIN	1.190	1.194	-0.010	1.186	1.194	-0.014	1.184	1.194	-0.016	1.176	1.194	-0.024	1.172	1.194	-0.028
1.6	MAX	1.600	1.600	0.006	1.596	1.600	0.002	1.594	1.600	0.000	1.586	1.600	-0.008	1.582	1.600	-0.012
	MIN	1.590	1.594	-0.010	1.586	1.594	-0.014	1.584	1.594	-0.016	1.576	1.594	-0.024	1.572	1.594	-0.028
2	MAX	2.000	2.000	0.006	1.996	2.000	0.002	1.994	2.000	0.000	1.986	2.000	-0.008	1.982	2.000	-0.012
	MIN	1.990	1.994	-0.010	1.986	1.994	-0.014	1.984	1.994	-0.016	1.976	1.994	-0.024	1.972	1.994	-0.028
2.5	MAX	2.500	2.500	0.006	2.496	2.500	0.002	2.494	2.500	0.000	2.486	2.500	-0.008	2.482	2.500	-0.012
	MIN	2.490	2.494	-0.010	2.486	2.494	-0.014	2.484	2.494	-0.016	2.476	2.494	-0.024	2.472	2.494	-0.028
3	MAX	3.000	3.000	0.006	2.996	3.000	0.002	2.994	3.000	0.000	2.986	3.000	-0.008	2.982	3.000	-0.012
	MIN	2.990	2.994	-0.010	2.986	2.994	-0.014	2.984	2.994	-0.016	2.976	2.994	-0.024	2.972	2.994	-0.028
4	MAX	4.003	4.000	0.011	3.996	4.000	0.004	3.992	4.000	0.000	3.985	4.000	-0.007	3.981	4.000	-0.011
	MIN	3.991	3.992	-0.009	3.984	3.992	-0.016	3.980	3.992	-0.020	3.973	3.992	-0.027	3.969	3.992	-0.031
5	MAX	5.003	5.000	0.011	4.996	5.000	0.004	4.992	5.000	0.000	4.985	5.000	-0.007	4.981	5.000	-0.011
	MIN	4.991	4.992	-0.009	4.984	4.992	-0.016	4.980	4.992	-0.020	4.973	4.992	-0.027	4.969	4.992	-0.031
6	MAX	6.003	6.000	0.011	5.996	6.000	0.004	5.992	6.000	0.000	5.985	6.000	-0.007	5.981	6.000	-0.011
	MIN	5.991	5.992	-0.009	5.984	5.992	-0.016	5.980	5.992	-0.020	5.973	5.992	-0.027	5.969	5.992	-0.031
8	MAX	8.005	8.000	0.014	7.996	8.000	0.005	7.991	8.000	0.000	7.983	8.000	-0.008	7.978	8.000	-0.013
	MIN	7.990	7.991	-0.010	7.981	7.991	-0.019	7.976	7.991	-0.024	7.968	7.991	-0.032	7.963	7.991	-0.037
10	MAX	10.005	10.000	0.014	9.996	10.000	0.005	9.991	10.000	0.000	9.983	10.000	-0.008	9.978	10.000	-0.013
	MIN	9.990	9.991	-0.010	9.981	9.991	-0.019	9.976	9.991	-0.024	9.968	9.991	-0.032	9.963	9.991	-0.037
12	MAX	12.006	12.000	0.017	11.995	12.000	0.006	11.989	12.000	0.000	11.979	12.000	-0.010	11.974	12.000	-0.015
	MIN	11.988	11.989	-0.012	11.977	11.989	-0.023	11.971	11.989	-0.029	11.961	11.989	-0.039	11.956	11.989	-0.044
16	MAX	16.006	16.000	0.017	15.995	16.000	0.006	15.989	16.000	0.000	15.979	16.000	-0.010	15.974	16.000	-0.015
	MIN	15.988	15.989	-0.012	15.977	15.989	-0.023	15.971	15.989	-0.029	15.961	15.989	-0.039	15.956	15.989	-0.044
20	MAX	20.006	20.000	0.019	19.993	20.000	0.006	19.986	20.000	-0.001	19.973	20.000	-0.014	19.967	20.000	-0.020
	MIN	19.985	19.987	-0.015	19.972	19.987	-0.028	19.965	19.987	-0.035	19.952	19.987	-0.048	19.946	19.987	-0.054
25	MAX	25.006	25.000	0.019	24.993	25.000	0.006	24.986	25.000	-0.001	24.973	25.000	-0.014	24.960	25.000	-0.027
	MIN	24.985	24.987	-0.015	24.972	24.987	-0.028	24.965	24.987	-0.035	24.952	24.987	-0.048	24.939	24.987	-0.061
30	MAX	30.006	30.000	0.019	29.993	30.000	0.006	29.986	30.000	-0.001	29.973	30.000	-0.014	29.960	30.000	-0.027
	MIN	29.985	29.987	-0.015	29.972	29.987	-0.028	29.965	29.987	-0.035	29.952	29.987	-0.048	29.939	29.987	-0.061

AMERICAN NATIONAL STANDARD PREFERRED METRIC LIMITS AND FITS   ANSI B4.2 - 1978

Dimensions in mm.

BASIC SIZE		LOCATIONAL TRANSN. Hole K7	Shaft h6	Fit	LOCATIONAL TRANSN. Hole N7	Shaft h6	Fit	LOCATIONAL INTERF. Hole P7	Shaft h6	Fit	MEDIUM DRIVE Hole S7	Shaft h6	Fit	FORCE Hole U7	Shaft h6	Fit
40	MAX	40.007	40.000	0.023	39.992	40.000	0.008	39.983	40.000	-0.001	39.966	40.000	-0.018	39.949	40.000	-0.035
	MIN	39.982	39.984	-0.018	39.967	39.984	-0.033	39.958	39.984	-0.042	39.941	39.984	-0.059	39.924	39.984	-0.076
50	MAX	50.007	50.000	0.023	49.992	50.000	0.008	49.983	50.000	-0.001	49.966	50.000	-0.018	49.939	50.000	-0.045
	MIN	49.982	49.984	-0.018	49.967	49.984	-0.033	49.958	49.984	-0.042	49.941	49.984	-0.059	49.914	49.984	-0.086
60	MAX	60.009	60.000	0.028	59.991	60.000	0.010	59.979	60.000	-0.002	59.958	60.000	-0.023	59.924	60.000	-0.057
	MIN	59.979	59.981	-0.021	59.961	59.981	-0.039	59.949	59.981	-0.051	59.928	59.981	-0.072	59.894	59.981	-0.106
80	MAX	80.009	80.000	0.028	79.991	80.000	0.010	79.979	80.000	-0.002	79.952	80.000	-0.029	79.909	80.000	-0.072
	MIN	79.979	79.981	-0.021	79.961	79.981	-0.039	79.949	79.981	-0.051	79.922	79.981	-0.078	79.879	79.981	-0.121
100	MAX	100.010	100.000	0.032	99.990	100.000	0.012	99.976	100.000	-0.002	99.942	100.000	-0.036	99.889	100.000	-0.089
	MIN	99.975	99.978	-0.025	99.955	99.978	-0.045	99.941	99.978	-0.059	99.907	99.978	-0.093	99.854	99.978	-0.146
120	MAX	120.010	120.000	0.032	119.990	120.000	0.012	119.976	120.000	-0.002	119.934	120.000	-0.044	119.869	120.000	-0.109
	MIN	119.975	119.978	-0.025	119.955	119.978	-0.045	119.941	119.978	-0.059	119.899	119.978	-0.101	119.834	119.978	-0.166
160	MAX	160.012	160.000	0.037	159.988	160.000	0.013	159.972	160.000	-0.003	159.915	160.000	-0.060	159.825	160.000	-0.150
	MIN	159.972	159.975	-0.028	159.948	159.975	-0.052	159.932	159.975	-0.068	159.875	159.975	-0.125	159.785	159.975	-0.215
200	MAX	200.013	200.000	0.042	199.986	200.000	0.015	199.967	200.000	-0.004	199.895	200.000	-0.076	199.781	200.000	-0.190
	MIN	199.967	199.971	-0.033	199.940	199.971	-0.060	199.921	199.971	-0.079	199.849	199.971	-0.151	199.735	199.971	-0.265
250	MAX	250.013	250.000	0.042	249.986	250.000	0.015	249.967	250.000	-0.004	249.877	250.000	-0.094	249.733	250.000	-0.238
	MIN	249.967	249.971	-0.033	249.940	249.971	-0.060	249.921	249.971	-0.079	249.831	249.971	-0.169	249.687	249.971	-0.313
300	MAX	300.016	300.000	0.048	299.986	300.000	0.018	299.964	300.000	-0.004	299.850	300.000	-0.118	299.670	300.000	-0.298
	MIN	299.964	299.968	-0.036	299.934	299.968	-0.066	299.912	299.968	-0.088	299.798	299.968	-0.202	299.618	299.968	-0.382
400	MAX	400.017	400.000	0.053	399.984	400.000	0.020	399.959	400.000	-0.005	399.813	400.000	-0.151	399.586	400.000	-0.378
	MIN	399.960	399.964	-0.040	399.927	399.964	-0.073	399.902	399.964	-0.098	399.756	399.964	-0.244	399.529	399.964	-0.471
500	MAX	500.018	500.000	0.058	499.983	500.000	0.023	499.955	500.000	-0.005	499.771	500.000	-0.189	499.483	500.000	-0.477
	MIN	499.955	499.960	-0.045	499.920	499.960	-0.080	499.892	499.960	-0.108	499.708	499.960	-0.292	499.420	499.960	-0.580

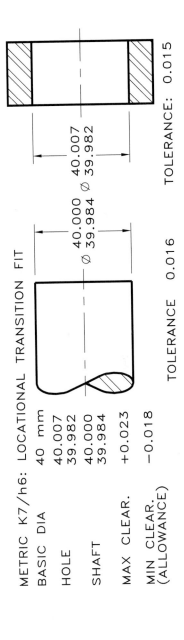

METRIC K7/h6: LOCATIONAL TRANSITION FIT

BASIC DIA   40 mm

HOLE   40.007
                 39.982   TOLERANCE 0.016

SHAFT   40.000
                 39.984   TOLERANCE: 0.015

MAX CLEAR.   +0.023

MIN CLEAR.   -0.018
(ALLOWANCE)

Ø 40.000   Ø 40.007
Ø 39.984   Ø 39.982

APPENDIX 44 • Hole Sizes for Non-Preferred Diameters (dimensions are in millimeters)

Basic Size		C11	D9	F8	G7	H7	H8	H9	H11	K7	N7	P7	S7	U7
OVER 0 TO 3		+0.120 +0.060	+0.045 +0.020	+0.020 +0.006	+0.012 +0.002	+0.010 0.000	+0.014 0.000	+0.025 0.000	+0.060 0.000	0.000 -0.010	-0.004 -0.014	-0.006 -0.016	-0.014 -0.024	-0.018 -0.028
OVER 3 TO 6		+0.145 +0.070	+0.060 +0.030	+0.028 +0.010	+0.016 +0.004	+0.012 0.000	+0.018 0.000	+0.030 0.000	+0.075 0.000	+0.003 -0.009	-0.004 -0.016	-0.008 -0.020	-0.015 -0.027	-0.019 -0.031
OVER 6 TO 10		+0.170 +0.080	+0.076 +0.040	+0.035 +0.013	+0.020 +0.005	+0.015 0.000	+0.022 0.000	+0.036 0.000	+0.090 0.000	+0.005 -0.010	-0.004 -0.019	-0.009 -0.024	-0.017 -0.032	-0.022 -0.037
OVER 10 TO 14		+0.205 +0.095	+0.093 +0.050	+0.043 +0.016	+0.024 +0.006	+0.018 0.000	+0.027 0.000	+0.043 0.000	+0.110 0.000	+0.006 -0.012	-0.005 -0.023	-0.011 -0.029	-0.021 -0.039	-0.026 -0.044
OVER 14 TO 18		+0.205 +0.095	+0.093 +0.050	+0.043 +0.016	+0.024 +0.006	+0.018 0.000	+0.027 0.000	+0.043 0.000	+0.110 0.000	+0.006 -0.012	-0.005 -0.023	-0.011 -0.029	-0.021 -0.039	-0.026 -0.044
OVER 18 TO 24		+0.240 +0.110	+0.117 +0.065	+0.053 +0.020	+0.028 +0.007	+0.021 0.000	+0.033 0.000	+0.052 0.000	+0.130 0.000	+0.006 -0.015	-0.007 -0.028	-0.014 -0.035	-0.027 -0.048	-0.033 -0.054
OVER 24 TO 30		+0.240 +0.110	+0.117 +0.065	+0.053 +0.020	+0.028 +0.007	+0.021 0.000	+0.033 0.000	+0.052 0.000	+0.130 0.000	+0.006 -0.015	-0.007 -0.028	-0.014 -0.035	-0.027 -0.048	-0.040 -0.061
OVER 30 TO 40		+0.280 +0.120	+0.142 +0.080	+0.064 +0.025	+0.034 +0.009	+0.025 0.000	+0.039 0.000	+0.062 0.000	+0.160 0.000	+0.007 -0.018	-0.008 -0.033	-0.017 -0.042	-0.034 -0.059	-0.051 -0.076
OVER 40 TO 50		+0.290 +0.130	+0.142 +0.080	+0.064 +0.025	+0.034 +0.009	+0.025 0.000	+0.039 0.000	+0.062 0.000	+0.160 0.000	+0.007 -0.018	-0.008 -0.033	-0.017 -0.042	-0.034 -0.059	-0.061 -0.086
OVER 50 TO 65		+0.330 +0.140	+0.174 +0.100	+0.076 +0.030	+0.040 +0.010	+0.030 0.000	+0.046 0.000	+0.074 0.000	+0.190 0.000	+0.009 -0.021	-0.009 -0.039	-0.021 -0.051	-0.042 -0.072	-0.076 -0.106
OVER 65 TO 80		+0.340 +0.150	+0.174 +0.100	+0.076 +0.030	+0.040 +0.010	+0.030 0.000	+0.046 0.000	+0.074 0.000	+0.190 0.000	+0.009 -0.021	-0.009 -0.039	-0.021 -0.051	-0.048 -0.078	-0.091 -0.121
OVER 80 TO 100		+0.390 +0.170	+0.207 +0.120	+0.090 +0.036	+0.047 +0.012	+0.035 0.000	+0.054 0.000	+0.087 0.000	+0.220 0.000	+0.010 -0.025	-0.010 -0.045	-0.024 -0.059	-0.058 -0.093	-0.111 -0.146

Cont.

APPENDIX 44 • Hole sizes for Non-Preferred Diameters (cont.)

Basic Size	c11	d9	f8	g7	h7	h8	h9	h11	k7	n7	p7	s7	u7
OVER 100 TO 120	+0.400 / +0.180	+0.207 / +0.120	+0.090 / +0.036	+0.047 / +0.012	+0.035 / 0.000	+0.054 / 0.000	+0.087 / 0.000	+0.220 / 0.000	+0.010 / −0.025	−0.010 / −0.045	−0.024 / −0.059	−0.066 / −0.101	−0.131 / −0.166
OVER 120 TO 140	+0.450 / +0.200	+0.245 / +0.145	+0.106 / +0.043	+0.054 / +0.014	+0.040 / 0.000	+0.063 / 0.000	+0.100 / 0.000	+0.250 / 0.000	+0.012 / −0.028	−0.012 / −0.052	−0.028 / −0.068	−0.077 / −0.117	−0.155 / −0.195
OVER 140 TO 160	+0.460 / +0.210	+0.245 / +0.145	+0.106 / +0.043	+0.054 / +0.014	+0.040 / 0.000	+0.063 / 0.000	+0.100 / 0.000	+0.250 / 0.000	+0.012 / −0.028	−0.012 / −0.052	−0.028 / −0.068	−0.085 / −0.125	−0.175 / −0.215
OVER 160 TO 180	+0.480 / +0.230	+0.245 / +0.145	+0.106 / +0.043	+0.054 / +0.014	+0.040 / 0.000	+0.063 / 0.000	+0.100 / 0.000	+0.250 / 0.000	+0.012 / −0.028	−0.012 / −0.052	−0.028 / −0.068	−0.093 / −0.133	−0.195 / −0.235
OVER 180 TO 200	+0.530 / +0.240	+0.285 / +0.170	+0.122 / +0.050	+0.061 / +0.015	+0.046 / 0.000	+0.072 / 0.000	+0.115 / 0.000	+0.290 / 0.000	−0.013 / −0.033	−0.014 / −0.060	−0.033 / −0.079	−0.105 / −0.151	−0.219 / −0.265
OVER 200 TO 225	+0.550 / +0.260	+0.285 / +0.170	+0.122 / +0.050	+0.061 / +0.015	+0.046 / 0.000	+0.072 / 0.000	+0.115 / 0.000	+0.290 / 0.000	+0.013 / −0.033	−0.014 / −0.060	−0.033 / −0.079	−0.113 / −0.159	−0.241 / −0.287
OVER 225 TO 250	+0.570 / +0.280	+0.285 / +0.170	+0.122 / +0.050	+0.061 / +0.015	+0.046 / 0.000	+0.072 / 0.000	+0.115 / 0.000	+0.290 / 0.000	+0.013 / −0.033	−0.014 / −0.060	−0.033 / −0.079	−0.123 / −0.169	−0.267 / −0.313
OVER 250 TO 280	+0.620 / +0.300	+0.320 / +0.190	+0.137 / +0.056	+0.069 / +0.017	+0.052 / 0.000	+0.081 / 0.000	+0.130 / 0.000	+0.320 / 0.000	+0.016 / −0.036	−0.014 / −0.066	−0.036 / −0.088	−0.138 / −0.190	−0.295 / −0.347
OVER 280 TO 315	+0.650 / +0.330	+0.320 / +0.190	+0.137 / +0.056	+0.069 / 0.017	+0.052 / 0.000	+0.081 / 0.000	+0.130 / 0.000	+0.320 / 0.000	+0.016 / −0.036	−0.014 / −0.066	−0.036 / −0.088	−0.150 / −0.202	−0.330 / −0.382
OVER 315 TO 355	+0.720 / +0.360	+0.350 / +0.210	+0.151 / +0.062	+0.075 / +0.018	+0.057 / 0.000	+0.089 / 0.000	+0.140 / 0.000	+0.360 / 0.000	+0.017 / −0.040	−0.016 / −0.073	−0.041 / −0.058	−0.169 / −0.226	−0.369 / −0.426
OVER 355 TO 400	+0.760 / +0.400	+0.350 / +0.210	+0.151 / +0.062	+0.075 / +0.018	+0.057 / 0.000	+0.089 / 0.000	+0.140 / 0.000	+0.360 / 0.000	+0.017 / −0.040	−0.016 / −0.073	−0.041 / −0.058	−0.187 / −0.244	−0.414 / −0.471
OVER 400 TO 450	+0.840 / +0.440	+0.385 / +0.230	+0.165 / +0.068	+0.083 / +0.020	+0.063 / 0.000	+0.097 / 0.000	+0.155 / 0.000	+0.400 / 0.000	+0.018 / −0.045	−0.017 / −0.080	−0.045 / −0.108	−0.209 / −0.272	−0.467 / −0.530
OVER 450 TO 500	+0.880 / +0.480	+0.385 / +0.230	+0.165 / +0.068	+0.083 / +0.020	+0.063 / 0.000	+0.097 / 0.000	+0.155 / 0.000	+0.400 / 0.000	+0.018 / −0.045	−0.017 / −0.080	−0.045 / −0.108	−0.229 / −0.292	−0.517 / −0.580

APPENDIX 45 • Shaft Sizes for Non-Preferred Diameters (dimensions are in millimeters)

Basic Size	c11	d9	f7	g6	h6	h7	h9	h11	k6	n6	p6	s6	u6
OVER 0 TO 3	−0.060 / −0.120	−0.020 / −0.045	−0.006 / −0.016	−0.002 / −0.008	0.000 / −0.006	0.000 / −0.010	0.000 / −0.025	0.000 / −0.060	+0.006 / 0.000	+0.010 / +0.004	+0.012 / +0.006	+0.020 / +0.014	+0.024 / +0.018
OVER 3 TO 6	−0.070 / −0.145	−0.030 / −0.060	−0.010 / −0.022	−0.004 / −0.012	0.000 / −0.008	0.000 / −0.012	0.000 / −0.030	0.000 / −0.075	+0.009 / +0.001	+0.016 / +0.008	+0.020 / +0.012	+0.027 / +0.019	+0.031 / +0.023
OVER 6 TO 10	−0.080 / −0.170	−0.040 / −0.076	−0.013 / −0.028	−0.005 / −0.014	0.000 / −0.009	0.000 / −0.015	0.000 / −0.036	0.000 / −0.090	+0.010 / +0.001	+0.019 / +0.010	+0.024 / +0.015	+0.032 / +0.023	+0.037 / +0.028
OVER 10 TO 14	−0.095 / −0.205	−0.050 / −0.093	−0.016 / −0.034	−0.006 / −0.017	0.000 / −0.011	0.000 / −0.018	0.000 / −0.043	0.000 / −0.110	+0.012 / +0.001	+0.023 / +0.012	+0.029 / +0.018	+0.039 / +0.028	+0.044 / +0.033
OVER 14 TO 18	−0.095 / −0.205	−0.050 / −0.093	−0.016 / −0.034	−0.006 / −0.017	0.000 / −0.011	0.000 / −0.018	0.000 / −0.043	0.000 / −0.110	+0.012 / +0.001	+0.023 / +0.012	+0.029 / +0.018	+0.039 / +0.028	+0.044 / +0.033
OVER 18 TO 24	−0.110 / −0.240	−0.065 / −0.117	−0.020 / −0.041	−0.007 / −0.020	0.000 / −0.013	0.000 / −0.021	0.000 / −0.052	0.000 / −0.130	+0.015 / +0.002	+0.028 / +0.015	+0.035 / +0.022	+0.048 / +0.035	+0.054 / +0.041
OVER 24 TO 30	−0.110 / −0.240	−0.065 / −0.117	−0.020 / −0.041	−0.007 / −0.020	0.000 / −0.013	0.000 / −0.021	0.000 / −0.052	0.000 / −0.130	+0.015 / +0.002	+0.028 / +0.015	+0.035 / +0.022	+0.048 / +0.035	+0.061 / +0.048
OVER 30 TO 40	−0.120 / −0.280	−0.080 / −0.142	−0.025 / −0.050	−0.009 / −0.025	0.000 / −0.016	0.000 / −0.025	0.000 / −0.062	0.000 / −0.160	+0.018 / +0.002	+0.033 / +0.017	+0.042 / +0.026	+0.059 / +0.043	+0.076 / +0.060
OVER 40 TO 50	−0.130 / −0.290	−0.080 / −0.142	−0.025 / −0.050	−0.009 / −0.025	0.000 / −0.016	0.000 / −0.025	0.000 / −0.062	0.000 / −0.160	+0.018 / +0.002	+0.033 / +0.017	+0.042 / +0.026	+0.059 / +0.043	+0.086 / +0.070
OVER 50 TO 65	−0.140 / −0.330	−0.100 / −0.174	−0.030 / −0.060	−0.010 / −0.029	0.000 / −0.019	0.000 / −0.030	0.000 / −0.074	0.000 / −0.190	+0.021 / +0.002	+0.039 / +0.020	+0.051 / +0.032	+0.072 / +0.053	+0.106 / +0.087
OVER 65 TO 80	−0.150 / −0.340	−0.100 / −0.174	−0.030 / −0.060	−0.010 / −0.029	0.000 / −0.019	0.000 / −0.030	0.000 / −0.074	0.000 / −0.190	+0.021 / +0.002	+0.039 / +0.020	+0.051 / +0.032	+0.078 / +0.059	+0.121 / +0.102
OVER 80 TO 100	−0.170 / −0.390	−0.120 / −0.207	−0.036 / −0.071	−0.012 / −0.034	0.000 / −0.022	0.000 / −0.035	0.000 / −0.087	0.000 / −0.220	+0.025 / +0.003	+0.045 / +0.023	+0.059 / +0.037	+0.093 / +0.071	+0.146 / +0.124

Cont.

Basic Size	c11	d9	f7	g6	h6	h7	h9	h11	k6	n6	p6	s6	u6
OVER 100 TO 120	−0.180 / −0.400	−0.120 / −0.207	−0.036 / −0.071	−0.012 / −0.034	0.000 / −0.022	0.000 / −0.035	0.000 / −0.087	0.000 / −0.220	+0.025 / +0.003	+0.045 / +0.023	+0.059 / +0.037	+0.101 / +0.079	+0.166 / +0.144
OVER 120 TO 140	−0.200 / −0.450	−0.145 / −0.245	−0.043 / −0.083	−0.014 / −0.039	0.000 / −0.025	0.000 / −0.040	0.000 / −0.100	0.000 / −0.250	+0.028 / +0.003	+0.052 / +0.027	+0.068 / +0.043	+0.117 / +0.092	+0.195 / +0.170
OVER 140 TO 160	−0.210 / −0.460	−0.145 / −0.245	−0.043 / −0.083	−0.014 / −0.039	0.000 / −0.025	0.000 / −0.040	0.000 / −0.100	0.000 / −0.250	+0.028 / +0.003	+0.052 / +0.027	+0.068 / +0.043	+0.125 / +0.100	+0.215 / +0.190
OVER 160 TO 180	−0.230 / −0.480	−0.145 / −0.245	−0.043 / −0.083	−0.014 / −0.039	0.000 / −0.025	0.000 / −0.040	0.000 / −0.100	0.000 / −0.250	+0.028 / +0.003	+0.052 / +0.027	+0.068 / +0.043	+0.133 / +0.108	+0.235 / +0.210
OVER 180 TO 200	−0.240 / −0.530	−0.170 / −0.285	−0.050 / −0.096	−0.015 / −0.044	0.000 / −0.029	0.000 / −0.046	0.000 / −0.115	0.000 / −0.290	+0.033 / +0.004	+0.060 / +0.031	+0.079 / +0.050	+0.151 / +0.122	+0.265 / +0.236
OVER 200 TO 225	−0.260 / −0.550	−0.170 / −0.285	−0.050 / −0.096	−0.015 / −0.044	0.000 / −0.029	0.000 / −0.046	0.000 / −0.115	0.000 / −0.290	+0.033 / +0.004	+0.060 / +0.031	+0.079 / +0.050	+0.159 / +0.130	+0.287 / +0.258
OVER 225 TO 250	−0.280 / −0.570	−0.170 / −0.285	−0.050 / −0.096	−0.015 / −0.044	0.000 / −0.029	0.000 / −0.046	0.000 / −0.115	0.000 / −0.290	+0.033 / +0.004	+0.060 / +0.031	+0.079 / +0.050	+0.169 / +0.140	+0.313 / +0.284
OVER 250 TO 280	−0.300 / −0.620	−0.190 / −0.320	−0.056 / −0.108	−0.017 / −0.049	0.000 / −0.032	0.000 / −0.052	0.000 / −0.130	0.000 / −0.320	+0.036 / +0.004	+0.066 / +0.034	+0.088 / +0.056	+0.190 / +0.158	+0.347 / +0.315
OVER 280 TO 315	−0.330 / −0.650	−0.190 / −0.320	−0.056 / −0.108	−0.017 / −0.049	0.000 / −0.032	0.000 / −0.052	0.000 / −0.130	0.000 / −0.320	+0.036 / +0.004	+0.066 / +0.034	+0.088 / +0.056	+0.202 / +0.170	+0.382 / +0.350
OVER 315 TO 355	−0.360 / −0.720	−0.210 / −0.350	−0.062 / −0.119	−0.018 / −0.054	0.000 / −0.036	0.000 / −0.057	0.000 / −0.140	0.000 / −0.360	+0.040 / +0.004	+0.073 / +0.037	+0.098 / +0.062	+0.226 / +0.190	+0.426 / +0.390
OVER 355 TO 400	−0.400 / −0.760	−0.210 / −0.350	−0.062 / −0.119	−0.018 / −0.054	0.000 / −0.036	0.000 / −0.057	0.000 / −0.140	0.000 / −0.360	+0.040 / +0.004	+0.073 / +0.037	+0.098 / +0.062	+0.244 / +0.208	+0.471 / +0.435
OVER 400 TO 450	−0.440 / −0.840	−0.230 / −0.385	−0.068 / −0.131	−0.020 / −0.060	0.000 / −0.040	0.000 / −0.063	0.000 / −0.155	0.000 / −0.400	+0.045 / +0.005	+0.080 / +0.040	+0.108 / +0.068	+0.272 / +0.232	+0.530 / +0.490
OVER 450 TO 500	−0.480 / −0.880	−0.230 / −0.385	−0.068 / −0.131	−0.020 / −0.060	0.000 / −0.040	0.000 / −0.063	0.000 / −0.155	0.000 / −0.400	+0.045 / +0.005	+0.080 / +0.040	+0.108 / +0.068	+0.292 / +0.252	+0.580 / +0.540

Grading graph

This graph can be used to determine the individual grades of members of a team and to compute grade averages for those who do extra assignments.

The percent participation of each team member should be determined by the team as a whole (see Chapter 2 problems).

Example: written or oral report grades

Overall team grade: 82

Team members $N = 5$	Contribution $C = \%$	$F = CN$	Grade (graph)
J. Doe	20%	100	82.0
H. Brown	16%	80	76.4
L. Smith	24%	120	86.0
R. Black	20%	100	82.0
T. Jones	20%	100	82.0
	100%		

Example: quiz or problem sheet grades

Number assigned:　30
Number extra:　　　6
　　　　　　　　　　——
Total　　　　　　　36

Average grade for total (36): 82

$$F = \frac{\text{No. completed} \times 100}{\text{No. assigned}} = \frac{36 \times 100}{30} = 120$$

Final grade (from graph): 86.0

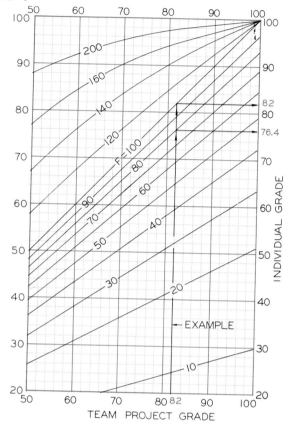

Fig. A50.–1. Grading graph.

The following programs were written by Professor Leendert Kersten of the University of Nebraska at Lincoln and they have been reproduced here with his permission. These programs were introduced in Chapter 27, in which the principles of descriptive geometry are covered. These very valuable programs can be duplicated and added as supplements to your AutoCAD software.

```
(defun C:PARALLEL ( )
  (setvar "aperture" 5)
  (setq sp (getpoint "\nSelect START point of parallel line:"))
  (setq ep (getpoint "\nSelect END point of parallel line:"))
  (setvar "osmode" 1)
  (setq sl (getpoint "\nSelect 1st point on line for parallelism:"))
  (setq el (getpoint "\nSelect 2nd point on line for parallelism:"))
  (setvar "osmode" 0)
  (setq pa (angle sl el))
  (setq la (angle sp ep))
  (setq ll (distance sp ep))
  (setq m -1)
  (setq d 0)
  (if (> pa d) (setq m 1))
  (if (> la d) (setq d 1))
  (if (/= m d) (setq pa (+ pa 3.141593)))
  (setq ep (polar sp pa ll))
  (setvar "cmdecho" 0)
  (command "line" sp ep "")
  (restore)
)
```

```
(defun C:PERPLINE ( )
  (setvar "aperture" 5) (setvar "cmdecho" 0)
  (setq sp (getpoint "\nSelect START point of perpendicular line:"))
  (setvar "osmode" 128)
  (setq cc (getpoint sp "\nSelect ANY point on line to which perp'lr:"))
  (setq beta (angle sp cc)) (setvar "osmode" 0)
  (setq ep
  (getpoint "\nSelect END point of desired perpendicular (for length only): "")
  (setq length (distance sp ep))
  (setq ep (polar sp beta length))
  (command "line" sp ep "")
  (restore)
)
```

```
(defun C: TRANSFER ( )
    (setvar "aperture" 5) (setvar "cmdecho" 0)
    (setvar "osmode" 1)
    (setq aa (getpoint "\nSelect start of transfer distance:"))
    (setvar "osmode" 128)
    (setq bb (getpoint aa "\nSelect the reference plane:"))
    (setq length (distance aa bb))
    (setvar "osmode" 1)
    (setq cc (getpoint "\nSelect point to be projected:"))
    (setvar "osmode" 128)
    (setq dd (getpoint cc "\nSelect other reference plane:"))
(setvar "osmode" 0)
    (setq alpha (angle cc dd))
    (setq ep (polar dd alpha length))
    (COMMAND "CIRCLE" EP 0.05)
    (restore)
)

(defun RESTORE ( )
    (setvar "aperture" 10)
    (setvar "cmdecho" 1)
    (setvar "osmode" 0)
(defun command: ( )
(defun *error* (st)
    (setvar "osmode" 0)
    (princ))
    (quit))
)

(defun C:COPYDIST ( )
    (setvar "aperture"5)
    (setvar "cmdecho" 0)
    (setvar "osmode" 1)
    (setq p1 (getpoint "\nSelect start point of line distance to be copied:"))
    (setq p2 (getpoint "\nEnd point?:"))
    (setvar "osmode" 0)
    (setq dist (distance p1 p2))
    (setq p1 (getpoint "\nStart point of new distance location:"))
    (setq ang (getangle p1 "\nWhich direction?: "))
    (setq p2 (polar p1 ang dist))
    (setvar "osmode" 0)
    (command "circle" p2 0.05)
    (restore)
)
```

```lisp
(defun C:BISECT ( )
  (setvar "aperture" 5)
  (setvar "osmode" 32)
  (setq sp (getpoint "\nSelect Corner of angle:"))
  (setvar "osmode" 2)
  (setq aa (getpoint "\nSelect first side (remember CCW):"))
  (setq alpha (angle sp aa))
  (setq bb (getpoint "\nSelect other side:"))
  (setvar "osmode" 0)
  (setq beta (angle sp bb))
  (setq m (/ (+ alpha beta) 2))
  (if (< alpha beta) (setq ang (+ pi m)) (setq ang m))

(setq ep
  (getpoint "\nSelect endpoint of bisecting line (for length only): "))
  (setq length (distance sp ep))
  (setq ep (polar sp ang length))
  (setvar "cmdecho" 0)
  (command "line" sp ep "")
  (restore)
)
```

Index

VPORTS command, 730
VSLIDE command, 716

W

Washers, 277
   lock, 277
   plain, 277
Waviness, 378
   height, 378
   width, 378
WBLOCK command, 696
WCS. *See* World coordinate system
Wedge, by computer, 738
Weld
   built-up, 392
   concave, 393
   flush, 393
   groove, 391
   joints, 387
   seam, 391
   types of, 387
Welding, 385
   processes, 385
   symbols, 388
      application of, 389
Window option, 672
Working drawings, 78, 185, 395
   dual dimensions, 399
   forged parts and castings, 410
   freehand, 410
   to lay out
      by computer, 399
      by hand, 401
   metric units, 399
World coordinates, 666
World coordinate system, 732, 734
Worm gears. *See* Gears, worm
Written reports, 74–76

X

XREF command, 696
Xerography, 448
XYZ filters, 747

Z

ZOOM command, 671

The following supplements are compatible with the following textbooks from Addison-Wesley Publishing Company:

Drafting Technology

Engineering Design Graphics

Design Drafting

Graphics for Engineers

Geometry for Engineers

DRAFTING TECHNOLOGY PROBLEMS (With Computer Graphics) is a problem book designed to cover basic graphics, descriptive geometry, and specialty drafting areas. It is designed to accompany *Drafting Technology*, which is available from Addison-Wesley Publishing Company, Reading, MA, 01867. 131 pages.

BASIC DRAFTING (With Computer Graphics) is a problem book that covers the basics for a one-semester high school drafting course. 67 pages.

CREATIVE DRAFTING (With Computer Graphics) is a problem book that covers mechanical drawing and architectural drafting for the high school. 106 pages.

DRAFTING & DESIGN (With Computer Graphics) is a problem book for mechanical drawing for a high school or college course. 106 pages.

DRAFTING FUNDAMENTALS 1 is a problem book for a one-year high school course in mechanical drawing. 94 pages.

DRAFTING FUNDAMENTALS 2 is a second version of **DRAFTING FUNDAMENTALS 1** for the same level and content. 94 pages.

TECHNICAL ILLUSTRATION is a problem book for a course in pictorial drawing for the college or high school. 70 pages.

ARCHITECTURAL DRAFTING is a problem book for a first course in architectural drafting. 71 pages.

GRAPHICS FOR ENGINEERS 1 (With Computer Graphics) is a problem book for a first course in engineering graphics for the college student. 100 pages.

GRAPHICS FOR ENGINEERS 2 (With Computer Graphics) is a problem book for a first course in engineering graphics for the college student. 100 pages.

GRAPHICS FOR ENGINEERS 3 (With Computer Graphics) is a problem book for a first course in engineering graphics for the college student. 108 pages.

GEOMETRY FOR ENGINEERS 1 is a problem book for a college-level descriptive geometry course. 100 pages.

GEOMETRY FOR ENGINEERS 2 is a problem book for a college-level descriptive geometry course. 100 pages.

GEOMETRY FOR ENGINEERS 3 is a problem book for a college-level descriptive geometry course. 100 pages.

GRAPHICS & GEOMETRY 1 (With Computer Graphics) is a problem book for a college course in graphics and descriptive geometry. 119 pages.

GRAPHICS & GEOMETRY 2 (With Computer Graphics) is a problem book for a college course in graphics and descriptive geometry. 121 pages.

GRAPHICS & GEOMETRY 3 (With Computer Graphics) is a problem book for a college course in graphics and descriptive geometry. 138 pages.

DESIGN GRAPHICS 1 (With Computer Graphics) is a problem book for a college course in graphics and descriptive geometry. 140 pages.

DESIGN GRAPHICS 2 (With Computer Graphics) is a problem book for a college course in graphics and descriptive geometry. 140 pages.

CREATIVE PUBLISHING CO.
Box 9292   College Station, Texas 77840
Phone 409-775-6047